The elements

Name	Symbol	Atomic Number	Atomic Mass	Name	Symbol	Atomic Number	Atomic Mass
Actinium	Ac	89	(227)	Neon	Ne	10	20.1797
Aluminum	Al	13	26.981539	Neptunium	Np	93	(237)
Americium	Am	95	(243)	Nickel	Ni	28	58.6934
Antimony	Sb	51	121.757	Niobium	Nb	41	92.90638
Argon	Ar	18	39.948	Nitrogen	N	7	14.00674
Arsenic	As	33	74.92159	Nobelium	No	102	(259)
Astatine	At	85	(210)	Osmium	Os	76	190.2
Barium	Ba	56	137.327	Oxygen	O	8	15.9994
Berkelium	Bk	97	(247)	Palladium	Pd	46	106.42
Beryllium	Be	4	9.012182	Phosphorus	P	15	30.973762
Bismuth	Bi	83	208.98037	Platinum	Pt	78	195.08
Boron	B	5	10.811	Plutonium	Pu	94	(244)
Bromine	Br	35	79.904	Polonium	Po	84	(209)
Cadmium	Cd	48	112.411	Potassium	K	19	39.0983
Calcium	Ca	20	40.078	Praseodymium	Pr	59	140.90765
Californium	Cf	98	(251)	Promethium	Pm	61	(145)
Carbon	C	6	12.011	Protactinium	Pa	91	(231)
Cerium	Ce	58	140.115	Radium	Ra	88	(226)
Cesium	Cs	55	132.90543	Radon	Rn	86	(222)
Chlorine	Cl	17	35.4527	Rhenium	Re	75	186.207
Chromium	Cr	24	51.9961	Rhodium	Rh	45	102.90550
Cobalt	Co	27	58.93320	Rubidium	Rb	37	85.4678
Copper	Cu	29	63.546	Ruthenium	Ru	44	101.07
Curium	Cm	96	(247)	Samarium	Sm	62	150.36
Dysprosium	Dy	66	162.50	Scandium	Sc	21	44.955910
Einsteinium	Es	99	(252)	Selenium	Se	34	78.96
Erbium	Er	68	167.26	Silicon	Si	14	28.0855
Europium	Eu	63	151.965	Silver	Ag	47	107.8682
Fermium	Fm	100	(257)	Sodium	Na	11	22.989768
Fluorine	F	9	18.9984032	Strontium	Sr	38	87.62
Francium	Fr	87	(223)	Sulfur	S	16	32.066
Gadolinium	Gd	64	157.25	Tantalum	Ta	73	180.9479
Gallium	Ga	31	69.723	Technetium	Tc	43	(98)
Germanium	Ge	32	72.61	Tellurium	Te	52	127.60
Gold	Au	79	196.96654	Terbium	Tb	65	158.92534
Hafnium	Hf	72	178.49	Thallium	Tl	81	204.3833
Helium	He	2	4.002602	Thorium	Th	90	232.0381
Holmium	Ho	67	164.93032	Thulium	Tm	69	168.93421
Hydrogen	H	1	1.00794	Tin	Sn	50	118.710
Indium	In	49	114.82	Titanium	Ti	22	47.88
Iodine	I	53	126.90447	Tungsten	W	74	183.85
Iridium	Ir	77	192.22	Unnilennium	Une	109	(267)
Iron	Fe	26	55.847	Unnilhexium	Unh	106	(263)
Krypton	Kr	36	83.80	Unniloctium	Uno	108	(265)
Lanthanum	La	57	138.9055	Unnilpentium	Unp	105	(262)
Lawrencium	Lr	103	(262)	Unnilquadium	Unq	104	(261)
Lead	Pb	82	207.2	Unnilseptium	Uns	107	(262)
Lithium	Li	3	6.941	Uranium	U	92	238.0289
Lutetium	Lu	71	174.967	Vanadium	V	23	50.9415
Magnesium	Mg	12	24.3050	Xenon	Xe	54	131.29
Manganese	Mn	25	54.93805	Ytterbium	Yb	70	173.04
Mendelevium	Md	101	(258)	Yttrium	Y	39	88.90585
Mercury	Hg	80	200.59	Zinc	Zn	30	65.39
Molybdenum	Mo	42	95.94	Zirconium	Zr	40	91.224
Neodymium	Nd	60	144.24				

The mass number of the longest-lived isotope of an element appears in parentheses.

Aqueous Environmental Geochemistry

Donald Langmuir

PRENTICE HALL
Upper Saddle River, New Jersey 07458

Library of Congress Cataloging-in-Publication Data

Langmuir, Donald.
    Aqueous environmental geochemistry / by Donald Langmuir.
        p.   cm.
    Includes index.
    ISBN 0-02-367412-1
    1. Water chemistry.   2. Environmental geochemistry.   I. Title.
    GB855.L36   1997
    551.48--dc21   96-37614
                                       CIP

Executive editor: Robert McConnin
Production: ETP Harrison
Copy editor: ETP Harrison
Cover director: Jayne Conte
Manufacturing manager: Trudy Pisciotti

© 1997 by Prentice-Hall, Inc.
Simon & Schuster / A Viacom Company
Upper Saddle River, New Jersey 07458

Printed in the United States of America

10 9 8 7 6 5 4 3 2 1

ISBN 0-02-367412-1

Prentice-Hall International (UK) Limited, *London*
Prentice-Hall of Australia Pty. Limited, *Sydney*
Prentice-Hall Canada Inc., *Toronto*
Prentice-Hall Hispanoamericana, S.A., *Mexico*
Prentice-Hall of India Private Limited, *New Delhi*
Prentice-Hall of Japan, Inc., *Tokyo*
Simon & Schuster Asia Pte Ltd., *Singapore*
Editora Prentice-Hall do Brasil, Ltda., *Rio de Janeiro*

Contents

Preface

A chief goal of this book is to help the reader understand controls on the chemical quality of surface- and subsurface-waters, both pristine and polluted. The focus is on inorganic processes and on the chemistry of soil and groundwaters, with less said about the chemistry of precipitation, surface-waters, or the ocean. The book leans heavily on the principles of chemical thermodynamics and the concept of chemical equilibrium. Chemical equilibrium, whether attainable or not, represents the reference state for purposes of explaining the concentrations of aqueous species in the hydrosphere. Concepts of chemical kinetics are introduced when they are known and seem applicable.

The book is intended for students who have taken at least one course dealing with chemical thermodynamics and solution chemistry, as well as an introductory course in physical geology that includes basic mineralogy. Professionals who have absorbed equivalent prerequisite knowledge should also benefit from the book. The full book contents were designed to comprise the background reading for two, three-credit courses. Many chapters have been written in such a way that their initial sections can serve as reading for the first course, with later chapter topics to be covered in the second course.

At the end of each chapter are study questions intended to examine a reader's understanding of important chapter concepts. These can generally be answered without performing calculations. If the book is used as a text, the study questions can be considered course objectives. In each chapter the study questions are followed by a collection of problems designed to illustrate the use and application of chapter materials. Detailed solutions to these problems, many of which require the use of geochemical computer models such as MINTEQA2 (Allison et al. 1991), PHREEQC (Parkhurst 1995),[†] or SOLMINEQ.88 (Kharaka et al. 1988), are available from the publisher in a companion volume to this text, titled *Worked Problems in Aqueous Environmental Geochemistry*. The worked problems should be especially useful for teaching purposes.

The book is an outgrowth of the author's 30 plus years of experience in basic research, teaching, and consulting in the aqueous geochemical-environmental arena. Examples and worked problems in the text and the worked problems book deal with subjects including acid mine drainage and mine tailings, oil-field brines, deep-well injection of wastes, toxic metal contamination, and radioactive-waste disposal. The book also contains extensive updated and critically evaluated thermodynamic data for numerous aqueous species and minerals, including those of the actinide elements.

[†]Instructions for obtaining the latest DOS versions of MINTEQA2 and PHREEQE, and of other U.S. Environmental Protection Agency and U.S. Geological Survey geochemical and hydrologic computer models are given in Obtaining Geochemical Software, at the end of the book. The model software may also be accessed through the author's web page at: http://www.igginc.com/iggi/langmuir/don.htm

The book is organized roughly into three parts. Chapters 1 and 2 on thermodynamics and kinetics introduce the laws and principles that govern chemical reactions. Chapters 3, 4, and 5 on complexation, activity coefficients, and acids and bases emphasize the occurrence and behavior of substances within aqueous solutions. Chapters 6 through 13 are chiefly devoted to controls on natural water chemistry in surface and subsurface systems. These controls involve heterogeneous reactions (reactions among gases, aqueous solutions, and/or solids). Chapters 6 through 12 consider carbonate systems, chemical weathering, clays, adsorption-desorption, redox reactions generally, and iron and sulfur geochemistry. The final chapter examines the geochemistry of uranium and other naturally occurring radioactive elements in water-rock systems and considers the radionuclides Tc, Np, Pu, and Am, which are of environmental concern in nuclear wastes.

A purpose of this book has been to introduce readers to the special language and assumptions of geochemical computer models so that they can confidently examine their output and know both what it means and what it does not mean (in spite of what it seems to be saying!). Recent years have seen a rapid growth in the development and availability of such models including PHREEQE (Parkhurst et al. 1990), PHRQPITZ (Plummer et al. 1988), PHREEQC (Parkhurst 1995), NETPATH (Plummer et al. 1991, 1994), SOLMINEQ.88 (Kharaka et al. 1988), EQ3/6 (Wolery et al. 1992a, 1992b), MINTEQA2 (Allison et al. 1991), WATEQ4F (Ball and Nordstrom 1991), MINEQL$^+$ (Schecher and McAvoy 1991), and The Geochemist's Workbench (Bethke 1994). These models base their analysis of aqueous systems on chemical equilibrium and, except for EQ3/6 and The Geochemist's Workbench, cannot consider reaction rates. The geochemical models have been reviewed and compared and their different capabilities discussed by Nordstrom et al. (1979), Mangold and Tsang (1991), van der Heijde and Elnawawy (1993), and Nordstrom and Munoz (1994). The attributes and capabilities of some geochemical models are also briefly summarized in Some Example Geochemical Computer Models at the end of the book.

Conclusions derived from the application of geochemical computer models are more and more frequently being offered as key evidence in environmental litigation. The use of geochemical models such as MINTEQA2 and PHREEQE has also become standard consulting and corporate practice in the United States as part of mine-site characterization and in the preparation of environmental impact statements for state and federal agencies. All the models are available in versions that can be run on a personal computer, although the largest, EQ3/6 and The Geochemist's Workbench, require a 486 or preferably a Pentium-based system. Several of these codes have simplified, "friendly" input files and so are readily accessible to users. Codes with an easy-to-use input file include SOLMINEQ.88, MINTEQA2, MINEQL$^+$, and The Geochemist's Workbench. A friendly input file for PHREEQC should become available in 1997.

Many individuals made valuable contibutions to the book. My thanks, to Richard B. Wanty, who wrote portions of Chapter 13, Actinides and their Daughter and Fission Products. Students in my classes at the Colorado School of Mines got to suffer through early versions of many of the problems and to ferret out errors in my solutions. Visiting scientist Shi Weijun patiently word-processed some of the worked problems. Ron Schmiermund tabulated and graphed much of the stability constant data in Chapter 1. I am particularly beholden to the reviewers, whose standards and expectations led to a much-improved final draft. Highly constructive and useful reviews were provided by Everett Jenne, who graciously reviewed the entire book, and by Ron Klusman, Don Rimstidt, Art Rose, Jim Conca, Janet Herman, Randy Bassett, David Parkhurst, Randy Arthur, and Cindy Palmer, who took on portions for review. I owe a sincere debt of gratitude to all of these generous colleagues.

Donald Langmuir

1

Thermochemical

Principles

1.1 SOME BASIC DEFINITIONS AND CONCEPTS

The influence of chemical equilibrium and/or kinetics on the progress of chemical reactions often determines the abundance, distribution, and fate of substances in the environment. An understanding of the basic concepts of chemical equilibrium and chemical kinetics, therefore, may help us to explain and predict the environmental concentrations of inorganic and organic species in aqueous systems, whether these species are present naturally or have been introduced by humans. In this chapter we will examine chemical equilibrium. The following chapter considers chemical kinetics or the study of rates of chemical reactions.

Given sufficient time, chemical substances in contact with each other tend to come to chemical equilibrium. Chemical equilibrium is the time-invariant, most stable state of a closed system (the state of minimum Gibbs free energy). We study chemical equilibrium concepts so as to learn the direction of spontaneous change of chemical reactions in any system, especially for conditions of constant temperature and pressure. We want to be able to compute the hypothetical equilibrium state of a system. We would like to predict the conditions for equilibrium in different systems and at different temperatures and pressures without having to measure them.

The statements above have introduced several concepts that need to be defined and expanded upon. A system is a grouping of atoms, minerals, rocks, and/or gases and waters under consideration within a single volume of space, the boundaries of which can be defined as is convenient. A system could be one mineral grain, a drop of rain, a water-logged soil, a well-mixed lake, or a regional groundwater/rock system tens of kilometers in diameter.

Among systems a closed system can exchange energy, but not matter, with its surroundings. An open system can exchange both energy and matter with its surroundings. A system may be closed in terms of some substances, but open in terms of others. For example, the ferric iron present in oxide minerals in an aerated soil will generally be immobile, in which case any volume of the soil is a closed system with respect to ferric iron. At the same time the soil is an open system with respect to gases such as CO_2 and O_2.

Whether a system can be considered open or closed depends not only on the specific substances under study, but also on both the rates of flux of matter in and out of the system and the time scale of interest. For slow rates of flux and/or short time scales, systems tend to be closed with respect to many substances. Given fast rates of flux and/or long time scales, systems will behave as if they were open with respect to many substances. Also, if reaction rates are much faster than flux rates of related components in and out of the system, we can assume the system is closed (and vice versa).

A phase is a restricted part of a system with distinct physical and chemical properties (Wood and Frazer 1976). A phase can also be defined as a physically and chemically homogeneous portion of a system with definite boundaries (Brownlow 1979). These attributes mean that a phase should be mechanically separable from a system. Example phases are minerals and well-mixed gases and liquids. Not true phases, because they comprise more than one mineral, are rocks such as granite or minerals such as the feldspars when they are chemically zoned and have spatially variable compositions.

Phases are made up of one or more components. Components are simple chemical units that can be combined to describe the chemical composition of the species or substances in a system. Our choice of components may depend upon the problem being addressed, although it is usual to identify the least number of components that will fully describe a system of interest.

If the system is a single mineral, there are often several ways to define the components that can be combined to form the mineral. For example, the clay mineral kaolinite [$Al_2Si_2O_5(OH)_4$] can be formed from the three oxides Al_2O_3, SiO_2, and H_2O as components, or from the four ion components Al^{3+}, Si^{4+}, H^+, and O^{2-}.

Selecting the least components (also called master species) is one of the fundamental and essential input decisions made in geochemical computer codes such as PHREEQE (Parkhurst et al. 1990), WATEQF (Ball and Nordstrom 1991), and MINTEQA2 (Allison et al. 1991), for example.

The Gibbs phase rule relates the number of components (C) and phases (P) that can exist in a system at equilibrium. The rule is

$$F = C - P + 2 \qquad\qquad (1.1)$$

where F is the number of independent variables or degrees of freedom (Stumm and Morgan 1981). In general $P \geq C$.

The simplest and perhaps most familiar phase rule example deals with the triple point of water, for which the equilibrium reaction is

$$H_2O(ice) = H_2O(liquid) = H_2O(vapor) \qquad\qquad (1.2)$$

At the triple point, $C = 1$ and $P = 3$, so that $F = 1 - 3 + 2 = 0$; thus, the triple point of water occurs at fixed temperature and pressure ($T = 0.01°C$, and $P = 0.006112$ bar).

If the temperature and pressure are fixed, as in aqueous systems in contact with the atmosphere at constant temperature, then two degrees of freedom are lost, and the phase rule becomes simply: $F = C - P$. Thus, for such a system to be invariant (have no degrees of freedom), the number of components will equal the number of phases.

Application of the phase rule to a system that includes a dissolving mineral, solution, and gas phase is a complex undertaking, which can be approached using a modified statement of the phase rule

$$F = C' + 2 - P - R \qquad\qquad (1.3)$$

in which $C = C' - R$. In this expression C' is the number of different chemical species in the system, and R equals the number of auxiliary restrictions. The latter includes the various chemical equilibria involving the species present, the equation of electroneutrality or charge balance, and stoichiometric relations (such as total analytical calcium equals the sum of all the dissolved calcium species) (cf. Reid 1990).

We will consider as an example of phase-rule calculation, calcite at saturation with an aqueous solution in equilibrium with gaseous CO_2. All species and substances in the system can be derived from a minimum of three components, which we will arbitrarily choose to be CaO, CO_2, and H_2O. (Other component choices would have been equally as valid; cf. Stumm and Morgan 1981.) The system has three phases, or $P = 3$. Ignoring complexes, nine different chemical species are present, or $C' = 9$. These include $CaCO_3$ (calcite), CO_2, H_2O, H^+, OH^-, Ca^{2+}, $H_2CO_3^\circ$, HCO_3^-, and CO_3^{2-}. There are six equations representing auxiliary restrictions ($R = 6$). These include a charge balance equation (every solution will have a charge balance or electroneutrality equation). In this case the charge-balance equation is

$$2mCa^{2+} + mH^+ = mHCO_3^- + 2mCO_3^{2-} + mOH^- \tag{1.4}$$

where m denotes molal concentrations that have units of mol/kg of solution. The five remaining auxiliary restrictions are equilibrium constant expressions for the five reactions

1) $CaCO_3(calcite) = Ca^{2+} + CO_3^{2-}$
2) $CO_2(g) + H_2O = H_2CO_3^\circ$
3) $H_2CO_3^\circ = H^+ + HCO_3^-$
4) $HCO_3^- = H^+ + CO_3^{2-}$
5) $H_2O = H^+ + OH^-$

The superscript $\circ$ in $H_2CO_3^\circ$ indicates that the species is in aqueous solution.

That $R = 6$ is also obvious from the fact that $R = C' - C = 9 - 3 = 6$. We can now compute $F = 9 + 2 - 3 - 6 = 2$. Two degrees of freedom indicate that at equilibrium, concentrations of the above species will be fixed at a given temperature and pressure.

Other important definitions include those of homogeneous, heterogeneous, and irreversible reactions. A homogeneous reaction occurs within a single phase, as in a stream; that is, $H^+ + HCO_3^- = H_2CO_3^\circ$. Heterogeneous reactions occur between phases, as gas-water, water-mineral, or, rarely, gas-mineral, as illustrated by the following reactions

$$[\text{gas-water}] \; CO_2(g) + H_2O(l) = H_2CO_3^\circ \tag{1.5}$$

$$[\text{water-mineral}] \; SiO_2(quartz) + 2H_2O(l) = H_4SiO_4^\circ \tag{1.6}$$

$$[\text{gas-mineral}] \; CaSO_4 \cdot 2H_2O(gypsum) = CaSO_4(anhydrite) + 2H_2O(g) \tag{1.7}$$

Irreversible reactions can go one way only. Equilibrium is not possible in general, or for existing conditions of temperature and pressure. The status of such reactions can often be described using the concepts of chemical kinetics. Radioactive decay is generally such an irreversible reaction. For example

$$^{226}Ra_{88} \rightarrow {}^{222}Rn_{86} + \alpha^- \tag{1.8}$$

In this expression subscripts are each element's atomic number or number of protons, superscripts are each element's atomic weight, or its sum of protons and neutrons, and α^- is an alpha particle or helium nucleus ($^4He_2$).

Oxidation of hydrogen sulfide at 25°C is also irreversible in the absence of sulfate-reducing bacteria.

$$H_2S(aq) + 2O_2(aq) \rightarrow 2H^+ + SO_4^{2-} \tag{1.9}$$

where (aq) indicates that the H_2S and oxygen are dissolved in the aqueous solution. Finally, many weathering reactions are irreversible, especially those involving the breakdown of high-temperature igneous minerals, such as olivine.

$$Mg_2SiO_4(\text{olivine}) + 4H^+ \rightarrow 2Mg^{2+} + SiO_2(\text{quartz}) + 2H_2O(l) \tag{1.10}$$

Another useful concept is that of congruent and incongruent reactions. These terms describe reactions involving the dissolution of minerals. If all the products of a dissolution reaction are soluble, the reaction is called congruent, as in the case of the quartz dissolution reaction (1.6) described above. Because, as written, the olivine weathering reaction leads to quartz precipitation; it is an incongruent reaction.

1.2 ENTHALPY AND ENTROPY

We need to understand the concepts of Gibbs free energy (G) and chemical potential (μ) in order to know the direction of spontaneous change of a reaction or system. These concepts can also be used to define or predict the most stable (equilibrium) assemblage and gas, fluid, or rock compositions expected in a system at a given pressure and temperature. Some phases and aqueous species in a system may be out of equilibrium with that system. Free-energy calculations permit us to decide which substances are out of equilibrium, and, therefore, which concentrations may be governed by chemical kinetics.

The concepts of enthalpy (H) and entropy (S) are basic to the definitions of Gibbs free energy, chemical potential, and the equilibrium constant (K_{eq}), and allow us to predict the effect of temperature on G and K_{eq}. We will also see that entropy is a measure of the state of order-disorder in solids and solid solutions, and helps explain the stabilities of aqueous complexes.

In sections to follow we make no attempt to present a rigorous, detailed development of thermodynamic principles. Such a rigorous approach can be found in most textbooks on physical chemistry and thermodynamics (cf. Lewis and Randall 1961; Reid 1990; Anderson and Crerar 1993; Nordstrom and Munoz 1994). We will limit ourselves to principles essential to an understanding of equilibrium concepts.

Enthalpy (H) is defined as the heat content of a substance at constant pressure. We cannot know absolute values of H, only differences in the enthalpies of substances. The enthalpy of formation, ΔH_f° of a substance at 25°C (298.15 K) and 1 bar pressure is its heat of formation from the elements in their most stable forms at that temperature and pressure. Here, and generally, the superscript degree symbol to the right of the state function denotes that the function is for 1 bar pressure. ΔH_f° for the elements in their most stable forms is taken as zero by definition at any temperature and 1 bar pressure. For example, $\Delta H_f^\circ = 0$ for rhombic sulfur, the most stable form of sulfur at 1 bar and 25°C. Monoclinic sulfur, with $\Delta H_f^\circ = 0.071$ kcal/mol at 1 bar and 25°C is unstable relative to the rhombic form.

The enthalpy of a reaction, ΔH_r°, is the heat transfer between a system and its surroundings for a process at constant pressure, but not at constant temperature and volume (V). For example, consider the formation of liquid water from gaseous hydrogen and oxygen at 25°C, which, with respective volumes and ΔH_f° values given beneath it is written

$$H_2(g) + \tfrac{1}{2} O_2(g) = H_2O(l) \tag{1.11}$$

$$
\begin{array}{ccc}
24.5\ \text{L} & 12.3\ \text{L} & 18\ \text{ml (0.018 L)} \\
0 & 0 & -68.315\ \text{kcal/mol}
\end{array}
$$

The large volume change for the reaction $\Delta V_r^\circ = 0.018 - 24 - 12.3 = -36.8$ L means that the reaction is explosive. In this case $\Delta H_r^\circ = \Delta H_f^\circ$ $(H_2O)(l) = -68.315 - 0 - 0 = -68.315$ kcal/mol. The negative sign of ΔH_r° indicates that heat is given off and the reaction is exothermic.

For the dissolution of potassium nitrate in water, with respective ΔH_f° (kcal/mol) values given below, we may write

$$KNO_3(c) \;\; = \;\; K^+ + NO_3^- \tag{1.12}$$
$$\quad -118.22 \quad -60.27 \quad -49.0$$

from which $\Delta H_r^\circ = +8.95$ kcal/mol. The reaction is endothermic, and the solution cools.

How would these reactions be affected by increases in temperature or pressure? The application of Le Chatelier's principle helps us answer this question. As restated by Glasstone (1946) Le Chatelier's principle is the following:

> If a change occurs in one of the factors, such as temperature or pressure, (or a species concentration) under which a system (or reaction) is in equilibrium, the system (or reaction) will tend to adjust itself so as to annul, as far as possible, the effect of that change.

Thus, because the reaction of hydrogen and oxygen to form water produces heat as a product, an increase in temperature that corresponds to an increase in system heat will favor the reverse reaction or the dissociation of water. Similarly, because the dissolution of potassium nitrate consumes heat (is endothermic, as is the dissolution of most minerals), it is favored by an increase in temperature. Further, Le Chatelier's principle tells us that because of the large reduction in volume ($\Delta V_r^\circ = -36.8$ L) that accompanies the formation of liquid water from gaseous hydrogen and oxygen, an increase in pressure favors the forward reaction.

All calculations involving energy here and in the following text are written in terms of calories and kilocalories. The current preference of many scientists (but not this one) is instead to use joules (J) and kilojoules (kJ). Dividing joules or kilojoules by exactly 4.184 yields calories and kilocalories, respectively.

In any reversible process, the change in entropy (ΔS) of any system, or part of it, equals the heat it absorbs (Q), divided by the absolute temperature: $\Delta S = Q/T$. In terms of usually tabulated entropy values for solids, liquids, or gases, at 1 bar pressure, $S_{298}^\circ = $ (total heat absorbed)$/T$(K), from $T = 0$ (K) to 298.15 K. In general, the value of S_T includes the heats involved in all state or phase transformations, first order (solid to liquid to gas), or higher order (changes in polymorphic form of solids, etc.), from absolute zero to the temperature and pressure conditions of interest. This is also called the absolute or third-law entropy and is determined calorimetrically for solids, liquids, or gases by measuring the heat capacity of the substance at constant pressure (Cp) from 0 (K) to temperature, T. At any temperature, Cp is the number of calories needed to raise the temperature of one mole of the substance by 1 (K). In differential terms: $dS = (Cp/T)\, dT$. Partial integration leads to

$$S_{T(2)} - S_{T(1)} = \int_{T(1)}^{T(2)} \frac{Cp}{T}\, dT \tag{1.13}$$

If $T(1) = 0$ (K), and $T(2) = T$, then we can write

$$S_T = S_0 + \int_0^T \frac{Cp}{T}\, dT \tag{1.14}$$

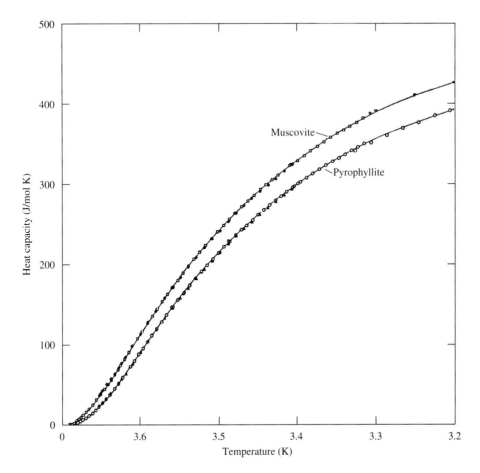

Figure 1.1 Low temperature heat capacities of muscovite [KAl$_2$(AlSi$_3$O$_{10}$)(OH)$_2$] and pyrophyllite [Al$_2$Si$_4$O$_{10}$(OH)$_2$]. The heat capacity in joules per mole degree (J/mol K) divided by 4.184 equals the heat capacity in calories per mole degree (cal/mol K). From Robie et al. (1976).

The third-law entropy of substances usually equals zero at $T(K) = 0$. A plot of the measured heat capacity of the minerals muscovite and pyrophyllite is shown in Fig. 1.1. To determine ΔS_T° for either mineral, we integrate the above function below the empirical curve from 0 to $T(K)$. (See Wood and Frazer 1976; Nordstrom and Munoz 1994.)

The entropy of dissolved species is approached quite differently. We cannot know (and do not need to know) their absolute entropies. Instead, we consider the differences in S (or ΔS) between an aqueous species and H$^+$ ion, assuming S_T for the H$^+$(aq) $= 0$. Treatment of the thermodynamic properties of aqueous species is addressed in more detail later.

Entropy is a measure of the degree of randomness or disorder of a phase (thermal, statistical, or the geometric arrangement of the atoms, for example). The most order exists at 0 (K), where $S = 0$ for most substances. With changes of state from solid to liquid to gas, the entropy increases discontinuously with increasing disorder of the phase. Thus, for H$_2$O at 25°C

$$S^\circ \text{ (ice } I\text{) (metastable)} < S^\circ \text{ (water)} < S^\circ \text{ (steam)} \qquad (1.15)$$

$$10.68 \qquad\qquad < 16.72 \qquad < 45.11 \text{ cal/mol K}$$

(For further useful entropy references see Krauskopf 1967; Lewis and Randall 1961; Wood and Fraser 1976; Stumm and Morgan 1981; Henderson 1982; Mason and Moore 1982.)

1.3 GIBBS FREE ENERGY, CHEMICAL POTENTIAL, AND THE EQUILIBRIUM CONSTANT

The Gibbs free energy (G_i) of a substance i is defined as $G_i = H_i - TS_i$. Free energy cannot be measured directly. As with enthalpy, we consider the difference (ΔG) between the free energies of substances in a reaction. The ΔG for a reaction is the maximum energy change for that reaction as useful work, measured at constant temperature and pressure. For a reaction we can write

$$\Delta G_r^\circ = \Delta H_r^\circ - T\Delta S_r^\circ \tag{1.16}$$

This is a most important, fundamental relationship.

Next we will show how ΔG_r° is related to the equilibrium constant (K_{eq}) for a reaction. Consider the general reaction

$$aA + bB = cC + dD \tag{1.17}$$

where the lowercase letters denote the number of moles of reactants A and B and products C and D. The Gibbs free energy of a mole of A at some pressure and temperature is G_A. This is also the definition of the chemical potential of A, or $G_A = \mu_A$. We can also write

$$G_A = \Delta G_A^\circ + RT \ln[A] \tag{1.18}$$

where T is in degrees K, and ΔG_A° is the standard molar Gibbs free energy of A. ΔG_A° is the Gibbs free energy of A at unit activity of A (when $[A] = 1$). The activity of A can be roughly thought of as the fraction of its total concentration that participates in reactions. For ions the activity of A is usually less than its concentration. (See Chap. 3).

For a moles of A, $aG_A = a\Delta G_A^\circ + RT \ln[A]^a$. Now ΔG_r for the general reaction equals the difference in the sum of values for the products minus that for the reactants

$$\Delta G_r = cG_C + dG_D - aG_A - bG_B \tag{1.19}$$

Introducing expressions for the other reacting substances similar to that for A, we obtain

$$\Delta G_r = [c\Delta G_C^\circ + d\Delta G_D^\circ - a\Delta G_A^\circ - b\Delta G_B^\circ] + RT \ln[C]^c + RT \ln[D]^d$$
$$- RT \ln[A]^a - RT \ln[B]^b \tag{1.20}$$

Collecting and combining terms

$$\Delta G_r = \Delta G_r^\circ + RT \ln \frac{[C]^c[D]^d}{[A]^a[B]^b} \tag{1.21}$$

where ΔG_r°, which equals the bracketed difference in terms in the previous equation, is the standard Gibbs free energy of the reaction. The ratio of concentrations on the right may be called the reaction quotient, Q. At equilibrium $\Delta G_r = 0$ and $Q = K_{eq}$ or

$$\Delta G_r^\circ = -RT \ln \frac{[C]^c[D]^d}{[A]^a[B]^b} \tag{1.22}$$

or simply

$$\Delta G_r^\circ = -RT \ln K_{eq} \tag{1.23}$$

In this important expression, R, the gas constant equals 1.9872 cal/mol K. Most free-energy data is published for 25°C (298.15 K), and given that $\ln K = 2.3026 \log K$, we can show for 25°C and 1 bar pressure:

$$\Delta G_r^\circ \text{(kcal/mol K)} = -1.3642 \log K_{eq} = -0.59248 \ln K_{eq} \tag{1.24}$$

We have defined the standard-state Gibbs free energy of a reaction in terms of the standard-state Gibbs free energies of reactants and products. However, as was true of enthalpy, absolute free energies cannot be measured. In any case, we are interested only in ΔG_r°, which is the difference in free energies of products and reactants. Therefore, we introduce the concept of the standard-state Gibbs free energy of formation from the elements, ΔG_f°. Consistent with the definition of ΔH_f°, ΔG_f° is the free energy of formation of a substance from its elements in their most stable forms at the temperature and pressure of interest. The value of ΔG_r° is then computed using tabulated ΔG_f° values for products and reactants, which are most often reported for 1 bar pressure and 25°C. Thus, $\Delta G_r^\circ = c\Delta G_f^\circ(C) + d\Delta G_f^\circ(D) - a\Delta G_f^\circ(A) - b\Delta G_f^\circ(B)$. As with ΔH_f°, ΔG_f° for aqueous species is based on the assumption that ΔG_f° for $H^+(aq) = 0$.

Our earlier general expression for the Gibbs free energy of a reaction can be restated as

$$\Delta G_r = \Delta G_r^\circ + RT \ln[Q] \tag{1.25}$$

where Q is the reaction quotient. Because $\Delta G_r^\circ = -RT \ln[K_{eq}]$, we find, in general, that

$$\Delta G_r = RT \ln \frac{Q}{K_{eq}} \tag{1.26}$$

At equilibrium, of course, $Q = K_{eq}$ and $G_r = 0$. When $Q < K_{eq}$ and $\Delta G_r < 0$, the reaction tends to go from left to right. Conversely, when $Q > K_{eq}$ and $\Delta G_r > 0$, the reaction tends to go from right to left.

Sometimes the state of a reaction relative to equilibrium is described in terms of reaction affinity (see Stumm and Morgan 1981). The affinity, A, simply equals $-\Delta G_r$.

In studies of the state of saturation of minerals in natural waters and in most of the geochemical computer codes, the saturation index (SI) is used. The index is defined as $SI = \log_{10}(Q/K_{eq})$, so that $SI = 0$ at equilibrium (at saturation) of the mineral with the solution. The saturation index and ΔG_r are related through $SI = \Delta G_r/(2.3026\,RT)$. If the reaction is written with the mineral as the reactant, then when SI and ΔG_r are both negative, the mineral is undersaturated and so will tend to dissolve. When both are positive, the mineral is supersaturated and will tend to precipitate from solution.

The sign of ΔG_r indicates which direction a reaction will spontaneously go, assuming the reaction is not rate-limited, with negative values favoring the forward reaction and vice versa. A common error made is to assume that the sign of ΔG_r° has the same significance as the sign of ΔG_r. Remember, $\Delta G_r = \Delta G_r^\circ$ only when all reactants and products have activities equal to one. This will only be true generally, when all reactants and products are pure phases and, therefore, does not apply to reactions that involve gas mixtures or dissolved species.

Whether or not kinetics limits reaction progress is sometimes (but not always) related to the magnitude of ΔG_r. Reactions with Gibbs free energies of tens of kilocalories are more likely to go spontaneously than are those with free energies of a few hundred calories or less.

The chemical potential concept provides a useful way to think about the tendency for spontaneous chemical change in complex environmental systems involving gases, liquids, and solids (cf. Wood and Fraser 1976; Stumm and Morgan 1981). In a particular phase, the chemical potential, μ_i of component i is related to the activity of i through the expression

$$\mu_i = \mu_i^\circ + RT \ln a_i \tag{1.27}$$

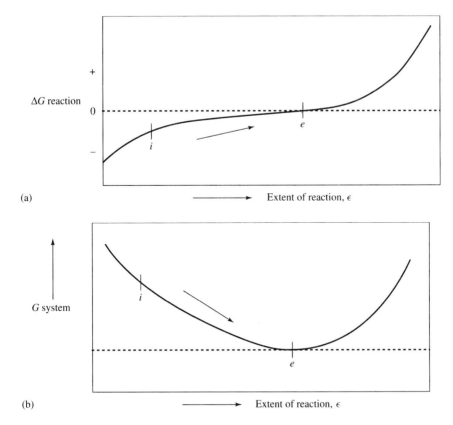

Figure 1.2 **(a)** Schematic plot of the change in Gibbs free energy of an individual reaction ΔG_r, as that reaction proceeds from a nonzero free-energy value at i, toward equilibrium, and $\Delta G_r = 0$ at point e. **(b)** Schematic plot showing the Gibbs free energy of a system as a function of the extent of the reaction described in (a). The system free energy will tend from a more positive value at some point i, to a minimum value e, as the ΔG_r for the individual reaction goes to zero.

where μ_i° is the chemical potential of i in some standard state. The chemical potential of component i will be equal in all phases in a system at equilibrium. Otherwise, components will tend to flow or diffuse from phases of high μ_i to those of lower μ_i.

Figure 1.2(a) shows schematically how ΔG_r might change as a function of reaction progress or extent of reaction ϵ (cf. Stumm and Morgan 1981). The reaction might start with a negative ΔG_r at point i, moving toward $\Delta G_r = 0$ at equilibrium (point e). At the same time, the Gibbs free energy, G, of the system, which includes all the reactants and products and in which the reaction is taking place, might move from point i toward a minimum value as shown in Fig. 1.2(b). In fact some computer models of complex solution equilibria use an approach called free-energy minimization, involving all reactions in the system to calculate the system's hypothetical equilibrium state (cf. Westall et al. 1976). This condition corresponds to the minimum value of G, where $G = \Sigma\mu_i n_i$, the sum of the chemical potentials of all components i in the system times the number of moles of each present (n_i).

We will close this discussion with a brief reminder of the conventions used in the reporting of most available entropy, heat capacity, enthalpy, and free-energy data. First, the heat capacity and entropy data for gases, liquids, and solids are absolute values and are always positive. Tabulated enthalpy (ΔH_f°) and Gibbs free-energy (ΔG_f°) values for gases, liquids, and solids are values for their formation from the most stable forms of their component elements at the same T and P (usually 25°C and 1 bar pressure). The ΔH_f° and ΔG_f° values for the most stable forms of the elements are then assumed equal to zero by convention. The thermodynamic properties of gases, liquids, and solids are for 1 mole (gram formula weight) of the substance. At 1 bar pressure, a mole of a gas occupies 22.41 L at 0°C and 24.46 L at 25°C.

The thermodynamic properties of aqueous species are for a one-molal concentration of the species (1 mol/kg of solvent water), which, for relatively insoluble species, may be strictly hypothetical. The thermodynamic properties of dissolved ionic species are based on the assumption that the heat capacity, entropy, ΔH_f° and ΔG_f° of the hydrogen ion [$H^+(aq)$], all equal zero at all temperatures and pressures; in other words, it is assumed that $\Delta G_f^\circ = \Delta H_f^\circ = S^\circ = 0$ for the hydrogen ion and that $\Delta G_r^\circ = \Delta H_r^\circ = \Delta S_r^\circ = 0$ for the reaction

$$\tfrac{1}{2}H_2(g) = H^+ + e^- \tag{1.28}$$

This also means, of course, that $\Delta G_f^\circ = \Delta H_f^\circ = 0$ for the electron. However, because the third law of entropy of H_2 gas equals 31.207 cal/mol K, we are left with the curious result that S° for the electron must be assumed equal to 15.604 cal/mol K. Because the entropy of the proton is arbitrarily set equal to zero, the entropies of other ions can be either positive or negative in sign.

1.4 EQUILIBRIUM CALCULATIONS

1.4.1 Pure Solids and Liquids and Their Mixtures

Henry's, Raoult's, and Nernst's laws.   If always present during a reaction (always in excess), the activities of pure solids and liquids may be assumed equal to unity, or $a_i = 1$. For solid or liquid mixtures we can define ideal solutions for which $a_i = N_i$, the mole fraction of i in the mixture. In a binary solution, for example, the mole fraction of component 1 in a solution with component 2 is given by

$$N_1 = \frac{n_1}{n_1 + n_2} \tag{1.29}$$

where n_1 and n_2 are the number of moles of the respective components in a given volume of their solution. Ideal solutions are said to obey Raoult's law. The activities of minor components or solutes in dilute solid or liquid solutions can generally be described using Henry's law. Thus, as N_i goes to zero $a_i = K_H N_i$, where K_H is the Henry's law constant. In general Raoult's law is obeyed by the major component, while Henry's law is obeyed by minor components in the same solution. Figure 1.3 shows how Raoult's and Henry's laws apply to an hypothetical binary solid solution. In the plot as drawn, $K_H < 1$ for the minor component. This is called a negative deviation from ideality. Such behavior is exhibited by many sulfide mineral solid solutions and other solid solutions that show complete miscibility (complete solid solution between endmembers) for the conditions of P and T of interest. When $K_H > 1$, it is called a positive deviation from ideality and corresponds to a miscibility (solubility) gap, or incomplete solid solution, between the two end members. This behavior is found for many solid solutions involving oxide, carbonate, or silicate minerals, in which the substituent cations are appreciably different in size and/or different in charge.

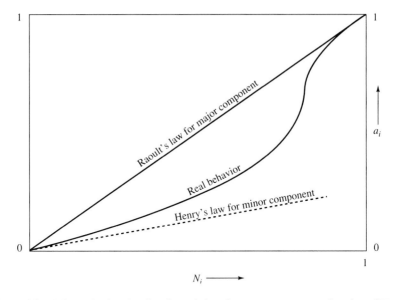

Figure 1.3   Schematic plot showing the activity of component i, a_i, as a function of N_i, the mole fraction of i in a solid solution. As a minor component i obeys Henry's law, whereas as a major component it obeys Raoult's law. The line describing Henry's law behavior is tangent to the curve that describes the solution's real behavior as N_i approaches zero.

Henry's law also applies to trace constituents or components in two coexisting solutions at equilibrium, for example, to two minerals in a rock or to a mineral in an aqueous solution. In general, the chemical potential of component i, μ_i must be the same in two phases (solutions) at equilibrium. For example, in a fluid phase such as groundwater (phase x), we may write

$$\mu_i = \mu_{ix}^{\circ} + RT \ln a_{ix} \tag{1.30}$$

In a coexisting mineral (phase y) at equilibrium the corresponding expression is

$$\mu_i = \mu_{iy}^{\circ} + RT \ln a_{iy} \tag{1.31}$$

But Henry's law applies to component i in each phase, or, $a_{ix} = K_{ix}N_{ix}$, and $a_{iy} = K_{iy}N_{iy}$. Because μ_i is the same in both phases, we can set the chemical potential expressions equal, and substituting for a_{ix} and a_{iy} obtain

$$\ln \left[\frac{K_{iy}N_{iy}}{K_{ix}N_{ix}} \right] = \frac{(\mu_{ix}^{\circ} - \mu_{iy}^{\circ})}{RT} \tag{1.32}$$

or

$$\frac{N_{iy}}{N_{ix}} = \frac{K_{ix}}{K_{iy}} \exp\left(\frac{\mu_{ix}^{\circ} - \mu_{iy}^{\circ}}{RT} \right) \tag{1.33}$$

At a particular temperature and pressure the exponential term is a constant, and the mole fraction ratio in the two coexisting phases will equal a constant, or

$$\frac{N_{iy}}{N_{ix}} = \frac{K_{ix}'}{K_{iy}'} = K \tag{1.34}$$

In other words, the mole fraction ratio of i in the coexisting phases at equilibrium for a given T and P should be constant. This is Nernst's law (cf. Lewis and Randall 1961). K is also called the distribution coefficient, often symbolized by D, and is used in the study of trace element partitioning between coexisting mineral solid solutions.

The distribution coefficient, K_d, which for a substance is the ratio of its amount sorbed on the surface of a solid to its aqueous concentration, is also an expression of Nernst's law. (See Chap. 10.)

In general for solid solutions we can write

$$a_i = \lambda_i N_i \tag{1.35}$$

where λ_i is called the rational activity coefficient of component i. When $N_i \to 1$ (Raoult's law) then $\lambda_i = 1$. Also $\lambda_i = K_H$ as $N_i \to 0$.

An ideal solid solution is one in which all the components, both major and minor, are assumed to obey Raoult's law. By definition, an ideal solid solution has the same enthalpy and volume (per mole of mixture) as the sum of the enthalpies and volumes of its components. In other words, $\Delta H_{\mathrm{mix}} = \Delta V_{\mathrm{mix}} = 0$ for an ideal solid solution. The entropy of mixing of components 1 and 2 (etc.) at one bar is given by

$$\Delta S^\circ_{\mathrm{mix}} = -R(N_1 \ln N_1 + N_2 \ln N_2 \cdots) \tag{1.36}$$

Because $\Delta G^\circ_{\mathrm{mix}} = \Delta H^\circ_{\mathrm{mix}} - T\Delta S^\circ_{\mathrm{mix}}$, and the enthalpy of mixing is zero

$$\Delta G^\circ_{\mathrm{mix}} = RT(N_1 \ln N_1 + N_2 \ln N_2 \cdots) \tag{1.37}$$

For the molar Gibbs free energy of the solid solution (G_{ss}) this mixing term is added to the contribution from mechanical mixing of the components to form the solid solution, or

$$G_{ss} = \underbrace{N_1 G_1^\circ + N_2 G_2^\circ + \cdots}_{\text{mechanical mixing}} + \underbrace{\Delta G_{\mathrm{mix}}}_{\text{ideal mixing}} \tag{1.38}$$

Regular solid solutions.   Solid solutions may behave ideally if the ions involved have the same charge and similar size and experience similar intermolecular forces. If such is not the case, but substituent ions are still mixed nearly randomly in the solid, their thermodynamic behavior and that of the solid solution may be described using regular solution concepts. Hildebrand proposed regular solutions in 1929 (cf. Hildebrand et al. 1970; Darken and Gurrey 1953). Since mixing is nearly random in a regular solution, the entropy of mixing is assumed equal to that of an ideal solution. However, both the molar volume and enthalpy of mixing in a regular solution are nonzero.

If both ions in a binary regular solution are of the same charge and of similar size and electronic configuration, they may have roughly equal solubilities at low mole fractions, but not show complete miscibility in a solid solution. For such conditions their solubilities may obey a *symmetrical regular solution*; the Gibbs free energy is given by

$$\Delta G^\circ_{ss} = N_A G_A^\circ + N_B G_B^\circ + \Delta G^\circ_{\mathrm{mix}} \text{ (ideal)} + \omega N_A N_B \tag{1.39}$$

The first two terms describe mechanical mixing of endmembers A and B, the third is the ideal solution mixing term, and the last term is the regular solution contribution, in which ω is a constant, presumably independent of temperature, pressure, or solid solution composition. (If ω is not so independent, the solution is not regular). The term $\omega N_A N_B$ in Eq. (1.39) is also sometimes called the excess Gibbs free energy of mixing, or $\Delta G^\circ_{\mathrm{mix}}$ (excess).

Ideal and excess mixing terms may together be described as the mixing terms, and set equal to ΔG°_{mix}. But because we can write

$$\Delta G^{\circ}_{mix} = \Delta H^{\circ}_{mix} - T\Delta S^{\circ}_{mix} \tag{1.40}$$

and

$$\Delta G^{\circ}_{mix} = RT(N_A \ln N_A + N_B \ln N_B) + \omega N_A N_B \tag{1.41}$$

$$\Delta S^{\circ}_{mix} = -R(N_A \ln N_A + N_B \ln N_B) \tag{1.42}$$

it is easy to show

$$\Delta H^{\circ}_{mix} = \omega N_A N_B \tag{1.43}$$

The enthalpy of formation of the overall solid solution thus equals

$$H^{\circ}_{ss} = N_A H^{\circ}_A + N_B H^{\circ}_B + \omega N_A N_B \tag{1.44}$$

where H°_A and H°_B are the enthalpies of formation of A and B from the elements at temperature and 1 bar pressure. The first two terms in Eq. (1.44) give the contribution of mechanical mixing to the enthalpy of the solid solution.

The chemical potentials of each component in the solid solution may be written

$$\mu^s_A = \mu^{\circ s}_A + RT \ln N_A + \omega N_B^2 \tag{1.45}$$

$$\mu^s_B = \mu^{\circ s}_B + RT \ln N_B + \omega N_A^2 \tag{1.46}$$

Because the ωN_B^2 term must equal $RT \ln \lambda_A$, and $\omega N_A^2 = RT \ln \lambda_B$, we find

$$\lambda_A = \exp(\omega N_B^2/RT) \tag{1.47}$$

$$\lambda_B = \exp(\omega N_A^2/RT) \tag{1.48}$$

In an aqueous solution at equilibrium with a solid solution the chemical potentials of components must be equal, thus we can write

$$\mu^a_A = \mu^{\circ a}_A + RT \ln \gamma_A m_A = \mu^{\circ s}_A + RT \ln N_A + \omega N_B^2 \tag{1.49}$$

$$\mu^a_B = \mu^{\circ a}_B + RT \ln \gamma_B m_B = \mu^{\circ s}_B + RT \ln N_B + \omega N_A^2 \tag{1.50}$$

in which $\mu^{\circ a}_A$ and $\mu^{\circ a}_B$ are the chemical potentials of hypothetical one-molal solutions of the pure components. Subtracting the second expression from the first gives

$$(\mu^{\circ a}_A - \mu^{\circ a}_B) + RT \ln \frac{\gamma_A m_A}{\gamma_B m_B} = (\mu^{\circ s}_A - \mu^{\circ s}_B) + RT \ln \left(\frac{N_A}{N_B}\right) + \omega(N_B^2 - N_A^2) \tag{1.51}$$

Rearrangement leads to

$$(\mu^{\circ a}_A - \mu^{\circ a}_B) - (\mu^{\circ s}_A - \mu^{\circ s}_B) = RT \ln \left(\frac{\gamma_B m_B}{\gamma_A m_A}\right)\left(\frac{N_A}{N_B}\right) + \omega(N_B^2 - N_A^2) \tag{1.52}$$

The chemical potential difference on the left is seen to equal ΔG°_{ex}, the free energy of the replacement reaction

$$B + AX = A + BX \tag{1.53}$$

where A and B are the dissolved components and AX and BX the components in the solid solution. Thus

$$\Delta G^{\circ}_{ex} = -RT \ln \left(\frac{\gamma_A m_A}{\gamma_B m_B}\right)\left(\frac{N_B}{N_A}\right) + \omega(N_B^2 - N_A^2) \tag{1.54}$$

Now $-\Delta G^{\circ}_{ex}/RT = \ln K_{ex}$, and $K_{ex} = K_{AX}/K_{BX}$, where K_{AX} and K_{BX} are known solubility products for the reactions $AX = A + B$, and $BX = B + X$. The value of ω can be determined from a water analysis and coexisting solid-solution analysis at equilibrium. Of course more than one such pair of analyses are needed to prove that the system obeys a regular solution.

At a given temperature the quotient ω/RT (unitless) is a constant, often symbolized as a_o. Thus, based on Eqs. (1.47) and (1.48) we have

$$\ln \lambda_A = \frac{\omega}{RT} N_B^2 = a_o N_B^2 \tag{1.55}$$

$$\ln \lambda_B = \frac{\omega}{RT} N_A^2 = a_o N_A^2 \tag{1.56}$$

Glynn (1990) lists values of a_o that range from <2 to 6 between 25 and 150°C for a large number of binary metal sulfate and carbonate solid solutions. In a broader survey of regular solutions, Garrels and Christ (1965) reported that a_o ranged from 0 to ± 11.5 "for most systems of geological interest." It is important to remember that $a_o = 0$ for ideal solutions and that a_o values decrease with increasing temperature, which favors more complete solid solution among components.

As suggested by Eqs. (1.53) and (1.54) a binary solid solution can be treated as an exchange reaction. Accordingly, the K_{ex} expression for reaction (1.53) is written

$$K_{ex} = \frac{K_{AX}}{K_{BX}} = \left(\frac{\gamma_A m_A}{\gamma_B m_B}\right)\left(\frac{\lambda_B N_B}{\lambda_A N_A}\right) \tag{1.57a}$$

If applicable, the regular solution model provides values for λ_A and λ_B. The following example calculation involving regular solutions is based on Appelo and Postma (1993).

Example 1.1

Cadmium in groundwater tends to be rapidly sorbed by any calcite present. The resultant solid then behaves like a $(Ca,Cd)CO_3$ symmetrical regular solid solution with an interaction parameter of $a_o = -0.8$. Assume the calcite in a freshwater aquifer contains 1.0 mole percent $CdCO_3$ and that $mCa^{2+} = 3 \times 10^{-3}$ M in the groundwater. Given also the solubility products $K_{Ca}(\text{calcite}) = 10^{-8.48}$ and $K_{Cd}(\text{otavite}, CdCO_3) = 10^{-11.31}$:

a) Calculate the activity coefficients of $CaCO_3$ and $CdCO_3$ in the solid solution.
b) Assuming stoichiometric (congruent) solubility of the solid solution, what is mCd^{2+} in the groundwater at equilibrium with the solid solution?
c) Compare this result to the solubility of pure otavite at pH = 7.0 and $[HCO_3^-] = 10^{-3.00}$ and comment.

First, write formation of the solid solution as an exchange reaction.

$$Ca^{2+} + CdCO_3 = Cd^{2+} + CaCO_3$$

The corresponding K_{ex} expression is

$$K_{ex} = \frac{K_{Cd}}{K_{Ca}} = \left(\frac{\gamma_{Cd} \, mCd}{\gamma_{Ca} \, mCa}\right)\left(\frac{\lambda_{Ca} N_{Ca}}{\lambda_{Cd} N_{Cd}}\right) \tag{1.57b}$$

To a good approximation $\gamma_{Ca} = \gamma_{Cd}$, and the values cancel out. According to the regular solution model $\lambda_{Ca} = \exp(a_o N_{Cd}^2)$ and $\lambda_{Cd} = \exp(a_o N_{Ca}^2)$. With $a_o = -0.8$, $N_{Ca} = 0.99$, and $N_{Cd} = 0.01$, we obtain $\lambda_{Cd} = 0.46$ and $\lambda_{Ca} = 1.00$. Substituting these values, K_{Ca}, K_{Cd} and $mCa^{2+} = 3 \times 10^{-3}$ into the K_{ex} expression and solving, we obtain $mCd^{2+} = 2.06 \times 10^{-8}$, a very small concentration.

To determine the solubility of otavite at pH = 7.00 and for $[HCO_3^-] = 10^{-3.00}$ we first compute the activity of carbonate ion via the expression for K_2(carbonic acid). Thus $[CO_3^{2-}] =$

$K_2 \times [HCO_3]/[H^+] = 10^{-10.33} \times 10^{-3.00}/10^{-7.00} = 10^{-6.33}$. Now $[Cd^{2+}] = K_{Cd}/[CO_3^{2-}] = 10^{-11.31}/10^{-6.33}$ $= 1.05 \times 10^{-5}$. The solid solution limits dissolved cadmium concentrations to a level 500 times less than the solubility of otavite.

This example leads to an important conclusion. Namely, major-minor species solid solutions are often capable of limiting concentrations of the minor species to levels well below saturation with respect to pure minerals of minor species. Solid solutions can thus provide a useful sink for removal of contaminant trace species. In complex natural systems the adsorption of trace species also limits their concentrations to very low levels. In such systems it is often impossible to decide whether the concentration of a trace species is limited by formation of a particular solid solution or adsorption or by both processes.

1.4.2 Gases

The following discussion is limited to the thermodynamics of gases and gas mixtures at ambient temperatures and total pressures near 1 bar. A perfect gas, i, by definition is one that obeys the gas law: $PV_i = n_iRT$, where V_i is the volume of the gas and n_i the number of moles of the gas in that volume (cf. Garrels and Christ 1965). In the simple case where a specific number of moles of the gas i, is subject to a change in P or T under conditions for which the perfect gas law applies, we can determine the resultant change in V_i through the expression: $P_1V_1/T_1 = P_2V_2/T_2$, where the superscripts 1 and 2 denote initial and final conditions. We can use this expression to show that the volume of one mole of a perfect gas, which is 22.41 L at 0°C (273.15 K) becomes 24.46 L at 25°C (298.15 K).

By definition all gases in a perfect gas mixture obey the perfect gas law such that $PV_T = RT$ Σn_i, where V_T is the total volume and Σn_i denotes the sum of the moles of gases within V_T. Also, for a perfect gas

$$a_i = N_iP \tag{1.58}$$

or, in other words, the activity of the gas equals its mole fraction times the total gas pressure. Because, in general, the partial pressure of a gas P_i also equals N_iP, then $a_i = P_i$.

At low temperatures and total pressures near 1 bar, gases tend to obey the perfect gas law, and the activity of a gas equals its partial pressure (1 bar = 10^5 pascals = 0.986923 atmosphere = 760 mm Hg at 0°C). Garrels and Christ (1965) offer a more exhaustive discussion of gases. (See also Stumm and Morgan 1981; Wood and Fraser 1976; Henderson 1982).

In normal dry air at 25°C and 1 bar total pressure (at sea level), we find the following partial pressures:

Gas	P_i (bar)	log P_i (bar)
O_2	0.21	-0.68
N_2	0.78	-0.11
Ar	0.0093	-2.03
CO_2	0.00033	-3.48

In water-saturated air at 25°C and sea level $P_{H_2O} = 0.031$ or $10^{-1.5}$ bar. By its presence water vapor proportionately reduces the partial pressures of other gases (P_{gas}), an effect that must be taken into account in exact calculations involving gas reactions. Correction for water vapor pressure (P_{H_2O}) may be made through the expression

$$P_{gas} = X_{gas} (P_T - P_{H_2O}) \tag{1.59}$$

where X_{gas} is the mole fraction or volume fraction of the gas, and P_T is the total pressure (cf. Pagen-kopf 1978). The partial pressure of water vapor from 0 to 50°C a 1 bar total pressure is given here.

$T(°C)$	P_{H_2O} (bar)	T(°C)	P_{H_2O} (bar)
0	0.00611	25	0.03169
5	0.00873	30	0.04246
10	0.01228	35	0.05627
15	0.01706	40	0.07381
20	0.02339	50	0.1234

Total pressure and atmospheric partial pressures decrease proportionately with increasing elevation. An approximate relationship between the total pressure and land-surface elevation in meters is $\log P$ (bars) $= -6.7 \times 10^{-5} \times$ elevation (m) (see also Manahan 1994). Correction for elevation is necessary when making field dissolved-oxygen (DO) measurements with a DO meter.

The solubilities of gases are described by Henry's law, which states that at constant T, the solubility of a gas in a liquid is proportional to the partial pressure of the gas. Henry's law gives the exact total solubility of gases that are not involved in further reactions in solution, but underestimates the total solubilities of reactive gases such as CO_2 and SO_2 (cf. Raiswell et al. 1980; Manahan 1994). Some Henry's law constants for gases at 25°C are given here. A more extensive listing of Henry's law constants from 0 to 50°C is given in Table 1.1. Obviously, the larger the value of K_H, the more soluble the gas. As will be noted later, the large K_H value for SO_2, combined with its reactivity with water, greatly enhances the importance of SO_2 as a trace gas in the formation of acid rain.

Gas	Henry's law constant, K_H (M/bar)
O_2	1.26×10^{-3}
CO_2	3.34×10^{-2}
H_2	7.80×10^{-4}
CH_4	1.32×10^{-3}
N_2	6.40×10^{-4}
SO_2	1.18
NH_3	58

TABLE 1.1 Henry's law constants for some gases at 1 bar total pressure in M/bar

$T(°C)$	O_2	N_2	CO_2	CO	H_2S	SO_2	NO
0	2.18×10^{-3}	1.05×10^{-3}	7.64×10^{-2}	1.57×10^{-3}	2.08×10^{-1}	3.56	3.29×10^{-3}
5	1.91×10^{-3}	9.31×10^{-4}	6.35×10^{-2}	1.40×10^{-3}	1.77×10^{-1}	3.01	2.88×10^{-3}
10	1.70×10^{-3}	8.30×10^{-4}	5.33×10^{-2}	1.26×10^{-3}	1.52×10^{-1}	2.53	2.55×10^{-3}
15	1.52×10^{-3}	7.52×10^{-4}	4.55×10^{-2}	1.13×10^{-3}	1.31×10^{-1}	2.11	2.30×10^{-3}
20	1.38×10^{-3}	6.89×10^{-4}	3.92×10^{-2}	1.03×10^{-3}	1.15×10^{-1}	1.76	2.10×10^{-3}
25	1.26×10^{-3}	6.40×10^{-4}	3.39×10^{-2}	9.56×10^{-4}	1.02×10^{-1}	1.46	1.92×10^{-3}
30	1.16×10^{-3}	5.99×10^{-4}	2.97×10^{-2}	8.91×10^{-4}	9.09×10^{-2}	1.21	1.79×10^{-3}
35	1.09×10^{-3}	5.60×10^{-4}	2.64×10^{-2}	8.37×10^{-4}	8.17×10^{-2}	1.00	1.67×10^{-3}
40	1.03×10^{-3}	5.28×10^{-4}	2.36×10^{-2}	7.92×10^{-4}	7.41×10^{-2}	0.837	1.56×10^{-3}
50	9.32×10^{-4}	4.85×10^{-4}	1.95×10^{-2}	7.21×10^{-4}	6.21×10^{-2}	—	1.41×10^{-3}

Source: Based on data tabulated by Pagenkopf (1978).

As an example calculation involving Henry's law, the dissolution of oxygen gas may be written as $O_2(g) = O_2(aq)$. The Henry's law expression is

$$K_H \text{ (M/bar)} = \frac{[O_2(aq)]}{P_{O_2}} = 1.26 \times 10^{-3} \qquad (1.60)$$

with P_{O_2} in bars. We might now ask, what is the solubility of atmospheric oxygen at 25°C in mg/L? At sea level the partial pressure of the oxygen in air is 0.21 bars, from which we compute via Henry's law

$$O_2(aq) = 0.21 \text{ (bar)} \times 1.26 \times 10^{-3} \text{ (M/bar)} = 2.65 \times 10^{-4} \text{ M} \qquad (1.61)$$

Multiplying this value by 32.0 g O_2/mol $\times 10^3$ mg/g, we obtain 8.5 mg/L dissolved oxygen (DO). In groundwater and most deeper soil systems the air is water saturated. Correcting for this effect the oxygen partial pressure is reduced to

$$P_{O_2} = 0.21 \ (1.00 - 0.0317) = 0.203 \text{ bar} \qquad (1.61)$$

The solubility of DO is then 8.2 mg/L.

Gas solubilities in water are also reported as Bunsen adsorption coefficients (cf. Pagenkopf 1978), or in terms of mole fractions dissolved (cf. Bodek et al. 1988). The Bunsen adsorption coefficient, α_B, is related to the Henry's law constant through the expression $K_H = \alpha_H/22.414$.

The solubility of oxygen in water at 25°C and 1 atm pressure as a mole fraction dissolved is 2.30×10^{-5}. This value can be related to K_H for oxygen in M/bar as follows. Given that 1.00 kg of water (GFW H_2O = 18.015 g/mol) contains 55.51 moles of H_2O and that by comparison to water we can neglect the mole fractions of dissolved species in the calculation, the moles of DO per kg/atm are $2.30 \times 10^{-5} \times 55.51 = 1.28 \times 10^{-3}$. Because 1.00 kg $H_2O \approx 1.00$ L H_2O from 0 to 30°C (1.00 kg = 1.01 L H_2O at 50°C), we will neglect the difference here. The DO solubility in mol/kg atm is converted to M/bar by dividing it by 1.01325 bar/atm, with the result $K_H = 1.26 \times 10^{-3}$ M/bar, as tabulated above.

1.4.3 Solutes

For solutes, which include all dissolved aqueous species (including ions and neutral or molecular species), the activity of the solute (a_i) is related to the solute's molal concentration (m_i) (moles of i in 1 kg of water) by the expression

$$a_i = \gamma_i m_i \qquad (1.62)$$

where γ_i is the activity coefficient of the species. We will detail approaches to determining the activity coefficients of solutes in Chap. 4. In general, activity coefficients are less than unity for ions and greater than unity for molecular or neutral species such as $CO_2(aq)$ and silicic acid ($H_4SiO_4^\circ$).

The degree superscript, used here after the species formula and used frequently in subsequent text, is equivalent to the abbreviation (aq) for aqueous, and indicates that the species is dissolved in aqueous solution.

Molar concentration units (moles of i in one liter of solution) practically equal molal concentrations for natural waters holding up to about 7000 mg/L of total dissolved solids (TDS). Below this TDS level, the difference between ppm (parts per million) and mg/L, and between molar (M) and molal (m or mol/kg) concentrations can be ignored. Molal concentrations are independent of temperature, whereas molar concentrations are not.

1.4.4 Solvent Water

The activity of solvent water (a_{H_2O} or $[H_2O]$) can be obtained exactly at low total pressures through water vapor pressure (P_{H_2O}) measurements over the solution, and the expression

$$a_{H_2O} = \frac{P_{H_2O}}{P^\circ_{H_2O}} \tag{1.63}$$

where $P^\circ_{H_2O}$ is the vapor pressure of H_2O over pure water measured at the same T and total P. An approximate measure of the activity of water, assuming an ideal solution, is its mole fraction (N_{H_2O}), which equals

$$N_{H_2O} = \frac{n_{H_2O}/\text{L solution}}{n_{H_2O}/\text{L solution} + \Sigma n \text{ solutes/L solution}} \tag{1.64}$$

or is simply $N_{H_2O} = n_{H_2O}/\Sigma n_i$ in a given volume of solution (usually 1 L) where n is the number of moles of water (n_{H_2O}) or solutes (n), and Σn_i is the total number of moles of water plus solutes in the same volume. A comparison of these approaches as applied to seawater and a halite (NaCl)-saturated brine is given below,

Solution	$a_{H_2O} = \dfrac{P_{H_2O}}{P^\circ_{H_2O}}$	$a_{H_2O} = N_{H_2O}$
Seawater (TDS = 35,000 ppm)	0.98	$\dfrac{55.51}{55.51 + 1.14} = 0.98$
Halite-saturated brine (TDS = 380,000 ppm)	0.75	$\dfrac{55.51}{55.51 + 13.0} = 0.81$

where 1.14 and 13.0 are the number of moles of salts in a liter of seawater and halite-saturated brine, respectively. That the mole fraction of water equals its activity in seawater shows that this approximation is accurate, in general, for potable waters. In fact, in waters with TDS levels (as NaCl) below about 9000 ppm, the activity of water exceeds 0.995 and so can be assumed equal to unity for most purposes (cf. Robinson and Stokes 1970). MINTEQA2 (Allison et al. 1991) uses a different ideal solution-based model that leads to $a_{H_2O} = 0.98$ and 0.78 respectively, for sea water and the halite-saturated brine.

For equilibria calculations involving waters more saline than seawater (saline groundwaters or evaporation brines, for example), an accurate value for the activity of water is necessary (cf. Pitzer 1987). We will discuss the treatment of equilibria in saline waters in Chap. 4.

1.5 SUMMATION OF REACTION-THERMODYNAMIC PROPERTIES

Not infrequently thermodynamic data are lacking for a reaction of particular interest. Sometimes such data does exist for related reactions which can then be combined to solve our problem. Combination may be either by addition or subtraction of their log K_{eq} (or ln K_{eq}) values as appropriate, leading to the cancellation of undesired species in the resulting summation. For example, assume we need K_{eq} at 25°C and 1 bar for reaction Eq. (1.65)

$$CaCO_3(calcite) + CO_2(g) + H_2O = Ca^{2+} + 2HCO_3^- \qquad \frac{\log K_{eq}}{?} \qquad (1.65)$$

and have been given log K_{eq} values for the reactions:

$$
\begin{array}{lll}
CaCO_3(c) = Ca^{2+} + CO_3^{2-} & (+\log K_c) & -8.48 \\
CO_2(g) + H_2O = H_2CO_3^\circ & (+\log K_{CO_2}) & -1.47 \\
H_2CO_3^\circ = H^+ + HCO_3^- & (+\log K_1) & -6.35 \\
H^+ + CO_3^{2-} = HCO_3^- & (-\log K_2) & +10.33 \\
\hline
CaCO_3(c) + CO_2(g) + H_2O = Ca^{2+} + 2HCO_3^- & (+\log K_{eq}) & -5.97
\end{array}
$$

We have produced our desired reaction and its K_{eq} value by writing the related reactions in such a way as to cancel out unwanted terms in the addition. Note that three of the four reactions have been written in their usual form, so their log K values are positive. The fourth and last is written in reverse so that log K_2 must be given a negative sign (log $1/K_2 = -\log K_2$). The same final answer for K_{eq} could have been obtained algebraically from the four K_{eq} expressions, combined to cancel out unwanted terms. Thus, it can be seen

$$K_{eq} = \frac{K_c K_{CO_2} K_1}{K_2} \qquad (1.66)$$

Note that because $\Delta G_r^\circ = -RT \ln K_{eq}$, we could also have added ΔG_r° values for the four reactions to obtain the corresponding ΔG_r° value. The same approach can also be employed to obtain ΔH_r° by summing the enthalpies of related reactions. This is called Hess's law. As a caution, it must be remembered that the thermodynamic properties of each reaction have uncertainties. Consequently, as we increase the number of reactions combined to obtain the properties of a desired reaction, we increase the uncertainty in our result.

1.6 THE EFFECT OF CHANGES IN TEMPERATURE AND PRESSURE ON THE EQUILIBRIUM CONSTANT

1.6.1 Introduction

Fortunately, many systems of environmental interest exist at low temperatures and pressures. The thermodynamic data published in standard references are usually given for 25°C and 1 bar pressure. Such data may often be used directly in calculations of equilibrium for Earth-surface conditions. However, when system conditions differ significantly from ambient conditions, it may become desirable to correct equilibrium constants for the effects of changes in temperature and pressure. Such corrections are more often needed to account for changes in temperature than changes in pressure.

A theoretical approach to both effects can be derived starting with an expression that combines the first and second laws of thermodynamics (cf. Darken and Gurry 1953), which for substance *i*, is

$$dG_i = V_i dP - S_i dT \qquad (1.67)$$

Similar differentials can be written for all substances in a chemical reaction. If we combine these expressions and assume all reactants and products are in their standard states, we find

$$d(\Delta G_r^\circ) = \Delta V_r^\circ dP - \Delta S_r^\circ dT \qquad (1.68)$$

which relates the Gibbs free-energy, volume, and entropy changes of the reaction (the sum of values for the products minus the sum for the reactants). This relationship shows that for conditions of constant pressure (say, 1 bar)

$$\left[\frac{\partial(\Delta G_r^\circ)}{\partial T}\right]_P = -\Delta S_r^\circ \tag{1.69}$$

and for constant temperature

$$\left[\frac{\partial(\Delta G_r^\circ)}{\partial P}\right]_T = \Delta V_r^\circ \tag{1.70}$$

1.6.2 Effect of Temperature

Starting with the fundamental equation $\Delta G_r^\circ = \Delta H_r^\circ - T\Delta S_r^\circ$, it is easy to show that

$$-\Delta S_r^\circ = \frac{\Delta G_r^\circ - \Delta H_r^\circ}{T} \tag{1.71}$$

We can eliminate ΔS_r° by equating this expression and the appropriate differential given above, to obtain for constant pressure

$$T\, d(\Delta G_r^\circ) - \Delta G_r^\circ\, dT = -\Delta H_r^\circ\, dT \tag{1.72}$$

Dividing this expression by T^2, it can be seen than the terms on the left then equal $d(\Delta G_r^\circ/T)$, and so we find

$$\frac{d(\Delta G_r^\circ/T)}{dT} = \frac{-\Delta H_r^\circ}{T^2} \tag{1.73}$$

but because $\Delta G_r^\circ = -RT \ln K_{eq}$, substituting into Eq. (1.73) gives

$$\frac{d(\ln K_{eq})}{dT} = \frac{\Delta H_r^\circ}{RT^2} \tag{1.74}$$

The last two expressions are statements of the van't Hoff equation. Partial integration of the second leads to

$$\ln \frac{K_2}{K_1} = \int_{T_1}^{T_2} \frac{\Delta H_r^\circ}{RT^2}\, dT \tag{1.75}$$

where K_1 and K_2 are the equilibrium constant values at T_1 and T_2. Clearly then, if we know the temperature dependence of the enthalpy of the reaction, we can determine the effect of temperature on K_{eq}.

The heat capacity of the reaction at constant pressure, ΔCp_r°, is related to the enthalpy of the reaction through

$$\left|\frac{\partial(\Delta H_r^\circ)}{\partial T}\right|_P = \Delta Cp_r^\circ \tag{1.76}$$

The van't Hoff equation is easily integrated if ΔH_r° is a constant, independent of temperature. This is equivalent to the assumption that $\Delta Cp_r^\circ = 0$. With a conversion to common logs, the integrated result is

$$-\log K_2 = -\log K_1 + \frac{\Delta H_r^\circ}{4.576}\left(\frac{1}{T_2} - \frac{1}{T_1}\right) \tag{1.77}$$

Subscript 1 usually refers to K_{eq} at the reference temperature T_1, which is most often 25°C. When this expression is obeyed, a plot of log K_{eq} versus $1/T(K)$ gives a straight line with a slope of $[-\Delta H_r^\circ/4.576]$ and intercept of $[\log K_1 + (\Delta H_r^\circ/4.576 T_1)]$. The linearity of such a plot is thus evidence that ΔH_r° is constant, independent of temperature. The value of ΔH_r° can then be evaluated from the slope of the plot.

The equilibrium constant at T_2 for reactions involving only condensed phases may be fairly accurately predicted using the integrated van't Hoff equation. K_2 for reactions with aqueous ions may also be predicted accurately in most cases using Eq. (1.77), as long as T_2 is within ± 10 to 15°C of 25°C.

In the more general case the enthalpy of the reaction is not constant and $\Delta Cp_r^\circ \neq 0$. If, however, ΔCp_r° is constant, then,

$$\Delta H_2^\circ = \Delta H_1^\circ + \Delta Cp_r^\circ (T_2 - T_1) \tag{1.78}$$

and we can integrate the van't Hoff expression to give

$$-\log K_2 = -\log K_1 + \frac{\Delta H_r^\circ}{4.576}\left[\frac{1}{T_2} - \frac{1}{T_1}\right] - \frac{\Delta C_p^\circ}{1.987}\left[\frac{1}{2.303}\left(\frac{T_1}{T_2} - 1\right) - \log \frac{T_1}{T_2}\right] \tag{1.79}$$

In these expressions T is in degrees kelvin, ΔH_r° is given in cal/mol, and ΔCp_r° is in cal/mol K.

Heat capacity data for pure phases is often available or can be estimated with some accuracy. However, such data is limited for aqueous species, particularly for complexes at elevated temperatures. In any case, when T_2 is 50 or more degrees from reference temperature T_1, and especially when we are considering reactions in solution, predictions of K_2 through the van't Hoff or even the extended van't Hoff equation can lead to values that are seriously in error.

Prediction of reliable K_{eq} values at elevated temperatures (above 50 to 100°C) requires accurate data for the enthalpy and heat capacity of reactants and products. Such data have been published by Helgeson et al. (1978) and Shock and Helgeson (1988). Methods of calculating reaction equilibria at elevated temperatures and pressures are described in these references and in Nordstrom and Munoz (1985) and Kharaka et al. (1988).

When such data are limited, it is sometimes possible to use what has been called *the isocoulombic approach*. This is based on the discovery that when reactions involving ionic species are written in such a way that there are an equal number of species on each side of the reaction having the same total valence, the heat capacity is often nearly constant, independent of temperature (cf. Murray and Cobble 1980; Langmuir and Mahoney 1985; Plyasunov and Grenthe 1994). For example, instead of writing

$$HCO_3^- = H^+ + CO_3^{2-} \tag{1.80}$$

we write $\qquad\qquad HCO_3^- + OH^- = CO_3^{2-} + H_2O \tag{1.81}$

so that with the well-known data for the dissociation of water above 25°C and a value of K_2 for the first reaction at that temperature, we can estimate values of K_2 at elevated temperatures.

For a detailed calculation involving this approach assume that we have reliable K_{sp} data for the dissolution of barite ($BaSO_4$)

$$BaSO_4 = Ba^{2+} + SO_4^{2-} \tag{1.82}$$

as a function of temperature from Blount (1977), and that we need such data for the K_{sp} of the mineral anhydrite ($CaSO_4$). We have accurate thermodynamic data for both reactions at a common reference temperature, which we will take as 56°C, at which log K_{sp}(anhydrite) $= -4.64$ (Langmuir and

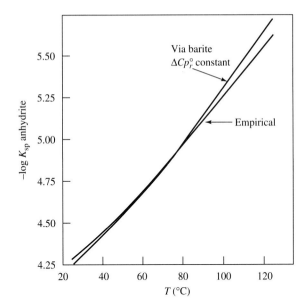

Figure 1.4 $-\log K_{sp}$ for anhydrite ($CaSO_4$) as a function of temperature based on empirical measurements and estimated assuming ΔCp_r° = constant for the reaction $Ca^{2+} + BaSO_4$ (barite) = $Ba^{2+} + CaSO_4$. After Langmuir and Mahoney (1985). Reprinted from National Water Well Assoc. Used by permission.

Melchior 1985). Our isocoulombic equation, which is obtained by subtracting the K_{sp} expression for anhydrite from that for barite, is

$$Ca^{2+} + BaSO_4 = Ba^{2+} + CaSO_4 \qquad (1.83)$$

for which at 56°C, $\Delta G_r^\circ = 7.518$ kcal/mol, $\Delta S_r^\circ = 7.98$ cal/mol K, and $\Delta Cp_r^\circ = -5.2$ cal/mol K. Several additional steps lead to $-\log K_{sp}$ for anhydrite as a function of temperature based on the temperature function for barite. In Fig. 1.4 are plotted $-\log K_{sp}$ for anhydrite based on this approach and from empirical measurements (cf. Langmuir and Melchior 1985). The agreement is clearly excellent up to about 100°C.

More recently Gu et al. (1994) have shown that ΔS_r° and ΔCp_r° are often small and of opposite sign in a well-balanced isocoulombic reaction and so tend to cancel each other out. For such conditions ΔG_r° is independent of temperature.

The variation of K_{eq} with temperature has been measured accurately for a number of important reactions. A general equation that can describe the temperature dependence of K_{eq} is

$$\log K_{eq} = A + BT + \frac{C}{T} + D \log T + ET^2 + \frac{F}{T^2} + GT^{1/2} \qquad (1.84)$$

where A through G are empirical constants, and T is in degrees kelvin. This function is used in the program MINTEQA2 (Allison et al. 1991) when the empirical constants are known. Of the 1000 species in the MINTEQA2 data base, however, temperature functions for K_{eq} are given for only about 25. In the absence of such a function, MINTEQA2 reverts to the simple van't Hoff equation to compute K_{eq} away from 25°C. Ball and Nordstrom (1991) give temperature functions for 37 of the nearly 650 reactions in the thermodynamic database for WATEQ4F. They use the van't Hoff equation to correct K_{eq} values of the other reactions for temperature. Rarely are more than the first four or five terms in Eq. (1.66) known or necessary to accurately fit empirical K_{eq} versus T data.

Plummer and Busenberg (1982) exactly modeled empirical K_{sp}/temperature data for the reaction

$$CaCO_3(calcite) = Ca^{2+} + CO_3^{2-} \qquad (1.85)$$

from 0 to 90°C with an equation of the form

$$\log K_{sp}(\text{calcite}) = A + BT + \frac{C}{T} + D \log T \tag{1.86}$$

where $A = -171.9065$, $B = -0.077993$, $C = 2839.319$, and $D = 71.595$.

Starting with this temperature function, Plummer and Busenberg show that it is straightforward to derive the other thermodynamic properties of the reaction, including ΔG_r°, ΔH_r°, ΔCp_r°, and ΔS_r°. Thus, in general, assuming this form of the log K_{sp} equation, and given that $\ln K_{sp} = 2.3026 \log K_{sp}$, the reaction free energy is given by

$$\Delta G_r^\circ = -RT \ln K_{sp} = \alpha R \left(AT + BT^2 + C + \frac{DT \ln T}{\alpha} \right) \tag{1.87}$$

where $\alpha = 2.302585$ and the gas constant $R = 1.987165$ cal/mol K. Transposing the van't Hoff equation we can write

$$\Delta H_r^\circ = \frac{RT^2 d \ln K_{sp}}{dT} = \alpha R \left(BT^2 - C + \frac{DT}{\alpha} \right) \tag{1.88}$$

Because the heat capacity is the temperature derivative of the reaction enthalpy, we find

$$\Delta Cp_r^\circ = \frac{\partial \Delta H_r^\circ}{\partial T} = \alpha R \left(2BT + \frac{D}{\alpha} \right) \tag{1.89}$$

Finally, the entropy of the reaction can be obtained from

$$\Delta S_r^\circ = -\frac{\partial \Delta G_r^\circ}{\partial T} = \alpha R \left(A + 2BT + \frac{D}{\alpha}(1 + \ln T) \right) \tag{1.90}$$

Pursuing these calculations for $K_{sp}(\text{calcite})$ yields the following thermodynamic properties for the reaction at 25°C (Plummer and Busenberg 1982): log $K_{sp} = -8.480 \pm 0.020$, $\Delta G_r^\circ = 11568 \pm 27$ cal/mol, $\Delta H_r^\circ = -2297 \pm 300$ cal/mol, $\Delta Cp_r^\circ = -70.5 \pm 4.0$ cal/mol K, and $\Delta S_r^\circ = -46.5 \pm 2.0$ cal/mol K, where the uncertainties are as given by the authors. Note that uncertainties are about $\pm 0.2\%$ for log K_{sp} and ΔG_r°, but range from 4% to 13% for the other properties. This reflects that the reaction entropy and enthalpy are derivatives of the K_{sp} expression, and so measure the *slope* of a log K_{sp} versus $1/T(\text{K})$ plot. (See Figs. 1.5 through 1.7.) As the second derivative of such a function, ΔCp_r° is the *curvature* of the plot and, therefore, is even less well defined. Generally more accurate values of the reaction heat capacity are obtained by direct calorimetric measurement of the enthalpy of a reaction as a function of temperature.

The computer code SOLMINEQ.88 (Kharaka et al. 1988) has been designed to model water-rock systems at high temperatures and pressures. Given the lack of thermodynamic data for such conditions, the code contains functions more complex than the integrated simple or extended van't Hoff equations that permit the estimation of K_{eq} values with reasonable accuracy up to about 150°C. These functions take into account such variables as the dielectric constant of water and the radii and valences of aquo-ions. Much of their theoretical basis has been developed by Helgeson and his coworkers (cf. Helgeson 1967, 1985; Helgeson et al. 1978, 1981; Schock and Helgeson 1988).

Empirical K_{eq}/temperature functions for 63 reactions are given in Table A1.1 in the Chapter Appendix. Several examples of these functions are plotted in Figs. 1.5 through 1.7 in terms of log K versus $1/T(\text{K})$. The log K_{dissoc} versus $1/T(\text{K})$ plots in these figures, which appear as straight or nearly straight lines, are those of ammonia (Fig. 1.5), the alkaline earth carbonate complexes (Fig. 1.6),

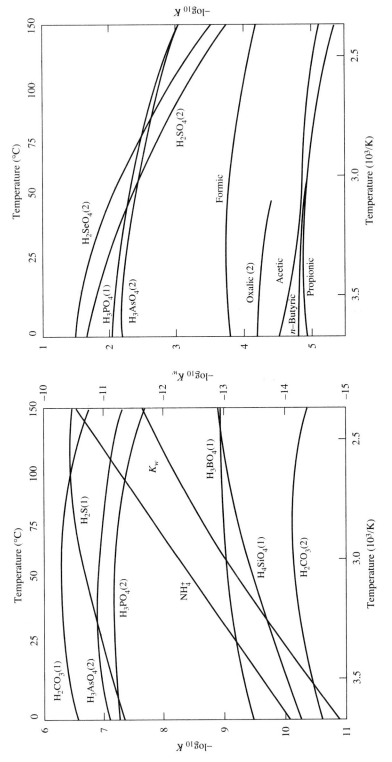

Figure 1.5 Plots of some log K_{diss} (stepwise) versus $1/T(K)$ functions for acids given in Table A1.1. Numbers in parentheses following the chemical formulae of the acids indicate whether the dissociation step is the first or second. Where no number is given, a single dissociation step is indicated. Right-side ordinant of figure on left applies to K_w curve only.

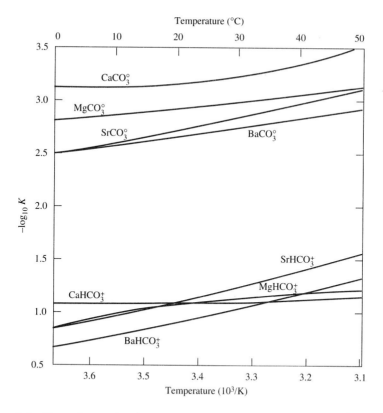

Figure 1.6 Plots of some log K_{dissoc} versus $1/T(K)$ functions of some carbonate and bicarbonate complexes given in Table A1.1.

melaterite ($FeSO_4 \cdot 7H_2O$), and the polymorphs of SiO_2 [SiO_2(am), chalcedony, and quartz] (Fig. 1.7). For these reactions, ΔH_r° is practically constant and independent of T, and the van't Hoff equation accurately predicts the change in K_{eq} values with *T*. For many of the other reactions, accurate prediction of K_{eq} with the van't Hoff equation is limited to within ± 10 to 20 degrees of the reference temperature of 25°C.

Earlier, in our introduction to enthalpy concepts, we defined *exothermic* (negative ΔH_r°) and *endothermic* (positive ΔH_r°) reactions. Exothermic reactions give off heat. This means that the reverse reaction is favored by an increase in temperature (reactant concentrations increase at the expense of product concentrations), thus K_{eq} values for the reaction *decrease* as the temperature rises. Conversely, endothermic reactions consume heat, so the forward reaction is more and more favored over the reverse reaction with increasing temperature. In other words, K_{eq} values increase with *T* for endothermic reactions.

As noted previously in the discussion of gas solubilities and Henry's law, the solubility of most gases decreases with increasing temperature (see Table 1.1). Consistent with this behavior, we can largely sweep the dissolved gases out of a water by boiling it. In other words, the dissolution of gases in water is exothermic. For example $\Delta H_r^\circ = -4.0$ kcal/mol for the reaction $CO_2(g) = CO_2(aq)$ and -2.9 kcal/mol for $O_2(g) = O_2(aq)$. The exothermic nature of oxygen solubility corresponds to sea-level solubilities of atmospheric oxygen (at 21% of air) of 14.74, 8.32, and 7.03 mg/L at 0, 25, and 35°C (Manahan 1991).

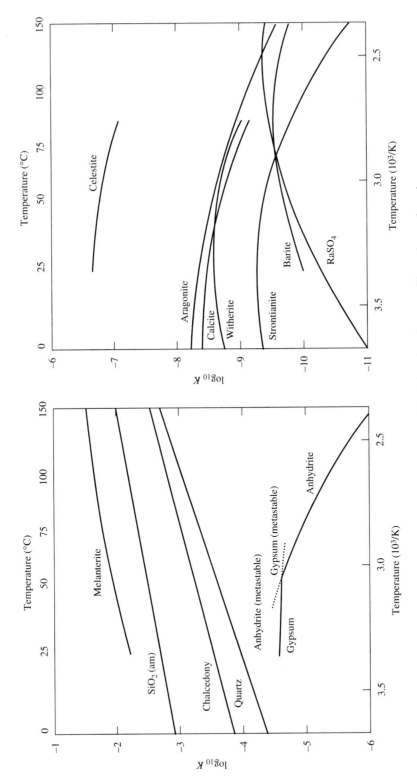

Figure 1.7 Plots of some log K_{sp} versus $1/T(K)$ functions for some silica polymorphs, sulfates, and carbonates given in Table 1.2.

Inspection of the van't Hoff equation shows that the effect of a temperature change on K_{eq} is proportional to the magnitude of the enthalpy of the reaction. The larger the ΔH_r°, the stronger the temperature dependence of K_{eq}. For example, Fig. 1.5 shows the steep temperature dependence of K_w, the dissociation constant of water ($H_2O = H^+ + OH^-$) for which $\Delta H_r^\circ = +13.362$ kcal/mol. This may be compared with the slight temperature dependence of K_1, the first dissociation constant of carbonic acid ($H_2CO_3 = H^+ + HCO_3^-$) for which $\Delta H_r^\circ = +1.84$ kcal/mol.

Weak acids (including carbonic acid) often have small reaction enthalpies (<1 to 2 kcal/mol), which makes their dissociation constants relatively temperature independent. Also they often exhibit maxima in their K_{eq} values when the latter are plotted against temperature. The dissociation constants for several weak acids are plotted against temperature in Fig. 1.5. Below the K_{eq} maxima for these acids, the dissociation reactions are endothermic ($\Delta H_r^\circ > 0$), and acid dissociation is favored with increasing temperature. At temperatures of the maxima, $\Delta H_r^\circ = 0$. At higher temperatures ΔH_r° values become increasingly negative, and the undissociated acid species become more and more important.

Most solution reactions that lead to the formation of aqueous inorganic complexes are endothermic, thus the complexes increase in stability with increasing temperature (see Fig. 1.6). This applies to the formation of most metal complexes with sulfate, fluoride, bicarbonate, carbonate, and phosphate ions (cf. Christensen et al. 1975; Smith and Martell 1976; Nordstrom et al. 1990). In other words the hotter the water is, the more the metal species in it will be present as complexes with these ligands rather than as free ions. Complexation thus tends to increase the mobilities of most metals at elevated temperatures. For example, the reaction

$$Fe^{3+} + F^- = FeF^{2+} \tag{1.91}$$

(log $K_{eq} = 6.2$) has $\Delta H_r^\circ = 2.7$ kcal/mol.

The stability of complexes formed with chloride, citrate, oxalate, and many metal-organic complexes, however, may increase ($\Delta H_r^\circ > 0$) or decrease ($\Delta H_r^\circ < 0$) with increasing temperature. Complexes formed with OH^- and EDTA are generally less stable at elevated temperatures. For example, the complex formed between EDTA and radioactive cobalt is

$$2(^{90}Co^{2+}) + EDTA^{2-} = (^{90}Co)_2EDTA \tag{1.92}$$

for which log $K_{eq} = 15.7$, has $\Delta H_r^\circ = -4.2$ kcal/mol (Sillen and Martell 1964; Christensen et al. 1975). This complex is found in subsurface waters beneath radionuclide research facilities at Oak Ridge, Tennessee, where EDTA was used as a laboratory cleansing agent (Manahan 1994).

Because it takes energy to break the bonds in most minerals (and in salts, generally) mineral dissolution is usually an endothermic process. In other words, most minerals (silicates, aluminosilicates, oxides, and sufides) increase in solubility (thus in their K_{sp} values) with increasing temperature (Fig. 1.7). This explains the relatively high concentrations of dissolved silica found in geothermal waters and why amorphous silica [SiO_2(am)] tends to precipitate, thus clogging heat exchangers when geothermal waters cool at the land surface. Silica precipitation upon cooling reflects $\Delta H_r^\circ = +1.55$ kcal/mol for the reaction SiO_2(am) $= SiO_2$(aq).

In contrast to the silicates, most carbonate and many sulfate minerals dissolve exothermically ($\Delta H_r^\circ < 0$). This makes them most soluble at low temperatures and gives them a tendency to precipitate as temperature increases (Fig. 1.7). (Compare log K_{sp} versus $1/T$ curves for calcite and quartz in Fig. 1.7.) Thus, calcite ($\Delta H_r^\circ = -2.30$ kcal/mol) can be a troublesome precipitate on the walls of boiler systems where its thermal insulating properties prevent heat exchange and can seriously damage boiler performance.

Example 1.2

The second dissociation step for carbonic acid corresponds to the reaction $HCO_3^- = H^+ + CO_3^{2-}$, for which $K_2 = 10^{-10.33}$ at 25°C. Using thermodynamic data tables we compute $\Delta H_r^\circ = 3550$ cal/mol, and from Shock and Helgeson (1988) we find $\Delta Cp_r^\circ = -61.0$ cal/mol K. Given this information, we ask the question, how does the value of K_2 at 50°C (323.15 K), predicted from the integrated van't Hoff equation, compare to the empirical value at 50°C? The van't Hoff equation is

$$-\log K_2 = 10.33 + \frac{3550}{4.576}\left(\frac{1}{323.15} - \frac{1}{298.15}\right) \tag{1.93}$$

from which we find

$$-\log K_2 = 10.33 - 0.20 = 10.13$$

This is in fair agreement with $-\log K_2 = 10.18$ at 50°C computed from the empirical temperature function in Table A1.1. The disagreement reflects the fact that the heat capacity of the reaction is not zero, as assumed in the calculation. If we refine our estimate by including the heat capacity correction term of the expanded integrated van't Hoff equation, the predicted value for K_2 at 50°C is

$$-\log K_2 = 10.33 - 0.20 + 0.04 = 10.17$$

in excellent agreement with the empirical value.

Example 1.3

Consider the reaction

$$AgCl(c) = Ag^+ + Cl^- \tag{1.94}$$

for which $K_{sp} = 10^{-9.75}$ and $\Delta H_r^\circ = 15,650$ cal/mol at 25°C. Silver chloride is also known as the mineral cerargyrite. This reaction takes place in the internal electrolyte of glass pH electrodes. The insolubility of cerargyrite also limits maximum concentrations of silver in many natural waters. We will ask the question, what is K_{sp} at 0°C? The van't Hoff equation leads to

$$-\log K_{sp} = 9.75 - \frac{15,650}{4.576}\left(\frac{1}{273.15} - \frac{1}{298.15}\right) \tag{1.95}$$

and

$$-\log K_{sp} = 9.75 + 1.05 = 10.80$$

This is in exact agreement with the empirical value, indicating that the assumptions inherent in the van't Hoff equation are valid (ΔH_r° is constant and $\Delta Cp_r^\circ = 0$).

1.6.3 Effect of Pressure

Substituting the quantity $(-RT \ln K_{eq})$ for ΔG_r° in the expression

$$\left(\frac{\partial(\Delta G_r^\circ)}{\partial P}\right)_T = \Delta V_r^\circ \tag{1.96}$$

derived previously, we obtain

$$\left(\frac{\partial \ln K_{\text{eq}}}{\partial P}\right)_T = \frac{\Delta V_r^\circ}{RT} \tag{1.97}$$

where ΔV_r° is the molar volume change of the reaction with all reactants and products in their standard states (cf. Millero 1982). Note the similarity of this expression to the van't Hoff equation. Equation (1.97) shows that the effect of a change in pressure on K_{eq} is proportional to the magnitude of the molar volume change for the reaction. Consistent with Le Chatelier's principle, an increase in pressure favors reaction products when ΔV_r° is negative and reactants when ΔV_r° is positive. When a gas is among the reactants or products, the effect of pressure is accounted for by writing the activity of the gas as its partial pressure. For reactions without gases, the effect of pressure can usually be ignored at depths less than 300 m (1000 ft). In following discussion, we will consider the pressure effect on reactions in more detail. Important environmental applications are to reactions in the deep ocean (average deep-ocean pressure is 200 bars (Stumm and Morgan 1981) and to geologic disposal of industrial wastes by deep-well injection. Deep-well injection is typically into saline groundwaters in sedimentary formations at temperatures of 20 to 150°C, with pressures from 50 to 300 bars (Environmental Protection Agency 1990a, 1990b). We will limit discussion in this section to the effect of pressure on reactions, which is usually less important than the coincident temperature effect as discussed previously.

If the molar volume of a reaction is independent of pressure (does not change with pressure in the *P* range of interest), Eq. (1.97) may be integrated to yield

$$\ln \frac{K_p}{K_1} = -\frac{\Delta V_r^\circ(P-1)}{RT} \tag{1.98}$$

where K_p and K_1 are the equilibrium constant at *P* and at the reference pressure, which is usually 1 bar.

If instead ΔV_r° is a function of pressure, we may define the standard partial molar compressibility which for single species *i* equals

$$\overline{K}_i^\circ = \left(\frac{\partial V}{\partial P}\right)_T \tag{1.99}$$

For a reaction we have

$$\Delta \overline{K}_r^\circ = \left(\frac{\partial \Delta V_r^\circ}{\partial P}\right)_T \tag{1.100}$$

Integration of Eq. (1.97) now leads to

$$\ln \frac{K_p}{K_1} = -\frac{\Delta V_r^\circ(P-1)}{RT} + \frac{\Delta \overline{K}_r^\circ(P-1)^2}{2RT} \tag{1.101}$$

in which $\Delta \overline{K}_r^\circ$ is independent of pressure. At elevated pressures (>100 bars) this simplifies to

$$\ln \frac{K_p}{K_1} = -\frac{\Delta V_r^\circ P}{RT} + \frac{\Delta \overline{K}_r^\circ P^2}{2RT} = -\frac{P}{RT}\left(\Delta V_r^\circ - \frac{\Delta \overline{K}_r^\circ P}{2}\right) \tag{1.102}$$

In these expressions ΔV_r° is in cm³/mol, $R = 83.15$ cm³ bar/mol K, and $\Delta \overline{K}_r^\circ$ is in cm³/bar mol.

Molar volumes of minerals are proportional to the size of molar formulas and may be computed by dividing a mineral's gram formula weight (g/mol) by its density (g/cm³). Values for minerals are tabulated by Naumov et al. (1974), Robie et al. (1978), and Helgeson et al. (1978), and may also be found in any edition of the *Handbook of Chemistry and Physics* (CRC Press).

Naumov et al. (1974), Helgeson et al. (1981), Millero (1982), and Tanger and Helgeson (1988) report molar volume data for aqueous species at 25°C. The last two references also give $V°$ data for aqueous species from 0°C up to 50°C and 350°C, respectively. Molar compressibilities of minerals are available in Birch (1966), Mathieson and Conway (1974), and Millero (1982). Millero (1982) also lists values of $\Delta V_r°$ and $\Delta \bar{K}_r°$ for aqueous species from 0 to 50°C. Molar volumes and compressibilities vary as a function of temperature and ionic strength (Millero 1982). Such effects may be important for aqueous species and especially for gases.

The molar volume of ions is based on the convention that the molar volume of H^+ ion equals zero at all temperatures. This assumption leads to molar volumes of -0.4 cm³/mol for Li^+ and 36 cm³/mol for I^-, for example. Molar volumes of ions increase with increasing temperature. $\Delta V_r°$ is negative for most ionization and dissolution reactions so that pressure generally increases the progress of ionization reactions and the solubility of minerals.

Molar compressibilities for minerals range from 0 to -4×10^{-9} cm³/bar mol and can be assumed equal to zero up to about 1 kb pressure. Values for ions $\bar{K}°$ are generally in the range -5 to -10×10^{-3} cm³/bar mol. $\Delta \bar{K}_r°$ for completely ionic reactions is small, but for dissolving minerals it may be large.

In a column of freshwater the pressure increases above atmospheric pressure by about 1 bar for every 33.5 ft of depth, or by 30 bars for 1000 ft (305 m) of depth. Pressure gradients are higher in saline than in fresh groundwaters. Pressures measured in groundwater at depth generally range between hydrostatic pressure (that of a column of water) and lithostatic pressure (that of the rock alone). Measured hydrostatic pressure gradients in sediments range from 1 bar/15 ft (roughly equal to the lithostatic pressure gradient) to 1 bar/70 ft of depth. In sediments of the Palo Duro basin of north Texas for example, the pressure gradient is 1 bar/42 to 58 ft to a depth of 5500 ft, where hydrostatic pressure is 130 bars and the temperature 38°C (Langmuir and Melchior 1985).

The effect of pressure on reactions involving only solid phases can be assumed to be negligible. For example, consider the reaction

$$Al_2O_3 + SiO_2 = Al_2SiO_5 \tag{1.103}$$

	(corundum)	(quartz)	(sillimanite)
$V°$ (cm³/mol)	25.57	22.69	49.92

The molar volumes lead to $(\partial \Delta G_r°/\partial P)_T = \Delta V_r° = 1.66$ cm³/mol, from which we conclude that an increase in pressure slightly favors corundum and quartz over sillimanite. Given the conversion factor 1 cm³ bar = 0.02390 cal, we obtain

$$(\partial \Delta G_r°/\partial P)_T = 1.66 \text{ cm}^3/\text{mol} \times 0.02390 \text{ cal/cm}^3 \text{ bar}$$
$$= 0.0397 \text{ cal/mol bar}$$

Thus, an increase in P of 100 bars increases $\Delta G_r°$ by only ~4 cal/mol. This is a negligible effect.

When a gas is among the reactants, changes in pressure are generally important. For example, given the reaction at 25°C,

$$CaCO_3(\text{calcite}) = CaO(\text{lime}) + CO_2(g) \tag{1.104}$$

$V°$ (cm³/mol)	36.934	16.764	24,460

TABLE 1.2 Effects of pressure and ionic strength on equilibrium constants for some reactions as a function of the molar volumes of the reactions

Reaction	ΔV_r° (cm³/mol)	K_p/K_1 at 25°C and 1000 bars
H_2O (l) $= H^+ + OH^-$	-22.2	2.36
$CH_3COOH = H^+ + CH_3COO^-$ (K_a acetic acid)	-11.2	1.4
$H_2CO_3 = H^+ + HCO_3^-$ (K_1 carbonic acid)	-27.2	3.2
$HCO_3^- = H^+ + CO_3^{2-}$ (K_2 carbonic acid)	-25.2	2.7
$CaSO_4$ (anhydrite) $= Ca^{2+} + SO_4^{2-}$	-48.9	5.8

Source: Based in part on Stumm and Morgan (1981).

we find $\Delta V_r^\circ = 24{,}440$ cm³/mol, or 584 cal/mol bar. The $(+)$ sign and large magnitude of ΔV_r° show that increasing pressure strongly favors calcite formation relative to lime and $CO_2(g)$.

The effect of a pressure increase to 1000 bars on some dissociation reactions is shown in Table 1.2. Increased pressure favors dissociation in every case. The effect is proportional to the size of $-\Delta V_r^\circ$. The solubility of calcite increases with increasing pressure as indicated by Table 1.3. The effect is reduced at the ionic strength of seawater.

Pressure generally increases the solubility of minerals. An accurate evaluation of the pressure effect on solubility may require that we consider both molar volumes and compressibilies of reactants and products. In the following problem we compare the solubility of celestite ($SrSO_4$) in a surface water at 1 bar pressure and 25°C to its solubility in groundwater at 6000 ft depth at a temperature of 75°C and pressure of 180 bars. The reaction of interest is $SrSO_4$(celestite) $= Sr^{2+} + SO_4^{2-}$. To solve the problem we need molar volumes and compressibilies for celestite and the ions. We adapt V° (celestite) $= 46.25$ cm³/mol from Robie et al. (1978). A nearly identical molar volume of 46.27 cm³/mol derives from the molecular weight (183.678 g/mol) divided by the density of celestite (3.97 g cm³; Palache et al. 1951). V°(celestite) can be assumed to be independent of pressure and temperature. Molar volumes for the ions at 75°C are obtained from Tanger and Helgeson (1988). The compressibility of celestite is read from a plot given by Millero (1982). $\overline{K}^\circ$ values for the ions are less readily available. We will use those given by Millero (1982) for 50°C. The comparatively small compressibility of the mineral (about 1% as large as values for the ions) suggests that its value is relatively unimportant in the calculation. The compressibility correction to the solubility will turn out

TABLE 1.3 Effects of increases in pressure and ionic strength at 25°C on the solubility product of calcite ($CaCO_3$) at P versus at 1 bar pressure, expressed as K_p/K_1

P (bar)	Pure water	0.725 M NaCl (seawater)
1	1	1
500	3.2	2.8
1000	8.1	6.7

Note: $\Delta V_r^\circ = -58.3$ cm³/mol for the reaction $CaCO_3 = Ca^{2+} + CO_3^{2-}$ at 25°C.
Source: Based in part on Stumm and Morgan (1981).

to be small, thus our use of 50°C $\overline{K}^\circ$ values for Sr^{2+} and SO_4^{2-} is a good approximation. Chosen V° and $\overline{K}^\circ$ values are given below.

$$SrSO_4(\text{celestite}) \quad = \quad Sr^{2+} \quad + \quad SO_4^{2-} \qquad (1.105)$$

V° (cm³/mol)	46.25	−18.27	+14.09
$\overline{K}^\circ$ (cm³/mol bar)	-0.08×10^{-3}	-8.35×10^{-3}	-6.87×10^{-3}

The V° and $\overline{K}^\circ$ data lead to $\Delta V_r^\circ = -50.43$ cm³/mol, and $\Delta \overline{K}_r^\circ = -1.514 \times 10^{-3}$ cm³/mol bar. Substituting into Eq. (1.101) or (1.102), we can determine the effect of pressure on celestite solubility. The result is log $(K_p/K_1) = 0.1364 - 0.0037 = 0.133$, where the first term in the difference reflects the contribution of ΔV_r° and the second the small, opposite effect of $\Delta \overline{K}_r^\circ$. At 1 bar pressure and 25°C, $K_{sp}(\text{celestite}) = 2.32 \times 10^{-7}$. For 75°C, $K_{sp}(\text{celestite}) = 7.45 \times 10^{-7}$, computed from the temperature function given by Langmuir and Melchior (1985). Overall results are summarized here.

T(°C)	P (bar)	$K_{sp}(\text{celestite})$	Overall percent increase above 1 bar and 25°C
25	1	2.32×10^{-7}	0
75	1	7.45×10^{-7}	221%
75	180 (ΔV_r°)	1.02×10^{-6}	340%
75	180 $(\Delta V_r^\circ$ and $\Delta \overline{K}_r^\circ)$	1.01×10^{-6}	335%

As shown, the increase in temperature from 25 to 75°C has more of an effect on solubility (+221%) than does the increase in pressure from 1 to 180 bars, which separately increases solubility by 36%. As evident from this example, and in general, the effect of a temperature increase on solubility greatly exceeds that of the concomitant pressure increase. The example also shows that the pressure correction due to $\Delta \overline{K}_r^\circ$ is minor and usually of an opposite sign to that due to ΔV_r°.

Aggarwal et al. (1990) have proposed a simpler model given by

$$\ln\left(\frac{K_{P,T}}{K_1^\circ}\right) = -\left(\frac{\Delta V^\circ}{RT\,\overline{K}_{H_2O}^\circ}\right)\ln\left(\frac{\rho_{H_2O(P,T)}}{\rho_{H_2O}^\circ}\right) \qquad (1.106)$$

to estimate the effect of a change in P and T on the equilibrium constant. Their equation requires only ΔV_r° for the reaction, the compressibility $(\overline{K}_{H_2O}^\circ)$ and density $(\rho_{H_2O}^\circ)$ of water for reference conditions (usually at 1 bar pressure and 25°C), and water density at P and T $(\rho_{H_2O(P,T)}^\circ)$. Reactions are written in isocoulombic form so that reaction heat capacity and the term $(\Delta V^\circ/RT\,\overline{K}_{H_2O}^\circ)$ are both constant, independent of temperature. The necessary properties of water have been published by Haar et al. (1984). The reader is referred to Aggarwal et al. (1990) for details of this approach.

STUDY QUESTIONS

1. Know the following definitions and how they relate to each other:
   (a) open and closed systems as they relate to mobile and less mobile substances
   (b) phases, components, master species, and the phase rule
   (c) homogeneous, heterogeneous and irreversible, congruent and incongruent reactions
   (d) Gibbs free energy, chemical potential, enthalpy, and entropy, as they relate to components, substances, reactions, and systems

2. Understand the relationships among Gibbs free energy, chemical potential, reaction quotients (Q), the equilibrium constant, and the saturation index (SI).

3. Learn the conventions used to define units of entropy, heat capacity, enthalpy, and Gibbs free energy of elements, solids, liquids, and ions, including the proton.

4. Know the concept of activity versus concentration and how activity is defined for gases, solutes, pure solids, and liquids.

5. Define ideal solutions, Raoult's and Henry's laws, and Nernst's law, with an application of each.

6. Describe perfect gases, perfect gas mixtures, and partial pressure. How does Henry's law apply to gas solubilities?

7. Know how the activities of solutes and solvent water are defined.

8. Be able to obtain the thermodynamic properties of reactions by summation.

9. Define an ideal solid solution. Discuss how you would determine the Gibbs free energy, enthalpy, molar volume, and entropy of formation of a binary ideal solid solution given the appropriate properties of its pure end members.

10. What are the equations for computing the Gibbs free energy, enthalpy, and entropy of formation of a binary symmetrical regular solution? How are the rational activity coefficients (λ values for the solid components) related to their mole fractions in such a solid solution?

11. Given a value of ω or a_0 for a binary, symmetrical, regular solid solution and the solubility products for the pure end-member components, be able to compute the dissolved concentrations of either component at equilibrium with the solid solution, assuming the concentration of the other dissolved component is known.

12. Why is it that the concentration of a trace component in a solid solution is unlikely to be controlled by equilibrium with the solid solution unless the pure end-member phase of the major component is at or near saturation with respect to the water?

13. Understand the following:
    (a) the effect of temperature changes on K_{eq} for exothermic and endothermic reactions
    (b) the significance of the size of ΔH_r° in the behavior of K_{eq} versus T
    (c) the use of the van't Hoff equation and its integrated form when $\Delta Cp_r^\circ \neq 0$
    (d) the isocoulombic approach
    (e) empirical temperature functions and their relationship to the thermodynamic properties of reactions

14. How can we calculate the molar volume of a mineral from its density?

15. The molar volume of ions may be positive or negative. Explain.

16. Explain how the effect of pressure on K_{eq} for a reaction relates to the magnitude and sign of the molar volume change of the reaction. What is the differential equation that expresses the effect of pressure on K_{eq} at constant T?

17. The effect of pressure on K_{eq} is of least importance for reactions involving only solid phases, of intermediate importance for aqueous-aqueous and mineral-aqueous reactions, and of major importance for reactions involving gases. Explain this statement with examples.

18. What is the partial molar isothermal compressibility and when is it an important variable in calculations of the effect of P on equilibrium?

19. An increase in pressure always leads to an increase in the solubility of minerals. Why? In what natural water-rock environments is the effect of pressure on K_{eq} important?

PROBLEMS

In all calculations related to this chapter concentrations of dissolved species will be considered equal to their activities. In other words activity coefficients of dissolved species are assumed equal to unity.

1. With Gibbs free-energy data from a single reference source (why one source?), calculate the difference in the free energies of diamond and graphite (polymorphs of pure carbon) at 25°C. What is the value of the equilibrium constant of the reaction between them? Which phase is stable and which metastable at 25°C and 1 bar total pressure? Discuss.

2. (a) Given the following thermodynamic data for 25°C and 1 bar pressure, calculate the dissociation constant of water, $K_w = [H^+][OH^-]$.

Species	ΔH_f° (kcal/mol)	S° (cal/mol K)
H^+	0	0
OH^-	-54.977	-2.560
$H_2O(l)$	-68.315	16.7

   (b) Calculate K_w given that $\Delta G_f^\circ[H_2O(l)] = -56.678$ kcal/mol, and $\Delta G_f^\circ(OH^-) = -37.604$ kcal/mol. Compare this result to K_w obtained in part (a). Are all the thermodynamic data internally consistent?
   (c) Does the pH of neutrality increase or decrease as the temperature increases?

3. Starting with published free-energy data, calculate the pH at which $[HF] = [F^-]$ at 25°C and 1 bar pressure.

4. Lime (CaO) reacts with water to form portlandite $[Ca(OH)_2]$ in cements and lime mortars.
   (a) Calculate the free energy of the reaction that forms portlandite in liquid water and discuss.
   (b) If the water is present as water vapor, what is the water vapor pressure (P_{H_2O}) in bars at which lime and portlandite can coexist at equilibrium?
   (c) Does the reaction produce heat or consume it?

5. Thermodynamic data that may be useful in solving this problem are tabulated below. Values are for 25°C and 1 bar pressure.

Species	ΔH_f° (kcal/mol)	ΔG_f° (kcal/mol)	S° (cal/mol K)
$O_2(g)$	0	0	49.003
C (graphite)	0	0	1.372
$CO_2(g)$	-94.051	-94.254	51.07
H_2CO_3	-167.22	-148.94	44.8
HCO_3^-	-165.39	-140.26	21.8
CO_3^{2-}	-161.84	-126.17	-13.6
$Ca(c)$	0	0	9.90
Ca^{2+}	-129.74	-132.30	-12.7
$CaCO_3$ (calcite)	-288.48	-269.79	22.2

   (a) Write the reaction that represents the formation of calcite ($CaCO_3$) from the elements in their most stable states at 25°C and 1 bar pressure, and compute ΔS_r° for the reaction.
   (b) How are ΔH_f° and ΔG_f° for calcite related to ΔH_r° and ΔG_r° for this reaction? -269.78 kcal/mol
   (c) Compute ΔG_f°(calcite) from its solubility product, given that $K_{sp} = 10^{-8.48}$.
   (d) Calculate ΔH_f°(calcite) from the entropy and free energy of formation data determined above.

6. (a) Write the reaction that describes the formation of sulfate ion from the elements. (The "elements" can include electrons if appropriate.) Calculate the entropy of formation of SO_4^{2-} from the elements at 25°C and 1 bar pressure.
   (b) Compute ΔG_f° for sulfate ion from the entropy of the above formation reaction and the fact that $\Delta H_f^\circ(SO_4^{2-}) = -217.40$ kcal/mol.

7. The breakdown of K-feldspar to form kaolinite clay is an important weathering reaction, particularly in humid climate soils. The reaction may be written:

$$KAlSi_3O_8 + H^+ + \tfrac{9}{2}H_2O(l) = \tfrac{1}{2}Al_2Si_2O_5(OH)_4 + K^+ + 2H_4SiO_4^\circ$$

K-feldspar   kaolinite

(a) The equilibrium constant at 25°C and 1 bar pressure is $K_{eq} = 10^{-0.9}$. Write the corresponding equilibrium constant expression.

(b) Assuming $[K^+] = 2 \times 10^{-4}$ mol/kg and the dissolved silica is $[H_4SiO_4^\circ] = 2 \times 10^{-4}$ mol/kg in soil moisture, at what pH would the feldspar be in equilibrium with the kaolinite and, therefore, not weather? Given that the pH of most humid climate soils is below 6.0, discuss the significance of your answer.

8. You are given the following free-energy data in kcal/mol for 25°C and 1 bar total pressure.

Species	ΔG_f°	Species	ΔG_f°
Ca^{2+}	-132.3	$CaCO_3$ (calcite)	-270.01
Mn^{2+}	-54.5	$MnCO_3$ (rhodocrosite)	-195.2
CO_3^{2-}	-126.17		

(a) Assuming manganoan calcite with a composition $Ca_{0.95}Mn_{0.05}CO_3$ is an ideal solid solution, calculate its free energy of formation.

(b) Write the dissolution reaction for the solid solution and calculate its solubility product.

(c) The drinking water standard for dissolved Mn^{2+} is 0.05 mg/L. Assuming dissolved $Ca^{2+} = 10^{-3.00}$ M and $CO_3^{2-} = 10^{-5.48}$ M, compute $Mn^{2+}(aq)$ at equilibrium with this solid solution. Compare this result to $Mn^{2+}(aq)$ at saturation with respect to pure rhodocrosite. How do both of these concentrations compare to the drinking water standard?

9. Assume that Henry's law applies to minor component B and Raoult's law to major component A of binary solid solution, $A_{1-N}B_NX$, and that A and B have the same charge. Given also that $K_{Henry} = 10$, and that the solubility products of pure AX and BX are $K_{sp}(AX) = 10^{-8.00}$ and $K_{sp}(BX) = 10^{-10.00}$, and that the composition of the solid solution is $A_{0.97}B_{0.03}X$.

(a) Compute the activities of components AX and BX in the solid solution.

(b) Express and calculate D, the distribution coefficient for the exchange reaction: $A + BX = B + AX$.

(c) Assuming that $A = 10^{-3.00}$ mol/kg, calculate B at equilibrium with the solid solution.

10. The curve in the figure at the end of this problem, which is based on Busenberg and Plummer (1989), shows the general trend of measured solubility products of relatively well-crystallized natural and synthetic magnesian calcites. (Biogenic calcites are typically 0.1 to 0.15 log K_{sp} units more soluble at a given mole fraction of Mg.) The Gibbs free energies of Ca^{2+}, Mg^{2+}, and CO_3^{2-} at 25°C and 1 bar pressure are -132.30, -108.7, and -126.17 kcal/mol, respectively.

Assuming that magnesian calcites are an ideal solid solution of calcite ($K_{sp} = 10^{-8.48}$) and magnesite ($K_{sp} = 10^{-8.03}$), calculate the K_{sp} for magnesian calcites up to a mole fraction of 0.16 $MgCO_3$ for $CaCO_3$ ($N_{MgCO_3} = 0.16$). Make your calculations for N_{MgCO_3} values of 0.02, 0.04, 0.08, 0.12, and 0.16. Compare your results to the empirical K_{sp} curve in the figure, and discuss the applicability of the ideal solution model to this binary solid solution. What does any departure of the empirical K_{sp} curve from ideal solution behavior tell you about the thermodynamic stability of high-Mg^{2+} calcites? Note that the ionic radii of Ca^{2+} and Mg^{2+} in six-fold coordination are very different at 1.00 and 0.72 Å, respectively.

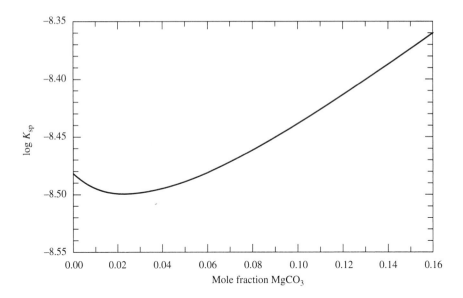

General trend of solubility products of magnesian calcites as a function of $MgCO_3$ mole fraction. Based on Busenberg and Plummer (1989).

11. Given the reaction

$$CaCO_3(\text{calcite}) + CO_2(g) + H_2O = Ca^{2+} + 2HCO_3^-$$

and the thermodynamic data table:

Substance	ΔH_f° (kcal/mol)	ΔG_f° (kcal/mol)	S° (cal/mol K)
$CaCO_3(\text{calcite})$	-288.46	-269.80	22.2
$CO_2(g)$	-94.051	-94.254	51.06
$H_2O(l)$	-68.315	-56.687	16.71
Ca^{2+}	-129.74	-132.30	-12.7
HCO_3^-	-165.39	-140.26	21.8

compute the following.

(a) The Gibbs free energy of the reaction (ΔG_r°) and its K_{eq}; express the free energy in kcal/mol and kJ/mol.

(b) The ΔH_r° of the reaction from the tabulated enthalpy data, also expressing the result in kcal/mol and kJ/mol.

(c) The ΔG_r° of the reaction from the tabulated enthalpy and entropy data. Compare your answer to ΔG_r° computed via the approach used in (a).

(d) K_{eq} at 10°C from thermodynamic data at 25°C using the van't Hoff equation.

12. The solubility of quartz below pH 8 can be written

$$SiO_2 + 2H_2O = H_4SiO_4^\circ$$

(a) What is the equilibrium constant expression for quartz dissolution?

Based on the literature and solubility measurements at 50°C and above, Rimstidt and Barnes (1980) obtained the following temperature function for the equilibrium constant of this reaction at 1 bar total pressure from 0 to 360°C:

$$\log K_{eq} = 1.8814 - 2.028 \times 10^{-3}T - 1560.46/T(\text{K})$$

(b) Compute the concentration of dissolved silica as SiO_2 (ppm) you would expect at equilibrium with quartz in water at 25, 100, and 300°C. Assume $\gamma = 1$ for silicic acid $H_4SiO_4^\circ$.

(c) Compute the expression for the enthalpy of the dissolution reaction as a function of $T(K)$ and its value in kcal/mol at 25°C. Is the dissolution reaction endothermic or exothermic?

(d) What is the heat capacity (ΔC_p°) of the reaction at 25°C?

(e) Could the simple van't Hoff equation ($\Delta H_r^\circ = $ constant) or its expanded version ($\Delta Cp_r^\circ = $ constant) have been used to predict the solubility of quartz at 300°C from the values for K_{eq}, ΔH_r°, and ΔCp_r° at 25°C? Compare the results of such calculations to the value computed above in (b), and explain any differences.

(f) Compute ΔS_r°, the entropy of the reaction at 25°C from the K_{eq} and ΔH_r° values.

(g) Is there any other way you could have determined this entropy from the experimental solubility data? Do so and compare the results.

(h) Given that the entropies of quartz and liquid water at 25°C are 9.909 and 16.718 cal/mol K, compute S° for silicic acid.

13. Palmer and Hyde (1993) measured association constants along the liquid water–water vapor curve from 25 to 300°C for the reaction

$$Fe^{2+} + Cl^- = FeCl^+$$

The experimental results are tabulated here.

$T°C$	$\log K_{assoc}$	$T°C$	$\log K_{assoc}$
25	-0.125	200	1.09
50	-0.026	250	1.58
100	0.266	300	2.10
150	0.648		

Their study led to the following expression for the association constant of $FeCl^+$, as a function of degrees K:

$$\log K_{assoc} = -7.1783 + \frac{911.13}{T} + 0.013407\,T$$

The following thermodynamic data are available from the literature for 25°C and 1 bar total pressure.

Species	ΔH_f° (kcal/mol)	ΔG_f° (kcal/mol)	S° (cal/mol K)
Fe^{2+}	-21.3	-18.85	-32.9
Cl^-	-39.933	-31.379	13.56

(a) Compute ΔG_f° for the $FeCl^+$ complex at 25°C.

(b) From the K_{assoc} versus temperature (K) equation, derive the equation that describes the dependence of ΔH_r° on $T(K)$, and compute the value of ΔH_r° in kcal/mol for the complexation (association) reaction at 25°C.

(c) Compute the entropy of the $FeCl^+$ complex at 25°C.

(d) Seawater contains $Cl^- = 0.56$ mol/kg. What percentage of $\Sigma Fe(II)$ in a shallow marine mud will be present as $FeCl^+$ at 25°C. Ignore activity coefficients in this problem and in (e).

(e) Answer question (d) assuming the seawater is present interstitially in a marine mud at 4°C and discuss.

14. **(a)** Calculate the Henry's law constant (K_H) for dissolution of H_2S gas in water at 25°C and 1 bar total pressure from free-energy data.

(b) Does an increase in temperature favor dissolution or exsolution of hydrogen sulfide?

15. The dissolution of amorphous silica in water can be written:

$$SiO_2(am) + 2H_2O = H_4SiO_4^\circ$$

According to Millero (1982), $\Delta V_r^\circ = -8.7$ cm³/mol and $\Delta K_r^\circ = -2.6 \times 10^{-3}$ cm³/mol bar for this reaction.

(a) Assuming that ΔV_r° is independent of pressure and given that $K_{sp} = 1.95 \times 10^{-3}$ at 25°C and 1 bar pressure, calculate K_{sp} at 500 and 1000 bars pressure.

(b) Millero (1982) plots the empirical solubility measurements for $SiO_2(am)$ at 25°C as a function of pressure. The plot indicates that at 500 and 1000 bars pressure, the solubility is 18% and 35% higher than at 1 bar pressure. Taking the isothermal compressibility into account, compute K_{sp} at 500 and 1000 bars. Compare your results to those computed in part (a) and to the pressure effect suggested by Millero's plot. Discuss any differences.

CHAPTER 1 APPENDIX

Following are the temperature functions for some equilibrium constants at 1 bar pressure below 100°C and at H_2O vapor-liquid saturation pressures above 100°C (Table A1.1), and equilibrium constants and enthalpies of selected reactions at 25°C and 1 bar pressure (Table A1.2).

$$\left(\frac{\partial H}{\partial T}\right)_V = \left(1 - \frac{\alpha \mu}{K_J}\right) C_p$$

$$K_T = -\frac{1}{V}\left(\frac{\partial V}{\partial P}\right)_T$$

TABLE A1.1 Temperature (K) functions for some equilibrium constants at 1 bar pressure below 100°C and at H_2O vapor-liquid saturation pressures above 100°C

$$\log K_T = a + bT + \frac{c}{T} + d\log_{10}T + \frac{e}{T^2}$$

Substance	K_n n	a	b	c	d	e	$\log K$ at 298.15 K (25°C)	T range (°C)	Source
H_2O	1	-283.9710	-0.05069842	13323.00	102.24447	-1119669	-14.000	0-200	1
H_3BO_3 (boric acid)	1	-255.3651	-0.03862042	14896.21	89.01867	-1119669	-9.242	0-200	2
H_2CO_3 (carbonic acid)	1	-356.3094	-0.06091964	21834.37	126.8339	-1684915	-6.352	0-250	3
H_2CO_3	2	-107.8871	-0.03252849	5151.79	38.92561	-563713.9	-10.329	0-250	3
CO_2 (Henry's law)		108.3865	0.01985076	-6919.53	-40.45154	669365	-1.468	0-250	3
HNO_3 (nitric acid)	1	-832.5472	-0.119	50736.3499	297.4825	-3275558.33	1.396	0-300	4
HNO_2 (nitrous acid)	1	-380.4716	-0.0635	21024.1734	137.9557	-1388701.86	-3.148	0-300	5
NH_4^+ (ammonium ion)	1	-2.797	.006358	-975.9	-1.141	-199500	-9.242	0-300	6
H_4SiO_4 (orthosilicic acid)	1	-624.921	-0.09761	33632.7	224.623	-2171000	-9.826	0-300	4
H_4SiO_4	2	8.424	-0.02196	-4465.2			-13.100	0-300	7
H_3PO_4 (orthophosphoric acid)	1	4.5535	-0.013486	-799.31			-2.148	0-300	4
H_3PO_4	2	5.3541	-0.01984	-1979.5			-7.200	0-300	4

TABLE A1.1 (continued)

$$\log K_T = a + bT + \frac{c}{T} + d\log_{10}T + \frac{e}{T^2}$$

Substance	K_n n	a	b	c	d	e	log K at 298.15 K (25°C)	T range (°C)	Source
H$_2$S (hydrogen sulfide)	1	-32.55	-0.02722	-1519.44	15.672		-6.983	0–300	8
H$_2$S$_2$O$_3$ (thiosulfuric acid)	1	-541.5804	-0.0838	32191.9934	194.2305	-2009986.85	-0.593	0–300	5
H$_2$S$_2$O$_3$	2	-1730.6936	-0.2362	105766.2837	614.9098	-6837046.11	-1.733	0–300	5
H$_2$SO$_3$ (sulfurous acid)	1	-1262.6927	-0.1794	77739.3993	449.3378	-5181506.81	-1.873	0–300	5
H$_2$SO$_3$	2	-1910.071	-0.2581	115644.769	677.7901	-7578788.5	-7.259	0–300	5
H$_2$SO$_4$ (sulfuric acid)	2	-854.2056	-0.1309	50019.7772	306.4789	-3098920.52	-1.965	0–300	4
HF (hydrofluoric acid)	1	-2.033	0.012645	429.01			3.18	0–100	1
HClO (hypochlorous acid)	1	-564.1484	-0.0862	31799.3262	201.7785	-2100605.68	-7.536	0–300	5
HClO$_2$ (chlorous acid)	1	-672.9805	-0.0987	40612.7057	239.9064	-2612786.88	-1.952	0–300	5
H$_2$CrO$_4$ (chromic acid)	2	-1819.2811	-0.247	110026.4085	646.0095	-7211873.02	-6.515	0–300	5
H$_3$AsO$_3$ (arsenious acid)	1	-459.1769	-0.0732	24684.7051	164.8478	-1687780.5	-9.290	0–300	5

40

TABLE A1.1 (continued)

$$\log K_T = a + bT + \frac{c}{T} + d\log_{10}T + \frac{e}{T^2}$$

Substance	K_n n	a	b	c	d	e	log K at 298.15 K (25°C)	T range (°C)	Source
H_3AsO_4 (arsenic acid)	1	−534.3496	−0.0826	31240.5266	191.7448	−2001838.69	−2.255	0–300	5
H_3AsO_4	2	−1395.4222	−0.1956	82683.1875	497.0172	−5364516.32	−6.931	0–300	5
H_2SeO_3 (selenious acid)	1	−553.0896	−0.0855	32151.2938	198.4306	−2046559.24	−2.765	0–300	5
H_2SeO_3	2	−1677.3859	−0.229	100856.6346	595.5211	−6640817.56	−8.515	0–300	5
H_2SeO_4 (selenic acid)	2	−1680.5823	−0.2294	102759.7564	597.2403	−6684221.24	−1.684	0–300	5
H_6TeO_6 (orthotelluric acid)	1	−497.955	−0.0772	27622.6829	178.6103	−1893918.66	−7.671	0–300	5
H_6TeO_6	2	−1442.9657	−0.2015	84031.7106	514.5146	−5593344.29	−10.988	0–300	5
HIO_3 (iodic acid)	1	−482.4423	−0.0765	27908.0585	174.2042	−1797340.39	−0.809	0–300	5
HCOOH (formic acid)	1	−547.0803	−0.0842	31649.6009	196.1681	−2055792	−3.753	0–300	5
CH_3COOH (acetic acid)	1	10.5850	−0.0372	−11175.3188	7.1388	1386117.6836	−4.731	0–300	5
HOOC–COOH (oxalic acid)	1	−15.44	0.05616	13220	−12.39	−1443000	−1.247	0–45	9
HOOC–COOH	2	6.5007	−0.00295	−1423.8			−4.266	0–50	9

41

TABLE A1.1 (continued)

$$\log K_T = a + bT + \frac{c}{T} + d\log_{10}T + \frac{e}{T^2}$$

Substance	K_n (n)	a	b	c	d	e	log K at 298.15 K (25°C)	T range (°C)	Source
CH₃CH₂COOH (propionic acid)	1	−490.8356	−0.0766	28233.9044	175.7014	−1835460.0845	−4.863	0–300	5
CH₃(CH₂)₂COOH (n-butyric acid)	1	2.6215	−0.013334	−1033.39			−4.820	0–60	10
CaHCO₃⁺	d	1209.120	0.31294	−34765.05	−478.782		1.106	0–90	3
MgHCO₃⁺	d	−59.215		2537.455	20.92298		1.07	0–50	11
SrHCO₃⁺	d	−3.248	0.014867				1.18	5–80	12
BaHCO₃⁺	d	−3.0938	0.013669				0.982	5–80	13
CaCO₃°	d	−1228.732	−0.299444	35512.75	485.818		3.224	0–90	3
MgCO₃°	d	0.9910	0.00667				2.98	0–50	14
SrCO₃°	d	−1.019	0.012826				2.81	5–80	12
BaCO₃°	d	0.113	0.008721				2.71	5–80	13
CaCO₃ (calcite)	sp	−171.9065	−0.077993	2839.319	71.595		−8.480	0–90	3
CaCO₃ (aragonite)	sp	−171.9773	−0.077993	2903.293	71.595		−8.336	0–90	3
SrCO₃ (strontianite)	sp	155.0305		−7239.594	−56.58638		−9.271	0–90	12

$$\log K_T = a + bT + \frac{c}{T} + d\log_{10}T + \frac{e}{T^2}$$

Substance	K_n n	a	b	c	d	e	log K at 298.15 K (25°C)	T range (°C)	Source
$BaCO_3$ (witherite)	sp	607.642	0.121098	-20011.25	-236.4948		-8.562	0-90	13
$CaSO_4$ (anhydrite)	sp	197.52		-8669.8	-69.835		-4.361	25-56	15
$CaSO_4$ (anhydrite)	sp	87.805		-3210.8	-32.8461		-4.24	70-150	15
$CaSO_4 \cdot 2H_2O$ (gypsum)	sp	68.2401		-3221.51	-25.0627		-4.581	25-90	15
$BaSO_4$ (barite)	sp	136.035		-7680.41	-48.595		-9.970	22-280	15
$SrSO_4$ (celestite)	sp	137.555		-6530.75	-49.419		-6.633	25-100	15
$PbSO_4$ (anglesite)	sp	0.0725	-0.0151	127.0938	-0.1472	-303524.2964	-7.782	0-60	16
$RaSO_4$	sp	137.98		-8346.87	-48.595		-10.261		17
$FeSO_4 \cdot 7H_2O$ (melanterite)	sp	1.447	-0.004153			-214949	-2.209	22-280	18
CaF_2 (fluorite)	sp	66.348		-4298.2	-25.271		-10.60	0-350	1
$AlOH^{2+}$	1	-38.253		-656.27	14.327		-5.00		19
$AlOH_2^+$	2	88.500		-9391.6	-27.121		-10.1		19

TABLE A1.1 (continued)

Substance	K_n n	$\log K_T = a + bT + \dfrac{c}{T} + d\log_{10}T + \dfrac{e}{T^2}$ a	b	c	d	e	$\log K$ at 298.15 K (25°C)	T range (°C)	Source
$AlOH_3^\circ$	3	226.374		-18247.8	-14.865		-16.9	50–300	19
$AlOH_4^-$	4	51.578		-11168.9	-14.865		-22.7	50–300	19
SiO_2 (quartz)	sp	1.8814	-0.002028	-1560.46			-3.96	50–300	20
SiO_2 (chalcedony)	sp	-0.09		-1032			-3.55	50–250	20
SiO_2 (amorphous)	sp	0.338037	-0.00078896	-840.075			-2.71	25–200	20
$Mg_3Si_2O_5(OH)_4$ (chrysotile)	sp	13.248		10217.1	-6.1894		32.20		21

Note: Sources 4, 5, 8, 9, and 16 report empirical $\log K$ values measured at different temperatures. R. Schmiermund employed Sigma-Plot (Jandell Scientific, Sausalito, CA) and a least-sum-of-squares curve-fitting routine to generate functions of the general form used in this table from the empirical data.

First (K_1) and second (K_2) stepwise acid dissociation constants (indicated by 1 and 2 in the "K_n" column, respectively) are for the following general reactions and dissociation constant expressions for the acid H_nA:

$$H_nA \rightleftarrows H^+ + H_{n-1}A^- \qquad H_{n-1}A^- \rightleftarrows H^+ + H_{n-2}A^{2-}$$

$$K_1 = \frac{[H^+][H_{n-1}A^-]}{[H_nA]} \qquad K_2 = \frac{[H^+][H_{n-2}A^{2-}]}{[H_{n-1}A^-]}$$

The "d" in the "K_n" column indicates that K values are for the ion pair *dissociation* reactions.

Solubility products (indicated by "sp" in the "K_n" column) are of the form $K_{sp} = [M]^a[L]^b$ for the salt M_aL_b, in equilibrium aqueous species M and L, except as noted in source 20.

Sources:

[1]Nordstrom et al. (1990) and references cited therein. The function for $\log K_w$ gives $\log K_w$ values that agree with those measured by Sweeton et al. (1974), and those model-predicted by Tanger and Helgeson (1988), to within ± 0.005 units up to $100°C$ and ± 0.007 units from 100 to $200°C$.

[2]The function has been obtained by combining the temperature function given by Mesmer et al. (1972) and the $K_w(H_2O)$ temperature function from Nordstrom et al. (1990) given in this table.

[3]Plummer and Busenberg (1982). The Henry's law constant (K_{CO_2}) for CO_2 is of the form: $K_{CO_2} = [H_2CO_3^\circ]/P_{CO_2}$.

[4]Parameters and K values for $25°C$ were obtained by fitting experimental $\log K_T$ data from various sources as provided by Smith et al. (1986, Tables 1 and 7).

[5]Parameters and K values at $25°C$ were obtained by fitting calculated values provided by Smith et al. (1986, Table 7), which were themselves obtained using the equation of Helgeson (1967, 1969).

[6]The temperature function for NH_4^+ has been obtained by fitting values given by Henley (1984), which were apparently derived from Hitch and Mesmer (1976).

[7]Baes and Mesmer (1976) for the value at $25°C$ only. The slope of the function above $25°C$ is from Volosov et al. (1972).

[8]Barberro et al. (1982). See also Hersey et al. (1988).

[9]K_1 data for oxalic acid is from Robinson and Stokes (1959). The K_2 value for $25°C$ is from McAuley and Nancollas (1961). The temperature function for K_2 has been obtained by fitting measured values reported by McAuley and Nancollas (op. cit.).

[10]Harned and Owen (1958).

[11]Siebert and Hostetler (1977a).

[12]Busenberg et al. (1984).

[13]Busenberg and Plummer (1986).

[14]Siebert and Hostetler (1977b).

[15]Langmuir and Melchior (1985). $K_{sp}(\text{gypsum}) = K_{sp}(\text{anhydrite}) = 10^{-4.64}$ at $56°C$. Anhydrite is metastable below about $56°C$ at unit activity of water.

[16]Parameters and K values for $25°C$ were obtained by fitting experimental $\log K_{sp}$ data given by Paige et al. (1992).

[17]Langmuir and Riese (1985).

[18]Reardon and Beckie (1987).

[19]The temperature functions are from Nordstrom and May (1989). K values are for reactions written in the form: $Al^{3+} + nH_2O = Al(OH)_n^{3-n} + nH^+$. Values in the "$K_n$" column are successive n values in this reaction. See also Pokrovskii and Helgeson (1995) for stability constants up to $250–350°C$.

[20]K values are for the dissolution reaction written in the form: $SiO_2 + 2H_2O \rightleftharpoons H_4SiO_4^\circ$. The K_{sp} function for chalcedony is from Nordstrom et al. (1990). The K_{sp} functions for quartz and amorphous silica are from Rimstidt and Barnes (1980).

[21]Nordstrom et al. (1990). The K value is K_{eq} for the dissolution reaction written: $Mg_3Si_2O_5(OH)_4 + 6H^+ = 3Mg^{2+} + 2H_4SiO_4^\circ + H_2O$.

TABLE A1.2　Equilibrium constants and enthalpies of selected reactions at 25°C and 1 bar pressure

I Fluoride and Chloride Species

Reaction	ΔH_r° (kcal/mol)	$\log K$	Reaction	ΔH_r° (kcal/mol)	$\log K$
$H^+ + F^- = HF^\circ$	3.18	3.18	$Al^{3+} + F^- = AlF^{2+}$	1.06	7.0
$H^+ + 2F^- = HF_2^-$	4.55	3.76	$Al^{3+} + 2F^- = AlF_2^+$	1.98	12.7
$Na^+ + F^- = NaF^\circ$	—	−0.24	$Al^{3+} + 3F^- = AlF_3^\circ$	2.16	16.8
$Ca^{2+} + F^- = CaF^+$	4.12	0.94	$Al^{3+} + 4F^- = AlF_4^-$	2.20	19.4
$Mg^{2+} + F^- = MgF^+$	3.2	1.82	$Al^{3+} + 5F^- = AlF_5^{2-}$	1.84	20.6
$Mn^{2+} + F^- = MnF^+$	—	0.84	$Al^{3+} + 6F^- = AlF_6^{3-}$	−1.67	20.6
$Fe^{2+} + F^- = FeF^+$	—	1.0			
$Fe^{3+} + F^- = FeF^{2+}$	2.7	6.2	$Si(OH)_4 + 4H^+ + 6F^-$	−16.26	30.18
$Fe^{3+} + 2F^- = FeF_2^+$	4.8	10.8	$= SiF_6^{2-} + 4H_2O$		
$Fe^{3+} + 3F^- = FeF_3^\circ$	5.4	14.0	$Fe^{2+} + Cl^- = FeCl^+$	—	0.14
$Mn^{2+} + Cl^- = MnCl^+$	—	0.61	$Fe^{3+} + Cl^- = FeCl^{2+}$	5.6	1.48
$Mn^{2+} + 2Cl^- = MnCl_2^\circ$	—	0.25	$Fe^{3+} + 2Cl^- = FeCl_2^+$	—	2.13
$Mn^{2+} + 3Cl^- = MnCl_3^-$	—	−0.31	$Fe^{3+} + 3Cl^- = FeCl_3^\circ$	—	1.13

Mineral	Reaction	ΔH_r° (kcal/mol)	$\log K$
Cryolite	$Na_3AlF_6 = 3Na^+ + Al^{3+} + 6F^-$	9.09	−33.84
Fluorite	$CaF_2 = Ca^{2+} + 2F^-$	4.69	−10.6

Redox potentials	ΔH_r° (kcal/mol)	E° (volts)	$\log K$
$Fe^{2+} = Fe^{3+} + e^-$	9.68	−0.770	−13.02
$Mn^{2+} = Mn^{3+} + e^-$	25.8	−1.51	−25.51

TABLE A1.2   (continued)

	II Oxide and Hydroxide Species				
Reaction	ΔH_r° (kcal/mol)	log K	Reaction	ΔH_r° (kcal/mol)	log K
$H_2O = H^+ + OH^-$	13.362	−14.000	$Fe^{3+} + H_2O = FeOH^{2+} + H^+$	10.4	−2.19
$Li^+ + H_2O = LiOH^\circ + H^+$	0.0	−13.64	$Fe^{3+} + 2H_2O = Fe(OH)_2^+ + 2H^+$	17.1	−5.67
$Na^+ + H_2O = NaOH^\circ + H^+$	0.0	−14.18	$Fe^{3+} + 3H_2O = Fe(OH)_3^\circ + 3H^+$	24.8	−12.56
$K^+ + H_2O = KOH^\circ + H^+$	—	−14.46	$Fe^{3+} + 4H_2O = Fe(OH)_4^- + 4H^+$	31.9	−21.6
$Ca^{2+} + H_2O = CaOH^+ + H^+$	—	−12.78	$2Fe^{3+} + 2H_2O = Fe2(OH)_2^{4+} + 2H^+$	13.5	−2.95
$Mg^{2+} + H_2O = MgOH^+ + H^+$	—	−11.44	$3Fe^{3+} + 4H_2O = Fe3(OH)_4^{5+} + 4H^+$	14.3	−6.3
$Sr^{2+} + H_2O = SrOH^+ + H^+$	—	−13.29	$Al^{3+} + H_2O = AlOH^{2+} + H^+$	11.49	−5.00
$Ba^{2+} + H_2O = BaOH^+ + H^+$	—	−13.47	$Al^{3+} + 2H_2O = Al(OH)^{2+} + 2H^+$	26.90	−10.1
$Ra^{2+} + H_2O = RaOH^+ + H^+$	—	−13.49	$Al^{3+} + 3H_2O = Al(OH)_3^\circ + 3H^+$	39.89	−16.9
$Fe^{2+} + H_2O = FeOH^+ + H^+$	13.2	−9.5	$Al^{3+} + 4H_2O = Al(OH)_4^- + 4H^+$	42.30	−22.7
$Mn^{2+} + H_2O = MnOH^+ + H^+$	14.4	−10.59			

Mineral	Reaction	ΔH_r° (kcal/mol)	log K
Portlandite	$Ca(OH)_2 + 2H^+ = Ca^{2+} + 2H_2O$	−31.0	22.8
Brucite	$Mg(OH)_2 + 2H^+ = Mg^{2+} + 2H_2O$	−27.1	16.84
Pyrolusite	$MnO_2 + 4H^+ + 2e^- = Mn^{2+} + 2H_2O$	−65.11	41.38
Hausmanite	$Mn3O_4 + 8H^+ + 2e^- = 3Mn^{2+} + 4H_3O$	−100.64	61.03
Manganite	$MnOOH + 3H^+ + e^- = Mn^{2+} + 2H_2O$	—	25.34
Pyrochroite	$Mn(OH)_2 + 2H^+ = Mn^{2+} + 2H_2O$	—	15.2
Gibbsite (crystalline)	$Al(OH)_3 + 3H^+ = Al^{3+} + 3H_2O$	−22.8	8.11
Gibbsite (microcrystalline)	$Al(OH)_3 + 3H^+ = Al^{3+} + 3H_2O$	(−24.5)	9.35
$Al(OH)_3$ (amorphous)	$Al(OH)_3 + 3H^+ = Al^{3+} + 3H_2O$	(−26.5)	10.8
Goethite	$FeOOH + 3H^+ = Fe^{3+} + 2H_2O$	—	−1.0
Ferrihydrite (amorphous to microcrystalline)	$Fe(OH)_3 + 3H^+ = Fe^{3+} + 3H_2O$	—	3.0 to 5.0

TABLE A1.2　(continued)

III Carbonate Species

Reaction	ΔH_r° (kcal/mol)	log K	Reaction	ΔH_r° (kcal/mol)	log K
$CO_2(g) = CO_2(aq)$	−4.776	−1.468	$Ca^{2+} + CO_3^{2-} = CaCO3^\circ$	3.545	3.224
$CO_2(aq) + H_2O = H^+ + HCO_3^-$	2.177	−6.352	$Mg^{2+} + CO_3^{2-} = MgCO_3^\circ$	2.713	2.98
$HCO_3^- = H^+ + CO_3^{2-}$	3.561	−10.329	$Sr^{2+} + CO_3^{2-} = SrCO_3^\circ$	5.22	2.81
$Ca^{2+} + HCO_3^- = CaHCO_3^+$	2.69	1.106	$Ba^{2+} + CO_3^{2-} = BaCO_3^\circ$	3.55	2.71
$Mg^{2+} + HCO_3^- = MgHCO_3^+$	0.79	1.07	$Mn^{2+} + CO_3^{2-} = MnCO_3^\circ$	—	4.90
$Sr^{2+} + HCO_3^- = SrHCO_3^+$	6.05	1.18	$Fe^{2+} + CO_3^{2-} = FeCO_3^\circ$	—	4.38
$Ba^{2+} + HCO_3^- = BaHCO_3^+$	5.56	0.982	$Na^+ + CO_3^{2-} = NaCO_3^-$	8.91	1.27
$Mn^{2+} + HCO_3^- = MnHCO_3^+$	—	1.95	$Na^+ + HCO_3^- = NaHCO_3^-$	—	−0.25
$Fe^{2+} + HCO_3^- = FeHCO_3^+$	—	2.0	$Ra^{2+} + CO_3^{2-} = RaCO_3^\circ$	1.07	2.5

Mineral	Reaction	ΔH_r° (kcal/mol)	log K
Calcite	$CaCO_3 = Ca^{2+} + CO_3^{2-}$	−2.297	−8.480
Aragonite	$CaCO_3 = Ca^{2+} + CO_3^{2-}$	−2.589	−8.336
Dolomite (ordered)	$CaMg(CO_3)_2 = Ca^{2+} + Mg^{2+} + 2CO_3^{2-}$	−9.436	−17.09
Dolomite (disordered)	$CaMg(CO_3)_2 = Ca^{2+} + Mg^{2+} + 2CO_3^{2-}$	−11.09	−16.54
Strontianite	$SrCO_3 = Sr^{2+} + COCO_3^{2-}$	−0.40	−9.271
Siderite (crystalline)	$FeCO_3 = Fe^{2+} + CO_3^{2-}$	−2.48	−10.89
Siderite (precipitated)	$FeCO_3 = Fe^{2+} + CO_3^{2-}$	—	−10.45
Witherite	$BaCO_3 = Ba^{2+} + CO_3^{2-}$	0.703	−8.562
Rodocrosite (crystalline)	$MnCO_3 = Mn^{2+} + CO_3^{2-}$	−1.43	−11.13
Rhodocrosite (synthetic)	$MnCO_3 = Mn^{2+} + CO_3^{2-}$	—	−10.39

IV Silicate Species

Reaction	ΔH_r° (kcal/mol)	log K	Reaction	ΔH_r° (kcal/mol)	log K
$Si(OH)_4^\circ = SiO(OH)_3^- + H^+$	6.12	−9.83	$Si(OH)_4^\circ = SiO_2(OH)_2^{2-} + 2H^+$	17.6	−23.0

Mineral	Reaction	ΔH_r° (kcal/mol)	log K
Kaolinite	$Al_2Si_2O_5(OH)_4 + 6H^+ = 2Al^{3+} + 2Si(OH)_4^\circ + H_2O$	−35.3	7.435
Chrysotile	$Mg_3Si_2O_5(OH)_4 + 6H^+ = 3Mg^{2+} + 2Si(OH)_4^\circ + H_2O$	−46.8	32.20
Sepiolite	$Mg_2Si_3O_{7.5}(OH) \cdot 3H_2O + 4H^+ + 0.5H_2O = 2Mg^{2+} + 3Si(OH)_4^\circ$	−10.7	15.76
Kerolite	$Mg_3Si_4O_{10}(OH)_2 \cdot H_2O + 6H^+ + 3H_2O = 3Mg^{2+} + 4Si(OH)_4^\circ$	—	25.79
Quartz	$SiO_2 + 2H_2O = Si(OH)_4^\circ$	5.99	−3.98
Chalcedony	$SiO_2 + 2H_2O = Si(OH)_4^\circ$	4.72	−3.55
Amorphous silica	$SiO_2 + 2H_2O = Si(OH)_4^\circ$	3.34	−2.71

TABLE A1.2 (continued)

V Sulfate Species

Reaction	ΔH_r° (kcal/mol)	log K	Reaction	ΔH_r° (kcal/mol)	log K
$H^+ + SO_4^{2-} = HSO_4^-$	3.85	1.988	$Mn^{2+} + SO_4^{2-} = MnSO_4^\circ$	3.37	2.25
$Li^+ + SO_4^{2-} = LiSO_4^-$	—	0.64	$Fe^{2+} + SO_4^{2-} = FeSO_4^\circ$	3.23	2.25
$Na^+ + SO_4^{2-} = NaSO_4^-$	1.12	0.70	$Fe^{2+} + HSO_4^- = FeHSO_4^+$	—	1.08
$K^+ + SO_4^{2-} = KSO_4^-$	2.25	0.85	$Fe^{3+} + SO_4^{2-} = FeSO_4^+$	3.91	4.04
$Ca^{2+} + SO_4^{2-} = CaSO_4^\circ$	1.65	2.30	$Fe^{3+} + 2SO_4^{2-} = Fe(SO_4)_2^-$	4.60	5.38
$Mg^{2+} + SO_4^{2-} = MgSO_4^\circ$	4.55	2.37	$Fe^{3+} + HSO_4^- = FeHSO_4^{2+}$	—	2.48
$Sr^{2+} + SO_4^{2-} = SrSO_4^\circ$	2.08	2.29	$Al^{3+} + SO_4^{2-} = AlSO_4^+$	2.15	3.02
$Ba^{2+} + SO_4^{2-} = BaSO_4^\circ$	—	2.7	$Al^{3+} + 2SO_4^{2-} = Al(SO_4)_2^-$	2.84	4.92
$Ra^{2+} + SO_4^{2-} = RaSO_4^\circ$	1.3	2.75	$Al^{3+} + HSO_4^- = AlHSO_4^{2+}$	—	0.46

Mineral	Reaction	ΔH_r° (kcal/mol)	log K
Gypsum	$CaSO_4 \cdot 2H_2O = Ca^{2+} + SO_4^{2-} + 2H_2O$	−0.109	−4.58
Anhydrite	$CaSO_4 = Ca^{2+} + SO_4^{2-}$	−1.71	−4.36
Celestite	$SrSO_4 = Sr^{2+} + SO_4^{2-}$	−1.037	−6.63
Barite	$BaSO_4 = Ba^{2+} + SO_4^{2-}$	6.35	−9.97
Radium sulfate	$RaSO_4 = Ra^{2+} + SO_4^{2-}$	9.40	−10.26
Melanterite	$FeSO_4 \cdot 7H_2O = Fe^{2+} + SO_4^{2-} + 7H_2O$	4.91	−2.209
Alunite	$KAl_3(SO_4)_2(OH)_6 + 6H^+ = K^+ + 3Al^{3+} + 2SO_4^{2-} + 6H_2O$	−50.25	−1.4

Source: Reprinted with permission from Nordstrom et al. *Chemical Modeling of Aqueous Systems II.* © 1990 American Chemical Society.

2

Chemical Kinetics

2.1 CHEMICAL EQUILIBRIUM AND CHEMICAL KINETIC CONCEPTS

We will study the reactions that occur in gas/water/rock systems, so that we may understand, model, and predict them. The state of reactions can be described in terms of concepts of equilibrium or kinetics. Equilibrium models describe boundary conditions defined by assuming attainment of equilibrium. They give no information regarding the pathways or time it takes to reach equilibrium. Kinetic models may describe reaction pathways toward equilibrium and reaction position and times along those pathways. Overall reactions provide no detailed kinetic information, although the magnitude and sign of the reaction free energy indicates the tendency for a reaction to proceed in a particular direction.

Several key questions must be answered initially in a study of reaction chemistry. First, is the reaction sufficiently fast and reversible so that it can be regarded as chemical-equilibrium controlled? Second, is the reaction homogeneous (occurring wholly within a gas or liquid phase) or heterogeneous (involving reactants or products in a gas and a liquid, or liquid and a solid phase)? Slow reversible, irreversible, and heterogeneous (often slow) reactions are those most likely to require interpretation using kinetic models. Third, is there a useful volume of the water-rock system in which chemical equilibrium can be assumed to have been attained for many possible reactions? This may be called the *local equilibrium assumption.*

The following example should serve to illustrate some of these ideas. We will assume an elementary first-order reaction:

$$A \rightleftarrows B \tag{2.1}$$

(see kinetic discussion below). In general, rates of the forward reaction (rate +) and reverse reaction (rate −) are given by

$$\text{rate} + = k_+ (A), \quad \text{and} \quad \text{rate} - = k_- (B) \tag{2.2}$$

Here and later in this chapter the parentheses denote concentrations of the enclosed species. In (2.2), k_+ and k_- are proportionality constants and are termed the forward ($A \rightarrow B$) and reverse ($B \rightarrow A$) reaction rate constants. The overall rate, R, equals (rate +) − (rate −). At equilibrium the forward and reverse rates are equal so that these expressions may be combined to give

$$k_+ (A) = k_- (B) \tag{2.3}$$

The equilibrium constant (K_{eq}) then equals

$$K_{eq} = \frac{k_+}{k_-} = \frac{(B)}{(A)} \tag{2.4}$$

Remember that at equilibrium, not only are forward and reverse reaction rates equal, but also $\Delta G_r = 0$.

In a system that is open and its volume (V) completely mixed, with the amounts of substances supplied or removed per unit time equal, we may write

$$t_R = \frac{V}{Q} \tag{2.5}$$

where t_R is the residence time of the water, and Q is the volume rate of flow. For our example it can be shown that

$$t_{1/2} = \frac{\ln 2}{k_+} = \frac{0.693}{k_+} \tag{2.6}$$

(see Table 2.2 in Section 2.3) where $t_{1/2}$ is the half-life or half-time of the reaction (the time it takes for half of A to go to B, assuming that initially B = 0). If the initial concentration of B = 0, we can also show:

$$\frac{(A)}{(B)} = \frac{1}{K_{eq}} + \frac{t_{1/2}}{0.693 \, t_R} \tag{2.7}$$

When $t_R \gg t_{1/2}$ the last term vanishes, we are left with the expression for K_{eq}, and equilibrium concepts can be used (cf. Hoffman, 1981; Stumm and Morgan 1981). Conversely, when $t_R \ll t_{1/2}$ for a particular reaction, then kinetic concepts are necessary to explain the state of that reaction.

The dissolved species composition of most deep, confined groundwater and of some deep unconfined groundwater may remain constant for periods of years or even thousands of years. Such constancy suggests that reactions involving those species have come to thermodynamic equilibrium. When this is the case, the water/rock system involved can be considered a closed system with respect to those species and their controlling reactions in a thermodynamic sense (the system is open to the flow of energy, but closed to the flow of matter; remember that equilibrium is the time-invariant state of a closed system). For these conditions equilibrium concepts can be used to explain concentrations of the aqueous species involved.

Most geologic formations are comprised of variable amounts of minerals that are thermodynamically stable at low temperatures, such as quartz, calcite, and ferric oxyhydroxides, along with minerals that formed at high temperatures, such as the olivines and pyroxenes and calcic feldspars. These high-temperature phases are thermodynamically unstable in most low-temperature weathering environments and so rarely equilibrate with the water. (See Chap. 7.) Groundwaters in such systems may equilibrate with low-temperature secondary minerals that have formed by breakdown of the high-temperature phases.

In some shallow confined or unconfined water/rock systems, and in large surface-water bodies, the chemistry may appear constant in time without the attainment of equilibrium. Such systems

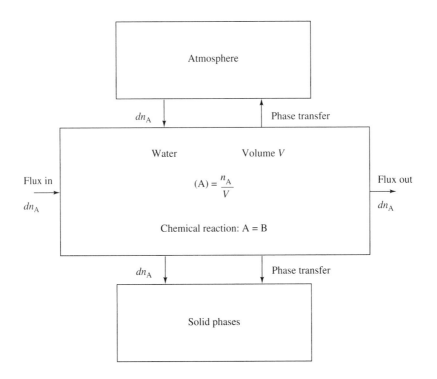

Figure 2.1   Schematic representation of an open groundwater system and a reaction A = B occurring in the aqueous phase, with a flux in A of dn_A taking place between the atmosphere, aqueous, and solid phases. (A) is the concentration of A, n_A the number of moles of A, and V the volume of the aqueous phase. After Langmuir and Mahoney (1985). Reprinted from the National Water Well Assoc. Used by permission.

may be described as open in a thermodynamic sense (open to the flow of both energy and matter). Their constancy would then be described as reflecting *steady-state conditions.*

Figure 2.1 depicts schematically how such concepts apply for a reaction A = B occurring in a volume of water in contact with both gas and solid phases. The volume of water and its temperature and pressure must be constant. For equilibrium concepts to apply to reaction A = B, the total fluxes of A (Σdn_A) and of B (Σdn_B) must equal zero. The number of moles of A (n_A) and of B (n_B) must be constant and homogeneously distributed throughout the water. Given that the initial concentration of A is (A)$_o$ and assuming that the concentration of B is zero to start, we find (B) = (A)$_o$ − (A). For equilibrium conditions where K_{eq} = (B)/(A), we obtain (A) = (A)$_o$/(1 + K_{eq}).

The mathematical description of a steady-state (also called stationary-state) open system in which the concentrations of A and B are governed instead by reaction rates is far more complicated. A simplified approach to such a system has been described by Stumm and Morgan (1981) (see also Morel and Hering 1993) and will only be highlighted here. The status of reaction A = B now depends on rates of the forward and reverse reactions, which, in the simplest case (assuming the rates are elementary, as discussed in Section 2.2), are described by the rate constants k_+ and k_-, respectively, thus

$$R_+ = k_+(A), \quad \text{and} \quad R_- = k_-(B) \tag{2.8}$$

where R_+ and R_- are rates of forward and reverse reactions. The rate constants themselves are functions of temperature and pressure. The time rates of change of (A) and (B) are

$$d(A)/dt = -k_+(A) + k_-(B) \tag{2.9}$$

and

$$d(B)/dt = +k_+(A) - k_-(B) \tag{2.10}$$

For the open system we must consider the material balance to the system of steady mole fluxes of A and B per unit volume, V. These fluxes can be designated r_A and r_B, where $r = Q/V$, and Q is again the overall volume rate of flux in units such as liters per second. In this context $(A)_o$ and $(B)_o$ are the inflow concentrations. For steady state, by definition $d(A)/dt = d(B)/dt = 0$, and inflow and outflow rates must be equal. Skipping several steps we find that for steady-state conditions

$$(A) = \frac{r_A + k_-[(A)_o + (B)_o]}{k_+ + k_- + r} \tag{2.11}$$

and

$$(B) = \frac{r_B + k_+[(A)_o + (B)_o]}{k_+ + k_- + r} \tag{2.12}$$

Obviously, the mathematics is considerably more cumbersome than was true of equilibrium conditions, requiring as it does both rate-constant data and flux information. We will say no more about steady-state conditions in this book. This discussion is intended only to suggest that we cannot assume equilibrium exists when a system exhibits constant composition over time, although the assumption will usually be correct, particularly when we are dealing with groundwaters having long residence times.

Returning to our earlier discussion we can generalize by stating that when a reaction is reversible and its rate is fast compared to the residence time of the aqueous system of interest ($t_R \gg t_{1/2}$) then equilibrium models may be used to describe the state of that reaction. When a reaction is irreversible, or its rate comparable to or slower than the system residence time ($t_{1/2} \geq t_R$) a kinetic model is needed to describe the state of such a reaction.

Regardless of the residence time of the aqueous system we are studying, there will always be some reactions fast enough to have attained equilibrium within it. There will also be other reactions that cannot or will not reach equilibrium within that system, regardless of the length of its residence time. In other words the typical system will be in *partial equilibrium*. We must be able to identify and distinguish among probable equilibrium and nonequilibrium reactions. Most geochemical computer models compare the chemical state of a system to its hypothetical equilibrium state. We should be able to examine the computer model output and, from our knowledge of rates, residence times, and the aqueous species and minerals involved, judge whether the apparent equilibrium computed for a particular reaction is even possible in the system under study.

Let us now assume that the residence time of a system is equivalent to the period of time that the system behaves as a closed system thermodynamically. With this assumption it is useful to qualitatively compare the residence times of different aqueous systems in the hydrosphere to the half-times of some example reactions and reaction types. This has been done schematically in Fig. 2.2. In essence, as we examine the diagram, we can assume reactions are at equilibrium in waters whose residence times significantly exceed the half-times of reactions of interest. Note that the half-times of some solute-solute and solute-water reactions (these include some complexation and acid-base reactions [see Chaps. 3 and 5]) are shorter than the "residence times" of raindrops and so can be assumed to be at equilibrium in rain. These are *homogeneous reactions*. However, the other types of reactions shown, including atmospheric gas exchange, which is *heterogeneous*, are too slow to have

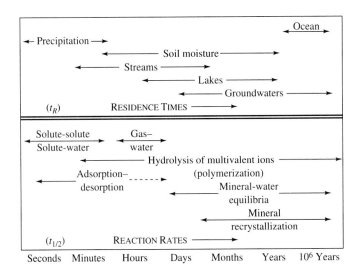

Figure 2.2 Schematic comparison between the half-times of some reaction types ($t_{1/2}$), and the residence times (t_R) of some waters in the hydrosphere. After Langmuir and Mahoney (1985). Reprinted from the National Water Well Assoc. Used by permission.

attained true equilibrium with the rain. The oxygen and CO_2 content of streams is often out of equilibrium with the atmosphere, consistent with the slow rates of the gas-exchange reactions relative to faster rates of processes that affect O_2 and CO_2 within streams.

The adsorption and desorption of metals onto the surface-exposed sorption sites of clay particles can approach equilibrium in seconds to minutes in well-mixed streamwaters. Similarly, the response of a pH electrode, which often stabilizes in a few seconds in stirred solutions, reflects the rapid rate of equilibration of aqueous protons with protons adsorbed on the hydrated surface of the glass electrode.

Longer times described by the dashed portion of the adsorption-desorption line represent the slower rates of metal and ligand adsorption found in some laboratory experiments and observed in most soil and groundwater systems (cf. Jenne 1995). Adsorption of metal ions generally involves concurrent desorption of protons or other metals. When adsorbed metals are strongly surface-bound (see Chap. 10) both adsorption and desorption reactions may be slow and desorption partially irreversible, requiring adsorption or desorption equilibration times that may approach 10 to 100 hours. Based on an analysis of published laboratory and field studies, Jenne (1995) concludes that metal adsorption and desorption reactions have initial fast and then slower reaction steps and are usually diffusion-rate limited.

The rates of metal and ligand adsorption and desorption involving surface sites within rock matrix pores are also diffusion limited. Thus, when ions must diffuse into the matrix of rock forming the walls of a groundwater-filled fracture, or into sand or gravel-sized stream sediments before adsorption can begin, slow ion diffusion and counter ion diffusion processes will limit the rate of attainment of equilibrium.

When concentrations of multivalent ions such as Al^{3+} and Fe^{3+} slightly exceed their saturation levels with respect to precipitation of the metal oxyhydroxides because of a pH increase for example, the solids may nucleate and gradually precipitate. At low temperatures both processes tend to be slow, requiring hours to much longer times for precipitation to go to completion.

TABLE 2.1   General reaction types and example reactions at low temperatures and pressures, with approximate reaction half-times.

Reaction type and example[†]		Half-times
Solute-solute		
$H_2CO_3^{\circ} = H^+ + HCO_3^-$	(acid-base)	$\sim 10^{-6}$ s
Solute-water		
$CO_2(aq) + H_2O = H_2CO_3^{\circ}$	(hydration/hydrolysis)	~ 0.1 s
$Cu^{2+} + H_2O = CuOH^+ + H^+$	(hydrolysis/complexation)	$\sim 10^{-10}$ s
$Fe(H_2O)_6^{2+} = Fe(H_2O)_5^{2+} + H_2O$	(hydrolysis/complexation)	$\sim 10^{-7}$ s
Adsorption-desorption		
$Cd^{2+} + CaX = Ca^{2+} + CdX$	(X^{-2} is the surface site)	$\sim$s–hr
Gas-water or gas solution-exsolution		
$CO_2(g) = CO_2(aq)$		$\sim$min
Oxidation-reduction		
$Fe^{2+} + \frac{1}{4}O_2(g) + \frac{5}{2}H_2O = Fe(OH)_3(ppt) + 2H^+$		min–hr
Hydrolysis of multivalent ions		
$Al_{n+m}(OH)_{3n+2m}^{+m} + mH_2O \rightarrow (n+m)Al(OH)_3(s) + mH^+$		hr–y
Mineral-water equilibria		
$Ca^{2+} + HCO_3^- = CaCO_3 + H^+$		week–y
Isotopic exchange		
$^{34}SO_4^{2-} + H^{32}S^- = H^{34}S^- + {}^{32}SO_4^{2-}$		y
Mineral recrystallization		
$Fe(OH)_3 \cdot nH_2O(am) \rightarrow \alpha\text{-}FeOOH(goethite) + (n+1)H_2O$		y
Radioactive decay		
$^{14}C \rightarrow {}^{14}N + e^-$		5570 y

[†]Note: Other descriptions or explanations of the reactions are given parenthetically.

Mineral recrystallization from relatively amorphous initial precipitates to more crystalline forms and mineral-water equilibria that are heterogeneous reactions are among the slowest reactions to equilibrate. Only in some ocean environments, large lakes, or older groundwaters can such reactions be expected to have attained equilibrium, and then only when they involve minerals that are thermodynamically stable in low-temperature environments.

To further illustrate the general concepts, Table 2.1 gives some examples of these and other reaction types and specific reactions with their reaction half-times. Ranges of reaction half-times have been given in this table where reaction rates in natural systems depend on other factors, such as changes in temperature, pH, or catalysis, or the involvement of micro-organisms. Mineral recrystallization and radioactive decay reactions have been written with arrows from reactants to products to emphasize that these reactions are irreversible; in other words, the reactants and products are incapable of equilibrating, regardless of system-residence time. Obviously then, the status of such reactions must always be described using kinetic models.

So far we have limited discussion to general concepts of chemical kinetics and chemical equilibrium. In this section we will take a more quantitative look at chemical kinetics, starting with some important definitions and concepts. The reader who wishes a more in-depth treatment is referred to the works of Berner (1978, 1981a), Hoffman (1981), Pankow and Morgan (1981), Lasaga (1981a, 1981b), Helgeson et al. (1984), Stumm (1990), Morel and Herring (1993), and Lasaga et al. (1994).

Compared to chemical equilibrium concepts, which have led to the quantitative understanding of chemistry in many environmental/ geochemical systems, chemical kinetic concepts are less simply applied to natural systems. This is particularly true of reactions in surface environments that involve biological activity. General rate information for such reactions may be known, although large uncertainties can be expected in predicted behavior (cf. Bowie et al. 1985).

The rates of reactions between minerals and groundwater are also difficult to predict because of their dependence on the surface characteristics of mineral grains, adsorbed trace substances on mineral surfaces, and often on the activities of organisms (Berner 1978). Laboratory rates of mineral dissolution may be orders of magnitude faster than observed in nature because of enhanced reactivities of the laboratory-prepared mineral grains and the adsorption of dissolution-inhibiting trace species such as phosphate in the natural system. Other laboratory rates, such as that of Fe(II) oxidation, are much slower than naturally observed oxidation rates when the latter are catalyzed by microorganisms.

A number of researchers have attempted to apply kinetic concepts and, in particular, the kinetics of mineral/precipitation/dissolution reactions to ground- or soil-water systems, or have compared the laboratory and field rates (cf. Claassen 1981; Paces 1983; Wanty et al. 1992; Swoboda-Colberg and Drever 1992; Stumm 1992). Rates of mineral solids/solution reactions are proportional to the surface area of reacting minerals in contact with the water. Comparison of laboratory and field rates has shown that they may agree within one or more orders of magnitude. However, our inability to measure the wetted surface areas of reacting minerals in aquifers or soils (cf. White and Peterson 1990), or to account for complex mineral-surface effects, including catalysis in the natural systems, will probably preclude the rigorous application of kinetic concepts to groundwater studies.

2.2 ELEMENTARY AND OVERALL REACTIONS

In kinetics a fundamental distinction is made between elementary and overall reactions. An elementary reaction describes an exact reaction mechanism or pathway. Three example elementary reactions are

$$H^+ + OH^- = H_2O \tag{2.13}$$

$$CO_2(aq) + OH^- = HCO_3^- \tag{2.14}$$

and
$$H_4SiO_4^\circ = SiO_2(quartz) + 2H_2O \tag{2.15}$$

The reactants and products of elementary reactions may be solids, ions, or molecular species, including gases, radicals, or free atoms. For example, the breakdown of ozone to oxygen, which is described by the overall reaction $2O_3 \rightarrow 3O_2$, probably involves the elementary reactions

$$O_3 \underset{k_{-1}}{\overset{k_{+1}}{\rightleftarrows}} O_2 + O, \quad \text{and} \quad O + O_3 \overset{k_{+2}}{\rightarrow} 2O_2 \tag{2.16}$$

As written, an overall reaction does not indicate the reaction mechanism or pathway. An example of an overall reaction is

$$CaCO_3(calcite) + CO_2(g) + H_2O = Ca^{2+} + 2HCO_3^- \tag{2.17}$$

Rates of overall reactions can be predicted only if the rates of component elementary reactions are known. Rates of elementary reactions are proportional to the concentrations of reactants. This may or may not be the case for overall reactions.

2.3 RATE LAWS

For a hypothetical elementary reaction A = B, the rate of the forward reaction A $\rightarrow$ B is given by $R_+ = dA/dt = k_+(A)$; and the rate of the reverse reaction B $\rightarrow$ A by $R_- = dB/dt = k_-(B)$, where k_+ and k_- (both functions of temperature) are the rate constants for the forward and reverse reactions, and the parentheses enclose concentrations of the species A and B. Values of R_+ and R_- can be positive or equal to zero. However, the differentials dA/dt and dB/dt may be positive or negative. Thus, if (B) = 0 at $t = 0$, (i.e., B$_0$ = 0) then (A) decreases with time, dA/dt is negative, and R_+ remains positive.

For elementary reactions in general [see Eq. (2.8)], we can write for the forward and reverse rates

$$R_+ = k_+ \prod_{\text{reactants}} (A_i)^{v_i} \tag{2.18}$$

and

$$R_- = k_- \prod_{\text{products}} (A_i)^{v_i} \tag{2.19}$$

where the A_i terms and their exponents are the products of the concentrations of reactants and products raised to the power of their stoichiometric coefficients, (v_i). The latter are positive for products and negative for reactants. At chemical equilibrium $R_+ = R_-$ and we can write

$$\frac{k_+}{k_-} = K_{\text{eq}} = \prod_{\substack{\text{reactants} \\ + \text{ products}}} (A_i)_{\text{eq}}^{v_i} \tag{2.20}$$

where K_{eq} is the equilibrium constant. For our simple elementary reaction A = B, this corresponds to

$$K_{\text{eq}} = \frac{k_+ (A)}{k_- (B)} \tag{2.21}$$

Equations (2.20) and (2.21) relate the rate constants of elementary reactions to their equilibrium constants, and show that if K_{eq} and a single rate constant are known, the second rate constant need not be measured. Rimstidt and Barnes (1980) used this relationship, called the principle of detailed balancing, to obtain the rate constant for dissolution of several silica polymorphs (k_+) given their empirical solubilities (K_{eq} values) and the precipitation rate constant (k_-). (See Section 2.7.8.)

The *order* of an elementary reaction is defined by the number of individual atoms or molecules involved in the reaction. The concept of *overall reaction order* can only be applied to single-term, simple, product rate equations such as

$$R_+ = k_+ (A)^{n_A}(B)^{n_B}(C)^{n_C} \cdots \tag{2.22}$$

where the exponents are whole numbers. For example, the rate of the forward reaction:

$$A + 2B \rightarrow C \tag{2.23}$$

can be written

$$-\frac{d(C)}{dt} = k_+(A) (B)^2 \tag{2.24}$$

This reaction is *first-order* with respect to A and C, *second-order* in B, and *third-order* overall. Benson (1960), Gardiner (1972), and Lasaga (1981a) discuss the rate expressions for third- and higher-order reactions. Such reactions are uncommon in natural systems and will not be considered here.

TABLE 2.2 Simple rate laws, their integrated forms, and
expressions for reaction half-times ($t_{1/2}$) for some simple
reaction orders (cf. Gardiner 1972; Lasaga 1981a)

	Integrated rate equation	Half-time
Zeroth-order		
(A as reactant)		
$\dfrac{d(A)}{dt} = -k$	$A = A_o - kt$	$+0.5A_o/k$
(A as product)		
$\dfrac{d(A)}{dt} = k$	$A = A_o + kt$	$-0.5A_o/k$
First-order		
(A $\rightarrow$ B)		
$\dfrac{d(A)}{dt} = -kA$	$\ln A = \ln A_o - kt$	$0.693/k$
(B $\rightleftarrows$ A)		
$\dfrac{d(A)}{dt} = k(A_s - A)$	$\ln\dfrac{(A_s - A)}{(A_s - A_o)} = -kt$	
Second-order		
(2A $\rightarrow$ B)		
$\dfrac{d(A)}{dt} = -kA^2$	$\dfrac{1}{A} = \dfrac{1}{A_o} - kt$	$-\dfrac{1}{kA_o}$
(A + B $\rightarrow$ C)		
$\dfrac{d(A)}{dt} = -k(A)(B)$	$\ln\dfrac{(A_o)(B)}{(B_o)(A)} = (B_o - A_o)kt$	

Note: General reaction types or conditions that correspond to the differential
rate equations are given parenthetically. Some reactions are irreversible
(denoted by $\rightarrow$) and others reversible (denoted by double arrows). Note that
the rate constant, k is always positive. In the integrated rate expressions the
concentration of $A = A_o$, at $t = 0$, and $A = A_o/2$ at half-time ($t_{1/2}$). A_s denotes
the equilibrium, mineral saturation or steady state concentration of species A.

Table 2.2 is a summary of the differential and integrated forms of some simple rate laws. Also
listed are expressions for the reaction half-times corresponding to several of the integrated rate expressions. These are the times at which the concentration of a reactant is half its initial value.

Differential rate equations are particularly useful when we wish to assess the rate of a specific
reaction for different conditions, or to compare the rates of several reactions. However, an integrated
rate equation is necessary if our goal is to determine concentrations of individual reactants at some
time t.

The rate constants of reactions without gaseous reactants have the following (or comparable)
units: mol/cm s (zeroth-order), 1/s or s^{-1} (first-order), and cm^3/mol s (second-order). Shown in
Fig. 2.3 for zeroth-, first-, and second-order reactions is the time-dependence of the concentration of
a reactant A, and its linear and log rate of change with concentration. The appearance of the plots,
and particularly the log rate versus log (A) plots, differs uniquely for each reaction order. This sug-

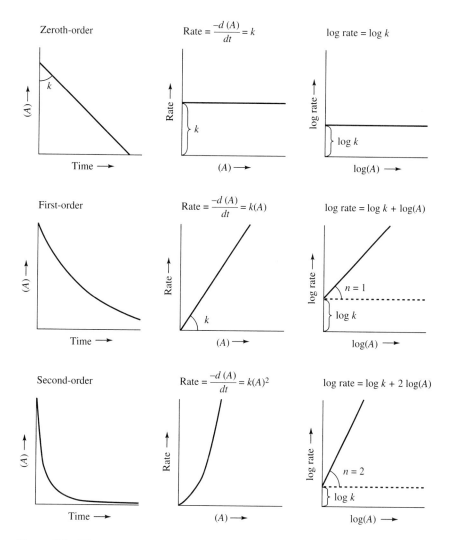

Figure 2.3 The general appearance of concentration time, rate concentration, and log rate-log concentration plots for simple rate laws involving reactant A. Reprinted from: Appelo, C. A. J. & D. Postma, Geochemistry, groundwater and pollution. 1993. 500 pp., Hfl.150/US\$85.00. Student edition: Hfl.90/US\$55.00. Computer program: Hfl.75/US\$45.00. Please order from: A. A. Balkema, Old Post Road, Brookfield, Vermont 05036 (telephone: 802-276-3162; telefax: 802-276-3837). The computer program is available directly from A. A. Balkema, P.O. Box 1674, Rotterdam, Netherlands.

gests a general procedure for determining the rate law and rate constant for a specific reaction, which involves plotting the empirical rate versus concentration data in log format as shown in Fig. 2.3.

     Zeroth-order rates are proportional to k_+ or k_-, independent of the concentrations of reactants or products. The rates of solution of quartz and many other silicate minerals are zeroth-order (see Sections 2.7.8 and 2.7.9).

First-order rates are proportional to the concentrations of a reactant or product. An important example is radioactive decay (Section 2.7.2), which generally is irreversible and has only a forward rate and rate constant, for which

$$\frac{dn}{dt} \rightarrow -k_+ n \tag{2.25}$$

Integration yields: $n = n_o \exp(-k_+ t)$, where n and n_o are the total number of atoms present at t and $t = 0$. The half-time of radioactive decay is defined by $t_{1/2} = 0.693/k_+$, where $n = 0.5$, and $n_o = 1$. Other examples of first-order reactions (see Section 2.7) are the oxidation of organic matter and sulfate reduction, gypsum ($CaSO_4 \cdot 2H_2O$) dissolution, and the oxidation of pyrite and marcasite (FeS_2).

Second-order rates depend on the concentrations of two reactants or products, and are thus *bimolecular* reactions. For example, the *overall reaction* of Fe^{2+} oxidation below pH 2.2 is

$$Fe^{2+} + \tfrac{1}{4}O_2 + H^+ = Fe^{3+} + \tfrac{1}{2}H_2O \tag{2.26}$$

for which the differential rate expression based on experimental measurements is

$$\frac{d(Fe(II))}{dt} = -k_+(Fe(II)) P_{O_2} \tag{2.27}$$

Clearly, the reaction order and form of the empirical rate law could not have been predicted from the overall reaction. If we hold P_{O_2} constant, for example assuming oxidation is in a turbulent, aerated stream, then the reaction is said to be *pseudo first-order*. Another example of a second-order rate is formation of the $FeSO_4^+$ complex (see Section 2.7.1).

A *complex composite rate law* may describe the rate of an overall reaction as the sum of rates of its constituent elementary reactions. Such a rate law may be written for calcite dissolution and precipitation, for which the overall rate equals

$$R(\text{mmol/cm s}) = k_1[H^+] + k_2[H_2CO_3] + k_3[H_2O] - k_4[Ca^{2+}][HCO_3^-] \tag{2.28}$$

(Plummer et al. 1979; and Section 2.7.6). The brackets denote activities of the enclosed species, which roughly equal their concentrations. (See Chap. 4.) The three positive terms on the right describe the forward reaction rate. The final, negative term defines the reverse rate as the water approaches saturation with respect to calcite. A graphic statement of the conditions of dominance of individual terms in this rate equation as a function of pH and P_{CO_2} is given in Fig. 2.4. If we are concerned about the weathering of carbonate stone (limestone and marble monuments and buildings) by acid rain with a pH below 5, then this expression is reduced to simply $R_+ = k_1[H^+]$. This term then describes the weathering process far from equilibrium with calcite. In air, the carbon dioxide pressure is about 0.0003 bar. In soils and groundwaters it rarely exceeds 0.1 bar. Thus the forward rate can usually be described by $R_+ = k_1[H^+] + k_3[H_2O]$, where the first term dominates below about pH 5.7 and the second at higher pH's. The activity of water in the k_3 term can usually be assumed equal to unity and ignored. (See Section 2.7.7.)

2.4   EFFECT OF TEMPERATURE ON REACTION RATES

Reaction rates are observed to increase (half-times to decrease) with increasing temperature. The reaction rate constant can be related to $T(K)$ through the Arrhenius expression

$$k = A_F \exp(-E_a/RT) \tag{2.29}$$

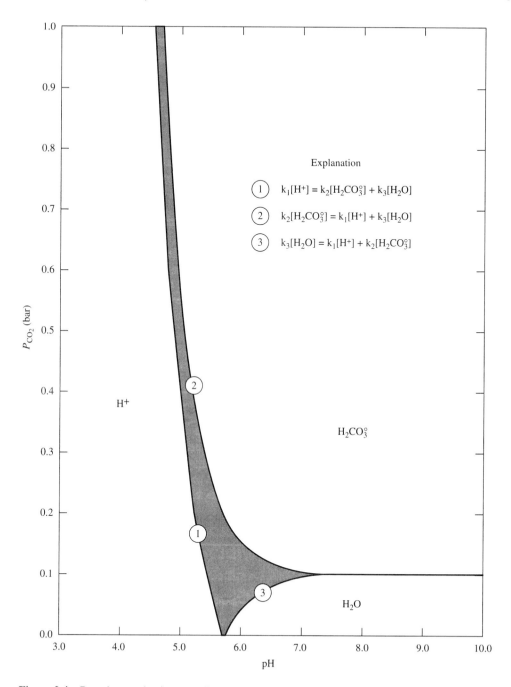

The explanation labels within the figure:

1. $k_1[H^+] = k_2[H_2CO_3^\circ] + k_3[H_2O]$

2. $k_2[H_2CO_3^\circ] = k_1[H^+] + k_3[H_2O]$

3. $k_3[H_2O] = k_1[H^+] + k_2[H_2CO_3^\circ]$

Figure 2.4   Reaction mechanism contributions to the rate of calcite dissolution as a function of pH and P_{CO_2} at 25°C. Although H^+, carbonic acid, and water reaction with calcite occur simultaneously throughout (far from equilibrium, as well as at equilibrium), the forward reaction is dominated by reaction with single species in the fields shown. More than one species contributes significantly to the forward rate in the stippled area. Along the lines labeled 1, 2, and 3, the forward rate attributable to one species balances that of the other two. After Plummer et al. (1979). Reprinted with permission from *Chemical modeling in aqueous systems,* Am. Chem. Soc. Symp. 93. © 1979, American Chemical Society.

Or $\ln k = \ln A_F - E_a/RT$, in which k is the rate constant, A_F the "A" factor, relatively independent of T, and E_a is the activation energy, which is always positive (+). Differentiation of the above expression and converting to common logs

$$\frac{d \log k}{dT} = \frac{E_a}{2.303 \, RT^2} \tag{2.30}$$

Note that this relationship is similar to the van't Hoff equation. If the Arrhenius expression is obeyed, then a plot of $\log k$ versus $1/T$ is a straight line, with a slope of $-E_a/2.303 \, R$. Such plots are shown in Figs. 2.5(a) and (b), which describe the effect of increasing temperature on the rates of dissolution of some silicate rocks and minerals.

An often-quoted rule of thumb is that the rates of reactions roughly double for every 10°C increase in temperature. It is instructive to compute what activation energy would correspond to such a rate change. We will start by taking the logarithm of the Arrhenius expression. For temperatures T_1 and T_2 this gives us

$$\log k_1 = \log A_F - E_a/(2.303 \, RT_1) \tag{2.31}$$

and $\qquad\qquad\qquad \log k_2 = \log A_F - E_a/(2.303 \, RT_2) \tag{2.32}$

Taking the difference we can eliminate the A_F terms. Then combining the equations we obtain

$$\log \frac{k_1}{k_2} = \frac{E_a}{2.303 \, R} \left[\frac{1}{T_2} - \frac{1}{T_1} \right] \tag{2.33}$$

Notice the similarity of this expression and the integrated van't Hoff equation. We know that the reaction is twice as fast at T_2 as at T_1. Thus $k_2 = 2 \times k_1$. Assuming that $T_1 = 298.15$ K, and $T_2 = 308.15$ K, we can substitute into the last expression and solve for E_a to obtain $E_a = 12.6$ kcal/mol.

The magnitude of E_a is, in fact, an important indicator of the reaction mechanism or process. This can be seen in the table below, much of which is based on Lasaga (1981b).

Reaction or process	Typical range of E_a values (kcal/mol)
Physical adsorption	2 to 6
Aqueous diffusion	< 5
Cellular and life-related reactions	5 to 20
Mineral dissolution or precipitation	8 to 36
Mineral dissolution via surface reaction control	10 to 20
Ion exchange	> 20
Isotopic exchange in solution	18 to 48
Solid-state diffusion in minerals at low temperatures	20 to 120

In general it is found that reactions that involve the least making and breaking of strong bonds have the lowest activation energies. Such reactions can proceed readily at low temperatures, and their rates are relatively independent of temperature. Having as it does the highest activation energies listed, solid-state diffusion is a very slow process at low temperatures.

As shown in the table, aqueous diffusion-controlled reactions typically have activation energies of less than 5 to 6 kcal/mol (Berner 1978; Rimstidt and Barnes 1980). Minerals dissolving and precipitating via surface reaction control usually have E_a values between 10 to 20 kcal/mol (Lasaga 1981b). Activation energies for solid-state diffusion in minerals at low temperature range

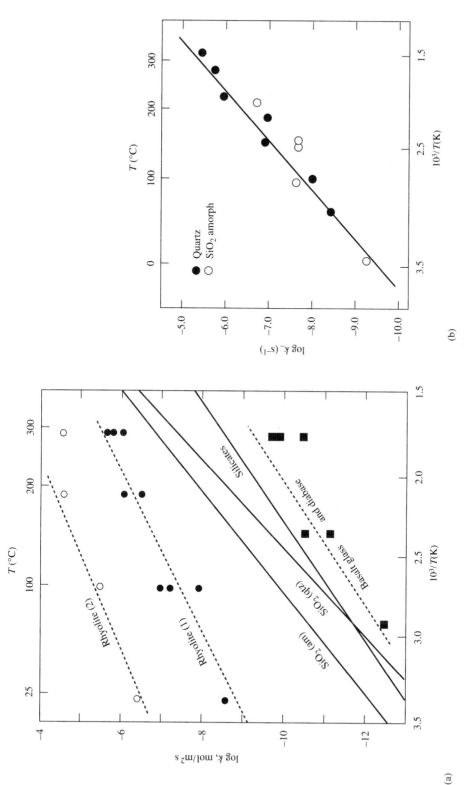

Figure 2.5 (a) An Arrhenius plot of log k versus $1/T$(K) for the dissolution rates of various silicate rocks and minerals. The data points and curves for rhyolite, basalt glass, and diabase are from Apps (1983), as is the curve labeled "silicates," which Apps computed from the results of Wood and Walther (1983). Curves for the SiO$_2$ polymorphs are based on Rimstidt and Barnes (1980). Modified from Langmuir and Mahoney (1985). Reprinted from the National Well Water Assoc. Used by permission. (b) An Arrhenius plot of log k versus $1/T$(K) for the precipitation of quartz and amorphous silica based on Rimstidt and Barnes (1980). Reprinted from *Geochim. Cosmochim. Acta*, 44, J.D. Rimstidt and H.L. Barnes, The kinetics of silica water reactions, 1683–99, © 1980, with permission from Elsevier Science Ltd, The Boulevard, Langford Lane, Kidlington OX5 1GB, U.K.

up to 120 kcal/mol, but are generally from 20 to 80 kcal/mol. Apps (1983) found that alteration (hydration) of obsidian and basalt glasses by diffusion had activation energies of about 22 to 23 kcal/mol. Lasaga (1981b) points out that the bond-breaking and bond-making associated with surface reaction–controlled mineral dissolution and precipitation would be expected to have activation energies similar to those of solid-state diffusion. That the former are so low may reflect that chemisorption/desorption occurs prior to mineral precipitation or dissolution. In other words, the heat of chemisorption may reduce the activation energy required for surface–controlled mineral reaction. It is worth remembering that the activation energies for inorganic processes given above, are for reactions in the absence of catalysis, which can greatly alter reaction rates (cf. Hoffman 1990).

2.5 MINERAL PRECIPITATION/DISSOLUTION REACTION KINETICS

Although the empirical rate laws for a number of geochemically important overall reactions are known, only rarely do we know their component elementary reactions (i.e., the reaction mechanism). This is particularly true of heterogeneous reactions such as mineral precipitation/dissolution. There are several different types of processes that can control the rates of mineral precipitation and dissolution. In moving groundwater these include mass transport, diffusion control, and surface-reaction control. Berner (1978, 1980, 1981a) notes that silicates including quartz, amorphous silica, and feldspars, and calcite and apatite all dissolve according to surface-reaction control. On the other hand, the dissolution-rate mechanism of more soluble minerals such as gypsum and halite is usually transport-controlled.

The mechanism of mineral dissolution/precipitation can, however, also depend on the rate of groundwater flow. Berner (1978) relates transport and surface-reaction rates in the expression

$$\frac{dC}{dt} = R - k_f C \tag{2.34}$$

where dC/dt is the rate of change of concentration in a fixed volume of the system. R is the rate of dissolution (assuming first-order kinetics, $R = k_+\{C_s - C\}$); and k_f is the flushing frequency (rate of flow/volume of system). C_s is the saturation concentration of the species of interest. For steady-state conditions (when $dC/dt = 0$) we find

$$C = k_+ C_s/(k_+ + k_f) \qquad \text{and} \qquad R = k_+ k_f C_s/(k_+ + k_f) \tag{2.35}$$

At high-groundwater flow rates ($k_f \gg k_+$), these expressions reduce to $C = k_+ C_s/k_f$, and $R = k_+ C_s$, so that a maximum solution rate is reached, independent of flow rate. At the opposite extreme of slow groundwater flow ($k_f \to 0$), $C = C_s$, and $R = k_f C_s$. Saturation is attained and the rate of dissolution is controlled by the groundwater flow rate. In other words, at high flow the dissolution rate is surface-reaction controlled. The slower process is rate limiting.

The slowest transport-controlled dissolution/precipitation is that governed by aqueous diffusion. Diffusion rates can be estimated (cf. Bodek et al. 1988; Fetter 1988), thus we can estimate the lower limit of rates attributable to transport control. Berner (1978) suggests that the rate of diffusion-controlled dissolution (R_d) is given by

$$R_d = D\rho A(C_s - C)/r \tag{2.36}$$

where R_d is in mass/volume/time, D is the diffusion coefficient, ρ the porosity, A is the surface area of dissolving crystals per unit volume of solution, and r is the spherical radius of dissolving crystals. It is interesting to compare the rate of dissolution that can result from diffusion to the rate from surface reaction. As a convenient example, James and Lupton (1978) report that the dissolution rate of gypsum via surface reaction (R_{sr}) is given by

$$R_{sr} = k_+ A(C_s - C) \qquad (2.37)$$

The ratio of the rates R_d/R_{sr} then equals $D\rho/rk_+$. Diffusion coefficients of ions typically range from 3×10^{-6} to 2×10^{-5} cm^2/s (Lerman 1979). Assuming $D = 10^{-6}$ cm^2/s, a porosity of 0.2, crystal radius of 1 cm, and $k_+ = 2 \times 10^{-4}$ cm/s at 25°C (based on James and Lupton 1978), we find $R_d/R_{sr} = 1/1000$. In other words $R_{sr} \gg R_d$, and the slower rate, diffusion (a transport process), will control the dissolution rate.

2.6 ABSOLUTE RATE (TRANSITION STATE) THEORY AND THE ACTIVATED COMPLEX

Considerable insight into reaction mechanisms has been derived from the concept of absolute-rate theory. The theory is based on two assumptions: first, that there is an energy maximum (barrier) between products and reactants in a reaction, and that an activated complex exists at that maximum; and second, that chemical equilibrium always exists among reactants, products, and the activated complex, $C^\pm$. Thus for the reaction

$$A + B \rightleftarrows C^\pm \rightleftarrows AB \qquad (2.38)$$

The equilibrium constant involving reactants A and B and the activated complex (see Fig. 2.6) is

$$K^\pm = \frac{[C^\pm]}{[A][B]} = \frac{(C^\pm)}{(A)(B)} \frac{\gamma^\pm}{\gamma_A \gamma_B} \qquad (2.39)$$

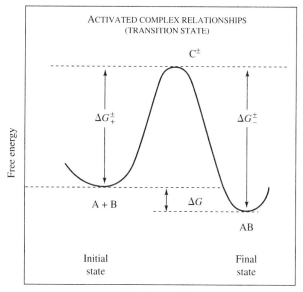

Figure 2.6   Schematic diagram of free-energy relationships for the reaction: $A + B = C^\pm = AB$, where $C^\pm$ is the activated complex. $\Delta G_+^\pm$ and $\Delta G_-^\pm$ are free energies of reactions $A + B = C^\pm$ and $C^\pm = AB$. For the overall reaction $A + B = AB$, $\Delta G_r = \Delta G_-^\pm - \Delta G_+^\pm$. After Langmuir and Mahoney (1985). Reprinted from the National Water Well Assoc. Used by permission.

where the brackets and parentheses enclose the activities and concentrations of the species, respectively, and $\gamma^{\pm}$, γ_A, and γ_B are their respective activity coefficients. (See Chap. 4.) Transition state (TS) theory assumes that the reaction $C^{\pm} \rightleftarrows AB$ is spontaneous (fast), so that $A + B \rightleftarrows C^{\pm}$ is rate limiting. According to TS theory (Wadsworth 1979, Stone and Morgan 1990), the reaction rate R is then given by

$$R = -\frac{d(A)}{dt} = -\frac{d(B)}{dt} = \frac{k_B T}{h}(C^{\pm}) \tag{2.40}$$

where k_B and h are Boltzmann and Planck constants. Eliminating $(C^{\pm})$ in Eq. (2.40) by substitution of Eq. (2.39) leads to

$$R = \frac{k_B T}{h}\left(\frac{\gamma_A \gamma_B}{\gamma^{\pm}}\right)K^{\pm}(A)(B) \tag{2.41}$$

But the rate also equals

$$R = k(A)(B) \tag{2.42}$$

where k is the second-order rate constant. The right side of Eqs. (2.41) and (2.42) are equal, from which we find for the rate constant, k

$$k = \frac{k_B T}{h}\left(\frac{\gamma_A \gamma_B}{\gamma^{\pm}}\right)K^{\pm} \tag{2.43}$$

The effect of ionic strength on the second-order rate constant may be obtained as follows. We define a specific rate constant k_o, for conditions of infinite dilution when the activity coefficients equal unity, so that

$$k_o = \frac{k_b T}{h}K^{\pm} \tag{2.44}$$

and

$$k = k_o\left(\frac{\gamma_A \gamma_B}{\gamma^{\pm}}\right) \tag{2.45}$$

Taking logs and rearranging gives

$$\log\frac{k}{k_o} = \log\left(\frac{\gamma_A \gamma_B}{\gamma^{\pm}}\right) \tag{2.46}$$

We assume applicability of the Debye-Hückel limiting law for computing the activity coefficients

$$\log \gamma_i = -A z_i^2 \sqrt{I} \tag{2.47}$$

where z_i is the charge of species i, and I is the molal ionic strength. The charge of the activated complex, $z^{\pm}$ must equal the sum of z_A and z_B. Substituting, and expanding Eq. (2.46) gives

$$\log\left(\frac{k}{k_o}\right) = -A\sqrt{I}\left[z_A^2 + z_B^2 - (z_A + z_B)^2\right] \tag{2.48}$$

which reduces to

$$\log\frac{k}{k_o} = 2A\sqrt{I}\ (z_A z_B) \tag{2.49}$$

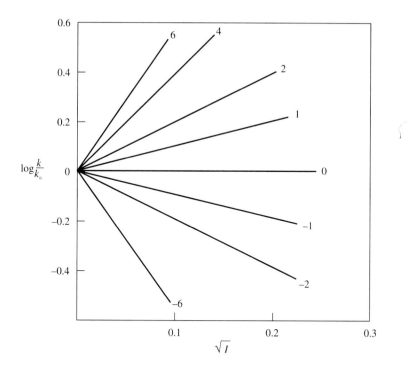

Figure 2.7   The predicted effect of ionic strength on the second-order rate constant k, for different values of the product of valences of reactants A and B (i.e., $z_A z_B$). $k = k_o$ at $I = 0$.

(Lasaga 1981b). Substitution shows that if A and B have the same sign, k increases with ionic strength; if they are of opposite sign, k decreases with ionic strength. Eq. (2.49) is plotted in Fig. 2.7, which shows, for example, that at $I = 6 \times 10^{-3}$ m the rate of a second-order reaction involving divalent ions A and B is double its value in pure water.

We next define a Gibbs free energy of activation, $\Delta G^\pm = -RT \ln K^\pm$. Substituting for $K^\pm$ in Eq. (2.44) yields

$$k_o = \frac{k_b T}{h} \exp \left(\frac{-\Delta G^\pm}{RT} \right) \tag{2.50}$$

Or replacing the free energy with the corresponding enthalpy and entropy of activation we may write

$$k_o = \frac{k_B T}{h} \exp \left(-\frac{\Delta H^\pm}{RT} + \frac{\Delta S^\pm}{R} \right) \tag{2.51}$$

Taking the logarithm gives

$$\ln k_o = \ln \left(\frac{k_B}{h} \right) + \ln T - \frac{\Delta H^\pm}{RT} + \frac{\Delta S^\pm}{R} \tag{2.52}$$

Assuming $\Delta H^\pm$ and $\Delta S^\pm$ are independent of temperature, differentiating we obtain

$$\frac{d \ln k_o}{dT} = \frac{1}{T} + \frac{\Delta H^\pm}{RT^2} \tag{2.53}$$

But the Arrhenius relationship in differential form is

$$\frac{d \ln k_o}{dT} = \frac{E_a}{RT^2} \tag{2.54}$$

so we find

$$E_a = RT + \Delta H^{\pm} \tag{2.55}$$

Because the value of RT is less than 1 kcal/mol up to about 500 K, this last expression reduces to $E_a = \Delta H^{\pm}$. We now have a tie-in between the empirical Arrhenius equation and transition-state theory and a way to evaluate the enthalpy of activation of a reaction.

2.7 SOME KINETIC EXAMPLES OF GEOCHEMICAL INTEREST

Following are some rate studies and models of geochemical interest. They have been chosen to illustrate the concepts presented in this chapter and to suggest where we are in the development and application of kinetic principles to low-temperature water/rock systems.

There is reason to believe that such rate information for simple homogeneous reactions and for the least complex heterogeneous reactions, including mineral dissolution/precipitation, can be more usefully applied to studies of natural systems than has so far been the case. The laboratory rate data and theoretical models often do not apply directly to complex natural systems. However, they at least provide guidelines and direct our study toward a better understanding of natural rates.

2.7.1 The $FeSO_4^+$ Complex

Formation of the $FeSO_4^+$ complex is described by the elementary reaction

$$Fe^{3+} + SO_4^{2-} \underset{k_-}{\overset{k_+}{\rightleftarrows}} FeSO_4^+ \tag{2.56}$$

(Pagenkopf 1978), for which the forward reaction (second-order) rate function is

$$R_+ = \frac{d(Fe^{3+})}{dt} = -\frac{d(SO_4^{2-})}{dt} = k_+(Fe^{3+})(SO_4^{2-}) \tag{2.57}$$

In this expression $k_+ = 6.37 \times 10^3$ L/mol s at 25°C. The reverse reaction rate (first-order) is given by $R_- = k_-(FeSO_4^+)$. The equilibrium constant is $K_{eq} = 205$/mol, from which (because $K_{eq} = k_+/k_-$ and assuming the principle of detailed balancing) we compute $k_- = 31$/s. Calculated rates and concentrations for this reaction as a function of time are given in Fig. 2.8. Note that the overall rate (R_{net} in Fig. 2.8[b]) equals the difference ($R_+ - R_-$).

2.7.2 Radioactive Decay: The Example of ^{14}C

Radioactive decay is a first-order, usually irreversible reaction with a rate given by

$$\frac{dn}{dt} = -k_+n \tag{2.58}$$

which, upon integration, yields $n = n_o \exp(-k_+t)$, where n_o and n are the total number of molecules present at $t = 0$ and t. The half time of radioactive decay is $t_{1/2} = 0.693/k_+$.

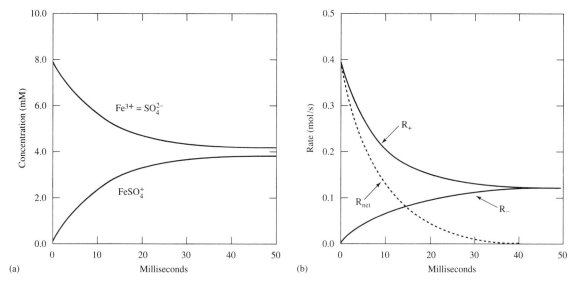

Figure 2.8 (a) Calculated concentrations of Fe^{3+}, SO_4^{2-}, and $FeSO_4^+$ as a function of time for the reaction $Fe^{3+} + SO_4^{2-} = FeSO_4^+$. Initially $(Fe^{3+}) = (SO_4^{2-}) = 8$ mM and $(FeSO_4^+) = 0$. (b) Calculated forward (R_+), reverse (R_-), and net rates (R_{net}) for the reaction $Fe^{3+} + SO_4^{2-} = FeSO_4^+$, $R_{net} = R_+ - R_-$. From G. K. Pagenkopf, *Introduction to natural water chemistry*, © 1978. Reprinted by courtesy of Marcel Dekker, Inc.

Natural carbon contains chiefly the stable isotopes ^{12}C (~98.89%) and ^{13}C (~1.10%) (Rosler and Lange 1972). Radioactive dating with ^{14}C depends upon the fact that trace amounts of radioactive ^{14}C (~10^{-10}% of total C) are created from nitrogen gas in the upper atmosphere through the reaction

$$^{14}N_7 + {}^1n_o \rightarrow {}^{14}C_6 + {}^1H_1 \tag{2.59}$$

which describes the formation of ^{14}C when the nitrogen is bombarded by neutrons produced by cosmic rays. The ^{14}C is then introduced into living organisms as $^{14}CO_2$. Upon their death, if no new ^{14}C is introduced

$$^{14}C_6 \xrightarrow{k_+} {}^{14}N_7 + {}^0e_{-1} \tag{2.60}$$

the ^{14}C decays back to ^{14}N plus an electron. The integrated form of Eq. (2.58) then becomes

$$^{14}C = ({}^{14}C_o) \exp(-k_+t) \tag{2.61}$$

or, in natural logs, $\ln({}^{14}C) = \ln({}^{14}C_o) - k_+t$. Transposing leads to

$$t = \frac{1}{k_+} \ln\left(\frac{{}^{14}C_o}{{}^{14}C}\right) = \frac{1}{k_+} \ln\left(\frac{d_o}{d}\right) \tag{2.62}$$

where d_o and d are the disintegration rates per minute per gram of carbon (dpm/g) at t_o and t. Because $t_{1/2} = 5570$ y, and $k_+ = 0.693/t_{1/2}$, then $k_+ = 1.24 \times 10^{-4}$/y. Substituting back into Eq. (2.62), and converting to common logs, with $d_o = 13.56 \pm 0.07$ dpm/g, the prehuman disintegration rate for freshly buried organic matter (cf. Faure 1991), we obtain

$$t \text{ (y)} = 18,500 \log\left(\frac{13.56}{d}\right) \tag{2.63}$$

The radioactivity of atmospheric ^{14}C may, in fact, have varied from this disintegration rate by more than $\pm 10\%$ in the last 6000 years (Henderson 1982). Age dating with ^{14}C has been found useful for samples up to about 50,000 years old (Mason and Moore 1982).

2.7.3 Oxidation of Organic Matter and Sulfate Reduction

There are several groups of organic substances in any sediment. Each group has its own decay rate (Westrich and Berner 1984; see also Middleburg 1989; Tarutis 1993). Decay rates, whether oxic or anoxic, obey first-order kinetics. For each group, G_i, $dG_i/dt = -k_iG_i$. For G_T, the total decomposable organic material present,

$$\frac{dG_T}{dt} = \sum_o^n k_iG_i \tag{2.64}$$

Integration of the initial-rate equation gives G_i as a function of time

$$G_i(t) = G_{oi}[\exp(-k_it)] \tag{2.65}$$

where G_{oi} is the initial amount of G_i. Experimental observation of the decay rate of organic matter in sediments leads to

$$G_T(t) = G_{01}[\exp(-k_1t)] + G_{02}[\exp(-k_2t)] + G_{nr} \tag{2.66}$$

where G_T is some measure of the total organic content (TOC) in the sediment at time t, the first term on the right is the highly reactive fraction, the second term the less reactive fraction, and G_{nr} denotes organic matter that was nonreactive during the study of Westrich and Berner.

In their study of modern marine sediments, Westrich and Berner (1984) found $G_{01} = 50\%$, $G_{02} = 16\%$, and $G_{nr} = 34\%$ of TOC. For oxic decay their empirical rate constants were: $k_{-1} = 18/y$ ($t_{1/2} = 0.039$ y $= 14$ days), and $k_2 = 2.3/y$ ($t_{1/2} = 0.3$ y $= 110$ days). For anoxic decay of the same organic matter, which follows the same rate equations, they obtained $k_1 = 4.4/y$, and $k_2 = 0.84/y$. All of this, of course, assumes G_{nr} is inert.

If similar rates apply to natural water/rock systems, which will generally have much longer residence times than considered in this study, we can expect that organic matter similar to G_1 will have disappeared ($t_{1/2} = 14$ days). G_2 kinetics will still be important. However, rates of decomposition of more recalcitrant organics (G_{nr}), because they include perhaps most of toxic organics, will still be important.

If sulfate reduction is organic-carbon limited in the sediment described by Westrich and Berner (1984), then the same rates of organic carbon oxidation apply. For sulfate reduction, assuming that the organic carbon may be generalized as CH_2O, the reaction may be written

$$2CH_2O + SO_4^{2-} \rightarrow H_2S + 2HCO_3^- \tag{2.67}$$

Taking into account the reaction stoichiometry, the rate of sulfate reduction, $R(t)$ as a function of time is

$$R(t) = \tfrac{1}{2}kG_o[\exp(-kt)] \tag{2.68}$$

But $R_o = \tfrac{1}{2}kG_o$ so that the overall rate is

$$\Sigma R(t) = R_{01}[\exp(-k_1t)] + R_{02}[\exp(-k_2t)] \tag{2.69}$$

From their measurements Westrich and Berner (1984) obtained $k_1 = 7.2/y$ and $k_2 = 1.0/y$; rate constants nearly equal to those for anoxic decay of organic carbon, as expected. These constants may be compared to mean values based on the work of others of $k_1 = 8.0 \pm 1/y$, and $k_2 = 0.94 \pm 0.25/y$.

Middleburg (1989) questions the approach of Westrich and Berner (1984) and suggests that all organic matter will oxidize, given sufficient time. He proposes a single rate law for the oxidation of organic matter, whether aerobic or anaerobic, with a rate constant that is itself a function of time as given by the empirical equation

$$\log_{10} k(1/y) = -0.95 \log_{10} t(y) - 0.81 \qquad (r = 0.987) \tag{2.70}$$

2.7.4 Gypsum Dissolution

The dissolution of gypsum according to the overall reaction

$$CaSO_4 \cdot 2H_2O \text{ (gypsum)} = Ca^{2+} + SO_4^{2-} + 2H_2O \tag{2.71}$$

follows first-order kinetics as expressed by the equation

$$\frac{d(C)}{dt} = k_+ A_w (C_s - C) \tag{2.72}$$

where $d(C)/dt$ is in mol/m^3 s, A_w in m^2/m^3 is the wetted mineral surface area exposed to a volume of solution at time t, and $C_s = 15.5$ mol/m^3 in pure water at 25°C (Langmuir and Melchior 1985). The solution rate constant is proportional to the diffusion coefficient, D, and inversely proportional to, the wetted film thickness on the mineral surface (Liu and Nancollas 1971). The film thickness decreases with increasing groundwater flow rate, thus increasing the gypsum dissolution rate. The approximate rate constant at 25°C and zero flow is $k_+ = 2 \times 10^{-6}$ m/s (Claassen 1981; estimated from James and Lupton 1978). James and Lupton's measurements at 5, 15, and 23°C lead to $E_a = 10.7$ kcal/mol, and a temperature dependence of the rate constant at groundwater-flow rates ($<2 \times 10^{-3}$ m/s) of

$$\log k_+ = 4.14 - 2338/T \tag{2.73}$$

where k_+ is in cm/s. In agreement, Liu and Nancollas (1971) suggest $E_a = 10 \pm 1.5$ kcal/mol. However, Karshin and Grigoryan (1970) obtained activation energies of 7.1 and 3.7 kcal/mol for dissolution of the {010} gypsum crystal face and faces perpendicular to {010}, respectively. Most authors consider gypsum dissolution to be chiefly diffusion-controlled. Calculated diffusion coefficients range from 4×10^{-4} to 8×10^{-4} m^2/s (Barton and Wilde 1971; Christoffersen and Christoffersen 1976).

James and Lupton (1978) found that gypsum solution rate increased with ionic strength. Their rate data, measured at a flow velocity of 0.15 m/s in up to 1.7 molal NaCl solutions, fit the equation $k/k_o = 1 + 2.2\sqrt{I}$. The rate constant for gypsum dissolution increases with flow velocity v, at 25°C according to

$$k_+ \text{ (m/s)} = 2 \times 10^{-6} + 5.8 \times 10^{-5} v \tag{2.74}$$

(James and Lupton 1978).

2.7.5 Oxidation of Ferrous Iron

There are two overall oxidation reactions for ferrous ion below pH 3.5 (cf. Stumm and Morgan 1981; Morel 1983; Morel and Hering 1993). Below pH = 2.2

$$Fe^{2+} + \tfrac{1}{4}O_2 + H^+ \rightarrow Fe^{3+} + \tfrac{1}{2}H_2O \tag{2.75}$$

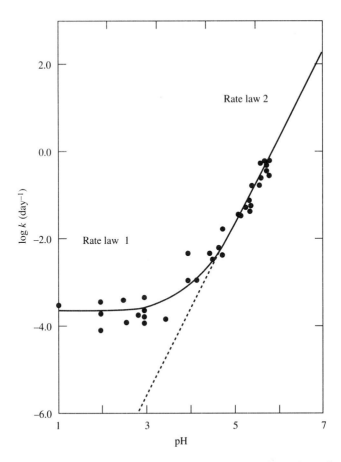

Figure 2.9   Oxidation rate of ferrous iron at 25°C as a function of pH. The ordinate is: $k = -d \log(Fe^{2+})/dt$. Data points are empirical results. Rate laws 1 and 2 are described in text. Modified after Stumm and Morgan (1981).

Above pH 2.2, but below about pH 3.5, the overall reaction is

$$Fe^{2+} + \tfrac{1}{4}O_2 + \tfrac{1}{2}H_2O \rightarrow FeOH^{2+} \tag{2.76}$$

The empirical rate law for Fe^{2+} oxidation under these acid conditions (rate law 1, Fig. 2.9) is

$$\frac{d(Fe(II))}{dt} = -k_+(Fe(II))\,P_{O_2} \tag{2.77}$$

which is seen to be second-order and pH independent. Empirically, at 20°C, $k_+ = 10^{-3.2}$/bar day. Under atmospheric conditions, with $P_{O_2} = 0.2$ bar (constant), the rate becomes pseudo first-order, and $t_{1/2} = \ln 2/0.2k_+ = 5500$ days (15 years). Studies of the same rate with bacterial mediation show it to be as much as 10^6 times faster than the inorganic rate, so that $t_{1/2} \approx 8$ minutes.

Above pH = 4, the rate of Fe^{2+} oxidation is related to the overall reaction

$$Fe^{2+} + \tfrac{1}{4}O_2 + \tfrac{5}{2}H_2O \rightarrow Fe(OH)_3 + 2H^+ \tag{2.78}$$

The empirical rate law (rate law 2, Fig. 2.9) is given by

$$\frac{d(\text{Fe(II)})}{dt} = - k_+ \frac{(\text{Fe}^{2+})}{(\text{H}^+)^2} P_{O_2} \tag{2.79}$$

for which $k_+ = 1.2 \times 10^{-11}$ mol^2/bar day at 20°C. If $P_{O_2} = 0.2$ bar, and pH = 6, the rate law becomes pseudo first-order, with $k'_+ = 2.4$/day. For these conditions $t_{1/2} = 0.693/2.4 = 0.29$ days = 7 hours. At pH = 7 and $P_{O_2} = 0.2$ bar, $k'_+ = 240$/day, and $t_{1/2} = 4.2$ min. The activation energy for the rate law above pH 4 is about 23 kcal/mol.

2.7.6 Pyrite and Marcasite Oxidation by Ferric Ion

Wiersma and Rimstidt (1984) measured the rate of the overall reaction

$$\text{FeS}_2 + 14\text{Fe}^{3+} + 8\text{H}_2\text{O} \rightarrow 15\text{Fe}^{2+} + 2\text{SO}_4^{2-} + 16\text{H}^+ \tag{2.80}$$

at pH 2.0 and 25°C, where k_+ is the rate constant, and FeS$_2$ is the mineral pyrite or marcasite. (See also Moses et al. 1987.) The reaction is first-order and follows the empirical rate law

$$\frac{d(\text{Fe}^{3+})}{dt} = - k_+ \frac{A_w}{M} (\text{Fe}^{3+}) \tag{2.81}$$

where (Fe^{3+}) is the molal concentration of free (uncomplexed) ferric iron and A_w/M the wetted surface area of reacting mineral per mass of solution. k_+ values range from 1.0×10^{-4} to 2.7×10^{-4}/s, and are not significantly different for pyrite and marcasite. The activation energy is 22 kcal/mol, based on rate measurements between 25 and 50°C. The rate-determining step is not known (see also Wadsworth 1979). Williamson and Rimstidt (1994) have published further analysis of the rate of pyrite oxidation by Fe^{3+} and by O$_2$, which is discussed in some detail in Chap. 12.

When molecular oxygen is the oxidizing agent, oxidation of sulfides and sulfur compounds by thiobacteria, which are all aerobic, can speed up oxidation rates by many orders of magnitude over the inorganic rates (Karamenko 1969).

2.7.7 Calcite Dissolution and Precipitation

According to Plummer et al. (1979) the rate of solution or precipitation of calcite is given by

$$R = k_1 [\text{H}^+] + k_2 [\text{H}_2\text{CO}_3^\circ] + k_3[\text{H}_2\text{O}] - k_4 [\text{Ca}^+] [\text{HCO}_3^-] \tag{2.82}$$

where the brackets denote activities, R is in mmol/cm^2 s and rate constants k_1 through k_3 correspond to the elementary reactions

$$\text{CaCO}_3 + \text{H}^+ = \text{Ca}^{2+} + \text{HCO}_3^- \qquad\qquad k_1$$

$$\text{CaCO}_3 + \text{H}_2\text{CO}_3^\circ = \text{Ca}^{2+} + \text{HCO}_3^- \qquad\qquad k_2$$

$$\text{CaCO}_3 + \text{H}_2\text{O} = \text{Ca}^{2+} + \text{HCO}_3^- + \text{OH}^- \qquad\qquad k_3$$

At concentrations well below saturation with calcite, the k_1 term dominates the rate equation up to pHs between about 4.5 and 5.5. At higher pHs when $P_{CO_2} > 0.1$ bar, the k_2 term dominates rates. At higher pHs and lower CO$_2$ pressures, the k_3 term dominates above pH 5.5. In freshwater [H$_2$O] = 1, and because $[\text{H}_2\text{CO}_3^\circ] = K_{CO_2} P_{CO_2}$, the rate law may be written

$$R = k_1 [\text{H}^+] + k_2 K_{CO_2} P_{CO_2} + k_3 - k_4 [\text{Ca}^{2+}] [\text{HCO}_3^-] \tag{2.83}$$

Temperature (K) functions for the rate constants are

$$\log k_1 = 0.198 - 444/T \tag{2.84}$$

$$\log k_2 = 2.84 - 2177/T \tag{2.85}$$

$$\log k_3 = -5.86 - 317/T \text{ (for } T < 25°C) \tag{2.86}$$

$$\log k_3 = 1.10 - 1737/T \text{ (for } T > 25°C) \tag{2.87}$$

At 25°C, values of the rate constants are: $k_1 = 10^{-1.29}$ s^{-1}; $k_2 = 10^{-4.46}$ s^{-1}; $k_3 = 10^{-6.92}$ s^{-1} ($T < 25°C$); and $k_3 = 10^{-4.73}$ s^{-1} ($T > 25°C$). Corresponding activation energies are: $E_a(k_1) = 2.0$ kcal/mol; $E_a(k_2) = 10.0$ kcal/mol; $E_a(k_3) = 1.5$ kcal/mol ($T < 25°C$) and $E_a(k_3) = 7.9$ kcal/mol ($T > 25°C$).

The low activation energy for k_1 suggests H$^+$ diffusion control is rate limiting at low pH's. Sjoberg and Rickard (1984) have suggested that for acid pHs (where k_1 is dominant), the rate of calcite dissolution is given by

$$R = k_1 [H^+]^{0.90} \tag{2.88}$$

reflecting that the rate-controlling process is not only H$^+$ diffusion to the surface, but also reaction-product diffusion away from it. The rate of dissolution at higher pH's is both surface-reaction and solution-product–diffusion controlled (Sjoberg and Rickard 1984).

Based on the rate equations of Plummer et al. (1979), we can compute that for groundwater pH's above 6 and P_{CO_2} values less than 0.1 bar at 25°C and below, the solution rate of calcite far from equilibrium reduces to $R = k_3$. In other words, for these conditions (which are typical of many shallow groundwaters) the reaction is zero-order, as long as the surface area of the calcite is constant. This assumes no catalysis or inhibition of the rate by adsorbed substances. (Sc, Cu, and PO$_4$ are strong inhibitors according to Sjoberg and Rickard 1984.) As equilibrium is approached, the rate equation becomes

$$R = k_3 - k_4 [Ca^{2+}] [HCO_3^-] \tag{2.89}$$

The value of k_4 is a complex function of temperature and P_{CO_2}. Based on Fig. 4 in Plummer et al. (1979), we can derive the approximate function

$$\log k_4 = -7.56 + 0.016 \, T - 0.64 \log P_{CO_2} \tag{2.90}$$

where T is in kelvin and P_{CO_2} in bars. This function permits the rough calculation of k_4 values for $P_{CO_2} < 10^{-1.5}$ bar. The reader is referred to Berner (1980) for a discussion of calcite dissolution/precipitation rate laws near saturation in the presence or absence of rate-inhibiting substances.

2.7.8 Silica Polymorphs, Dissolution and Precipitation

The reaction (which is elementary) describing the dissolution and precipitation of silica polymorphs is

$$SiO_2(s) + 2H_2O \underset{k_-}{\overset{k_+}{\rightleftarrows}} H_4SiO_4^° \tag{2.91}$$

The rate equals the change in dissolved silica concentration with time, or $d(H_4SiO_4^°)/dt$. Multiplying both sides of the differential rate equation by the activity coefficient γ H$_4$SiO$_4^°$ (see Chap. 4), leads to

$$\frac{d[H_4SiO_4^°]}{dt} = \frac{A_w}{M} (\gamma \, H_4SiO_4^°) (k_+[SiO_2] [H_2O]^2 - k_-[H_4SiO_4^°]) \tag{2.92}$$

(Rimstidt and Barnes 1980), where brackets enclose activities of the species, A_w/M (m^2/kg) is the wetted surface area of solid exposed to solution, divided by the mass of water. In most cases the activities of solid silica and water equal unity, and the rate equation reduces to

$$\frac{d[H_4SiO_4^\circ]}{dt} = \frac{A_w}{M}(k_+ - k_-[H_4SiO_4^\circ]) \tag{2.93}$$

Thus the dissolution reaction is zero-order and the precipitation reaction first-order.

The precipitation rate constant for quartz and α and β cristobalite up to about 300°C, and of amorphous silica to about 200°C, with the reaction written as in Eq. (2.91) is given by

$$\log k_-(s^{-1}) = -0.707 - 2598/T \tag{2.94}$$

where T is in kelvin (see Fig. 2.5).

Because the solubilities of the four polymorphs are known as a function of temperature (see Chap. 7, Section 7.5) the forward rate constants, k_+ (M/s) can be computed from the relationship $K_{eq} = k_+/k_-$ for each elementary reaction. Thus the K_{eq} versus T(K) expressions for quartz and amorphous silica written as in Eq. (2.91), are

$$\log K_{eq}(\text{quartz}) = 1.881 - 2.028 \times 10^{-3}T - 1560/T \tag{2.95}$$

and
$$\log K_{eq}(\text{SiO}_2\text{-amorph}) = 0.3380 - 7.8890 \times 10^{-4}T - 840.1/T \tag{2.96}$$

from which we find

$$\log k_+(\text{quartz}) = 1.174 - 2.028 \times 10^{-3}T - 4158/T \tag{2.97}$$

and
$$\log k_+(\text{SiO}_2\text{-amorph}) = -0.369 - 7890 \times 10^{-4}T - 3438/T \tag{2.98}$$

Corresponding activation energies are: E_a(quartz) = 16.1 to 18.3 kcal/mol, and E_a(SiO$_2$-amorph) = 14.6 to 15.5 kcal/mol. Rimstidt and Barnes (1980) examine the kinetics in terms of the activated complex (SiO$_2$ · 2H$_2$O)$^\pm$.

Recently, Dove (1994) (cf. Dove and Rimstidt [1995]) has developed a single function that successfully describes the dissolution kinetics of quartz from 25 to 300°C, pH 2 to 12, and in up to 0.3 molal Na solutions. The function takes into account the reactivity of neutral and negatively charged sites on the quartz surface.

2.7.9 Silicates Including Feldspars, Dissolution and Precipitation

Empirical studies of silicate rock or mineral solution rates at low temperatures, under conditions where the water is far from equilibrium with the solid, obey zero-order kinetics (cf. Apps 1983; Paces 1983, Bodek et al. 1988), also called linear kinetics (White and Claassen 1979). The best example of such behavior is the dissolution of SiO$_2$ polymorphs (see Rimstidt and Barnes 1980; and Section 2.7.8). Linear or zero-order kinetics is observed when the area of reacting mineral exposed to a volume of solution or volume of the water-rock system (also called the specific wetted surface, A_w, in cm^2 or m^2/m^3) may be considered constant with time. The general form of the empirical rate law is

$$R = \frac{dC}{dt} = A_w k_+ \tag{2.99}$$

where C is the aqueous concentration of a chemical species such as sodium or silica in the mineral. Paces (1983) relates this rate constant to the properties of a water-bearing formation with the equation

$$k_+ = (C - Ft)/A_w Ct \tag{2.100}$$

where k_+ is in mol/m^2 s, $C = 0$ at $t = 0$. F is the net input of the chemical species from the surroundings to groundwater, and A_w is defined as above (m^2/m^3). Helgeson et al. (1984) suggest $k_+ = 3 \times 10^{-12}$ mol/m^2 s at 25°C for K-feldspar, albite, and anorthite.

Based on laboratory dissolution-rate studies of feldspars, obsidian, and volcanic glass, White and Claassen (1979) observe that the initial solution rate of silicates can either obey linear (zero-order) kinetics as in Eq. (2.99), or parabolic kinetics, where the rate is given by

$$R = \frac{dC}{dt} = \left(\frac{A_w k_+}{2}\right) t^{-1/2} \tag{2.101}$$

Equation (2.101) corresponds to transport-controlled kinetics (cf. Stumm 1990). White and Claassen conclude that after long times in natural water/rock systems parabolic rates tend to become linear. Helgeson et al. (1984) show that feldspar dissolution rates are linear if the feldspar is pretreated to remove ultrafine reactive particles. In other words initial parabolic rates are probably an artifact of sample preparation. It seems likely that, in general, the dissolution or weathering of most silicates in natural water/rock systems obeys zero-order kinetics.

Silicate mineral dissolution is usually incongruent, with precipitation of relatively amorphous metastable products that may crystallize with time to form minerals such as gibbsite, kaolinite, illite, and montmorillonite (Helgeson et al. 1984). The incongruency means that the net release rates of individual components from a silicate mineral into the water may not be equal (cf. White and Claassen 1979; Helgeson et al. 1984).

Aagaard and Helgeson (1982) and Murphy and Helgeson (1989) propose general-rate expressions for the dissolution of silicate minerals. They suggest that the solution rate in acid water is determined by the decomposition rate of a critical activated surface complex. The rate is given by

$$R = k_+ [H^+]^n [1 - \exp(\Delta G/\sigma RT)] \tag{2.102}$$

where $[H^+]$ is the hydrogen activity in solution, n a constant, and σ the average stoichiometric number of the reaction (Helgeson et al. 1984). $\Delta G = -RT \ln (K_{eq}/Q)$, where K_{eq} is the equilibrium constant, and Q the corresponding activity quotient in solution. In other words, Eq. (2.102) may be written

$$R = k_+ [H^+]^n (1 - Q/K_{eq}) \tag{2.103}$$

Far from equilibrium ($Q/K_{eq} \leq 0.05$) Eqs. (2.102) and (2.103) reduce to

$$R = k_+ [H^+]^n \tag{2.104}$$

which might correspond to feldspar or mica weathering (cf. Kalinowsci and Schweda 1996) in humid climate soils or shallow groundwaters. Close to equilibrium ($Q/K_{eq} \geq 0.8$) Eq. (2.103) becomes

$$R = k_+ [H^+]^n [-\ln (Q/K_{eq})] \tag{2.105}$$

The solution rate of feldspars is pH-independent between about pH 4.5 and 8 at 25°C (Helgeson et al. 1984; Lasaga et al. 1994; Drever 1994). For these conditions the general rate equation may be written

$$R = k_+ \left(\frac{1 - Q}{K_{eq}}\right) \tag{2.106}$$

where k_+ is the pH-independent rate constant. Far from equilibrium this becomes

$$R = k_+ \tag{2.107}$$

Near equilibrium the rate is given by

$$R = k_+ [-\ln (Q/K_{eq})] \tag{2.108}$$

(At equilibrium $Q = K$ and $R = 0$.) Eq. (2.107) perhaps applies to the rate of feldspar dissolution in a fresh, unconfined groundwater, whereas Eq. (2.108) is more consistent with the near-equilibrium conditions we might expect in older confined groundwaters. Far from equilibrium at pH values above 7 to 8, feldspar dissolution rates again increase, and the rate equation has the general form

$$R = k_+ [OH^-]^m \tag{2.109}$$

In general, dissolution rates of other silicate and aluminosilicate minerals also increase under acid and alkaline conditions and are relatively independent of pH in the near-neutral pH range. Drever (1994) summarizes this behavior with a general equation for the forward rate far from equilibrium

$$R = k_H [H^+]^n + k_n + k_{OH} [OH^-]^m + k_L [ligand]^p \tag{2.110}$$

where the terms from left to right are the acid, neutral, and alkaline pH terms, and a less important term (usually) that describes the effect of complexing ligands on the rate. The general dissolution rate of silicates and aluminosilicates according to Drever (1994) is shown schematically in Fig. 2.10. (See also Fig. 2.11.) Drever indicates that n ranges from 0.3 to 1 and equals 0.9 for forsterite, 0.8 for amphiboles, 0.4 for biotite, and 0.5 for the Na- and K-feldspars. The pH of the important transition from acid to neutral rates is between pH 4.5 and 5.5 for most of the silicates, and from 4.5 to 5 for the feldspars. Because soil pH is usually close to or exceeds these values, weathering in soils will generally be defined by the neutral rate. The less important transition from neutral to alkaline rates may be above pH 7 to 8, with values of m in Eq. (2.110) ranging from 0.3 to 0.5 (Drever 1994).

In recent years rapid advances have been made in understanding mechanisms and controls on the dissolution and precipitation rates of silicate and aluminosilicate minerals in general. Much of this work has focused on rates of chemical weathering (cf. Steefel and Van Cappellen 1990; Lasaga

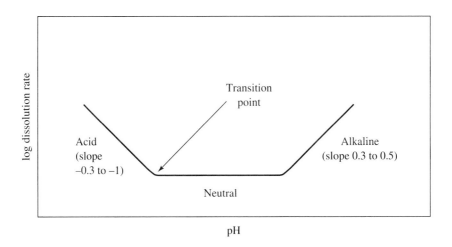

Figure 2.10 Schematic relationship between pH and dissolution rate for silicate and aluminosilicate minerals. Reprinted from *Geochim. et Cosmochim. Acta*, 58(10), J.T. Drever, The effect of land plants on weathering rates of silicate materials, 2325–32, © 1994, with permission from Elsevier Science, Ltd, The Boulevard, Langford Lane, Kidlington, OX5 1GB, U.K.

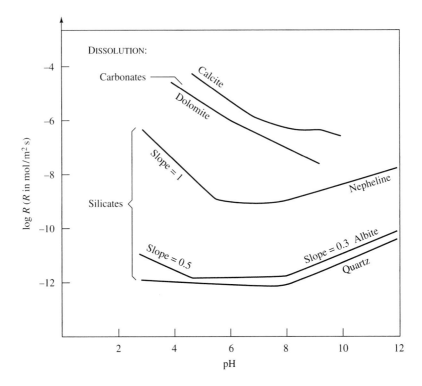

Figure 2.11   Generalized dissolution rates (R, mol/m s) of carbonate and silicate minerals in water near 25°C as a function of pH based on laboratory measurements. After A. Lerman, Transport and kinetics in surficial processes. In *Aquatic chemical kinetics,* ed. W. Stumm. Copyright © 1990 by John Wiley & Sons. Reprinted by permission of John Wiley & Sons, Inc.

et al. 1994). Among the more successful modeling efforts have been those relating the rates to concentrations and reactions of protonated and deprotonated mineral surface sites (cf. Wieland et al. 1988; Carrol-Webb and Walther 1988; Xie and Walther 1992; Dove 1994).

2.8 SUMMARY OBSERVATIONS

The rates of dissolution of carbonates and aluminosilicates as a function of pH are generalized in Fig. 2.11. Calcite and dolomite dissolution rates are generally 10^2 to 10^3-fold faster than rates for the silicates and decrease with pH up to saturation with the carbonates, usually between pH 8 and 10. Dissolution rates among the silicates range widely and are greatest for rapidly weathered minerals such as nepheline and olivine and slowest for quartz, muscovite (illite) and kaolinite, important products of chemical weathering in soils, discussed in more detail in Chap. 7.

The kinetics of reactions is very important in low-temperature environments, where rates tend to be slow, particularly for solid-solid or heterogeneous reactions with large activation energies. The rates of reactions involving the oxidation or reduction of species of Fe, S, N, and C are often very slow if governed strictly by inorganic rate laws. However, where biological activity is involved, rates of the same reactions may be orders of magnitude faster, decreasing from years to minutes or even seconds. Theoretical rates at low temperatures may also be invalid because of the adsorption of surface active trace substances such as metal species and organics, that may either inhibit or catalyze reactions.

STUDY QUESTIONS

1. Contrast the applicability of equilibrium and kinetic models to some environmental problems. Explain local and partial equilibrium assumptions.
2. Understand the definitions of system (or substance) residence time and the half-time of reactions in a system and how these concepts can be used to decide the applicability of equilibrium and kinetic concepts.
3. Explain why we need to know the relative mobilities (flux rates) of the substances involved in a reaction (e.g., solids, liquids, solutes, gases) to decide on the applicability of equilibrium versus kinetic models. This relates to the concepts of open and closed systems and residence time.
4. Know the difference between a system in steady state and one in equilibrium and how you might distinguish them.
5. Learn the approximate half-times of important types of reactions and the approximate residence times of important aqueous systems (e.g., streams, rain, the ocean, groundwater).
6. Contrast elementary and overall reactions with some examples. What is a rate-limiting step?
7. Define a zero-, first-, and second-order reaction, and a pseudo first- or second-order reaction, and discuss an example of each from the last section of this chapter.
8. Contrast the Arrhenius and van't Hoff equations and their integrated forms. Know how the magnitude of the activation energy, E_a, provides clues as to the reaction mechanism or process.
9. The rate of mineral dissolution may be mass-transport or surface-reaction controlled. Explain how the applicability of these controls relates to mineral solubility, with examples.
10. Define the activated complex. Explain how the activation energy of a reaction relates to the enthalpy of activation.
11. Study examples of reaction rates associated with formation of the $FeSO_4^+$ complex, radioactive decay, oxidation of organic matter, and ferrous iron, and the dissolution and precipitation of calcite and quartz.

PROBLEMS

1. Define and discuss an example of a pseudo first-order reaction.
2. You are given the elementary reaction, $H_2S = H^+ + HS^-$.
   (a) Write the overall rate equation that relates the changes in concentrations of the three species.
   (b) Given that the rate constant, $k_+ = 8.3 \times 10^3$/s for this reaction, and that $k_1(\text{dissoc.}) = 9.33 \times 10^{-8}$, calculate the reverse rate constant, k_-.
3. The rate of a reaction doubles between 25 and 50°C. What is its activation energy? What kind of a reaction might have an activation energy of this magnitude?
4. The dating of ^{14}C involves counting beta particles (electrons) emitted from its decay. Researchers interested in dating archaeological finds have found that their best counting equipment is limited to dating materials no older than about 50,000 years. Given that the half-time of radioactive decay of ^{14}C is $t_{1/2} = 5589$ y, calculate the number of decays per minute per gram (dpm/g) that a counter must detect to allow the dating of a sample of 50,000 years old. What is the length of time it takes for the emission of 1 beta particle from 50,000-year-old carbonaceous material?
5. Radon gas as Rn-222 is formed by the radioactive decay of U-238, the chief natural isotope of uranium, found in soils and especially in granitic rocks. (See Chap. 13.) As Rn-222, radon decays through several daughter-product isotopes to stable Pb-206. The gas occurs in air at natural levels of 0.1 to 0.2 pCi/L, and

is considered a health hazard (cancer risk) by the U.S. Environmental Protection Agency at or above 4 pCi/L in air. Rn-222 decays by the emission of alpha particles (helium nuclei) with a half-life or decay half-time of 3.823 days.

(a) What is its decay rate constant in s^{-1}?

(b) The mole fraction solubility of pure radon gas in water (N_{Rn}) from 3 to 97°C at 1 atm radon gas pressure, is given by the equation

$$R \ln N_{Rn} = -499.309 + \frac{25864.2}{T} + 69.3241 \ln T + 0.00101227\, T$$

(Bodek et al. 1988), where $R = 1.9872$ cal/mol K, and T is in degrees K. Calculate the Henry's law constant for Rn gas at 25°C in M/bar?

(c) Well waters pumped from granitic rocks or uraniferous sandstones may be high in radon. Exsolution of the gas into the air in a shower stall may result in air levels in excess of 4 pCi/L. What M concentration of dissolved Rn in a groundwater would be in equilibrium with 4 pCi/L of radon gas in bathroom air at 25°C?

6. Westrich and Berner (1984) suggest that the oxidation of organic matter can be considered a first-order reaction, with the differential rate equation written as: $dG_i/dt = -k_i G_i$. They assume three forms of organic matter, behaving with three different and decreasing rates of oxidation. (See Section 2.7). The third form is assumed recalcitrant and unreactive. Middleburg (1989) questions this approach and suggests that all organic matter will oxidize, given sufficient time. He proposes a single rate law for the oxidation of organic matter, whether aerobic or anaerobic, with a rate constant that is itself a function of time as given by the equation:

$$\log_{10} k \;(y^{-1}) = -0.95 \log_{10} t \;(y) - 0.81$$

for which $r = 0.987$. Assuming Middleburg's approach applies to the oxidative breakdown of sewage sludge, calculate how long it should take the fresh sewage sludge at a disposal site to decompose to 1% of its original weight.

7. Below about pH 3.5, ferrous iron oxidizes in streams according to the overall reaction

$$Fe^{2+} + \tfrac{1}{4}O_2 + H^+ \rightarrow Fe^{3+} + \tfrac{1}{2}H_2O$$

The rate law for the inorganic oxidation of ferrous iron under these conditions is given by:

$$\frac{d(Fe(II))}{dt} = -k_+(Fe(II))\, P_{O_2}$$

at 20°C, where $k_+ = 10^{-3.2}$/bar day.

Nordstrom (1985) measured the oxidation rate of ferrous iron in an acid mine drainage stream in which the initial Fe^{2+} concentration was 300 mg/L. The stream had a practically constant pH of about 2.5. The ferrous iron concentration dropped to about 5 mg/L after the stream had flowed for about 24 hours at about 0.2 m/s. Nordstrom concluded that the oxidation process was independent of the ferrous iron concentration, but was instead proportional to the concentration of the iron-oxidizing bacteria, *T. ferrooxidans*, in the stream.

(a) Calculate the reduction in ferrous iron concentration expected in the stream during this same 24-hour period due only to inorganic oxidation as expressed in the above rate law and compare it to the reduction in Fe^{2+} reported by Nordstrom.

(b) Using Nordstrom's empirical data for the rate of bacterial oxidation of Fe(II) in the stream, calculate the apparent rate constant, k_+^b, of the pseudo first-order reaction assuming $P_{O_2} = 0.21$ bar, and compare it to the apparent rate constant for the inorganic oxidation reaction.

8. It is often necessary to remove ferrous iron from groundwater so that the water complies with drinking water standards (Fe(II) < 0.3 mg/L). A New Jersey groundwater contains Fe(II) = 10 mg/L at a pH of 5. A water-treament plant operator has a large tank to which he can add lime (calcium oxide) to raise the pH, while aerating the water. He is interested in how long the water must be held in the tank at different pH val-

ues to remove the iron to values below 0.3 mg/L. To answer his question, a study is run where one-liter samples of the groundwater are aerated in bottles for periods of 15 and 30 minutes, at constant, but increasing pH values, with the pH adjusted by lime addition. Test results are given below.

pH	Fe(II) (mg/L) after 15 minutes	Fe(II) (mg/L) after 30 minutes
5	9.0	9.0
5.5	5.5	4.6
6.5	2.8	1.8
7.0	1.4	0.5
7.5	0.1	<0.1
8.0	<0.1	<0.1

You are asked to calculate the theoretical concentration of Fe(II) after 15 minutes at pH 6.5 and 7.0 and compare it to the empirical data.

9. Rates of dissolution or weathering of rocks and minerals are proportional to their surface areas exposed to a given volume of solution, or rate α (A/V). Exposed areas can vary widely, depending on the occurrence of the rock. In this problem you are asked to compare and contrast the rates of rock dissolution under attack by acid rain and in the pores of a rock. We will assume that the rock, temperature, pressure, and solution chemistry are the same in both cases.

   (a) Calculate the value of A/V (cm^{-1}) for a stone monument exposed to acid rain attack. Assume the water contacting the monument is running down the stone surface in a layer 1 mm thick.

   (b) Calculate A/V (cm^{-1}) for dissolution of the same rock by water in the rock pores. In this calculation assume 18% rock porosity, a rock density of 2.7 g/cm^3, and an internal rock surface area of 1 m^2/g.

   (c) Compare the rates of dissolution in each case and discuss.

10. Morey et al. (1964) collected samples of silica-rich hot spring waters from Yellowstone National Park and stored them in the laboratory at about 25°C. Silica concentrations in their samples stored for 2 years showed a strong pH dependence, from about 250 ppm as SiO_2(aq) at pH 1.5, to 150 ppm at pH 3, and 115 ppm at pH 6. Silica levels in the pH 3 sample had dropped from about 450 to 150 ppm in 2 years.

    Estimate how long it should take for silica concentrations in the pH 3 sample to drop to saturation with respect to amorphous silica (about 115 ppm as SiO_2) if the sample is stored at 25°C. Can silica kinetics explain the observed pH dependence of silica concentrations after 2 years?

3

Aqueous Complexes

3.1. INTRODUCTION AND OVERVIEW

Most trace metals and many major elements are transported in surface and groundwaters chiefly in complexed form. A *complex* is a dissolved species that exists because of the association of a cation and an anion or neutral molecule. A *ligand* is an anion or neutral molecule that can combine with a cation to form a complex.

Complexes are important for at least four reasons.

(1) *Complexing of a dissolved species that also occurs in a mineral tends to increase the solubility of that mineral over its solubility in the absence of the aqueous complexing.* This leads to a higher total concentration of the species that forms the complex in solutions saturated with respect to the mineral. For example, assume concentrations and activities are equal (see Chap. 1), pH is constant, and that the total calcium concentration (ΣmCa) is controlled by calcite solubility, that is, by $K_{sp} = (mCa^{2+})(mCO_3^{2-})$. If calcium ion is not complexed, then

$$\Sigma mCa = mCa^{2+} = K_{sp}/mCO_3^{2-} \tag{3.1}$$

But if sulfate and bicarbonate concentrations are high enough so that significant amounts of the complexes $CaSO_4^{\circ}$ and $CaHCO_3^{+}$ are formed, the total calcium concentration in equilibrium with calcite is increased, because

$$\Sigma mCa = mCa^{2+} + mCaHCO_3^{+} + mCaSO_4^{\circ} \tag{3.2}$$

where $mCa^{2+} = K_{sp}/mCO_3^{2-}$ and is unchanged. A problem calculation based on this example is given at the end of this section.

(2) *Some elements occur in solution more often in complexes than as free ions.* For example, the cations Cu^{2+}, Hg^{2+}, Pb^{2+}, Fe^{3+}, and U^{4+} are found chiefly in complexes rather than as the free ions. The oxycation UO_2^{2+} and oxyanions such as AsO_4^{2-} and SO_4^{2-} are complexes in which the uranium, arsenic, and sulfur are always complexed with oxygen. The mobilities and other environmental properties of such elements are, therefore, chiefly those of their complexes. This topic is discussed in some detail in this chapter.

(3) *Adsorption of cations or anions may be greatly favored or inhibited when they occur as complexes rather than as free (uncomplexed) ions.* For example, the hydroxide complexes of uranyl ion (UO_2^{2+}) are strongly adsorbed by oxide and hydroxide minerals, whereas uranyl carbonate complexes are poorly adsorbed by these minerals (Hsi and Langmuir 1985). In fact, carbonate, sulfate, and fluoride complexes of metals are often poorly adsorbed in general, whereas OH and phosphate complexes are usually readily adsorbed, particularly by oxide and hydroxide solids. Metal adsorption is considered at some length in Chap. 10.

(4) *The toxicity and bioavailability of metals in natural waters depends on the aqueous speciation or complexation of those metals.* The toxicities to aquatic life of Cu^{2+}, Cd^{2+}, Zn^{2+}, Ni^{2+}, Hg^{2+}, and Pb^{2+} are a function of the activities of the metal ions and their complexes, not of total metal concentrations (cf. Morel and Hering 1993; Manahan 1994). For example, monomethyl mercury ion (CH_3Hg^+) and Cu^{2+} are toxic to fish, but some other Hg and Cu complexes (such as $CuCO_3^o$) are far less so. The bioavailability of essential metals such as Fe, Mn, Zn, and Cu to plants is also a function of their metal speciation (Morel and Hering 1993). Until recently the U.S. Environmental Protection Agency did not recognize the importance of metal speciation in its water quality assessments (cf. Hall and Raider 1993). Metal toxicity is considered briefly near the end of this chapter.

Example 3.1

When the concentration of a dissolved species is controlled or limited by the solubility of a mineral containing that species, the total concentration of the species will be higher in solutions in which it is complexed than in those in which it is not. The following problem illustrates this point. Assuming the groundwater is at saturation with calcite ($CaCO_3$), $K_{sp} = 10^{-8.5}$ at 25°C and 1 bar pressure:

a) Compute the solubility of calcite as dissolved total calcium (ΣmCa) if the groundwater is pure water; and

b) Compute calcite solubility as ΣmCa in a water in which signficant complexing of the calcium occurs.

Assume, in general, that the pH is fixed and that $mCO_3^{2-} = 10^{-5.0}$ and is constant. In the complexed system (b) also assume $mSO_4^{2-} = 10^{-2.0}$ and $mHCO_3^- = 10^{-3.0}$. We are also told that $K_{assoc}(CaSO_4^o) = 10^{2.3}$, and $K_{assoc}(CaHCO_3^+) = 10^{1.1}$. Because activities are assumed equal to concentrations in our calculations, exponents will be rounded to one decimal place.

Solution

a) $K_{sp}(calcite) = 10^{-8.5} = (mCa^{2+})(mCO_3^{2-})$, so that the concentration of free calcium equals

$$mCa^{2+} = 10^{-8.5}/10^{-5.0} = 10^{-3.5} = 3.2 \times 10^{-4} \tag{3.3}$$

The mass-balance equation for the pure water system is simply

$$\Sigma mCa = mCa^{2+} \tag{3.4}$$

b) When calcium reacts to form complexes, decreasing the amount of uncomplexed calcium in solution, more calcite must dissolve to return the water to saturation with respect to calcite. The calcium mass-balance equation is now

$$\Sigma mCa = mCa^{2+} + mCaSO_4^o + mCaHCO_3^+ \tag{3.5}$$

To determine the quantitative effect, we calculate the amounts of the complexes that are formed. This is done by solving simultaneously the mass-balance equation, the two association constant expressions for the complexes, and the K_{sp}(calcite) expression. The association constant expressions are

$$K_{assoc}(CaSO_4^\circ) = 10^{2.3} = \frac{(mCaSO_4^\circ)}{(mCa^{2+})(mSO_4^{2-})} \tag{3.6}$$

$$K_{assoc}(CaHCO_3^+) = 10^{1.1} = \frac{(mCaHCO_3^+)}{(mCa^{2+})(mHCO_3^-)} \tag{3.7}$$

We solve these expressions for concentrations of the complexes and substitute into the mass-balance equation to eliminate the complexes, with the result

$$\Sigma mCa = mCa^{2+} + 10^{0.3}mCa^{2+} + 10^{-1.9}mCa^{2+}$$
$$= mCa^{2+}[1.0 + 2.0 + 0.01] = 3.01\ mCa^{2+} \tag{3.8}$$

but $mCa^{2+} = 3.2 \times 10^{-4}$, so $\Sigma mCa = 9.6 \times 10^{-4}$

The first, second, and third terms on the right side of the total calcium equation are the concentrations of mCa^{2+} and the $CaSO_4^\circ$ and $CaHCO_3^+$ complexes. The calculation shows that complexing has led to an increase in the total concentration of calcium and thus the solubility of calcite by roughly 300%.

3.1.1 Outer- and Inner-Sphere Complexes

Water molecules have a unique dipolar structure that results in locally unsatisfied negative and positive charges associated with the oxygen atom and hydrogen atoms in the molecule (see Fig. 3.1). Because of this dipolar charge distribution, water molecules in contact with cations tend to be oriented with their oxygens toward the cation and their protons away from it, as shown schematically below for Ca^{2+} and a single water molecule.

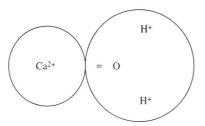

Ca^{2+} with radius = 1.0 nm
H_2O with a radius = 1.4 nm
H–O distance in H_2O = 1.1 nm
H–O–H angle = 104.5°

The six water molecules in the spherical envelope surrounding and in direct contact with a Ca^{2+} ion are all oriented with their protons away from the Ca^{2+}. (See Fig. 3.2.) The hydrated cation is also called a calcium aquocomplex.

Outer-sphere complexes, also described as *ion pairs,* involve the association of a hydrated cation and an anion, held by long-range electrostatic forces. The association is transient and not strong enough for the anion to displace any of the water molecules in immediate contact with the cation. Ion pairs are most often formed between major ($>10^{-4}$ m) monovalent and divalent metal

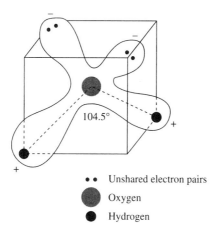

Figure 3.1 Electron cloud of the water molecule showing hydrogen and oxygen nucleii and unshared electrons. The "side of the jack" with hydrogens is positive and the other "side" is negative, giving rise to the dipolar character of water. The H-O-H bond angle is 104.5°. From E. K. Berner and R. A. Berner. *The global water cycle, geochemistry and environment.* Copyright © 1987. Used by permission of Prentice Hall, Inc., Upper Saddle River, NJ.

cations and major anions such as Cl^-, HCO_3^-, SO_4^{2-}, and CO_3^{2-}. Example metal ions involved include Na^+, K^+ (forms very weak pairs), Ca^{2+}, Mg^{2+}, and Sr^{2+}. A typical ion pair is $CaSO_4^\circ$, created by the reaction

$$Ca(H_2O)_6^{2+} + SO_4^{2-} = Ca(H_2O)_6SO_4^\circ \tag{3.9}$$

Ignoring the water molecules as is conventional, we have

$$K_{assoc} = \frac{[CaSO_4^\circ]}{[Ca^{2+}] \, [SO_4^{2-}]} \tag{3.10}$$

where the brackets denote activities of the enclosed species.

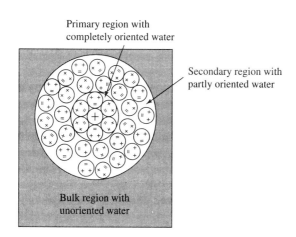

Primary region with completely oriented water

Secondary region with partly oriented water

Bulk region with unoriented water

Figure 3.2 A cation aquocomplex, showing the primary region with oriented water molecules in contact with the cation, typically in six- or fourfold coordination. The plot is schematic in that the water molecules in the primary region actually occur in a spherical envelope surrounding the cation. A secondary spherical envelope of partially oriented water molecules lies farther away from the cation. Water molecules in the bulk solution beyond this secondary region are unoriented. From J. O. Bockris and A. K. N. Reddy, 1973, *Modern electrochemistry,* vol. 1. © 1973, Plenum Pub. Corp. Used by permission.

TABLE 3.1　Radii of some cations in solids (chiefly oxides) and their ionic potentials (Ip = z/r)

Cation	Radius (Å)	Ip	Cation	Radius (Å)	Ip
Cs^+	1.88(12)		Ni^{2+}	0.69(6)	2.90
	1.67(6)	0.60	Be^{2+}	0.45(6)	4.44
Rb^+	1.72(12)		U^{4+}	0.89(6)	4.49
	1.52(6)	0.66	Fe^{3+}	0.645(6)	4.65
K^+	1.51(8)		V^{3+}	0.64(6)	4.69
	1.38(6)	0.72	Cr^{3+}	0.615(6)	4.88
Hg^+	1.19(6)	0.84	Co^{3+}	0.61(6)	4.92
Ag^+	1.00(4)		Pb^{4+}	0.775(6)	5.16
	1.15(6)	0.87	Al^{3+}	0.39(4)	
Na^+	1.02(6)	0.98		0.535(6)	5.61
Cu^+	0.60(4)		Ti^{4+}	0.605(6)	6.61
	0.77(6)	1.30	V^{4+}	0.58(6)	6.90
Li^+	0.76(6)	1.32	Mn^{4+}	0.530(6)	7.55
Ba^{2+}	1.35(6)	1.48	U^{6+}	0.73(6)	8.22
	1.42(8)		V^{5+}	0.54(6)	9.26
Pb^{2+}	1.19(6)	1.68	Si^{4+}	0.26(4)	
Sr^{2+}	1.18(6)	1.69		0.400(6)	10.0
	1.26(8)		W^{6+}	0.60(6)	10.0
Hg^{2+}	1.02(6)	1.96	Mo^{6+}	0.59(6)	10.2
Ca^{2+}	1.00(6)	2.00	As^{5+}	0.46(6)	10.9
Cd^{2+}	0.95(6)	2.11	B^{3+}	0.27(6)	11.1
Mn^{2+}	0.83(6)	2.41	P^{5+}	0.38(6)	13.2
Fe^{2+}	0.78(6)	2.56	Cr^{6+}	0.44(6)	13.6
Co^{2+}	0.745(6)	2.68	Se^{6+}	0.42(6)	14.3
Zn^{2+}	0.74(6)	2.70	S^{6+}	0.29(6)	20.7
Cu^{2+}	0.73(6)	2.74	C^{4+}	0.16(6)	25.0
Mg^{2+}	0.72(6)	2.78	N^{5+}	0.13(6)	38.5

Note: Radii, which are from Shannon (1976), are for the coordination numbers (CN's) given in parentheses. Where more than one CN is given, the table has been arbitrarily ordered with increasing Ip values for the six-fold coordinated cations. Radii and Ip values for C^{4+} and N^{5+} are from Ahrens (1952). Radii and Ip values for cations having Ip's greater than about 5 are of questionable meaning, given their tendency to form strongly covalent rather than ionic bonds with oxygen.

Many researchers prefer to define all complexes in which the bonding is chiefly electrostatic as ion pairs, regardless of whether waters of hydration surround the cation or it is in direct contact with the ligand (cf. Nancollas 1966). For example, the larger alkaline metals Cs^+ and Rb^+ and larger alkaline earths Ra^{2+} and Ba^{2+} have such low charge densities (as reflected by their low ionic potentials, see Table 3.1) that they are relatively unhydrated in solution, and so tend to form "contact" ion pairs.

The association or formation constants of ion pairs are generally less than 10^4 and increase with the increasing charges of cation and ligand. Summarized here are association constants at 25°C for ion pairs as a function of cation and ligand charge.

Charge of cation-ligand	$\log K_{\text{assoc}}$
1-1	0–1
1-2 or 2-1	0.7–1.3
2-2	2.3–3.2

Example pairs are $NaHCO_3^\circ$, $CaCO_3^\circ$, $CaHCO_3^+$, and $MgSO_4^\circ$. Because the cation and ligand in an ion pair are far apart (separated by waters of hydration around the cation) the electron configuration of the cation is "not seen" by the ligand. Evidence for this dominantly long range, electrostatic bonding is the remarkable constancy of K_{assoc} values for pairs formed by metals of the same charge and a given ligand. For example, $-\log K_{\text{assoc}} = 2.28$ to 2.35 for the 1:1 sulfate ion pairs of the divalent cations Ca, Mg, Ni, Zn, Co, Cd, Mn, Fe, and Cu.

In seawater, important amounts of sulfate, bicarbonate, and carbonate are ion-paired with calcium, magnesium and sodium ions (cf. Garrels and Christ 1965; Stumm and Morgan 1981). Percentage distributions of free and ion-paired species in seawater, according to Pytkowicz and Hawley (1974), are given here, where L denotes the respective anion in the ion pair.

Anion	Molality	% free ion	% CaL	% MgL	% NaL	% KL	% other ion pairs
SO_4	0.0291	39.0	4.0	19.5	37.1	0.4	—
HCO_3	0.00213	81.3	1.5	6.5	10.7	—	—
CO_3	0.000171	8.0	21.0	43.9	16.0	—	11.1

In high-salinity waters such as seawater, both ion-pairing and activity-coefficient effects (see Chap. 4) increase the concentrations of species limited by the solubility of minerals. For example, in pure water saturated with respect to calcite, the molal solubility product $\Sigma mCa^{2+} \times \Sigma mCO_3^{2-} = 10^{-8.5}$, whereas in seawater this product equals $10^{-6.1}$. If the concentration of carbonate is constant, this corresponds to a 250-fold increase in the concentration of dissolved calcium in seawater relative to that in pure water.

As with *outer-sphere complexes*, *inner-sphere complexes* are an ideal case. Actual complexes tend to have a component of both types of bonding. As noted above, most multivalent cations are aquocomplexes, with four to six ligand water molecules bonded to the cation. Accordingly, aquocomplexes are themselves inner-sphere complexes. For another ligand, L, to form an inner-sphere complex, it must displace one or more coordinating water molecules, forming a bond usually with some covalent character. This process may be written

$$M(H_2O)_n + L = ML(H_2O)_{n-1} + H_2O \tag{3.11}$$

Spectroscopic, conductimetric, and kinetic evidence have been used to establish the degree to which a complex is of outer- or inner-sphere character (cf. Nancollas 1966; Shaw et al. 1991). As noted above, the formation constants of many divalent metal-sulfate complexes are practically identical, consistent with a dominantly outer-sphere character. In fact, the sulfate pairs of divalent Mg, Zn, Ni, Co, Mn, and Be ($\log K_{\text{assoc}}(BeSO_4^\circ) = 1.95$) have been shown to have about 10% inner-sphere character (Nancollas 1966). The greater stabilities of $PbSO_4^\circ$ ($\log K_{\text{assoc}} = 2.69$) and $UO_2SO_4^\circ$ ($\log K_{\text{assoc}} =$

2.72) suggest that their percentages of inner sphere bonding exceed 10%. Higher-charged cations form bonds with ligands such as sulfate that are even more covalent. The formation constant of the $CrSO_4^+$ complex is $10^{4.8}$ (Sillen and Martell 1964). Nancollas (1966) notes that this complex is about 27% outer and 73% inner sphere.

In general, cations form increasingly inner-sphere complexes with a given ligand as their charge (z) increases and radius (r) decreases. In other words, the higher the ionic potential (Ip = z/r) of the cation (Table 3.1), the more covalent its bonding in a complex, and the stronger and more inner sphere the complex.

Examples of inner-sphere complexes include: HCO_3^-, AgS^-, HgI^+, $CdSH^+$, and UOH^{3+}. The cations Ag^+, Cd^{2+}, Zn^{2+}, and Hg^{2+} form strong inner-sphere complexes with Se^{2-}, Te^{2-}, S^{2-}, and SH^-. Trivalent and higher valent ions tend to form predominantly inner-sphere complexes in general. The tendency of metal cations and ligands to form inner-sphere complexes and the relative strengths of those complexes, are most readily explained and understood in terms of the concepts of electronegativity and of hard and soft acids and bases that are considered later in this chapter.

3.1.2 General Observations on Complexation

- The stability of complexes usually increases with increasing charge and/or decreasing radius of cations for a given ligand, or of ligands for a given cation. Ion pairs or outer-sphere complexes are weak and electrostatically bonded. Inner-sphere complexes involve important covalent bonding between the cation and ligand, and are generally stronger (have larger K_{assoc} values) than outer-sphere complexes.

- Cations and ligands that form strong complexes also tend to form minerals with low solubilities (low K_{sp} values).

- Complexing tends to increase the solubility of minerals that contain the species being complexed. Complexing of a species may also enhance or inhibit its adsorption and will usually affect its toxicity and bioavailability.

- The more saline a water, the more ions in it (especially multivalent ions) exist as complexes.

- The more saline a water, the more soluble minerals tend to be in it, both because of complex formation and activity-coefficient effects (see Chap. 4).

This introductory section has touched only briefly on the nature of complexes and their behavior in aqueous environments. The rest of the chapter expands on such topics, with particular emphasis on how to at least qualitatively predict the thermodynamic stabilities of complexes. This is a meaningful topic in that the stabilities of many potentially important complexes are poorly known.

3.2 METAL CATION–LIGAND RELATIONS IN COMPLEXES

As a general rule, metal cations are appreciably smaller than the ligands with which they form complexes. Thus the ionic radii of F^-, Cl^-, S^{2-}, O^{2-}, and all oxyanions, exceed about 1.3 to 1.4 Å, whereas (except for Ba^{2+}, Cs^+, Rb^+, and K^+) all monatomic cations are smaller than this. Typically, the absolute charge of metal cations exceeds that of coordinating ligands. These size and charge differences usually cause complexes to consist of one or more cations surrounded by several larger coordinating anions or neutral molecules. The maximum possible number of ligands surrounding and bonded to a given cation in a complex equals the maximum coordination number of the cation with respect to that ligand. Because of the long-range nature of coulombic forces of attraction, structuring of the ligands around a cation in aqueous solution extends to ligands somewhat beyond those in immediate contact with the cation. Nevertheless, we generally restrict our concerns in discussions of coordination to the ligands in direct contact with a core cation, or within that cation's inner-coordination sphere. The most commonly observed maximum-coordination numbers (CN's) for metal cations in aqueous complexes are 4, and 6, although CN values of 2, 3, 5, 7, 8, and 12 are also known.

In the simplest case, assuming that both cation and ligand are rigid spheres, complexation can be viewed as an exercise in closest packing. Listed here are theoretical coordination numbers and packing arrangements for given radius ratios (r_{cation}/r_{ligand}), assuming both cation and ligand are contact rigid spheres (Mason and Moore 1982).

Radius ratio	Arrangement of ligands around cation	CN of cation
0.14–0.22	corners of an equilateral triangle	3
0.22–0.41	corners of a tetrahedron	4
0.41–0.73	corners of an octahedron	6
0.73–1.00	corners of a cube	8
1.00	closest packing	12

The radius of oxygen (O^{2-}) (1.4 nm) roughly equals that of hydroxyl ion and the water molecule, so that similar maximum coordination numbers are found for all three. Of course, when the ligand is an oxyanion and contains multiple oxygens and/or larger atoms such as Cl^- or S^{2-} ($r = 1.8$ Å for both), fewer ligands can associate with a given cation, so its maximum possible coordination number will be less.

In any case, except for aquocation complexes in which water molecules occupy all possible coordination sites (typically six sites), the maximum observed number of coordinating ligands in complexes rarely exceeds three or four, so is usually less than a cation's maximum possible coordination number. This is because the coordination number we observe is proportional to the ligand concentration, which in natural waters is usually too low to stabilize higher numbers of ligands. This also reflects the fact that ligands must compete with and displace bonded water molecules in aquo-complexes in order to form their own complex. A typical complexation reaction in which M is the cation and L the ligand may be written

$$M(H_2O)_6 + L = ML(H_2O)_5 + H_2O \tag{3.12}$$

for which

$$K_{eq} = \frac{[ML(H_2O)_5]\,[H_2O]}{[M(H_2O)_6]\,[L]} \tag{3.13}$$

When we consider that the concentration of water molecules exceeds 55.6 mol/kg in fresh natural waters, whereas that of competing ligands is usually less than 0.001 to 0.0001 mol/kg, it is clear why aquocomplexes are important.

Except for the halogens and a few other monatomic anions, most ligands are not spherical. Summarized here, based chiefly on crystallographic evidence for solids, is the geometry of some common inorganic ligands (cf. Evans 1952), which generally retain the same geometry in solution. Note that apart from the monatomic ions, these ligands are themselves complexes. Under "Type formula," B denotes the cation and X the anion within the ligand.

Geometric arrangement	Type formula	Example ligands
Plane equilateral triangle	BX_3	BO_3^{3-}, CO_3^{2-}, NO_3^-
Regular trigonal low pyramidal:. Oxygens at corners of equilateral triangle, with central atom ~ 0.5 Å above this plane.	BX_3	AsO_3^{3-}, BrO_3^-, PO_3^{3-}, SbO_3^{3-}, SeO_3^{2-}, SO_3^{2-}
Regular tetrahedral: The most stable ligand group	BX_4	SO_4^{2-}, SeO_4^{2-}, CrO_4^{2-}, MnO_4^-, $Al(OH)_4^-$, PO_4^{3-}, ClO_4^-, $Si(OH)_4^0$
Distorted tetrahedron	BX_4	MoO_4^{2-}, WO_4^{2-}, ReO_4^-, IO_4^-
Spherical	X BX_n	F^-, Cl^-, Br^-, I^-, O^{2-}, S^{2-}, Se^{2-} OH^-, NH_4^+

G. N. Lewis defined an acid as an electron pair acceptor and a base as an electron pair donor (cf. Stumm and Morgan 1981). In this sense cations are acids and complexing ligands are bases. If a single pair of electrons is shared in a complex, the bond is said to be *monodentate,* where the word *dentate* means "having toothlike projections." Shown in Table 3.2 is the dentate nature of bonding of some common ligands in complexes. Clearly, most inorganic ligands form monodentate complexes. *Multidentate* ligands tend to be organic. Those, such as ethylenediaminetetraacetate (EDTA), which form multiple bonds with cations in a cagelike structure, are called *chelates* and are among the strongest of metal complex formers. Humic and fulvic acids are also thought to form chelate-type complexes with metal cations. The exceptional strength of metal-chelate bonding has led to the design of organic chelators that can selectively complex toxic elements such as plutonium, cadmium, or chromium to facilitate their removal from waste streams (cf. Hart 1993).

3.3 COMPLEXATION MASS-BALANCE AND EQUILIBRIA EQUATIONS

In a solution in which a metal M occurs in several successive mononulcear complexes of M (complexes containing one M atom) with ligand L, the total metal concentration is defined by the mass-balance equation

$$\sum M = M + ML + \cdots + ML_N = L + \sum_{i=0}^{N} ML_i \tag{3.14}$$

where M is the free metal ion concentration, and N is the maximum number of ligand groups that can coordinate with M. N is usually 2 to 4, but sometimes equals 6. The corresponding mass-balance equation for ligand L is

$$\sum L = L + ML + 2ML_2 + \cdots + (N)ML_N = L + \sum_{i=1}^{N} iML_i \tag{3.15}$$

or

$$\sum L - L = \sum_{i=1}^{N} iML_i \tag{3.16}$$

ΣL and ΣM are related through a concept called *the extent of complex formation,* or *average ligand number,* $\bar{n},$ first suggested by J. Bjerrum (cf. Beck 1970). This equals

$$\bar{n} = \frac{\Sigma L - L}{\Sigma M} \tag{3.17}$$

TABLE 3.2 Some important ligands and the nature of their bonding in complexes

Monodentate

H_2O, OH^-, NH_3, Cl^-, F^-, Br^-, HPO_4^{2-}, PO_4^{3-}, SO_4^{2-}, CO_3^{2-}, HCO_3^-,

$CH_3C\overset{O}{\underset{O^{(-)}}{}}$, $HC\overset{O}{\underset{O^{(-)}}{}}$, O^{2-}, I^- CN^-, H_2S

Bidentate

Glycinate $NH_2CH_2C\overset{O}{\underset{O^{(-)}}{}}$

Oxalate $\underset{(-)O}{\overset{O}{}}C-C\overset{O}{\underset{O^{(-)}}{}}$

1,10–orthophenanthroline

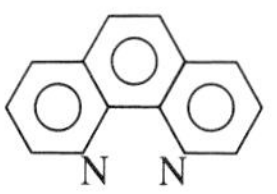

Salicylate

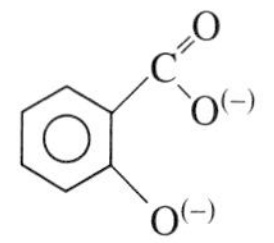

Ethylenediamine $NH_2CH_2CH_2NH_2$

Tridentate

Citrate $H_2CCO_2^-$
       |
       $HOCCO_2^-$
       |
       $H_2CCO_2^-$

Tetradentate

Nitrilotriacetate $N(CH_2CO_2^-)_3$

Hexadentate

Ethylenediaminetetraacetate (EDTA) anion

$\overset{-O_2CCH_2}{\underset{-O_2CCH_2}{}}NCH_2CH_2N\overset{CH_2CO_2^-}{\underset{CH_2CO_2^-}{}}$

Note: Inorganic ligands are chiefly monodentate, whereas organic ligands are most often multidentate in complexes.
Source: Modified after Pagenkopf (1978) and Phillips and Williams (1965).

or the ratio of the complexed ligand to the total metal concentration. The average ligand number is a measure of the fraction of M complexed by L. In typical natural water we will often have several important complexing ligands and cations, and thus several such equations. The total metal cation and total ligand equations, which are also called mass-balance equations, are fundamental to geochemical computer models. The average ligand number, however, is rarely used in such codes.

To parameterize and solve the total metal and total ligand equations for a given solution, we also need to know the formation constants of the complexes. The *stepwise formation constants* of mononuclear complexes are usually defined as follows (cf. Beck, 1970)

$$K_1 = \frac{[ML]}{[M]\,[L]}$$

$$K_2 = \frac{[ML_2]}{[ML]\,[L]} \tag{3.18}$$

$$K_N = \frac{[ML_N]}{[ML_{N-1}]\,[L]}$$

where the brackets denote activities of the enclosed species. The *overall or cumulative formation constants,* or β values, for ML_N complexes are given by

$$\beta_1 = K_1 = \frac{[ML]}{[M]\,[L]}$$

$$\beta_2 = K_1 K_2 = \frac{[ML_2]}{[M]\,[L]^2} \tag{3.19}$$

Or in general,

$$\beta_N = K_1 K_2 \cdots K_N = \frac{[ML_N]}{[M]\,[L]^N} \tag{3.20}$$

which can be abbreviated as,

$$\beta_N = \prod_{i=1}^{N} K_i \tag{3.21}$$

Finally, assuming activities equal concentrations, the cumulative formation constants are related to total metal and total ligand concentrations through the expressions

$$\Sigma M = [M] \sum_{i=0}^{N} \beta_i\,[L]^i \tag{3.22}$$

$$\Sigma L = [L] + [M] \sum_{i=0}^{N} \beta_i\,[L]^i \tag{3.23}$$

Additional equations relating stepwise and cumulative formation constants are suggested by Beck (1970) and Sillen and Martell (1964).

Other conventional notation refers to cumulative complex formation reactions that involve proton-metal exchange. Thus, for the general reaction

$$M + i\mathrm{HL} = ML_i + i\mathrm{H}^+ \tag{3.24}$$

the equilibrium constant expression is written

$$*\beta_i = \frac{[ML_i]\,[\mathrm{H}^+]^i}{[M]\,[\mathrm{HL}]^i} \tag{3.25}$$

If a complexation reaction is written as a *dissociation,* the corresponding complexation constant is called an *instability or dissociation constant.* Association constants generally have positive exponents, whereas dissociation constants most often have negative exponents.

Polynuclear complexes are those in which more than one metal cation is present. Such complexes are uncommon in natural waters, because they are only stabilized by relatively high concentrations of metal cations and ligands. (See Table 3.3.) Formation of the polynuclear complex M_mL_n may be written

$$mM + nL = M_mL_n \tag{3.26}$$

for which

$$\beta_{mn} = \frac{[M_mL_n]}{[M]^m[L]^n} \tag{3.27}$$

If the reaction is written as a proton-metal exchange, expressed in general terms as

$$mM + nHL = M_mL_n + nH^+ \tag{3.28}$$

an asterisk may be used in the formation constant notation, thus

$$*\beta_{mn} = \frac{[M_mL_n]\,[H^+]^n}{[M]^m[HL]^n} \tag{3.29}$$

(Sillen and Martell 1964), where in this case, m and n are the number of metal ions and protons involved in the reaction, respectively. For proton-cation exchange, where $m = 1$, the value of m is omitted from the subscript notation for β and $*\beta$, which is then of the form given above for mononuclear complexation reactions.

Two illustrative complexation problems follow. They are but slightly modified from their original format in Butler (1964). For simplicity, activities are assumed equal to concentrations in both problems.

Example 3.2

Calculate the concentration of all species present in a solution containing 1.00 M and 0.010 M $Cd(NO_3)_2$ at 25°C. Because the solution is strongly acidic, Cd-OH complexing need not be considered. Solving the problem requires the simultaneous solution of equilibrium constant expressions and mass-balance equations involving the aqueous species. We are given stepwise formation (equilibrium) constant expressions for the Cd-Cl complexes that can be reformatted to give

$$(CdCl^+) = 21(Cd^{2+})(Cl^-) \tag{3.30}$$

$$(CdCl_2^0) = 7.9(CdCl^+)(Cl^-) \tag{3.31}$$

$$(CdCl_3^-) = 1.23(CdCl_2^0)(Cl^-) \tag{3.32}$$

$$(CdCl_4^{2-}) = 0.35(CdCl_3^-)(Cl^-) \tag{3.33}$$

where the parentheses denote molar concentrations of the enclosed species. Mass-balance equations for total chloride and total cadmium are:

$$\Sigma Cl = (Cl^-) + (CdCl^+) + 2(CdCl_2^0) + 3(CdCl_3^-) + 4(CdCl_4^{2-}) = 1.00 \tag{3.34}$$

$$\Sigma Cd = (Cd^{2+}) + (CdCl^+) + (CdCl_2^0) + (CdCl_3^-) + (CdCl_4^{2-}) = 0.010 \tag{3.35}$$

We thus have six equations and six unknowns. Lacking a computer, the simplest approach is to first look for an approximate solution. Because total chloride exceeds total Cd by 100-fold, we can assume as a reasonable *first approximation* that nearly all of the chloride is uncomplexed, or $(Cl^-) = 1.00$ M. Introducing $(Cl^-) = 1.00$ in Eqs. (3.30) to (3.33), we can solve for the ratios of the complexes to (Cd^{2+}) to obtain: $(CdCl^+) = 21(Cd^{2+})$, $(CdCl_2^0) = 165(Cd^{2+})$,

$(CdCl_3^-) = 204(Cd^{2+})$, and $(CdCl_4^{2+}) = 71(Cd^{2+})$. Substituting for the complexes in Eq. (3.35) we find $(Cd^{2+}) = 2.16 \times 10^{-5}$. Concentrations of the complexes are now: $(CdCl^+) = 4.55 \times 10^{-4}$; $(CdCl_2^0) = 3.58 \times 10^{-3}$, $(CdCl_3^-) = 4.40 \times 10^{-3}$, and $(CdCl_4^{2-}) = 1.54 \times 10^{-3}$. Relative importance of the Cd species is: $CdCl_3^- > CdCl_2^0 > CdCl_4^{2-} > CdCl^+ > Cd^{2+}$. Now check our initial approximation by substituting these values into Eq. (3.34), which gives $(Cl^-) = 0.973$. Our approximation that $(Cl^-) = 1.00$ was off by only 3%. In other words, only 3% of the chloride is complexed by Cd. However, $[(\Sigma Cd - Cd^{2+})/\Sigma Cd] \times 100 = 99.8\%$ so practically all of the cadmium is complexed by chloride. We could repeat the entire calculation starting with $(Cl^-) = 0.973$ M in a *second approximation*. The final result would, however, differ little from results of the first approximation.

Example 3.3

It is useful to construct diagrams that show changes in the relative importance of complexes as a function of pH or the concentration of a metal or ligand. In such a problem, Butler (1964) describes how to calculate as a function of uncomplexed chloride ion, (Cl^-), the fraction of Cd present as Cd^{2+} ion and as the several Cd-Cl complexes. The general approach involves defining and solving simultaneously a series of equilibria, mass-balance, and charge-balance equations.

  First, Cd-Cl complexing is described in terms of the cumulative (overall) constant expressions. These are:

$$(CdCl^+) = \beta_1(Cd^{2+})(Cl^-), \qquad \beta_1 = K = 21 \tag{3.36}$$

$$(CdCl_2^0) = \beta_2(Cd^{2+})(Cl^-)^2, \qquad \beta_2 = K_1K_2 = 166 \tag{3.37}$$

$$(CdCl_3^-) = \beta_3(Cd^{2+})(Cl^-)^3, \qquad \beta_3 = K_1K_2K_3 = 204 \tag{3.38}$$

$$(CdCl_4^{2-}) = \beta_4(Cd^{2+})(Cl^-)^4, \qquad \beta_4 = K_1K_2K_3K_4 = 71.5 \tag{3.39}$$

For simplicity, total cadmium (ΣCd) is symbolized as C in following calculations. The mass-balance equation for cadmium is then

$$\Sigma Cd = C = (Cd^{2+}) + (CdCl^+) + (CdCl_2^0) + (CdCl_3^-) + (CdCl_4^{2-}) \tag{3.40}$$

Eliminating the complexes with the cumulative constant expressions in Eqs. (3.36) to (3.39) we have

$$C = (Cd^{2+})[1 + \beta_1(Cl^-) + \beta_2(Cl^-)^2 + \beta_3(Cl^-)^3 + \beta_4(Cl^-)^4] \tag{3.41}$$

The fraction of cadmium present as each species (α_n) can thus be expressed as a function of the constants and (Cl^-) alone.

$$\alpha_0 = \frac{Cd^{2+}}{C} = \frac{1}{[1 + \beta_1(Cl^-) + \beta_2(Cl^-)^2 + \beta_3(Cl^-)^3 + \beta_4(Cl^-)^4]} \tag{3.42}$$

Similar expressions may be written for each of the Cd-Cl complexes, which, after substitution, are also seen to be functions of α_0. The simplest way to obtain these expressions is to divide both sides of Eqs. (3.36) to (3.39) by C, and then to substitute $\alpha_0 C$ for (Cd^{2+}) in each equation. Fractions of each complex then equal

$$\alpha_1 = \frac{(CdCl^+)}{C} = \beta_1(Cl^-)\,\alpha_0 \tag{3.43}$$

$$\alpha_2 = \frac{(CdCl_2^0)}{C} = \beta_2(Cl^-)^2\alpha_0 \tag{3.44}$$

$$\alpha_3 = \frac{(CdCl_3^-)}{C} = \beta_3 (Cl^-)^3 \alpha_o \tag{3.45}$$

$$\alpha_4 = \frac{(CdCl_4^{2-})}{C} = \beta_4 (Cl^-)^4 \alpha_o \tag{3.46}$$

Two interesting conclusions derive from Eqs. (3.42) to (3.46). First, the fraction of Cd present as each species is independent of the total Cd concentration. (This is true when only *mononuclear* Cd complexes are present, but would not be true in the presence of *polynuclear* Cd complexes, that is, Cd complexes that contained more than one atom of Cd). The second noteworthy conclusion is that higher chloride complexes increase rapidly in relative importance with increasing total chloride. This is because the concentration of each complex is proportional to $(Cl^-)^n$, where n is the number of Cl^- groups in the complex.

In preceding expressions (Cl^-) denotes the concentration of free or uncomplexed chloride. Total chloride is given by the mass-balance equation

$$\Sigma Cl = (Cl^-) + (CdCl^+) + 2(CdCl_2^o) + 3(CdCl_3^-) + 4(CdCl_4^{2-}) \tag{3.47}$$

Substituting successive (Cl^-) values into Eqs. (3.42) to (3.46) we can solve for the fractions of each Cd species as a function of increasing chloride concentration. The results are plotted in Fig. 3.3, which shows that when chloride is less than 10^{-4} M, free Cd^{2+} dominates and Cd-Cl complexing is negligible. As chloride concentrations increase (moving from right to left in the figure), successively higher chloride complexes become important. In seawater, for example, where $pCl \approx 0.25$, most of the Cd is complexed, and all four complexes occur in significant amounts.

The answer to Example 3.2 can be read directly from Fig. 3.3 or found by substitution into Eqs. (3.42) to (3.46). First, given that $(Cl^-) = 1.00$ M ($pCl = 0$), we solve for the fractional values. These are then multiplied times $\Sigma Cd = 0.010$ M, the total analytical Cd concentration to obtain amounts of the individual species.

$$
\begin{aligned}
\alpha_0 &= 0.2\%, & (Cd^{2+}) &= 2 \times 10^{-5} \\
\alpha_1 &= 4.6\%, & (CdCl^+) &= 4.6 \times 10^{-4} \\
\alpha_2 &= 35.8\%, & (CdCl_2^o) &= 3.58 \times 10^{-3} \\
\alpha_3 &= 44.0\%, & (CdCl_3^-) &= 4.40 \times 10^{-3} \\
\alpha_4 &= 15.4\%, & (CdCl_4^{2-}) &= 1.54 \times 10^{-3}
\end{aligned}
\tag{3.48}
$$

They equal the same values determined in Example 3.2. Usually fractional species curves such as are shown in Fig. 3.3 are plotted together in one figure to save space and for more direct comparison of the relative importances of the species.

3.4 HYDROLYSIS OF CATIONS IN WATER AND IONIC POTENTIAL

The dipolar nature of the water molecule (see Fig. 3.1) is an important property of water that influences its interactions with cations. Because of its dipolar character, charge is unevenly distributed at the surface of each water molecule, and the protons of one molecule attract the oxygens of adjacent water molecules. This attractive force is called hydrogen bonding and relates to ϵ, the dielectric constant of water. The dielectric constant is a measure of a solvent's ability to dissolve ionic solids and

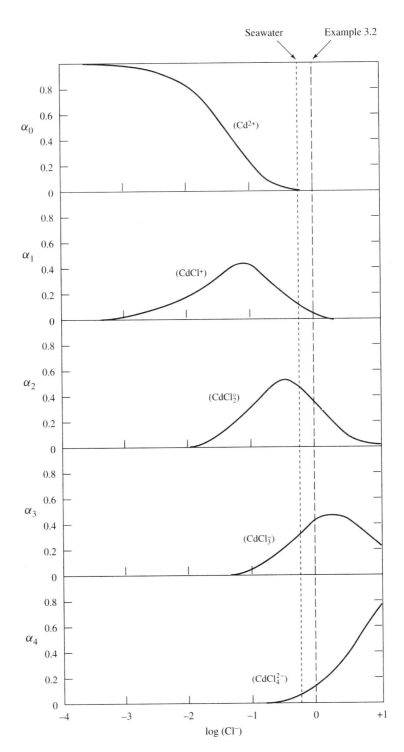

Figure 3.3 Fraction of Cd present as the free ion and as chloride complexes as a function of chloride ion concentration (M) for a total Cd concentration of 0.010 M at 25°C. Also shown are the distribution of species as computed in Example 3.2, and in seawater. Modified after J. N. Butler, 1964. *Ionic equilibrium, a mathematical approach.* Used by permission.

is higher for water than for any other common inorganic liquid. (For H_2O, $\epsilon = 78.54$ at 25°C.) Note that the bonding between monovalent and divalent cations and the Si-O and Al-O groups in minerals is mostly ionic. This helps to explain the effectiveness of water as the agent of chemical weathering of alumino-silicate minerals such as the feldspars. (See Chap. 7).

As noted above, multivalent cations in solution tend to be hydrated or coordinated with water molecules and thus may be called aquocomplexes. The extent of hydration of a cation is proportional to its ion size parameter (a_i). (This parameter is used in the Debye-Hückel equation for ion activity coefficients; see Chap. 4). For strongly hydrated Li^+, $a_i = 6$ Å, whereas for unhydrated Cs^+, $a_i = 2.5$ Å. This compares with the unhydrated sixfold coordination radii of 0.76 Å for Li^+ and 1.76 Å for Cs^+ in minerals (Shannon 1976).

The number of surrounding water molecules bonded directly to a cation (the coordination number of the cation with respect to H_2O) is a measure of the cation's surface-charge density. A cation's surface-charge density is proportional to the charge of the ion z and inversely proportional to ion size or radius r measured in solids. Thus we may define the concept of ionic potential, Ip, where Ip $= z/r$. (See Table 3.1.) Note that assuming sixfold radii for the cations, the Ip for strongly hydrated $Li^+ = 1/0.76 = 1.32$, whereas Ip for practically unhydrated $Cs^+ = 1/1.67 = 0.60$.

A plot of ionic radius versus cation charge is instructive. Such a plot is shown in Fig. 3.4, with dashed lines drawn to roughly separate cation-hydrolysis products by their nature. The following discussion considers the behavior of the cations at pH = 7. The plot in Fig. 3.4 shows that most of the "soluble" monovalent and divalent cations found in carbonate and evaporate minerals (especially sulfates and chlorides of Na^+, K^+, Ca^{2+}, and Mg^{2+}) have ionic potentials less than about 3 (the approximate slope of the dashed straight line). The larger ions in this part of the diagram (e.g., K^+ and Ra^{2+}) and even larger Cs^+ and Rb^+, which are not shown, are unhydrated or weakly hydrated, whereas those with higher ionic potentials form aquocations (e.g., Li^+, Mg^{2+}, and Fe^{2+}).

As ionic potentials increase above 3 (core cation charge densities increase), the positive charge densities of the cations become great enough so that they repulse protons from one or more coordinating water molecules, and core cations can only exist in solution if bonded in OH^- and O^{2-} species. In this region of the diagram between the dashed lines, core cations may occur in soluble oxycations (e.g., VO^{2+}, UO_2^+, and UO_2^{2+}), hydroxycations (e.g., $FeOH^{2+}$, $AlOH^{2+}$, and $Th(OH)_2^{2+}$), and hydroxyanions (e.g., $Al(OH)_4^-$, $Fe(OH)_4^-$, $Sb(OH)_6^-$, and $TeO(OH)_5^-$) (see Table 3.3). Also, in this region, particularly among trivalent and tetravalent cations, the number of repulsed protons may equal the positive charge on the cation, resulting in the formation of neutral complexes (e.g., $Fe(OH)_3^\circ$) and/or oxide or hydroxide solids. When thermodynamically stable, most of such solids are highly insoluble in natural waters at pH = 7. They include gibbsite [$Al(OH)_3$], goethite [$FeOOH$], hematite [Fe_2O_3], pyrolusite [MnO_2], thorianite [ThO_2], rutile [TiO_2], uraninite [UO_2], zircon [$ZrSiO_4$], and cassiterite [SnO_2], among others.

As Ip increases above values as low as 8.5, and we enter the upper-left portion of Fig. 3.4, the bonding between core cation and associated oxygen or hydroxyl is even stronger and largely covalent. The result is the formation of oxyanionic species such as silicate, selenate, borate, carbonate, arsenate, and sulfate, which, because of their relatively low charge densities *as oxyanions*, form rather weak bonds with cations and are soluble.

As a complication to Ip concepts, several of the above elements are redox-sensitive, and thus can occur in more than one valence state. These include As, Cr, Fe, Mn, Mo, and U among many others. The solubilities of these elements' compounds depend on which oxidation state of the element is stable in the compound and in solution for the given conditions. Thus, minerals of reduced Fe^{2+} and Mn^{2+}, and of UO_2^{2+}, the oxidized form of uranium, are comparatively soluble, and these species occur as soluble aquocations. In contrast, Fe^{3+} and Mn^{4+} (oxidized) and U^{4+} (reduced) most often occur in insoluble oxides and hydroxides.

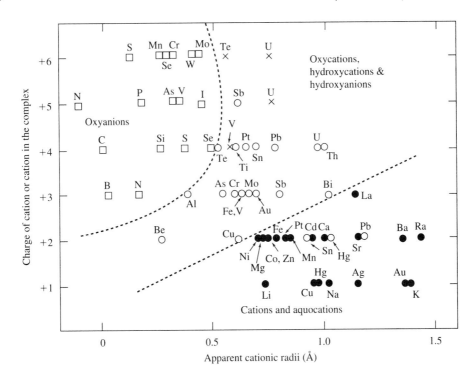

Figure 3.4   Charge of core cations in their aquocomplexes plotted against apparent crystal radii of the cations in solids (1 Å = 1 nm). The radii are mostly from Shannon and Prewitt (1969). Dashed curves roughly divide species by their behavior. Radii of the cations were computed assuming the radius of the oxygen atom equals 1.4 Å and is constant. However, it is less than 1.4 Å in the oxyanions because of strong covalent bonding between oxygen and multivalent cations of N, C, S, and B, for example. Consequently, the apparent radii of these cations shown in the figure are only qualitatively meaningful. (●) cations and aquocations; (○) hydroxycations and hydroxyanions; (×) oxycations; (□) oxyanions. Reprinted with permission from Techniques of estimating thermodynamic properties for some aqueous complexes of geochemical interest, D. Langmuir. In *Chemical modeling in aqueous systems*, ed. E. A. Jenne, Am. Chem. Soc. Symp. Ser. 93. Copyright 1979 American Chemical Society.

Given in Table 3.3 are the species formed by hydrolysis of cations in water at 25°C and between pH 2 to 12. The nature of these species is broadly related to ionic potentials of the core cations in Table 3.1, and more qualitatively a function of core cation charge as apparent from the listing in Table 3.3. Complex formation in natural waters will involve the replacement of one or more coordinating hydroxyls or water molecules in these species by other ligands.

As might be expected from the preceding discussion, the tendency of cations to form OH complexes, thus the association constants of such complexes, are roughly proportional to ionic potentials of the cations. We would then expect a plot of log K_{assoc} versus Ip to have a positive slope for cations that form chiefly electrostatic bonds with OH. A related and theoretically better measure of electrostatic bonding than ionic potential is the coulombic function $z_M z_L/(r_M + r_L)$, which considers the charge and radius of both the metal cation (M) and the complexing ligand (L). Shown in Fig. 3.5 is a plot of log K_{assoc} versus this function for a number of 1:1 cation-hydroxy complexes. (Remember $\Delta G°_{assoc} = -RT \ln K_{assoc}$.) The plot shows that the weakly OH-complexed alkali metal and alkaline earth cations plot along a straight line. The line has a slope of 3.1, predicted by the simple electro-

TABLE 3.3 Species formed by hydrolysis of cations in water at 25°C and pH 2 to 12

Valence of core cation	Cations and complexes
1+	Ag, Au, Cu, Hg, K, Li, Na
2+	Ba, Ca, Cd, Co, Fe, Mg, Mn, Ni, Pt, Ra, Sr, Zn; $Be(OH)_n^{2-n}[0-4]$, $Cu(OH)_n^{2-n}[0-4]$, $Cu_2(OH)_2^{2+}$, $Hg(OH)_n^{2-n}[0-3]$, $Pb(OH)_n^{2-n}[0-3]$, $Pb_3(OH)_4^{2+}$, $Sn(OH)_n^{2-n}[0-3]$
3+	$Al(OH)_n^{3-n}[0-4]$, AsO^+, $As(OH)_n^{3-n}[3,4]$, $Au(OH)_n^{3-n}[2-4]$, $B(OH)_n^{4-n}[3,4]$, $Bi(OH)_n^{3-n}[0-4]$, $Bi_6(OH)_{12}^{6+}$, $Bi_9(OH)_n^{27-n}[20-22]$, $Cr(OH)_n^{3-n}[0-4]$, $Cr_3(OH)_4^{5+}$, $Fe(OH)_n^{3-n}[0-4]$, $H_nNO_2^{3-n}[0,1]$, La^{3+}, Mo^{3+}, $Sb(OH)_n^{3-n}[2-4]$, $V(OH)_n^{3-n}[0-3]$
4+	$H_nCO_3^{2-n}[0-2]$, $H_nSO_3^{2-n}[0,1]$, $H_nSeO_3^{2-n}[0-2]$, $H_nSiO_4^{4-n}[2-4]$, $Pb(OH)_n^{4-n}[3-6?]$, $Pt(OH)_n^{4-n}[1-4?]$, $Sn(OH)_4^{\circ}?$, $Te(OH)_n^{4-n}[3,4]$, $TeO(OH)_4^{4-n}[2,3]$, $VO(OH)_n^{2-n}[0,1]$, $Th(OH)_n^{4-n}[0-4]$, $Th_2(OH)_2^{6+}$, $Ti(OH)_n^{4-n}[2-4]$, $U(OH)_n^{4-n}[1-4]$
5+	$H_nAsO_4^{n-3}[0-3]$, $H_nPO_4^{n-3}[0-3]$, $H_nVO_4^{n-3}[1-4]$, IO_3^-, NO_3^-, $Sb(OH)_n^{5-n}[5-6]$, UO_2^+
6+	$H_nCrO_4^{2-n}[0,1]$, $H_nMoO_4^{2-n}[0-2]$, $H_nSO_4^{2-n}[0,1]$, $H_nSeO_4^{2-n}[0,1]$, $H_nWO_4^{2-n}[0-2]$, MnO_4^{2-}, $TeO_n(OH)_m^{6-2n-m}[nm = 06, 15, 24]$, $(UO_2)_n(OH)_m^{2n-m}[nm = 10, 11, 22, 35]$, $W_{12}O_{39}^{6-}$

Note: Only species that may exceed 10% of the total core cation concentration are listed. Cations are written as free species if they occur chiefly as such in the pH range 2 to 12. When a cation is less stable than its hydroxy or oxycomplex at a pH level below 7, the hydroxy or oxycomplex is listed instead of the free cation. Most of the free cations occur as aquocations. Elements that form polynuclear complexes are assumed to occur at $\leq 10^{-5}$ molal total concentrations. Brackets enclose possible values of n and/or m.

Source: Most of the tabulated species are discussed by Baes and Mesmer (1976, 1981). Table is modified after Langmuir (1979). Reprinted with permission from Techniques of estimating thermodynamic properties for some aqueous complexes of geochemical interest, D. Langmuir. In *Chemical modeling in aqueous systems*, ed. E. A. Jenne, Am. Chem. Soc. Symp. Ser. 93. Copyright 1979 American Chemical Society.

static model (Langmuir 1979), which assumes metal cation-OH bonding is purely electrostatic and, therefore, proportional to the coulombic function. Metal-OH complexing is stronger and becomes increasingly covalent (inner sphere) as $z_M z_L/(r_M + r_L)$ values increase for species plotting above the line (cf. Baes and Mesmer 1976, 1981).

3.5 ELECTRONEGATIVITY AND THE STABILITIES OF INNER-SPHERE COMPLEXES

In this and following sections we will consider other approaches that have been used to understand the stabilities of inner- and outer-sphere complexes. The concept of electronegativity (EN) helps to explain the stabilities of complexes that have some inner-sphere character. Electronegativity is defined as the power of an atom or ion to attract electrons (cf. Pauling 1960). Electronegativity data for monatomic cations and ligands can be found in Pauling (1960), Gordy and Thomas (1956), Allred (1961), Wells (1962), Fyfe (1964), Rosler and Lange (1972), Clifford (1959), and Moskvin (1975). Unfortunately, such data are lacking for polyatomic cations and ligands, although EN may be estimated or computed for such species from mineral solubility products (Clifford 1959).

Atoms with high EN's (especially above 2) are Lewis bases. These are chiefly nonmetal ions or potential ligands. Atoms with EN values less than about 2 are generally metal cations and Lewis acids (see Table 3.4). Bonding in inner-sphere complexes depends (in part) on ΔEN, the difference

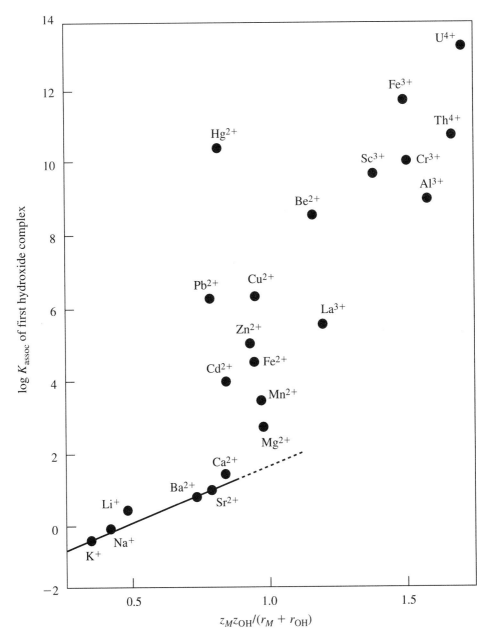

Figure 3.5 Plot of the association constant of some 1:1 metal cation-hydroxy complexes at zero ionic strength (see Chap. 4) versus the electrostatic function $z_M z_{OH}/(r_M + r_{OH})$, where the association reaction is written $M^z + OH^- = MOH^{z-1}$, and z and r are the charge and radius in nanometers (nm) or angstroms (Å) (1 nm = 1 Å) of cation M and OH ($r_{OH} = 1.40$ nm). Cation radii are from Shannon and Prewitt (1969), log K_{assoc} values from Baes and Mesmer (1981). The slope of the straight line suggests the contribution of electrostatic (ionic) bonding to the stability of the complexes. The extent to which species plot above this line presumably reflects the increased contribution of covalency to their stabilities.

TABLE 3.4 Electronegativities of some ions

Atomic number	Species	EN	Atomic number	Species	EN
1	H^+	2.20	29	Cu^+	1.8
3	Li^+	0.98		Cu^{2+}	2.0
4	Be^{2+}	1.57	30	Zn^{2+}	1.65
5	B^{3+}	2.04	33	As^{3+}	2.0
6	C^{4+}	2.55		As^{5+}	2.2
7	N^{3+}	3.04	35	Br^-	2.96
8	O^{2-}	3.44	37	Rb^+	0.82
9	F^-	3.98	38	Sr^{2+}	0.95
11	Na^+	0.93	42	Mo^{4+}	1.6
12	Mg^{2+}	1.31		Mo^{6+}	2.1
13	Al^{3+}	1.61	47	Ag^+	1.93
14	Si^{4+}	1.90	48	Cd^{2+}	1.69
15	P^{3+}	2.19	50	Sn^{2+}	1.7
16	S^{2-}	2.58		Sn^{4+}	1.96
17	Cl^-	3.16	53	I^-	2.66
19	K^+	0.82	55	Cs^+	0.79
20	Ca^{2+}	1.00	56	Ba^{2+}	0.89
21	Sc^{3+}	1.36	57	La^{3+}	1.10
22	Ti^{4+}	1.54	79	Au^+	2.54
23	V^{3+}	1.35		Au^{3+}	2.9
	V^{4+}	1.6	80	Hg^+	1.8
	V^{5+}	1.8		Hg^{2+}	2.00
24	Cr^{3+}	1.6	81	Tl^+	1.5
	Cr^{6+}	2.1		Tl^{3+}	2.04
25	Mn^{2+}	1.4	82	Pb^{2+}	1.6
	Mn^{7+}	2.3		Pb^{4+}	1.8
26	Fe^{2+}	1.7	88	Ra^{2+}	(0.83)
	Fe^{3+}	1.8	90	Th^{4+}	1.1
27	Co^{2+}	1.88	92	U^{4+}	1.3
28	Ni^{2+}	1.91		U^{6+}	1.9
				UO_2^{2+}	(1.8)

Complex ligands	Coord. no.	EN	Complex ligands	Coord. no.	EN
OH^-	1	3.1	HPO_4^{2-}		2.8
	2	2.75	CO_3^{2-}		(2.5)
	3 & 4	2.15	HCO_3^-		(~4)
NO_3^-		3.5	HS^-		(2.33)
$H_2PO_4^-$		3.15	SO_4^{2-}		3.7

Source: From or based on Pauling (1960), Wells (1962), Allred (1961), Fyfe (1964), Rosler and Lange (1972), and Moskvin (1975). Parenthetic values are estimates based chiefly on Clifford (1959).

in electronegativities between the cation and ligand in the complex. When $\Delta EN = 0$ (as for the carbon-carbon bond in diamond), the bonding is purely covalent. Pauling (1966) suggests that when $\Delta EN < 1.7$, covalency predominates over ionicity. Larger ΔEN values indicate chiefly electrostatic or ionic bonding between cation and ligand.

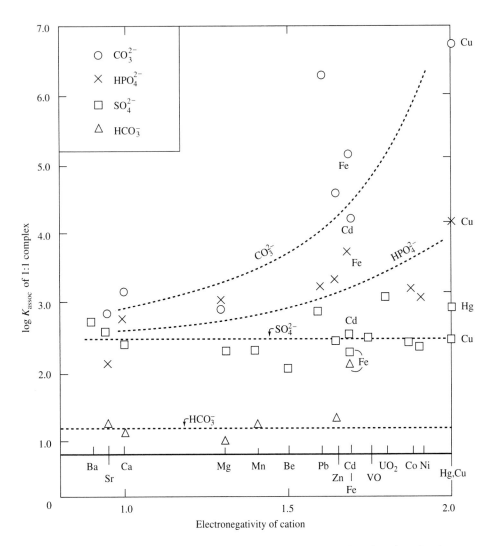

Figure 3.6 Stabilities of some 1:1 oxyanion complexes are plotted against the electro-negativity (EN) of the cation. Literature data have been corrected to zero ionic strength when necessary. Lines through the data for HCO_3^- and SO_4^{2-} complexes are for mean val-ues. Curves drawn through the HPO_4^{2-} and CO_3^{2-} data have no statistical significance. Assuming the following approximate EN values for HCO_3^- (4 estimated), SO_4^{2-} (3.7), HPO_4^{2-} (2.8), and CO_3^{2-} (2.5 estimated), ΔEN differences between cations and ligands suggest that bonding is largely electrostatic in the bicarbonate complexes, but becomes increasingly covalent in the phosphate and carbonate complexes. Reprinted with per-mission from Techniques of estimating thermodynamic properties for some aqueous com-plexes of geochemical interest, D. Langmuir. In *Chemical modeling in aqueous systems*, ed. E. A. Jenne, Am. Chem. Soc. Symp. Ser. 93. Copyright 1979 American Chemical Society.

Plotted in Fig. 3.6 are the association constants of some 1:1 complexes of divalent metal ions versus their electronegativities. The figure shows that association constants for the bicarbonate and most of the sulfate complexes are practically independent of EN or ΔEN. For the alkaline earths, bonding with all four ligands is also largely ionic (i.e., independent of EN). These all must be, therefore, chiefly ion pairs. Increases in K_{assoc} values for the bicarbonate and sulfate pairs from Be (Ip = 4.44) to Ba (Ip = 1.48) may reflect the decreasing strength of hydration of the cations, which allows the closer approach of ligands and a corresponding increase in the strength of ionic bonding in the complex.

Figure 3.6 shows that the carbonate and biphosphate complexes increase in stability with increasing EN of the cation (decreasing ΔEN between cation and ligand), especially for cation EN values above 1.5. Based on the relative solubilities of their salts, EN values for CO_3^{2-}, HPO_4^{2-}, and SO_4^{2-} should decrease from near 4 to about 2.5 in the order $SO_4^{2-} > HPO_4^{2-} > CO_3^{2-}$. Thus, covalency of bonding should be least for the sulfate complexes, and greatest for the carbonate complexes that may be chiefly of inner-sphere character for cation EN values above 1.5.

3.6. SCHWARZENBACH'S CLASSES A, B, AND C, AND PEARSON'S HARD AND SOFT ACIDS AND BASES

There are no simple rules involving strictly electrostatic bonding or covalent bonding that can explain all the observed stabilities of complexes. To develop a more comprehensive understanding of complex stabilities we will now discuss two schemes that consider the following: (1) the detailed electronic structures of individual cations (Schwarzenbach's class A, B, and C cations); and (2) the electronic structures of both cations and ligands as well as their electronic interactions (the concept of hard and soft acids and bases [HSAB]).

Pearson (1973) and Ahrland (1973) (see also Brimhall and Crerar 1987) classify cations and ligands as *hard* or *soft* acids or bases. The cations are Lewis acids and the ligands Lewis bases, with the metal cation and ligand acting as electron acceptor and donor, respectively. Soft implies that the species' electron cloud is deformable or polarizable with the electrons mobile and easily moved. Such species prefer to participate in covalent bonding. Hard species are comparatively rigid and nondeformable, have low polarizability, hold their electrons firmly, and prefer to participate in ionic bonds. Hard acids form strong, chiefly ionic bonds with hard bases, whereas soft acids form strong, chiefly covalent bonds with soft bases. In contrast, the bonds formed between hard-soft or soft-hard acids and bases are weak, and such complexes tend to be rare. Table 3.5 summarizes the relationship between the cation and ligand classifications of Schwarzenbach (1961) and the HSAB classification for a number of substances.

Schwarzenbach (1961) proposed three general classes of metal cations (A, B, and C) based upon their electronic configurations (cf. Phillips and Williams 1965; Nancollas 1966). Class A metal cations, which have noble gas configurations, are listed here.

Cation (increasing covalency →)	Atomic configuration
Li^+, Be^{2+}	He
Na^+, Mg^{2+}, Al^{3+}	Ne
K^+, Ca^{2+}, Sc^{3+}, Ti^{4+}	Ar
Rb^+, Sr^{2+}, Y^{3+}, Zr^{4+}	Kr
Cs^+, Ba^{2+}, La^{3+}	Xe

Note that the bonding of class A cations becomes increasing covalent as the cations increase in charge and decrease in size, or in other words, as their ionic potentials increase. Class A cations have spherical symmetry and low polarizability and thus are *hard spheres* (cf. Stumm and Morgan 1981). They are included among the list of hard acids in Table 3.5.

TABLE 3.5. Relationship (where there is one) between Schwarzenbach's (1961) cation classifications (A, B, and C) and the hard and soft acids and bases (HSAB) classification of Pearson (1973)

Ion	Schwarzenbach's classification	HSAB classification
Li^+, Be^{2+}, Na^+, Mg^{2+}, Al^{3+}, Si^{4+}, K^+, Ca^{2+}, Sc^{3+}, Ti^{4+}, Rb^+, Sr^{2+}, Y^{3+}, Zr^{4+}, Cs^+, Ba^{2+}, La^{3+}	Class A	Hard acids
Ga^{3+}, In^{3+}, Sn^{4+}	Class B	
VO^{2+}, Cr^{3+}, Mn^{2+}, Fe^{3+}, Co^{3+}	Class C	
UO_2^{2+}, U^{4+}, Th^{4+}, Pu^{4+} (Actinides)	—	
Zn^{2+}	Class B	Borderline acids (between hard and soft)
Fe^{2+}, Co^{2+}, Ni^{2+}, Cu^{2+}	Class C	
Pb^{2+}, Sn^{2+}, Bi^{3+}	—	
Cu^+, Ag^+, Au^+, Au^{3+}, Pd^{2+}, Cd^{2+}, Hg^{2+}, Tl^{3+}	Class B	Soft acids
Tl^+, Hg^+, CH_3Hg^+	—	
F^-, H_2O, OH^- Oxyanions: SO_4^{2-}, CO_3^{2-}, HCO_3^-, $C_2O_4^{2-}$, $H_nPO_4^{n-3}$, $H_nAsO_4^{n-3}$, etc.	—	Hard bases
Cl^-, Br^-, NO_2^-, SO_3^{2-}	—	Borderline bases (between hard and soft)
I^-, HS^-, S^{2-}, CN^-, SCN^-, Se^{2-}, Te^{2-}, $S_2O_3^{2-}$, $-SH$, $-SCH_3$, $-NH_2$	—	Soft bases

Hard cations tend to form largely electrostatic bonds with ligands, especially when the ligands are hard and the cations monovalent or divalent. Important complexes are formed between hard di- and higher valent cations and the hard ligands F^-, H_2O, and OH^-. The stabilities of complexes generally decrease in the order: $OH^- > F^-$, $CO_3^{2-} > HPO_4^{2-} \gg NO_3^-$, and $PO_4^{3-} \gg HPO_4^{2-} > SO_4^{2-}$. Complexes are usually not formed with S, N, C, Cl^-, Br^-, or I^-, because these species cannot compete with H_2O or OH^-, except in high chloride brines where Cl^- complexes with tri- and higher valent hard cations can be important. Complexes probably decrease generally in stability in the order of decreasing ligand hardness: $F^- > Cl^- > Br^- > I^-$, although data are mostly lacking for the extremely weak I^- and Br^- complexes.

Because bonding is in large part electrostatic, when a single ligand is considered, K_{assoc} values are usually proportional to z_M/r_M (= Ip) or to $z_M z_L/(r_M + r_L)$, where the M and L subscripts denote cation and ligand charge and radii, respectively (Phillips and Williams 1965; Cotton and Wilkinson 1988). As z_M/r_M values increase, however, covalent bonding becomes important. For example, important covalency and cation deformation occurs when complexes are formed with Be^{2+} or species such as Fe^{3+}, Al^{3+}, and U^{4+} (see Fig. 3.5).

Schwarzenbach (1961) defines class B metal cations as those with electron configurations of $Ni°$, $Pd°$, or $Pt°$ (cf. Stumm and Morgan 1981). The group includes Cu^+, **Ag^+**, Au^+, **Zn^{2+}**, **Cd^{2+}**, **Hg^{2+}**, Ga^{3+}, In^{3+}, Tl^{3+}, and **Sn^{4+}**, where the species in boldface here and below are the most important , geo-

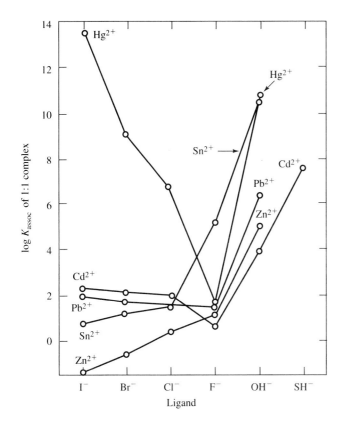

Figure 3.7 Stabilities of some class B divalent cation complexes are shown at zero ionic strength, based on published data. Ligand EN values are I^- (2.66), Br^- (2.96), Cl^- (3.16), F^- (3.98), OH^- (3.1), and SH^- (2.33). Cation EN values all range from 1.6 to 1.7, except EN = 2.00 for Hg^{2+}. Reprinted with permission from Techniques of estimating thermodynamic properties for some aqueous complexes of geochemical interest, D. Langmuir. In *Chemical modeling in aqueous systems,* ed. E. A. Jenne, Am. Chem. Soc. Symp. Ser. 93. Copyright 1979 American Chemical Society.

chemically. Mono- and divalent class B cations are considered soft acids and are highly polarizable, whereas the trivalent cations and Sn^{4+} are hard acids. Sometimes also included in the soft acid list are Tl^+, Sn^{2+}, **Pb^{2+}**, and Bi^{3+}. However, Pearson (1968) prefers to treat Zn^{2+}, Pb^{2+}, Sn^{2+}, and Bi^{3+} as "borderline," rather than as truly soft or hard cations (see Table 3.5).

Class B cations form largely covalent bonds in complexation, so that the stability of their inner-sphere complexes tends to decrease with increasing ΔEN values between the cation and ligand. Thus stability constants for the monovalent cation complexes and Pb^{2+}, Cd^{2+}, and Hg^{2+} complexes decrease in the orders $I^- > Br^- > Cl^- > F^-$, and $S^{2-} > SH^- > OH^- > F^-$ (see Figs. 3.7 and 3.8). For Sn^{2+}, Zn^{2+}, and In^{3+}, complex stabilities are reversed for the halogens and decrease in the order: $OH^- > F^- > Cl^- > Br^- > I^-$. The latter three cations thus exhibit hard-acid behavior with respect to OH^- and the halogens. As shown in Fig. 3.6, the divalent class B cations form SO_4^{2-} complexes, and probably also HPO_4^{2-} complexes. Soft-acid cations, including those among the class B group, generally form their strongest complexes when they associate with the soft ligands I^-, SH^-, S^{2-}, Se^{2-}, and Te^{2-}, among others (Table 3.5).

Zn^{2+}, Cd^{2+}, and especially Hg^{2+} are strong complexers, although the mercuric complexes are typically orders of magnitude stronger than those of Zn^{2+} or Cd^{2+} (Fig. 3.7). Unlike Zn^{2+} and Cd^{2+}, Hg^{2+} also forms strong organometallic complexes (e.g., monomethylmercury ion, CH_3Hg^+), which are very important environmentally. Mercury also occurs as the Hg_2^{2+} ion, which is a weak complexer.

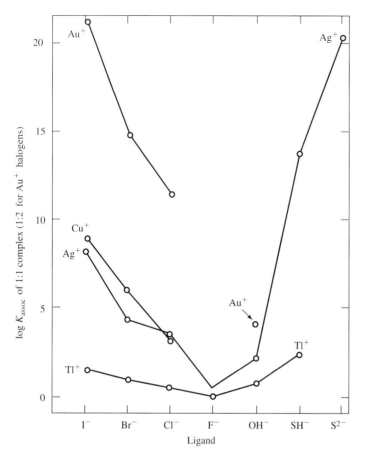

Figure 3.8 Stabilities of some class B monovalent cation complexes and Tl⁺ complexes, based on published data. Cation EN values are: Tl⁺ (1.5), Ag⁺ (1.93), Cu⁺ (1.8), Au⁺ (2.54). Reprinted with permission from Techniques of estimating thermodynamic properties for some aqueous complexes of geochemical interest, D. Langmuir. In *Chemical modeling in aqueous systems*, ed. E. A. Jenne, Am. Chem. Soc. Symp. Ser. 93. Copyright 1979 American Chemical Society.

For the reaction $2Cu^+ = Cu^\circ + Cu^{2+}$, $K = 10^6 = [Cu^{2+}]/[Cu^+]^2$, so that Cu^+ is relatively unimportant in solution. It occurs, however, in sulfides such as Cu_2S (chalcocite) with $K_{sp} = 10^{-48.5}$ (Smith and Martell 1976). In solution, Cu^+ and Ag^+ usually occur in fourfold coordinated complexes. Au^+ is unimportant in natural waters.

The transition element cations, which comprise Schwarzenbach's class C, have 0 to 10 d subshell electrons in the M shell (first series) or N shell (second series), etc. Examination of Table 3.5 shows that the class C cations are generally considered either hard or borderline hard-soft acids. These cations have partially filled $3d$ subshells either in the ground state or when ionized (Table 3.6). Moving across the periodic chart from Sc to Cu, protons are added to the nucleus and electrons to the unfilled inner $3d$ subshell. Attraction of these inner electrons to the nucleus leads to an overall decrease in cation radii (Table 3.7). The divalent ions are generally sixfold, coordinated in complexes. Cu^{2+} is an exception and, because of its small size and unique electronic configuration, tends to occur in distorted four- or sixfold coordination. Table 3.7 also lists Ip, EN values, and thermodynamic data for divalent cations of the first transition series. The near-constancy of S° for the transition metal cations (except for Mn^{2+}) indicates that their aquoion stability differences reflect chiefly ΔH_f° differences. Typically, the stability order for inner-sphere M^{2+} complexes (the K_{assoc} values) follows the ΔG_f°, ΔH_f° and EN sequences for the aquocations shown in Table 3.7 (cf. Yatsimirskii and Vasil'ev 1966). Thus for the M^{2+} cations of geochemical interest, complex stabilities with a given ligand increase in the order Ca < Mn < Fe < Co < Ni < Cu > Zn. This is the same as the order of in-

TABLE 3.6. Number of 3d electrons and common valence states in natural waters of elements with atomic numbers 20 to 30.

Element	Number of 3d electrons	(+) Valence states	Comments
Ca	0	2	Ion is class A
Sc	1	3	Ion is class A
Ti	2	(2), (3), 4	4+ is class A
V	3	2, 3, 4, 5	—
Cr	5	(2), 3, 4	—
Mn	5	2, (3), 4	3+ in minerals only
Fe	6	2, 3	—
Co	7	2, (3)	3+ in minerals only
Ni	8	2	—
Cu	10	1, 2	1+ is class B
Zn	10	2	2+ is class B

Note: Parentheses indicate ion is rare or does not exist in natural waters as a complex former.
Source: Reprinted with permission from Techniques of estimating thermodynamic properties for some aqueous complexes of geochemical interest, D. Langmuir. In *Chemical modeling in aqueous systems*, ed. E. A. Jenne, Am. Chem. Soc. Symp. Ser. 93. Copyright 1979 American Chemical Society.

creasing Ip, except for Cu^{2+}, which is preferentially stabilized in complexes because of Jahn-Teller effects (Cotton and Wilkinson 1988). The above sequence is called *the Irving-Williams order*. It is followed by the stability constants of inner-sphere complexes formed with ligands that include bisulfide, oxalate, citrate, glycinate, NTA, and EDTA (Nancollas 1966; Yatsimirskii and Vasil'ev 1966; Ringbom 1963; Stumm 1992; Luther et al. 1996), soil fulvic-acid complexes (Schnitzer 1971), and also by the stability constants of the sulfide, telluride, and hydroxide minerals (Smith and Martell

TABLE 3.7. Some properties of the divalent aquocations of elements with atomic numbers 20 to 30, at 25°C and 1 bar total pressure

Cation	Ca	Ti	V	Cr	Mn
Radius (nm)	1.00	0.86	0.79	0.80	0.83
Ip	2.00	2.33	2.53	2.50	2.41
EN	1.00	1.54	1.63	1.66	1.55
ΔG_f° (kcal/mol)	−132.24	(−75.1)	−54.7	−42.1	−54.5
ΔH_f° (kcal/mol)	−129.71	—	—	−33.2	−52.76
S° (cal/mol K)	−12.79	—	—	—	−17.6

Cation	Fe	Co	Ni	Cu	Zn
Radius (nm)	0.78	0.745	0.69	0.73	0.74
Ip	2.56	2.68	2.90	2.74	2.70
EN	1.83	1.88	1.91	1.90	1.65
ΔG_f° (kcal/mol)	−21.8	−13.0	−10.9	+15.68	−35.19
ΔH_f° (kcal/mol)	−22.1	−13.9	−12.9	+15.70	−36.66
S° (cal/mol K)	−25.6	−27	−30.8	−23.21	−26.2

Note: Ti^{2+}, V^{2+}, and Cr^{2+} are unstable in water.
Source: Crystal radii are for sixfold coordination (Shannon 1976). EN values are from Allred (1961). Thermodynamic data for Ca, Cu, Co, Mn, Ni, and Zn are from Wagman et al. (1982); Ti, V, and Cr from Latimer (1952); and Fe from Tremaine et al. (1977). Reprinted with permission from Techniques of estimating thermodynamic properties for some aqueous complexes of geochemical interest, D. Langmuir. In *Chemical modeling in aqueous systems,* ed. E.A. Jenne, Chem. Soc. Symp. Ser. 93. Copyright 1979 American Chemical Society.

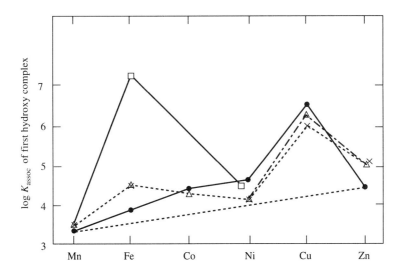

Figure 3.9   Association constants of the 1:1 hydroxycomplexes of divalent metal cations of the first transition series based on the following sources: (□) Wagman et al. (1982), (×) Baes and Mesmer (1976), (●) Yatsimirskii and Vasil'ev (1966), (△) Smith and Martell (1976). From Langmuir (1979). Reprinted with permission from Techniques of estimating thermodynamic properties for some aqueous complexes of geochemical interest, D. Langmuir. In *Chemical modeling in aqueous systems,* ed. E. A. Jenne, Am. Chem. Soc. Symp. Ser. 93. Copyright 1979 American Chemical Society.

1976; Naumov et al. 1974). The tendency for the divalent transition metal cations to be adsorbed by hydrous ferric oxides follows this order (see Chap. 10). Although data are incomplete, available stability constants for $M^{2+}\text{-}CO_3^{2-}$, and HPO_4^{2-} complexes (Fig. 3.6) also appear to follow the Irving-Williams order, suggesting that they have significant inner-sphere character.

The Irving-Williams order results from an enhancement of stabilities for complexes formed with Fe^{2+}, Co^{2+}, Ni^{2+}, and Cu^{2+}, because of electronic interaction between the cation and ligand, that is, ligand field stabilization energies, LFSEs (Cotton and Wilkinson 1988). Such effects do not occur with Mn^{2+} or Zn^{2+}, so that their complexes are generally less stable than those formed with Fe^{2+} and Cu^{2+}, respectively. Accordingly, a plot of $\Delta G°$, $\Delta H°$, or log K_{assoc} data for a particular ligand against atomic number will usually show minima at Ca^{2+}, Mn^{2+}, and Zn^{2+}. LFSE effects will raise the stabilities of intermediate complexes above the line connecting Ca^{2+}, Mn^{2+}, and Zn^{2+}. The Mn^{2+} to Zn^{2+} portion of such a plot is shown in Fig. 3.9 for 1:1 divalent metal-OH^- complexes. Stability data for these complexes reported by Yatsimirskii and Vasil'ev (1966) follow the Irving-Williams order. These authors suggest log K_{assoc} values of 3.9 and 3.7, respectively, for the reaction $Fe^{2+} + OH^- = FeOH^+$. Data in Wagman et al., (1982) yield log $K_{assoc} = 7.2$, with Baes and Mesmer (1976) and Smith and Martell (1976) both suggesting log $K_{assoc} = 4.5$. These latter three data sets violate the Irving-Williams order, and, in addition, the sets of Baes and Mesmer and Smith and Martell suggest no LFSE for $NiOH^+$, which seems unlikely. Whichever of these data are correct, it is clear that $K_{assoc} = 7.2$ for $FeOH^+$ from Wagman et al. (1982) is too large, making this complex an important species in most groundwaters.

Based on LFSE arguments, the stability constants of trivalent cation transition metal complexes often follow the Irving-Williams order, which is Sc < Ti < V < Cr < Mn > Fe < Co < Ni < Cu > Ga. For the trivalent cations stable in natural waters, ΔG_f°, ΔH_f°, and EN data suggest the order of increasing complex stabilities should be V < Cr << Fe < Co >> Ga. Among the trivalents, Sc^{3+}, Fe^{3+}, and Ga^{3+} form the sequence of minimal complex stabilities, reflecting the absence of LFSE effects, with intermediate cation complexes having greater stabilities because of such effects.

Exceptions to the two Irving-Williams orders for inner-sphere complexes occur particularly when complexes with more than four ligands or large, asymmetric ligands are compared (Ringbom 1963). Stability-constant sequences for weak outer-sphere complexes, such as those formed with chloride (Libus and Tialowskia 1975) and fluoride (Smith and Martell 1976; Hancock and Marsicano 1978) do not obey the Irving-Williams orders.

As noted earlier, Pearson (1973) lists cations Mn^{2+}, Ga^{3+}, Cr^{3+}, Co^{3+}, and Fe^{3+} as hard acids. The divalent cations Fe, Co, Ni, and Cu are described as "borderline" between the hard and soft acids (Table 3.5).

3.7 MODEL-PREDICTION OF THE STABILITIES OF COMPLEXES

The stability constants of ion pairs (their log K_{assoc} values) have been shown to be proportional to the electrostatic function $z_M z_L / d$, where z_M and z_L are the charge of metal cation and ligand, and $d = r_M + r_L$, the sum of their crystal radii (cf. Fig. 3.5). Mathematical models for predicting ion pair stabilities generally assume this proportionality and include the simple electrostatic model, the Bjerrum model, and the Fuoss model (cf. Langmuir 1979). Such models can predict stabilities in fair agreement with empirical data for monovalent and divalent cation ion pairs.

A number of researchers have developed useful models for predicting the stability of complexes that involve partially ionic (electrostatic) and partially covalent bonding. The models weight the influence of each bond type. The weighting may proportionalize the ionic contribution with some variation of the electrostatic function for the complex and account for the covalent bond contribution via the EN difference between the cation and ligand (cf. Nieboer and McBryde 1973). Weighting may involve assigning numerical values to both cation and ligand to describe their degree of hardness or softness (cf. Pearson, 1973). The two approaches have been shown to be closely related (cf. Hancock and Marsicano 1978). More recently, Brown and Sylva (1987) introduced the concept of *electronicity* and a single numerical scale of electronicity values for both cations and ligands, with success predicting the stabilities of complexes. Electronicity is also related to the hard and soft character of cation and ligand in the complex.

Langmuir (1979) showed the usefulness of a diversity of graphic methods for predicting the stabilities of complexes. For example, a plot that compares the log K_{assoc} values of complexes formed by a variety of cations and two similar ligands (e.g., the hard ligands CO_3^{2-} and $C_2O_4^{2-}$) is often a straight line (cf. Fig. 3.10). In the same way, a plot of the log K_{assoc} values of complexes formed by different ligands and two similar cations (e.g., the hard cations Al^{3+} and Fe^{3+}) will often be close to a straight line. Such comparison plots are most useful when the cations and ligands being compared have the same charge, similar size and geometry, similar electron configurations and bonding properties, and form complexes having the same structures.

Another useful graphic approach is to plot the number of ligands (n) in monomeric complexes with a given ligand (L) against log β_n, where β_n is the cumulative formation constant of each complex with cation M ($\beta_n = [ML_n]/[M][L]^n$). Such a plot is given in Fig. 3.11 for uranium complexes.

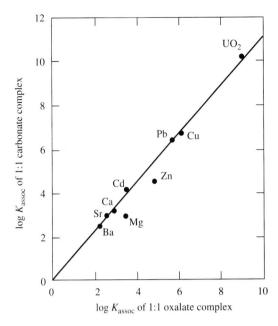

Figure 3.10   Stability constants of some 1:1 divalent metal carbonate complexes plotted against stability constants for the corresponding oxalate complexes. Data are from the literature and are for zero ionic strength. The equation of the line is $\log K_{\mathrm{assoc}}(\mathrm{MCO_3^\circ}) = 1.11 \times \log K_{\mathrm{assoc}}(\mathrm{MC_2O_4^\circ})$. Based on Martell and Smith (1977), $\log K_{\mathrm{assoc}}(\mathrm{FeC_2O_4^\circ}) = 4.52$, from which we predict $\log K_{\mathrm{assoc}}(\mathrm{FeCO_3^\circ}) = 5.02$.

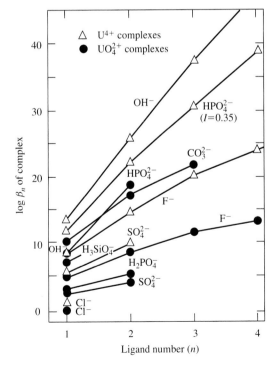

Figure 3.11   Cumulative formation constants of monomeric U^{4+} and UO_2^{2+} complexes plotted against their ligand numbers. All the data are for zero ionic strength (I) unless otherwise indicated (see Chap. 4). Reprinted with permission from Techniques of estimating thermodynamic properties for some aqueous complexes of geochemical interest, D. Langmuir. In *Chemical modeling in aqueous systems,* ed. E. A. Jenne, Am. Chem. Soc. Symp. Ser. 93. Copyright 1979 American Chemical Society.

When lines connecting successive log β_n values for different ligands cross on such a plot, values of the cumulative constants involved may be suspect. In this instance plotted β_n values for $UO_2HPO_4^\circ$ and $UO_2(HPO_4)_2^{2-}$ are suspect (Langmuir 1978). In fact, Grenthe et al. (1992) have since questioned the existence of the 1:2 complex.

3.8 THE THERMODYNAMICS OF COMPLEXATION

The stability of a complex is usually described by its stability constant, which is also known as its formation or association constant. The stability constant may refer to a stepwise or a cumulative (overall) complex-formation reaction. If K_{assoc}, applies to the simple reaction

$$M + L = ML \tag{3.49}$$

where M and L are the cation and ligand and ML the complex, we also know that for the complexation reaction

$$\Delta G^\circ = -RT \ln K_{assoc} \tag{3.50}$$

and

$$\Delta G^\circ = \Delta H^\circ - T\Delta S^\circ \tag{3.51}$$

The stability of the complex depends on the relative and absolute contributions of ΔH° and ΔS°. For most complexation reactions a positive ΔS° stabilizes the complex. This net positive entropy change reflects structural changes in the solution consequent to complexation. The total entropy change that attends complexation may be written:

$$\Delta S^\circ = \Delta S^\circ_{net\,chg} + \Delta S^\circ_{tr} + \Delta S^\circ_{rot} + \Delta S^\circ_{vibr} + \Delta S^\circ_{dehydr} \tag{3.52}$$

$$(+) \qquad\quad (-) \qquad\quad (-) \qquad\quad (+) \qquad\quad (+) \qquad\quad (+)$$

where the signs in parentheses indicate the usual contribution of each term. There is an entropy reduction ($\Delta S^\circ_{net\,chg}$) due to the decrease in number of charged particles. Some of the translational entropy (ΔS°_{tr}) of the separate cation and ligand is converted to vibrational (ΔS°_{vibr}) and rotational entropy (ΔS°_{rot}) in the complex. However, the chief contribution is usually the $+\Delta S^\circ_{dehydr}$ term that reflects partial dehydration of the cation and/or ligand. Complexation leads to a breakdown of the structured water of hydration, especially around the cation, with a resultant decrease in the order of the solution. Accordingly, the ΔS°_{dehydr} term is proportional to the number of water molecules displaced by the ligand. Thus ΔS°_{dehydr} (and $+\Delta S^\circ$ total) is greatest for multivalent ion complexation, since these species are initially the most hydrated. If little or no coordinating water is eliminated in complex formation, a net $-\Delta S^\circ$ may result because of the contributions of $-\Delta S^\circ_{net\,chg}$ and $-\Delta S^\circ_{tr}$. This is the case with cation-neutral ligand complexes, which consequently are weak. For the same reason ΔS° is less positive in ion pair formation than in inner-sphere complexation.

The $+\Delta S^\circ$ term is usually greatest for addition of the first ligand to the cation, decreasing for each successive ligand. Consistent with this fact, stepwise complexation constants tend to decrease with increasing ligand number.

Because chelates (such as EDTA) bond a cation via more than one electron-donating atom, they tend to free several cation-coordinating waters upon complexation. The result is an unusually large $+\Delta S^\circ$ contribution to the formation of chelates as opposed to that of monodentate complexes (cf. Cotton and Wilkinson 1988).

Complexes are stabilized more often by $+\Delta S^\circ_{\text{assoc}}$ than by $-\Delta H^\circ_{\text{assoc}}$. In fact for many complexes $\Delta H^\circ_{\text{assoc}}$ is positive and their formation is endothermic (see Chap. 1, Table A1.2), which means that their stabilities increase with increasing temperature.

3.9 DISTRIBUTION OF COMPLEX SPECIES AS A FUNCTION OF pH

Early in the chapter we showed how to construct a distribution plot that described the relative concentrations of Cd^{2+} and its chloride complexes as a function of chloride concentration (see Fig. 3.3). A similar approach is used to create a diagram showing the fractional distribution of an acid and its dissociation products with pH (see Chap. 5, Acids and Bases). Butler (1964) provides example calculations for the related but more intricate problem of constructing a plot to describe the relative amounts of metal-ligand complexes with pH when one metal cation and several ligands are present. In present practice such calculations are conveniently performed with a computer spreadsheet and the figure drawn by a computer-graphics program. They may also be performed using the sweep option of MINTEQA2, usually assuming an ionic strength of zero and ignoring activity coefficients.

In this section we will study examples of several distribution plots that offer a convenient and powerful way to examine changes in the speciation of complex-forming metal cations with pH. Such plots are particularly useful when studying the complexation of multivalent hard cations, such as the actinides that tend to form important hydroxyl complexes. Other ligands must displace OH in order to complex the cation. They have difficulty in so doing except in acid waters where competition with OH is minimal.

A logical way to approach metal cation complexing in natural waters is to first study hydroxyl complexing of the cation as a function of pH. For example, Fig. 3.12(a) is a distribution plot of thorium hydroxy complexes from pH 2 to 8 in pure water at a typical ΣTh concentration of 0.01 μg/L (4.3×10^{-11} M). As is usually observed for trivalent and quadrivalent cations, the free ion, in this case Th^{4+}, predominates only in acid waters. With increasing pH, thus with increasing OH concentrations, successive Th-OH complexes become predominant. The most notable characteristics of Fig. 3.12(a) are that Th^{4+} and the $Th(OH)_4^\circ$ complex dominate between pH 2 and 8, and that the intermediate complexes rarely reach 50% of Th. $Th(OH)_4^\circ$ is clearly a strong complex and the most important of these species at pHs typical of natural waters. The general stepwise formation reaction for the complexes may be written

$$Th(OH)_n^{4-n} + H_2O = Th(OH)_{n+1}^{3-n} + H^+ \qquad (3.53)$$

where $n = 0$ to 3. For the first stepwise reaction the equilibrium expression is

$$*K_1 = \frac{[ThOH^{3+}]\,[H^+]}{[Th^{4+}]} = 10^{-3.2} \qquad (3.54)$$

Examination of this relationship shows that concentrations of Th^{4+} and $ThOH^{3+}$ are equal when $p*K_1 = \text{pH} = 3.2$. This is the point in Fig. 3.12(a) where the mole percent curves for the two species cross. Similarly, the curves for $ThOH^{3+}$ and $Th(OH)_2^{2+}$ cross and the species are equal when $p*K_2 = \text{pH} = 3.87$. Note that $Th(OH)_3^+$ never predominates. The last important crossover point is the pH at which $Th(OH)_2^{2+}$ and $Th(OH)_4^\circ$ are equal. This corresponds to the reaction:

$$Th(OH)_2^{2+} + 2H_2O = Th(OH)_4^\circ + 2H^+ \qquad (3.55)$$

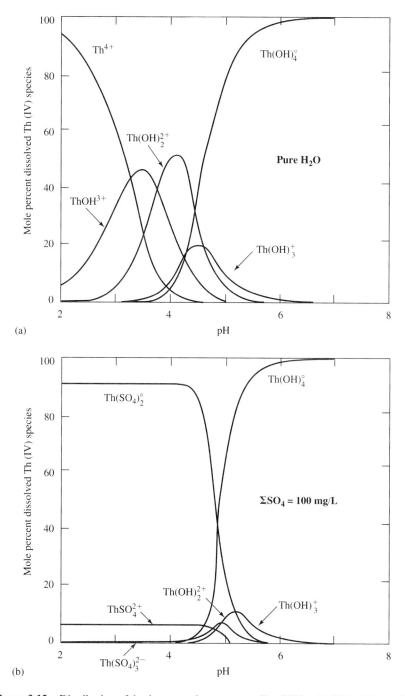

Figure 3.12 Distribution of thorium complexes versus pH at 25°C with $\Sigma Th = 0.01\ \mu g/L$: **(a)** in pure water; **(b)** with $\Sigma SO_4 = 100$ mg/L. Reprinted from *Geochim. et Cosmochim. Acta* 44(11), D. Langmuir and J. S. Herman, The mobility of thorium in natural waters at low temperatures, 1753–1766, © 1980, with permission from Elsevier Science Ltd., The Boulevard, Langford Lane, Kidlington OX5 1GB, U.K.

The equilibrium constant K_{eq} for this reaction can be shown equal to

$$K_{eq} = {}^*K_3 {}^*K_4 = (10^{-4.77}) \times (10^{-4.2}) = 10^{-8.97} = \frac{[Th(OH)_4^\circ] [H^+]^2}{[Th(OH)_2^{2+}]} \qquad (3.56)$$

The complexes are, therefore, equal when pH = 4.49 as seen in Fig. 3.12(a).

The addition of another complexing ligand, such as sulfate changes thorium speciation, especially at low pH. Sulfate readily complexes with Th^{4+} and effectively competes with and displaces the weaker Th-OH complexes. However, the stability field of the strong $Th(OH)_4^\circ$ complex is barely affected as shown in Fig. 3.12(b). When we add other complexing ligands, the picture becomes more involved as in Fig. 3.13(a), which is drawn for some typical inorganic ligand concentrations in natural waters. Clearly, all the soluble thorium is complexed. The ligands compete with each other to form Th complexes. With increasing pH the dominant complexes are formed with sulfate, fluoride, orthophosphate, and hydroxyl. As shown in Fig. 3.13(b), the speciation is drastically altered by the addition of trace amounts of a chelating organic substance such as EDTA to the same water. Th-EDTA and Th-oxalate complexes are much stronger than complexes formed with the inorganic ligands, so the organic complexes dominate thorium chemistry.

It was mentioned earlier that complexing increases the solubility of minerals and thus the mobility of the species involved. This is evident from Fig. 3.14, which shows the solubility of thorianite (ThO_2) as a function of pH in pure water and in the presence of inorganic and organic complexing ligands. The lower, crosshatched curve is the solubility in pure water. At low pH the curve corresponds to the reaction

$$ThO_2 + 4H^+ = Th^{4+} + 2H_2O \qquad (3.57)$$

Above roughly pH = 5 the dissolution reaction is

$$ThO_2 + 2H_2O = Th(OH)_4^\circ \qquad (3.58)$$

and so above pH 5 the solubility becomes pH-independent. Between about pH 3 and 5, the pure-water solubility-curve corresponds to ThO_2 dissolution to form the intermediate Th-OH complexes. Note that in pure water, thorianite is highly insoluble (thus Th is immobile), particularly above pH 3. However, in the presence of typical levels of inorganic ligands, the solubility is increased by up to 1000 times because of complexing [Fig. 3.14(a)]. Addition of trace organics such as oxalate or EDTA has an even greater effect. Fig. 3.14(c) shows up to a 10^7 increase in ThO_2 solubility due to the organic complexers. These results are summarized in Fig. 3.14(c). Such strong complexing is common among the multivalent actinides, which consequently become highly mobile. High actinide solubility and mobility, presumably caused by complexing, has been observed in some groundwaters contaminated by nuclear wastes.

3.10. TOXICITY AND THE ROLE OF SOFT-ACID METAL CATIONS

According to Manahan (1994), a toxic substance or toxicant "is harmful to living organisms because of its detrimental effects on tissues, organs, or biological processes." Toxicants occur in air, water, and solids. We will limit ourselves here to an overview of inorganic toxicants in water, with a focus on the metals. Readers will find more detailed general discussion of toxicological chemistry in Moore and Moore (1976), Morel and Hering (1993), and especially in Manahan (1994). A detailed accounting of organic chemical toxicants and their sources, occurrences, and fate is available in Howard (1990a, 1990b). (See also Manahan 1994.)

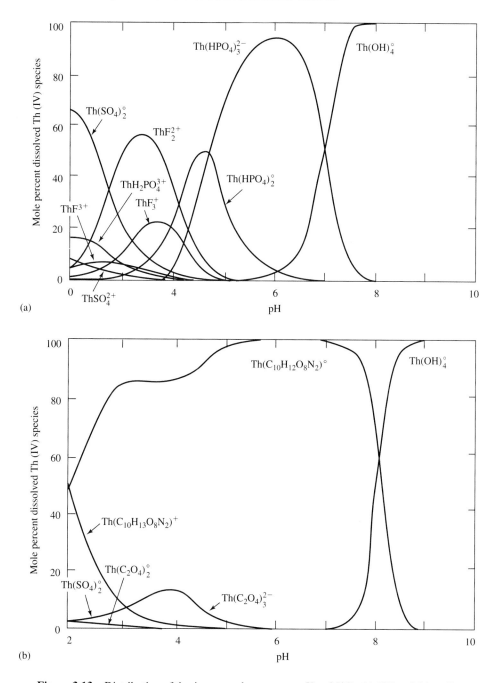

Figure 3.13   Distribution of thorium complexes versus pH at 25°C with $\Sigma Th = 0.01\ \mu g/L$ for some typical ligand concentrations in groundwater. **(a)** Inorganic ligands at total (mg/L) concentrations of $\Sigma F = 0.3$, $\Sigma Cl = 10$, $\Sigma PO_4 = 0.1$, and $\Sigma SO_4 = 100$. **(b)** Inorganic ligand concentrations as in Fig. 3.10, plus the following organic ligands at total (mg/L) concentrations: $\Sigma C_2O_4 = 1$, Σcitrate $= 0.1$, $\Sigma EDTA = 0.1$. Reprinted from *Geochim. et Cosmochim. Acta* 44(11), D. Langmuir and J. S. Herman, The mobility of thorium in natural waters at low temperatures, 1753–1766, © 1980, with permission from Elsevier Science Ltd., The Boulevard, Langford Lane, Kidlington OX5 1GB, U.K.

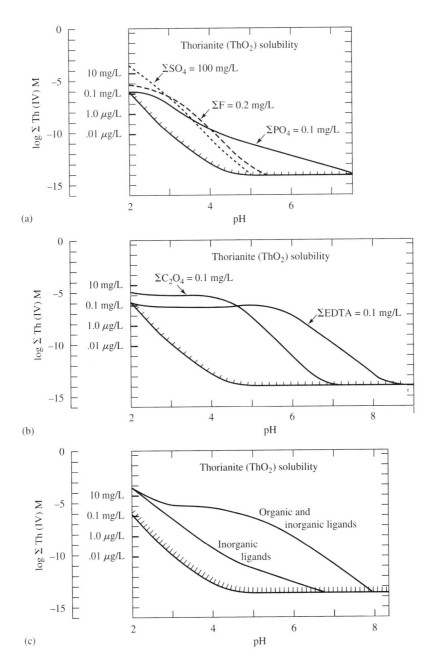

Figure 3.14 The effect of complexing on the solubility of thorianite (ThO_2) as a function of pH at 25°C. The cross-hatched curve denotes thorianite solubility in pure water and reflects only Th-OH complexing. Curves in (a) and (b) assume Th complexing due to each ligand present in the absence of the others at the concentrations shown. Part (c) shows the cumulative effects on ThO_2 solubility of Th complexing by inorganic ligands present together: ΣCl = 10 mg/L, ΣNO_3 = 2.5 mg/L, ΣSO_4 = 100 mg/L, ΣF = 0.3 mg/L, ΣPO_4 = 0.1 mg/L, and the organic ligands Σoxalate = 1 mg/L, Σcitrate = 0.1 mg/L and, $\Sigma EDTA$ = 0.1 mg/L. Reprinted from *Geochim. et Cosmochim. Acta* 44(11), D. Langmuir and J. S. Herman, The mobility of thorium in natural waters at low temperatures, 1753–66, © 1980, with permission from Elsevier Science Ltd., The Boulevard, Langford Lane, Kidlington OX5 1GB, U.K.

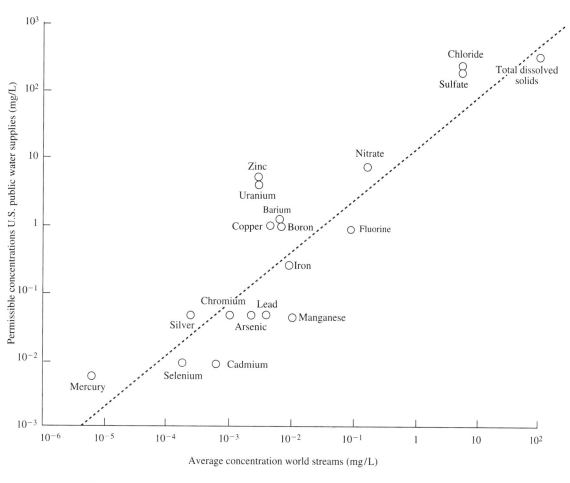

Figure 3.15   Correlation between trace-element concentrations in streams and permissible concentrations in water supplies. From R. M. Garrels et al. 1975. *Chemical cycles and the global environment.*

Garrels et al. (1975) identified an intriguing correlation between average species concentrations in world streams and permissible concentrations of the same species in U.S. public water supplies as shown in Fig. 3.15. The limits on a few of the species (e.g., Fe and Mn) have been set strictly for aesthetic reasons, although for most species the limits are a public health concern. Simple reasons for the strong general correlation shown by the plot are not obvious. A tentative explanation might relate to the fact that our ancestors evolved in a similar aquatic environment. They then adjusted to and flourished in the presence of the average concentrations in Fig. 3.15. Toxic effects would then be observed only at higher concentrations. Similar arguments may apply to average concentrations of background species in soil moisture versus the higher levels observed to cause plant toxicity (cf. Sposito 1989).

Macronutrients are major elements essential to life. According to Manahan (1994) for plants these include C, H, O, N, S, P, K, Ca, and Mg. Of these, C, H, and O are constitutents of biomass, N and P are in proteins, S in proteins and enzymes, and K, Mg, and Ca have metabolic functions. Plant

TABLE 3.8. Representative toxicity sequences for plants.

Organisms	Sequence of decreasing toxicity
Algae	**Hg** > *Cu* > **Cd** > *Fe* > Cr > *Zn* > *Co* > Mn
Flowering plants	**Hg** > *Pb*>*Cu* > **Cd** > Cr > *Ni* > *Zn*
Fungi	**Ag** > **Hg** > *Cu* > **Cd** > Cr > *Ni* > *Pb* > *Co* > *Zn* > *Fe*
Phytoplankton (freshwater)	**Hg** > *Cu* > **Cd** > *Zn*>*Pb*

Note: Elements in bold are soft acids. Italicized elements are borderline hard-soft acids. Cr(III) and
Mn(II) are hard acids. (See Table 3.5). Hg = Hg(II), Fe = Fe(II), Cr = Cr(III), Co = Co(II), Mn = Mn(II),
Pb = Pb(II).
Source: From The chemistry of soils, by G. Sposito. Copyright © 1989 by Oxford University Press,
Inc. Reprinted by permission.

micronutrients, or essential trace elements, include B, Cl, Co, Cu, Fe, Mn, Mo, Na, Si, V, and Zn.
The micronutrients have metabolic functions or play a role in enzymes. Enzymes are biological cat-
alysts that can accelerate essential metabolic reactions by 10^8 to 10^{11} times over uncatalyzed rates
(Moore and Moore 1976). Metals important in enzymes include Cu, Co, Fe, K, Mg, Mn, and Zn.

Shown in Table 3.8 are observed toxicity sequences for some plants. (Toxicity to plants is
termed *phytotoxicity*). Soft-acid cations Hg(II), Ag(I), and Cd(II) are generally the most toxic of the
metals. Close behind are the borderline hard-soft–acid cations. The only toxic hard cations listed are
Cr(III) and Mn(II). Although not listed in Table 3.8, in soils affected by acid rain (see Chap. 7, Section 7.7), hard acid Al(III) presents a serious toxicity problem to plants (Bohn et al. 1985).

Mechanisms by which toxic metals poison plants and animals relate to their tendency to form
strong complexes with the generally soft functional groups on biomolecules (cf. Sposito 1989; Morel
and Herring 1993). Sposito (1989) proposes several processes by which soft-metal cations cause
phytotoxicity. First, a soft metal such as Cd can displace an essential metal such as a Ca bond to a
bioligand. Also, complexation of a bioligand by a soft-metal cation can block that ligand from reacting normally or modify it structurally and thus interfere with its intended activity. Enzymes have
active or catalytic sites with which they bind to biological substrates and that facilitate enzyme function, and these are especially vulnerable to damage by soft-metal cations. The amino acids cysteine
and methionine present at the active sites in some enzymes contain –SH and SCH$_3$ groups (Manahan 1994). These sulfur-containing groups are soft ligands and form strong covalent bonds with
soft-metal cations such as Hg, Ag, Cd, Cu, and Pb. A result is the breakdown of normal enzyme function and a toxic reaction of the affected organism.

3.11. SUMMARY

Many environmentally important chemicals are transported as complexes in natural waters. Complexes may increase or decrease the toxicity and/or bioavailability of elements. Complexation increases the solubility of minerals and may increase or decrease the adsorption of elements. The major
monovalent and divalent cations and anions (especially $> 10^{-4}$ m) form outer-sphere complexes or
ion pairs, in which the bonding is chiefly long-range and electrostatic. Ion pairs are unimportant in
dilute fresh waters, but become important in saline waters such as seawater. Minor and trace ions
such as Cu^{2+}, Fe^{3+}, Pb^{2+}, and Hg^{2+} are usually complexed, and occur in inner-sphere complexes,
which are usually much stronger complexes than the ion pairs. Written in terms of Gibbs free energy,

the energy that stabilizes the association of a cation and ligand in a complex equals $\Delta G^\circ_{assoc} = \Delta H^\circ_{assoc} - T\Delta S^\circ_{assoc}$. Inner-sphere complexes are stabilized chiefly by the entropy increase ($+\Delta S^\circ_{assoc}$) that accompanies the release of water molecules to the bulk solution from their ordered positions in the primary hydration sphere around the cation, when they are replaced by ligands.

Insights as to the nature of aquospecies and some understanding of the stabilities of complexes, including their bonding, can be gained by considering concepts such as ionic potential, and for inner-sphere complexes, electronegativity, and classifications based on: (1) the electron configuration of cations (Schwarzenbach 1961), and (2) the "hard" and "soft" acid and base (HSAB) behavior of cations and ligands (Pearson 1973). The hard and soft approach has provided a particularly useful basis for predicting the stabilities of complexes (cf. Brown and Sylva 1987). Because biological ligands are often soft, soft-metal cations such as Hg^{2+} and Cd^{2+} are among the most toxic species in natural waters.

STUDY QUESTIONS

1. Show how complexation of a metal can increase the solubility of a mineral that contains the metal.

2. Why do ions that form strong complexes also tend to form insoluble minerals?

3. Complexation affects the toxicity, bioavailability and adsorption behavior of metals. Explain with examples.

4. Contrast outer-sphere complexes (ion pairs) and inner-sphere complexes, using examples, in terms of the character and strength of the metal-ligand bond in solution (the value of K_{assoc} for the complex) and its co-valency and ionicity.

5. Why does ion pairing increase in importance with the salinity of a water?

6. The ionic-potential concept is useful to describe why certain cations are soluble and hydrated, why other cations form insoluble oxides and hydroxides, and why still others are found only in soluble oxyanions or oxycations. Explain with examples.

7. Understand the conventions and algebraic relations among mass-balance equations and stepwise and cumulative formation constants of complexes.

8. Be able to derive the equations needed to construct diagrams showing the percent or fractional distribution of metal-ligand complexes as a function of the pH or ligand concentration.

9. The stabilities (K_{assoc} values) of inner-sphere complexes formed between metal cations and a given ligand increase as the difference in the electronegativities of the metals and the ligand decrease. Explain this statement with examples.

10. Schwarzenbach described cations as being of Class A, B, or C. Pearson proposed the concept of hard and soft acids and bases. Pearson's approach is generally more useful than Schwarzenbach's for systematizing and predicting the stabilities of inner-sphere complexes. Explain with examples.

11. What is the Irving-Williams order, and how and when can it be used to predict the stability of complexes? Does it apply to the stabilities of complexes formed with fulvic acid, for example?

12. Be able to look at a table of cations and ligands and predict:
    (a) the innner- or outer-sphere character of the complex that might form between any cation and ligand in the table, and
    (b) the approximate stability of such a complex.

13. Discuss the applicability of hard and soft acid and base concepts to phytotoxicity (toxicity to plants).

14. Explain why the stability of inner-sphere complexes is usually favored by a positive entropy of formation of the complex.

15. Be able to read and understand in general terms, distribution diagrams that depict metal complexation as a function of ligand concentration or pH.

16. After reading the pertinent discussion in Langmuir (1979), contrast the applicability of the electrostatic, Bjerrum, and Fuoss thermodynamic models for predicting the stabilities of ion pairs.

PROBLEMS

1. (a) The thermodynamic data base for the computer model MINTEQA2 (Allison et al. 1991) is found in the file *thermo.dbs*. Examine *thermo.dbs* and determine the manner in which the complexation reactions are written for cupric chloride and cupric hydroxide complexes. Make a table of these reactions as written in *thermo.dbs*, along with their corresponding ΔH_r° and log K values.
   (b) Some of the complexation reactions are written with negative protons among the reactants. Assuming that the activity of liquid water equals unity and that log $K_w = -14.00$ for the reaction $H_2O = H^+ + OH^-$, reformat the formation reactions for the cupric hydroxide complexes as cumulative formation reactions. Then recompute the corresponding cumulative formation constants (log β_i values). Tabulate your results.
   (c) Using the formation constants obtained in parts (a) and (b), hand calculate the aqueous speciation of a 0.01 M $CuCl_2$ solution at 25°C and pH = 4.00, assuming zero ionic strength (solute activities equal molar concentrations). Which cupric species occur at concentrations equal to or greater than 1% of the concentration of the most abundant cupric species?
   (d) Using the sweep option of MINTEQA2 (edit level IV) and assigning an ionic strength of zero, speciate the 0.01 M $CuCl_2$ solution at pH 4, 5, 6, 7, 8, 9, and 10. Compare your results at pH = 4 to what you have computed by hand in part (c). What happens to the charge balance of the solution as the pH is increased in sweep? Tabulate and plot the percent molal concentrations of the various cupric species from pH 4 to 10, as given in Part 4 of the output file of the sweep results for each pH, and discuss.

2. In the text we hand calculated the speciation of a solution containing 1.00 M HCl and 0.010 M $Cd(NO_3)_2$ at 25°C and zero ionic strength. Enter the same information into MINTEQA2 and let the program calculate the speciation. Compare your MINTEQA2 result to the result given in the book. Results may differ because of a different choice of Cd-Cl complexes and/or the use of different stability constants for the complexes in the computer model (listed in the file *thermo.dbs*). Tabulate and compare the two sets of stability constants and explain differences between the results.

3. A metal fabrication plant is discharging wastewaters that contain lead into the ocean. You are asked to determine what the resultant maximum lead concentration in the ocean water might be, assuming 25°C, and that the maximum concentration of free Pb^{2+} ion in the water is limited by the solubility of $PbSO_4$ (the mineral anglesite), which has a solubility product, $K_{sp} = 10^{-7.79}$. The pH of seawater is 8.1 and the following total concentrations (mol/kg) in seawater are $\Sigma mCl^- = 0.56$ and $\Sigma mSO_4^{2-} = 0.028$. From the literature you obtain dissociation constants at 25°C for the six Pb^{2+} complexes given here.

Dissociation reaction	log K_{dissoc}
$PbOH^+ = Pb^{2+} + OH^-$	$-6.29^\dagger$
$PbSO_4^\circ = Pb^{2+} + SO_4^{2-}$	$-2.75^\ddagger$
$PbCl^+ = Pb^{2+} + Cl^-$	$-1.59^\ddagger$
$PbCl_2^\circ = Pb^{2+} + 2Cl^-$	$-1.8^\ddagger$
$PbCl_3^- = Pb^{2+} + 3Cl^-$	$-1.7^\ddagger$
$PbCl_4^{2-} = Pb^{2+} + 4Cl^-$	$-1.4^\ddagger$

Source: †Baes & Mesmer (1976)
‡Smith & Martell (1976)

Ignoring activity-coefficient corrections (a very approximate approach for seawater), and assuming that the total concentrations of sulfate and chloride given above are so large relative to that of the lead that they can be assumed constant, derive an equation that describes the total molal concentration of lead in the seawater, as defined by the solubility of anglesite and the presence of the above complexes. Solve this equation to determine the molal concentrations of free Pb^{2+} and of each complex in seawater and calculate the percent of the total lead that each of these species represents. How much does the presence of complexes increase the concentration of total dissolved lead in seawater?

4. Using the sweep option in MINTEQA2, calculate and tabulate the percentage distribution of lead complexes as a function of the chloride concentration, from $mCl^- = 10^{-6.0}$ to $mCl^- = 0.56$. (Seawater contains 0.56 mol/kg chloride). Ignore activity coefficients in your calculation. Assume 25°C, $\Sigma mSO_4 = 0.028$, $\Sigma mPb(II) = 10^{-6.0}$, and pH = 8.1, and that all are constant. Explain the changes in speciation with increasing chloride concentration.

5. You are given the following analysis of a well adjacent to a uranium mill-tailings pond south of Moab, Utah. The high alkalinity and pH reflect the fact that the mill uses a sodium carbonate leach to extract uranium from the ore. Assume 25°C.

Parameter	Concentration (mg/L) except for pH and SpC	Parameter	Concentration (mg/L) except for pH and SpC
Cl^-	739	Ni	0.11
SO_4^{2-}	3362	Ba	0.10
Alkalinity (as $CaCO_3$)	2629	Cd	0.01
NO_3^-	22	F	4.1
pH	9.0	Mo	14
Pb	0.3	SpC (μmhos/cm)	13,994
Cr (as CrO_4^{2-})	0.01		

(a) The analysis is obviously missing information on major cation concentrations. Enter the tabulated data into MINTEQA2 and, from the unspeciated charge balance in the output and (assuming the charge imbalance is entirely as Na^+), estimate the sodium ion concentration. What is the danger of such an estimation approach? Hint: To solve this problem you may have to increase the number of program iterations in MINTEQA2 from 40 (standard) to 200.

(b) Now rerun MINTEQA2 with the above analysis as before, plus the estimated Na^+ concentration. Examine Part 4 of the output file, which lists the percentage distribution of components (aqueous species including complexes). Discuss the speciation of the water in terms of percentages of Cd species that equal or exceed 1% of total Cd. Assuming that saturation with otavite ($CdCO_3$) limits Cd concentrations, how much more soluble is otavite in the water because of Cd complexing that it would be in the absence of Cd complexing?

(c) Assuming that calcite is present in the rock and is at saturation with the wellwater (calcite occurs as a "fixed solid"), what is the total dissolved calcium concentration and what is the speciation of dissolved calcium in the water? Hint: You may have to enter calcite as an fixed-solid *before* you enter other species concentrations, or the program will not accept the fixed-solid entry.

6. Starting with the temperature functions for the Al^{3+}-OH hydrolysis reactions tabulated in Chap. 1, plot the fields of dominance of Al^{3+} and its hydroxide complexes from 0 to 100°C, and discuss. Remember that at boundaries between the fields the species are present in equal concentrations. Does Al-OH complexing increase or decrease with increasing T?

7. Using MINTEQA2 and the sweep option, calculate and plot the solubility of ferrihydrite ($Fe(OH)_3$) with a K_{sp} of $10^{-38.5}$ at 25°C from pH 2 to 10 at whole pH values, as molar total dissolved Fe:

(a) in water that contains 0.1 M NaCl; and

(b) in a 0.1 M NaCl solution that is also saturated with respect to gypsum, assuming log K_{sp} for gypsum equals -4.59.

Tabulate the speciation of ferric iron in mole percent read from the output file as computed in parts (a) and (b). Compare the results of the two calculations, including differences in the relative importances of Fe(III) species as a function of pH. Note that part (b) represents conditions typical of some acid mine waters.

8. Among the least soluble of lead minerals are its phosphates. However, complexing of the lead tends to make the lead phosphates more soluble than if complexing were absent. Given the following water analysis, compute the solubility of hydroxypyromorphite [$Pb_5(PO_4)OH$] as $\Sigma Pb(aq)$ at 25°C using MINTEQA2 and report the lead speciation. Assume the water has a pH of 7.00, Eh = 700 mV, $Na^+ = 2 \times 10^{-3}$ mol/kg, $Cl^- = 10^{-3}$ mol/kg, $HCO_3^- = 10^{-3}$ mol/kg, and $\Sigma PO_4 = 10^{-6}$ mol/kg.

9. The following cations (M^{n+}) and the association constants (K_1) of their MOH^{n-1} complexes as defined in the following general reaction and equilibrium expressions

$$M^{n+} + OH^- = MOH^{n-1}$$

$$K_1 = \frac{[MOH^{n-1}]}{[M^{n+}][OH^-]}$$

are tabulated here. Also listed are $-\log K_{sp}$ values for the hydroxide solids of these cations, where the dissolution reaction and K_{sp} expressions are:

$$M(OH)_n = M^{n+} + n(OH^-)$$

$$K_{sp} = [M^{n+}] + [OH^-]^n$$

The K_1 and K_{sp} values are mostly from Smith and Martell (1976).
   (a) Plot $-\log K_{sp}$ and $\log K_1$ versus EN for these cations.
   (b) Discuss the plots in terms of the covalency versus ionicity of the M-OH bond in the complexes.
   (c) Do the plots make sense in terms of concepts of hard and soft acids and bases?
   (d) Plot $-\log K_{sp}$ versus $\log K_1$ and comment on the suggestion that cations that form strong complexes with a ligand tend to form insoluble solids with the same ligand, and conversely.
   (e) Calculate and tabulate $\log {}^*K_1$ values for the complexes and compute the pH at which $[MOH^{n-1}] = [M^{n+}]$.
   (f) Discuss by cation valence groups, the importance of MOH^{n-1} complexing versus the importance of the uncomplexed cations in natural waters (pH especially 2 to 10).

Cation	EN	$\log K_1$	$-\log K_{sp}$	Cation	EN	$\log K_1$	$-\log K_{sp}$
Li^+	0.98	0.36		Sr^{2+}	0.95	0.8	
Be^{2+}	1.57	8.6	21.7	Y^{3+}	1.22	6.3	23.2
Na^+	0.93	−0.2		Zr^{4+}	1.33	14.3	(54.1)
Mg^{2+}	1.31	2.58	11.15	Pd^{2+}	2.20	13.0	28.5
Al^{3+}	1.61	9.01	33.5	Ag^+	1.93	2.0	
K^+	0.82	−0.5		Cd^{2+}	1.69	3.9	14.35
Ca^{2+}	1.00	1.3	5.19	Sn^{2+}	1.7	(10.4)	26.2
Sc^{3+}	1.36	9.7	32.7	Ba^{2+}	0.89	0.6	3.6
V^{3+}	1.35	11.7	34.4	La^{3+}	1.10	5.5	20.7
VO^{2+}		8.3	23.5	Hf^{4+}	(1.34)	13.7	(54.8)
Cr^{3+}	1.6	10.07	(29.8)	Hg^{2+}	2.00	10.6	(25.44)
Mn^{2+}	1.4	3.4	12.8	Tl^{3+}	2.04	13.4	(45.2)
Fe^{2+}	1.7	4.5	15.1	Pb^{2+}	1.87	6.3	15.3
Fe^{3+}	1.8	11.81	38.8	Ra^{2+}	(0.83)	0.5	
Co^{2+}	1.88	4.3	14.9	Th^{4+}	1.1	10.8	44.7
Ni^{2+}	1.91	4.1	15.2	U^{4+}	1.3	13.35	56.2
Cu^{2+}	2.0	6.3	19.32	UO_2^{2+}	(1.8)	8.2	22.4
Zn^{2+}	1.65	5.0	16.46	Pu^{4+}		(12.2)	47.3

4

Activity Coefficients of Dissolved Species

4.1 ACTIVITIES OF DISSOLVED SPECIES, IONIC STRENGTH

Reactions are written in terms of *effective concentrations* or *activities* of dissolved species (a_i), not their concentrations (m_i). By definition the extent that these differ is expressed by the activity coefficient (γ_i):

$$a_i = \gamma_i m_i \tag{4.1}$$

The activities and chemical potentials (μ_i) of individual dissolved species are also related through the expressions:

$$\mu_i = \mu_i^\circ + \text{RT ln } a_i = \mu_i^\circ + \text{RT ln } \gamma_i + \text{RT ln } m_i \tag{4.2}$$

(see Chap. 1) where m_i is the molal concentration (mol/kg, i.e., moles solute/1000 g H_2O) and μ_i° is the chemical potential of the species under reference conditions.

In nearly pure water, where all solute ions or molecules contact only water molecules, $a_i = m_i$, and the activity coefficients of all solutes equal unity. As salt concentrations increase, however, individual aqueous species must move closer together and are, therefore, more and more likely to come in contact. Because the interaction between adjacent ions is largely Coulombic, it is also proportional to the charge of the ions involved. These effects are embodied in the definition of ionic strength (I) which is given by

$$I = \tfrac{1}{2}\Sigma(m_i z_i^2) \tag{4.3}$$

where I is in molal (preferably) or molar units and z_i is the charge of ion i. The ionic strength is summed for all charged species in solution. According to Lewis and Randall (1921), "In dilute solutions, the activity coefficient of a given strong (completely dissociated) electrolyte is the same in all solutions of the same ionic strength." (Cf. Lewis and Randall 1961; Garrels and Christ 1965.)

Example calculations of ionic strength for some simple pure salt solutions are straightforward. Thus, for the solution of a monovalent-monovalent salt such as NaCl, we find from the charge balance that $mNa^+ = mCl^- = mNaCl$, so that $I = \tfrac{1}{2}[mNa^+ \times (+1)^2 + mCl^- \times (-1)^2] = \tfrac{1}{2}(2mNaCl) = mNaCl$.

For a divalent-monovalent salt solution in which the electrolyte is $CaCl_2$ (a $Ca(HCO_3)_2$ electrolyte would be similar), $2mCa^{2+} = mCl^- = 2mCaCl_2$. The ionic strength is then $I = \frac{1}{2}(mCa^{2+} \times 2^2 + mCl^- \times 1^2) = \frac{1}{2}(6mCaCl_2) = 3\ mCaCl_2$. For a divalent-divalent salt such as $MgSO_4$ it can similarly be shown that $I = 4\ mMgSO_4$.

The most accurate value of ionic strength is obtained from a total water analysis, which includes all ionic species. However, partial analyses are more commonly available than complete ones. Under such conditions the molal ionic strength can be estimated from the total dissolved solids or specific conductance of the water (cf. Langmuir and Mahoney 1985; Polemo et al. 1980). Expressions that permit such an approach are:

$$I \approx 2 \times 10^{-5} \times TDS \qquad \text{NaCl solutions} \qquad (4.4)$$

$$I \approx 2.5 \times 10^{-5} \times TDS \qquad \text{"average" water} \qquad (4.5)$$

$$I \approx 2.8 \times 10^{-5} \times TDS \qquad Ca(HCO_3)_2 \text{ waters} \qquad (4.6)$$

where the TDS is in milligrams per liter (mg/L) or parts per million (ppm). Approximations written in terms of specific conductance in micromhos (μmhos/cm) or microsiemens (μS/cm) are:

$$I \approx 0.8 \times 10^{-5} \times SpC \qquad \text{NaCl waters} \qquad (4.7)$$

$$I \approx 1.7 \times 10^{-5} \times SpC \qquad CaSO_4\text{–}MgSO_4 \text{ waters} \qquad (4.8)$$

$$I \approx 1.9 \times 10^{-5} \times SpC \qquad Ca(HCO_3)_2 \text{ waters} \qquad (4.9)$$

A specific-conductance measurement generally permits a more accurate estimate of the ionic strength than can a total-dissolved-solids measurement. This is because specific conductance and ionic strength are both measures of the total concentrations of ionic species, whereas the TDS also includes concentrations of uncharged species, such as dissolved silica and organic acids.

4.2 ACTIVITY COEFFICIENTS OF IONS

4.2.1. Mean Ion-Activity Coefficients

Because solutions must always contain both cations and anions, there is no simple, direct way to measure the activity or activity coefficient of a single ion without making some assumptions about these values for an oppositely charged ion in the same solution. What is usually measured, in fact, is the net effect of both cations and anions present in solution, or the *mean ion activity coefficient*. (See Robinson and Stokes 1970; Desnoyers 1979; Klotz and Rosenberg 1986). Mean ion activity coefficients for salts in aqueous solutions ($\gamma\pm$ values) have been obtained by measurements of: (a) freezing-point lowering; (b) boiling-point elevation; (c) water-vapor pressure; (d) osmotic pressure; (e) solubility and distribution; and (f) transport properties, including conductivity and diffusion. In methods (a) through (d) the properties of the salt are obtained indirectly from its effect on the properties of solvent water. Methods (e) and (f) provide a more direct measure of a salt's mean ion activity coefficient.

Following is an example calculation to illustrate the conventional mathematical treatment of mean activity coefficients. Let us assume we are interested in the solubility of the salt $K_2SO_4(c)$. Its dissolution can be written

$$K_2SO_4(c) = 2K^+ + SO_4^{2-} \qquad (4.10)$$

for which we obtain

$$K_{sp} = (\gamma_K^2)\,(\gamma_{SO_4})\,(mK^+)^2\,(mSO_4^{2-}) \qquad (4.11)$$

We cannot measure the individual ion activity coefficients here, only their total effect on K_{sp}. It is convenient to lump this total effect in the geometric mean of the product of the individual activity coefficients and to call this the mean ion activity coefficient of the salt. Thus for a K_2SO_4 solution, by definition

$$\gamma_{\pm K_2SO_4} = [(\gamma_K^2)\,(\gamma_{SO_4})]^{1/3} \tag{4.12}$$

and the K_{sp} expression may be rewritten

$$K_{sp} = \gamma_{\pm K_2SO_4}^2\,(mK^+)^2\,(mSO_4^{2-}) \tag{4.13}$$

In general, for an electrolyte that dissociates into a total of n ions, the mean ion activity coefficient equals, by definition,

$$\gamma_\pm = [(\gamma_+^{n+})\,(\gamma_-^{n-})]^{1/n} \tag{4.14}$$

where n_+ and n_- are the number of positive and negative ions in the formula of the salt, and $n = n_+ + n_-$.

The mean ion activity coefficients of several salts taken from tables in Robinson and Stokes (1970) are plotted against ionic strength in Fig. 4.1. For reasons discussed later in this chapter, $\gamma\pm$

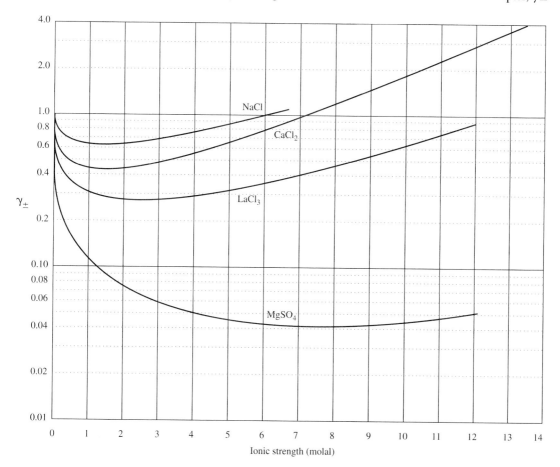

Figure 4.1 Mean ion activity coefficients of several salts as a function of ionic strength. Data are from Robinson and Stokes (1970).

values first decrease and then increase with ionic strength for all the salts shown. The decrease is far greater for multivalent salts, such as $MgSO_4$, than for monovalent NaCl. Mean ion activity coefficients for pure salts are of little use to us as such, except perhaps in some brines that are dominantly NaCl in character. Otherwise, natural waters, including seawater, tend to be mixtures of several electrolytes.

In order for us to have flexibility in our modeling of natural water chemistry we need a way to obtain individual ion activity coefficients from mean values. To do so requires that we make an assumption, called *the MacInnes convention* (MacInnes 1919), which states $\gamma_K = \gamma_{Cl}$. The convention is based on the observation that K^+ and Cl^- ions are of the same charge and nearly the same size, have similar electron structures (inert gas), and similar ionic mobilities. In support of this assumption, tracer diffusion coefficients, D°, of K^+ and Cl^- at infinite dilution are nearly equal at 19.6 and 20.3×10^{-6} cm^2/s (Lerman 1979). Also, limiting equivalent conductances, λ°, of K^+ and Cl^- are comparable at 73.50 and 76.35 cm^2/(ohm) (equiv.) at 25°C (Robinson and Stokes 1970).

The MacInnes convention leads to $\gamma_K = \gamma_{Cl} = \gamma_{\pm KCl}$. We can now compute individual ion activity coefficients from their mean values measured in solutions of *strong electrolytes* using $\gamma_{\pm KCl}$ values as our starting point. (In the ideal strong electrolyte, cations and anions are unassociated with each other and thus do not form complexes [see Chap. 3].) It is important to remember that all such calculations must be done with $\gamma_\pm$ values for KCl and other salts *measured at the same ionic strength*, which is not the same molality except for monovalent-monovalent salts.

In general we use $(\gamma_{\pm \text{cation}_1 Cl_{n-1}})$ values to obtain (γ_{cation}) values, and $(\gamma_{\pm K_{n-1} \text{anion}_1})$ values to compute (γ_{anion}) values. Again, $n = n_+ + n_-$. For the general cation case, where M is the cation,

$$\gamma_{\pm MCL_{n-1}} = [(\gamma_M)(\gamma_{Cl})]^{n-1}]^{1/n} \tag{4.15}$$

Substituting $\gamma_{\pm KCl}$ for γ_{Cl} and solving for γ_M, we obtain

$$\gamma_M = \frac{(\gamma_{\pm MCl_{n-1}})^n}{(\gamma_{\pm KCl})^{n-1}} \tag{4.16}$$

For example, for $\gamma_{Fe^{3+}}$ from $\gamma_{\pm FeCl_3}$,

$$\gamma_{\pm FeCl_3} = [(\gamma_{Fe^{3+}})(\gamma_{\pm KCl}^3)]^{1/4} \tag{4.17}$$

But since $\gamma_{Cl^-} = \gamma_{\pm KCl}$, we have,

$$\gamma_{Fe^{3+}} = \frac{\gamma_{\pm FeCl_3}^4}{\gamma_{\pm KCl}^3} \tag{4.18}$$

For γ_{anion} values, where A is the anion in a $K_{n-1}A$ solution,

$$\gamma_{\pm K_{n-1}A} = [(\gamma_k)^{n-1}(\gamma_A)]^{1/n} \tag{4.19}$$

and

$$\gamma_A = \frac{(\gamma_{\pm K_{n-1}A})^n}{(\gamma_{\pm KCl})^{n-1}} \tag{4.20}$$

Thus, to obtain γ_{SO_4} from $\gamma_{\pm K_2SO_4}$ we have

$$\gamma_{\pm K_2SO_4} = [(\gamma_K^2)(\gamma_{SO_4})]^{1/3} \tag{4.21}$$

and

$$\gamma_{SO_4} = \frac{\gamma_{\pm K_2SO_4}^3}{\gamma_{\pm KCl}^2} \tag{4.22}$$

Following are three example calculations of individual ion activity coefficients from mean salt data. Such calculations must always be performed using $\gamma_\pm$ values measured at the same ionic strengths of KCl and of the second salt. However, tables of mean activity coefficients are generally listed as a function of molal salt concentrations. The example calculations thus show the relationship between molalities of the two salts and the ionic strength of interest.

Example 4.1

a) Compute γ_{Na} from $\gamma_{\pm NaCl}$ at $I = 0.1$ mol/kg, using mean salt data from Hamer and Wu (1972). The appropriate expression is

$$\gamma_{Na} = \frac{\gamma^2_{\pm NaCl}}{\gamma_{\pm KCl}} \tag{4.23}$$

In this case $I = mNaCl = mKCl = 0.1$. From the tabulated mean salt data we obtain

$$\gamma_{Na} = \frac{(0.779)^2}{(0.768)} = 0.790 \tag{4.24}$$

b) Calculate γ_{Ca} at $I = 0.12$ mol/kg from the mean salt data for $CaCl_2$ given by Goldberg and Nuttall (1978). In this example, $I = 3mCaCl_2 = mKCl = 0.12$. Substitution gives

$$\gamma_{Ca} = \frac{(0.598)^3}{(0.754)^2} = 0.376 \tag{4.25}$$

c) Compute γ_{SO_4} at $I = 0.3$ mol/kg from mean activity coefficient data for K_2SO_4 listed by Robinson and Stokes (1970). $I = 3mK_2SO_4 = mKCl = 0.3$. The expression for γ_{SO_4} is

$$\gamma_{SO_4} = \frac{\gamma^3_{\pm K_2SO_4}}{\gamma^2_{\pm KCl}} \tag{4.26}$$

For $I = 0.3$ mol/kg substitution gives

$$\gamma_{SO_4} = \frac{(0.436)^3}{(0.687)^2} = 0.176 \tag{4.27}$$

Individual ion activity coefficients for Na^+, HCO_3^-, Ca^{2+}, SO_4^{2-}, and La^{3+} computed from mean salt data and, assuming the MacInnes convention, are plotted in Fig. 4.2. As before, the mean salt data for KCl are from Hamer and Wu (1972). The $\gamma\pm$ data for HCO_3^- and La^{3+} are from Roy et al. (1983) and Robinson and Stokes (1970), respectively. Sources for the other ions are as given above.

The accuracy of individual ion activity coefficients derived from mean salt data is limited by the fact that important ion pairing takes place in solutions of many "strong electrolytes" at ionic strengths as low as 0.1 mol/kg. For example, data given by Reardon (1975) suggests that formation of the pair KSO_4^- ($pK_{assoc} = 0.85$ at 25°C) makes the γ_{SO_4} values shown in Fig. 4.2 for $I > 0.3$ mol/kg (which are based on mean salt data for K_2SO_4 uncorrected for such pairing) too low by 60% or more. Millero and Schreiber (1982) show how to correct mean activity coefficients for ion pairing in electrolyte solutions with significant pairing (see also Parkhurst 1990).

4.2.2 Dilute Solutions and the Debye-Hückel Equation

In fresh, potable waters (TDS < 500 ppm), and even in brackish waters with TDS values up to about 5000 ppm ($I \approx 0.1$), activity coefficients of monovalent and divalent ions can be computed through

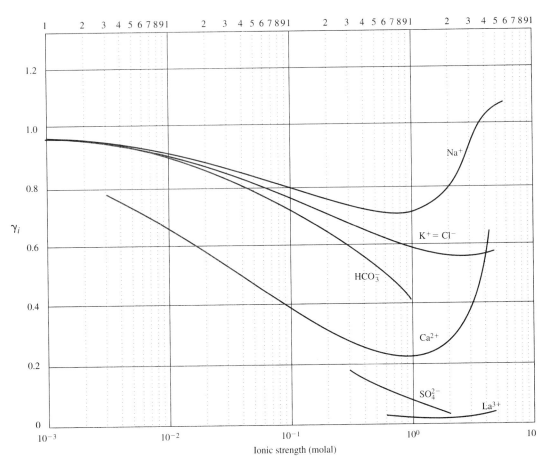

Figure 4.2 Some individual ion activity coefficients computed from mean salt data assuming the MacInnes convention.

use of the theoretical Debye-Hückel equation (c.f. Robinson and Stokes 1970). Assumptions inherent in this equation include that ion interactions are purely Coulombic, ion size does not vary with ionic strength, and ions of the same sign do not interact. For individual ions the equation is

$$\log \gamma_i = \frac{-A\,z_i^2\,\sqrt{I}}{1 + B\,a_i\sqrt{I}} \qquad (4.28)$$

where $A = 1.824928 \times 10^6\,\rho_o^{1/2}\,(\epsilon T)^{-3/2}$, $B = 50.3\,(\epsilon T)^{-1/2}$, with ρ_o the density of water, ϵ the dielectric constant of water, and T in kelvin. At 25°C, $\rho_o = 0.99707$, $\epsilon = 78.4528$, $A = 0.5092$, and $B = 0.3283$ (Helgeson and Kirkham 1974).

At other temperatures the dielectric constant of water can be obtained from the equation

$$\epsilon = 2727.586 + 0.6224107\,T - 466.9151\,\ln T - 52000.87/T \qquad (4.29)$$

where T is in kelvin. This function fits empirical values of ϵ to within ± 0.01 units up to 100°C (Nordstrom et al. 1990). The dielectric constant of water is a measure of the effect of water (versus a vacuum) in decreasing the force (F) of the electrical field between ionic species in solution, where $F = z_1 z_2/\epsilon r^2$ and z_1 and z_2 are the charges on ions 1 and 2, and r is the distance between them. The dielectric constant is determined by measuring the capacitance of water, $\mathbb{C}_{H_2O}$ versus that of a vacuum, $\mathbb{C}_o$, or $\epsilon = \mathbb{C}_{H_2O}/\mathbb{C}_o$.

The term a_i is the effective size of the hydrated ion in angstroms (1 Å = 1 nm = 10^{-8} cm) and is a function of the degree of hydration of the ion, which is roughly proportional to its ionic potential (its charge z over radius r [see below]. Given in Table 4.1 are ion size parameters for many inorganic and organic species based chiefly on Kielland (1937). According to Kielland (1937) a_i values for inorganic ions range from 2.5 to 4.5 for monovalent ions, (except for Li^+ and H^+), 4 to 8 for divalent ions, 4 to 9 for trivalent ions, and 5 to 11 for quadrivalent ions. The anomalous size of the proton ($a_i = 9$) reflects its occurrence in the large hydronium ion (H_3O^+) and in similar hydrated species.

Some ion activity coefficients at 25°C computed with the Debye-Hückel equation as a function of ionic strength, ion size, and charge, are shown in Table 4.2. Debye-Hückel ion activity coefficients up to 0.1 mol/kg ionic strength, are plotted in Fig. 4.3 for some monovalent and divalent ions. The Debye-Hückel equation can be used to compute accurate activity coefficients for monovalent ions up to about $I = 0.1$ mol/kg, for divalent ions to about $I = 0.01$ mol/kg, and for trivalent ions up to perhaps $I = 0.001$ mol/kg.

At ionic strengths between 0.001 and 0.0001 mol/kg the separate curves for individual ions of the same valence converge on a single curve. At these low ionic strengths, the term $B\, a_i \sqrt{I}$ approaches zero, and the Debye-Hückel equation is reduced to

$$\log \gamma_i = -A\, z_i^2 \sqrt{I} \tag{4.30}$$

This is called the Debye-Hückel limiting law. At higher ionic strengths, differences in ion sizes (a_i values) become important and cause the γ_i curves for individual ions to diverge.

The effect of temperature on ion activity coefficients is largely predicted by changes in the value of A, which is proportional to $-\log \gamma_i$ in the Debye-Hückel equation. The value of A increases from 0.492 to 0.534 between 0 and 50°C. Thus, activity coefficients become smaller with increasing temperature. Because A is multiplied by z^2 in the Debye-Hückel equation, the effect of temperature on activity coefficients is greatest for multivalent ions.

4.2.3 Intermediate Ionic Strengths: The Davies and Truesdell-Jones Equations, Specific Ion-Interaction Theory

Examination of Figs. 4.1 and 4.2 shows that individual and mean ion activity coefficients generally decrease to minimal values between 1 and 10 mol/kg, and then increase at higher ionic strengths. However, the form of the Debye-Hückel equation is such that it predicts γ values that decline continuously with ionic strength. Several assumptions incorporated in the Debye-Hückel equation become invalid and lead to its failure at high ionic strengths where ion activity coefficients increase (cf. Pytkowicz 1983). These assumptions include: ion interactions are purely coulombic, ion size does not vary with I, and ions of the same sign do not interact. Chemical equilibria calculations in seawater ($I = 0.7$ mol/kg) and in higher ionic strength brines require activity coefficient models without such limitations.

The increase in ion activity coefficients at elevated ionic strengths results from several effects, including the fact that an increasing fraction of water molecules are involved in hydration spheres around ions. This causes a proportionate decrease in the concentration of free water molecules in the solution. For example, assuming there are six water molecules associated with each pair of dissolved Na^+ and Cl^- ions in a 0.01 molar NaCl solution (see Bockris and Reddy 1973), the moles of free water molecules in a liter of solution are $55.5 - 6 \times (0.01) = 55.44$. On the other hand, if the solution

TABLE 4.1. Debye-Hückel ion size parameters (Å) for selected ions

	Inorganic ions, charge 1
9	H^+
6	Li^+
4–4.5	Na^+, $CdCl^+$, ClO_2^-, IO_3^-, HCO_3^-, $H_2PO_4^-$, HSO_3^-, $H_2AsO_4^-$, $Co(NH_3)_4(NO_2)_2^+$
3.5	OH^-, F^-, NCS^-, NCO^-, HS^-, ClO_3^-, ClO_4^-, BrO_3^-, IO_4^-, MnO_4^-
3	K^+, Cl^-, Br^-, I^-, CN^-, NO_2^-, NO_3^-
2.5	Rb^+, Cs^+, NH_4^+, Tl^+, Ag^+

	Inorganic ions, charge 2
8	Mg^{2+}, Be^{2+}
6	Ca^{2+}, Cu^{2+}, Zn^{2+}, Sn^{2+}, Mn^{2+}, Fe^{2+}, Ni^{2+}, Co^{2+}
5	Sr^{2+}, Ba^{2+}, Ra^{2+}, Cd^{2+}, Hg^{2+}, S^{2-}, $S_2O_4^{2-}$, WO_4^{2-}
4.5	Pb^{2+}, CO_3^{2-}, SO_3^{2-}, MoO_4^{2-}, $Co(NH_3)_5Cl^{2+}$, $Fe(CN)_5NO^{2-}$
4	Hg_2^{2+}, SO_4^{2-}, $S_2O_3^{2-}$, $S_2O_8^{2-}$, SeO_4^{2-}, CrO_4^{2-}, HPO_4^{2-}, $S_2O_6^{2-}$

	Inorganic ions, charge 3
9	Al^{3+}, Fe^{3+}, Cr^{3+}, Sc^{3+}, Y^{3+}, La^{3+}, In^{3+}, Ce^{3+}, Pr^{3+}, Nd^{3+}, Sm^{3+}
6	$Co(ethylenediamine)_3^{3+}$
4	PO_4^{3-}, $Fe(CN)_6^{3-}$, $Cr(NH_3)_6^{3+}$, $Co(NH_3)_6^{3+}$, $Co(NH_3)_5H_2O^{3+}$

	Inorganic ions, charge 4
11	Th^{4+}, Zn^{4+}, Ce^{4+}, Sn^{4+}
6	$Co(S_2O_3)(CN)_5^{4-}$
5	$Fe(CN)_6^{4-}$

	Inorganic ions, charge 5
9	$Co(SO_3)_2(CN)_4^{5-}$

	Organic ions, charge 1
8	$(C_6H_5)_2CHCOO^-$, $(C_3H_7)_4N^+$
7	$[OC_6H_2(NO_2)_3]^-$, $(C_3H_7)_3NH^+$, $CH_3OC_4H_4COO^-$
6	$C_6H_5COO^-$, $C_6H_4OHCOO^-$, $C_6H_4ClCOO^-$, $C_6H_5CH_2COO^-$, $CH_2=CHCH_2COO^-$, $(CH_3)_2CHCH_2COO^-$, $(C_2H_5)_4N^+$, $(C_3H_7)_3NH_2^+$
5	$CHCl_2COO^-$, CCl_3COO^-, $(C_2H_5)_3NH^+$, $(C_3H_7)NH_3^+$
4.5	CH_3COO^-, CH_2ClOO^-, $(CH_3)_4N^+$, $(C_2H_5)_2NH_2^+$, $NH_2CH_2COO^-$
4	$NH_3CH_3COOH^+$, $(CH_3)_3NH^+$, $C_2H_5NH_3^+$
3.5	$HCOO^-$, H_2-citrate$^-$, $CH_3NH_3^+$, $(CH_3)_2NH_2^+$

	Organic ions, charge 2
7	$OOC(CH_2)_5COO^{2-}$, $OOC(CH_2)_6COO^{2-}$, Congo red anion^{2-}
6	$C_6H_4(COO)_2^{2-}$, $H_2C(CH_2COO)_2^{2-}$, $(CH_2CH_2COO)_2^{2-}$
5	$H_2C(COO)_2^{2-}$, $(CH_2COO)_2^{2-}$, $CHOHCOO)_2^{2-}$
4.5	$(COO)_2^{2-}$, H-citrate^{2-}

	Organic ions, charge 3
5	Citrate^{3-}

Source: Reprinted with permission from J. Kielland, Individual activity coefficients of ions in aqueous solutions, *J. Am. Chem. Soc.* 59, copyright 1937 American Chemical Society.

TABLE 4.2. Individual ion activity coefficients at 25°C for different ion sizes (a_i) in angstroms (1 Å = 10^{-8} cm) as a function of ionic strength, computed using the extended Debye-Hückel equation with $A = 0.5091$ and $B = 0.3286$

Ion Size Parameter	Ionic Strength							
	0.0005	0.001	0.0025	0.005	0.01	0.025	0.05	0.1
				Ion charge 1				
9	0.975	0.967	0.950	0.933	0.914	0.88	0.86	0.83
8	0.975	0.966	0.949	0.931	0.912	0.88	0.85	0.82
7	0.975	0.965	0.948	0.930	0.909	0.875	0.845	0.81
6	0.975	0.965	0.948	0.929	0.907	0.87	0.835	0.80
5	0.975	0.964	0.947	0.928	0.904	0.865	0.83	0.79
4.5	0.975	0.964	0.947	0.928	0.902	0.86	0.82	0.775
4	0.975	0.964	0.947	0.927	0.901	0.855	0.815	0.77
3.5	0.975	0.964	0.946	0.926	0.900	0.855	0.81	0.76
3	0.975	0.964	0.945	0.925	0.899	0.85	0.805	0.755
2.5	0.975	0.964	0.945	0.924	0.898	0.85	0.80	0.75
				Ion charge 2				
8	0.906	0.872	0.813	0.755	0.69	0.595	0.52	0.45
7	0.906	0.872	0.812	0.755	0.685	0.58	0.50	0.425
6	0.905	0.870	0.809	0.749	0.675	0.57	0.485	0.405
5	0.903	0.868	0.805	0.744	0.67	0.555	0.465	0.38
4.5	0.903	0.868	0.805	0.742	0.665	0.55	0.455	0.37
4	0.903	0.867	0.803	0.740	0.660	0.545	0.445	0.355
				Ion charge 3				
9	0.802	0.738	0.632	0.54	0.445	0.325	0.245	0.18
6	0.798	0.731	0.620	0.52	0.415	0.28	0.195	0.13
5	0.796	0.728	0.616	0.51	0.405	0.27	0.18	0.115
4	0.796	0.725	0.612	0.505	0.395	0.25	0.16	0.095
				Ion charge 4				
11	0.678	0.588	0.455	0.35	0.255	0.155	0.10	0.065
6	0.670	0.575	0.43	0.315	0.21	0.105	0.055	0.027
5	0.668	0.57	0.425	0.31	0.20	0.10	0.048	0.021
				Ion charge 5				
9	0.542	0.43	0.28	0.18	0.105	0.045	0.020	0.009

Source: Reprinted with permission from J. Kielland, Individual activity coefficients of ions in aqueous solutions, *J. Am. Chem. Soc.* 59, copyright 1937 American Chemical Society.

is 5 molar NaCl, the moles of free water molecules are $55.5 - 6 \times 5 = 25.5$. In other words, more than half the water is not free in the solution. As a result the activity of water is reduced from near unity in the dilute solution to 0.807 in the 5 molar solution of NaCl. Such effects and important ion pairing of electrolyte ions cause the ion activity coefficient increase.

This increase in γ values with ionic strength can be modeled by adding positive terms to some form of the extended Debye-Hückel expression for log γ. Simple models using this approach have been proposed by Hückel (see Harned and Owen 1958), and more recently by others, including

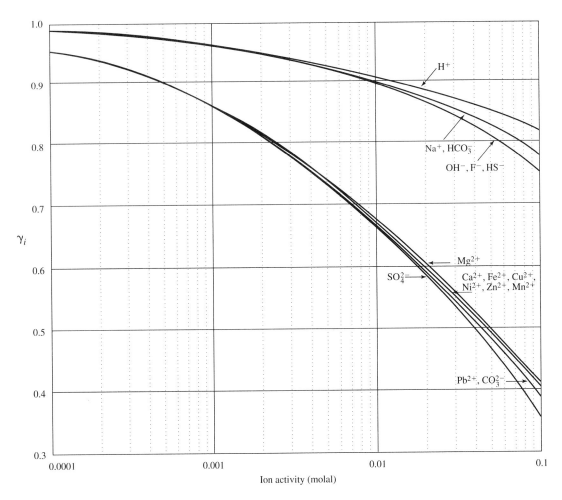

Figure 4.3   Some individual ion activity coefficients calculated using the extended Debye-Hückel equation.

Davies, Bronsted, and others, as well as Truesdell and Jones (1974) (cf. Nordstrom and Munoz 1994). Written in a form for single ion activity coefficients, the empirical Davies expression is

$$\log \gamma = -Az^2\left(\frac{\sqrt{I}}{1 + \sqrt{I}} - 0.3I\right) \tag{4.31}$$

Earlier versions of this equation used $0.2I$ instead of $0.3I$ as the add-on term (cf. Butler 1964). In MINTEQA2 the add-on term is assigned a value of $0.24I$. The add-on term accounts for effects that include lowering of the dielectric constant of water and increased ion pairing caused by the increase in dissolved ion concentrations (Harned and Owen 1958).

Because it lacks an ion size parameter, the Davies equation gives the same activity coefficient for all ions of the same charge at a given ionic strength in all electrolyte solutions. The Davies equation has been found to be accurate for monovalent salts up to ionic strengths of between 0.1 and about 0.7 mol/kg (the ionic strength of sea water), although its lack of an ion size parameter makes it less accurate than the Debye-Hückel equation at low ionic strengths.

TABLE 4.3   Ion size (a_i) and b values for the Truesdell-Jones equation for individual ion activity coefficients

	a_i	b	Source		a_i	b	Source
H^+	4.78	0.24	‡	Fe^{2+}	5.08	0.16	‡
Li^+	4.76	0.20	‡	Co^{2+}	6.17	0.22	‡
Na^+	4.0	0.075	†	Ni^{2+}	5.51	0.22	‡
	4.32	0.06	‡	Zn^{2+}	4.87	0.24	‡
K^+	3.5	0.015	†	Cd^{2+}	5.80	0.10	‡
	3.71	0.01	‡	Pb^{2+}	4.80	0.01	‡
Cs^+	1.81	0.01	‡	OH^-	10.65	0.21	‡
Mg^{2+}	5.5	0.20	†	F^-	3.46	0.08	‡
	5.46	0.22	‡	Cl^-	3.71	0.01	‡
Ca^{2+}	5.0	0.165	†	ClO_4^-	5.30	0.08	‡
	4.86	0.15	‡	HCO_3^-, CO_3^{2-}	5.4	0	†
Sr^{2+}	5.48	0.11	‡	SO_4^{2-}	5.0	−0.04	†
Ba^{2+}	4.55	0.09	‡		5.31	−0.07	‡
Al^{3+}	6.65	0.19	‡				
Mn^{2+}	7.04	0.22	‡				

Note: The equation, which applies to predominantly NaCl solutions, may be generally reliable up to an ionic strength of about 2 mol/kg (Parkhurst 1990).

Source: † From Truesdale and Jones (1974)
　　　　‡ From Parkhurst (1970)

　　　　The Truesdell-Jones (TJ) equation (Truesdell and Jones 1974), which was in fact proposed by Hückel in 1925 (see Pytkowicz 1983), is simply the extended Debye-Hückel equation plus an add-on term, bI, specific for each ion. The ion size parameter, a_i, and b in the add-on term are empirical, based on fitting the TJ equation to individual ion activity coefficients obtained from mean salt data and assuming the MacInnes convention as described above. Values of a_i and b from Truesdell and Jones (1974) and from a greatly expanded study by Parkhurst (1990) are listed and compared in Table 4.3. These values apply to chloride electrolyte solutions for the cations and to potassium and thus approximately to sodium electrolyte solutions for the anions. The Truesdell-Jones activity coefficient model is used by the geochemical computer models SOLMINEQ.88 (Kharaka et al. 1988) and PHREEQE (Parkhurst et al. 1990). Note that the ion size parameters in Table 4.3 generally disagree with the a_i values for the same ions suggested by Kielland (1937). Complete agreement would not be expected, however, in that the former have been selected so that the Truesdell-Jones function will optimally fit the full range of mean salt data. This author has also adjusted the Debye-Hückel a_i value to optimize the fit between Debye-Hückel and empirical mean salt data at the lowest mean salt values. The exercise suggests $a_i = 5.5$ for Ca^{2+} and 6.5 for Mg^{2+}; these values are intermediate between those of Kielland and Truesdell-Jones.

　　　　Bronsted-Guggenheim-Scatchard specific ion interaction theory (SIT) (cf. Grenthe and Wanner 1989; Giridhar and Langmuir 1991; Nordstrom and Munoz 1994) is an ion- and electrolyte-specific approach to activity coefficients, which is, therefore, theoretically capable of greater accuracy than the Davies equation. The general SIT equation for a single ion, i, can be written

$$\log{(\gamma_i)} = -z^2D + \sum_k \epsilon(i, j, I)m(j) \tag{4.32}$$

where the D, the Debye-Hückel term, equals

$$D = \frac{0.5901\sqrt{I}}{1 + 1.5\sqrt{I}}$$ (4.33)

In these expressions z is the charge of ion i, $m(j)$ is the molality of major electrolyte ion j, which is of opposite charge to ion i. Interaction parameters, $\epsilon(i, j, I)$ refer to the interaction between ion i and major electrolyte ion j. The value of ϵ is assumed zero for neutral species or between ions of like sign. As an approximation, the interaction parameters are often assumed ionic strength independent. This assumption may be correct for monovalent or divalent ions up to an ionic strength of about 3.5 mol/kg, but is questionable for more highly charged ions at such ionic strengths. Application of the SIT approach to studies of natural waters is somewhat limited by the availability of values for the interaction parameters, the most complete tabulation of which is probably that of Grenthe and Wanner (1989). Some interaction parameters of importance for natural waters are given in Table 4.4.

A similar model is presented by Baes and Mesmer (1976). Such models, which consider only interactions between two ions of opposite sign (so-called *binary interactions*) can give accurate activity coefficients for monovalent and divalent ions up to about $I = 2$ to 3.5 mol/kg (TDS about 100,000 to 175,000 ppm).

The activity coefficient of Ca^{2+} from 0.001 to 10 mol/kg ionic strength, computed using the mean salt approach and the equation

$$\gamma_{Ca^{2+}} = \frac{\gamma_{\pm CaCl_2}^3}{\gamma_{\pm KCl}^2}$$ (4.34)

is plotted in Fig. 4.4, where it is compared to the same coefficient obtained via the extended Debye-Hückel equation, the Davies equation, the SIT model equation, and the Truesdell-Jones (TJ) equation. Given that $CaCl_2$ is a strong electrolyte, with no measureable Ca-Cl ion pairing, the mean-salt derived coefficients are probably accurate. They are practically identical at all ionic strengths to the values obtained with the TJ equation using the a_i and b parameters from Truesdell and Jones in Table 4.3. This should come as no surprise, for the TJ parameters for Ca^{2+} are based on the same mean salt data. The extended Debye-Hückel and Davies equations are clearly inaccurate at ionic strengths above roughly 0.3 mol/kg, whereas the SIT model is accurate to within about ±5 to 10% up to $I = 2$ mol/kg. Because both Davies and SIT models ignore ion size, they are less accurate than

TABLE 4.4   Some SIT interaction parameters of importance for natural waters

Ion i	$\epsilon(i,j)$	Ion i	$\epsilon(i,j)$
Mg^{2+}	0.19	UO_2^{2+}	0.21
Ca^{2+}	0.14	Al^{3+}	0.33
Ba^{2+}	0.07	La^{3+}	0.22
Mn^{2+}	0.13	Th^{4+}	0.25
Co^{2+}	0.16	Cl^-	0.03
Ni^{2+}	0.17	HCO_3^-	−0.03
Cu^{2+}	0.08	SO_4^{2-}	−0.12

Note: For the cations, the major electrolyte ion, j, is chloride. For the anions it is sodium. Although a single parameter is given here for sulfate, the authors also present a function to describe its dependence on ionic strength.

Source: From Grenthe and Wanner (1989).

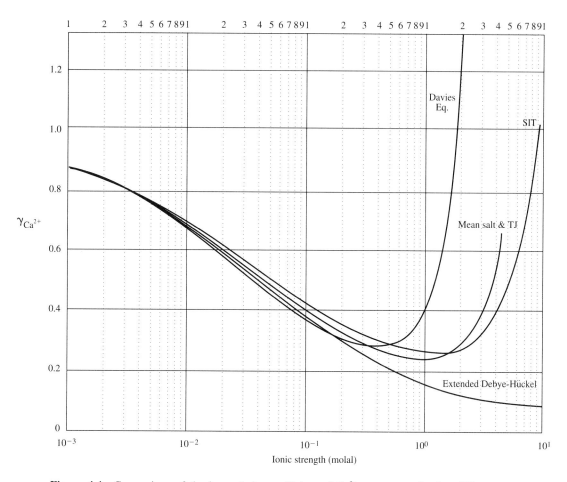

Figure 4.4   Comparison of the ion activity coefficient of Ca^{2+} as computed using different approaches. These include: (1) the Davies equation; (2) the mean salt approach using the MacInnes convention, and Truesdell-Jones equation (curve labeled Mean salt & TJ); (3) the specific ion interaction (SIT) equation; and (4) the extended Debye-Hückel equation.

the extended Debye-Hückel equation at low ionic strengths. SIT model interaction parameters nevertheless result in activity coefficients that are in general agreement with mean salt values over a relatively wide range of ionic strengths.

4.2.4 Stoichiometric and Effective Ionic Strength

When the total concentrations of dissolved species are used to compute ionic strength through its definition

$$I = \tfrac{1}{2}\Sigma(m_i z_i^2) \tag{4.3}$$

and the existence of ion pairs is ignored, the resultant ionic strength is said to be a total or stoichiometric (I_s) value. If, on the other hand, the presence of ion pairs is accounted for in the calculation of I, its value will be reduced. This is because the formation ion pairs (e.g., $NaCO_3^-$ or $CaSO_4^\circ$) removes charged species from the ionic strength calculation, and the pair always has a lower charge than its component ions (remember I is proportional to z^2). An ionic strength value so corrected is called an effective ionic strength (I_e). Because typically less than 1% of the ionic species in fresh-water occur in ion pairs, for such waters $I_s = I_e$. In seawater however, because of significant ion pairing, $I_s = 0.718$ mol/kg and $I_e = 0.668$ mol/kg (7% lower). The difference becomes even more significant at higher ionic strengths. Geochemical computer models such as WATEQ4F, MINTEQA2, and PHREEQE calculate and use the effective ionic strength. SOLMINEQ.88 and PHRQPITZ are capable of calculating the stoichiometric ionic strength. In computer codes, the Davies, SIT, and TJ equations and the Pitzer model are usually defined in terms of I_s.

Ion activities in a solution should be the same, regardless of the ionic-strength approach used. Remember that by definition $a_i = \gamma_i m_i$. Because the I_s approach ignores ion pairs, $I_s > I_e$ and m_i (I_s) is greater than m_i (I_e). The smaller ion activity coefficients computed from higher values of ionic strength in the stoichiometric approach, are compensated for by higher ion molalities in that approach. The result is an ion activity that is the same regardless of the ionic strength model used. The following worked problems illustrate this point. The first examines the solubility of gypsum ($CaSO_4 \cdot 2H_2O$) in pure water. The second considers calcite ($CaCO_3$) solubility in seawater.

Example 4.2

Compute the total or stoichiometric and effective ionic strengths, and corresponding total- and free-ion activity coefficients of Ca^{2+} and SO_4^{2-} in pure water saturated with gypsum at 25°C. It is given that the solubility product of gypsum, $K_{sp} = 10^{-4.59}$ and that at gypsum saturation $\Sigma mCa = \Sigma mSO_4 = 0.0154$.

The stoichiometric ionic strength equals simply

$$I_s = \tfrac{1}{2}(2 \times 0.0154 \times 4) = 0.616 \text{ mol/kg} \tag{4.35}$$

The effective ionic-strength approach assumes the existence of ion pairs, thus the mass-balance equations are

$$\Sigma mCa = mCa^{2+} + mCaSO_4^{\circ} \tag{4.36}$$

$$\Sigma mSO_4 = mSO_4^{2-} + mCaSO_4^{\circ} \tag{4.37}$$

The solubility reaction for gypsum is written

$$CaSO_4 \cdot 2H_2O = Ca^{2+} + SO_4^{2-} + 2H_2O \tag{4.38}$$

Calcium and sulfate ions form the $CaSO_4^{\circ}$ ion pair, with the association constant expression

$$K_{assoc} = 10^{2.31} = \frac{[CaSO_4^{\circ}]}{[Ca^{2+}][SO_4^{2-}]} \tag{4.39}$$

where the brackets denote activities of the species. But at saturation we know $[Ca^{2+}][SO_4^{2-}] = 10^{-4.59}$. Substituting for this product in Eq. (4.39) leads to $[CaSO_4^{\circ}] = 10^{-2.28} = 5.25 \times 10^{-3}$ mol/kg. Assuming for the ion pair that $\gamma_{CaSO_4} \approx 1$, we solve for mCa^{2+} and mSO_4^{2-} in Eqs. (4.36) and (4.37) and find $mCa^{2+} = mSO_4^{2-} = 1.02 \times 10^{-2}$ mol/kg. The effective ionic strength is now given by

$$I_e = \tfrac{1}{2}(2 \times 1.02 \times 10^{-2} \times 4) = 0.0408 \text{ mol/kg} \tag{4.40}$$

With the Debye-Hückel equation, and assuming I_e, we compute $\gamma_{Ca}^e = 0.511$ and $\gamma_{SO_4}^e = 0.477$, from which, since $a_1 = \gamma_i^e m_i^e$, the ion activities are $[Ca^{2+}] = 5.21 \times 10^{-3} = 10^{-2.28}$ mol/kg, and $[SO_4^{2-}] = 4.87 \times 10^{-3} = 10^{-2.31}$ mol/kg.

Note that I_s is about 50% greater than I_e. Calcium and sulfate ion activities must be the same, whether computed assuming effective or stoichiometric I. Because $\gamma_i^s = a_i/m_i^s$, we find that for I_s

$$\gamma_{Ca}^s = \frac{[Ca^{2+}]}{0.0154} = 0.333 \qquad \text{versus } \gamma_{Ca}^e = 0.511$$

and

$$\gamma_{SO_4}^s = \frac{[SO_4^{2-}]}{0.0154} = 0.311 \qquad \text{versus } \gamma_{SO_4}^e = 0.477$$

(4.41)

Example 4.3

The state of saturation of seawater with respect to calcite must be the same whether calculated using effective ionic strength ($I_e = 0.668$ mol/kg) and assuming both uncomplexed species and ion pairing, or computed assuming the stoichiometric or total ionic strength ($I_s = 0.718$ mol/kg), using total concentrations and neglecting ion pairing. This follows from the fact that the activities of calcium and carbonate ions must be identical regardless of which approach is taken to compute them. In the former case we consider molal concentrations of the free (unpaired) ions (m_i^e), and their activity coefficients (γ_i^e values). In the latter case the calculation is made in terms of total molar concentrations of the ions (m_i^s) and their total ion activity coefficients (γ_i^s values). Because ion activities are the same, we know that for a given ion

$$a_i = \gamma_i^s\, m_i^s = \gamma_i^e\, m_i^e \tag{4.42}$$

You are given the following information for seawater of 34.8% salinity at 25°C from Pytkowicz (1983).

$$m_{Ca}^s = 0.01063 \text{ mol/kg} \qquad m_{CO_3}^s = 0.000171 \text{ mol/kg}$$

$$\gamma_{Ca}^s = 0.228 \qquad \gamma_{CO_3}^s = 0.029$$

Tabulated here are concentrations of free species and ion pairs expressed as percentages of the total concentrations given above assuming the ion pair model and effective ionic strength.

	Percent of total Ca		Percent of total CO_3
Ca^{2+}	88.35	CO_3^{2-}	8.0
$CaSO_4^\circ$	10.87	$NaCO_3^-$	16.0
$CaHCO_3^+$	0.29	$MgCO_3^\circ$	43.9
$CaCO_3^\circ$	0.41	$CaCO_3^\circ$	21.0
$CaMgCO_3^{2+}$	0.07	$Mg_2CO_3^{2+}$	7.4
CaF^+	0.02	$CaMgCO_3^{2+}$	3.8

Note that Pytkowicz suggests the presence of three, *triplet* carbonate ion pairs. These have not been considered by most other observers.

Given the above information, compute the ion activities of Ca^{2+} and CO_3^{2-} ions using both approaches, and from these values calculate the ion activity product of calcite in seawater. Tabulate and compare your results and explain their agreement. What is the significance of the total ion-activity coefficients used in the I_s approach?

Ion activities, a_i, are the same whether we use the "free" or "total" values of activity coefficients and concentrations to obtain them. Solution to the problem is based on Eq. (4.42). From the total calcium and the percent free value, we obtain the free calcium concentration:

$$m_{Ca}^e = (0.8835)\,(0.01063) = 9.39 \times 10^{-3} \tag{4.43}$$

Similarly, the concentration of free carbonate ion is

$$m_{CO_3}^e = (0.08)\,(1.71 \times 10^{-4}) = 1.37 \times 10^{-5} \tag{4.44}$$

We can now solve for the ion activity coefficients of the free ions.

$$\gamma_{Ca}^e = \frac{(0.228)\,(0.01063)}{9.39 \times 10^{-3}} = 0.258 \tag{4.45}$$

$$\gamma_{CO_3}^e = \frac{(0.029)\,(1.71 \times 10^{-4})}{1.37 \times 10^{-5}} = 0.362 \tag{4.46}$$

The ion activity product of calcite (IAP_c) based on total activity coefficients and concentrations is

$$IAP_c = (0.228)\,(1.063 \times 10^{-2})\,(0.029)\,(1.71 \times 10^{-4}) = 1.20 \times 10^{-8} = 10^{-7.92} \tag{4.47}$$

Based on the free ion values, the product is

$$IAP_c = (0.258)\,(9.39 \times 10^{-3})\,(0.362)\,(1.37 \times 10^{-5}) = 1.20 \times 10^{-8} = 10^{-7.92} \tag{4.48}$$

As expected, we have obtained an identical result.

4.2.5 High Ionic Strengths and the Pitzer Model

At ionic strengths above 2 to 3.5 mol/kg, the high density of ions in solution can lead to binary interactions between species of like charge and *ternary interactions* (simultaneous interactions) between three or more ions, some of which will be of like sign. The likelihood of binary and ternary interactions is suggested by the relative number of water and Na^+ and Cl^- ions associated in a liter of solution. Thus there are an equal number of water molecules and NaCl pairs in a 0.06 molar solution. Further, with 55.4 moles of H_2O in a liter of water, in a 5 molar NaCl solution about 30 moles of water are directly associated with ions of the salt and only 25.4 moles of water are free. Obviously, in such systems some ions must be touching, and dilute solution concepts do not apply. The most accepted model for high ionic strengths is the Pitzer model, which has been shown to accurately model the behavior of electrolyte solutions up to 6 mol/kg (cf. Pitzer 1987). While the Debye-Hückel model considers long-range electrostatic effects in dilute solutions, the Pitzer model also takes into account short range interactions in concentrated solutions.

The Pitzer model requires so-called "interaction parameters" involving the aqueous species of interest and major species in the water. Such parameters have been measured for major ions, but are often unavailable for trace species, including strong complexes. Sometimes the missing parameters can be reasonably assumed to be equal to those for similar species, however (cf. Langmuir and Melchior 1985).

Weare (1987) (see also Millero 1983) suggested the following conceptual equation to describe the activity coefficient of an individual ion in the Pitzer model approach

$$\ln \gamma_i = z_i^2 f^\gamma + \sum_i D_{ij}(I)\,m_j + \sum_{ijk} E_{ijk}\,m_j m_k + \cdots \tag{4.49}$$

The first term on the right is a modified Debye-Hückel term where

$$f^\gamma = -0.392\left[\frac{I^{1/2}}{1 + 1.2I^{1/2}} + \frac{2}{1.2}\ln\left(1 + 1.2I^{1/2}\right)\right] \tag{4.50}$$

The second term, the binary term, is the sum of interactions that involve two solution species of opposite or the same sign. The parameters required to calculate the effect of the second term on the ion activity coefficient are called binary virial coefficients. Those describing the interaction of species of opposite sign (B terms) are functions of ionic strength. Like-like sign species interaction (θ) terms are assumed independent of I. Binary interaction B terms are obtained from measurements of osmotic or activity coefficients in single electrolyte solutions. The Debye-Hückel (DH) and binary terms taken together are roughly comparable to the SIT and Truesdell-Jones model equations, which consider only binary interactions among species of opposite charge. The DH and binary terms define the chief contributions to an ion activity coefficient, while ternary interaction terms in the Pitzer model (the E_{ijk} terms) can be viewed as refinements to the DH and binary term contributions, generally not needed for major ions at ionic strengths below about 3.5 mol/kg. Within the ternary terms are ternary (ψ) virial coefficients that account for interactions among two like-charged and a third unlike-charged species. Ternary coefficients are assumed independent of ionic strength. They are obtained from measurements in mixed electrolyte solutions or from mineral solubility measurements in electrolyte solutions. In practice, when Pitzer equation expressions are written out, binary and ternary virial coefficients are often collected in the same terms.

Pitzer model equations are linear algebraic functions of $\ln \gamma_i$. They are often extremely long and involve numerous individual parameters and substitutions (cf. Pitzer 1987; Weare 1987). Tables of binary and ternary virial coefficients have been published by Pitzer (1987) and Weare (1987). A larger and comprehensive list of published virial coefficients is given by Plummer et al. (1988). Solution of Pitzer model equations is best accomplished by computer. The Pitzer model is available in the computer codes SOLMINEQ.88 (Kharaka et al. 1988), PHRQPITZ (Plummer et al. 1988), and PHREEQC (Parkhurst 1995). The extensive Pitzer-parameter data base in PHRQPITZ updates earlier lists and adds virial coefficients to describe the interaction of alkali metals, alkaline earths, transition metal cations, and NH_4^+, Cd^{2+}, Pb^{2+}, and UO_2^{2+} with a variety of ligands, including nitrate, sulfate, and bisulfate, borate, orthophosphate, arsenate, and chromate. The Pitzer model thus becomes useful in studies of concentrated, acid mine waters and industrial wastewaters. The model approach works best, however, when cation and anion interaction produces a complex with a formation constant less than about 10^2 (Weare 1987). The interaction effect is then accurately accounted for with model virial coefficients. More stable complexes need to be considered as separate, interacting species, and they require their own virial coefficients. Application of the Pitzer model to concentrated, mixed wastewaters, for example, has been somewhat limited because of the lack of interaction-parameter data for strong trace-element complexes. As a strategy in such cases, it is useful to first determine activity coefficients and activities of major electrolyte ions using the Pitzer model. The activity coefficients of trace species (for which virial coefficients are lacking) can then sometimes be set equal to those of major ions of the same size and charge (cf. Langmuir and Melchior 1985).

Example 4.4

Langmuir and Melchior (1985) examined the solubility of Ca, Sr, and Ba sulfates in saline groundwaters of the Wolfcamp Formation of north Texas. The Wolfcamp is composed of limestone and dolomite. The groundwater, which is at a depth of about 970 m. was at 32°C and 67 bars pressure and had a pH of 6.1. Species total concentrations are listed here.

Species	mol/kg	Species	mol/kg
Na	1.88	Ba	7.64×10^{-7}
K	3.62×10^{-3}	Ra	4.65×10^{-12}
Ca	0.188	Cl	2.53
Mg	0.123	SO$_4$	2.35×10^{-2}
Sr	1.68×10^{-3}	Br	5.61×10^{-3}

The water is dominantly a NaCl brine with important amounts of Ca and Mg. The analysis indicates a stoichiometric ionic strength of 2.88 mol/kg, too high to justify use of the Davies equation. Because solubility of the akaline earth sulfates is directly proportional to the activity coefficients of Ca^{2+}, Sr^{2+}, Ba^{2+}, and SO_4^{2-}, we are interested in their values in the brine. We will compute the activity coefficients using the Truesdell-Jones (TJ) and SIT models.

The TJ equation is

$$\log \gamma_i = \frac{-A z_i^2 \sqrt{I}}{1 + B a_i \sqrt{I}} + bI \tag{4.51}$$

The constants A and B in the Debye-Hückel term equal 0.5154 and 0.3294 respectively at 32°C. Values of a_i and b are listed in Table 4.3, which is for dominantly NaCl-type waters. We will use Parkhurst's (1990) values in the calculation.

For cations and anions in a predominantly sodium chloride solution, the activity coefficient equations in the SIT model are

$$\log \gamma_{\text{cation}} = -z^2 D + \epsilon \, mCl^-$$
$$\log \gamma_{\text{anion}} = -z^2 D + \epsilon \, mNa^+ \tag{4.52}$$

where the Debye-Hückel type term is

$$D = \frac{A\sqrt{I}}{1 + 1.5\sqrt{I}} \tag{4.53}$$

At $I = 2.88$ mol/kg, $D = 0.2467$. According to Table 4.4, $\epsilon = 0.14$ for Ca^{2+}, 0.07 for Ba^{2+} and -0.12 for SO_4^{2-}. We will assume ϵ for Sr^{2+} is the same as for Ca^{2+}. (Because the radius of Sr^{2+} [1.18 Å] is between those of Ca^{2+} [1.00 Å] and Ba^{2+} [1.35 Å], a better choice might be $\epsilon = 0.1$ for Sr^{2+}.) Resultant activity coefficients assuming $\epsilon_{Sr,Cl} = 0.14$ are listed below, along with values computed with the Truesdell-Jones equation.

	Truesdell-Jones model	SIT model	Pitzer model
$\gamma_{Ca^{2+}}$	0.310	0.233	0.237
$\gamma_{Sr^{2+}}$	0.286	0.233	0.204
$\gamma_{Ba^{2+}}$	0.187	0.155	0.156
$\gamma_{SO_4^{2-}}$	0.101	0.049	0.037

Source: From Langmuir and Melchior 1985

Also listed for comparison purposes are coefficients from Langmuir and Melchior (1985) based on the Pitzer model, which are probably the most accurate at these high ionic strengths. The SIT model results are mostly in good agreement with those obtained using the Pitzer model. If $\epsilon = 0.1$ for Sr^{2+} instead of 0.14, we obtain $\gamma_{Sr} = 0.185$, in slightly better agreement with the Pitzer result. Parkhurst (1990) notes that the TJ model accurately fits mean salt data generally up to about $I \approx 2$ mol/kg. That the TJ derived activity coefficients are systematically too large suggests inapplicability of the TJ model at this more elevated ionic strength.

Example 4.5

Following is a relatively simple calculation using the Pitzer model to compute the activity coefficient of HCO_3^- in seawater. The exercise is based largely on Millero (1983). (See also Harvie et al. 1984; Pitzer 1987). The activity coefficient of a trace cation in NaCl electrolyte solution can be written

$$\ln \gamma_M^T = z_A^2 f^\gamma + 2mNa(B_{MCl} + mNa\ C_{MCl}) + mNa\ mCl(B_{NaCl}^1 + C_{NaCl}) \\ + mNa(2\theta_{MNa} + mNa\ \psi_{MNaCl}) \tag{4.54}$$

For a trace anion in NaCl solution the expression is

$$\ln \gamma_A^T = z_A^2 f^\gamma + 2mCl(B_{NaA} + mCl\ C_{NaA}) + mNa\ mCl(B_{NaCl}^1 + C_{NaCl}) \\ + mCl(2\theta_{Cl,A} + mCl\ \psi_{Cl,A,Na}) \tag{4.55}$$

For trace bicarbonate ion this becomes

$$\ln \gamma_{HCO_3}^T = f^\gamma + 2mCl(B_{NaHCO_3} + mCl\ C_{NaHCO_3}) + mNa\ mCl(B_{NaCl}^1 + C_{NaCl}) \\ + mCl(2\theta_{ClHCO_3} + mCl\ \psi_{ClHCO_3Na}) \tag{4.56}$$

Among the virial coefficients in these equations, B and B^1 are functions of ionic strength, while C, θ, and ψ are independent of I. Millero (1983) gives general equations for B, B^1, and C for 1-1 (same as 2-1) and 2-2 electrolytes. For salt MX, the expressions are:

$$B_{MX} = \beta_{MX}^\circ + (\beta_{MX}^1/2I)\ [1 - (1 + 2I^{1/2})\exp(-2I^{1/2})] \tag{4.57}$$

$$B_{MX}^1 = (\beta_{MX}^1/2I^2)\ [-1 + (1 + 2I^{1/2} + 2I)\exp(-2I^{1/2})] \tag{4.58}$$

$$C_{MX} = C_{MX}^\phi/(2\ |\ Z_M Z_X\ |^{1/2}) \tag{4.59}$$

Virial coefficients needed to solve these equations are tabulated below.

	β_{MX}°	β_{MX}^1	C_{MX}^ϕ	θ	ψ
Na, Cl	0.0765	0.2644	0.00127		
Na, HCO_3	0.0277	0.0411			
Cl, HCO_3				0.0359	
Cl, HCO_3, Na					−0.0143

Given $I = 0.718$ mol/kg, $mNa = 0.48$, and $mCl = 0.56$ mol/kg for seawater. From Eq. (4.52), $f^\gamma = -0.6230$. Also, $B_{NaCl}^1 = -0.06190$, $B_{NaHCO_3} = 0.04216$, and $C_{NaCl} = 0.00635$. Since a value for $C_{NaHCO_3}^\phi$ is lacking, we set it equal to zero and find $C_{NaHCO_3} = 0$. Solving for the four terms in Eq. (4.56) leads to

$$\ln \gamma_{HCO_3}^T = -0.6230 + 0.0472 - 0.1673 + 0.0357 = -0.7074 \tag{4.60}$$

The first (−) term on the right is the Debye-Hückel (DH) term for long-range electrostatic effects. The second (+) term defines electrolyte trace-ion (Na–HCO_3) interactions. The third term is a Na–Cl electrolyte term, and the fourth a small correction due to binary and ternary trace-ion (HCO_3)–NaCl mixing. Expressing Eq. (4.60) in linear format we have

$$\gamma^T_{HCO_3} = \gamma_{DH} \quad \gamma_{NaHCO_3} \quad \gamma_{NaCl} \quad \gamma_{mixing}$$

$$0.493 = 0.536 \times 1.048 \times 0.846 \times 1.036$$

(4.61)

Clearly, the chief contribution to $\gamma^T_{HCO_3}$ is electrostatic effects as described by the DH term. At the relatively low ionic strength of seawater the other effects combined change $\gamma^T_{HCO_3}$ by less than 10%.

4.3 OVERVIEW OF ACTIVITY COEFFICIENT MODELS FOR IONS

The algorithms that describe the change in ion activity coefficients (γ_i) with ionic strength all relate the logarithm of γ_i to a function of $I^{1/2}$ with or without additional terms. In generalizing the applicability of such models it is, therefore, helpful to do so with a schematic plot of log γ_i versus $I^{1/2}$. Such a plot is shown in Fig. 4.5 for hypothetical cation, M^{2+}.

Up to about 0.02 mol/kg ionic strength ions are so far apart in solution that their interaction is accurately measured by their charge only, and the Debye-Hückel (DH) limiting law (Eq. 4.30) applies. The limiting law appears as a tangent (dashed line) to the log γ_i versus $I^{1/2}$ curve in Fig. 4.5. As salt concentrations rise, ion size becomes significant, and the extended DH equation (Eq. 4.28) and a hydrated-ion size are needed to accurately model γ_i values. The size parameter increases the denominator of the extended DH equation, but maintains the negative slope of the plot. Between about 0.01 and 0.1 mol/kg ionic strength, increased solution concentrations result in significant interaction between anions and cations and the curvature of a γ_i versus log I plot turns upward (see Fig. 4.4). This produces a minimum followed by an increasingly positive slope in the plot in Fig. 4.5, neither of which are predicted by the extended DH equation.

The Davies equation (Eq. 4.31) generates the positive change in slope with an add-on term, bI, where b is the same constant for all ions. The denominator of the Davies equation equals $1 + \sqrt{I}$, which is equivalent to assigning a constant a_i value of about 3.0 to all ions in the extended DH equation. These simplifications make the Davies equation less accurate than the extended Debye-Hückel equation at low ionic strengths, and limit its use to ionic strengths below that of seawater (0.7 mol/kg).

The TJ equation (Eq. 4.51) improves on shortcomings of the Davies equation by retaining the extended DH equation with its a_i parameter and assigning different values to b in the add-on bI term based on the MacInnes convention and on fitting mean salt data for the ions. The TJ equation has been found reliable up to $I = 3.5$ mol/kg in some cases, but is conservatively limited to $I \leq 2$ mol/kg.

The denominator of the DH term in the SIT model (Eqs. 4.32 and 4.33) equals $1 + 1.5\sqrt{I}$, which is equivalent to assuming $a_i \approx 4.6$ and constant in the extended DH equation. The SIT model accounts for binary interactions of anions and cations with interaction parameters that are specific for individual ions in chloride or other media. Lacking an adjustable ion size parameter, the SIT model is less accurate than the extended DH model at low ionic strengths. It does, however, give accurate results in some solutions up to $I = 3.5$ mol/kg.

At ionic strengths between about 2 and 3.5 mol/kg the high salt concentrations lead to further interaction between anion and cation pairs, but also to important interaction between pairs of ions of the same charge and simultaneously between three ions where two have the same charge. The Pitzer model can accurately account for such effects with a complex series of interaction terms added to a

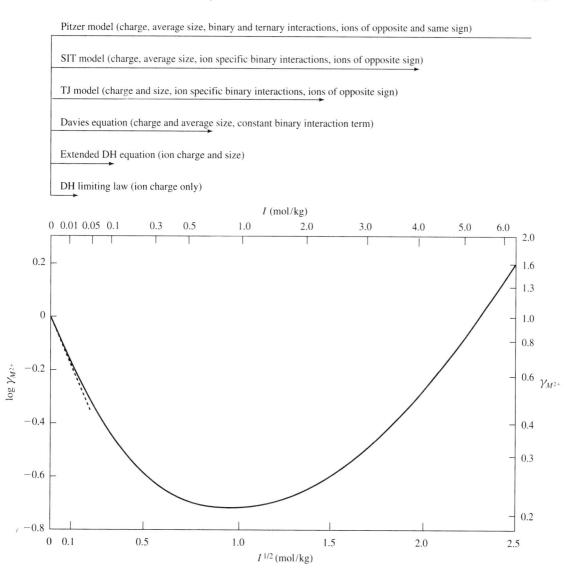

Figure 4.5 Schematic plot showing the general applicability of different activity coefficient models as a function of ionic strength for a divalent cation. The dashed tangent to the curve at its origin is a plot of the Debye-Hückel limiting law for the ion.

DH-type, low ionic-strength term. Because the DH-type term lacks an ion size parameter, the Pitzer model is also less accurate than the extended DH equation in dilute solutions. However, assuming the necessary interaction parameters (virial coefficients) have been measured in concentrated salt solutions, the model can accurately model ion activity coefficients and thus mineral solubilities in the most concentrated of brines.

4.4 ACTIVITY COEFFICIENTS OF MOLECULAR SPECIES

Activity coefficients for uncharged, molecular species generally obey the empirical Setchenow equation up to high ionic strengths (cf. Lewis and Randall 1961; Miller and Schreiber 1982). Such species include dissolved gases, weak acids, and molecular organic species. The Setchenow equation is

$$\log \gamma_i = K_i \, I \tag{4.62}$$

where K_i is a constant, generally ranging from 0.02 to 0.23 at 25°C. Aqueous species whose activity coefficients obey this equation include: CO_2, H_2S, N_2, O_2, CH_3COOH (acetic acid), NH_3, H_3PO_4, and $H_4SiO_4^\circ$. In NaCl solutions K_i equals 0.231 for $H_2CO_3^\circ$ (Millero 1983) and 0.080 for $H_4SiO_4^\circ$ (Marshall and Chen 1982). Thus, based on the Setchenow equation, in seawater ($I = 0.7$) the activity coefficients of carbonic and silicic acids equal 1.18 and 1.06, respectively.

Values of K_i for molecular species in NaCl solutions at 25°C are given in Table 4.5. Pytkowicz (1983) lists additional K_i values for seawater. MINTEQA2 assumes $K_i = 0.1$ for all uncharged species. The largest K_i values in Table 4.5 equal about 0.2 for several species. In fresh, potable waters (TDS < 500 ppm, $I < 0.01$ m), the activity coefficients of these species still equal 1.00. Even in brackish waters with TDS values of about 5000 ppm ($I \approx 0.1$ mol/kg), for $K_i = 0.2$, molecular species activity coefficients equal 1.02. Thus, to a good approximation the γ_i of such species can be taken equal to unity in fresh and brackish waters.

Plotted in Fig. 4.6 is the general behavior of activity coefficients of ions and molecular species with increasing ionic strength. Note that the former decrease and the latter increase with ionic strength. Remembering the definition, $a_i = \gamma_i m_i$, and given that activities are often fixed by a chemical reaction at equilibrium, it is clear that if activity coefficients increase, dissolved concentrations must decrease, and conversely. Thus, because the γ_i value of molecular species increases with I, m_i values must decline. For this reason molecular species, such as gases at a fixed partial pressure, become less soluble with increasing I. This is the *salting out* effect. Conversely, for a fixed ion activity, because γ_i values decrease with I, the m_i values for ions increase with I. Thus ionic species are

TABLE 4.5 Values of the salting out coefficient, K_i, where $\log \gamma_i = K_i I$, for some molecular species in NaCl solutions at 25°C based on various sources

Aqueous species	K_i	Source
H_2	0.094	†
O_2	0.132	†
C_2H_4(ethylene)	0.093	†
C_6H_5COOH(benzoic acid)	0.191	†
$C_6H_4(OH)COOH$(salicylic acid)	0.196	†
CH_3COOH(acetic acid)	0.066	†
CH_4	0.129	‡
N_2	0.20	‡
$H_2CO_3^\circ$	0.231	§
$H_4SiO_4^\circ$	0.080	‖
H_2S	0.020	§
$H_3PO_4^\circ$	0.052	§
NH_3	0.036	§
$H_3BO_3^\circ$	0.045	§

Source: †Harned and Owen (1958); ‡Millero and Schrieber (1982); §Millero (1983); ‖Marshall and Chen (1982).

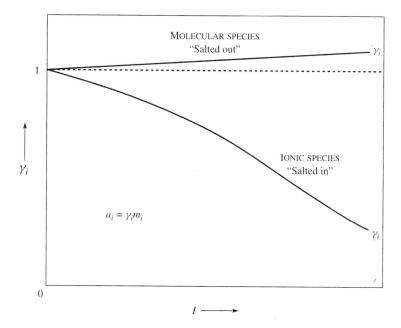

Figure 4.6   Schematic plot showing the "salting in" ($\gamma < 1$) and "salting out" ($\gamma > 1$) behavior of ionic and molecular species.

salted in. Because of this effect, the solubilities of minerals such as calcite that dissolve to form ionic species, increase with increasing ionic strengths, whereas a mineral such as quartz, with a molecular dissolution product ($H_4SiO_4^\circ$), becomes less soluble as ionic strength increases.

STUDY QUESTIONS

1. Define ionic strength. Explain how it is computed and how it may be estimated from the specific conductance or total dissolved solids content of a water.

2. Define the mean ion activity coefficient of a salt and comment on its significance in a weak versus a strong electrolyte solution.

3. What is the MacInnes convention, what is its justification, and how is it used to estimate individual ion activity coefficients from mean salt data?

4. Describe the general behavior of ion activity coefficients with ionic strength as a function of ion charge and size.

5. What are the Debye-Hückel limiting law and the extended Debye-Hückel equation and under what general conditions can they be used to compute ion activity coefficients? Discuss the meaning and use of the ion size parameter in the Debye-Hückel equation. How is it related to the ionic potential?

6. How can we estimate the ion size parameter in the Debye-Hückel equation from mean salt data?

7. Davies, Truesdell-Jones (TJ), specific ion interaction theory (SIT), and Pitzer equations are formulated to model ion activity coefficients, both in dilute solutions and at elevated ionic strengths where ionic species tend to be salted out (they become less soluble). Thus, each of these models for ln γ_i includes a Debye-Hückel term or terms of net negative sign (salting in terms) and an add-on term or terms that overall have a net positive sign. Discuss and compare the accuracy of the Davies and TJ equations at low and high ionic strengths.

8. The stoichiometric or total ionic strength (I_s) is computed ignoring solution ion pairs, whereas the effective ionic strength (I_e) takes ion pairs into account. I_s is greater than I_e. Why? Ion activities (a_i) must be the same, whichever ionic strength model is used, so that $a_i = \gamma_t m_t = \gamma_f m_f$, where the subscript t denotes total ion values that apply when I_s is used. Subscript f indicates free ion values should be used when I_e is employed. Explain and contrast the two approaches with an example.

9. The Pitzer model equation for an ion activity coefficient includes a Debye-Hückel–type term to describe long-range electrostatic effects, and a series of add-on terms to describe nearer-neighbor contributions to the activity coefficient. Some of the binary add-on terms for interactions between two ions of opposite sign (the B terms), are functions of ionic strength. Other binary interaction terms between ions of opposite sign (the C terms) and ions of the same sign (the θ terms) are constant and independent of ionic strength, as are the ternary interaction (ψ) terms. Comment on the form of the Pitzer model equation and its relevant interaction terms for the activity coefficients of Na and Cl in a pure NaCl solution and of trace HCO_3^- in a NaCl solution.

10. The stoichiometric activity coefficient of bicarbonate ion is less in a 0.7 mol/kg solution of $NaHCO_3$ than it is in a 0.7 mol/kg solution of NaCl. Why might this be the case?

11. Show how salting out and salting in apply to the solubility of a gas such as oxygen and a mineral such as calcite as ionic strength increases.

PROBLEMS

1. This problem deals with the general treatment of a water analysis, including the calculation of its charge balance, ionic strength, ion activity coefficients and activities, and saturation state with respect to calcite.

   The following analysis is of the Iowa River at Iowa City, Iowa. The specific conductance is 365 μmhos/cm. Assume the water temperature is 25°C. TDS is the solid residue on evaporation.

Species	Concentration (mg/L)	Species	Concentration (mg/L)
SiO_2	15	Cl	4.3
Ca	49	F	0.2
Mg	14	NO_3	14
Na	5.4	TDS	251
K	3.1	Hardness as $CaCO_3$	180
HCO_3	168	Noncarbonate hardness	42
SO_4	40		

(a) Compute the total equivalents per million (epm) of cations and anions and compare them by calculating the charge balance,

$$\text{charge balance (\%)} = \frac{|\Sigma\text{epm cations} - \Sigma\text{epm anions}|}{\Sigma\text{epm cations} + \Sigma\text{epm anions}} \times 100$$

where the numerator on the right side is always assumed positive.

(b) As an approximation, it has been found roughly that for a given water $100 \times (\Sigma\text{epm cations}) = $ Specific conductance (μmhos/cm at 25°C). How accurate is the approximation for this water?

(c) What is the molal ionic strength of the water and what factor (K) relates it to the specific conductance, where $I(\text{molal}) = K \times \text{SpC}$ (μmhos/cm)? Is the value of K you have determined consistent with such values given in this chapter?

(d) Compute Debye-Hückel ion activity coefficients for the ions Ca^{2+} and HCO_3^-.

(e) Given that $K_{sp} = 10^{1.85}$ at 25°C for the reaction: $CaCO_3 + H^+ = Ca^{2+} + HCO_3^-$, and assuming the river is at saturation with respect to calcite, compute what the pH of the water should be, using ion activities in your calculation but ignoring complexes.

(f) Assuming the pH computed in (e) is correct, what is the CO_2 pressure (in bars) that is in equilibrium with the water? Remember that $K_{CO_2} = 10^{-1.47}$ and, for carbonic acid, $K_1 = 10^{-6.35}$ and $K_2 = 10^{-10.33}$ at 25°C.

(g) Given that the Henry's law constant for the solubility of oxygen in the water at 25°C is $K_H = 1.26 \times 10^{-3}$ M/bar, compute the concentration of O_2 in the river in equilibrium with air at 25°C. Remember that the air is 21% O_2.

2. Goldberg (1979) reevaluated the published mean ion activity coefficient data for UO_2Cl_2, which has been measured at and above a salt concentration of 0.1 mol/kg (cf. Robinson and Stokes 1970). He then fit a correlation equaiton to the reevaluated data for $\gamma_{\pm UO_2Cl_2}$, and extrapolated the data down to a salt molality of 0.001 mol/kg. Values of $\gamma_{\pm UO_2Cl_2}$, as a function of mUO_2Cl_2 from Goldberg's (1979) study, are tabulated in the table that accompanies this problem. Also shown in the table are $\gamma_{\pm}$ values for KCl as a function of salt molality from Hamer and Wu (1972). These are based on a reevaluation of published mean salt data for KCl measured at and above 0.1 mol/kg, and a similar fitting of that data with a correlation equation that has been extrapolated down to mKCl = 0.001.

(a) Using the tabulated mean salt data and assuming the MacInnes convention, calculate and plot the individual ion activity coefficient of uranyl ion over the range of ionic strengths for which the mean activity coefficients of both salts are available. Do not forget that the mean salt data for KCl and UO_2Cl_2 must be compared at the *same ionic strength* for each salt, not at the same salt molalities. You may have to graph the $\gamma_{\pm}$ data for KCl and interpolate or extrapolate that data in order to obtain accurate values for both salts at the same ionic strengths. If you are plotting your resutls on graph paper, four-cycle log-linear paper is the most appropriate choice.

(b) Arthur E. Martell suggested that the ion size parameter (a_i) for uranyl ion (UO_2^{2+}) in the Debye-Hückel equation equals 7.35 Å. Assuming this value, calculate $\gamma_{UO_2^{2+}}$ for $I = 0.001$ and 0.003 mol/kg using the Debye-Hückel equation and plot your results on the graph showing $\gamma_{UO_2^{2+}}$ derived from the mean salt data.

(c) If the Debye-Hückel and mean salt derived curves for $\gamma_{UO_2^{2+}}$, do not smoothly connect, adjust the ion size paramet for uranyl ion until they do, emphasizing extrapolation of the Debye-Hückel curve to the lowest ionic-strength mean salt values. Report your result for the size parameter, replot the Debye-Hückel activity coefficient if necessary, and draw a single curve to make the Debye-Hückel values smoothly connect to the curve of the mean salt data.

(d) Assuming the ion size value estimated above, what value (or values) of (b) in the TJ equation gives the best fit to the mean salt derived activity coefficients at $I = 0.06$, 0.6, and 3.0 mol/kg?

mUO_2Cl_2	$\gamma_{\pm UO_2Cl_2}$	mUO_2Cl_2	$\gamma_{\pm UO_2Cl_2}$	mKCl	$\gamma_{\pm KCl}$	mKCl	$\gamma_{\pm KCl}$
0.001	0.8885	0.500	0.5009	0.001	0.9649	1.200	0.5937
0.003	0.8245	0.600	0.5156	0.002	0.9515	1.400	0.5861
0.005	0.7871	0.700	0.5335	0.005	0.9266	1.600	0.5803
0.010	0.7293	0.800	0.5540	0.010	0.9011	1.800	0.5759
0.020	0.6662	0.900	0.5766	0.020	0.8689	2.000	0.5726
0.030	0.6285	1.000	0.6013	0.050	0.8155	2.500	0.5684
0.040	0.6024	1.250	0.6712	0.100	0.7682	3.000	0.5682
0.050	0.5827	1.500	0.7522	0.200	0.7170	3.500	0.5712
0.060	0.5674	1.750	0.8443	0.300	0.6865	4.000	0.5765
0.070	0.5549	2.000	0.9479	0.400	0.6653	4.500	0.5838
0.080	0.5447	2.250	1.0634	0.500	0.6492	4.800	0.5880
0.090	0.5360	2.500	1.1913	0.600	0.6365		
0.100	0.5287	2.750	1.3320	0.700	0.6261		
0.200	0.4929	3.000	1.4868	0.800	0.6174		
0.300	0.4859	3.174	1.6026	0.900	0.6101		
0.400	0.4903			1.000	0.6038		

   (e) On the same plot draw the curve for $\gamma_{UO_2^{2+}}$ based on the specific ion interaction (SIT) approach in Table 4.4 and comment on the fit of this curve to the mean salt derived curve.

   (f) The uranyl ion forms complexes with chloride. Cumulative constants for the formation of UO_2Cl^+ and $UO_2Cl_2^0$ are $10^{0.17}$ and $10^{-1.1}$, respectively. Above what chloride concentration is 1% and 10% of the uranyl ion complexed? How does such complexation affect our calculation of the activity coefficient of uranyl ion?

3. Mean salt data indicate that $\gamma_{\pm FeCl_2} = 0.4427$ in a 0.5 molal $FeCl_2$ salt solution at 25°C and that $\gamma_{\pm KCl} = 0.5830$ in a KCl solution of the same ionic strength.

   (a) What is the ionic strength?

   (b) Assuming the MacInnes convention, what is $\gamma_{Fe^{2+}}$ at this ionic strength?

   (c) Estimate the interaction parameter $\epsilon_{Fe,Cl}$ for the SIT model from parameters listed for other divalent transition metals in Table 4.4, assuming the parameters are proportional to the ionic radii.

   (d) Using $\epsilon_{Fe,Cl}$ estimated in (c), compute $\gamma_{Fe^{2+}}$ with the SIT model for the same ionic strength. How does this value compare with the value computed in (b)?

   (e) Assuming $\gamma_{Fe^{2+}}$ computed in (b) is correct, what value of $\epsilon_{Fe,Cl}$ in the SIT model would have been required for the models to agree?

4. The solubility of halite (NaCl) in water at 25°C and 1 bar pressure is 6.15 mol/kg.

   (a) Using Pitzer equations from Millero (1983) and single electrolyte solution parameters from text Example 4.5, calculate the activity coefficients of Na^+ and Cl^- in a brine that contains only these ions and that is saturated with respect to halite. (In this simple system many of the virial coefficients equal zero.) Compare γ_{Na^+} and γ_{Cl^-} and discuss the relative contributions of Debye-Hückel and binary interaction terms to their values.

      The applicable Pitzer equations for the activity coefficients of cation M and anion X from Millero (1983) are given by

$$\ln \gamma_M = z_M^2 f^\gamma + 2\sum_a m_a(B_{Ma} + EC_{Ma}) + z_M^2 \sum_c \sum_a m_c m_a B'_{ca} + z_M \sum_c \sum_a m_c m_a C_{ca}$$

$$\ln \gamma_X = z_X^2 f^\gamma + 2\sum_c m_c(B_{cX} + EC_{cX}) + z_X^2 \sum_c \sum_a m_c m_a B'_{ca} + z_X \sum_c \sum_a m_c m_a C_{ca}$$

(where z_i is the charge, m_i the molality of cation (c) and anion (a) in the mixed solution, and $E = \frac{1}{2}\Sigma_i m_i |z_i|$. The Debye-Hückel term is defined in Eq. (4.50). The second and third virial coefficients for 1-1 and 2-1 electrolytes equal

$$B_{MX} = \beta_{MX}^\circ + (\beta_{MX}^1/2I)\left[1 - (1 + 2I^{1/2})\exp(-2I^{1/2})\right]$$

$$B'_{MX} = (\beta_{MX}^1/2I^2)\left[-1 + (1 + 2I^{1/2} + 2I)\exp(-2I^{1/2})\right]$$

$$C_{MX} = C_{MX}^\phi/(2 \mid z_M z_X \mid^{1/2})$$

The β and C interaction terms for NaCl are tabulated in the text.

   (b) Compute the solubility product (K_{sp}) of halite from the foregoing information. What is the total dissolved solids (TDS) content in ppm of a brine at saturation with respect to halite?

   (c) With the following Gibbs free-energy data, compute the solubility product of halite. Compare this result to the value you computed in (b) and comment on any difference.

Species	ΔG_f°(kcal/mol)	Species	ΔG_f°(kcal/mol)
Na^+	-62.59	NaCl(c) halite	-91.815
Cl^-	-31.379		

5

Acids and Bases

5.1. THE SIGNIFICANCE AND MEASUREMENT OF pH

The importance of water in aqueous environments reflects the behavior of hydrogen and hydroxyl ions, as well as that of substance water itself as a solvent and reactant. As a weak acid or base, the dissociation of water may be written:

$$H_2O = H^+ + OH^- \tag{5.1}$$

for which at 25°C

$$K_w = \frac{[H^+][OH^-]}{[H_2O]} = 10^{-14.00} \tag{5.2}$$

In most aqueous systems, except brines, the activity of water is close to unity and can be ignored. (See Chap. 1.)

The activity of a hydrogen ion is generally measured and reported in pH units. The pH is defined as the negative, base-10 logarithm of the hydrogen-ion activity, or

$$pH = -\log [H^+] \tag{5.3}$$

Most reactions in gas/water/rock systems involve or are controlled by the pH of the system. Among these are:

1. Aqueous acid-base equilibria, including hydrolysis and polymerization.
2. Adsorption, because protons compete with cations and hydroxyl ions compete with anions for adsorption sites. Also, the surface charge of most minerals is pH dependent.
3. The formation of metal-ligand complexes, because protons compete with metal ions to bond with weak-acid anions, and OH^- competes with other ligands that would form complexes.
4. Oxidation-reduction reactions, whether abiological or biologically mediated. Oxidation usually produces protons, whereas reduction consumes them.

5. The solubility and rate of dissolution of most minerals is strongly pH-dependent. Weathering of carbonate, silicate, and alumino-silicate minerals consumes protons and releases metal cations. (See Chaps. 7 and 9.)

The ultimate reference for pH (and Eh) measurement is the hydrogen electrode, which is formed by bubbling $H_2(g)$ over a platinum electrode surface (cf. Parsons 1985, and Chap 2). The electrode half-cell reaction is

$$\tfrac{1}{2}H_2(g) = H^+ + e^- \tag{5.4}$$

for which the measured electrode potential E, equals

$$E = E^\circ - \frac{RT}{nF} \ln \frac{[H^+]}{[P_{H_2}]^{1/2}} = E^\circ - \frac{2.303RT}{nF} \log \frac{[H^+]}{[P_{H_2}]^{1/2}} \tag{5.5}$$

where the gas constant $R = 1.9872$ cal/deg mol, the Faraday $(F) = 23,061$ cal/V g eq., $n = 1$ electron. E° is the standard electrode potential with $E^\circ = 0$ (by definition), thus $\Delta G_r^\circ = nFE^\circ = 0$. (See Chap. 11.) The quantity $(2.303RT/F)$ is called the Nernst factor and equals 54.2, 59.16, and 64.12 mV at 0, 25, and 50°C. This is the theoretical response of the electrode in volts to a one-unit increase in pH.

Although the hydrogen electrode is the ultimate reference for pH, it is unsuitable for general pH measurement. For convenience the pH of natural waters is measured using a glass electrode in conjunction with a reference electrode (Fig. 5.1). The reference electrode is most often a silver–silver chloride (Ag–AgCl) or calomel (Hg–Hg$_2$Cl$_2$) electrode. The unknown solution pH is determined by comparing the potential measured in the unknown to the potential measured in a solution of known pH, called a pH buffer solution. The pH of the unknown is then obtained from the relationship

$$pH_x = pH_b + \frac{(E_x - E_b)}{1.984 \times 10^{-4}\,T} \tag{5.6}$$

where pH_x and pH_b are the pH measured in the solution of interest, and in the pH buffer solution, and E_x and E_b are the corresponding voltages measured with the glass and reference electrode pair and an electrometer. The term $1.984 \times 10^{-4}\,T = 2.303RT/F$ is the Nernst factor. At 25°C, expression (5.6) reduces to

$$pH_x = pH_b + \frac{(E_x - E_b)}{0.05916} \tag{5.7}$$

Expression (5.7) is, in fact, built into the routine measurement of pH with a commercial pH meter. Such meters have a scale graduated in pH units, obviating the need to convert volts to pH units. Prior to pH measurement, the buffer solution and unknown solution are brought to the same temperature. The temperature compensation dial on the meter is then turned to that temperature. This sets the meter with the appropriate Nernst-factor value for pH response at the temperature of measurement. The pH and reference electrode pair (preferably available in a single, combination pH electrode) are then immersed in a buffer solution of known pH, and the meter is set to that pH. Immersion of the electrodes in the unknown solution then yields that solution's pH.

In a pH measurement, the silver chloride or calomel internal couple of the reference electrode contributes a voltage, E_{ref}, to the measured potential at the meter. The reference electrode makes electrical contact with the buffer and unknown solutions into which it is immersed via a salt bridge that leaks a concentrated KCl electrolyte solution through an outer electrode orifice into those solutions. This leaking salt bridge, which is called a liquid junction, creates a voltage called a liquid-junction

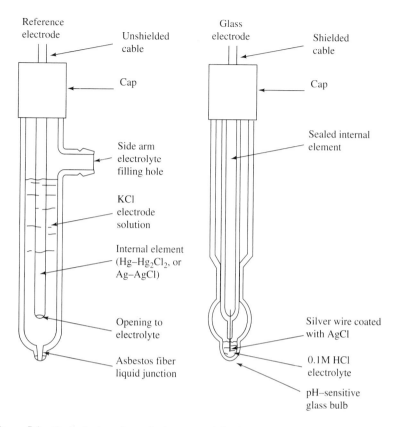

Figure 5.1   Basic design of a typical commercial reference electrode and glass electrode. After D. Langmuir, Eh-pH determination, in *Procedures in Sedimentary Petrology*, R. E. Carver, ed. Copyright © 1971 by John Wiley & Sons, Inc. Used by permission of John Wiley & Sons, Inc.

potential, E_j. Inherent in pH measurement is the assumption that voltages such as E_{ref} and E_j are the same when measurements are made in the unknown and buffer solutions, so that these voltages cancel out in Eq. (5.6). This assumption is particularly suspect for E_j, which depends on constant and nearly equal diffusion rates of K^+ and Cl^- ions and the dominance of these ions at the salt bridge. E_j can differ by tens of millivolts between a buffer and unknown solution. Such differences in E_j make accurate pH measurement difficult in brines in low pH solutions (cf. Bates 1964) and in clay suspensions due to adsorption of K^+, but not Cl^-, by the clay.

    Although the precision (repeatability) of pH measurements using modern equipment may be ±0.005 units in the laboratory, the accuracy of lab measurements can be no better than about ±0.02 pH units. This is because the buffer solutions, which have been calibrated by the U.S. National Bureau of Standards (NBS) using a hydrogen electrode, have an absolute accuracy in the range of ±0.01 to 0.02 pH units. Under ideal field conditions (similar air and water temperatures, measuring equipment unaffected by wind, sun, or electrical machinery) the accuracy of field pH measurements (as evidenced by the reproducibility of pH buffer checks and measurements) may be within ±0.02 pH units, but is generally no better than about ±0.05 pH units. For greatest accuracy in the laboratory or field, pH equipment should be calibrated in two NBS-certified buffers having nominal pH values above and below that of the unknown solution before determining the pH of the unknown.

5.2 ACIDS AND BASES: OVERVIEW

The ionic-potential (Ip) concept discussed in Chap. 3 best describes element behavior at intermediate pH's. Most of the metal oxides and hydroxide minerals formed by cations with Ip values between about 3 and 8.2 become significantly soluble in acid waters, where the high H^+ concentrations can break metal-O or metal-OH bonds to form water and release metal cations to solution. Some of these otherwise insoluble metal oxides and hydroxides (for example, those of Al^{3+} and Fe^{3+}) are also solubilized by high pH's. In other words, the mobilities of these metal cations depend on the acid-base properties of the water, as well as on their ionic potentials. The rate of chemical weathering is also greatly accelerated in strongly acid waters.

Some definitions are appropriate here. In 1928, Bronsted and Lowry defined an *acid* as a substance that can give or donate a proton and a *base* as a substance that can receive or accept a proton (Butler 1964). Strong acids release a greater proportion of their protons than do weak acids. For example, above pH 4, the reaction

$$H_2SO_4^\circ = 2H^+ + SO_4^{2-} \tag{5.8}$$

and, in general, the reaction

$$HCl^\circ = H^+ + Cl^- \tag{5.9}$$

have gone completely to the right, and essentially none of the neutral acid species remain. Sulfuric and hydrochloric thus are strong acids. In contrast, the reaction

$$H_2CO_3^\circ = H^+ + HCO_3^- \tag{5.10}$$

only advances part way to the right, so that some undissociated carbonic acid remains at equilibrium. Note that in the above examples, sulfuric, hydrochloric, and carbonic are acids, and sulfate, chloride, and bicarbonate ions are bases.

In general, we can write the equilibrium expression for the dissociation of an acid as

$$K_a = \frac{[H^+] [A^n]}{[HA^{n+1}]} \tag{5.11}$$

Values of $pK_a = -\log K_a$ at 25°C for some important dissolved acids and bases in natural waters are given in Table 5.1. In the table Fe^{3+} and Al^{3+} have been written as aquocomplexes, to show the role of water in their acid-base behavior. The convention is generally not to write the waters of hydration. Notice that the species HSO_4^-, $H_2PO_4^-$, $FeOH(H_2O)_5^{2+}$, HCO_3^-, HPO_4^{2-}, and $H_3SiO_4^-$ can act either as acids or as bases. Thus they are called *ampholytes* or *amphiprotic*.

In natural waters the principal *strong acids* (pH's < 4) are sulfuric and hydrochloric acids, bisulfate ion, ferric ion, and $FeOH^{2+}$ and $Fe(OH)_2^+$ ions. The pH of most natural waters lies between 4 and 8.5, under which conditions the acids present are *weak*, and include organic acids, Al^{3+}, carbonic acid, and bicarbonate ion.

Acetic acid is a common organic acid found in landfill leachates, sewage, and deep oil-field brines, for example. Formic acid is also found in groundwaters associated with hydrocarbons. The most common natural organic acids are called humic and fulvic acid. These acids are described in some detail in Section 5.4.

The pH of seawater, which is about 8.15, chiefly reflects the presence of bicarbonate and carbonate and, to less extent, boric acid species. (B = 4.5 ppm in seawater [Goldberg et al. 1971].)

Values of pH above 8.5 are rare in groundwaters, but may be found in a few springs discharging Mg–OH and Ca–OH (strong bases) waters from serpentine (hydrated Mg-silicate) rocks in the coastal mountain ranges of California and Alaska (cf. Barnes 1970), and from ultramafic rocks of

TABLE 5.1 Some important acids (HA^{n+1}) and their conjugate bases (A^n) in natural waters, and pK_a ($-\log K_a$) values for the acids, where $K_a = [H^+][A^n]/[HA^{n+1}]$

Acid	Base	pK_a
HCl°	Cl^-	~ -3
$H_2SO_4^{\circ}$	HSO_4^-	~ -3
HNO_3°	NO_3^-	~ -0
HSO_4^-	SO_4^{2-}	1.99
$H_3PO_4^{\circ}$	$H_2PO_4^-$	2.15
$Fe(H_2O)_6^{3+}$	$FeOH(H_2O)_5^{2+}$	2.19
HF°	F^-	3.18
$FeOH(H_2O)_5^{2+}$	$Fe(OH)_2(H_2O)_4^+$	3.48
$HCOOH^{\circ}$ (formic)	$COOH^-$ (formate)	3.75
CH_3COOH° (acetic)	CH_3COO^- (acetate)	4.76
$Al(H_2O)_6^{3+}$	$AlOH(H_2O)_5^{2+}$	5.00
$H_2CO_3^{\circ}$	HCO_3^-	6.35
H_2S°	HS^-	7.03
$H_2PO_4^-$	HPO_4^{2-}	7.20
NH_4^+	NH_3°	9.24
$H_3BO_3^{\circ}$	$H_2BO_3^-$	9.24
$H_4SiO_4^{\circ}$	$H_3SiO_4^-$	9.82
HCO_3^-	CO_3^{2-}	10.33
HPO_4^{2-}	PO_4^{3-}	12.35
$H_3SiO_4^-$	$H_2SiO_4^{2-}$	13.10
H_2O	OH^-	14.00
HS^-	S^{2-}	(18.51)

Note: A conjugate base is the base formed when an acid gives or donates a proton. Parenthetic values are estimates.

Northern Oman (Neal and Stanger 1985). pH's around 10 are also found in evaporitic lakes, high in Na carbonate and bicarbonate concentrations, and in streams and lakes clogged with active photo-synthetic aquatic plants (cf. Berner and Berner 1987). Detailed discussion of these and other processes that affect the $CO_2(aq)$ or carbonic-acid content of surface- and groundwaters, and thus the pH, is presented next.

5.3 CARBON DIOXIDE AND CARBONIC ACID SPECIES IN NATURAL WATERS

5.3.1 Theoretical Relationships

Carbonic acid is the most abundant acid in natural water systems and is the acid most responsible for rock weathering. Bicarbonate ion is generally the dominant anion in fresh surface- and ground-waters. Bicarbonate and carbonate ions are also the chief contributors to total alkalinity in natural waters (see below). For such reasons we will consider carbonate-solution chemistry in some detail.

The reaction

$$CO_2(aq) + H_2O = H_2CO_3^* \tag{5.12}$$

where $H_2CO_3^*$ is the *true* concentration of carbonic acid, has an equilibrium constant of

$$K_{eq} = \frac{[H_2CO_3^*]}{[CO_2(aq)]} = 2.6 \times 10^{-3} \tag{5.13}$$

at 25°C (Butler 1964). In other words, the concentration of true carbonic acid is less than 0.3% of the $CO_2(aq)$ present. Nevertheless, it is conventional to designate the dissolved carbon dioxide plus true carbonic acid as the species $H_2CO_3^\circ$. We can then write the dissolution of CO_2 gas in water as

$$CO_2(g) + H_2O = H_2CO_3^\circ \tag{5.14}$$

for which the Henry's law expression equals at 25°C

$$K_{CO_2} = \frac{[H_2CO_3^\circ]}{P_{CO_2}} = 10^{-1.47} \tag{5.15}$$

The first dissociation step of carbonic acid is written

$$H_2CO_3^\circ = H^+ + HCO_3^- \tag{5.16}$$

and has the equilibrium expression

$$K_1 = \frac{[H^+][H_2CO_3^-]}{[H_2CO_3^\circ]} = 10^{-6.35} \tag{5.17}$$

The second dissociation step and its equilibrium expression and constant at 25°C are

$$HCO_3^- = H^+ + CO_3^{2-} \tag{5.18}$$

and

$$K_2 = \frac{[H^+][CO_3^{2-}]}{[HCO_3^-]} = 10^{-10.33} \tag{5.19}$$

It is useful for some purposes to combine the expressions for K_{CO_2} and K_1, eliminating $H_2CO_3^\circ$. The result is

$$[H^+] = \frac{K_{CO_2}K_1P_{CO_2}}{[HCO_3^-]} \tag{5.20}$$

What are the pH ranges of dominance of the species of carbonic acid? The answer is simply related to the values of K_1 and K_2. Inspection of the equilibrium expressions for K_1 and K_2 shows that carbonic acid dominates below pH = pK_1 = 6.35, bicarbonate ion dominates between pH = pK_1 = 6.35, and pH = pK_2 = 10.33. Carbonate ion dominates above pH 10.33.

We will now take a more rigorous approach to this question and compute the distribution diagram for carbonic acid species in water as a function of pH, but ignore ion activity coefficients. First let us define the *total carbonate*, C_T, where

$$C_T = H_2CO_3^\circ + HCO_3^- + CO_3^{2-} \tag{5.21}$$

Also needed for the calculation is the value of K_w, the dissociation constant of water. The expressions for K_1, K_2, K_w, and C_T give us four equations and five unknowns. Our goal is to solve for CO_3^{2-}, HCO_3^-, and $H_2CO_3^\circ$ in terms of pH and the constants C_T, K_1, and K_2. The result is

$$(CO_3^{2-}) = \frac{C_T}{\alpha_H} \tag{5.22}$$

$$(HCO_3^-) = \frac{C_T (H^+)}{K_2 \alpha_H} \tag{5.23}$$

$$(H_2CO_3^\circ) = \frac{C_T (H^+)^2}{K_1 K_2 \alpha_H} \tag{5.24}$$

where α_H equals

$$\alpha_H = \left[\frac{(H^+)^2}{K_1 K_2} + \frac{(H^+)}{K_2} + 1 \right] \tag{5.25}$$

Assuming carbonic acid dissociation constants of $K_1 = 10^{-6.35}$ and $K_2 = 10^{-10.33}$ at 25°C, we can substitute in Eq. (5.25) to obtain

$$\alpha_H = (H^+)^2 10^{16.68} + (H^+) 10^{10.33} + 1 \tag{5.26}$$

Values of this function from pH 2 to 12 are listed in Table 5.2.

Given values for C_T, K_1, K_2 and α_H, we can solve Eqs. (5.22) to (5.24) for concentrations of the carbonate species as a function of pH. Carbonate species concentrations computed for $C_T = 10^{-3.0}$ are plotted in Fig. 5.2. Also shown are concentrations of H^+ and OH^-. Important features of the diagram include: (1) the pH regions of dominance of $H_2CO_3^\circ$, HCO_3^-, and CO_3^{2-} with increasing pH; (2) the fact that crossovers of their concentration curves occur where pH = pK for each dissociation step; (3) the pH dependence of the distribution plot is independent of C_T; and (4) concentrations of H^+ and OH^- are independent of C_T. We will return to this figure later in the chapter when we discuss acidity and alkalinity.

5.3.2 Carbonic Acid/pH Relations in Natural Waters

A variety of reactions and processes influence the CO_2/carbonic acid concentration and thus pH of surface- and groundwaters. Some of these are given in Table 5.3, along with their respective effect

TABLE 5.2 Values of α_H from pH 2 to 12 at 25°C

pH	log α_H	pH	log α_H
2	12.68	8	2.34
3	10.68	9	1.35
4	8.68	10	0.497
5	6.70	11	0.0841
6	4.84	12	0.00919
7	3.42		

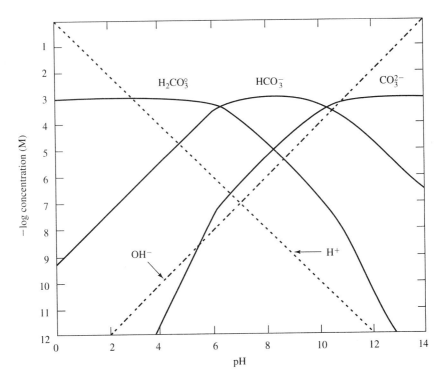

Figure 5.2   Distribution diagram for carbonate species as a function of pH, assuming $C_T =$ 10^{-3} M. Concentrations of H^+ and OH^-, which are independent of C_T, are shown as dashed straight lines.

on pH. Remember that when free protons are consumed in reactions that produce carbon dioxide and/or carbonic acid (a weak acid), the pH will rise, and conversely. Among the reactions and processes listed in the table, some operate in both surface-water and groundwater systems, whereas others are largely limited to one or the other.

     The carbon dioxide content of surface water tends toward a limiting value defined by Henry's law and the CO_2 pressure in the atmosphere (0.00033 bar or $10^{-3.5}$ bar). Differences from $10^{-3.5}$ bar CO_2 pressure reflect relative rates of reactions and processes, including (1) through (9) in Table 5.3.

     Some surface waters (deep, stratified, and poorly mixed, with little sunlight penetration, heavily polluted) are physically isolated from the atmosphere. In such waters, the apparent CO_2 pressure, which is defined by $P_{CO_2} = [H_2CO_3^\circ]/K_{CO_2}$ will usually exceed the atmospheric value. This disequilibrium partly reflects the fact that the rate of CO_2 degassing from water is slower than its rate of gaseous uptake from the atmosphere (cf. Schmiermund 1991; Morel and Hering 1993; Cole et al. 1994).

     In a study of lakes worldwide (4665 samples from 1835 lakes) Cole et al. (1994) found the pH ranged from about 4 to 10 and CO_2 pressures from about $10^{-1.7}$ to $10^{-6.0}$ bar. Sources of CO_2 in lakes include its inflow in surface waters and groundwaters and respiration and decay related to lake sediments and biota. The overall mean CO_2 pressure of 1.049×10^{-3} ($10^{-2.98}$) bar in lakes was three times greater than the atmospheric value. This reflects a CO_2 production and accumulation rate that exceeds its rates of consumption in the growth of aquatic plants and rate of degassing to the atmosphere.

TABLE 5.3 Some important processes and reactions that control the CO_2 content of surface- and groundwaters and, therefore, their pH

Processes and reactions	pH up or down
1. CO_2(g) dissolution $\rightarrow$, CO_2(aq) exsolution $\leftarrow$ CO_2(g) + H_2O $\leftrightarrow$ CO_2(aq) + H_2O $\leftrightarrow$ $H_2CO_3^\circ$	$\rightarrow$down/$\leftarrow$up
2. Photosynthesis[†] $\rightarrow$, respiration & aerobic decay $\leftarrow$ CO_2 + H_2O $\leftrightarrow$ 1/6 $C_6H_{12}O_6$ (glucose) + O_2	$\rightarrow$up/$\leftarrow$down
3. Methane fermentation (anaerobic decay) $\rightarrow$ $C_6H_{12}O_6$ (glucose) $\rightarrow$ $3CH_4$ + H_2O + CO_2	$\rightarrow$down
4. Nitrate uptake and reduction $\rightarrow$ NO_3^- + $2H^+$ + $2CH_2O$ $\rightarrow$ NH_4^+ + $2CO_2$ + H_2O	$\rightarrow$up
5. Denitrification $\rightarrow$ $5CH_2O$ + $4NO_3^-$ + $4H^+$ $\rightarrow$ $5CO_2$ + $2N_2$ + $7H_2O$	$\rightarrow$up
6. Sulfate reduction $\rightarrow$ $2CH_2O$ + SO_4^{2-} + H^+ $\rightarrow$ $2CO_2$ + HS^- + $2H_2O$	$\rightarrow$up
7. Carbonate mineral dissolution $\rightarrow$ or precipitation $\leftarrow$ $CaCO_3$(calcite) + H^+ = Ca^{2+} + H_2O + CO_2 $NaHCO_3$(nahcolite) + H^+ = Na^+ + H_2O + CO_2	$\rightarrow$up/$\leftarrow$down
8. Common-ion driven calcite precipitation $\rightarrow$ $CaSO_4 \cdot 2H_2O$ + $2HCO_3^-$ $\rightarrow$ $CaCO_3$ + SO_4^{2-} + $2H_2O$ + CO_2 gypsum calcite	$\rightarrow$down
9. Chemical weathering of Al-silicate minerals $2KAlSi_3O_8$ + $2CO_2$ + $11H_2O$ $\rightarrow$ $Al_2Si_2O_5(OH)_4$ + $2K^+$ + $2HCO_3^-$ K-feldspar kaolinite	$\rightarrow$up

Note: CH_2O represents organic matter.

[†]Other reactions that lead to carbon fixation may involve species of S and N (cf. Morel & Hering 1993; Stumm and Morgan 1996).

Some of the lowest apparent CO_2 pressures (and related highest pH's) found in lakes and other surface waters result from the photosynthetic activity of aquatic vegetation (reaction 2$\rightarrow$) around midday, removing CO_2 far more rapidly than it can be replenished from the atmosphere (reaction 1$\rightarrow$), from respiration and decay (reactions 2$\leftarrow$ and 3$\rightarrow$), or from groundwater inflow to the surface water. (Groundwaters usually contain more CO_2 than surface waters.) The highest pH's due to photosynthesis are found in streams, lakes, and reservoirs clogged with green algae (waters that have a high biomass/water ratio).

The highest apparent CO_2 pressures in surface waters may occur at night, when the dominant processes are respiration and aerobic decay (reaction 2$\leftarrow$), and perhaps groundwater inflow. Such behavior is reflected in the data from a study of Slab Cabin Run, near State College, Pennsylvania, given in Table 5.4 (Jacobson et al. 1971). Slab Cabin Run is a small stream fed by carbonate springs and often clogged with photosynthetic aquatic vegetation. The apparent CO_2 pressures for the stream in Table 5.4 have been computed from the total analysis using MINTEQA2. Measurements at 4:45 A.M. on a summer day showed a minimum stream pH of 7.76 and maximum apparent CO_2 pressure of $10^{-2.68}$ bar, probably reflecting chiefly the chemistry of groundwater inflow. At 2 P.M. that same day, active photosynthesis by aquatic vegetation produced a maximum stream pH of 9.42 and corresponding reduction in the apparent CO_2 pressure to $10^{-4.41}$ bar.

Explanation for the inverse relationship between pH and P_{CO_2} in carbonate-dominated systems such as Slab Cabin Run is most obvious if we recast Eq. (5.20) in log form

$$\log P_{CO_2} = -pH + \log \frac{(\gamma_{HCO_3} mHCO_3^-)}{K_{CO_2} K_1} \tag{5.27}$$

The data in Table 5.4 show that the total alkalinity (chiefly HCO_3^-) is practically constant at 2.80 ± 0.04 (SD) mM, regardless of pH. The product of $K_{CO_2} K_1$ is nearly constant at $10^{-7.73}$ between 8.4 and 16.5°C. Thus, $\Delta pH \approx \Delta(-\log P_{CO_2})$ in the stream. In essence, the stream has a constant background composition, perturbed diurnally by the effect of photosynthesis on its dissolved CO_2 content and thus its pH. This behavior of Slab Cabin Run has been generalized in Fig. 5.3, which is based on Eq. (5.28) and on MINTEQA2 calculations.

In some surface waters (and shallow groundwaters) depleted in dissolved oxygen because of their organic-matter content, there is an inverse correlation between the O_2 consumed by aerobic decay and respiration and the increase in CO_2 found in the water over its equilibrium atmospheric value. This would be expected if changes in the dissolved oxygen and carbon dioxide contents of the water are controlled chiefly by respiration and aerobic decay (reaction 2←). The ratio of the CO_2 produced to the O_2 consumed is called the *respiratory coefficient*, or,

$$R_c = \frac{(CO_2 \text{ moles produced})}{(O_2 \text{ moles consumed})} \tag{5.28}$$

(Hutchinson 1957). Typical values of R_c are between 0.7 and 0.9. Thus, somewhat more oxygen is consumed than can be accounted for by reaction 2←. This presumably reflects oxygen consumption by other processes, including oxidation of such species as Fe^{2+}, HS^-, and NH_4^+.

The apparent CO_2 pressure in shallow groundwaters is chiefly controlled by the production of CO_2 in the unsaturated zone. Most of this carbon dioxide is produced in the A horizon of soils by plant-root respiration and aerobic decay of organic matter, including humic (discussed in Section 5.4) and other plant materials. Thus, soil CO_2 pressures from 0.01 to 0.1 bar (10,000 to 100,000 parts per million by volume, or ppmv) may be found during the growing season in organic-rich soils. During warm months, soil CO_2 (grams per formula weight [GFW] 44 g) tends to diffuse upward and escape to the atmosphere, as well as move downward because of its relatively high density compared to N_2 (GFW 28 g) and O_2 (GFW 32 g). Solomon and Cerling (1987) found, however, that during the winter months, snow cover limited the atmospheric escape of CO_2 and so caused a severalfold increase in soil CO_2 values at shallow soil depths.

The concentration of natural, dissolved organic carbon (DOC) content in the soil moisture of shallow soils is often 5 to 10 ppm. This DOC may percolate downward in the unsaturated zone, perhaps in association with colloidal-sized clay particles. Its aerobic bacterial oxidation can produce increasing amounts of CO_2 with depth (cf. Wood and Petraitis 1984; Wood 1985).

Subsurface contamination by organic-rich wastes from landfills, leaky septic tanks, sewerage tile fields, and toxic-waste dumps, can also produce high CO_2 concentrations in both the unsaturated and saturated zones. For example, a shallow well 50 m from two septic tanks had an apparent CO_2 pressure of $10^{-1.32}$ bar (see Chap. 6 and Table 6.7, analyses 3 to 5 and 9).

As the groundwater moves downdip from the recharge zone of a sedimentary formation, it tends to deplete its CO_2 content by dissolving carbonate and alumino-silicate minerals in the formation (e.g., reactions 7 and 9 in Table 5.3). When such weathering reactions dominate the chemistry, the apparent P_{CO_2} of the water decreases and the pH increases with groundwater flow. Such behavior is typical of groundwaters in the silt, sand, and gravel aquifers of the U.S. coastal plain along the

TABLE 5.4 Some chemical analyses of waters from Slab Cabin Run, near State College, PA, May 28–29, 1971, and the corresponding apparent CO_2 pressure

Sample	Date	Time	$T°C$	Ca^{2+} (mM)	Mg^{2+} (mM)	Total alkalinity[†] (mM)	DO (mM)	pH	$-\log P_{CO_2}$ (bar)	SpC (μmhos/cm) 25°C	Solar radiation daily cumulative (langleys)
1	28th	1430	16.5	1.05	0.66	2.72	0.363	9.18	4.13	336	412
2	28th	1830	14.8	1.00	0.66	2.77	0.313	8.94	3.85	332	502
3	28th	2300	10.4	1.02	0.66	2.84	0.266	8.03	2.92	330	504
4	29th	0445	8.4	0.97	0.70	2.77	0.338	7.76	2.68	331	0
5	29th	0930	10.0	0.97	0.66	2.82	0.375	8.46	3.37	336	106
6	29th	1400	16.5	1.00	0.66	2.84	0.359	9.42	4.41	338	305
7	29th	1800	14.5	1.00	0.70	2.82	0.313	8.88	3.78	333	369
8	29th	2000	13.1	1.02	0.70	2.82	0.275	8.54	3.42	335	376
Average ($\pm$SD)				1.00 ± 0.02	0.675 ± 0.019	2.80 ± 0.04					

[†] C_B (total alkalinity) = $HCO_3^- + 2CO_3^{2-} + OH^- - H^+$ (see Section 5.7).

Source: Unpublished data of Jacobson et al. (1971).

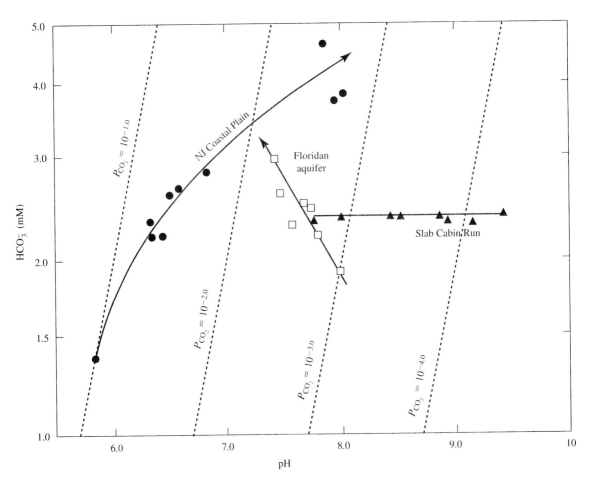

Figure 5.3   Example trends of the apparent CO_2 pressure in bars in a New Jersey coastal plain groundwater aquifer (Langmuir 1969), in the Floridan carbonate aquifer (Back and Hanshaw 1970), and in Slab Cabin Run, a small stream. Arrows denote the direction of groundwater flow. The data for Slab Cabin Run represent the diurnal cycle in May (Jacobson et al. 1971). See Table 5.4.

East and Gulf coasts. In the Potomac-Raritan-Magothy Formation of coastal plain New Jersey, for example, P_{CO_2} decreases in the direction of flow from about $10^{-1.0}$ to $10^{-2.6}$ bar while the pH rises from 5.8 to 8.0 over a 19 km distance downdip (Langmuir 1969). (See Fig. 5.3.) Chapelle (1983) and Chapelle and Knobel (1983) also note a pH increase and CO_2 decrease with groundwater flow in the coastal plain Aquia aquifer of Maryland. The aquifer is chiefly composed of quartz sand (~55%), glauconite clay (~30%), and carbonate shell material (~8%). Over a downgradient-flow distance of about 64 km, the groundwater pH increases from about 7.2 to 8.9, and the CO_2 pressure drops from about $10^{-2.5}$ to as low as $10^{-3.4}$ bar.

  If, instead, the production of CO_2 in the formation exceeds its consumption, the apparent CO_2 content will increase and pH may decrease with flow. Deeper groundwaters tend to become anaerobic because traces of organic matter and/or Fe(II) and sulfide in minerals at depth deplete the oxygen introduced in groundwater recharge. Reactions such as anaerobic decay, nitrate reduction, denitrification, and sulfate reduction then become progressively important. The apparent CO_2 (carbonic acid) content of the water may increase with groundwater flow because of such reactions.

In the Floridan aquifer, traces of gypsum are present in the carbonate rock. The high calcium concentration from gypsum dissolution exceeds its value at saturation with calcite, leading to precipitation of the carbonate and the production of additional CO_2 (reaction 8). Concurrently, anaerobic decay of buried organic matter (reaction 3) and sulfate reduction (reaction 6) take place. The combination of these processes has caused an increase in the CO_2 pressure of the groundwater from $10^{-2.9}$ bar in the recharge zone to $10^{-2.1}$ bar downdip as the pH decreases from 8.0 to 7.4 over a map distance of 115 km (Fig. 5.3) (cf. Back and Hanshaw 1970; Plummer et al. 1983).

5.4 HUMIC AND FULVIC ACIDS

Most of the organic material in soils and natural waters is described as *humic substances*, associated with smaller amounts of *nonhumic substances* (Schnitzer and Kahn 1978). The nonhumic fraction includes organic compounds having definite physical and chemical characteristics (e.g., a melting and/or boiling point, refractive index, defined chemical composition). Nonhumics include carbohydrates, proteins, peptides, amino acids, fats, and waxes (cf. Thurman, 1985). About 25% of the dissolved organic carbon (DOC) in seawater, groundwater, streams, and lakes is nonhumic, as is 10% of the DOC in wetlands (Thurman 1985). These compounds tend to be more rapidly biodegradable than are humic substances, which are biologically refractory (Sposito 1994).

Humic substances tend to be acidic, dark-colored, and partly aromatic and have molecular weights from a few hundred to several thousand grams per formula weight. They lack specific chemical and physical characteristics. They also have a high surface charge or cation exchange capacity (CEC) and so account for most of the CEC of organic-rich stream, estuarine, and lake muds, and of the A horizon of many humid-climate soils (see Chap. 10). Their negative charge is chiefly due to their carboxylic and phenolic OH structural groups.

Traditionally, *soil humic substances* have been defined by the fact they are soluble in 0.1 N NaOH. *Aquatic humic substances,* however, are operationally defined as polyelectrolytic acids that can be isolated from water by sorption onto XAD or weak base-ion exchange resins, for example (Thurman 1985). They are nonvolatile, have molecular weights from about 500 to 5000 g/mol, and a molar composition of about 50% C, 4 to 5% H, 35 to 40% O, and 1% N.

Measured in terms of the weight of their carbon, concentrations of aquatic humic substances range from about 0.03 to 0.1 mg C/L in groundwaters, to 0.06 to 0.60 mg C/L in seawater, 0.5 to 4.0 mg C/L in rivers and lakes, and 10 to 30 mg C/L in wetlands (Thurman 1985). Three fractions of aquatic humic substances are defined, based on their solublities in acids and bases. These include *fulvic acid,* which is soluble in both acid and alkaline solutions. Fulvic acids have the lowest molecular weights (about 500 to 1500 g/mol). *Humic acid* is soluble in alkaline solutions, but insoluble in acid below pH = 2, and has an intermediate molecular weight. *Humin* is insoluble in acid or alkaline solutions and has the highest molecular weights. The pK_a's of the humic and fulvic acids are generally above 3.6.

Fulvic acid is present in all soils where it makes up roughly 25 to 75% of the total organic matter. In natural waters fulvic acid ranges from about 12 to 60% of DOC (Malcolm 1985). The COOH/(total OH groups) ratio is about 1, and the COOH/(phenolic OH) ratio is about 3 in fulvic acid. (Phenolic OH is OH attached to a benzene ring, phenol is C_6H_5OH). The average composition of fulvic acid is $C_{135}H_{182}O_{95}N_5S_2$ (Sposito 1989). The acid is 100% water soluble; 60% as COOH, OH, and carbonyl (CO) groups (Schnitzer 1971).

Fulvic acid plays a major role in the transport and deposition of Fe, Al, and other metals in soils. The acid is produced by organic decay in the top of the soil's A horizon. Fulvic acid ligands can form soluble complexes with Fe^{3+} and Al^{3+} and other metals, which facilitates metal movement downward through the soil. As a rule of thumb, if the molar ratio of metals/fulvic acid is less than 1/1, the metals are water soluble and mobile (Schnitzer 1971). If that ratio exceeds 1/1, the metals become insoluble and immobile. Thus, as fulvic acids are destroyed by aerobic decay or other processes during downward percolation, the metals precipitate, typically in the soil's B horizon. Precipitation of Fe and Al (and also Mn) oxyhydroxides, in turn, leads to coprecipitation and concentration of trace metals such as Cu, Cd, Zn, Co, Ni, and Pb in the soil (cf. Suarez and Langmuir 1975).

5.5 SUMMARY OF CONTROLS ON THE pH OF NATURAL WATERS

Shown in Fig. 5.4 is a schematic plot of the frequency of occurrence of natural water pH's, suggesting that their pH's most often lie between about 4 and 9. This reflects the dynamic interplay between natural acids and bases, with extreme pHs found when either dominates. Lowest pH's (1 to 3) are usually associated with the products of ferrous sulfide (FeS_2) weathering, which generates the strong acid's sulfuric acid and bisulfate ion, and acid cations, especially Fe^{3+}. The pH of acid rain, which is sometimes below 3, may reflect the presence of sulfuric, nitric, and hydrochloric acids.

In natural waters having pH values between about 4.5 and 7 the acids are weak and usually include carbonic acid and smaller amounts of organic acids such as fulvic acid. In near-surface environments these acids derive chiefly from organic decay and, in the case of carbonic acid, from plant root respiration as well. The carbon dioxide in groundwater may also be generated by other reactions and processes (Table 5.4). In *water-dominated systems*, which include humid climate soils and stream sediments, the carbonic and organic acids in the water, although often dilute, may be frequently replenished by rainfall and fresh recharge. They then readily attack and cause the chemical breakdown of carbonate, silicate, and alumino-silicate minerals in a process called *chemical weathering*, which is discussed in more detail in Chap. 7. Intermediate natural-water pH's reflect a balance between the production of H^+ ions from the dissociation of these weak acids and H^+ (and CO_2) consumption by weathering reactions.

Elevated pH values occur in waters whose chemistry is dominated by minerals, most of which, as noted above, are salts of strong bases and weak acids. Thus, in the absence of sources of acidity, mineral carbonate and silicate and alumino-silicate minerals tend to raise the pH to values of 9 to 10 or even higher. Such high pH's are found in arid soils and in some deep groundwaters that are not exposed to fresh recharge and so can be said to be *rock dominated* as opposed to *water dominated*. In such systems the minerals present can exist stably, without weathering. Water pH values above 10 are exceptional and may reflect contamination by strong bases, such as NaOH or $Ca(OH)_2$.

High pH's (10 to 11) are caused by the dissolution of such minerals as nahcolite ($NaHCO_3$) and natron ($Na_2CO_3 \cdot 10H_2O$), which form in evaporative alkaline lakes and dissolve according to reactions such as

$$NaHCO_3(c) = Na^+ + HCO_3^- \tag{5.29}$$

and
$$Na_2CO_3 \cdot 10H_2O(c) = 2Na^+ + CO_3^{2-} + 10H_2O \tag{5.30}$$

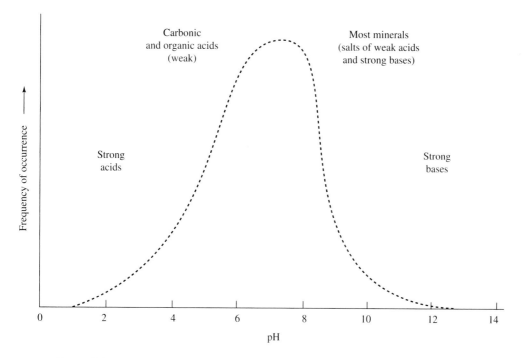

Figure 5.4   Schematic plot showing the frequency of occurrence of pH values and the major controls on natural-water pH's.

Assuming saturation with natron at atmospheric CO_2 pressure ($10^{-3.5}$ bar) and 25°C, MINTEQA2 computes pH = 10.14. Highly alkaline surface water pH's are also caused by photosynthetic depletion of dissolved carbonic acid.

Exceptionally, portlandite [$Ca(OH)_2$] and/or brucite [$Mg(OH)_2$] cause similarly high pH values in springs issuing from ultramafic rocks (cf. Neal and Stanger 1985; Bath et al. 1987). (At $P_{CO_2} = 10^{-3.5}$ bar and 25°C, MINTEQA2 finds pH = 12.43 at portlandite saturation and pH = 10.32 at brucite saturation.) Portlandite is also a component in concrete, which is used in the construction of water-supply wells, and may cause pH values above 12 in the groundwater from such wells for a time following their completion.

Example 5.1

What is the pH of an evaporating lake that contains 0.01 mol/kg dissolved Na^+ and HCO_3^-? Nahcolite ($NaHCO_3(c)$) is the salt of a strong base and weak acid, so pH > 7. In the calculation we will assume equilibrium with atmospheric CO_2 ($10^{-3.5}$ bar) at 25°C and, for simplicity, will ignore ion pairing and activity coefficients. The problem involves setting up mass-balance, charge-balance, and equilibrium expressions and solving them simultaneously. The only mass-balance equation is simply

$$\Sigma Na = mNa^+ = 0.01 \text{ mol/kg} \qquad (5.31)$$

We cannot assume a mass balance for total bicarbonate because the system is open to atmospheric CO_2. The charge-balance equation states that in any water the total equivalents of the cations must equal that of the anions. Here, the charge-balance equation is

$$mH^+ + mNa^+ = mOH^- + mHCO_3^- + 2mCO_3^{2-} \qquad (5.32)$$

Given the high concentrations of Na^+ and HCO_3^- relative to H^+ above pH 7, H^+ can be ignored. Combining the mass-balance and charge-balance equations gives

$$mNa^+ = mOH^- + mHCO_3^- + 2mCO_3^{2-} = 0.01 \qquad (5.33)$$

Our first goal is to replace the terms in Eq. (5.33) with terms that are only functions of $[H^+]$. This is described as putting Eq. (5.33) *in the proton condition*. The resultant equation can then be solved for the solution pH. We will use equilibrium expressions to develop an equation in the proton condition.

That the CO_2 pressure is given simplifies our first equilibrium calculation, which is

$$K_{CO_2} = 10^{-1.47} = \frac{[H_2CO_3^\circ]}{P_{CO_2}} = \frac{[H_2CO_3^\circ]}{10^{-3.5}} \qquad (5.34)$$

and $\qquad\qquad\qquad [H_2CO_3^\circ] = 10^{-4.97}$

Hydroxyl is eliminated with

$$[OH^-] = \frac{K_w}{[H^+]} = \frac{10^{-14.00}}{[H^+]} \qquad (5.35)$$

Now we can reformat Eq. (5.20) to obtain

$$[HCO_3^-] = \frac{K_{CO_2} K_1 P_{CO_2}}{[H^+]} = \frac{10^{-1.47} \times 10^{-6.35} \times 10^{-3.5}}{[H^+]} = \frac{10^{-11.32}}{[H^+]} \qquad (5.36)$$

Finally, $[CO_3^{2-}]$ is eliminated through the expression for $K_2[H_2CO_3^\circ]$.

$$K_2 = 10^{-10.33} = \frac{[H^+][CO_3^{2-}]}{[HCO_3^-]} \qquad (5.37)$$

which, after substitution for $[HCO_3^-]$, leads to

$$[CO_3^{2-}] = \frac{10^{-21.65}}{[H^+]^2} \qquad (5.38)$$

Substituting into Eq. (5.33) now provides us with our equation in the proton condition

$$\frac{10^{-14.00}}{[H^+]} + \frac{10^{-11.32}}{[H^+]} + \frac{10^{-21.35}}{[H^+]^2} = 10^{-2.00} \qquad (5.39)$$

Inspection shows that the $[OH^-]$ term will always be less than 1% of $[HCO_3^-]$ and so can be ignored. Reformatting the remaining terms as a quadratic equation gives

$$[H^+]^2 - 10^{-9.32}[H^+] - 10^{-19.35} = 0 \qquad (5.40)$$

This can be solved by completing the squares, for example, and we find pH = 9.25.

The problem can also be solved with MINTEQA2, assigning an ionic strength of zero and assuming a total alkalinity (defined below) of 0.01 equiv/L. The result is also pH = 9.25. If, instead, we allow for ionic strength, activity coefficients, and ion pairs, MINTEQA2 gives a pH of 9.19. The difference reflects the fact that 11% of the carbonate is present as the $NaCO_3^-$ pair and that $\gamma_{CO_3^{2-}} = 0.651$ and $\gamma_{HCO_3^-} = 0.898$.

5.6 ACIDITY

We have discussed the important individual acid and base species in natural waters. As (or more) important environmentally is the combined effect of such species, that is, the acidity and alkalinity of the system. This must include contributions from minerals and other solids present, as well as from the aqueous species.

By definition, the acidity is the capacity of a water to give or donate protons; the alkalinity is the capacity of a water to accept protons. The conventional mathematical definitions of these quantities consider only species present in solution, or the *immediate acidity and alkalinity*, which can be measured by titration with strong acids or bases. Solids, including minerals, contribute to the long-term acidity and alkalinity of natural waters via such reactions as ion exchange of protons and weathering, for example, and must especially be considered in studies involving soils or groundwaters.

We will focus first on *immediate contributions* to acidity and alkalinity. For the acidity of natural waters, the contributors are strong and weak acids, salts of strong acids and weak bases, hydrolysis of Fe^{3+} and Al^{3+}, and oxidation and hydrolysis of Fe^{2+} and Mn^{2+}.

Acidity is significant and undesirable because:

1. Acidity gives water a greater capacity to attack geological materials (it thus accelerates rock weathering) and so is usually accompanied by high total dissolved solids (TDS), including hardness (the sum of the concentrations of multivalent cations, chiefly Ca and Mg).
2. Acidity increases the solubilities of hazardous substances, such as heavy metals, and is corrosive and toxic to fish and other aquatic forms.
3. Acidity thus limits the use of water without its extensive treatment due to the high TDS, high metals (including calcium and magnesium), and low pH. This often requires that acid waters be diluted, as well as neutralized.

Acidity is measured by titrating water with a strong base, such as NaOH. Resultant reactions include, for example:

a) Strong acids
$$H^+ + OH^- = H_2O$$
$$HSO_4^- + OH^- = SO_4^{2-} + H_2O$$

b) Weak acids
$$H_2S(aq) + OH^- = HS^- + H_2O$$
$$H_2CO_3^\circ + OH^- = HCO_3^- + H_2O$$
$$HCO_3^- + OH^- = CO_3^{2-} + H_2O$$

c) Metal ions

    hydrolysis only:
$$Al^{3+} + 3OH^- = Al(OH)_3$$

    oxidation plus hydrolysis:
$$2Fe^{2+} + 4OH^- + H_2O + \tfrac{1}{2}O_2 = 2Fe(OH)_3$$

In dilute, potable water systems the acidity (C_A) is usually defined as

$$C_A = 2H_2CO_3^\circ + HCO_3^- + H^+ - OH^- \tag{5.41}$$

where C_A is given in equivalents per liter (eq/L) or milliequivalents per liter (meq/L). This equation corresponds to the acidity of a typical potable water, in which most of the acidity is caused by carbonate species.

For an acid-mine drainage (pH 2 to 3) the total acidity might instead be given by

$$C_A = H^+ + HSO_4^- + 2Fe^{2+} + 3Fe^{3+} + 2FeOH^{2+} + 3Al^{3+} \tag{5.42}$$

Note that the equivalents of acidity due to each cationic species, or the number of protons it can donate, equals the charge of that species.

5.7 ALKALINITY

Alkalinity is the capacity of a water to accept protons and is the sum effect of all bases present. In most potable waters the alkalinity is due chiefly to bicarbonate ion and, to a minor extent, carbonate, ion. *Carbonate alkalinity* is defined as $HCO_3^- + 2CO_3^{2-}$. Hydroxyl, when present (usually unimportant below about pH 10), is called *caustic alkalinity*. The conventional definition of total alkalinity (C_B) in eq/L or meq/L, is

$$C_B = HCO_3^- + 2CO_3^{2-} + OH^- - H^+ \tag{5.43}$$

or, in other words, C_B equals the total equivalents of bases minus those of acids. Other bases that can contribute to total alkalinity are the ligands of fulvic acid; organic acid anions such as formate, propionate, and acetate (cf. Lundergard and Kharaka 1990) and bisulfide, orthophosphates, ammonia, borate, and silicates. Alkalinity is usually reported as mg/L $CaCO_3$ or meq/L $CaCO_3$. A useful conversion is C_B (or C_A) as mg/L $CaCO_3 = C_B$ (or C_A) as meq/L $\times 50.04$.

In potable waters below about pH 8.3, bicarbonate is usually the only significant base. Bicarbonate alkalinity in water derives from two sources: (1) the weathering of silicate and carbonate minerals by carbonic acid, for example

$$2KAlSi_3O_8 + 2H_2CO_3^\circ + 9H_2O = Al_2Si_2O_5(OH)_4 + 2K^+ + 4H_4SiO_4^\circ + 2HCO_3^- \tag{5.44}$$

| K-feldspar | carbonic
acid | kaolinite
clay | silicic
acid |

and (2) the dissolution of carbonate minerals, which, if dissolved by carbonic acid, contributes twice as much bicarbonate alkalinity relative to $H_2CO_3^\circ$ consumed as does silicate weathering. For example,

$$CaMg(CO_3)_2 + 2H_2CO_3^\circ = Ca^{2+} + Mg^{2+} + 4HCO_3^- \tag{5.45}$$

dolomite

Bicarbonate ion is usually the chief anion in freshwaters. In and on silicate rocks, the HCO_3^- concentration is usually 50 to 200 mg/L, whereas in groundwaters that contact a few percent carbonate materials up to pure limestone and dolomite, bicarbonate levels are usually in the range of 200 to 400 ppm. Seawater contains 140 mg/L HCO_3^-. Carbonate alkalinity (CO_3^{2-}) rarely exceeds 10 mg/L. Why? The presence of caustic alkalinity (free OH^-) at pH's above 10 usually indicates artificial contamination of a water by, for example, $Ca(OH)_2$ (portlandite) from the setting of concrete at newly completed wells. C_B concentrations can reach 1000 ppm as HCO_3^- in sodium carbonate-bicarbonate brines found in evaporative, closed basin lakes.

For natural waters in which the only acids and bases are species of carbonic acid and strong acids or bases, there is a simple relationship between C_A, C_B, and the total carbonate, C_T. Remember that the simple definitions of these parameters are:

$$C_T = H_2CO_3^\circ + HCO_3^- + CO_3^{2-} \tag{5.21}$$

$$C_A = 2H_2CO_3^\circ + HCO_3^- + H^+ - OH^- \tag{5.41}$$

and

$$C_B = HCO_3^- + 2CO_3^{2-} + OH^- - H^+ \tag{5.43}$$

Algebraic manipulation shows that these expressions are related in the equation

$$2C_T = C_A + C_B \tag{5.46}$$

5.8 ACID-BASE PROPERTIES OF MINERALS AND ROCKS

Most minerals are salts of weak acids and strong bases. The weak acids include silicic and carbonic acids. The strong bases are, for example, NaOH, KOH, and $Ca(OH)_2$. Because of their weak acid–strong base character, most minerals are stable under alkaline conditions, but tend to dissolve under acid conditions. Consistent with this behavior, when ground-up silicate and aluminosilicate minerals are placed in pure water, the pH generally rises to alkaline values. Such behavior gives most important rocks and minerals (but not quartz, SiO_2) a substantial alkalinity that can neutralize natural or contaminant acidity. The rate of such neutralization is relatively fast for carbonate rocks, but slow for most silicate rocks, except the clays. The neutralization process is called chemical weathering and is dealt with in more detail in Chap. 7. Typical weathering reactions are the dissolution of calcite and the weathering of potassium feldspar to form kaolinite clay (see Table 5.3).

A few minerals produce acid when they contact water. These minerals can be described as salts of weak bases and strong acids. They chiefly result from weathering and oxidation of the pyrite or marcasite (FeS_2) exposed in the mining of mineral deposits and coal. Such acid minerals, which are dominantly Fe^{3+} sulfates and to a minor extent Al^{3+} sulfates, typically form from the evaporation of pooled acid-mine waters or of the moisture in unsaturated mine wastes or spoils that contain the sulfides. Acidity is produced when they are dissolved by fresh runoff or recharge. For example

$$Fe_2(SO_4)_3 \cdot nH_2O = 2Fe^{3+} + 3SO_4^{2-} + nH_2O \tag{5.47}$$

where $n = 6$ in lausenite, 7 in kornelite, 9 in paracoquimbite and coquimbite, and 10 in quenstedtite (Palache et al. 1951). Acidity is produced by hydrolysis of the ferric iron,

$$Fe^{3+} + 3H_2O = Fe(OH)_3(ppt) + 3H^+ \tag{5.48}$$

More common minerals in such a setting are jarosite and alunite, which often dissolve incongruently to form their oxyhydroxides, releasing protons. The jarosite reaction produces significant acidity, whereas alunite is a very weak acid. For the reaction

$$KFe_3(SO_4)_2(OH)_6 + 3H_2O = K^+ + 3Fe(OH)_3(ppt) + 2SO_4^{2-} + 3H^+ \tag{5.49}$$

jarosite

$$K_{eq} = [K^+][SO_4^{2-}]^2[H^+]^3 = 10^{-19.5} \tag{5.50}$$

Assuming that $K_{sp} = 10^{-39}$ for the ferric oxyhydroxide and $K_{sp}(\text{jarosite}) = 10^{-94.5}$, and given $[K^+] = 10^{-3}$ mol/kg and $[SO_4^{2-}] = 10^{-2.5}$ mol/kg, and ignoring activity coefficients, the pH for equilibrium with jarosite is 3.8.

The pH at equilibrium with the aluminum sulfate salts on the other hand, is near neutral. For alunite in equilibrium with amorphous aluminum hydroxide, the reaction is

$$KAl_3(SO_4)_2(OH)_6 + 3H_2O = K^+ + 3Al(OH)_3(ppt) + 2SO_4^{2-} + 3H^+ \tag{5.51}$$

alunite

for which $K_{eq} = 10^{-28.3}$, assuming $K_{sp} = 10^{-33}$ for the aluminum oxyhydroxide and $K_{sp}(\text{alunite}) = 10^{-85.3}$. Again, given $[K^+] = 10^{-3}$ mol/kg, and $[SO_4^{2-}] = 10^{-2.5}$ mol/kg, and ignoring activity coefficients, the equilibrium pH is 6.8.

Another common mineral under such conditions is basaluminite ($Al_4SO_4(OH)_{10}$), which has a solubility product of $10^{-121.3}$ and dissolves according to the reaction

$$Al_4SO_4(OH)_{10} + 2H_2O = 4Al(OH)_3(ppt) + SO_4^{2-} + 2H^+ \qquad (5.52)$$

for which $K_{eq} = 10^{-17.3}$, again assuming K_{sp} for the aluminum oxyhydroxide is 10^{-33}. With $[SO_4^{2-}] = 10^{-2.5}$ mol/kg, the pH at equilibrium is 7.4.

5.9 ACIDITY AND ALKALINITY DETERMINATION

Acidity and alkalinity indicate a water's ability to resist changes in pH if mixed with acid or alkaline waters or wastes. In this section we will examine the measurement of acidity and alkalinity by titration and prediction of the measured titration curves for acidity and alkalinity.

5.9.1 The Titration Mass Balance

The total equivalents of acidity (C_A) or alkalinity (C_B) of a solution is usually measured by titrating it with a standardized solution of base or acid, respectively. The titration is continued to an endpoint pH, which ideally may be identified as an inflection point in a plot of the measured pH versus the acidity or alkalinity of the solution, or versus the volume of added titrant. C_A or C_B in the unknown solution are then computed from the relationship

$$\text{Concentration (eq/L)} \times \text{Volume (L)} = \text{Concentration (eq/L)} \times \text{Volume (L)} \qquad (5.53)$$
$$\qquad\qquad \text{titrant acid or base} \qquad\qquad\qquad \text{solution unknown}$$

where the titrant is the standard solution being added to the solution under analysis. This equation represents a mass balance calculated at the titration endpoint and states that the number of equivalents of added titrant acid or base equals the corresponding number of equivalents of base or acid present in the unknown solution, respectively.

Theoretically, the acidity or alkalinity can be measured by titration as just described, or may be computed from a total analysis of the water, taking into account concentrations of all of the acidic and/or basic species present. Such a computation is simple enough for freshwaters, but may be difficult for high ionic-strength acid or alkaline wastewaters because it requires that we know the extent of complexation of the acids and bases present, as well as their ion activity coefficients.

5.9.2 Acidity Titration

If a water contains only strong acids (or strong bases), the titration curve generated by plotting pH, measured with a glass electrode, against C_B (which equals $-C_A$) or against milliliters of base added, will have an inflection point at pH = 7, where $mH^+ = mOH^-$ (see Fig. 5.5). If, on the other hand, the water contains a mixture of strong and weak acids (or strong and weak bases), the titration curve will show several inflection points, some of which may be ill-defined because of overlapping neutralization endpoints (cf. Figs. 5.6 and 5.7).

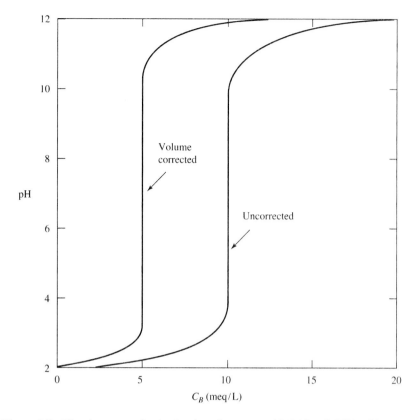

Figure 5.5   Titration curves for the titration of a strong acid (0.10 eq/L HCl) with a strong base (0.01 eq/L NaOH), drawn in terms of base added, C_B, with and without volume correction (see Section 5.9.4).

Figure 5.6 shows the acidity titration curve of a water (initial pH = 3.0) that contains a strong acid and $10^{-2.6}$ M carbonic acid. Below the titration curve is the corresponding distribution plot of carbonate species versus pH. (See also Fig. 5.2.) The total acidity of the water is given by

$$C_A = 2H_2CO_3^\circ + HCO_3^- + H^+ - OH^- \qquad (5.41)$$

Titrating the water with a strong base such as NaOH, the first inflection point around pH 4.5 is taken as the endpoint of the mineral acidity or strong-acid titration. It corresponds to completion of the reaction

$$H^+ + OH^- \rightarrow H_2O \qquad (5.54)$$

and at the endpoint $H^+ \approx HCO_3^-$.

The titration endpoint for the CO_2-acidity (i.e., $H_2CO_3^\circ$ content) of natural waters with a strong base such as NaOH is usually taken as pH = 8.3 (pOH = 5.7). This is the pH at which more than 99% of the $H_2CO_3^\circ$ has been converted to HCO_3^- based on the titration reaction

$$H_2CO_3^\circ + OH^- = H_2O + HCO_3^- \qquad (5.55)$$

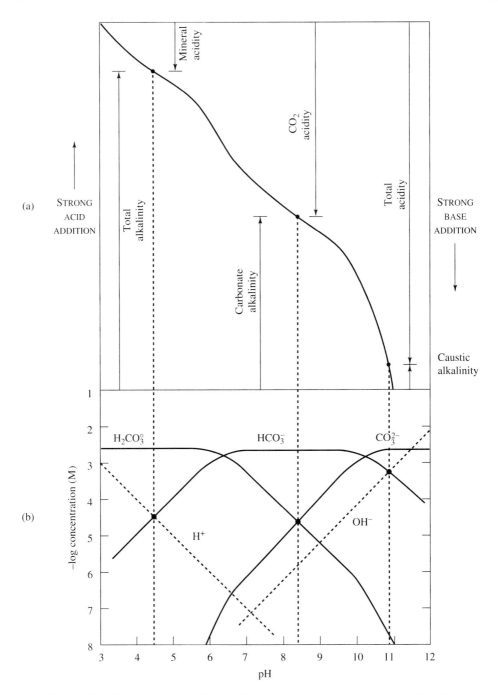

Figure 5.6 The carbonate distribution diagram of a solution with constant $C_T = 2.5 \times 10^{-3}$ M showing **(a)** the strong acid titration curve for the same solution from pH 12 to 3 and strong base titration curve between pH 3 and 12. Dashed straight lines in **(b)** indicate concentrations of H^+ and OH^-, which are independent of C_T. Modified after V. L. Snoeyink and D. Jenkins, *Water Chemistry*. Copyright © 1980 by John Wiley & Sons, Inc. Used by permission of John Wiley & Sons, Inc.

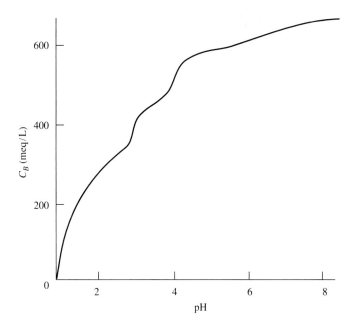

Figure 5.7   Titration of an acid uranium mill-tailings solution with a standard NaOH so-lution. The tailings solution had a pH of 0.9 and TDS = 112 g/L. A solution analysis showed that in addition to H^+, the important acid species present were HSO_4^-, Fe^{2+}, and Al^{3+}. Data and plot from Langmuir and Nordstrom (1995). See Problem 8 at the end of this chapter.

for which we can write

$$K_{eq} = \frac{K_1}{K_w} = \frac{10^{-6.35}}{10^{-14.00}} \approx 10^{7.7} = \frac{[HCO_3^-]}{[H_2CO_3^\circ]\,[OH^-]} \tag{5.56}$$

Substituting $[OH^-] = 10^{-5.7}$ (pH = 8.3) in this expression, we find

$$\frac{[HCO_3^-]}{[H_2CO_3^\circ]} = 10^{2.0} = \frac{100}{1} \tag{5.57}$$

Thus, for pH values above 8.3, the carbonic acid concentration is less than 1% of the bicarbonate concentration. We can then argue that above this pH the reaction has gone completely to the right and that all strong acids plus $H_2CO_3^\circ$ have been neutralized.

The exact endpoint pH of the CO_2-acidity titration may, in fact, differ from 8.3 by several tenths of a pH unit. Its value will depend on the total carbonate (C_T) concentration of the solution (Eq. [5.21]). The endpoint is defined by the equality

$$H^+ + H_2CO_3^\circ = CO_3^{2-} + OH^- \tag{5.58}$$

(Stumm and Morgan 1981). For C_T values above 10^{-4} mol/kg (6 ppm as HCO_3^-) one can ignore H^+ in this expression. We will assume total carbonate is constant in the acidity titration or, in other words, that the system is "closed" and no $CO_2(g)$ has been lost. Substituting the appropriate α_H values from Eqs. (5.22) and (5.24) into Eq. (5.58) and simplifying we obtain

$$(H^+) = \left(K_1 K_2 + \frac{K_1 K_w}{C_T} \right)^{1/2} \tag{5.59}$$

Introducing values for the constants at 25°C, Eq. (5.59) becomes

$$(H^+) = 10^{-3.18} \left(10^{-10.33} + \frac{10^{-14.00}}{C_T} \right)^{1/2} \tag{5.60}$$

Substituting values for C_T into Eq. (5.60) we can solve for the corresponding pH of the acidity endpoint. Some results are given below.

C_T (M)	Endpoint pH
10^{-4}	8.10
10^{-3}	8.30
10^{-2}	8.34

Problems with the determination of CO_2-acidity by titration, particularly that of concentrated acid-mine waters, may include:

1. Rates of metal (chiefly Fe^{2+}) oxidation and hydrolysis may be too slow to be completed during titration at room temperature, so the sample may have to be boiled to increase rates.
2. Boiling drives off $O_2(aq)$ needed for oxidation, so that hydrogen peroxide (H_2O_2) is sometimes added as an oxidant.
3. Boiling drives off the weak-acid gases H_2O and CO_2.
4. The H^+ produced by boiling the sample must be retitrated after the water is cooled to room temperature.

As evident from the definition of total acidity in Eq. (5.41), HCO_3^- is also an acid. At the endpoint of the total acidity titration near pH 11 (the third inflection point in Fig. 5.6) the reaction

$$HCO_3^- + OH^- \rightarrow CO_3^{-2} + H_2O \tag{5.61}$$

has gone to completion and $HCO_3^- \approx OH^-$.

Example 5.2

What is the acidity (C_A) of a pure water that contains only a strong acid at pH = 2.5? We will answer this question in terms of the several ways in which the acidity may be reported. Given the pH, and ignoring activity coefficients as usual, we know $mH^+ = 10^{-2.5} = 3 \times 10^{-3}$ mol/kg $\times 10^3$ mg/mol $H^+ = 3$ mg/kg $H^+ \approx 3$ mg/L H^+. Because the gram equivalent weight of hydrogen is 1.008, this is equivalent to 3 meq/L of acidity.

The gram equivalent weight of H_2SO_4 is 48.6 g, so that 48.6 mg of H_2SO_4 per liter has an acidity of 1 meq/L, and 3 meq/L $H^+ \times 48.6$ mg/meq $H_2SO_4 = 146$ mg/L acidity as H_2SO_4.

5.9.3 Alkalinity titration

In a simple carbonate system the definition of alkalinity is

$$C_B = HCO_3^- + 2CO_3^{2-} + OH^- - H^+ \tag{5.43}$$

The alkalinity is measured by titration with a standardized solution of a strong acid such as H_2SO_4, HNO_3, or HCl. The titration curve is identical, but inverse to the acidity titration curve in Fig. 5.6. The caustic alkalinity titration endpoint near pH 11 measures free OH^- from strong bases. At this endpoint $HCO_3^- \approx OH^-$. At the carbonate alkalinity endpoint (pH $\approx$ 8.3) conditions are identical to those defined above for CO_2-acidity. $H_2CO_3^\circ + OH^- = H_2O + HCO_3^-$

The important total alkalinity titration endpoint near pH 4.5 is defined exactly by

$$H^+ = HCO_3^- + 2CO_3^{2-} + OH^- \tag{5.62}$$

We will next compute the pH of the endpoint, which depends on C_T. Because the endpoint pH is generally below 5, carbonate and hydroxyl terms may be neglected so that $H^+ = HCO_3^-$. Substituting the expression given in Eq. (5.23) for HCO_3^- leads to

$$(H^+) = (HCO_3^-) = \frac{C_T (H^+)}{K_2 \alpha_H} \tag{5.63}$$

For pH values below 8 we can ignore 1 in the expression for α_H. Then, dividing by $(H^+)/K_2$ gives

$$(H^+) = \frac{C_T}{\dfrac{(H^+)}{K_1} + 1} \tag{5.64}$$

Solving for H^+,

$$(H^+) = \left(K_1 C_T + \frac{K_1^2}{4}\right)^{1/2} - \frac{K_1}{2} \tag{5.65}$$

If $C_T \geq 10^{-4}$ mol/L, then the second term in the parentheses in Eq. (5.65) is negligible and the expression simplifies to

$$(H^+) = (K_1 C_T)^{1/2} - \frac{K_1}{2} \tag{5.66}$$

Finally, introducing $K_1 = 10^{-6.35}$ for 25°C, Eq. (5.66) becomes

$$(H^+) = 10^{-3.18} C_T^{1/2} - 10^{-6.65} \tag{5.67}$$

We can now compute the pH of the alkalinity titration endpoint for different total carbonate concentrations. Some results are as follows

C_T (M)	Endpoint pH
10^{-4}	5.2
10^{-3}	4.7
10^{-2}	4.2
$10^{-1.5}$	3.9

The endpoint pH of 3.9 corresponds to a solution in equilibrium with 1 bar CO_2 pressure and an unusually high alkalinity as bicarbonate of 1950 mg/L. An assumption inherent in these calculations is that C_T is constant; thus no $CO_2(g)$ has been lost during the titration (cf. Barnes 1964). This is called a *closed-system assumption*. Accordingly, the endpoint pH is that of a solution of pure carbonic acid of concentration C_T.

5.9.4 Calculation of Titration Curves for Acid and Base Determination

If, by starting with a chemical analysis, we can mathematically model the acidity or alkalinity of a water as it is mixed with another water, we can then predict the change in pH of the mixture without the need to measure it. To accomplish such modeling we must derive equations that describe how a water's pH changes with the addition of an acid or a base. The simplest models assume that the mixing or addition of solutions occurs at constant volume. The more realistic but computationally difficult approach involves correcting solution concentrations for the volumetric changes that attend solution mixing or titration.

A number of geochemical computer codes can be used to simulate an acid or base titration. MINTEQA2 (Allison et al. 1991) makes such calculations assuming a constant volume. PHREEQE (Parkhurst et al. 1990) and PHRQPITZ (Plummer et al. 1988) can mix or titrate solutions with volume correction, but do not correct for changes in water mass. This deficiency is insignificant except when the mixing involves brines. The codes SOLMINEQ.88 (Kharaka et al. 1988), NETPATH (Plummer et al 1991,1994), EQ3/6 (Wolery et al. 1992a), and PHREEQC (Parkhurst 1995) can calculate mixing or titration results with both volume and water-mass corrections.

In following hand calculations of titration results we will ignore activity coefficients to avoid excessively complicating the calculations. Titration curves are typically plotted in two ways. The plot is usually either: (1) of the volume of a titrant acid or base added (V_a) versus the pH of the solution; or (2) of the concentration of acid or base added (C_A or C_B) versus the solution pH. The plots have a similar appearance. We will derive algebraic expressions for each type of plot. The advantage of C_A or C_B versus pH plots and mathematical expressions is that they can be used to derive the *buffer capacity* of a solution directly (see Section 5.10), whereas a plot of V_a versus pH cannot.

As a modeling exercise, consider the titration of a strong acid (HCl) with a strong base (NaOH). Derivation of the titration curve without volume correction must be in terms of C_A or C_B versus pH. For titration of an HCl solution with NaOH titrant, the amount of base added will equal the Na^+ in solution. In other words, $C_B = Na^+$. The charge-balance equation for the solution being titrated is

$$H^+ + Na^+ = OH^- + Cl^- \tag{5.68}$$

Substituting C_B for Na^+ and transposing we obtain

$$C_B = \frac{K_w}{H^+} + Cl^- - H^+ \tag{5.69}$$

This last equation is now in *the proton condition*. This means that we have eliminated all terms other than C_B, H^+ as pH, and known concentrations and stability constants, allowing us to plot the equation as C_B versus pH. Assuming $Cl = 0.01$ eq/L, we now introduce values of H^+ and solve for C_B. The results are plotted in Fig. 5.5. The figure shows the inflection point and endpoint at pH 7.0.

We will next derive an expression for the titration curve showing the change in pH with volume of added base (cf. Butler 1964; Pankow 1991). First, define V_o and V_a as the original volume of solution to be titrated and the volume of added titrant. Thus, V_o is known and fixed, and the final volume, V, is $V = V_o + V_a$. The original concentration of strong acid, HCl, will equal Cl_o, the original chloride present. Similarly, the concentration of titrant OH^- that has been added must equal the added sodium, Na_a. The total amount of acid before and during dilution by the titration (ignoring neutralization) must be equal, or

$$Cl_oV_o = Cl(V_o + V_a) \tag{5.70}$$

where Cl is the chloride concentration in the solution during the titration. Thus

$$Cl = \frac{Cl_oV_o}{V_o + V_a} \tag{5.71}$$

Similarly, all the sodium present during the titration has been added in titrant, so

$$Na_aV_a = Na(V_o + V_a) \tag{5.72}$$

or

$$Na = \frac{Na_aV_a}{V_o + V_a} \tag{5.73}$$

The titration calculation is begun with the charge-balance equation for the solution being titrated, which is as before

$$H^+ + Na^+ = OH^- + Cl^- \tag{5.68}$$

Next, substitute into the charge-balance equation to place it in the proton condition, its proper form for plotting. The equation thus becomes

$$H^+ + \frac{Na_aV_a}{(V_o + V_a)} = \frac{K_w}{H^+} + \frac{Cl_oV_o}{(V_o + V_a)} \tag{5.74}$$

and rearranging

$$\frac{Na_aV_a - Cl_oV_o}{(V_o + V_a)} = \frac{K_w}{H^+} - H^+ \tag{5.75}$$

At the inflection point of the titration curve, which is the endpoint of the titration,

$$Na_aV_a = Cl_oV_o \tag{5.76}$$

Thus, for the endpoint, the left-hand side of Eq. (5.75) disappears, and we calculate an endpoint pH of 7.0. In fact, all titrations of strong acids with strong bases, and vice versa, will have their endpoint at pH = 7.

Below pH = 6, OH^- is negligible and Eq. (5.75) simplifies to

$$H^+ = \frac{Cl_oV_o - Na_aV_a}{V_o + V_a} \tag{5.77}$$

Above pH = 8, with H^+ negligible, Eq. (5.75) reduces to

$$H^+ = \frac{K_w(V_o + V_a)}{(Na_aV_a - Cl_oV_o)} \tag{5.78}$$

Now let us assign values to terms in these equations and calculate the titration curve. First, assume $V_o = 1.0$ L, and that acid-solution and titrant base concentrations are both 0.01 eq/L as above or, in other words, that $Cl_o = Na_a = 0.01$ eq/L. With these substitutions into Eqs. (5.77) and (5.78) and for different volumes of added base titrant, V_a, we can solve for the solution pH. Now because $C_B = Na$ and Na is given by Eq. (5.73), we can calculate C_B. The resultant titration curve of C_B versus pH is shown in Fig. 5.5, where it may be compared with the curve generated without volume correction. The same endpoint pH of 7.0 is obtained in both cases. However, $C_B = 5.0$ meq/L at the endpoint of the volume-corrected titration versus 10.0 meq/L at the endpoint of the titration without volume correction.

Example 5.3

Calculate the volume-corrected titration curve for the titration of 0.01 N HCl with 0.01 N NaOH between pH 2 and 12 using a geochemical computer code and compare the result to the corresponding curve in Fig. 5.5. This problem can be solved using the mixing option in SOLMINEQ.88 (Kharaka et al. 1988). The titration is simulated by mixing different fractions of solution 1 (pH = 12.0, $Na^+ = 0.01$ mol/kg) with solution 2 (pH = 2.0, $Cl^- = 0.01$ mol/kg). In the output $C_B = Na^+$. Computed results are given here.

pH	C_B (meq/L)	pH	C_B (meq/L)
2.0	0	10.18	5.0
2.11	1.0	11.32	6.0
2.23	2.0	11.61	7.0
2.41	3.0	11.89	9.0
2.73	4.0		

Because SOLMINEQ computes ionic strength and activity coefficients by default, these results differ slightly from the volume-corrected curve in Fig. 5.5, particularly at low pH's, since that curve was derived assuming zero ionic strength.

Titrations involving weak acids and bases are far more complex to model, paticularly when the acids or bases are polyprotic and volume corrections are made. For detailed discussion of both simple and complex acid/base modeling with volume corrections, the reader is referred to Butler (1964), Pankow (1991), and Stumm and Morgan (1996).

We will now derive equations for the titration curve of carbonate alkalinity with a strong acid, HCl, first assuming the titration takes place at constant volume, then taking into account titration volume changes. To simplify our calculations the following assumptions will be made in both derivations:

1. Total carbonate, C_T, is constant (there is no CO_2 loss during titration, that is, *the system is closed*).
2. Activity coefficients can be ignored in the calculation.
3. Only carbonic acid species and strong acids or bases are present.

First, consider the titration assuming constant volume, which will yield an equation and plot of C_A versus pH. In the carbonate system the total acidity was defined in Eq. (5.41) as

$$C_A = 2H_2CO_3^\circ + HCO_3^- + H^+ - OH^- \tag{5.41}$$

Substituting into this equation the expressions for carbonic acid and bicarbonate ion given in Eqs. (5.24) and (5.23), we obtain

$$C_A = \frac{2C_T(H^+)^2}{K_1 K_2 \alpha_H} + \frac{C_T(H^+)}{K_2 \alpha_H} + H^+ - \frac{K_w}{H^+} \tag{5.79}$$

where α_H is defined in Eq. (5.25). With stability-constant values for 25°C, and assuming $C_T = 5 \times 10^{-3}$ mol/L, Eq. (5.79) becomes

$$C_A = \frac{10^{14.7}(H^+)^2}{\alpha_H} + \frac{10^{8.0}(H^+)}{\alpha_H} + H^+ - \frac{10^{-14.0}}{H^+} \tag{5.80}$$

The same result is obtained with somewhat more effort if we start with the charge-balance equation, which is

$$Na^+ + H^+ = HCO_3^- + 2CO_3^{2-} + Cl^- + OH^- \tag{5.81}$$

Now $C_A = Cl^-$, thus

$$C_A = Na^+ - HCO_3^- - 2CO_3^{2-} + H^+ - OH^- \tag{5.82}$$

Comparing Eqs. (5.41) and (5.82), which must be equal, it is easy to show that $Na^+ = 2C_T$, and so Eq. (5.82) becomes

$$C_A = 2C_T - HCO_3^- - 2CO_3^{2-} + H^+ - OH^- \tag{5.83}$$

which obviously must also equal Eq. (5.41). Substituting to put this expression in the proton condition we find

$$C_A = 2C_T - \frac{C_T(H^+)}{K_2\alpha_H} - \frac{2C_T}{\alpha_H} + H^+ - \frac{K_w}{H^+} \tag{5.84}$$

Introducing $C_T = 5 \times 10^{-3}$ mol/L, and stability constant values for 25°C, Eq. (5.84) becomes

$$C_A = 10^{-2} - \frac{10^{8.0}(H^+)}{\alpha_H} - \frac{10^{-2.0}}{\alpha_H} + (H^+) - \frac{10^{-14.0}}{(H^+)} \tag{5.85}$$

which equals Eq. (5.80). Values of α_H from pH 2 to 12 are given in Table 5.2. A plot of Eqs. (5.80) or (5.85) is shown in Fig. 5.8 and, as expected, shows the two inflection points at pH values near 8.3 and 4.5 corresponding to endpoints of the respective carbonate and bicarbonate alkalinity titrations.

The development of a mathematical model for the carbonate alkalinity titration with a standard acid with correction for volume change, can begin with the solution charge-balance equation, which is

$$Na^+ + H^+ = HCO_3^- + 2CO_3^{2-} + Cl^- + OH^- \tag{5.81}$$

The expression for total carbonate, C_T is also useful here, where

$$C_T = H_2CO_3^\circ + HCO_3^- + CO_3^{2-} \tag{5.21}$$

We will assume that except for dilution by titrant, total moles of C_T and Na^+ are fixed during the titration and that all Cl^- in the solution has come from the titrant acid HCl. With volume correction, concentrations of Na^+, Cl^- (which equals C_A, or acid added), and C_T in the solution being titrated will equal

$$Na^+ = \frac{Na_oV_o}{V_o + V_a} \qquad Cl^- = \frac{Cl_aV_a}{V_o + V_a} \qquad C_T = \frac{C_{T_o}V_o}{V_o + V_a} \tag{5.86}$$

As before, the subscripts o and a denote concentrations or volumes of the original solution and the added titrant, respectively. We can eliminate CO_3^{2-} from the charge-balance equation by subtracting twice the total carbonate equation from it. The result is

$$Na^+ + H^+ + 2H_2CO_3^\circ + HCO_3^- = 2C_T + Cl^- + OH^- \tag{5.87}$$

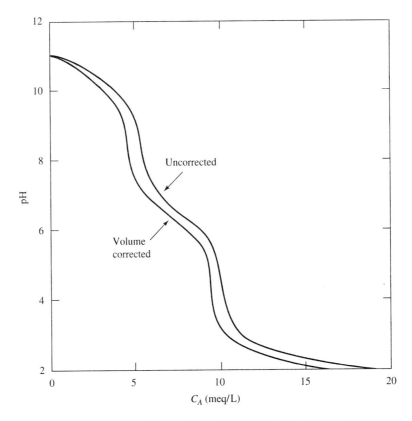

Figure 5.8 Titration of total carbonate ($C_{T_o} = 5 \times 10^{-3}$ M) with a strong acid (0.1 M HCl) with and without volume correction and assuming a closed system.

Substituting for Na^+, C_T, Cl^-, and the carbonate species in this equation, after some rearrangement we obtain

$$\frac{Na_o V_o}{(V_o + V_a)} + \frac{C_{T_o} V_o}{(V_o + V_a)}\left(\frac{2(H^+)^2}{K_1 K_2 \alpha_H} + \frac{(H^+)}{K_2 \alpha_H} - 2\right) = \frac{Cl_a V_a}{(V_o + V_a)} + \frac{K_w}{(H^+)} - (H^+) \qquad (5.88)$$

Let us assume $Na_o = 10^{-2}$ M, $Cl_a = 0.1$ M, $C_{T_o} = 5 \times 10^{-3}$ M, and $V_o = 1.00$ L. With these substitutions and values for K_1, K_2, and K_w at 25°C, the last expression "simplifies" to

$$\frac{10^{14.7}(H^+)^2}{\alpha_H} + \frac{10^{8.0}(H^+)}{\alpha_H} = 10^{-1.0}V_a + (1 + V_a)\left(\frac{10^{-14.0}}{(H^+)} - (H^+)\right) \qquad (5.89)$$

which can be solved with values of α_H from Table 5.2. The results are shown in Fig. 5.9.

Remembering that $C_A = Cl^-$, we can also create the more useful plot of pH versus C_A for the titration by substituting values of V_a into Eq. (5.86) for Cl^-. The resultant plot of C_A versus pH with volume correction is compared to the titration curve without volume correction in Fig. 5.8. As in the strong acid/base titration plot, the general features of the curves derived with and without volume correction are the same. The curves are close, indicating a minor correction for volume change. This reflects the fact that the acid titrant is 20 times more concentrated than the carbonate water, so a relatively small acid volume is consumed in the titration.

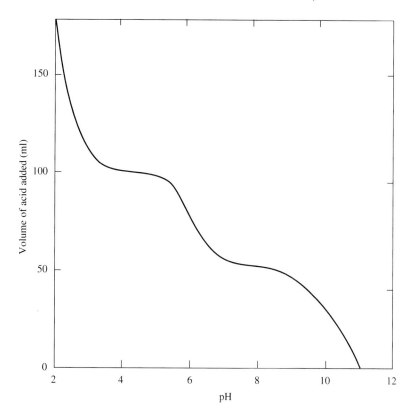

Figure 5.9   Plot of pH versus added volume (V_a) of titrant acid (0.1 M HCl) for the titration of total carbonate ($C_{T_0} = 5 \times 10^{-3}$ M), assuming a closed system (no CO_2 loss).

Example 5.4

Calculate the volume-corrected titration curve for the titration of 0.005 M total carbonate (C_T) with 0.1 eq/L HCl between pH 2 and 11 using the mixing option in SOLMINEQ.88 (Kharaka et al. 1988); compare the result to the corresponding curve in Fig. 5.8. The titration is simulated by mixing different fractions of solution 1 (pH = 1.0, Cl⁻ = 0.1 M) with solution 2 (pH = 11.0, Na⁺ = 0.01 M, $C_T = 5 \times 10^{-3}$ M as C). In the output $C_A = $ Cl⁻. Computed results given here are seen to be in generally good agreement with the volume-corrected curve in Fig. 5.8.

pH	C_A (meq/L)	pH	C_A (meq/L)
10.66	1.0	5.27	8.0
10.30	2.0	3.09	9.0
9.92	3.0	2.71	10.0
9.34	4.0	2.50	11.0
7.18	5.0	2.36	12.0
6.52	6.0	2.26	13.0
6.05	7.0	2.18	14.0

5.10 BUFFER CAPACITY OF AQUEOUS SPECIES AND MINERAL SYSTEMS

Acidity and alkalinity titrations determine the total capacity of natural waters to consume strong bases or acids as measured to specified pH values defined by the endpoints of titrations. Of more interest for many purposes is the ability of a water or water-rock system to resist pH change when mixed with a more acid or alkaline water or rock. This system property is called its buffer capacity. Buffer capacity is important in aqueous/environmental studies for reasons that include:

1. The buffer concept helps us understand what reactions control the pH of natural waters including soil waters (cf. Van Breemen and Wielemaker 1974) and wastewaters.
2. The principles of pH buffering can be used to understand the evolution of pH and CO_2 and thus the mineral alteration that takes place during the burial diagenesis of sediments (cf. Hutcheon et al. 1993).
3. The concept of pH buffer capacity as a measure of resistance to pH change can be applied to our thinking about the buffering of environmental systems with respect to their concentrations of other substances, including electrons (as defined by redox potential, cf. Nightingale 1958) or contaminant trace metals (cf. Pankow 1991).
4. The standard buffer solutions used in pH electrode calibration are designed compositionally to have maximal buffer capacities to assure that their advertised pH values are as constant as possible (cf. Bates 1964; Langmuir 1971a).

Here we restrict ourselves to pH buffers. Van Slyke (1922) defined pH buffers as "substances which by their presence in solution increase the amount of acid or alkali which must be added to cause a unit change in pH." In aqueous solutions pH buffering is especially due to the interaction of weak acids and bases and their salts with water. The quantification of this effect, the buffer capacity or buffer index, β, is by definition

$$\beta = -\frac{dC_B}{d\text{pH}} \tag{5.90}$$

where dC_B is the increment of strong base added in gm eq/L, and dpH is the corresponding change in solution pH. β equals the inverse slope of the acidity titration curve and is always positive in sign. Because C_A has the opposite effect on system pH, the buffer index in terms of added acid is defined as

$$\beta = -\frac{dC_A}{d\text{pH}} \tag{5.91}$$

In a solution of weak acid HA and conjugate base A^-, the dissociation reaction is

$$HA = H^+ + A^- \tag{5.92}$$

If protons are added, they combine with A^- to make more HA. Conversely, if protons are somehow consumed, the acid HA dissociates to produce more protons. Thus, the pH change is less than would take place in the absence of the weak acid and its base. We can write

$$K_a = \frac{(H^+)(A^-)}{(HA)} \tag{5.93}$$

After taking logs we can expand this last expression to obtain

$$\text{pH} = -\log K_a + \log \frac{(A)}{(HA)} \tag{5.94}$$

Now let us assume that the acid is carbonic acid and $K_a = K_1(H_2CO_3^\circ) = 10^{-6.35}$. Equation (5.94) then becomes

$$pH = 6.35 + \log \frac{(HCO_3^-)}{(H_2CO_3^\circ)} \tag{5.95}$$

Now, for example, assume that pH = 7.00 and $(HCO_3^-) = 10^{-3}$ mol/L. Through the expression for K_1 we calculate $(H_2CO_3^\circ) = 10^{-3.65} = 2.24 \times 10^{-4}$ mol/L. If we introduced 10^{-4} mol/L of H^+ ions, the pH would drop to 4 in pure water. However in the presence of carbonic acid species, the added protons participate in the reaction, $H^+ + HCO_3^- = H_2CO_3^\circ$. In terms of expression (5.95) the resulting change in pH is given by

$$pH = 6.35 + \log \frac{(10^{-3.00} - 10^{-4.0})}{(10^{-3.65} + 10^{-4.0})} = 6.35 + 0.44 = 6.79 \tag{5.96}$$

So the buffer capacity of the carbonate system limits the pH drop to about 0.2 pH units rather than 3 full units.

The following discussion and numerical calculations of buffer-capacity values will be approximate, in that for simplicity we will ignore ion activity coefficients and volume corrections.

5.10.1 Buffer Capacity of Water

To calculate the buffer index of an aqueous system we begin by deriving the equation for the acidity titration curve for that system. For pure water we will assume the titrant is NaOH. The charge-balance equation is then simply

$$Na^+ + H^+ = OH^- \tag{5.97}$$

and the mass-balance equation for added base is $C_B = Na^+$. Substituting into the charge-balance equation we can eliminate Na^+ and OH^- to obtain an equation in the proton condition

$$C_B = \frac{K_w}{H^+} - H^+ \tag{5.98}$$

Taking the derivative leads to

$$dC_B = \left[-\frac{K_w}{(H^+)^2} - 1 \right] dH^+ \tag{5.99}$$

But because pH $= -\ln(H^+)/2.3$, the derivative gives us

$$dH^+ = dpH[-2.3(H^+)] \tag{5.100}$$

Substituting for dH^+ results in the final expression

$$\frac{dC_B}{dpH} = \beta_{H_2O} = 2.3 \left[\frac{K_w}{H^+} + H^+ \right] = 2.3 \, [OH^- + H^+] \tag{5.101}$$

This equation is plotted in Figs. 5.10 and 5.11, which show that strong acids and bases have considerable buffer capacity below pH 4 and above pH 10. β_{H_2O} is negligible in natural waters at intermediate pH's. The units of the buffer capacity, as computed from Eq. (5.101) are equivalents of strong base per liter per pH unit. Thus at pH 7 we find $\beta_{H_2O} = 4.6 \times 10^{-7}$ eq/L pH or 4.6×10^{-4} meq/L pH. At pH 2, $\beta_{H_2O} = 2.3 \times 10^{-2}$ eq/L pH or 23 meq/L pH. Generally buffer index units of meq/L pH are preferred. It is important to remember that the buffer capacity contributions of all chemical species in a system are always positive and additive. In aqueous systems β_{H_2O} is always included.

5.10.2 Weak Monoprotic Acids

To derive a function for the buffer index of a weak monoprotic acid, HA, we begin as before with equilibrium, mass-balance, and charge-balance equations, and first derive an equation for the titration curve. We are given the following expressions:

$$K_a = \frac{(H^+)(A^-)}{(HA)} \tag{5.102}$$

$$C_A = HA + A^- \tag{5.103}$$

or

$$HA = C_A - A^- \tag{5.104}$$

and electroneutrality

$$Na^+ + H^+ = A^- + OH^- \tag{5.105}$$

where C_A is the total concentration of the acid species. But because $C_B = Na^+$, we obtain

$$C_B = A^- + OH^- - H^+ \tag{5.106}$$

Our goal is to rewrite Eq. (5.106) in the proton condition. First eliminate HA from Eq. (5.102) by substitution of Eq. (5.104). Solving for (A^-) gives

$$(A^-) = \frac{K_a C_A}{[K_a + (H^+)]} \tag{5.107}$$

Now, in the proton form Eq. (5.106) becomes

$$C_B = \frac{K_a C_A}{[K_a + (H^+)]} + \frac{K_w}{(H^+)} - (H^+) \tag{5.108}$$

Differentiating we find

$$-\frac{dC_B}{dH^+} = \left[\frac{K_a C_A}{[K_a + (H^+)]^2} + \frac{K_w}{(H^+)^2} + 1 \right] \tag{5.109}$$

But $dH^+ = -2.3(H^+)\,dpH$, so substitution gives the equation

$$\beta = 2.3 \left[(H^+) + \frac{K_w}{(H^+)} + \frac{K_a C_A (H^+)}{[K_a + (H^+)]^2} \right] \tag{5.110}$$

Note that the first two bracketed terms are the buffer capacity of water, and the third term is the buffer capacity due to the weak acid. For intermediate pH's the water terms are negligible and we have

$$\beta = 2.3 \, K_a C_A \left[\frac{(H^+)}{[K_a + (H^+)]^2} \right] \tag{5.111}$$

One can now determine the pH at which a weak monoprotic acid exhibits its maximum buffer capacity (β_{max}). This involves differentiating Eq. (5.111) with respect to pH and then setting the expression for $d\beta/dpH$ equal to zero. This leads to $K_a = [H^+]$ for the condition of maximum buffer capacity. Substituting this fact into Eq. (5.111) indicates that

$$\beta_{max} = 0.58 \, C_A \tag{5.112}$$

In other words, the pH for the maximum buffer capacity of a weak acid is defined by $pH = pK_a$, and the value of β_{max} is a function only of C_A, the total acid-species concentration. The same reasoning applies to the maximum buffer capacity of weak bases.

5.10.3 Polyprotic Acids

The buffer capacity or index of a polyprotic acid is the sum of the contributions from each successive acid species. Or, in general terms,

$$\beta = \beta_{H_2O} + \beta_{HA} + \beta_{HA'} + \beta_{HA''} + \cdots \tag{5.113}$$

Remembering that for a monoprotic weak acid the buffer index equals

$$\beta = 2.3 \left[(H^+) + \frac{K_w}{(H^+)} + \frac{K_a C_A (H^+)}{[K_a + (H^+)]^2} \right] \tag{5.110}$$

the buffer index for a polyprotic acid will equal the sum of similar weak-acid terms for each successive dissociation step. For a diprotic acid the β index expression is, therefore,

$$\beta = 2.3 \left[(H^+) + \frac{K_w}{(H^+)} + \frac{K_1 C_A (H^+)}{[K_1 + (H^+)]^2} + \frac{K_2 C_A (H^+)}{[K_2 + (H^+)]^2} \right] \tag{5.114}$$

Where C_A is the total concentration of acid species and K_1 and K_2 are the first and second stepwise dissociation constants of the acids. This equation can be used to compute the buffer index of a polyprotic acid as long as successive dissociation constants differ by at least 20 times (this assures a calculation error of 5% or less). In other words, for a diprotic acid K_2/K_1 should be less than 0.05 (cf. Butler 1964). Thus, for example, Eq. (5.114) may be used to compute the buffer index due to species of carbonic acid, for which $K_1 = 10^{-6.35}$ and $K_2 = 10^{-10.33}$, or β for species of silicic acid, for which $K_1 = 10^{-9.86}$ and $K_2 = 10^{-13.1}$.

5.10.4 Carbonic Acid

The buffer capacity due to carbonic acid species may be computed assuming either constant total carbonate, C_T, which implies a closed system, or for constant CO_2 pressure, which suggests an open system. We will first assume constant C_T and constant Ca^{2+} with an HCl titrant. The mass-balance equations are

$$C_T = H_2CO_3^\circ + HCO_3^- + CO_3^{2-} \tag{5.115}$$
$$C_A = Cl^-$$

The charge balance is

$$2Ca^{2+} + H^+ = HCO_3^- + 2CO_3^{2-} + Cl^- + OH^- \tag{5.116}$$

so that
$$C_A = 2Ca^{2+} + H^+ - OH^- - HCO_3^- - 2CO_3^{2-} \tag{5.117}$$

From Eqs. (5.22) and (5.23), and the expression for K_w we have

$$(OH^-) = \frac{K_w}{(H^+)} \qquad (CO_3^{2-}) = \frac{C_T}{\alpha_H} \qquad \text{and} \qquad (HCO_3^-) = \frac{C_T (H^+)}{K_2 \alpha_H} \tag{5.118}$$

where α_H is given by Eq. (5.25).

Substituting into Eq. (5.117) and combining terms gives us the equation of the titration curve,

$$C_A = 2Ca^{2+} + H^+ - \frac{K_w}{H^+} - \frac{C_T}{\alpha_H}\left[\frac{H^+}{K_2} + 2\right] \tag{5.119}$$

Differentiating we have

$$\frac{dC_A}{dH^+} = 1 + \frac{K_w}{H^+} + \frac{C_T}{(\alpha_H)^2}\left[\frac{(H^+)^2}{K_1K_2^2} + \frac{4(H^+)}{K_1K_2} + \frac{1}{K_2}\right] \tag{5.120}$$

But $dH^+ = -2.3(H^+)d$pH, and so

$$\beta = 2.3\left[(H^+) + (OH^-) + \frac{C_T(H^+)}{K_1K_2(\alpha_H)^2}\left(\frac{(H^+)^2}{K_2} + 4(H^+) + K_1\right)\right] \tag{5.121}$$

where the H^+ and OH^- terms are the contribution of β_{H_2O} and the final terms are the buffer capacity due to species of carbonic acid. With $K_1 = 10^{-6.4}$ and $K_2 = 10^{-10.3}$, substitution gives

$$\beta = 2.3\left[(H^+) + (OH^-) + \frac{C_T\,10^{10.3}\,(H^+)\,[10^{16.7}\,(H^+)^2 + 10^{7.0}(H^+) + 1]}{[10^{16.7}\,(H^+)^2 + 10^{10.3}\,(H^+) + 1]^2}\right] \tag{5.122}$$

where β is in (eq/L pH). This function is plotted in Fig. 5.10, which shows the overall buffer capacity and the separate contribution of β_{H_2O}. The β curve shows the features of a monoprotic acid with $\beta_{max} = 0.58C_T = 0.58$ meq/L pH and the maximum buffer capacity of carbonic acid at pH = pK_1. Note that a second β_{max} for carbonic acid at pH = $pK_2 = 10.3$ is masked by the rapid increase in β_{H_2O} above pH 10.

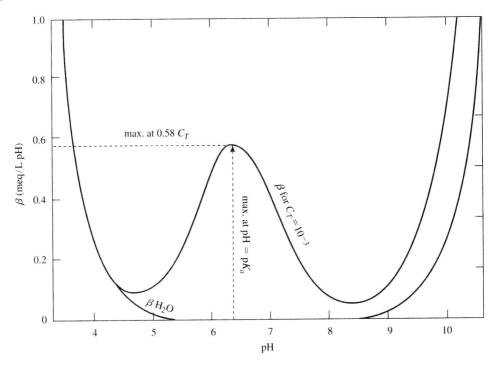

Figure 5.10 A linear plot of the buffer capacity of carbonic acid species as a function of pH for $C_T = 10^{-3.0}$ M showing that the maximum buffer capacity equals 0.58 C_T, and occurs at pH = $pK_1(H_2CO_3^\circ) = 6.35$. The lower curve is the buffer capacity of water, β_{H_2O}.

Next, consider the buffer capacity of carbonic acid species at constant CO_2 pressure. In the previous calculation β was obtained by taking the derivative of an equation for C_B in the proton condition that had been obtained from the solution charge-balance equation. We will instead base the calculation of β directly on the equation for C_B, differentiating in parts. Thus, we write

$$\beta = \frac{dC_B}{dpH} = \frac{d(HCO_3^-)}{dpH} + \frac{2d(CO_3^{2-})}{dpH} + \frac{d(OH^-)}{dpH} - \frac{d(H^+)}{dpH} \tag{5.123}$$

Equation (5.20) can be rewritten,

$$(HCO_3^-) = (K_1 K_{CO_2} P_{CO_2})10^{pH} \tag{5.124}$$

At a constant CO_2 pressure the expression in parentheses is set equal to a constant K'. With natural logs we have

$$\ln (HCO_3^-) = 2.3 \ln K' + 2.3 \, pH \tag{5.125}$$

Taking the derivative leads to

$$\frac{d(HCO_3^-)}{dpH} = 2.3 \, (HCO_3^-) \tag{5.126}$$

Combining carbonic acid constant expressions we can derive

$$(CO_3^{2-}) = (K_1 K_2 K_{CO_2} P_{CO_2})10^{2pH} \tag{5.127}$$

We set the constant term in parentheses equal to K'', take natural logs, and then the derivative to obtain

$$\frac{d(CO_3^{2-})}{dpH} = 4.6 \, (CO_3^{2-}) \tag{5.128}$$

Using a similar approach to the OH^- and H^+ terms gives

$$\frac{d(OH^-)}{dpH} = 2.3 \, (OH^-) \qquad \text{and} \qquad \frac{d(H^+)}{dpH} = -2.3 \, (H^+) \tag{5.129}$$

The separate terms now add up to the expression for total β,

$$\beta = 2.3 \, [(HCO_3^-) + 4(CO_3^{2-}) + (H^+) + (OH^-)] \tag{5.130}$$

This equation shows the β contribution of each species, but is not solved readily unless substitutions are made to put it in the proton condition.

5.10.5 Calcite-Carbonic Acid

When an acidic or basic water is mixed with a second water, the pH response of the mixture is immediate. Any resistance to pH change exhibited by the second water then reflects what could be termed its immediate buffer capacity. In soils and sediments the presence of reacting minerals can vastly increase the buffer capacity of associated waters. However, most reactions involving minerals are relatively slow to equilibrate (Chap. 2), thus the buffer capacity response of minerals can be viewed as long term. As salts of strong bases and weak acids, mineral buffer capacity generally resists a drop in pH. Among common minerals, calcite dissolves much more rapidly than the aluminosilicates. The presence of a few tenths of a percent or more calcite in soils and sediments is, therefore, extremely important in order to avoid environmental damage from acid rain or acid wastewaters.

We will compute β for pure water in equilibrium with calcite in a closed system with constant total carbonate, C_T. Dissolved Ca^{2+} is thus controlled by calcite saturation, as defined by $(Ca^{2+}) = K_c/(CO_3^{2-})$ where $K_c = 10^{-8.48}$. With an HCl titrant $C_A = Cl^-$, and the charge-balance equation is

$$2Ca^{2+} + H^+ = HCO_3^- + 2CO_3^{2-} + Cl^- + OH^- \tag{5.131}$$

Transposing after substitution for Cl^- we have

$$C_A = H^+ - OH^- + 2Ca^{2+} - HCO_3^- - 2CO_3^{2-} \tag{5.132}$$

Now substituting to attain the proton condition gives

$$C_A = (H^+) - \frac{K_w}{(H^+)} + \frac{2K_c\alpha_H}{C_T} - \frac{C_T(H^+)}{K_2\alpha_H} - \frac{2C_T}{\alpha_H} \tag{5.133}$$

Differentiation and introduction of β and equilibrium constants for 25°C result in

$$\beta = 2.3 \left[(H^+) + (OH^-) + \frac{C_T\, 10^{10.3}(H^+)\, [10^{16.7}\,(H^+)^2 + 10^{7.0}\,(H^+) + 1]}{[10^{16.7}\,(H^+)^2 + 10^{10.3}\,(H^+) + 1]^2} \right]$$

$$+ 2.3 \left[\frac{10^{8.5}\,(H^+)}{C_T}\, (2H^+ + 10^{-6.4}) \right] \tag{5.134}$$

The first bracketed term is the contribution of water and dissolved carbonate species. The second term in brackets is the calcite contribution. Equation (5.134) is plotted in Fig. 5.11 which shows that, in equilibrium with calcite below about pH 8.5 for $C_T = 10^{-3}$, system buffer capacity is 100 times or more greater than it is without calcite. Because calcite can both dissolve and precipitate under ambient conditions, its buffer capacity provides resistance to a pH change in either direction.

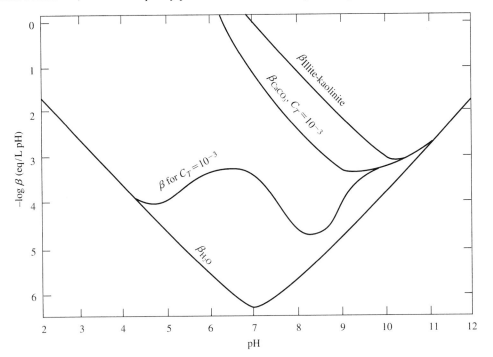

Figure 5.11   A log plot of the buffer capacity due to carbonic acid species for $C_T = 10^{-3.0}$ M (see Fig. 5.10); at saturation with respect to calcite for $C_T = 10^{-3.0}$ M; and for equilibrium between the clays illite and kaolinite. The lower curve is β_{H_2O}.

5.10.6 Clays

As a second example of mineral-controlled buffer capacity, consider the reaction in pure water between the clays kaolinite and illite (here assumed the same as muscovite), which may be written

$$2KAl_3Si_3O_{10}(OH)_2 + 2H^+ + 3H_2O = 3Al_2Si_2O_5(OH)_4 + 2K^+ \tag{5.135}$$

illite   kaolinite

for which
$$K_{eq} = 10^{6.5} = \frac{(K^+)}{(H^+)} \tag{5.136}$$

(Garrels and Christ 1965). With HCl as the titrant ($C_A = Cl^-$) charge balance is

$$H^+ + K^+ = OH^- + Cl^- \tag{5.137}$$

The total acidity equation and its proton form after substitution are then

$$C_A = K^+ + H^+ - OH^- = 10^{6.5} (H^+) + (H^+) = \frac{10^{-14.0}}{(H^+)} \tag{5.138}$$

Taking the derivative

$$\frac{dC_A}{dH^+} = \left[1 + 10^{6.5} + \frac{10^{-14.0}}{(H^+)^2} \right] \tag{5.139}$$

Introducing the definition of β

$$\beta = -\frac{dC_A}{dpH} = 2.3\,(H^+)\frac{dC_A}{dH^+} = 2.3 \left[(H^+) + \frac{K_w}{(H^+)} + 10^{6.5}\,(H^+) \right] \tag{5.140}$$

This equation is plotted in Fig. 5.11 which shows that the reaction at equilibrium has about 10 times more buffer capacity than calcite. Because this reaction will not often be at equilibrium, however, the buffer capacity is a maximum possible value. Further, the reaction is usually irreversible with kaolinite more often stable than illite in weathering environments. For this reason the reaction resists a pH decrease, but not an increase.

    Although clay reactions are slower than those involving carbonates, they are fast enough to be very important, particularly in acid-contaminated systems. The alkalinity and buffer capacity of the clays in clay liners and soils is often more important than that of small amounts of associated carbonates. A few percent of carbonates are rapidly dissolved by a drop in pH below about 4.5, whereas the clays provide substantial resistance to a further lowering of pH values. This is evident from a lab batch study in which a waste mill-tailings solution of pH = 1.2 and total acidity of 0.35 eq/L was added to rock pastes A and B to assess the ability of each rock to neutralize the tailings solution. The rocks were chiefly calcareous claystones (rock A, 66% clay and 9% calcite; rock B, 65% clay and 2.4% calcite). Titration curves in Fig. 5.12 show that the carbonates provide a limited alkalinity and buffer capacity above pH 4.5, which is rapidly consumed by the tailings solution. The clays offer about five times more resistance to a further drop from pH 4.5 to 2. Figure 5.13 gives the results of experiments in which the same acid tailings solution was added to pastes of 10 different rocks to maintain mixture pH's at 2.0 or 4.5 for 24 hours. The plot shows that calcite contributes about one-fourth of the rocks' total alkalinity, which is fully consumed above pH 4.5. Between pH 4.5 and 2, clays provide three times the alkalinity available from the calcite.

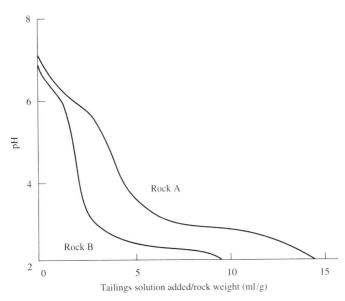

Figure 5.12   Titration of lightly crushed pastes of rock A (66% clay, 9% calcite) and rock B (65% clays, 2.4% calcite) with acid-tailings solution of pH 1.2, $C_A = 0.35$ eq/L (Langmuir and Nordstrom 1995).

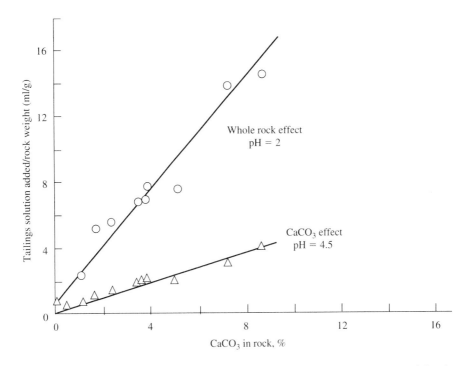

Figure 5.13   Volume of tailings solution needed (pH = 1.2, $C_A = 0.35$ eq/L) per weight of rock paste for 10 different rocks to bring their mixtures down to pH 4.5 and pH 2.0 in 24-hour tests. The pH = 4.5 line reflects the neutralizing capacity of calcite. The difference between this line and the whole rock line at pH = 2 indicates the neutralizing capacity of the clays (Langmuir and Nordstrom 1995).

STUDY QUESTIONS

1. Define pH. Explain briefly why pH is important to our understanding of acid-base equilibria, ion adsorption, metal-ligand complexation, oxidation-reduction, and the solubility of most minerals.

2. The hydrogen electrode is the standard method of measuring pH. How is the voltage measured by the electrode related to the pH? What is the Nernst factor?

3. What is a pH buffer solution? Explain how one calibrates the pH measured with a glass/reference electrode pair in an unknown solution. What is E_j, the liquid junction potential?

4. Define an acid and a base. Distinguish the behavior of strong and weak acids and bases with examples. Define amphiprotic and conjugate acid and base.

5. What are the most important strong acids and weak acids in natural waters, and under what pH conditions do they control pH?

6. Sketch and label important features of the percent distribution diagram for carbonate species as a function of pH.

7. Define total carbonate (C_T), total alkalinity (C_B), and total acidity (C_A) for a typical natural freshwater, and explain their significance from a water-use viewpoint.

8. Explain how the CO_2 content and pH of natural waters is affected by processes and reactions including the dissolution of $CO_2(g)$, photosynthesis and respiration, aerobic decay, anaerobic decay (fermentation), nitrate reduction, and denitrification and sulfate reduction.

9. Give reasons why the CO_2 content of soils and groundwaters is generally higher than that of surface waters.

10. What is the respiratory coefficient and how is it applied?

11. In some groundwaters (many coastal plain groundwaters), the pH increases and the CO_2 pressure decreases with groundwater flow. However, in a few groundwaters (as in the Floridan carbonate aquifer) the opposite occurs. Explain what reactions determine these contrasting trends in pH and CO_2 pressure.

12. Describe the origin and characteristics of humic and fulvic acids and discuss their role in metal transport in soils.

13. Explain why the pH of natural waters generally lies between about 5 and 9.

14. Write the equation for the total acidity of an acid mine water.

15. Most minerals are salts of weak acids and strong bases. Explain this statement and discuss its implications in terms of chemical weathering, controls on the pH of subsurface waters, and the capacity of most rocks to neutralize acid wastes.

16. What is the titration mass-balance equation?

17. The endpoints of titrations of carbonate and bicarbonate alkalinity with an acid depend on the total carbonate of the water. Explain.

18. The equation of a titration curve of an acid or base may be derived, taking into account the charge-balance equation for the solution, the total acidity (C_A) or alkalinity (C_B) equation, and deriving an equation in terms of stability constants, known concentrations and volumes, and the pH. This resultant equation, which can be solved, is said to be "in the proton condition." Explain this statement as it applies to a simple titration curve.

19. In the derivation of an equation for a titration curve, how are corrections made to account for the volume change on mixing of the titrant and unknown solution?

20. Be able to calculate the pH of a solution that contains a known concentration of the salt of a strong base and weak acid, or of a weak base and strong acid.

21. Define buffer capacity and explain how the buffer capacity of a water relates to the titration curve for its acidity and alkalinity.

22. What are the properties of a good pH buffer?

23. The buffer capacity is always high for very acid or alkaline natural waters and may also be high at intermediate pH's for a different reason. Explain.

24. One can compute the buffer capacity of a water from its chemical analysis or graphically from its strong acid or strong base titration curve. Explain.

25. Be able to set up the equations necessary to compute the buffer capacity of a water as a function of pH from its chemical analysis, given that reactions involving minerals contribute to the buffer capacity.

26. Know the meaning of *reversible* and *irreversible* and *immediate* and *long-term buffer capacity.*

PROBLEMS

Assume the activities of dissolved species are equal to their concentrations in the following calculations.

1. What is the pH of a 3×10^{-4} mol/L solution of HCl?

2. What is the pH of a 3×10^{-4} mol/L solution of acetic acid (CH_3COOH), given that the acidity constant $K_a = 1.75 \times 10^{-5}$? In your answer also report solution concentrations of CH_3COOH and CH_3COO^- (acetate ion). Compare your computed pH to the value obtained for the same HCl concentration in problem 1.

3. Given that the acidity constant, K_a, of hydrogen cyanide (HCN) is 6.17×10^{-10} at 25°C, what is the pH of a 10^{-3} mol/L sodium cyanide solution, and what are the concentrations of HCN and CN^- in that solution?

4. A natural water contains both carbonic acid and acetic acid (CH_3COOH) species. Dissociation constants for carbonic acid species and the procedure used to derive and plot a distribution diagram for carbonic acid species are given in this chapter. The dissociation constant of acetic acid is $K_a = 10^{-4.75}$ at 25°C. Assuming that the total carbonate concentration in the water is constant, and equals 2×10^{-3} M, and that the total acetic acid-acetate ion concentration is 10^{-4} M, compute and plot a distribution diagram that shows the molar concentrations of each of the carbonate and acetate species of H^+ and OH^- as function of pH between pH 2 and 12. The ordinant should be in terms of $\log_{10}$ M of each of the species, and the abscissa should be in pH units. Discuss the relative importances of the various species as a function of pH and how the plot relates to the dissociation constants of the acids.

5. What is the pH of a 5×10^{-3} molal Na_2CO_3 solution in a container isolated from the atmosphere? This problem is based on a calculation given by Drever (1988).

6. The dissociation constant K_a for acetic acid is $10^{-4.76}$. Given that a water has a total acetic acid concentration of 10^{-3} mol/kg, what is the pH at which the acid has its maximum buffer capacity, and what is the value of the buffer capacity at that maximum?

7. Derive an equation for the buffer capacity of a solution due to its carbonate alkalinity at constant CO_2 pressure. Base the calculation on the equation for total carbonate alkalinity (C_B), but do not convert that equation to the proton form before differentiation. This approach was suggested by van Breemen and Wielamaker (1974a; 1974b).

8. Below is the total analysis of a raffinate (tailings waste liquor) from the acid processing of uranium ore. The raffinate was filtered before analysis. Raffinate pH is 0.9.

Aqueous species	Total concentration (mg/L)	Aqueous species	Total concentration (mg/L)
Al^{3+}	3600	Mg^{2+}	1600
Ca^{2+}	750	Mn^{2+}	150
Cl^-	2900	Na^+	5300
Fe^{2+}	10,500	NH_4^+ (as N)	1100
K^+	270	SO_4^{2-}	81,000

The raffinate is to be released to compacted-clay-lined "evaporation" ponds constructed in local sandstone and silty-sand formations. These formations contain a few percent of clays such as kaolinite and smaller amounts of calcite cement. You are asked to evaluate the capacity of the rock to neutralize the raffinate acid-

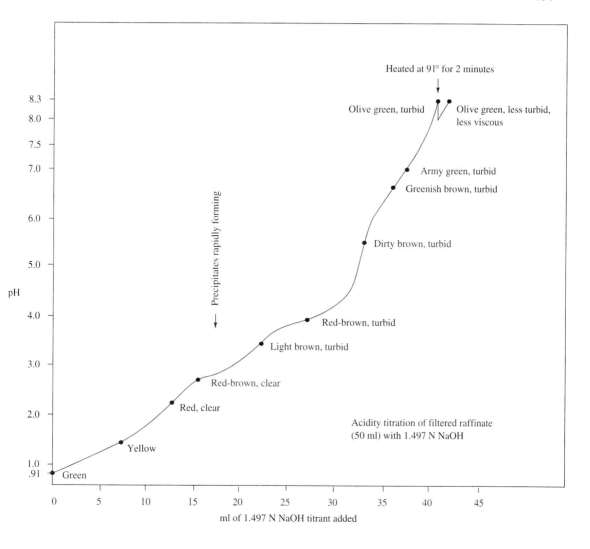

Acidity titration of filtered raffinate (50 ml) with 1.497 N NaOH

ity and so to protect local surface and groundwater quality. Your initial task is to measure raffinate acidity and buffer capacity. In tests to determine C_A, the total acidity (equivalents per liter) of the raffinate, three approaches are taken. In all three, atmospheric oxygen is either present or assumed present. First, the raffinate is neutralized with reagent grade calcite. It is found that 23 g of $CaCO_3$ neutralizes 150 ml of the raffinate to a pH of 7.66 (assume to pH 8.3). Second, 50 ml of the raffinate is neutralized by titration with 1.497 N NaOH solution to pH 8.3. The titration curve is plotted in the accompanying figure. Third, the acidity is computed from the total analysis tabulated above.

(a) Calculate C_A from the calcite neutralization experiment.

(b) Calculate C_A from the titration curve.

(c) Calculate C_A from the total analysis following the steps outlined below. First, write the equation that defines the total acidity. (Hint: Check the pK_a values (etc.) for the above species to see which are significant acids that will be consumed by base titrant up to pH 8.3 under oxidizing conditions.)

   A key issue in this part of the problem is how to determine the concentration of bisulfate ion. This can be done in two ways, although suggestion (1), which follows, is approximate.

(1) Ignoring complexing, compute the bisulfate and sulfate concentrations from the mass-balance equation for total sulfate and the charge-balance equation for the raffinate, assuming any charge in-balance can be explained by adjusting the relative amounts of sulfate and bisulfate.

(2) Input the raffinate analysis into MINTEQA2 and let the program speciate the water as a means of determining the bisulfate. Compare this result to bisulfate computed above, and discuss the difference. Now calculate C_A using the bisulfate and H^+-ion concentrations computed by MINTEQA2. Remember, the acidity is a sum of equivalent concentrations, not activities. Assume the raffinate will be completely oxidized during its titration.

(d) Compare and discuss the results of the three approaches to C_A.

(e) Compare the ionic strength computed by MINTEQA2 in part (c2) to the stoichiometric ionic strength you can determine from the raffinate chemical analysis given above and explain the difference.

(f) Remembering the definition of buffer capacity, β, determine the buffer capacity of the raffinate graph-ically from its titration curve in Fig. 5.7 and plot β from pH 0.9 to 8.3 in meq/L pH unit.

(g) This question asks you to explain the general shape of the titration and buffer capacity curves in terms of specific acid-base reactions as a function of pH. How might the acid-base and buffer capacity be-havior of the raffinate affect the detailed release and migration of raffinate liquids from a pond into and through surrounding geologic formations?

Some raffinate neutralization reactions are strictly aqueous (homogeneous) reactions, others in-volve mineral precipitation. To answer this question, you will need to titrate the raffinate with a base using the sweep option in MINTEQA2. Use 0.5 pH unit increments in the titration. Solids are likely to precipitate during your computer titration, which of course is performed at constant volume. First, run sweep with no possible solids. By examining the saturation indices output you can get an idea of what minerals might precipitate. Some will later dissolve as the pH further increases during neutralization. Solids that perhaps are not inhibited kinetically from precipitating during a titration include $AlOHSO_4$, gypsum, anhydrite, ferrihydrite, $Fe(OH)_{2.7}Cl_{0.3}$, birnessite, and basaluminite. Remember, you can de-termine the compositions of these minerals in the program documentation, or by examining their stoi-chiometry in the file *type6.dbs*. Now run sweep inputting these minerals as possible minerals. Your sweep run may not converge for the full pH range of the computer titration. Set the number of itera-tions allowed to 200 to assist convergence. (The program still may not converge for a full 20 sweep in-crements.) Convergence is assisted if you add an unreactive cation such as Rb^+ or anion such as ClO_4^- to maintain charge balance.

(h) In light of your answer to part (g) explain possible consequences and implications of the raffinate acid-base reactions if the raffinate is released into and migrates in a sandstone formation that contains traces of calcite.

(i) Could an equation in the proton form for the buffer capacity of the raffinate be derived starting with the acidity equation in part (c)? Derive such an equation showing first the acid-base terms that reflect ho-mogeneous reactions and discuss how you could write the terms that involve heterogeneous reactions.

6

Carbonate Chemistry

6.1 OCCURRENCE AND STABILITY OF THE CALCIUM-MAGNESIUM CARBONATES

Limestone (chiefly calcite, $CaCO_3$) and dolomite rocks (chiefly dolomite, $CaMg(CO_3)_2$) are exposed at about 20% of Earth's surface. Carbonate detritus, fossil shell materials, and carbonate cements are also common in noncarbonate sedimentary rocks and arid-climate soils. The carbonate minerals found in such occurrences, in decreasing order of importance, are: calcite, dolomite, magnesian calcites ($Ca_{1-x}Mg_xCO_3$ where x is usually <0.2), aragonite (a $CaCO_3$ polymorph) and, perhaps, magnesite. As a rule of thumb, when such materials are present in silicate or aluminosilicate rocks or soils at a level of about 1% or more, they will tend to dominate the chemistry of the soil or groundwater. This fact is extremely important when one is concerned about the ability of a rock to neutralize acid mine waters, other acid wastewaters, or acid rain.

Equilibrium constants for minerals and aqueous species in the system CaO-MgO-CO_2-SO_3–H_2O at 25°C are given in Tables 6.1 and 6.2. Solubility product data in Table 6.1 indicate that calcite is more stable (i.e. less soluble) than aragonite. The K_{sp} data correspond to $\Delta G_r^\circ = -191$ cal/mol K for the reaction aragonite $\rightleftarrows$ calcite. This is in excellent agreement with $\Delta G_r^\circ = -196 \pm 31$ cal/mol K proposed by Plummer and Busenberg (1982) and -198 ± 5 cal/mol K suggested by Konigsberger et al. (1989). Aragonite, the high-density polymorph of $CaCO_3$, is thermodynamically stable relative to calcite at 25°C only at pressures above 3.0 kb (Konigsberger et al. 1989). The abundance of aragonite in recent marine sediments reflects its often more favorable kinetics of nucleation and crystal growth in seawater than that of calcite (cf. Drever 1988) and the ability of some carbonate-mineralizing organisms to preferentially nucleate either aragonite or calcite from supersaturated waters (cf. Falini et al. 1996). When groundwaters near saturation with calcite at low temperatures are pumped to the surface and rapidly heated, aragonite, rather than calcite, may precipitate (cf. Jenne 1990).

Vaterite is a rare mineral that only forms from solutions highly supersaturated with respect to calcite and aragonite. Ogino et al. (1987) precipitated vaterite with calcite between 14 and 30°C from laboratory solutions in which the $[Ca^{2+}][CO_3^{2-}]$ product equaled about 10^{-6}. The vaterite was transformed to calcite in several hours.

TABLE 6.1 Solubility product expressions and K_{sp} values for some minerals in the system $CaO-MgO-CO_2-SO_3-H_2O$ at 25°C and 1 bar total pressure

Mineral	Formula	K_{sp} expression	$-\log K_{sp}$
portlandite[†]	$Ca(OH)_2$	$[Ca^{2+}][OH^-]^2$	5.23
brucite[‡]	$Mg(OH)_2$	$[Mg^{2+}][OH^-]^2$	11.16
calcite[‡]	$CaCO_3$	$[Ca^{2+}][CO_3^{2-}]$	8.48
aragonite[‡]	$CaCO_3$	$[Ca^{2+}][CO_3^{2-}]$	8.34
vaterite[§]	$CaCO_3$	$[Ca^{2+}][CO_3^{2-}]$	7.91
magnesite[‖]	$MgCO_3$	$[Mg^{2+}][CO_3^{2-}]$	4.9, 7.93
nesquehonite[#]	$MgCO_3 \cdot 3H_2O$	$[Mg^{2+}][CO_3^{2-}][H_2O]^3$	5.58
lansfordite[††]	$MgCO_3 \cdot 5H_2O$	$[Mg^{2+}][CO_3^{2-}][H_2O]^5$	5.45
artinite[‡‡]	$MgCO_3 \cdot Mg(OH)_2 \cdot 3H_2O$	$[Mg^{2+}][CO_3^{2-}][OH^-]^2[H_2O]^3$	15.4
hydromagnesite[§§]	$4MgCO_3 \cdot Mg(OH)_2 \cdot 4H_2O$	$[Mg^{2+}]^5[CO_3^{2-}]^4[OH^-]^2[H_2O]^4$	35.0
huntite[‖‖]	$CaMg_3(CO_3)_4$	$[Ca^{2+}][Mg^{2+}]^3[CO_3^{2-}]^4$	29.14
dolomite[‡]	$CaMg(CO_3)_2$	$[Ca^{2+}][Mg^{2+}][CO_3^{2-}]^2$ (ordered)	17.09
		(disordered)	16.54
anhydrite[‡]	$CaSO_4$	$[Ca^{2+}][SO_4^{2-}]$	4.36
gypsum[‡]	$CaSO_4 \cdot 2H_2O$	$[Ca^{2+}][SO_4^{2-}][H_2O]^2$	4.58

Sources: [†]Langmuir (1968). [‡]Nordstrom et al. (1990). [§]Plummer and Busenberg (1982). [‖]K_{sp}(nesquehonite) in this table, and other data from Langmuir (1965) suggest K_{sp}(magnesite) = $10^{-4.9}$; based on the literature and high temperature phase equilibria involving magnesite, Koziol and Newton (1995) propose $\Delta H_f^\circ = -265.70$ kcal/mol and $S^\circ = 15.557$ cal/mol K for magnesite at 25°C, which, with other thermodynamic data from Wagman et al. (1982), leads to K_{sp}(magnesite) = $10^{-7.93}$. [#]K_{sp} = $10^{-5.58 \pm 0.03}$ based on 15 reversed solubility runs in up to 5.5 mol/kg NaCl solutions (Ranville 1988). [††]Based on the nesquehonite-lansfordite equilibrium (Langmuir 1965), and K_{sp}(nesquehonite) this table. [‡‡]Recomputed from Langmuir (1965) (K_{sp} = $10^{-18.3}$ calculated from data of Hemingway and Robie (1973), makes artinite stable relative to brucite plus hydromagnesite, which seems doubtful). [§§]Formula from Robie and Hemingway (1973); K_{sp} recomputed from data given by Langmuir (1965). [‖‖]Based on electrochemical cell measurements, Walling et al. (1995) propose ΔG_f°(huntite) = -1002.8 ± 1.5 kcal/mol; with free energy data for the ions from Wagman et al. (1982) this gives $K_{sp} = 10^{-29.14}$; which is in good agreement with measured entropy and enthalpy data from Hemingway and Robie (1973), $\Delta G_f^\circ = -1004.6 \pm 0.4$ kcal/mol and $K_{sp} = 10^{-30.52}$.

TABLE 6.2 Some equilibrium constants in the system $CaO-MgO-CO_2-SO_3-H_2O$ at 25°C and 1 bar pressure

Equilibrium constant expression	$-\log K_{eq}$
$K_w = [H^+][OH^-]$	14.00
$K_{CO_2} = [H_2CO_3^\circ]/P_{CO_2}$	1.47
$K_1 = [H^+][HCO_3^-]/[H_2CO_3^\circ]$	6.35
$K_2 = [H^+][CO_3^{2-}]/[HCO_3^-]$	10.33
$K(CaOH^+) = [Ca^{2+}][OH^-]/[CaOH^+]$	1.22
$K(CaCO_3^\circ) = [Ca^{2+}][CO_3^{2-}]/[CaCO_3^\circ]$	3.22
$K(CaHCO_3^+) = [Ca^{2+}][HCO_3^-]/[CaHCO_3^+]$	1.11
$K(MgCO_3^\circ) = [Mg^{2+}][CO_3^{2-}]/[MgCO_3^\circ]$	2.98
$K(MgHCO_3^+) = [Mg^{2+}][HCO_3^-]/[MgHCO_3^+]$	1.07
$K(CaSO_4^\circ) = [Ca^{2+}][SO_4^{2-}]/[CaSO_4^\circ]$	2.30
$K(MgSO_4^\circ) = [Mg^{2+}][SO_4^{2-}]/[MgSO_4^\circ]$	2.37

Source: All values are from Nordstrom et al. (1990).

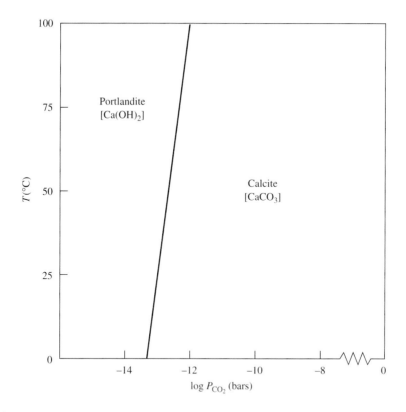

Figure 6.1   Stability relations among solid phases in the system $CaO–CO_2–H_2O$ from 0 to 100°C at 1 bar total pressure as a function of P_{CO_2} at a water activity of unity.

Phase relations in the system $CaO–CO_2–H_2O$ from 0 to 100°C are shown in Fig. 6.1. Given the fact that CO_2 pressures in natural environments are rarely less than about 10^{-5} bar, and that equilibrium between calcite and portlandite is at a CO_2 pressure of $10^{-13.10}$ bar at 25°C, calcite is clearly the stable phase in Earth-surface environments.

Magnesite is the most common magnesium carbonate mineral in geological environments. In spite of this fact, it is rarely observed precipitating from natural waters. Further, its low-temperature solubility has been extremely difficult to measure because of its very slow rate of dissolution in the laboratory (cf. Langmuir 1965). The $-\log K_{sp}$ of 4.9 for magnesite given in Table 6.1 is consistent with its general behavior at low temperatures and is the basis for the phase diagrams in Figs. 6.2 and 6.3. Nevertheless, recent solubility measurements reported by Khodakovsky (1993) support a $-\log K_{sp}$ for magnesite near 8, as does entropy and enthalpy data summarized by Koziol and Newton (1995), which suggests $K_{sp} = 10^{-7.93}$ (see Table 6.1 footnotes). The MINTEQA2 data base (Allison et al. 1991) lists $K_{sp}(\text{magnesite}) = 8.03$.

In seawater the ion activity product $[Mg^{2+}][CO_3^{2-}]$ equals about $10^{-7.0}$. This makes the oceans about 10-fold supersaturated with respect to magnesite if the lower solubilities just given are correct. However, magnesite has never been observed to precipitate from seawater. Figures 6.2 and 6.3 suggest that if the readily precipitated hydrates of magnesium carbonate are heated, dehydrated at

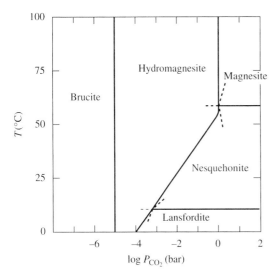

Figure 6.2 Stability relations among solid phases in the system MgO-CO$_2$-H$_2$O as a function of P_{CO_2} from 0 to 100°C, at a water activity of unity and 1 bar total pressure. The figure is drawn assuming K_{sp}(magnesite) = $10^{-4.9}$ at 25°C. If, instead, K_{sp}(magnesite) = $10^{-7.93}$, the hydrated Mg carbonates are metastable relative to magnesite, and the brucite/magnesite boundary is at P_{CO_2} = $10^{-6.62}$ bar. See Table 6.1. From D. Langmuir, Stability of carbonates in the system MgO-CO$_2$-H$_2$O, *J. Geol.* 73. © 1965 by *J. Geol.*, University of Chicago Press. Used by permission.

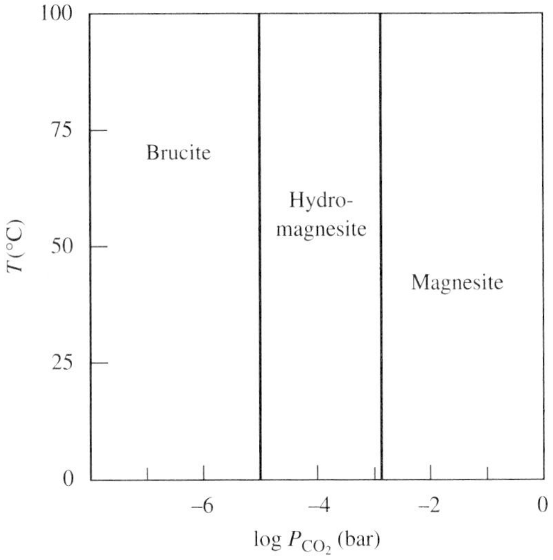

Figure 6.3 Stability relations among solid phases in the system MgO-CO$_2$-H$_2$O as a function of P_{CO_2} from 0 to 100°C, at a water activity of 0.2 (relative humidity of 20%) and 1 bar total pressure. The figure is drawn assuming K_{sp}(magnesite) = $10^{-4.9}$ at 25°C. If, instead, K_{sp}(magnesite) = $10^{-7.93}$, the hydrated Mg carbonates are metastable relative to magnesite, and the brucite/magnesite boundary is at P_{CO_2} = $10^{-7.32}$ bars. See Table 6.1. From D. Langmuir, Stability of carbonates in the system MgO-CO$_2$-H$_2$O, *J. Geol.* 73. © 1965 by *J. Geol.*, University of Chicago Press. Used by permission.

ambient temperatures by exposure to low relative humidities, or if precipitation is from high ionic strength brines (low water activities), particularly above 25°C, magnesite can form. Once formed, magnesite is very slow to dissolve, perhaps because traces of silica adsorb to its surface.

The minerals nesquehonite through huntite in Table 6.1 are comparatively rare. Among these only hydromagnesite can precipitate in water at 25°C and atmospheric CO$_2$ pressures (Fig. 6.2). Because of their high solubilities, the Mg-carbonate and hydroxy-carbonate minerals, are generally uncommon except in some evaporative cave pools, evaporating ocean embayments, arid soils, and closed-basin lakes (cf. Hostetler 1964). Nesquehonite is about 24 times as soluble as calcite at 25°C and a CO$_2$ pressure of 1 bar. Similarly, at atmospheric CO$_2$ pressure ($10^{-3.5}$ bar) the solubility of hydromagnesite is 5.5 mmol/kg, whereas that of calcite is only 0.52 mmol/kg.

There are several ways one can write the solubility of carbonate minerals in water. Most useful are reactions written in terms of CO$_2$ or H$^+$. As noted earlier, carbon dioxide is the chief source of acidity in most natural waters. The reaction between calcite and CO$_2$ is

$$CaCO_3 + CO_2(g) + H_2O = Ca^{2+} + 2HCO_3^- \tag{6.1}$$

for which, at 25°C,

$$K_{eq} = \frac{[Ca^{2+}]\,[HCO_3^-]^2}{P_{CO_2}} = 10^{-5.97} \tag{6.2}$$

Comparable reactions can be written for the other carbonate minerals.

In terms of H^+ the calcite reaction is

$$CaCO_3 + H^+ = Ca^{2+} + HCO_3^- \tag{6.3}$$

for which, at 25°C,

$$K_{eq} = \frac{[Ca^{2+}]\,[HCO_3^-]}{[H^+]} = 10^{+1.85} \tag{6.4}$$

These equations indicate that calcite (or other carbonate) dissolution is favored by an increase in CO_2 pressure and a decrease in pH, and that calcite precipitation can result from a loss of CO_2, or increase in pH for whatever reason.

6.2 CALCITE SOLUBILITY AS A FUNCTION OF CO_2 PRESSURE

It is useful to construct a graph relating carbonate mineral solubilities to CO_2 pressure. This can be done for calcite starting with equilibrium constant expression (6.2) above. If done rigorously, the derivation accounts for the effects of ion activity coefficients and the presence of $CaHCO_3^+$ and $CaCO_3^0$ ion pairs and of $CaOH^+$. Considering all complexation, the exact charge-balance equation for a pure water in which calcite is dissolving is

$$2mCa^{2+} + mCaHCO_3^+ + mCaOH^+ + mH^+ = mHCO_3^- + 2mCO_3^{2-} + mOH^- \tag{6.5}$$

Do we need to consider all of these species in our derivation? Most natural waters in which calcite saturation is observed have pH's between about 6 and 9. Also, at calcite saturation concentrations of Ca^{2+} and alkalinity are roughly at 10^{-3} mol/kg or higher. If we assume that all species having concentrations less than 1% of this amount can be ignored, we can drop mH^+, mOH^-, and $CaOH^+$ from the equation. Also, for pH's below 8, mCO_3^{2-} is less than 1% of $mHCO_3^-$. (How do we know this?) Equation (6.5) thus reduces to

$$2mCa^{2+} + mCaHCO_3^+ = mHCO_3^- \tag{6.6a}$$

Given that

$$K_{dissoc}\,(CaHCO_3^+) = \frac{[Ca^{2+}]\,[HCO_3^-]}{[CaHCO_3^+]} = 10^{-1.11} \tag{6.6b}$$

the $CaHCO_3^+$ concentration is significant at all CO_2 pressures greater than the atmospheric value of about $10^{-3.5}$ bar. This can be shown if we write calcite dissolution as

$$CaCO_3 + H^+ = CaHCO_3^+ \tag{6.7a}$$

The equilibrium constant for this expression at 25°C can be derived in several ways. One approach is to sum reactions (6.7b) through (6.7d) and their log K_{eq} values as shown below.

reaction	log K_{eq}	
$CaCO_3 = Ca^{2+} + CO_3^{2-}$	-8.48	(6.7b)
$H^+ + CO_3^{2-} = HCO_3^-$	$+10.33$	(6.7c)
$Ca^{2+} + HCO_3^- = CaHCO_3^+$	$+1.11$	(6.7d)
$CaCO_3 + H^+ = CaHCO_3^+$	$+2.96$	(6.7a)

Thus,

$$K_{eq} = \frac{[CaHCO_3^+]}{[H^+]} = 10^{+2.96} \tag{6.8}$$

We could also have added the ΔG_r° values for reactions (6.7b) to (6.7d) to obtain ΔG_r° for reaction (6.7a) and K_{eq} for reaction (6.8) through $\Delta G_r^\circ(kcal/mol) = -1.3642 \log K_{sp}$ at 25°C. A third approach is to write out the K_{sp} expressions for reactions (6.7b), (6.7c), and (6.7d), then through algebraic substitution eliminate the undesired terms to obtain K_{sp} and expression (6.8).

Examination of expression (6.8) and its K_{eq} value indicates that between pH 8 and 6 the concentration of the $CaHCO_3^+$ pair increases from about 10^{-5} to 10^{-3} mol/kg. Stated differently, the error in ignoring the pair in our calculations increases from about 1% to 10% between CO_2 pressures of $10^{-3.5}$ bar and 1 bar. Nevertheless, ignoring $CaHCO_3^+$ for the moment, the charge-balance equation is reduced to

$$2mCa^{2+} = mHCO_3^- \tag{6.9}$$

With this approximation and ignoring activity coefficients, we can eliminate HCO_3^- from expression (6.2), and solve for mCa^{2+} with the result

$$Ca^{2+} \, (mmol/kg) = 6.45 \times (P_{CO_2})^{0.33} \tag{6.10}$$

This explains why the solubility of calcite (and carbonate minerals generally) varies roughly as the cube root of the CO_2 pressure. The accurate, empirical equation for calcite solubility at 25°C that has been plotted in Fig. 6.1 is, in fact,

$$Ca^{2+} \, (mmol/kg) = 9.057 \times (P_{CO_2})^{0.3615} \tag{6.11}$$

The difference between expressions (6.10) and (6.11) reflects the consequences of our ignoring activity coefficients and the $CaHCO_3^+$ ion pair in our derivation of (6.10).

We have also ignored the potentially important $CaCO_3^\circ$ ion pair. At saturation with calcite we can write simply

$$CaCO_3(calcite) = CaCO_3^\circ \tag{6.12}$$

for which $K_{eq} = [CaCO_3^\circ]$. In other words the concentration of the calcium carbonate ion pair is fixed at saturation with calcite. The equilibrium constant for this reaction (and the $CaCO_3^\circ$ concentration) at 25°C can be obtained by combining the solubility-product expression for calcite

$$K_{sp}(calcite) = 10^{-8.48} = [Ca^{2+}][CO_3^{2-}] \tag{6.13}$$

and the dissociation-constant expression for the $CaCO_3^\circ$ ion pair, which is

$$K_{dissoc} = 10^{-3.22} = \frac{[Ca^{2+}] \, [CO_3^{2-}]}{[CaCO_3^\circ]} \tag{6.14}$$

Substituting for $[Ca^{2+}][CO_3^{2-}]$ we obtain

$$K_{eq} = 10^{-5.26} = [CaCO_3^\circ] \tag{6.15}$$

Given in Fig. 6.4, in addition to the solubility curve for calcite, are theoretical solubility curves for dolomite and for calcite plus dolomite. The plot shows that carbonate mineral solubilities increase with increasing CO_2 pressures and decrease with decreasing CO_2 pressures. You should be able to explain to yourself why the upper-left portion of the diagram describes conditions supersaturated with respect to the carbonates, whereas undersaturated conditions are found below and to the right of the solubility curves.

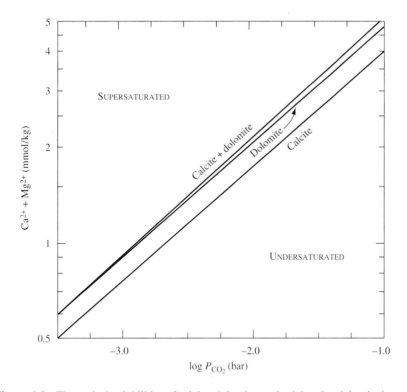

Figure 6.4   Theoretical solubilities of calcite, dolomite, and calcite plus dolomite in pure water at 25°C and 1 bar pressure in terms of CO_2 pressure and the total concentrations of calcium plus magnesium in solution. From D. Langmuir (1984), Physical and chemical characteristics of carbonate water. In *Guide to the hydrology of carbonate rocks*, P. E. La-moreaux, B. M. Wilson, and B. A. Memeon, eds. ©1984 by UNESCO. Used by permission.

The concentrations of individual aqueous species including the ion pairs at 25°C and at saturation with calcite as a function of CO_2 pressure are shown in Fig. 6.5. The plot shows the dominance of calcium and bicarbonate ions ($\sim 10^2$ times greater than the next most abundant species, or >99% of total species) at all CO_2 pressures above atmospheric, with the calcium bicarbonate ion pair increasing from less than 1% to about 10% of Ca^{2+} as the CO_2 pressure increases from $10^{-3.5}$ bar to about 1 bar.

6.3 CALCITE SOLUBILITY AS A FUNCTION OF pH AND CO₂ PRESSURE

We will next derive a general expression for the solubility of calcite as a function of pH and CO_2 pressure. First, write the solubility as a mass-balance equation

$$\Sigma mCa^{2+} = mCa^{2+} + mCaHCO_3^+ + mCaCO_3^\circ \tag{6.16}$$

Our goal is to put this equation in the proton condition. We will do so dealing with each successive right-hand term in Eq. (6.16).

The value of mCa^{2+} is a function of the solubility product of calcite, K_c, which equals

$$K_c = \gamma_{Ca^{2+}} mCa^{2+} [CO_3^{2-}] \tag{6.17}$$

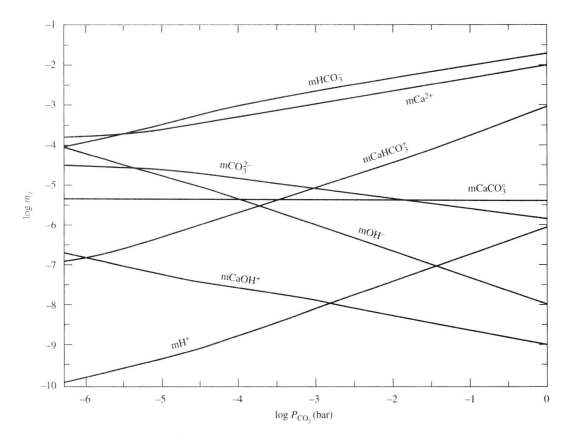

Figure 6.5   The solubility of calcite in pure water at 25°C and 1 bar total pressure as a function of P_{CO_2}, showing the molalities of dissolved species present at saturation. From D. Langmuir (1984), Physical and chemical characteristics of carbonate water. In *Guide to the hydrology of carbonate rocks*, P. E. Lamoreaux, B. M. Wilson, and B. A. Memeon, eds. © 1984 by UNESCO. Used by permission.

Solving for mCa^{2+} gives

$$mCa^{2+} = \frac{K_c}{\gamma_{Ca^{2+}} [CO_3^{2-}]}$$ (6.18)

The $[CO_3^{2-}]$ term can be eliminated by summing the following expressions

$$CO_2(g) + H_2O = H_2CO_3^{\circ} \qquad\qquad K_{CO_2}$$

$$H_2CO_3^{\circ} = H^+ + HCO_3^- \qquad\qquad K_1$$

$$\underline{HCO_3^- = H^+ + CO_3^{2-} \qquad\qquad K_2}$$

$$CO_2(g) + H_2O = 2H^+ + CO_3^{2-} \qquad\qquad K_1K_2K_{CO_2}$$

From this result we obtain

$$K_1 K_2 K_{CO_2} = \frac{[H^+]^2 \, [CO_3^{2-}]}{P_{CO_2}} \tag{6.19}$$

rearranging

$$[CO_3^{2-}] = \frac{K_1 K_1 K_{CO_2} P_{CO_2}}{[H^+]^2}$$

and

$$mCa^{2+} = \frac{K_c [H^+]^2}{\gamma_{Ca^{2+}} K_1 K_2 K_{CO_2} P_{CO_2}} \tag{6.20}$$

Equations (6.7) and (6.8) lead to

$$(\gamma_{CaHCO_3^+}) = \frac{K_c [H^+]}{(\gamma_{HCO_3^-}) K_2 K^d_{CaHCO_3^+}}$$

where $K^d_{CaHCO_3^+}$ is the dissociation constant of the complex. But assuming that $\gamma_{CaHCO_3^+} \approx \gamma_{HCO_3^-}$ gives

$$mCaHCO_3^+ = \frac{K_c [H^+]}{(\gamma_{HCO_3^-}) \, K_2 K^d_{CaHCO_3^+}} \tag{6.21}$$

Equations (6.13) and (6.14) indicate that

$$[CaCO_3^\circ] \approx mCaCO_3^\circ = \frac{K_c}{K^d_{CaCO_3^\circ}} \tag{6.22}$$

We now have our final expression, which is

$$\Sigma mCa^{2+} = \frac{K_c [H^+]^2}{\gamma_{Ca^{2+}} K_1 K_2 K_{CO_2} P_{CO_2}} + \frac{K_c [H^+]}{\gamma_{HCO_3^-} K_2 K^d_{CaHCO_3^+}} + \frac{K_c}{K^d_{CaCO_3^\circ}} \tag{6.23}$$

Introducing K values for 25°C and simplifying, this becomes

$$\Sigma mCa^{2+} = \frac{10^{9.67} \, [H^+]^2}{(\gamma_{Ca^{2+}}) \, P_{CO_2}} + \frac{10^{2.96} \, [H^+]}{\gamma_{HCO_3^-}} + 10^{-5.26} \tag{6.24}$$

$$\Sigma mCa^{2+} = \quad mCa^{2+} \quad + mCaHCO_3^+ + mCaCO_3^\circ$$

where the identity of each term is indicated by our original mass-balance expression.

Equation (6.24) shows that mCa^{2+} is a function of both pH and P_{CO_2}, whereas $mCaHCO_3^+$ depends on pH, but is independent of P_{CO_2}. The concentration of $mCaCO_3^\circ$ is independent of both pH and P_{CO_2}. This equation is plotted for CO_2 pressures from 1 to 10^{-5} bar in Fig. 6.6. At first glance the plot seems illogical, for it suggests that at a given pH calcite solubility increases with decreasing CO_2 pressures. The diagram makes sense, however, if we remember Eq. (6.20), which states that for a given pH, mCa^{2+} is proportional to $1/P_{CO_2}$. Equation (6.24) and Figs. 6.5 and 6.6 show that mCa^{2+}, which is a function of both pH and CO_2 pressure, dominates calcite solubility for ΣmCa^{2+} greater than about 10^{-4} to 10^{-5} mol/kg (4 to 0.4 ppm). The $mCaHCO_3^+$ term is never more than about 10% of mCa^{2+} and becomes negligible at CO_2 pressures below atmospheric ($10^{-3.5}$ bar) and pH > 8.5 (Fig. 6.5). The minimum solubility of calcite corresponds to $\Sigma mCa^{2+} = mCaCO_3^\circ$ and is 5.5×10^{-6} mol/kg (0.22 ppm as Ca) at 25°C.

Figure 6.6 may be used to estimate calcite solubility in surface versus groundwaters. Thus, for $\Sigma mCa^{2+} = 10^{-3}$ mol/kg (40 ppm) and a relatively high groundwater CO_2 pressure of 10^{-1} bar, calcite is saturated at about pH = 7.1. At atmospheric CO_2 pressure ($10^{-3.5}$ bar) for $mCa^{2+} = 10^{-3.3}$ mol/kg (20 ppm) the saturation pH is 8.3, a value typical of many surface-waters.

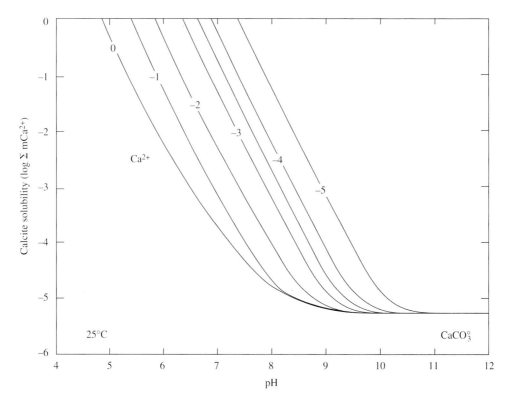

Figure 6.6   The solubility of calcite at 25°C and 1 bar total pressure as a function of P_{CO_2} and pH. Solubility curves for different CO_2 pressures are labeled with corresponding $\log P_{CO_2}$ values from 0 to −5, except for the curve for $P_{CO_2} = 10^{-3.5}$ bar, which is unlabeled. With increasing pH, calcite solubility depends successively on its dissolution to form Ca^{2+} and $CaHCO_3^+$ (negative-sloping solubility curves) and $CaCO_3^\circ$ at the solubility minimum (horizontal curve).

6.4 INFLUENCES ON THE SOLUBILITY AND SATURATION STATE OF CARBONATE MINERALS

We will next examine the kinds of natural processes that can influence the dissolution or precipitation of carbonate minerals. The following discussion revolves around Fig. 6.7, which is a simplified, schematic version of Fig. 6.4, but considers only Ca^{2+} and P_{CO_2} and their relationship to calcite precipitation or dissolution. The same arguments apply to other carbonate minerals, as well.

6.4.1 Dissolution and Exsolution of CO₂

Groundwaters usually contain more dissolved CO_2 than surface waters, especially because of the CO_2 produced by decay and plant-root respiration in soil zones, which is picked up by percolating groundwater recharge. Surface waters generally contain CO_2 concentrations near the atmospheric CO_2 pressure of $10^{-3.5}$ bar, or slightly greater amounts, with the highest surface-water CO_2 concentrations associated with decaying organic matter (including that in sewage) or caused by the inflow of higher-CO_2 groundwaters. When surface waters infiltrate through soils and sediments, they tend

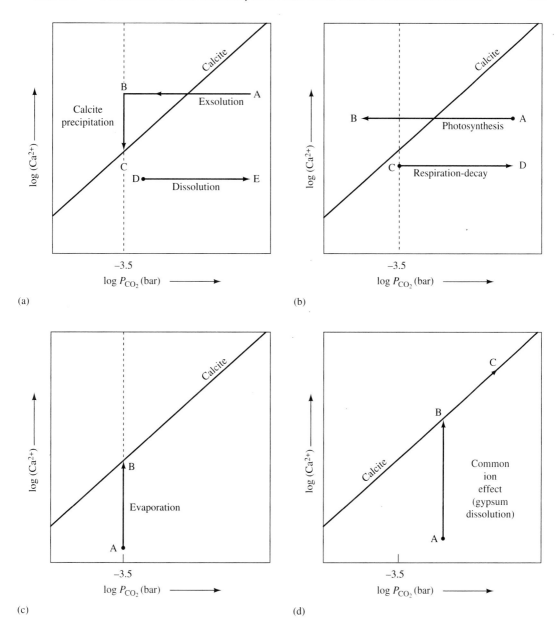

Figure 6.7   Schematic plot showing the general effects of various processes on the saturation state of natural waters with respect to calcite. Waters are supersaturated with respect to calcite above and to the left of the calcite line and undersaturated with respect to calcite below and to the right of the calcite line.

to pick up additional CO_2, making them less saturated with respect to calcite. However, groundwaters discharged into caves or streams generally exsolve CO_2 to the atmosphere, which may lead to the precipitation of carbonate minerals. This CO_2 loss is aided by stream turbulence (cf. Herman and Lorah 1987).

Decreasing and increasing the CO_2 content of a water is shown schematically by horizontal arrows pointing to the left (line A–B) and right (line D–E), respectively, in Fig. 6.7(a). The CO_2 exsolution pathway in Fig. 6.7(a) shows the water becoming supersaturated with respect to calcite (line A–B), followed by calcite precipitation under atmospheric CO_2 pressures until equilibrium with calcite is reached at C.

6.4.2 Photosynthesis, Aerobic Decay, and Respiration

The photosynthesis reaction may be written in simplified form as

$$CO_2 + H_2O \rightarrow \tfrac{1}{6}C_6H_{12}O_6(glucose) + O_2 \tag{6.25}$$

Aerobic decay and respiration are the reverse reaction, although the reaction is not reversible in a chemical sense. (See Chap. 5, Section 5.3.2). Photosynthesis occurs in streams, lakes, and the shallow ocean where the water contains abundant aquatic vegetation. The process operates on a diurnal cycle, with maximum photosynthesis in the early afternoon during warm months. Because of CO_2 removal by aquatic vegetation, the pH may rise from about 6 or 7 to 10. The effect is most pronounced in slow-moving streams clogged with vegetation, where there are high N, P, and organic loadings derived from pollution. Such waters may become greatly supersaturated with calcite, as suggested by line A–B in Fig. 6.7(b). However, the kinetics of calcite precipitation is generally too slow for it to precipitate during a short diurnal cycle. Calcite supersaturation and precipitation, due in part to photosynthesis by aquatic vegetation, is common in lakes, especially in carbonate rock terranes. Calcite supersaturation in lakes may persist throughout the year (cf. Driscoll et al. 1994).

Aerobic and anaerobic decay and respiration dominate the nighttime chemistry of surface waters, leading to an increase in the CO_2 pressure, such as described by line C–D.

6.4.3 Evaporation

Evaporation causing carbonate precipitation occurs in soils and surface waters. Line A–B in Fig. 6.7(c), positioned at atmospheric CO_2 pressure ($10^{-3.5}$ bar), describes the increase in Ca^{2+} with evaporation. If evaporation continues to point B, calcite precipitation takes place according to the reaction

$$Ca^{2+} + 2HCO_3^- \rightarrow CaCO_3 + CO_2(g)\uparrow + H_2O \tag{6.26}$$

which produces CO_2 gas. Evaporation of all liquid water, as in a soil or a closed-basin lake reduces [H_2O] activity to values below unity and so further drives calcite precipitation.

In the southwestern United States and elsewhere in arid or semiarid climates, rainfall may largely evaporate at the land surface, or it may infiltrate into shallow soil horizons, evaporating there or as it returns toward the surface because of capillary forces. Resultant carbonate mineral layers are called *caliche* and may occur in and on soils, along with layers of chert (impure, precipitated SiO_2) and iron(III) oxides, similarly formed, which are called *silcrete* and *ferricrete*.

Evaporative lakes are common in midcontinent regions in Australia, the western United States, southwest Canada, and Africa, for example. Depending upon the extent of evaporation and the amounts and compositions of inflowing waters, the first precipitates are calcite and, perhaps, dolomite, which may be followed by gypsum/anhydrite, then halite, and finally more complex sulfate, carbonate and halide salts (cf. Holland 1978; Berner and Berner 1987; Faure 1991). Similar sequences of mineral precipitates may accumulate in restricted evaporite basins in equatorial areas (Berner and Berner 1987).

6.4.4 The Common Ion Effect

Increases in the concentrations of calcium, bicarbonate or carbonate from sources other than the dissolution of calcite may supersaturate a water with respect to calcite, causing it to precipitate. In the Floridan carbonate aquifer, groundwaters attain calcite saturation by contact with limestone, but then may become supersaturated with respect to calcite because of gypsum dissolution (Back and Hanshaw 1970; Wicks and Herman 1996). Calcium from the gypsum, which is far more soluble than calcite, drives the common ion effect reaction

$$CaSO_4 \cdot 2H_2O(gypsum) + 2HCO_3^- \rightarrow CaCO_3 + SO_4^{2-} + 2H_2O + CO_2 \qquad (6.27)$$

If the common ion is calcium, then the reaction is described by a vertical arrow pointing upward in Fig. 6.7(d) (line A–B). With further dissolution of gypsum and Ca^{2+} increases, groundwater composition then moves upward along the calcite saturation curve (line B–C) in Fig. 6.7(d) to higher CO_2 pressures. Point C will correspond to saturation with respect to gypsum, if the mineral is not completely leached from the rock.

6.4.5 Temperature

Unlike the aluminosilicates and most other minerals, the carbonates have an exothermic heat of dissolution, which means that their solubilities decrease with increasing temperature. For example, K_{sp} for calcite decreases from $10^{-8.38}$ at 0°C to $10^{-8.51}$ at 30°C. The effect of temperature on the solubility products of aragonite, calcite, and ordered dolomite is plotted in Fig. 6.8. The figure shows that between 0 and 90°C solubilities of the carbonates decrease by about 6-fold for aragonite and calcite and 14-fold for dolomite. This decreasing solubility with temperature is magnified by the fact that the solubility of CO_2 gas also declines with temperature. K_{CO_2} decreases from $10^{-1.11}$ to $10^{-1.52}$ between 0 and 30°C.

The combined effects of changes in temperature and in CO_2 pressure on calcite and dolomite solubility are summarized in Table 6.3. Values in the table have been computed from the thermodynamic data in previous tables using the computer model PHREEQE (Parkhurst et al. 1990). The temperatures of 25 and 10°C are typical of shallow groundwaters in Florida and in the northeastern United States, respectively. The table shows the effect on solubility of a CO_2 increase from the atmospheric value of $10^{-3.5}$ bar, to a typical groundwater value of $10^{-2.5}$ bar, and a somewhat high but not unusual groundwater value of $10^{-1.5}$ bar (3.2% CO_2). Solubilities can be seen to increase 5-fold with this 100-fold increase in CO_2. (Remember the increase in solubility is roughly proportional to the cube root of the ratio of the CO_2 pressures, or $(10^{-1.5}/10^{-3.5})^{1/3} = 4.6$ times). At the same time the pH at saturation with the carbonates drops from about 8.3 or 8.4 (typical of a surface water at calcite saturation) to about 7.0 or 7.1, reflecting the higher level of carbonic acid present at saturation at the lower pHs, which is typical of some groundwaters. Ordered dolomite is more soluble than calcite, although its rate of solution (and precipitation) is much slower (cf. Herman and White 1985). The temperature effect is important, but less pronounced than the CO_2 effect. Calcite solubility thus drops by about 20% when the water temperature rises from 10 to 25°C.

6.4.6 Mixing

Mixing of two waters can lead to either carbonate mineral supersaturation or undersaturation of their mixture. The effect can be simply understood for calcite from a plot of Ca^{2+} versus $CO_2(aq)$ (which, by convention, equals $H_2CO_3^\circ$). On such a plot, shown in Fig. 6.9, saturation with respect to calcite is defined by a curve, whereas mixtures of two waters lie on a straight line connecting their compositions. If we mix saturated waters of compositions A and B in Fig. 6.9, regardless of their

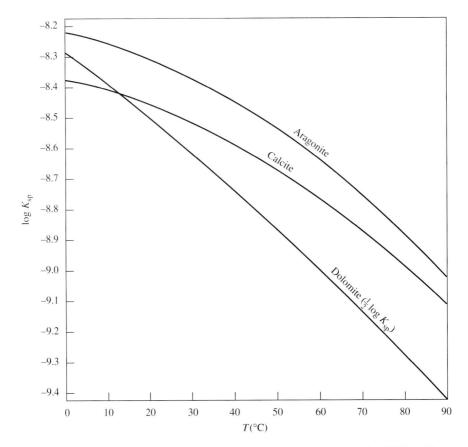

Figure 6.8   Solubility products of aragonite, calcite, and dolomite from 0 to 90°C at 1 bar pressure. The curves for aragonite and calcite are based on the empirical solubility data of Plummer and Busenberg (1982). The curve for dolomite is a plot of $-\log K_{sp}(\text{dolomite}) = -175.90 + 6835.4/T(\text{K}) + 68.727 \log T(\text{K})$. This function is based on $K_{sp}(\text{dolomite}) = 10^{-17.09}$ and $\Delta H_r^\circ = 9436$ cal/mol K at 25°C (Nordstrom et al. 1990) and $\Delta C_p^\circ = -136.6$ cal/mol K at 25°C for the reaction $\text{CaMg(CO}_3)_2 = \text{Ca}^{2+} + \text{Mg}^{2+} + 2\text{CO}_3^{2-}$, which is assumed independent of temperature. ΔC_p° for the dolomite dissolution reaction has been computed from the following C_p° data: C_p° values for dolomite and calcite (Robie et al. 1978), C_p° for Ca^{2+} and Mg^{2+} (Tanger and Helgeson 1988), and $C_p^\circ(\text{CO}_3^{2-}) = -43.0$ cal/mol K based on $\Delta C_p^\circ = -70.5$ cal/mol K for $\text{CaCO}_3 = \text{Ca}^{2+} + \text{CO}_3^{2-}$ (Plummer and Busenberg 1982).

proportions, in a closed system all mixtures will be undersaturated and have compositions on the line A–B. If, for example, undersaturated mixture *m* results, it can further dissolve calcite until saturation is reached at point S. This is shown in Fig. 6.9 which is based on modeling mixed waters using the program PHREEQE (Parkhurst et al. 1990). It is also possible to mix undersaturated waters to create mixtures supersaturated with respect to the carbonate minerals.

Mixing effects on carbonate mineral saturation are important in coastal regions, where meteoric or brackish carbonate groundwaters mix with seawater, which is generally supersaturated with respect to calcite (cf. Fig. 6.10; and Folk and Land 1975; Back et al. 1986). Musgrove and Banner (1993) have also documented calcite supersaturation caused by mixing of regional fresh and saline groundwaters in Missouri, Kansas, and Oklahoma. Mixing may also be important in uncased wells that tap dissimilar carbonate groundwaters with depth, as in the Floridan aquifer.

TABLE 6.3   The theoretical solubilities of calcite and dolomite in pure water at 25°C and 10°C for three CO_2 pressures in bars

25°C

	Calcite			Dolomite		
$\log P_{CO_2}$	−3.5	−2.5	−1.5	−3.5	−2.5	−1.5
Ca^{2+}	20	44	100	12	27	63
Mg^{2+}	0	0	0	7.5	17	38
HCO_3^-	58	131	298	72	164	376
pH	8.29	7.62	6.97	8.37	7.72	7.07

10°C

	Calcite			Dolomite		
$\log P_{CO_2}$	−3.5	−2.5	−1.5	−3.5	−2.5	−1.5
Ca^{2+}	25	55	127	17	37	87
Mg^{2+}	0	0	0	10	23	53
HCO_3^-	74	166	380	98	223	514
pH	8.31	7.65	7.00	8.43	7.78	7.12

Note: Concentrations (in mg/L) have been calculated using the computer model PHREEQE (Parkhurst et al. 1990).

Although we have discussed separately the natural processes affecting carbonate mineral solubilities, in fact such processes often act together. For example, the calcite supersaturation observed in lakes and streams and shallow ocean embayments may result from the simultaneous operation of three processes. These include evaporation, CO_2 removal by photosynthesis, and increases in temperature that reduce the solubilities of both CO_2 and the carbonate minerals.

Another example is the effect of mixing sewage with groundwater on carbonate mineral solubilities. Sewage is relatively high in TDS (including Ca^{2+}) and in dissolved and suspended organic matter that, upon oxidation, produces CO_2. When groundwater is contaminated by sewage (leaked from sewage disposal systems, sewage lagoons, septic tanks, sewage tile fields or injection wells, etc.), the result thus may be an increase in Ca^{2+} and in CO_2 pressure in the mixture (see causative reactions in Table 5.3, Chap. 5). The consequent CO_2 pressure rise in the groundwater, which will generally be the more important effect, increases the solubility of calcite and dolomite. (See carbonate groundwater analyses numbers 4, 5, 9, and 10, Table 6.7).

A third example is the increasing solubility of calcite with depth in the ocean. Surface seawater is generally supersaturated with respect to calcite, whereas at depths below about 500 m the oceans become undersaturated with calcite (cf. Drever 1988). For this reason there is a tendency for carbonate materials to dissolve and disappear with increasing depth as they settle into ocean bottom sediments. Reasons for the shift to undersaturation with increasing ocean depth include declining temperatures, increased pressure (see Chap. 1, Section 1.6.2), and increasing CO_2 concentrations. The depth at which a pronounced reduction in the carbonate content of ocean sediments is observed is termed the calcite compensation depth (CCD). Because of the slow rate at which the calcite dissolves, the CCD is found well below the approximately 500 m depth at which calcite becomes undersaturated. The CCD is thus at about 3.5 km depth in the Pacific Ocean and 5 km depth in the Atlantic Ocean (Drever 1988).

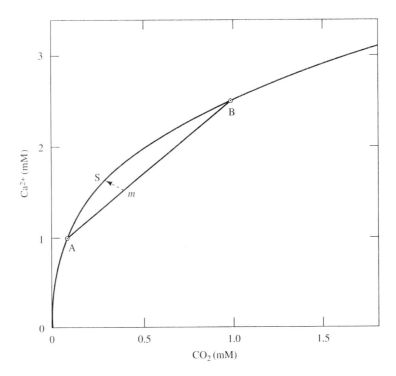

Figure 6.9   Solubility of calcite in pure water at 25°C and 1 bar total pressure as a function of Ca^{2+} and $CO_2(aq)$ concentrations. Waters with compositions below the curve are undersaturated and those above the curve are supersaturated with respect to calcite. The line
A–B is the locus of compositions of mixtures of waters A and B. Mixture m resaturates with
calcite at point S. From D. Langmuir, Physical and chemical characteristics of carbonate
water. In *Guide to the hydrology of carbonate rocks*, P. E. Lamoreaux, B. M. Wilson, and
B. A. Memeon, eds. © 1984 by UNESCO. Used by permission.

6.5 THE SOLUBILITY OF DOLOMITE

Dolomite is the second most abundant carbonate mineral after calcite. In their occurrences, dolomitic
rocks are usually associated with limestones. In the crystal structure of ideal (ordered) dolomite,
which is the thermodynamically most stable phase, layers of carbonate groups are separated by and
coordinated with alternating layers of calcium and then magnesium ions. In disordered dolomite,
which is less stable than the ordered form, a significant number of calcium and magnesium ions are
mixed throughout the cation layers. Recently formed dolomite tends to be disordered, whereas the
dolomite found in older rocks, such as those of Paleozoic age, is usually well-ordered. The molar
Ca^{2+}/Mg^{2+} ratio in ordered dolomites tends to be close to unity, whereas that ratio in disordered
dolomites is usually several percent enriched in calcium. In Table 6.1, solubility products are given
for both ordered and disordered dolomite. As expected, ordered dolomite is less soluble than its disordered form.

Dissolution of ideal, stoichiometric dolomite may be written

$$CaMg(CO_3)_2 + 2CO_2 + 2H_2O = Ca^{2+} + Mg^{2+} + 4HCO_3^- \qquad (6.28)$$

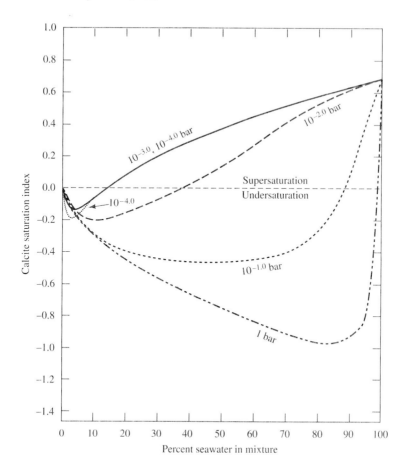

Figure 6.10 Saturation index of calcite in mixtures of seawater and freshwater in equilibrium with calcite at 25°C and different CO_2 pressures. From L. N. Plummer, Mixing of seawater with calcium carbonate water, *Geol. Soc. Am. Memoir* 142. © 1975 by The Geological Society of America. Used by permission.

which shows that a groundwater in equilibrium with dolomite should have a Ca^{2+}/Mg^{2+} molar ratio close to unity. However, because of the usual presence of limestones with dolomitic rocks, the Ca^{2+}/Mg^{2+} ratio of groundwaters in carbonates is generally greater than unity. Assuming that equilibrium can be attained between calcite and dolomite, we can write

$$2CaCO_3 + Mg^{2+} = CaMg(CO_3)_2 + Ca^{2+} \tag{6.29}$$

for which, assuming $\gamma_{Ca^{2+}} \approx \gamma_{Mg^{2+}}$

$$K_{eq} = \frac{[Ca^{2+}]}{[Mg^{2+}]} = \frac{mCa^{2+}}{mMg^{2+}} \tag{6.30}$$

By algebraic manipulation of the calcite and dolomite K_{sp} expressions, we obtain

$$K_{eq} = \frac{mCa^{2+}}{mMg^{2+}} = \frac{[K_{sp}(\text{calcite})]^2}{[K_{sp}(\text{dolomite})]} \tag{6.31}$$

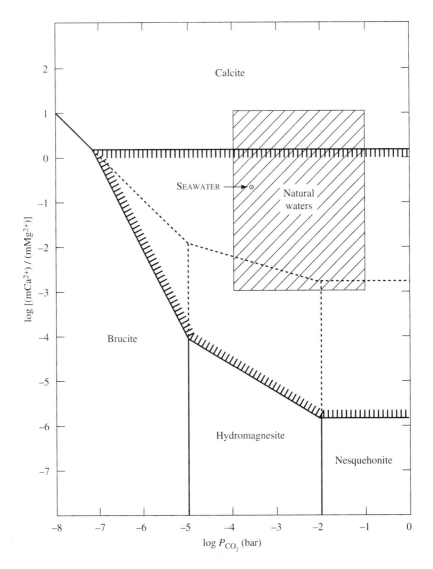

Figure 6.11 Stability relations among the calcium-magnesium carbonates at 25°C and 1 bar total pressure as a function of the mCa^{2+}/mMg^{2+} ratio and CO_2 pressure. The stippled field is that of ordered, well-crystallized dolomite. The plot shows that when dolomite is kinetically inhibited from precipitating, nesquehonite, hydromagnesite, or brucite can co-exist with calcite under metastable conditions. Huntite (not shown) is stable relative to crystalline dolomite for $mCa^{2+}/mMg^{2+} < 10^{-5.0}$. From D. Langmuir, Physical and chemical characteristics of carbonate water. In *Guide to the hydrology of carbonate rocks,* P. E. Lamoreaux, B. M. Wilson, and B. A. Memeon, eds. © 1984 by UNESCO. Used by permission.

With K_{sp} values for ordered and then disordered dolomite from Table 6.1, we find $(mCa^{2+})/(mMg^{2+}) =$ 1.35 and 0.38, respectively, at equilibrium between dolomite and calcite.

Equilibrium reactions among all the calcium-magnesium carbonates can be written in terms of the CO_2 pressure and mCa^{2+}/mMg^{2+} ratio in water ($[H_2O] = 1$). A diagram in which the phase boundaries correspond to these reactions is shown in Fig. 6.11. The crosshatched rectangle includes the compositions of nearly all carbonate groundwaters and surface waters, with the latter having the lower CO_2 pressures. The figure shows that because dolomite cannot easily precipitate for kinetic reasons, most natural waters lie within the calcite field. If, in fact, dolomite precipitated readily from waters having compositions within its stability field, then the common precipitation of calcite (and aragonite) from seawater and their coprecipitation with nesquehonite or hydromagnesite in cave deposits and surface sediments would not be observed.

   Although a huntite stability field is not shown, for $P_{CO_2} > 10^{-2}$ bar we compute that huntite is thermodynamically stable relative to crystalline dolomite and nesquehonite for mCa^{2+}/mMg^{2+} values between $10^{-5.03}$ and $10^{-7.47}$, respectively. If huntite is capable of precipitating, but for kinetic reasons dolomite is not, the stability field of huntite for $P_{CO_2} > 10^{-2}$ bar lies between $mCa^{2+}/mMg^{2+} = 10^{-1.59}$ (the calcite/huntite boundary) and $= 10^{-7.47}$ (the huntite/nesquehonite boundary). Because mCa^{2+}/mMg^{2+} ratios below 10^{-2} are rare in natural waters, huntite precipitation must generally occur metastably, under conditions when dolomite is the stable phase. Similarly, the magnesium carbonates are nearly always precipitated under conditions when they are metastable relative to huntite and dolomite.

6.6 OPEN AND CLOSED CARBONATE SYSTEMS

In an open carbonate system the CO_2 gas pressure is assumed constant, and unaffected by changes in gaseous-aqueous CO_2 exchange rates or by the differential consumption or production of CO_2 by reactions in the groundwater. Such conditions imply the presence of a large gaseous CO_2 reservoir. Examples may include the atmosphere in contact with surface waters or the subsurface gas phase in contact with soil moisture in porous and permeable soils or with groundwaters at the water table in porous and permeable geological materials, such as sands and gravels or cavernous carbonate rocks.

   Because the CO_2 pressure is fixed, $H_2CO_3^\circ$ is also fixed. Given $H_2CO_3^\circ$ and pH and ignoring ion pairing and activity coefficients, corresponding concentrations of C_T, HCO_3^-, and CO_3^{2-} can be obtained using Eqs. (5.17) to (5.21) in Chap. 5. This approach has been used to construct Fig. 6.12, which assumes open-system conditions and a fixed CO_2 pressure of $10^{-1.5}$ bar. The line denoting the Ca^{2+} concentration at calcite saturation has been computed from K_{sp}(calcite) and the value for CO_3^{2-}. The plot shows a rapid increase in bicarbonate and carbonate species concentrations above pH 6 in the open system where the CO_2 consumed in the dissolution of calcite is continuously replenished by gaseous exchange. At the same time calcite solubility as Ca^{2+} decreases rapidly with increasing pH.

   A closed carbonate system is defined as one in which the carbon dioxide (carbonic acid) initially present in the water is not replenished as it is consumed in carbonate mineral dissolution. This may simply reflect that soil moisture/infiltration is charged with CO_2 chiefly in the A horizon of the soil, whereas carbonate mineral dissolution by $H_2CO_3^\circ$ takes place at greater depths in the soil C horizon or below the water table in the absence of further sources of carbon dioxide. (There may be, however, sources of additional CO_2 at depth, including pollution. See Chap. 5.)

   The equation describing the change in bicarbonate concentration with pH during the closed-system dissolution of calcite may be derived as follows (cf. Langmuir 1971c). First, the dissolution reaction is written

$$CaCO_3(\text{calcite}) + H_2CO_3^\circ = Ca^{2+} + 2HCO_3^- \qquad (6.32)$$

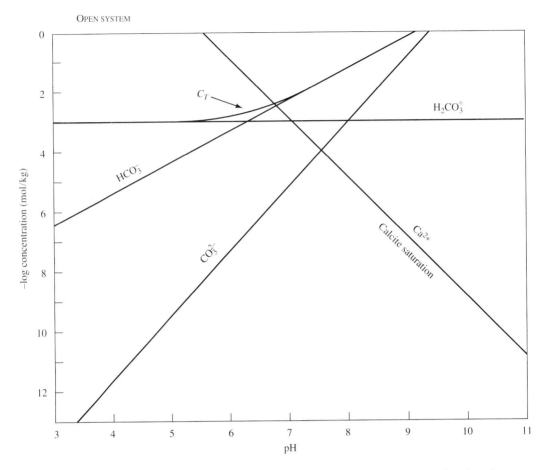

Figure 6.12 Concentrations of total carbonate (C_T) and carbonate species as a function of pH for the open-system dissolution of calcite in pure water at 25°C and 1 bar total pressure. P_{CO_2} is assumed constant at $10^{-1.5}$ bar, which corresponds to $[H_2CO_3^\circ] = 10^{-3.0}$ mol/kg. The curve showing Ca^{2+} concentrations at calcite saturation is also shown.

This equation indicates that for every mole of carbonic acid consumed, two moles of bicarbonate are produced. In differential terms this statement corresponds to

$$d(H_2CO_3^\circ) = -\tfrac{1}{2}d(HCO_3^-) \tag{6.33}$$

Next we differentiate the expression for $K_1(H_2CO_3^\circ)$ to obtain

$$d(H_2CO_3^\circ) = \frac{1}{K_1}(H^+)\ d(HCO_3^-) + \frac{1}{K_1}(HCO_3^-)\ dH^+ \tag{6.34}$$

Eliminating $d(H_2CO_3^\circ)$ between Eqs. (6.33) and (6.34), and rearranging gives

$$\frac{d(HCO_3^-)}{(HCO_3^-)} = \frac{-2\ dH^+}{K_1 + 2(H^+)} \tag{6.35}$$

This can be integrated with the result

$$(HCO_3^-) = \frac{C}{K_1 + 2(H^+)} \tag{6.36}$$

where C is the constant of integration, constant for a given closed-system water.

If we assume, in this case, that initially $(H_2CO_3^\circ) = 10^{-3.00}$ mol/kg at pH = 3.00, we can determine (HCO_3^-) for these same conditions through the expression for $K_1(H_2CO_3^\circ)$ to find $(HCO_3^-) = 10^{-6.36}$ mol/kg. Substituting this value into Eq. (6.36) gives $C = 10^{-9.05}$ mol/kg. This allows us to solve for values of bicarbonate as a function of pH for the initial conditions just given in this closed-system case. Corresponding concentrations of total carbonate, C_T, $H_2CO_3^\circ$, and CO_3^{2-} obtained ignoring ion pairing and activity coefficients, can be computed from the HCO_3^- value and the pH using the dissociation constant expressions for carbonic acid. The results of these calculations are plotted in Fig. 6.13, along with the Ca^{2+} concentration curve at calcite saturation, computed as in the

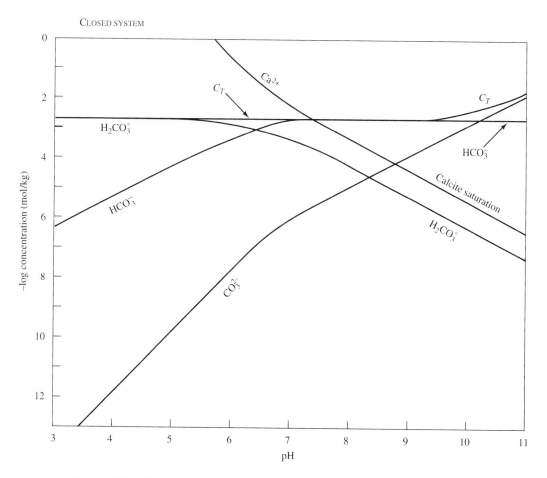

CLOSED SYSTEM

Figure 6.13   Concentrations of total carbonate (C_T) and carbonate species as a function of pH for the closed-system dissolution of calcite in pure water at 25°C and 1 bar total pressure. The initial concentration of $[H_2CO_3^\circ]$ is assumed equal to $10^{-3.0}$ mol/kg at pH = 3. The curve showing Ca^{2+} concentrations at calcite saturation is also shown.

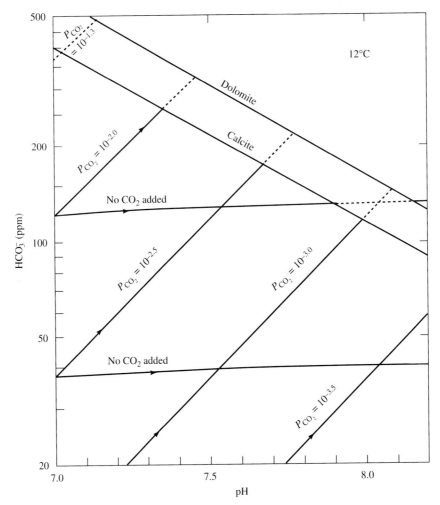

Figure 6.14 Idealized, end-member approaches to mineral saturation of groundwater, dissolving calcite or dolomite at 12°C and 1 bar total pressure. Open-system dissolution pathways are labeled with fixed values of P_{CO_2}. Closed-system pathways are labeled "no CO_2 added." Reprinted from *Geochim. et Cosmochim. Acta,* 35, D. Langmuir, The geochemistry of some carbonate ground waters in central Pennsylvania, 1023–45, © 1971, with permission from Elsevier Science Ltd, The Boulevard, Langford Lane, Kidlington OX5 1GB, U.K.

open-system case. Comparison of Figs. 6.12 and 6.13 shows that closed-system dissolution of carbonate minerals may lead to lower concentrations of C_T, Ca^{2+}, Mg^{2+} and carbonate alkalinity in subsurface waters at saturation with calcite and dolomite than observed in the open-system case.

The same equations and general approach have been used to construct Fig. 6.14 in which the "no CO_2 added" curves show the pathways of carbonate mineral dissolution for closed-system conditions, and the fixed CO_2 pressure lines indicate the trend of open-system dissolution of the carbonates.

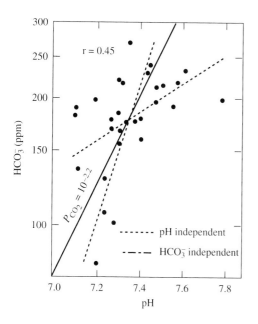

Figure 6.15 Plot of pH versus HCO_3^- values for 29 springwaters from carbonate rocks in central Pennsylvania. Regression lines are drawn assuming HCO_3^- or pH as the independent variable. Also shown is the line for $P_{CO_2} = 10^{-2.2}$ bars at 12°C. Reprinted from *Geochim. et Cosmochim. Acta,* 35, D. Langmuir, The geochemistry of some carbonate ground waters in central Pennsylvania, 1023–45, © 1971, with permission from Elsevier Science Ltd, The Boulevard, Langford Lane, Kidlington OX5 1GB, U.K.

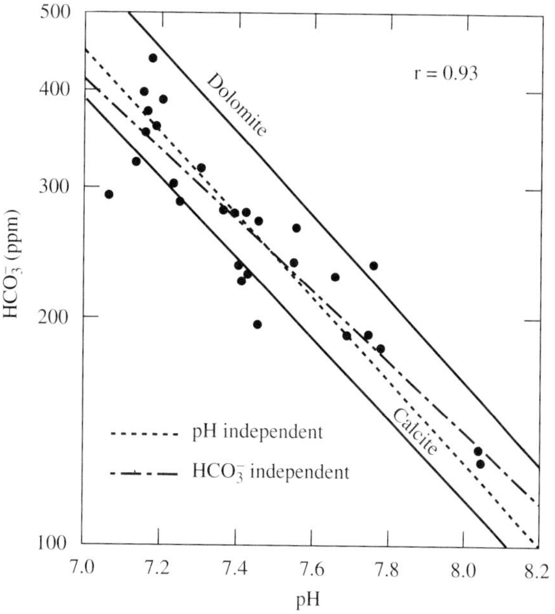

Figure 6.16 Plot of pH versus HCO_3^- values for 29 well waters from carbonate rocks in central Pennsylvania. Regression lines through the data are drawn assuming HCO_3^- or pH as the independent variable. Also shown are theoretical solubility curves for calcite and dolomite. Reprinted from *Geochim. et Cosmochim. Acta,* 35, D. Langmuir, The geochemistry of some carbonate ground waters in central Pennsylvania, 1023–45, © 1971, with permission from Elsevier Science Ltd, The Boulevard, Langford Lane, Kidlington OX5 1GB, U.K.

Most natural waters, and especially subsurface waters, evolve under conditions intermediate between open and closed. Thus, between the points of recharge and discharge of a groundwater, carbonate solution may occur first under open conditions and later under closed conditions. A given spring or well water may then be a mixture of waters that have flowed variously under open- or closed-system conditions.

Figure 6.15 is a plot of the pH versus bicarbonate content of 29 springwaters issuing for carbonate rocks in central Pennsylvania. Most are undersaturated with respect to calcite. Although the correlation is weak, the trend in Fig. 6.15 suggests that the springwaters, which have subsurface ages ranging from several days to years, may have evolved under open-system conditions at a CO_2 pressure of roughly $10^{-2.2}$ bar. A similar plot in Fig. 6.16 for 29 well waters from the same area indicates (as does Fig. 6.17) that most of the well waters are at or near saturation with respect to calcite and dolomite.

Based solely on the groundwater chemistry of springwaters or well waters, there is no way to establish whether groundwaters have evolved under open- or closed-system regimes. However, differences in the stable carbon-isotope ratios of the carbon in (a) local soil gas CO_2 in the zone of

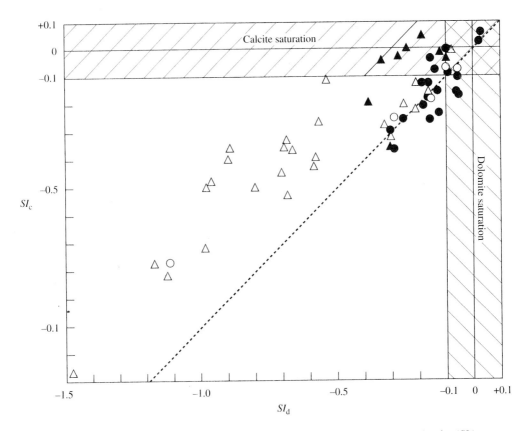

Figure 6.17 Saturation index of calcite (SI_c) plotted against the index for dolomite (SI_d), for 29 springwaters and 29 well waters from carbonate rocks in central Pennsylvania. Springwaters are denoted by open symbols; well waters by solid symbols. Triangles denote a limestone (especially calcite) source rock; circles denote a dolomite source rock. $SI_c = SI_d$ along the dashed line. Crosshatched area shows the limits of uncertainty in the saturation indices. Reprinted from *Geochim. et Cosmochim. Acta,* 35, D. Langmuir, The geochemistry of some carbonate ground waters in central Pennsylvania, 1023–45, © 1971, with permission from Elsevier Science Ltd, The Boulevard, Langford Lane, Kidlington OX5 1GB, U.K.

groundwater recharge, (b) carbonate rock CO_3^{2-}, and (c) dissolved carbonate species in the groundwater, provide unique, additional information. This isotopic ($\delta^{13}C$) data indicate that most of the same 58 springwaters and well waters have evolved under dominantly closed-system conditions (Deines et al. 1974).

6.7. THE SATURATION STATE OF NATURAL WATERS WITH RESPECT TO CALCITE AND DOLOMITE

Laboratory and field studies indicate that calcite dissolves more rapidly than dolomite and that approaching equilibrium with respect to these minerals by dissolution, starting with fresh, undersaturated waters (the only way one can approach dolomite saturation), takes from weeks to months (cf. Plummer et al 1979; Herman and White 1985). A rapid increase in temperature, mixing of waters, the addition of a common ion, evaporation, or CO_2 loss by exsolution or through photosynthesis can all create supersaturated conditions or, at least, accelerate the approach to saturation with the carbonates.

In Chap. 1 we introduced the concept of the saturation index (SI), which is defined as $SI = \log_{10} (IAP/K_{sp})$. In this expression IAP is the ion activity product of the mineral computed from the water analysis and K_{sp} its solubility product for the same temperature and pressure. For a water at saturation, $IAP = K_{sp}$ and $SI = 0$. $IAP > K_{sp}$ and $SI > 0$ indicates the water is supersaturated with the mineral, with $IAP < K_{sp}$ and $SI < 0$ denoting undersaturated conditions.

In order to make the saturation indices of carbonates roughly comparable, one can normalize their SI values to the same mole number of carbonate groups or of cations. Thus, if SI_d is the saturation index of dolomite, and K_d and IAP_d its solubility and ion activity products, we write

$$SI_d = \log (IAP_d/K_d)^{1/2} \tag{6.37}$$

Similarly, for calcite,

$$SI_c = \log (IAP_c/K_c) \tag{6.38}$$

Taking into account probable uncertainties in the field-measured pH (usually about ± 0.05 pH units) lab-analyzed Ca^{2+}, Mg^{2+}, and HCO_3^- values, ionic strength and equilibrium constants involved, SI values for the carbonates so defined are probably known to ± 0.1 units at best (Langmuir 1971c).

Figure 6.17 is a plot of computed SI_c versus SI_d values for some springwaters and well waters in carbonate rocks in central Pennsylvania. First, it is clear that groundwater concentrations of calcium, magnesium, and carbonate species are limited by the solubilities of calcite and dolomite. Second, the well waters are closer to saturation with the carbonates than are the springwaters. This reflects the fact that wells are most often drilled in convenient locations for users and only fortuitously tap into the high-yielding cavernous or weathering-enlarged fractured zones in carbonate rocks that feed springs. For this reason most well waters are relatively isolated from fresh recharge and have longer residence times to dissolve the rock than do most springwaters. The residence times of springwaters are as short as a few days to weeks, versus well-water residence times of months to years (Langmuir 1971c; Jacobson and Langmuir 1974). Also evident from Fig. 6.17 is the fact that springwaters are closer to saturation with respect to calcite than with dolomite. This is because most of the springs flow in limestone not dolomite, since the limestone is more readily weathered to create the high-permeability zones characteristic of spring systems. Such studies also show that even in carbonate groundwaters highly polluted with sewage or landfill leachates, concentrations are limited by saturation with respect to calcite and dolomite. In fact, many of the data points plotted in Fig. 6.17 are for such polluted wells and springs. (See Table 6.7).

6.8 SOLUBILITIES OF OTHER CARBONATE MINERALS

Although the Ca–Mg carbonates constitute the most frequently encountered carbonate minerals, other metal carbonates can also be important. Of these, the most abundant are siderite ($FeCO_3$) and rhodocrosite ($MnCO_3$). The divalent metal carbonates typically occur as solid solutions and so contain trace to major amounts of one or more secondary metal cations (cf. Reeder 1983). Solubilities of some binary metal-carbonate solid solutions are known (cf. Glynn 1990).

Table 6.4 lists formulas and solubility products of the pure carbonate minerals. The 1:1 carbonates of divalent Fe, Mn, Zn, Cd, Co, Sr, Pb, UO_2, are all less soluble than calcite and, in fact, the

TABLE 6.4   Solubility products of some carbonate minerals at 25°C and 1 bar pressure

Mineral	Formula	pK_{sp} ($-\log K_{sp}$)	Source
calcite	$CaCO_3$	8.33–8.48	1,2
magnesite	$MgCO_3$	4.9?, 7.93, 8.03	3,3,4
siderite	$FeCO_3$	10.45–10.89	2
rhodocrosite	$MnCO_3$	10.39–11.13	2
smithsonite	$ZnCO_3$	9.92, 10.8	5,6
	$NiCO_3$	6.87	7
otavite	$CdCO_3$	11.21, 11.49, 12.14	5,8,9
cobaltocalcite	$CoCO_3$	9.98	7
aragonite	$CaCO_3$	8.34	2
witherite	$BaCO_3$	8.56, 8.69	2,7
strontianite	$SrCO_3$	9.27	2
cerrusite	$PbCO_3$	12.80, 13.13	10,7
rutherfordine	UO_2CO_3	14.44	11
malachite	$Cu_2CO_3(OH)_2$	33.78	12,13
azurite	$Cu_3(CO_3)_2(OH)_2$	45.96	12,13
hydrozincite	$Zn_5(CO_3)_2(OH)_6$	74.3	6
hydrocerrusite	$Pb_3(CO_3)_2(OH)_2$	46.9	14
hydromagnesite	$Mg_5(CO_3)_4(OH)_2 \cdot 4H_2O$	35.0	3
dolomite (ordered)	$CaMg(CO_3)_2$	17.09	2
dolomite (disordered)	$CaMg(CO_3)_2$	16.54	2
huntite	$CaMg_3(CO_3)_4$	29.15	3
ankerite	$CaFe(CO_3)_2$	20.5	15
kutnahorite	$CaMn(CO_3)_2$	19.83, 21.81	16,17
alstonite	$BaCa(CO_3)_2$	17.99	5
barytocalcite	$BaCa(CO_3)_2$	17.84	5
artinite	$Mg_2CO_3(OH)_2 \cdot 3H_2O$	15.4	3
zaratite	$Ni_3CO_3(OH)_4 \cdot 4H_2O$	>33.2	15
nesquehonite	$MgCO_3 \cdot 3H_2O$	5.58	3
lansfordite	$MgCO_3 \cdot 5H_2O$	5.45	3

Note: The more common rock-forming minerals are italicized (cf. Deer et al. 1992). Single pK_{sp} values are presumably for well-crystallized, least soluble forms. Where pK_{sp} ranges are given, they reflect the solubility range between relatively amorphous and well-crystallized forms. pK_{sp} values have been computed assuming specific solution speciation models (e.g., specific complexes and complex stability constants), which must also be used when these constants are employed in mineral saturation index calculations.

Source: (1) Busenberg and Plummer (1989); (2) Nordstrom et al. (1990); (3) See Table 6.1, footnotes; (4) Allison et al. (1991); (5) Wagman et al. (1982); (6) Schindler (1967); (7) Smith and Martell (1976); (8) Rock et al. (1994); (9) Stipp et al. (1992); (10) Robie et al. (1978); (11) Grenthe et al. (1992); (12) Silman (1958); (13) Duby (1977); (14) Based on information in Baes and Mesmer (1976); (15) Estimated by Woods and Garrels (1992). See also Davidson et al. (1994); (16) Garrels et al. (1960); (17) Mucci,(1991); (18) Based on K_{sp}($NiCO_3$) this table and free energy data for related substances from Wagman et al. (1982), and the argument that zaratite (a mineral) is at least as thermodynamically stable as $NiCO_3$ (not reported as a mineral) for $P_{CO_2} = 10^{-3.5}$ bar and [H_2O] = 1 (see Palache et al. 1957).

Fe, Mn, Zn, Cd, Pb, and UO_2 carbonates are over 100 times less soluble. In most cases (except in some ore deposits), the Zn, Cd, Co, Sr, Pb, and UO_2 carbonates come to equilibrium in systems that are first saturated with respect to rock-forming carbonates (e.g., calcite, dolomite, rhodocrosite, or siderite). For these conditions, maximum concentrations of the trace divalent metals are limited by solubilities of the rock-forming carbonates. Such control may be evident from metal cation concentration ratios in groundwater. For example, at saturation of the water with respect to calcite and rhodocrosite we may write

$$CaCO_3 + Mn^{2+} = MnCO_3 + Ca^{2+} \qquad (6.39)$$

for which, assuming equal ion activity coefficients for the cations, it is easy to show

$$K_{eq} = \frac{K_c}{K_r} = \frac{[Ca^{2+}]}{[Mn^{2+}]} \approx \frac{mCa^{2+}}{mMn^{2+}} \qquad (6.40)$$

where K_c and K_r are solubility products of calcite and rhodocrosite, and the ratio is of the free (uncomplexed) metal cations. Now, assuming $K_c = 10^{-8.48}$ and $K_r = 10^{-11.13}$, we find $mCa^{2+}/mMn^{2+} = 447/1$. Thus, if $Ca^{2+} = 40$ mg/kg, Mn^{2+} is limited to about 0.12 mg/kg at equilibrium.

Example 6.1

Healing Spring is a thermal (27.5°C) artesian water of prevalent chemical character Ca > Mg: HCO_3 > SO_4 issuing from limestone in northwest Virginia (Helz and Sinex 1974). The water has an ionic strength of 1.76×10^{-2} mol/kg, and a field-measured pH of 6.84. The chemical analysis (number 6 in Table 6.7) shows $\Sigma Ca = 123$ mg/kg, $\Sigma Fe(II) = 1.30$ mg/kg, and $\Sigma HCO_3 = 460$ mg/kg. We will address the following questions:

(a) Are Ca and Fe(II) concentrations contolled by saturation with respect to calcite and siderite? Is the mCa^{2+}/mFe^{2+} ratio limited by saturation with both minerals? Inputting the complete water analysis from Table 6.7 into MINTEQA2 we obtain $SI_c = 0.072$ and $SI_s = 0.065$, with a probable uncertainty of about ±0.1 SI units, indicating perhaps slightly supersaturated conditions for both carbonate minerals. MINTEQA2 assumes $K_c = 10^{-8.49}$ and $K_s = 10^{-10.58}$ at 25°C. The ratio $mCa^{2+}/mFe^{2+} = 2.723 \times 10^{-3}/2.180 \times 10^{-5} = 125/1$. In excellent agreement, $K_c/K_s = 123/1$.

(b) Might the slight supersaturations obtained for calcite and siderite have resulted from an error in the measured pH? Given the computed apparent CO_2 pressure of $10^{-1.18}$ bar, CO_2 gas would tend to exsolve during sampling to give a positive error in the field-measured pH (cf. Pearson 1978). Assuming the pH should have been 0.07 units lower (pH = 6.77), we obtain $SI_c = 0.002$ and $SI_s = -0.005$.

Siderite and rhodocrosite may approach saturation when acid mine waters are neutralized by contact with carbonate rocks. Figure 6.18 shows the trend toward saturation with respect to carbonate minerals of some surface waters and groundwaters affected by bituminous coal mining (Gang and Langmuir 1974). Plotted are the saturation indices of various metal carbonates versus the index for calcite. The plots show that siderite and rhodocrosite reach equilibrium in some streams (probably because of CO_2 loss), whereas calcite and the other carbonates remain undersaturated in all samples.

6.9 EXPLANATION FOR CARBONATE MINERAL SUPERSATURATION

There are a variety of potential explanations for carbonate mineral supersaturations that exceed uncertainties in the saturation indices of the pure, well-crystallized carbonates (about ±0.1 SI units for calcite). As discussed earlier in this chapter, calcite supersaturation in surface-waters may result from a temperature increase, evaporation, and/or a loss in CO_2 to photosynthesis or by exsolution to the atmosphere. Mixing of surface-waters or groundwaters can also produce a supersaturated mixture. The dissolution of more soluble gypsum causes calcite supersaturation in some groundwaters of the Floridan aquifer.

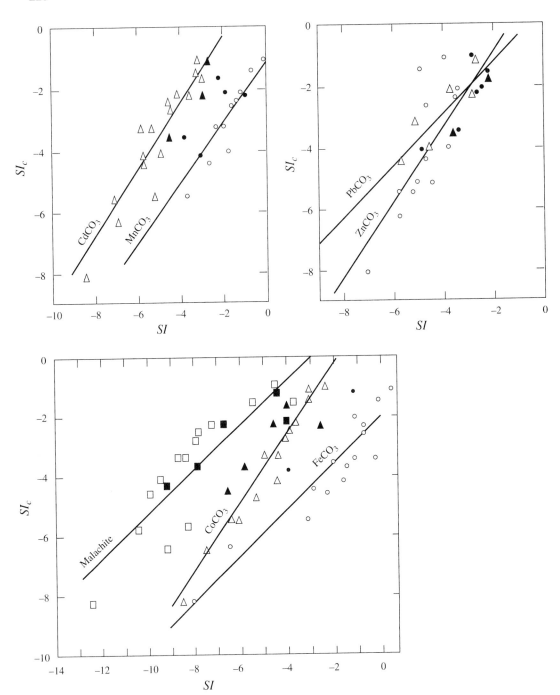

Figure 6.18 Saturation index of calcite (SI_c) plotted against the indices of other carbonate minerals computed for surface- and groundwaters from bituminous coal mining areas in northeastern Pennsynvania. Regression lines are drawn through the data. Open symbols denote surface-waters, closed symbols groundwaters. Malachite is $Cu_2CO_3(OH)_2$. From Gang and Langmuir (1974). Copyright 1974 by Natl. Coal Assoc. Used by permission.

There are other possible explanations when a model calculation indicates a water is supersaturated with respect to one or more carbonate minerals. They include: (1) the use of inaccurate, inconsistent, or incomplete thermodynamic data for carbonate minerals and aqueous complexes; (2) nonstoichiometry (i.e., solid solution) and/or small (submicron) particle sizes of the carbonates, making them more soluble than the well-crystallized pure phases assumed in the calculation (cf. Busenberg and Plummer 1989); (3) different solution models used to define the mineral K_{sp} and in the calculation of saturation state in a natural water; (4) inhibition of carbonate nucleation by adsorbed substances (cf. Inskeep and Bloom 1986); and (5) slow nucleation and precipitation rates that require times exceeding residence times of the water in the water-rock system (cf. Herman and Lorah 1987). The same possible explanations apply to model-computed supersaturations obtained for noncarbonate minerals.

When attempting to explain a mineral saturation state that appears unreasonable, one should first consider whether the mineral of interest is known to be present in contact with the natural water, and has the same composition and stability as the phase listed in the geochemical model's thermodynamic data base. Data bases often include only the stability constants of pure, well-crystallized phases, whereas the natural phases are often impure solid solutions with enhanced solubilities because of poor crystallinity and small particle sizes. Table 6.4 gives a range of K_{sp} values for chemically pure calcite, siderite, rhodocrosite, otavite, and dolomite. The greatest solubilities are presumably for materials of smallest particle size and poorest crystallinity (e.g. protodolomite). The plot in Fig. 6.19 shows that the solubility range for a carbonate is roughly proportional to its pK_{sp}. Thus, poorly crystalline calcite and otavite are roughly 1.4 times and 8.5 times more soluble than their most stable forms. A similar pK_{sp} versus δpK_{sp} correlation is found for metal oxide and hydroxide minerals and can be expected for the other carbonates, although the data are incomplete.

Precipitated modern marine carbonates often contain several percent Mg substituting for Ca and/or small amounts of sulfate substituting for carbonate, both of which may decrease calcite solubility if present in small amounts (below 1 to 2%) but make it more soluble at higher concentrations of the minor constituents (cf. Mackenzie et al. 1983; Busenberg and Plummer 1985). Recent marine carbonates also tend to be of small particle size and may include aragonite and protodolomite (disordered dolomite). Combined with inhibition effects (see below) such factors can create waters that are supersaturated with respect to well-crystallized calcite and dolomite, as in seawater and the Floridan and Yucatan (Mexico) aquifers (see Table 6.7 in Section 6.11, analyses 3, 7, 10, and 11; Back and Hanshaw 1970; Plummer 1977). The dense, Paleozoic-aged carbonate rocks in contact with most fresh continental groundwaters, however, have been recrystallized and approach the stoichiometry and solubilities of pure calcite and dolomite (cf. Langmuir 1971c).

Jacobson and Langmuir (1974) pointed out that if one ignored the ion pairs $CaHCO_3^+$ and $CaCO_3^\circ$, the solubility product of calcite at 25°C was $10^{-8.42}$. Including the pairs in the calculation of K_c (K_{sp}[calcite]) led to $K_c = 10^{-8.47}$ (Plummer and Busenberg [1982] suggest $K_c = 10^{-8.48}$ with the pairs present). Thus, to a good approximation one would expect to compute the same calcite-saturation index for a water assuming the higher solubility and no pairs as with the lower solubility assuming the pairs are present. The key issue is what does the computer model we are using assume? Further, even assuming ion pairs are present, the choice of pairs and their stabilities may differ. For example, MINTEQA2 assumes for well-crystallized siderite $K_s = 10^{-10.55}$ at 25°C, and ignores ferrous carbonate and bicarbonate ion pairs. Estimation and literature analysis suggest association constants of $K(FeHCO_3^+) = 10^{2.0}$ (Nordstrom et al. 1990), and $K(FeCO_3^\circ) = 10^{5.1}$, consistent with a significantly smaller K_{sp}. Bruno et al. (1992) adapt a different solution model with no bicarbonate ion pair, but assume 1:1 and 1:2 ferrous carbonate pairs, and so obtain $K_s = 10^{-10.8}$. All of this suggests the need to ascertain that the K_{sp} value we have chosen was computed using the same solution model that is

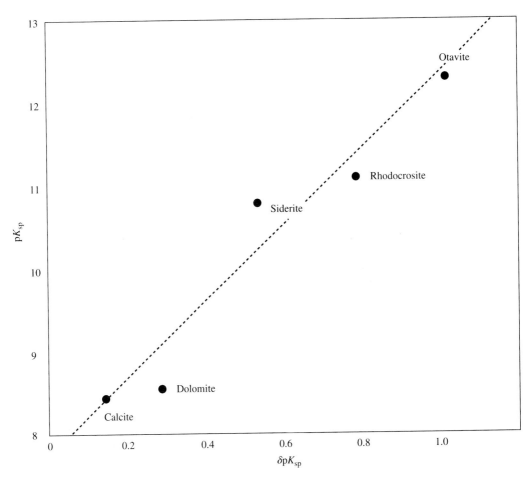

Figure 6.19 Plot showing the increasing difference in pK_{sp} values (δpK_{sp}) between well-crystallized and relatively amorphous forms of carbonate minerals and the pK_{sp} value of the crystallized form. Data are from Table 6.4.

employed in a code such as MINTEQA2 if we would obtain meaningful speciation and mineral saturation results from geochemical computer modeling.

It has long been known that the magnesium in seawater inhibits calcite nucleation, resulting in the precipitation of more soluble and fine-grained aragonite and Mg calcites (cf. Plummer and Busenberg 1982). Adsorption of other substances can also slow dissolution, prevent nucleation, and inhibit carbonate crystal growth (cf. Hooghart and Posthumus 1990). Known inhibitors include Mg and also PO_4, SO_4, Sr, Ba, Pb, and organics, with the latter being a more effective inhibitor than the inorganics at lower concentrations. Hooghart and Posthumus (1990) note, in fact, that a few mg/L of organics are added to process waters to prevent carbonate-scale formation in industrial water treatment and in the petroleum industry.

At some level of calcite supersaturation, depending on the concentration and reactivity of inhibitors and other kinetic factors, calcite tends to nucleate and precipitate. Herman and Lorah (1987) noted previous field studies in which calcite precipitation did not occur until CO_2 outgassing had raised calcite supersaturation levels to 5 or 10 times above equilibrium. In their own work on Falling

Spring Creek, Virginia, they observed some calcite precipitation at SI_c (calcite saturation index) values between +0.1 and +0.7. Major precipitation of calcite as travertine took place at a waterfall where the local turbulence facilitated CO_2 outgassing. The CO_2 loss caused the water to exceed calcite saturation by up to 20 times ($SI_c = +1.3$).

Jenne (1992) suggests that in the presence of a few mg/L of inhibiting DOC, which is common in many groundwaters, SI_c values may have to exceed +0.3 or more for calcite nucleation and precipitation. (See also Appelo and Postma 1993.) Jacobson et al. (1971) reported no carbonate precipitation in a stream in which SI_c values averaged 0.74 diurnally and were as high as 1.5 (see below). These experiences, presumably reflect both nucleation-inhibition effects and the relatively slow rate of calcite precipitation compared to the residence times of streams.

6.10 A SURFACE-WATER CARBONATE SYSTEM: SLAB CABIN RUN

The data in Table 6.5 are for Slab Cabin Run, a small stream (~60 L/s) in central Pennsylvania. The stream flows over alluvial muds in a valley underlain by limestone and dolomite rocks. Its waters originate as springs issuing from the carbonate rocks about 3 km upstream. The small standard deviations in the averages for Ca^{2+}, Mg^{2+}, C_B (total alkalinity), and specific conductance in Table 6.5 indicate that overall stream chemistry does not change during the day. These values presumably reflect the constant composition of springs feeding Slab Cabin Run. At the time of sampling (May 28 and 29) the stream was clogged with green aquatic vegetation. The following discussion is intended to help explain controls on the carbonate chemistry, including the calcite saturation state of the stream and its carbon isotopic composition ($\delta^{13}C$) over the 30-hour period.

First, we must define $\delta^{13}C$ and the concept of isotope fractionation (cf. Deines et al. 1974; Faure 1991; Stumm and Morgan 1996). The atomic weight of natural carbon is 12.011. Of this, about 98.9% occurs as the stable isotope ^{12}C with a nominal atomic weight of exactly 12 atomic mass units (amu) and 1.1% as stable ^{13}C with an atomic weight of 13 amu. A variety of natural processes can lead to small changes in these proportions, which are conventionally expressed in terms of $\delta^{13}C$ where

$$\delta^{13}C\ (\permil) = \left[\frac{^{13}C/^{12}C(\text{sample}) - \,^{13}C/^{12}C(\text{standard})}{^{13}C/^{12}C(\text{standard})} \right] \times 10^3 \qquad (6.41)$$

The standard is a marine carbonate fossil (calcite–PDB belemnite; or PDB) for which, by definition, $\delta^{13}C = 0\permil$ (zero per mil). When a reaction or process leads to a change in $\delta^{13}C$, the change is termed *isotope fractionation*. Negative $\delta^{13}C$ values, which reflect enrichment in ^{12}C, are described as isotopically light. Positive (or less negative) values identify ^{13}C-enriched carbon and are termed isotopically heavy. Some typical $\delta^{13}C$ values for natural carbon are: marine limestone and dolomite, +5 to −3‰; ocean water bicarbonate, −2‰; atmospheric CO_2, −7‰; freshwaters, −5 to −10‰; coal, land plants, soil organic matter, and respired CO_2, −22 to −29‰ (cf. Mason and Moore 1982). Aquatic plants preferentially extract ^{12}C over ^{13}C for growth during daylight hours for photosynthesis, leaving behind isotopically heavier dissolved carbonate species. Plants respire light carbon throughout the day, but the comparatively small contribution of respiration to $\delta^{13}C$ values may be unimportant, except at night.

The data in Table 6.5 have been input into MINTEQA2 to compute the concentrations of HCO_3^-, CO_3^{2-}, OH^-, $H_2CO_3^\circ$, C_T, and in bars, and the saturation indices of calcite and dolomite. These values (except for SI_d) are listed in Table 6.6. Measured and computed study results are also plotted in Fig. 6.20. Most variables have been plotted against the computed apparent CO_2 pressure. Given the constancy of Ca^{2+}, Mg^{2+}, C_B, and specific conductance as shown in Table 6.5, it is clear that

TABLE 6.5 Chemical, carbon isotopic, and solar radiation analyses for Slab Cabin Run, Nittany Valley, State College, PA

Sample	May date	Time	T°C	Ca²⁺ (mM)	Mg²⁺ (mM)	Total alkalinity (mM)	DO (mM)	pH	SpC (μmhos/cm) 25°C	δ¹³C (‰)	Solar rad. daily cum. (langleys)
1	28	1430	16.5	1.05	0.66	2.72	0.363	9.18	336	−11.5	412
2	28	1830	14.8	1.00	0.66	2.77	0.313	8.94	332	−12.2	502
3	28	2300	10.4	1.02	0.66	2.84	0.266	8.03	330	−12.5	504
4	29	0445	8.4	0.97	0.70	2.77	0.338	7.76	331	−12.9	0
5	29	0930	10.0	0.97	0.66	2.82	0.375	8.46	336	−11.7	106
6	29	1400	16.5	1.00	0.66	2.84	0.359	9.42	338	−10.9	305
7	29	1800	14.5	1.00	0.70	2.82	0.313	8.88	333	−11.5	369
8	29	2000	13.1	1.02	0.70	2.82	0.275	8.54	335	−12.0	376
Average ± 1 SD			13.0 ± 3.1	1.00 ± 0.02	0.675 ± 0.019	2.80 ± 0.04			334 ± 3		

Source: Jacobson et al. (1971).

TABLE 6.6   Computed carbonate speciation, CO_2 pressure, total carbonate C_T, and calcite saturation index, SI_c, of samples 1–8 from Slab Cabin Run

Sample	HCO_3^- (mM)	CO_3^{2-} (mM)	OH^- (mM)	$H_2CO_3^\circ$ (mM)	$-\log P_{CO_2}$ (bars)	C_T (mM)	SI_c
1	2.103	0.152	0.0084	0.0027	4.13	2.421	1.34
2	2.380	0.095	0.0042	0.0052	3.85	2.590	1.12
3	2.766	0.012	0.0004	0.0460	2.92	2.863	0.24
4	2.720	0.006	0.0002	0.0818	2.68	2.840	-0.088
5	2.678	0.031	0.0009	0.0165	3.37	2.778	0.62
6	1.913	0.240	0.0146	0.0014	4.41	2.380	1.50
7	2.464	0.085	0.0036	0.0061	3.78	2.659	1.07
8	2.635	0.040	0.0015	0.0140	3.43	2.755	0.76

Source: Based on the data in Table 6.5.

changes in pH, $\delta^{13}C$, P_{CO_2}, SI_c and SI_d are a consequence of diurnal changes in the dissolved CO_2 content of the stream. The maximum CO_2 pressure of $10^{-2.68}$ bar measured at 4:45 A.M. is close to the CO_2 level in springs at the stream source, perhaps slightly elevated by respiration and decay. Lowest CO_2 pressures of $10^{-4.13}$ and $10^{-4.41}$ bar are recorded in early afternoon on both days at times of maximum CO_2 removal during photosynthesis. The inverse dependence of pH and P_{CO_2} at constant HCO_3^- is described by the equation

$$\log P_{CO_2} = -\mathrm{pH} + \log\left(\frac{\gamma_{HCO_3}\mathrm{m}HCO_3^-}{K_{CO_2}K_1}\right) \tag{6.42}$$

(Eq. [5.27], Chap. 5) in which the right term is practically constant during the study. That the saturation indices and $\log P_{CO_2}$ are inversely related should be obvious from earlier discussion in this chapter.

We noted above that green plants preferentially extract light carbon for growth. This explains why the heaviest $\delta^{13}C$ values and lowest CO_2 pressures in the stream are measured in early afternoon. Lightest $\delta^{13}C$ values and maximum CO_2 pressures are observed at night, when respiration and decay are dominant processes in the stream.

6.11 THE CARBONATE MINERAL SATURATION STATE OF SOME REPRESENTATIVE GROUNDWATERS AND SEAWATER

Chemical analyses of 10 groundwaters from springs and wells in carbonate rocks are shown in Table 6.7, along with their apparent CO_2 pressures and saturation indices with respect to calcite and dolomite, which have been calculated using the computer model SOLMINEQ.88 (Kharaka et al. 1988). The composition of seawater and its modeled carbonate-mineral saturation state is also shown. SOLMINEQ.88 calculates the concentrations of ion pairs, such as $CaHCO_3^+$ and $MgSO_4^\circ$, and uses the Truesdell-Jones equation to compute ion activity coefficients. (See Chap. 4.)

These waters are derived from both pristine and polluted carbonate systems. The analyses are ordered according to their increasing TDS contents. The freshest waters have a prevalent chemical character (PCC) of the calcium-bicarbonate type. As TDS values increase, the waters become relatively more enriched in Na, Cl, and/or SO_4. The high Na and Cl content of analysis number 9 in Table 6.7 (TDS = 1269 mg/L) is derived from its pollution by sewage. The even higher Na and Cl concentrations of analysis number 10 reflect the fact that waters in the X-Can well, which is in coastal Yucatan, have been mixed with seawater.

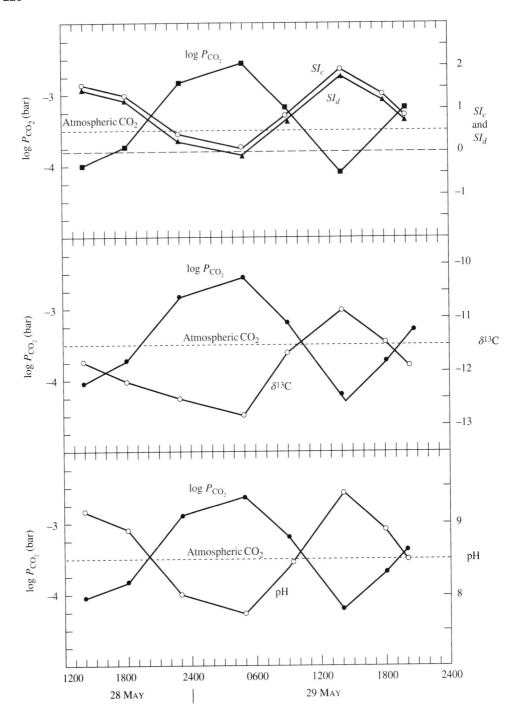

Figure 6.20 Variations in log P_{CO_2}, pH, $\delta^{13}C$ (per mil units, PDB), and the saturation indices of calcite (SI_c) and dolomite (SI_d) in Slab Cabin Run, near State College, Pennsylvania. Based on Jacobson et al. (1971).

TABLE 6.7 Chemical analyses of carbonate groundwaters from the United States and Mexico and of seawater

Analysis number	1[†]	2[‡]	3[§]	4[‖]	5[#]	6[††]	7[§§]	8[‖‖]	9[##]	10[†††]	11[‡‡‡]
SiO_2	12		5.1				31			0.4	4.4
Fe^{2+}						1.30		0.023			
Ca	34	54.9	94	60.8	71	123	106	276	133	145	404.2
Mg	5.6	23	28	57.4	61	36.9	60	58.4	54.2	166	1269
Na	3.2	6	16	25.1	9.9	6.9	21	17.5	233	1060	10565
K	0.5	1.4	0.6	3.5	1.2	13.9	3.7	23	29	41	391.4
HCO_3	124	238	382	443	292	460	206	691	447	513	142.3
SO_4	2.4	16.5	5.9	34	87	128	344	387	67.5	259	2660
Cl	4.5	13.5	34	63	110	6.0	28	29.5	480	1920	18971
F	0.1	0.06	0.2	0.02			2.2		0.05	1.2	1.3
NO_3	0.1	22	14	14	49		0		52	11	
TDS sum	111	254	386	476	553	543	698	1135	1269	3856	34410
TDS residue	138		418				726			4340	
SpC (μS/cm)	218	458	685	814	945	935	1000	1580	2150	6390	
T (°C)	23.8	10.3	24.1	10.6	12.0	27.5	26.3	23	12.5	27.5	25
pH	8.00	7.36	7.23	7.01	7.06	6.84	7.44	6.46	6.88	7.31	8.15
DO (mg/l)		7.5		0.9		2.0		1.4	3.0		
DO % sat.	(100)		(100)		38		(100)			(100)	(100)
Eh (mV)	3.7					389		423			
Ca/Mg (molar)	−2.92	1.45	2.0	0.64	0.71	2.02	1.1	2.9	1.5	0.53	0.193
log P_{CO_2} (bars)	−2.09	−2.04	−1.68	−1.43	−1.66	−1.21	−2.16	−0.68	−1.32	−1.69	−3.31
SI_c	0.47	−0.20	0.27	−0.29	−0.35	0.07	0.19	0.05	−0.13	0.44	0.70
SI_d(ordered)	−0.30	0.23	0.67	0.31	0.24	0.48	0.73	0.37	0.30	1.07	1.65
SI_d(disordered)		−0.60	−0.10	−0.51	−0.58	−0.28	−0.04	−0.41	−0.52	0.39	0.88
Rock type	limestone	limestone	limestone (?)	dolomite	dolomite	limestone	limestone & dolomite?	limestone	dolomite	?	
Sampling date	3/15/62	6/17/72	5/1/66	6/17/72	8/70	7/73	5/10/62	7/73	8/72	5/1/66	

Note: Concentrations are in mg/L except for seawater, which is in mg/kg. All temperatures, pH, Eh, and dissolved oxygen (DO) values have been measured in the field. The groundwater analyses are from Langmuir (1984). Saturation indices (SI values) and apparent CO_2 pressures (bars) have been calculated from the analyses using the computer program SOLMINEQ.88 (Kharaka et al. 1988; Debraal and Kharaka 1989). Specific conductance (SpC) is reported for 25°C. Parentheses indicate estimated values. As shown, $SI_d = \log (IAP_d/K_d)^{1/2}$.

Source: [†]Polk City Well, central FL, USA (Back and Hanshaw 1970). [‡]Thompson Spring, State College, PA, USA (Back and Hanshaw 1970). [§]X-Can Well, Yucatan, Mexico (Jacobson and Langmuir 1974). [‖]Well beneath the municipal solid-waste landfill in State College, PA, USA. TDS had increased to 708 mg/L in 1973, and was accompanied by an E. coli count of 500/100 ml. [#]Well beneath a sewage lagoon near State College, PA, USA (Langmuir 1971). [††]Healing Spring in Warm Springs Valley, NW VA, USA. Water also has 8.3 mg/L Cu, 0.1 mg/L Cd, and 1.2 mg/L Sr (Helz and Sinex 1974). [§§]Arcadia Well, central FL, USA (Back and Hanshaw 1970). [‖‖]Sweet Spring in Sweet Spring Valley, NW VA, USA. Water also has 1.2 mg/L Cd, 0.7 mg/L Cu, and 7.7 mg/L Zn (Helz and Sinex 1974). [##]Well within 50 m of two septic tanks, near State College, PA, USA. Total coliform count 46/100 ml, fecal coliforms 4/100 ml (Jacobson, 1973). [†††]Isla Mujeres Well, Yucatan, Mexico (Back and Hanshaw 1970). SI(aragonite) = 0.304. TDS was 1398 mg/L in 1973, with an E. coli count of 2200/100 mg (Doehring and Butler 1974). [‡‡‡]Seawater analysis from Pytkowicz (1983). Concentrations are in mg/kg. The pH is from Berner (1971).

Computed, apparent CO_2 pressures for the ground waters range from a high of $10^{-0.68}$ bar for Sweet Spring, a thermal spring in Virginia (analysis number 8) to $10^{-2.92}$ bar for the Polk City well in central Florida (analysis number 1), which samples shallow groundwaters in a recharge zone, mantled with soils poor in organic matter. Where groundwaters have probably approached equilibrium with respect to calcite and dolomite from undersaturation (analyses 2, 4, 5, 6, 8, and 9), SI_c values are at or below saturation and SI_d values average +0.32 and −0.48 for ordered and disordered dolomite, respectively. This suggests that a partially ordered dolomite may limit groundwater concentrations. These groundwaters are from Paleozoic-age carbonate rocks. Analyses 1, 3, 7, and 10 are of groundwaters from the younger carbonate rocks of Florida and the Yucatan Peninsula. Except for Polk City well waters (analysis number 1), which are undersaturated with respect to calcite, the other waters have SI_c values ranging from 0.19 to 0.44. SI_d values for the four waters average 0.74 and −0.01 for ordered and disordered dolomite, respectively, perhaps reflecting control of the water chemistry by protodolomite (disordered dolomite) solubility. Significant levels of supersaturation of these waters with respect to ordered dolomite and calcite result from kinetic inhibition of their precipitation. (See above discussion of inhibition.) Carbonate mineral supersaturation in these waters may also be caused by groundwater mixing with seawater (analysis number 10), or by such processes as sulfate reduction (which produces carbonate species) or dissolution of gypsum (which adds Ca^{2+}, the common ion effect) (analyses number 3 and 7).

The analysis of surface seawater shows it to be moderately supersaturated with respect to both calcite ($SI_c = 0.70$) and disordered dolomite ($SI_d = 0.88$) and more supersaturated with respect to ordered dolomite ($SI_d = 1.65$). It is also supersaturated with respect to magnesite ($SI_m = 0.76$ assuming $K_m = 10^{-7.93}$), and aragonite ($SI_a = 0.56$). Equilibration of seawater with respect to ordered dolomite, magnesite, and pure calcite is kinetically inhibited (high Mg^{2+} may inhibit calcite nucleation in seawater), although aragonite does readily precipitate from seawater.

STUDY QUESTIONS

1. What are the relative stabilities and typical occurrences of the common carbonate minerals?

2. Contrast the stabilities of the calcium versus the magnesium carbonates and comment on the possible metastability of the magnesium carbonate hydrates.

3. Why is the solubility of carbonate minerals roughly proportional to the cube root of the CO_2 pressure.

4. The solubility of calcite at 25°C decreases with increasing pH up to about pH 10, but is constant, independent of pH, above this pH. Explain.

5. Understand the general effects of CO_2 exsolution and dissolution, photosynthesis and respiration or decay, evaporation, common ion effects, mixing of waters, and temperature changes on the saturation state of waters with respect to calcite. Give examples of the environmental conditions when each of these processes or effects might control calcite-saturation state.

6. How might increases in groundwater pollution affect the solubility of carbonate minerals?

7. For the Ca^{2+}/Mg^{2+} ratio of seawater, ideal, stoichiometric and well-crystallized dolomite is more thermodynamically stable than calcite. Why then do calcite and aragonite, rather than dolomite, precipitate from seawater?

8. Compare and discuss the evolution of groundwater in contact with calcite under open- versus closed-system conditions.

9. How is calcite solubility affected by the formation of ion pairs and increases in ionic strength? Explain.

10. The maximum concentrations of trace metals may be limited in groundwaters at equilibrium with minerals such as calcite, dolomite, siderite, and rhodocrosite. Explain.

11. One should compute about the same saturation index (SI value) for a water with respect to a carbonate mineral when the computation is done ignoring ion pairs as when considering them, as long as the respective K_{sp} values used in the two calculations were also determined by one ignoring and the other assuming the same ion pairs. Explain how this statement can be true.

12. Why do some surface waters remain supersaturated with respect to calcite without calcite precipitation, even at SI_c values as high as 1.0?

13. How and why is the $\delta^{13}C$ content of a stream water related to its pH and apparent CO_2 pressure during the day and at night?

14. Read the paper by Deines et al. (1974). How did the authors use $\delta^{13}C$ data to conclude that the carbonate groundwaters were moving under largely closed-system conditions?

15. Carbonate mineral saturation can limit metal and carbonate concentrations even in sewage and landfill leachate polluted groundwaters. What would you expect the apparent CO_2 pressure to be in such waters and why? What about the solubility of carbonates under such conditions?

16. Explain why the $CaCO_3$ tests (shells) from planktonic (floating) marine organisms dissolve when they sink into the deep ocean.

PROBLEMS

1. A well water sampled from a depth of 53.3 m in dolomite rock, has a temperature of 9.8°C, a field-measured pH of 7.17, a dissolved oxygen concentration of 65% of saturation, specific conductance of 720 μmhos/cm at 25°C, and the following chemical analysis, where concentrations are given in mmol/kg: $Ca^{2+} = 1.70$, $Mg^{2+} = 2.30$, $Na^+ = 0.048$, $K^+ = 0.023$, $HCO_3^- = 7.18$, $SO_4^{2-} = 0.30$, $Cl^- = 0.09$, and $NO_3^- = 0.124$. The well is adjacent to a farmhouse septic tank.

    (a) Compute the charge balance and total ionic strength.

    (b) Calculate Debye-Hückel ion activity coefficients for Ca^{2+}, Mg^{2+}, and HCO_3^- in the well water at 9.8°C (assume 10°C). Assume the respective ion size (a_i) parameters for these species are: 5.5, 6.5, and 4.5 Å. At 10°C, A(molal) = 0.4976 and B(molal) = 0.3261 (Helgeson et al. 1981).

    To answer questions (c), (e), and (f) below, you are given the following equilibrium constants for 9.8°C, $K_{CO_2} = 10^{-1.27}$, $K_1(H_2CO_3^\circ) = 10^{-6.47}$, $K_2(H_2CO_3^\circ) = 10^{-10.49}$, $K_{sp}(\text{calcite}) = 10^{-8.41}$, and $K_{sp}(\text{dolomite}) = 10^{-16.71}$. Henceforth we will abbreviate these respective K_{sp} values as K_c and K_d. In questions (c) and (e) below, ignore ion pairing, but use the γ values for the ions computed in part (b).

    (c) Compute the apparent CO_2 pressure in bars in equilibrium with the water. How does it compare with the atmospheric carbon dioxide pressure? Discuss reasons for any difference.

    (d) What chemical properties of the water might reflect its proximity to the septic tank, and given that some pollution is apparent, what unreported substances of concern might also be present in the water? This question is related to question (c).

    (e) Compute the saturation index of calcite, SI_c, where $SI_c = \log(\text{IAP}_c/K_c)$, and of dolomite, SI_d, where $SI_d = \log[(\text{IAP}_d)/K_d]^{1/2}$, and comment on the significance of the values.

    (f) How different would your answers have been in question (e) if you had ignored ion activity coefficients and assumed all $\gamma = 1$?

    (g) So far we have also ignored ion pairs in our calculations. Now input the above groundwater analysis into the computer program MINTEQA2 and tabulate for the free ions and ion pairs of Ca and Mg their output concentrations in mmol/kg and percents of total Ca and Mg. Also report the computed ionic strength and SI values of calcite and dolomite. How do the ionic strength and saturation indices obtained with MINTEQA2 compare to the values calculated in parts (a), (e), and (f) above? Explain the differences.

    (h) It is instructive to hand-calculate a few ion pair concentrations in order to understand how a computer model such as MINTEQA2 operates. Such an exercise gives us a glimpse inside "the black box" of such a model. Hand calculation can be extremely tedious if several ion pairs are present. We will, therefore, simplify the water analysis so that we need consider only the bicarbonate ion pairs in our calculations. This will be done by assuming that the concentration of K^+ is replaced by an equivalent amount of Na^+, and that SO_4^{2-} and NO_3^- are replaced by equivalent amounts of Cl^-. Our "simplified" analysis is tabulated below.

Cations	mmol/kg	meq/kg	Anions	mmol/kg	meq/kg
Ca^{2+}	1.70	3.40	HCO_3^-	7.18	7.18
Mg^{2+}	2.30	4.60	Cl^-	0.81	0.81
Na^+	0.07	0.07			

Given this analysis and that the dissociation constants for the ion pairs at 9.8°C are $K_{dissoc}(CaHCO_3^+) = 10^{-0.97}$, and $K_{dissoc}(MgHCO_3^+) = 10^{-1.05}$, hand calculate the molal concentrations of free Ca^{2+}, Mg^{2+}, and HCO_3^- and of the bicarbonate ion pairs and compute their percent values. Use the ion activity coefficients computed in part (b) in this calculation, assuming a constant ionic strength, and that the activity coefficients of both bicarbonate ion pairs equal γ_{HCO_3}. Also compute the apparent CO_2 pressure and the saturation indices of calcite and dolomite.

(i) Finally, input this analysis into MINTEQA2. Discuss and compare the results of the hand calculation and the MINTEQA2 run.

2. What is the pH, apparent CO_2 pressure in bars, and speciation (as percent concentrations) in solution when pure water equilibrates with calcite in a closed system at 25°C? What are the total concentrations of major species? Solve the problem using MINTEQA2.

3. Tabulated below in mg/L is the composition of an average drinking water supply, and average effluent from secondary treatment of municipal wastewater (Weinberger et al. 1966). The difference in these compositions is listed as the increment of substances added by use. High SiO_2 and PO_4 concentrations in the sewage are probably chiefly in suspended solids. To the extent possible, solve the problem using MINTEQA2. Assume 25°C.

(a) In this problem the sewage is leaking from a sewage lagoon into a water-table groundwater in shallow limestone rock. The apparent CO_2 pressure of the uncontaminated groundwater is $10^{-2.2}$ bar. Assume that initially, the sewage pH is 6.8. Neglecting species of N and P except for NO_3^-, use MINTEQA2 to compute the apparent CO_2 pressure of the sewage and its saturation index with respect to calcite (SI_c).

(b) If all of the sewage BOD (25 mg/L as C) is oxidized to CO_2, what will happen to the apparent CO_2 pressure of the sewage assuming its alkalinity remains unchanged?

(c) If the sewage now equilibrates with calcite in the limestone at the CO_2 pressure computed in part (b) (assumed constant), what are the pH and concentrations of total Ca^{2+} and total HCO_3^- at calcite saturation?

(d) How much more soluble is calcite in the sewage-affected groundwater than in uncontaminated groundwater?

Species	Water supply	Secondary sewage	Increment added	Species	Water supply	Secondary sewage	Increment added
$BOD^†$	0	25	25	NO_3	5	15	10
Na	65	135	70	HCO_3	200	300	100
K	5	15	10	SO_4	70	100	30
NH_3	0	20	20	SiO_2	35	50	15
Ca	45	60	15	PO_4	0	25	25
Mg	18	25	7	TDS	410	730	320
Cl	55	130	75				

†BOD is biodegradable organics or biochemical oxygen demand.

7

Chemical Weathering

7.1 GENERAL OBSERVATIONS

Among the most important processes affecting the chemistry of waters in streams, soils, and shallow groundwaters is *chemical weathering*. Chemical weathering takes place because essentially all minerals are undersaturated in contact with atmospheric precipitation and in the shallow soil moisture in humid climates (see Chapter 8). Further, many of the minerals present in subaerial environments were formed at elevated temperatures and pressures and so are thermodynamically unstable at Earth's surface. Chemical weathering involves the dissolution, alteration, and sometimes replacement of such minerals and the precipitation of new minerals, such as clays and ferric and aluminum oxyhydroxides, which are more nearly at chemical equilibrium with their environment under Earth-surface conditions. Chemical weathering chiefly takes place in soils, but also occurs in streams and stream sediments. In soils, the products of chemical weathering are a function of the nature of the parent material, climate, vegetation (biological activity), drainage, and temperature. Further observations include the following:

1. *Most chemical weathering occurs in the vadose zone (zone of aeration) of soils.* It is most active in deeper soil horizons which contain partially weathered or unweathered materials.
2. *The rate of chemical weathering is usually increased if the rate of physical or mechanical weathering is increased.* Physical weathering (the breakdown of rock due to physical processes), which is most active in mountainous areas, can repeatedly expose fresh mineral surfaces to chemical weathering.
3. *The rate of chemical weathering is proportional to soil organic/biological activity, including the abundance of vegetation.* Such activity produces CO_2, carbonic acid, and also organic acids that actively react with and break down many minerals.
4. *The chemical weathering rate of a given soil is proportional to the infiltration rate (I) of water through it.* $I = P - R - ET$, where P, R, and ET are the precipitation, runoff and evapotranspiration rates (see Chap. 8, Section 8.2).

5. *The rate of chemical weathering increases with increasing temperature.* This reflects the fact that reaction rates increase (half-times decrease) with increasing temperature. (See Chap. 2, Section 2.4).

6. *Chemical weathering generally involves the attack of* H_2O *and associated acidity, and atmospheric* O_2 *(oxidation) on parent materials.*

7. *Compared with the parent rock, mineral products of weathering are usually more hydrated (as in the clays), have lower metal cation concentrations relative to* Al *and* Si *(as in the clays and quartz), are oxidized (as in the ferric oxyhydroxides), or are chemical precipitates such as the carbonates or evaporite minerals, stable at earth-surface temperatures.*

8. *Dissolved products of weathering include the major cations and anions in natural waters, and dissolved silica.*

Silicate and aluminosilicate mineral-weathering reactions involve both H^+-ion attack on the mineral and *hydration* or *hydrolysis* of the product minerals (clays and metal oxyhydroxides) or product aqueous species. If the primary mineral contains aluminum, then clays may result from the weathering. If the primary mineral contains Fe(II) or other reduced metal ions, weathering may involve *oxidation* as well as hydrolysis.

Some examples of chemical weathering reactions are:

$$Mg_2SiO_4 + 4H^+ = 2Mg^{2+} + H_4SiO_4^\circ \tag{7.1}$$

forsterite
olivine

$$Fe_2SiO_4 + \tfrac{1}{2}O_2 + 3H_2O = 2(\alpha\text{-FeOOH}) + H_4SiO_4^\circ \tag{7.2}$$

fayalite   goethite
olivine

$$CaAl_2Si_2O_8 + H_2O + 2H^+ = Ca^{2+} + Al_2Si_2O_5(OH)_4 \tag{7.3}$$

Ca-feldspar   kaolinite
(anorthite)   clay

(ΔG_f°) −960.15   −56.687   −132.3   −908.07

Note that because olivine contains no Al, no clays result from reactions (7.1) or (7.2). Also, reaction (7.1) is *congruent*, whereas reactions (7.2) and (7.3) are *incongruent*. In other words, in the first reaction all the weathering products are soluble, whereas in the second and third reactions, weathering results in precipitation of new solid phases. Reaction (7.2) involves oxidation of Fe(II) to Fe(III), with precipitation of the ferric iron in a mineral such as goethite. Among the three examples, only the feldspar contains Al and so can result in a clay. In reaction (7.3) we assume that all the Al moves directly from the feldspar to the kaolinite. However, if the pH is below about 4 or 5, the kaolinite itself becomes soluble and appreciable Al also goes into solution.

Are there Earth-surface conditions for which the above minerals are thermodynamically stable and so would not weather? The answer is rarely for the olivine minerals (cf. Bath et al. 1987), which rapidly disappear in the volcanic soils and beach sands of Hawaii, for example. What about the stability of the plagioclase feldspar? Gibbs free energies of the reactants and products (in kcal/mol) given under weathering reaction (7.3) indicate that $\Delta G_r^\circ = -23.53$ kcal/mol, and $K_{eq} = 10^{17.25} = [Ca^{2+}]/[H^+]^2$. If we assume a typical soil pH of 6, then we can compute that Ca^{2+} would have to equal $10^{5.25}$ M for the Ca-feldspar to be thermodynamically stable relative to kaolinite. This is obviously an impossibly high concentration.

One could ask, is there any pH at which the Ca-feldspar would be stable relative to kaolinite? We must assume a calcium concentration to make this calculation. If we set $Ca^{2+} = 10^{-2}$ M (roughly 400 mg/L), which is a bit higher than likely at this pH, we can compute that the pH at which the feldspar would be stable is about 9.6, a pH found only in unusual environments such as highly alkaline lakes and groundwaters.

7.2 WEATHERING RATES OF SOME ROCK-FORMING MINERALS

In Chap. 2 we discussed rates of mineral dissolution reactions. An important application of such data is to estimate absolute and relative rates of chemical weathering (cf. Lasaga et al. 1994; Drever 1994). Based on his field observations, S. S. Goldich (1938) suggested the weathering sequence of common rock-forming minerals shown in Fig. 7.1. The diagram shows that in surface environments, olivine and Ca-plagioclase feldspar are highly vulnerable to chemical weathering, whereas quartz and muscovite mica are practically inert. Lasaga et al. (1994) used laboratory dissolution rate data far from equilibrium to estimate the lifetime of a 1-mm-diameter cube of a number of common minerals at pH = 5. Data from their paper and related sources are ordered in Table 7.1 with the most weatherable minerals listed first. The theoretical lifetimes are also shown in parentheses in Fig. 7.1. Clearly, the Goldich weatherability sequence is consistent with the rate data. Quartz, kaolinite, and muscovite (related to the clay illite), are common products of chemical weathering. Their persistence in mature, weathered soils and sediments is consistent with their crystal lifetimes in Table 7.1, all of which exceed 10^6 years. The relative dissolution rates are also generally consistent with instabilities of the minerals computed from free-energy data, as in Section 7.1. Although their solution rates may be slow, it is important to remember that all minerals, whether formed at high or low temperatures (e.g., quartz, muscovite, and the clays) will eventually dissolve and be leached away if placed long enough in contact with precipitation or with typical (acid) soil moisture.

TABLE 7.1 Log dissolution rates and corresponding theoretical mean lifetimes of a 1-mm-diameter cube of some silicate and aluminosilicate minerals at pH = 5 and 25°C

Mineral	−log rate (mol/m²/s)	Lifetime (y)	Mineral	−log rate (mol/m²/s)	Lifetime (y)
Wollastonite	8.00	7.9×10^1	Tremolite[‡]	11.7	5.8×10^4
Anorthite	8.55	1.12×10^2	Sanidine	12.00	2.91×10^5
Nepheline	8.55	2.11×10^2	Albite	12.26	5.75×10^5
Forsterite	9.5	2.3×10^3	Microcline	12.50	9.21×10^5
Diopside	10.15	6.8×10^3	Muscovite	13.07	2.6×10^6
Enstatite	10.00	1.01×10^4	Kaolinite	13.28	6.0×10^6
Biotite[†]	11.25	3.8×10^4	Quartz	13.39	2.4×10^7
Gibbsite	11.45	2.76×10^5			

Note: Minerals are listed in order from most weatherable to least weatherable.

Idealized formulas: wollastonite, $CaSiO_3$; anorthite, $CaAl_2Si_2O_8$; nepheline, $NaAlSiO_4$; forsterite, Mg_2SiO_4; diopside, $CaMgSi_2O_6$; enstatite, $MgSiO_3$; biotite, $K(Mg,Fe)_3(AlSi_3O_{10})(OH)_2$; gibbsite, $Al(OH)_3$; tremolite, $Ca_2Mg_5Si_8O_{22}(OH)_2$; sanidine, $KAlSi_3O_8$; albite, $NaAlSi_3O_8$; microcline, $KAlSi_3O_8$; muscovite, $KAl_2[Si_3AlO_{10}](OH)_2$; kaolinite, $Al_2Si_2O_5(OH)_4$; quartz, SiO_2 (Deer et al. 1992).

Source: Most of the data are from Lasaga et al. (1994). [†]Rate is average from Sverdrup (1989) as reported by Acker and Bricker (1992). Lifetime computed assuming molar volume of phlogopite (149.91 cm³). [‡]Rate from Mast and Drever (1987). Lifetime computed assuming molar volume of 272.92 cm³.

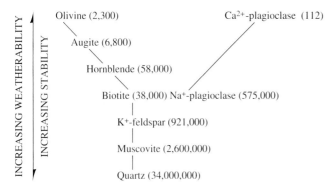

Figure 7.1 Goldich's sequence of increasing weatherability of common minerals (cf. Loughnan 1969; Faure 1991). In parentheses are the lifetimes in years from Table 7.1, assuming olivine = forsterite, augite = diopside, hornblende = tremolite, Ca-plagioclase = anorthite, Na-plagioclase = albite, K-feldspar = microcline, and the stability of muscovite is comparable to the related clay, illite.

7.3 A WEATHERING EXAMPLE

Many of the above observations can be illustrated through discussion of an example of weathering taken from Goldich (1938) (see also Krauskopf 1967). The parent rock in this humid, temperate climate example is a quartz-feldspar-biotite gneiss. The mineralogic and oxide composition of this rock and its weathered product soil is given in Table 7.2.

In order to quantify changes in the element oxides caused by chemical weathering we need to identify a "tracer" element that can be assumed nearly inert (immobile) during weathering. Among the oxides listed, Al_2O_3 (alumina) is probably the best choice, in that Al is relatively insoluble in natural waters above pH 5.5 (near pH = 7 Al(aq) is usually <0.1 mg/L). (The greater insolubilities of ZrO_2 or TiO_2, compared to Al_2O_3, may make them even better tracers [cf. Nesbitt and Wilson 1992]). Assuming Al_2O_3 is our tracer and that initially 100 wt % ≡ 100 g of soil, we can normalize the concentrations of other metal oxides to 14.61 g of alumina in the weathered material, as shown in the third column at the bottom of Table 7.2. Now, a comparison of the fresh rock (wt %) and remaining rock (wt %) columns, subtracting out the water added during soil formation, indicates that 25% of the rock has been removed by weathering.

Comparison of the mineralogy of the rock and product soil shows that the amount of quartz has increased from 30 to 43%,[†] potassium feldspar has decreased from 19 to 13%, biotite and hornblende have disappeared, and resistant trace (assessory) minerals in the rock, such as magnetite (Fe_3O_4) and ilmenite ($FeTiO_3$) have been concentrated (remember the Ip concept and insoluble, multivalent metal oxides in Chapter 3). Most conspicuously, the plagioclase feldspar (composition between $NaAlSi_3O_8$ and $CaAl_2Si_2O_8$) has practically disappeared. Kaolinite has formed from the Al in the plagioclase and K-feldspars and the biotite.

Obvious conclusions are:

1. Of total silica in the rock, 22% has been lost to the soil solution, chiefly from weathering of the plagioclase feldspar, biotite, and hornblende. All or most of the quartz has evidently been unaffected by weathering.

[†]There is no proof, however, that the weathered material has derived from the exact parent rock given in the table. This may explain the excess in soil vol % quartz, which, assuming quartz is not dissolved or precipitated during soil formation and that volume and weight percents are roughly equal, should not exceed about (30/75) × 100 = 40 wt % of the soil. (Densities of quartz, K-feldspar, plagioclase feldspar, and kaolinite are all 2.60 ± 0.05 g/cm³).

TABLE 7.2 The composition of a granitic-type rock and of its weathering products[†]

Approximate Mineral Composition (Volume %)

	Fresh rock	Weathered material
Quartz	30	43
K-feldspar	19	13
Plagioclase feldspar	40	1
Biotite (+ chlorite)	7	trace
Hornblende	1	none
Magnetite, ilmenite, other secondary oxides	1.5	2
Kaolinite	none	40

Approximate Oxide Composition (Weight %)

	Fresh rock (wt %)	Weathered material (wt %)	Remaining[‡] (wt %)	Percent loss or gain
SiO_2	71.48	70.51	55.99	−22
Al_2O_3	14.61	18.40	14.61 (assumed)	0
Fe_2O_3	0.69 ⎫	1.55 ⎫	1.23	+78
FeO	1.64 ⎭ 2.3	0.22 ⎭ 1.8	0.17	−90
MgO	0.77	0.21	0.17	−78
CaO	2.08	0.10	0.08	−96
Na_2O	3.84	0.09	0.07	−98
K_2O	3.92	2.48	1.97	−50
H_2O	0.32	5.90	4.68	+1,360
Others[§]	0.70	0.54	0.43	−39
Total	100.00	100.00	79.40	

[†]For compositions of the minerals listed in this table, see also Table 7.1 footnotes. Other minerals are: K-feldspar, $KAlSi_3O_8$; plagioclase (feldspar), a solid solution with a composition between albite and anorthite; chlorite, a fine-grained mica of composition $(Mg,Fe^{2+},Fe^{3+},Mn,Al)_{12}([Si,Al]_8O_{20})(OH)_{16}$; hornblende, a black prismatic mineral with the general composition $(Na,K)_{0-1}Ca_2(Mg,Fe^{2+},Fe^{3+},Al)_5Si_{6-7.5}Al_{2-0.5}O_{22}(OH)_2$; magnetite, Fe_3O_4; ilmenite, $FeTiO_3$ (Deer et al. 1992).

[‡]Assuming all Al_2O_3 remains.

[§]TiO_2, Mn oxides, BaO, P_2O_5, CO_2, and S oxides.

Source: From Goldich (1938) and Krauskopf (1967).

2. Weathering of the feldspars, biotite, and hornblende has also released their cations to the soil solution, as apparent from the high percentage losses of MgO, CaO, Na_2O, and K_2O.

3. Weathering has led to oxidative loss of FeO in the biotite and hornblende and a relative increase in Fe_2O_3 in the soil. A comparison of the total iron as Fe in the parent rock and the soil shows that a 44% loss of Fe has occurred on weathering.

4. The most profound change in composition on weathering is the addition of a large amount of water (+1360%), chiefly present as water of hydration in the kaolinite clay.

5. Practically all the Na^+ and Ca^{2+}, most of the Mg^{2+}, and half of the K^+ have been removed from the rock. The 25% weight loss on weathering has released dissolved cations and silica in the soil solution. This is consistent with the domination of Na^+, Ca^{2+}, K^+, and Mg^{2+} as the major cations in natural waters. Bicarbonate is the major anion produced by weathering. This reflects the fact that carbonic acid, the chief weathering agent in the soil, forms HCO_3^- according to the reaction $H_2CO_3^\circ \rightarrow H^+ + HCO_3^-$, as protons are consumed (and the pH rises) during weathering.

7.4 SOIL CLASSIFICATION AND PROCESSES

Soils are extremely important environmentally, for they provide the medium and the essential nutrients for plant growth and strongly influence the chemical composition of surface- and groundwaters. Soils are also called upon to remove or attenuate contaminants derived from the atmosphere or released from the surface or near-surface disposal of wastes. We will limit our discussion of soils to a few basic principles. For a more thorough treatment of soil geochemistry the reader is referred to books by Loughnan (1969), Greenland and Hayes (1978), Lindsay (1979), Rose et al. (1979), Bolt (1979), Drever (1985), and Sposito (1989).

Soils are the products of chemical weathering of parent rocks and sediments and, usually to a lesser extent, the physical breakdown of those materials. When the parent material for soil formation is underlying bedrock, the soil so formed in place is called a residual soil. Soils may also develop in materials that have been transported and deposited by wind (loess), streams (alluvium), ice (till), or by landslides (colluvium).

In arctic and in high-altitude temperate climates, the freezing and thawing of water in rock pores and fractures (a physical process called frost wedging or riving) may assist in the rupture of rocks and diminution of rock fragments (the molar volume of ice is 8.99% greater than that of liquid water at 0°C and 1 bar pressure). Diminution of rock materials, especially of sand size or larger ($> \frac{1}{16}$ mm) is also caused by their abrasion and fragmentation through the action of wind, streams, landslides, and glaciers. Although physical processes may be the source of materials that become soils, the development of mature (more weathered) soils chiefly results from processes related to chemical weathering.

Young or immature soils are those that have been little weathered and so are in a changing, unstable state. Profiles of such soils may show minimal physical or chemical differentiation from the surface down to bedrock. Older or more mature soils tend toward steady-state conditions. The soil profiles of mature soils have developed subhorizontal layers called soil horizons. From the surface downward these are termed O, A, E, B, C, and R horizons (Table 7.3). Frequently, soil scientists further subdivide individual horizons on a basis of textural and color differences within them. Figure 7.2 is an example of such subdivision for a soil developed on loess.

As soil is formed from the weathering of underlying bedrock, the O, A, E, and B (together called the *solum*), and especially C horizons tend to thicken with time unless surface soil is eroded away, in which case O, A, E, and B horizons will migrate downward as underlying materials near the land surface.

Immature soils may lack one or more soil horizons and will compositionally resemble their parent materials. The well-developed horizons of mature soils result from long-term, relatively constant, climatic and drainage conditions with minimal physical erosion. Mature soils then reflect the development of steady-state conditions among organic-matter accumulation and breakdown in the O and A horizons, transport and migration of dissolved and suspended weathering products to the B horizon, chemical weathering (which is most pronounced in the C horizon), and the leaching of dissolved weathering products out of the soil profile into nearby surface or groundwaters. In mature soils the nature of parent materials may be erased above the C horizon. Soil composition will then chiefly reflect the effects of climate, vegetation, and drainage. The above description applies to soil systems in which there is a net downward movement of soil moisture as would be expected generally in moist, humid climates.

Low rainfall and high moisture losses due to evapotranspiration are typical features of arid regions. Thus, following precipitation events and resultant infiltration of moisture, arid soils often experience net moisture movement upward toward the land surface driven by capillary forces. For such

TABLE 7.3   General description of soil horizons

Soil horizon	Description
O	A surface layer composed primarily of organic matter, black or dark brown in color.
A	Mineral particles mixed with finely divided organic matter that produces a dark grey color.
E	Primarily mineral grains resistant to chemical weathering and too large to have been translocated downward by percolating soil water. Low in organic content and so light grey in color.
O, A, E	General Comments: O, A, and E horizons may together be a few cm to 1.2 m thick. Humic and fulvic acids and CO_2 produced in the O and A horizons are the chief weathering agents. The A horizon overall is strongly chemically weathered and so depleted of materials that have been carried downward as suspended or dissolved substances.
B	Horizon in which substances leached and transported from the A and E horizons (such as Fe, Al, and Mn oxyhydroxides, and organic matter as humus) are deposited. Deposition of Fe, Al, and Mn oxyhydroxides is favored by the breakdown of metal complexing organic acids produced in the O and A horizons. Horizon is enriched in clay minerals, may also include precipitated calcite and gypsum, and contains highly weathered, altered materials relative to underlying bedrock. B horizon may be a few cm to 2 m thick.
C	Parent materials largely unaffected by chemical weathering. May be partially weathered underlying bedrock (saprolite) or sediment (e.g., alluvium or glacial till), or volcanic tuff, for example. Source of materials for horizons O, A, E, and B. Maximum thicknesses of C horizon (e.g., ~30 m) are found in tropical soils.
R	Unaltered or unweathered parent materials.

Source: Modified after Foth (1984).

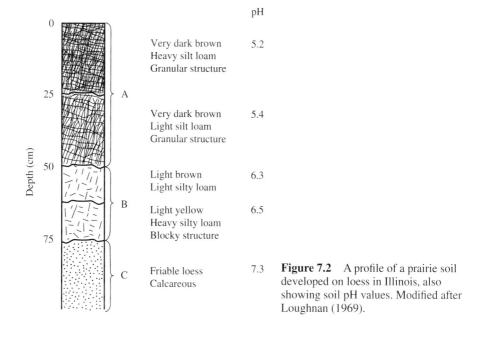

Figure 7.2   A profile of a prairie soil developed on loess in Illinois, also showing soil pH values. Modified after Loughnan (1969).

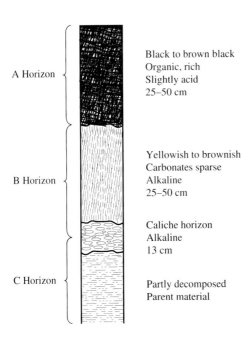

A Horizon

Black to brown black
Organic, rich
Slightly acid
25–50 cm

B Horizon

Yellowish to brownish
Carbonates sparse
Alkaline
25–50 cm

Caliche horizon
Alkaline
13 cm

C Horizon

Partly decomposed
Parent material

Figure 7.3   A generalized profile of a forested soil developed in a humid climate with a moderate temperature, showing a caliche ($CaCO_3$) layer between the B and C horizons. From *Fundamentals of Soil Science*, 3d ed. C. E. Millar, L. M. Turk, and H. D. Foth, Copyright © 1958 by John Wiley & Sons, Inc. Reprinted by permission of John Wiley & Sons, Inc.

conditions, the soluble or colloidal species that normally precipitate in the B horizon or are leached from the soil altogether in wet, humid climates, may instead be concentrated and precipitated in the solum. In this way, layers or concretions of silica, carbonates (sometimes called caliche), and sulfate salts can form in the B and C horizons (see Fig. 7.3). They form at depths that increase with rainfall (Fig. 7.4). However, in desert soils such precipitates accumulate at or just below the land surface.

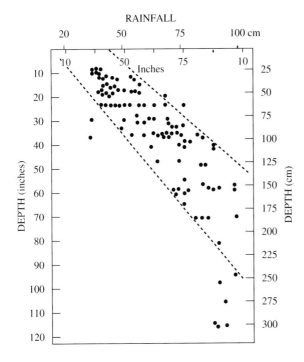

Figure 7.4   Plot of the relation between rainfall and depth to the top of carbonate mineral (caliche) accumulation in soils derived from loess. From H. Jenny and C. D. Leonard, *Soil Sci.* 38:363–81. © 1934 by Soil Science Society of America. Used by permission.

Developed at the surface as a hardpan layer, they have been described as caliche or calcrete (chiefly $CaCO_3$), silcrete (chiefly SiO_2), or ferricrete (ferric oxyhydroxides).

According to Loughnan (1969), "Soils are generally classified as *zonal, intrazonal*, or *azonal*. Zonal soils . . . possess mature characteristics and . . . have developed over wide geographic areas under reasonably uniform climatic, drainage, and vegetational influences . . . Intrazonal soils also have mature profiles, but differ from the prevailing zonal soils type" because of controlling local conditions of soil formation, such as drainage or relief. Azonal soils are immature soils that "bear little relation to the prevailing zonal type."

Zonal soils have been broadly classified overall as *pedocals* or *pedalfers* (obsolete terms). Their occurrence is qualitatively a function of rainfall amount. Pedocals (pedon is Greek for ground, cal is short for calcium) are found in relatively arid regions with rainfall amounts less than about 25 in (64 cm). Pedocals have carbonate accumulations (e.g., caliche) in their A and B horizons (the solum) because of incomplete leaching of calcium and bicarbonate ions. According to Loughnan (1969) such carbonate layers are leached out of and absent from temperate climate soils that experience 100 cm or more of precipitation. Pedalfers are aluminum- and iron-rich soils developed in humid climates where rainfall amounts are greater than about 25 in. (64 cm). Carbonate mineral accumulations are lacking, but aluminum and ferric oxyhydroxide precipitates are concentrated in the B horizon.

There are many terms used to define and distinguish mature zonal soils. Among the most familiar are the great soil groups. These are groups "of soils having a wide distribution and a number of common fundamental internal characteristics" (Bunting 1965). The great soil groups have developed under different but relatively uniform conditions of climate, drainage, and vegetative cover. Their occurrences are generalized as a function of temperature and precipitation in Fig. 7.5.

Among the great soil groups, those in wet climates (pedalfers) have been called *podzols* and *latosols* (Fig. 2.5). *Podzols* include a wide range of leached soils having excellent profile development. They are found in cool, humid, forested areas, but also in the tropics and subtropics. Podzols are characterized by a relative abundance of organic matter and intense leaching of the A horizon consistent with its strong acidity. Because of high rainfall, carbonates and sulfate salts are generally absent in podzols. Clays and ferric oxyhydroxides are concentrated in the B horizon. Minerals found in the B horizon of podzols may include hematite, goethite, kaolinite, illite, gibbsite, and quartz.

Latosols and *laterites* are formed in tropical and subtropical areas of high rainfall, where the soils are well drained, and physical erosion is minimal. Such conditions lead to intensely leached and oxidized soils. Organic decay rates are rapid so that little organic matter accumulates in the A horizon. Soluble weathering products are leached away, leaving insoluble Al and Fe oxyhydroxides (also TiO_2) to accumulate in the B horizon. If the soil is seasonally waterlogged, then reducing conditions may develop, leading to the reduction of ferric iron and its removal from the soil as soluble ferrous iron. The result is an alumina-rich soil called *bauxite,* which is the chief ore of aluminum metal.

Culminating a long-term international effort, the Soil Survey staff of the Soil Conservation Service (U.S. Department of Agriculture) developed a Comprehensive Soil Classification System (CSCS) for world soils (Soil Survey Staff 1975). The CSCS defines soil classes strictly in terms of soil morphology, rather than based on soil genesis. A brief explanation of the 10 soil orders in the CSCS is given in Table 7.4 (see also Bodek et al. 1988). Their temporal relationships are considered in Fig. 7.6.

An understanding of the properties and behavior of soil systems is essential if we are to understand and predict the behavior of contaminant species released into and on soils. A variety of physical, geochemical, and biological processes operate in soils. Soils can physically filter out particulate contaminants, including bacteria and viruses. Filtration is most effective in A, E, and B horizons,

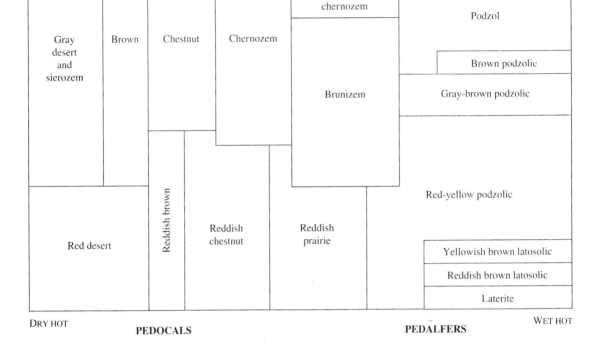

Figure 7.5 Temperature-moisture (climate) relations favorable to the formation of some of the great soil groups. From *Fundamentals of Soil Science,* 3d ed. C. E. Millar, L. M. Turk, and H. D. Foth, Copyright © 1958 by John Wiley & Sons, Inc. Reprinted by permission of John Wiley & Sons, Inc.

because of their sand- and especially silt- and clay-size materials. Filtration may also occur in the C horizon of parent alluvial sediments, but is less effective when the C horizon is made up of the coarser materials formed on crystalline igneous or metamorphic rocks.

The removal of dissolved or particulate organic contaminants is most effective where they can be aerobically broken down. Aerobic decay is most rapid in well-aerated, unsaturated soils and is most complete in thick, unsaturated soils. Organic decay in water-saturated soils tends to be anaerobic, which is much slower and produces more noxious products than aerobic decay (see Chap. 5). Because the O and A horizons of a soil are usually relatively acid (cf. Figs. 7.2 and 7.3), the alkalinity of soils chiefly resides in the clays and carbonates within B and C horizons. Because of their carbonate content, mollisols and aridisols can neutralize acid wastes more rapidly and completely than can oxisols and spodosols, for example.

Soil organic matter in the A horizon and clays and metal oxyhydroxides in the B and C horizons (also humus in the B horizon) have high sorptive capacities for trace organic and metal contaminants (see Chap. 10). Such sorbent materials are much less abundant in the C horizon of soils, which is, therefore, less effective as a sorbent zone for such contaminants, unless it is relatively thick. These observations suggest that there are obvious ways to predict and maximize (or minimize) the ability of soil systems to remove or attenuate contaminant substances.

TABLE 7.4 A simplified description of the 10 soil orders as defined in the Comprehensive Soil Classification System (CSCS)

Order	Description
Soils with poorly developed horizons or no horizon and capable of further mineral alteration	
Entisols	Soils at an early stage of development and lacking horizons, due to short time available or location on slopes subject to constant removal by physical erosion.
Inseptisols	Young soils with poorly developed horizons. May have some leaching in the A horizon, but has only a weakly developed B horizon. Found in moderately humid upland areas.
Soils with a large proportion of organic matter	
Histosols	Organic-rich soils with thick, peaty horizons; formed in low-lying permanently water–logged areas. Typical of cool swamp environments.
Soils with well-developed horizons or with fully weathered minerals, resulting from long-continued adjustment to prevailing soil-temperature and soil-water regimes	
Alfisols	Forest soils of humid and subhumid climates with strongly leached acidic A horizon containing decomposing organic matter over a base-rich (smectitic) clayey and well-developed B horizon. High content of primary weatherable minerals (e.g., feldspars). One kind of pedalfer soil.
Spodosols	Highly acidic surface organic-rich layer over strongly leached and quartz-rich E horizon, on base-poor and Al and Fe-oxide/hydroxide rich B horizon. Typical under coniferous forests in temperate climates. Formerly called podzols.
Ultisols	Strongly-leached acidic A and E horizons over well-developed, kaolinite-rich (base-poor) B horizon. Low content of primary minerals due to extensive weathering. Midlatitude and warmer temperature deciduous forest soils, common in piedmont and coastal plain, southeastern United States.
Oxisols	Reddish soils of low latitudes, deeply weathered like ultisols. Strongly leached A horizon. Clays such as kaolinite largely leached from B horizon, where oxides of Al and Fe abundant. Almost no remaining primary minerals. Formerly called laterites.
Mollisols	Thick, well-developed and base-rich A horizon with much organic matter. B horizon clay-rich with abundant Ca^{2+} and Mg^{2+}. Clayey, calcareous or gypsiferous subsurface accumulations. Found under grassland and steppe vegetation in subhumid to semiarid climates.
Aridisols	Soils of dry climates and deserts. Often contain wind-blown dust. A and B horizons thin with little organic material. Calcium carbonate (caliche) accumulations generally present, sometimes with gypsiferous or saline horizons.
Vertisols	Old weathered soils with uniform, thick, clay-rich profiles. Deeply cracked and hummocky topography produced by intense seasonal drying of expandable clay minerals such as smectites. Soil profile poorly developed because of vertical mixing caused by the seasonal cracking. High in exchangeable cations (base-rich).

Note: Base-rich and base-poor clays are clays rich or poor in adsorbed Ca^{2+}, Mg^{2+}, Na^+, and K^+.

Source: Modified after Retallack (1990), Faure (1991), and Strahler and Strahler (1992).

7.5. AQUEOUS SILICA SPECIES AND THE SOLUBILITIES OF QUARTZ, AMORPHOUS SILICA, AND OTHER SILICA POLYMORPHS

As noted above, a major fraction of the silica released to natural waters from the weathering of silicate and aluminosilicate minerals, such as the feldspars, remains in solution, where it occurs chiefly

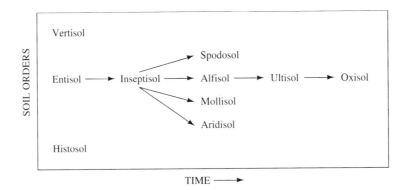

Figure 7.6 Approximate sequence of development of different soil orders. Entisols and inseptisols are early stages of soil development leading to spodosols, alfisols, mollisols, and aridisols, depending on the mineral composition of the parent material and climatic conditions. Utisols and oxisols occur in humid tropical regions as end stages of soil development enriched in oxides of Al and Fe. Vertisols develop only from parent material composed of swelling clay in regions with alternating wet and dry seasons. Histosols consist of partly decomposed plant material in places that are deprived of O_2, because they are waterlogged. From H. D. Foth, *Fundamentals of soil science,* 7th ed. Copyright © 1984 by John Wiley & Sons, Inc. Used by permission of John Wiley & Sons, Inc.

as monomeric silicic acid, $H_4SiO_4^\circ$. As a molecular species, the activity coefficient of silicic acid may be modeled with the Setchenow equation, which in NaCl solutions is

$$\log \gamma_{H_4SiO_4^\circ} = 0.0803I \tag{7.4}$$

(Marshall and Chen 1982). In seawater ($I = 0.7$ mol/kg) this gives $\gamma_{H_4SiO_4^\circ} = 1.14$. Thus, because $a_i = \gamma_i m_i$, the solubility of silica is lowered by 14% in seawater. The activity coefficient of silicic acid is sufficiently close to unity to be ignored, however, in potable waters.

Based on the weathering example, we concluded that some quartz may have formed as a product of weathering. Laboratory studies of quartz dissolution and precipitation kinetics (cf. Rimstidt and Barnes 1980) indicate that quartz dissolution and precipitation are extremely slow at low temperatures. In fact, no measurable dissolution of quartz was observed in three years at 25°C (Siever 1962). The first successful quartz precipitation experiment at low temperatures was performed by Mackenzie and Gees (1971), who grew quartz in one year at 20°C on preexisting quartz grains in seawater. At equilibrium they obtained a concentration of 4.4 ppm dissolved silica as SiO_2(aq). Given that $a_i = \gamma_i m_i$, and with $\gamma_{H_4SiO_4^\circ} = 1.14$ in seawater, this corresponds to SiO_2(aq) = 5.0 ppm in pure water.

The dissolution of quartz below about pH 7.8 can be written

$$SiO_2(\text{quartz}) + H_2O = H_4SiO_4^\circ \tag{7.5}$$

or

$$SiO_2(\text{quartz}) = SiO_2(\text{aq}) \tag{7.6}$$

for which $K_{sp} = [H_4SiO_4^\circ] = [SiO_2(\text{aq})]$. Based on measurements from 50 to 300°C, Rimstidt and Barnes (1980) give for quartz solubility

$$\log K_{sp} = 1.8814 - 2.028 \times 10^{-3}\, T - \frac{1560.46}{T} \tag{7.7}$$

where T is in kelvin. Solution of this equation yields

$T°C$	$SiO_2(aq)$ (ppm)
0	2.4
25	6.6
50	15
100	53

Richet et al. (1982) present thermodynamic data for silica species equivalent to a quartz solubility of 6 ppm at 25°C, which corresponds to a concentration of $10^{-4.00}$ mol/kg. The same solubility is supported by Fournier, who gives $\log K_{sp}$ (quartz) $= 0.41 - 1309/T(K)$ (see Nordstrom et al. 1990). We will assume a quartz solubility value of 6 ppm at 25°C in future calculations and discussions.

The most soluble form of SiO_2 is amorphous silica. The equilibrium reaction for $SiO_2(am)$ in water can be written as for quartz

$$SiO_2(am) + 2H_2O = H_4SiO_4°$$ (7.8)

or

$$SiO_2(am) = SiO_2(aq)$$ (7.9)

for which $K_{sp} = [H_4SiO_4°] = [SiO_2(aq)]$. Rimstidt and Barnes give the following K_{sp}-temperature function for $SiO_2(am)$ solubility, which applies up to about 200°C:

$$\log K_{sp} = 0.338037 - 7.8896 \times 10^{-4}\, T - \frac{840.075}{T}$$ (7.10)

where T is in kelvin. According to Ellis and Mahon (1964), amorphous silica alters to quartz at temperatures roughly above 200°C, for which conditions the nucleation energy for quartz is sufficiently low that it readily precipitates instead of $SiO_2(am)$. Equation (7.10) leads to $K_{sp} = 10^{-2.71}$ at 25°C and the following ppm solubilities for $SiO_2(am)$ with temperature:

$T°C$	$SiO_2(aq)$ (ppm)
0	67
25	116
50	183
100	372

In good agreement, the temperature function

$$\log K_{sp} = -0.26 - \frac{731}{T(K)}$$ (7.11)

for $SiO_2(am)$ (proposed by Fournier [1985]), which applies between about 25 and 200°C, yields $SiO_2(aq)$ solubilities of 117 ppm at 25°C and 363 ppm at 100°C. Such concentrations are common in hot springs (cf. Morey et al. 1964) and in groundwaters having temperatures above 50°C.

Other crystalline polymorphs of SiO_2 have solubilities greater than that of quartz, the thermodynamically most stable (and thus least soluble) form of SiO_2, and less than that of amorphous silica, which is the least stable form of silica (cf. Dove and Rimstidt 1994). The only crystalline non-quartz silica polymorphs found in significant amounts at low temperatures are the minerals cristobalite and tridymite, which are both unstable relative to quartz at such temperatures (cf. Brown et al. 1978). The low- and high-temperature structural forms of the silica polymorphs are designated their α and β forms, respectively. At 1 bar pressure, α-quartz is stable up to 573°C and β-quartz from that temperature up to 870°C. At 1 bar pressure, β-tridymite is thermodynamically stable above 870°C relative to β-quartz, and β-cristobalite is stable relative to β-tridymite above 1470°C (cf. Deer et al. 1992). This indicates that α-cristobalite is probably less stable (more soluble) than α-tridymite at 25°C and 1 bar. Published thermodynamic data and reported solubilities for α-cristobalite and α-tridymite are in poor agreement. The solubilities suggested here are consistent with the likelihood that α-cristobalite is more soluble than α-tridymite at 25°C.

The solubility of α-cristobalite measured between 85 and 250°C by Fournier and Rowe (1962) corresponds to the fortuitously simple function

$$\log K_{sp}(\alpha\text{-cristobalite}) = -\frac{1000}{T(K)} \tag{7.12}$$

This extrapolates to an α-cristobalite solubility of 27 ppm as $SiO_2(aq)$ at 25°C. Thermodynamic data in Robie et al. (1978) suggest a 25°C solubility of 16 ppm for α-tridymite.

The silica in soils occurs in both crystalline and relatively amorphous forms. High dissolved silica concentrations from the weathering of mafic minerals in volcanic rocks (e.g., basalts and tuffs) may lead to the precipitation of cristobalite, tridymite, and/or opal (microcrystalline $SiO_2 \cdot nH_2O$), particularly in the unsaturated zone (cf. Deer et al. 1992; Murphy and Palaban 1994). The relatively amorphous cristobalite in soils, which is sometimes called opal or opaline silica, may largely result from organic processes. Thus, opaline silica is formed in many grasses and in the leaves of some deciduous trees. It is also precipitated in lakes by diatoms (microscopic algae) and siliceous sponges and in the ocean by diatoms, sponges, and radiolaria (a large group of one-celled sea animals).

The common silica rocks, chert and chalcedony, are made up of fine-grained quartz and microcrystalline silica. Accordingly, their solubility exceeds that of well-crystallized quartz. Fournier (1985) suggests the temperature function

$$\log K_{sp}(\text{chalcedony}) = -0.09 - \frac{1032}{T(K)} \tag{7.13}$$

which corresponds to a chalcedony solubility of 17 ppm as SiO_2 at 25°C.

As a weak acid, silicic acid dissociates in two steps, according to the reactions with the stepwise constants given below for 25°C:

$$H_4SiO_4^\circ = H^+ + H_3SiO_4^- \qquad\qquad \beta_1 = K_1 = 10^{-9.82} \tag{7.14}$$

$$H_3SiO_4^- = H^+ + H_2SiO_4^{2-} \qquad\qquad K_2 = 10^{-13.10} \tag{7.15}$$

Here, K_1 is from Busey and Mesmer (1977), and K_2 is computed from K_1 and the cumulative constant for the reaction:

$$H_4SiO_4^\circ = 2H^+ + H_2SiO_4^{2-} \qquad\qquad \beta_2 = K_1 \times K_2 = 10^{-22.92} \tag{7.16}$$

from Baes and Mesmer (1976). Given the value for β_1, it is clear that the $H_4SiO_4^\circ$ concentration exceeds that of the anionic silica species up to pH = 9.82 and that the latter are less than 1% of total silica and so can be ignored up to about pH 7.8.

Although the solubilities of quartz, its polymorphs, and amorphous silica are fixed and pH-independent in most natural waters, the dissociation of silicic acid at alkaline pH's leads to substantial increases in their solubilities above pH 9 to 10. The following calculations show how we can predict this effect. First, the solubility of any silica solid must equal the sum of the concentrations of all species of silica in solution at equilibrium. This summation is given by the mass-balance equation

$$\Sigma SiO_2(\text{molal}) = H_4SiO_4^\circ + H_3SiO_4^- + H_2SiO_4^{2-} \tag{7.17}$$

Using the silicic acid dissociation constant expressions for β_1 and β_2 for reactions (7.14) and (7.16) above, we can replace the last two terms in Eq. (7.17) with terms that contain only the species $H_4SiO_4^\circ$. This leads to

$$\Sigma SiO_2 (\text{molal}) = H_4SiO_4^\circ + \frac{K_1[H_4SiO_4^\circ]}{[H^+]} + \frac{K_1 K_2 [H_4SiO_4^\circ]}{[H^+]^2} \tag{7.18}$$

Factoring out $H_4SiO_4^\circ$, we are left with the simple expression

$$\Sigma SiO_2 (\text{molal}) = [H_4SiO_4^\circ]\left(1 + \frac{K_1}{[H^+]} + \frac{K_1 K_2}{[H^+]^2}\right) \tag{7.19}$$

As shown previously, the dissolution of silica solids can be written

$$SiO_2(s) + 2H_2O = H_4SiO_4^\circ$$

or

$$SiO_2(s) = SiO_2(aq)$$

Given the above solubility data for silica solids and values for K_1 and K_2, we can substitute into expression (7.19) for different pH values to compute solubilities of the solids. At 25°C we will assume $K_{sp}(\text{quartz}) = 10^{-4.00}$ and $K_{sp}[SiO_2(\text{am})] = 10^{-2.71}$. Substituting these values for $H_4SiO_4^\circ$ in Eq. (7.19) and ignoring activity coefficients leads to the solubilities tabulated below, where concentrations have been converted from molal to ppm units.

pH	Quartz $\Sigma SiO_2(aq)$ ppm	SiO_2(am) $\Sigma SiO_2(aq)$ ppm
7	6.0	117
8	6.1	122
9	7.0	139
10	15.5	310
11	102	2,020
12	1,028	20,500

These values are plotted in Fig. 7.7. Consistent with the thermodynamic results, silica concentrations can be very high in evaporative alkali lakes.

7.6 SILICA IN NATURAL WATERS

Silica concentrations are summarized in Table 7.5 for a variety of natural waters. Diatoms can lower lake silica levels to below 3 ppm and to as low as 0.5 to 0.8 ppm (Hutchinson 1957). Some freshwater sponges can reduce lake silica values to 1.2 ppm.

Streams pick up most of their dissolved silica in roughly three days or fewer following storm runoff (cf. Kennedy 1971), probably because of stream recharge from water in adjacent soils. Silica levels in soil moisture appear to be buffered by silica associated with the surfaces of fine-grained soil materials, including clays (cf. Mackin and Swider 1987) and oxyhydroxides (cf. Jones and Handreck 1963), which can be rapidly released or adsorbed. Bricker et al. (1968) found, for example, that when a soil column was treated with percolating distilled water or with a solution containing 50 ppm dissolved silica, the column effluent in either case contained 9 ppm dissolved silica, even after the equivalent of six years of rainfall had been added to the column. Soil mineralogy included quartz,

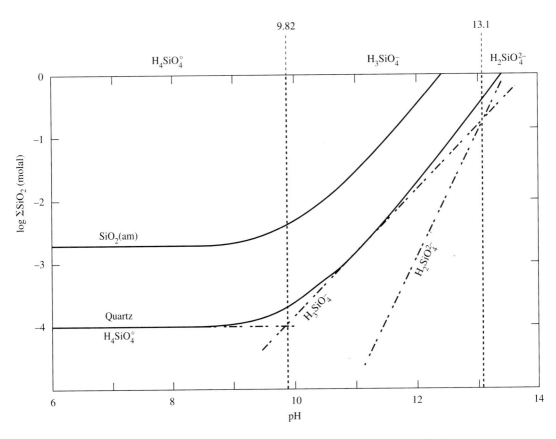

Figure 7.7 Solubilities of quartz and amorphous silica as a function of pH. Also shown are lines indicating the solubility of quartz due to the individual dissolved silica species $H_4SiO_4^\circ$, $H_3SiO_4^-$, $H_2SiO_4^{2-}$, and the fields of dominance of $H_4SiO_4^\circ$, $H_3SiO_4^-$, and $H_2SiO_4^{2-}$ as a function of pH.

TABLE 7.5 Silica concentrations as $SiO_2(aq)$ in some natural waters

Type of water	Approximate range (ppm)	Median (ppm)	Comments
Streams, rivers, & lakes	5–25	14	Lowest in lakes (diatoms)
Seawater	0.01–7	1	Lowest values, especially diatoms
Soil moisture[†]	<1–117	54	Concentrated by evaporation
Unsaturated volcanic rocks[‡]	78–102	90(?)	Concentrated by evaporation, unstable mafic minerals and glasses
Groundwater	5–85	17	See below
Oil field brines	5–60		Especially <30 ppm
Hot springs & geysers	100–600		Especially temperature effect

Source: After Davis (1964), Kennedy (1971), and Anderson (1972) unless otherwise indicated.
[†]Sears and Langmuir (1982). [‡]Murphy and Palaban (1994).

muscovite, and the clays kaolinite and illite. (See also Garrels and Mackenzie 1971; and Mackin and Swider 1987.)

The silica in soil moisture in temperate humid climates may reach saturation with amorphous silica, probably because of evaporative concentration (Sears and Langmuir 1982). Mean or median silica concentrations in soil moisture are poorly known. The median value of 54 ppm in Table 7.4 is from a study of C-horizon (deep) soils in central Pennsylvania (Sears and Langmuir 1982).

The lowest soil moisture values probably reflect dilution of silica concentrations in highly leached soils. The reaction

$$\tfrac{1}{2}Al_2Si_2O_5(OH)_4 + \tfrac{5}{2}H_2O = Al(OH)_3 + H_4SiO_4^\circ \qquad (7.20)$$

$$\text{kaolinite} \qquad\qquad\qquad \text{gibbsite}$$

has $K_{eq} = [H_4SiO_4^\circ] = 10^{-4.7}$, which is equivalent to a low concentration of ≈ 1.2 ppm dissolved silica. However, this reaction and other clay mineral equilibria probably take years to reach equilibrium at low temperatures (cf. May et al. 1986) and so cannot control short-term or rapid-response silica concentrations in soils, groundwaters or surface-waters, generally.

Groundwater silica concentrations are remarkably uniform for most aquifers, with trends relatively minor or nonexistent (Davis 1964). Silica levels in groundwater are a function of rock type, as shown in Table 7.6. The highest silica concentrations derive from the weathering of high-temperature mafic minerals, such as the olivines and pyroxenes, which are abundant in pyroclastic volcanic and basaltic rocks. Weathering of the high-temperature plagioclase feldspars that are present in some plutonic rocks and in arkose and greywacke sediments can also produce high dissolved-silica values. The lowest silica concentrations are found in rocks that contain chiefly quartz and/or minor amounts of clays, such as quartz-rich sandstones and carbonate rocks.

In summary, the lowest SiO_2 concentrations in natural waters (below a few ppm) reflect: (1) in lakes or ocean water, the action of silica-extracting organisms such as diatoms; (2) in soil moisture, dilution by fresh infiltration in highly leached soils; and (3) in soil- or groundwater systems, generally, the absence of silicate minerals other than quartz or, perhaps, kaolinite. The surprisingly constant and similar median silica values of 14 ppm in streams and 17 ppm in typical groundwaters probably reflects silica buffering by its association with the surfaces of fine-grained materials, including clays and aluminum and ferric oxyhydroxides.

TABLE 7.6   Typical silica concentrations (as ppm SiO_2(aq)) in the groundwater from different rock types

Rock type	Concentration	Remarks
Pyroclastic volcanics[†]	~85	Abundant mafic minerals & high rock surface areas
Fresh basalts[‡]	50	Mafic minerals
Plutonic felsic rocks, arkoses and greywackes	~50	Plagioclase feldspars, etc.
Marine sandstones	~15	Weathered quartz
Limestones & dolomites	5–14	Trace clays & quartz present

[†]SiO_2(aq) = 36–78 ppm in groundwaters in saturated volcanic tuffs under Yucca Mountain (Murphy and Palaban 1994).

[‡]SiO_2(aq) = 44 ppm (average) for nine groundwaters from basalt reported by White et al. (1963).

Source: After Davis (1964).

Highest silica concentrations in groundwater (≥ 50 ppm) result from the weathering of rocks that contain high-temperature silicate minerals. Elevated dissolved-silica values are also found in thermal groundwaters and alkaline lakes, because of the increased solubility of silica at high temperatures and high pH's.

7.7 SOLUBILITY OF THE ALUMINUM OXYHYDROXIDES, KAOLINITE, AND THE FERRIC OXYHYDROXIDES AS A FUNCTION OF pH

An assumption inherent in the weathering calculation above was that the aluminum is insoluble during weathering. As we will see, this assumption is a good approximation around pH 6 to 7, but becomes less valid at lower and higher pH's where aluminum oxides and hydroxides are more soluble. Solids that become increasing soluble at both low and high pH's are said to exhibit *amphoteric* behavior. Such behavior reflects the fact that the solids dissolve at low pH's to form cationic species and at high pH's to form anionic species. Both the aluminum and ferric-iron oxyhydroxides are amphoteric.

Let us calculate the solubility of aluminum hydroxide in pure water at 25°C as a function of pH to explain this behavior. In our calculations we will ignore activity coefficients so that activities equal concentrations.

We start with the general solubility product expression $K_{sp}[Al(OH)_3] = [Al^{3+}][OH^-]^3$. Given also that $K_w = [H^+][OH^-]$, we can eliminate OH^- from the K_{sp} expression, and solve for $[Al^{3+}]$ to obtain

$$[Al^{3+}] = \frac{K_{sp} [H^+]^3}{(K_w)^3} \tag{7.21}$$

But the total solubility of $Al(OH)_3$ ($\Sigma Al[aq]$) is the sum of concentrations of all the Al-OH complex species *plus* the free Al^{3+} concentration. This is written:

$$\Sigma Al(aq) = [Al^{3+}] + [AlOH^{2+}] + [Al(OH)_2^+] + [Al(OH)_3^\circ] + [Al(OH)_4^-] \tag{7.22}$$

Concentrations of each Al-OH complex can be related to that of Al^{3+} through the following *cumulative complexation* reactions:

$$Al^{3+} + H_2O = AlOH^{2+} + H^+ \tag{7.23}$$

$$Al^{3+} + 2H_2O = Al(OH)_2^+ + 2H^+ \tag{7.24}$$

$$Al^{3+} + 3H_2O = Al(OH)_3^\circ + 3H^+ \tag{7.25}$$

$$Al^{3+} + 4H_2O = Al(OH)_4^- + 4H^+ \tag{7.26}$$

The complexation constant expressions for these reactions are:

$$*\beta_1 = \frac{[AlOH^{2+}] [H^+]}{[Al^{3+}]} \qquad *\beta_2 = \frac{[Al(OH)_2^+] [H^+]^2}{[Al^{3+}]} \tag{7.27), (7.28}$$

$$*\beta_3 = \frac{[Al(OH)_3^\circ] [H^+]^3}{[Al^{3+}]} \qquad *\beta_4 = \frac{[Al(OH)_4^-] [H^+]^4}{[Al^{3+}]} \tag{7.29), (7.30}$$

Our goal is to derive an equation in terms of $\Sigma Al(aq)$, pH, and the complexation constants, so that we can plot the solubility of $Al(OH)_3$ (as $\Sigma Al[aq]$) versus pH. We begin the derivation by solving each complexation constant expression for its Al-OH complex and then substituting into the $\Sigma Al(aq)$ expression to eliminate all the Al-OH species. The result is

$$\Sigma Al(aq) = [Al^{3+}] + \frac{*\beta_1 [Al^{3+}]}{[H^+]} + \frac{*\beta_2 [Al^{3+}]}{[H^+]^2} + \frac{*\beta_3 [Al^{3+}]}{[H^+]^3} + \frac{*\beta_4 [Al^{3+}]}{[H^+]^4} \qquad (7.31)$$

Notice that individual terms in this sum give the solubility contributions from each aluminum species. Factoring out $[Al^{3+}]$ and remembering that it is controlled by K_{sp} through the expression $[Al^{3+}] = K_{sp}[H^+]^3/(K_w)^3$, we eliminate $[Al^{3+}]$ and obtain the final mass-balance expression in the proton form

$$\Sigma Al(aq) = \frac{K_{sp}}{(K_w)^3}\left([H^+]^3 + *\beta_1 [H^+]^2 + *\beta_2 [H^+] + *\beta_3 + \frac{*\beta_4}{[H^+]}\right) \qquad (7.32)$$

Examination of this equation shows that $Al(OH)_3$ solubility depends on the value of K_{sp}, but the shape of the solubility curve depends only on the $*\beta_n$ values.

According to Nordstrom et al. (1990), at 25°C: $*\beta_1 = 10^{-5.00}$, $*\beta_2 = 10^{-10.1}$, $*\beta_3 = 10^{-16.9}$, and $*\beta_4 = 10^{-22.7}$. Also, $K_{sp} = 10^{-31.2}$ for amorphous $Al(OH)_3$ and $10^{-33.9}$ for gibbsite, a common crystalline form of $Al(OH)_3$.[†] We can substitute these constants into the expression for $\Sigma Al(aq)$, and derive solubility curves for the amorphous and crystalline solids as a function of pH. The result is shown in Fig. 7.8.

In our weathering example (Table 7.2) kaolinite, rather than the aluminum oxyhydroxides, was the chief weathering product of the feldspars. This reflects the fact that the silica present in soil moisture and natural waters, generally, is high enough to stabilize kaolinite relative to the aluminum oxyhydroxides. This observation is better understood if we write the kaolinite dissolution reaction:

$$\tfrac{1}{2}Al_2Si_2O_5(OH)_4 + \tfrac{5}{2}H_2O = \alpha\text{-}Al(OH)_3 + H_4SiO_4^\circ \qquad (7.33)$$
$$\text{kaolinite} \qquad\qquad\qquad \text{gibbsite}$$

for which

$$K_{sp} = [H_4SiO_4^\circ] \qquad (7.34)$$

The value of K_{sp} and $[H_4SiO_4^\circ]$ will depend on the stabilities we assign to kaolinite and gibbsite. The natural behavior of these minerals in weathering environments suggests that the equilibrium silica concentration for this reaction is about 2×10^{-5} mol/kg, or 1.2 ppm silica as SiO_2 (cf. Hess 1966). In good agreement, stability constant data given by Nordstrom et al. (1990) correspond to $K_{eq} = [H_4SiO_4^\circ] = 4.05 \times 10^{-5}$ mol/kg, or 2.4 ppm silica for kaolinite-gibbsite equilibrium. At higher silica concentrations, kaolinite is more stable than gibbsite.

The solubility of kaolinite in terms of $\Sigma Al(aq)$ versus pH, can be derived and graphed in the same way we have developed the $Al(OH)_3$ solubility plot, except that we must fix the concentration of dissolved silica. Thus, for example, we write:

$$\tfrac{1}{2}Al_2Si_2O_5(OH)_4 + 3H^+ = Al^{3+} + H_4SiO_4^\circ + \tfrac{1}{2}H_2O \qquad (7.35)$$
$$\text{kaolinite}$$

[†]In excellent agreement, Pokrovskii and Helgeson (1995) list free energy data that corresponds to the following values for these constants: $*\beta_1 = 10^{-4.96}$, $*\beta_2 = 10^{-10.59}$, $*\beta_3 = 10^{-16.43}$, $*\beta_4 = 10^{-22.88}$, and $K_{sp}(\text{gibbsite}) = 10^{-34.2}$.

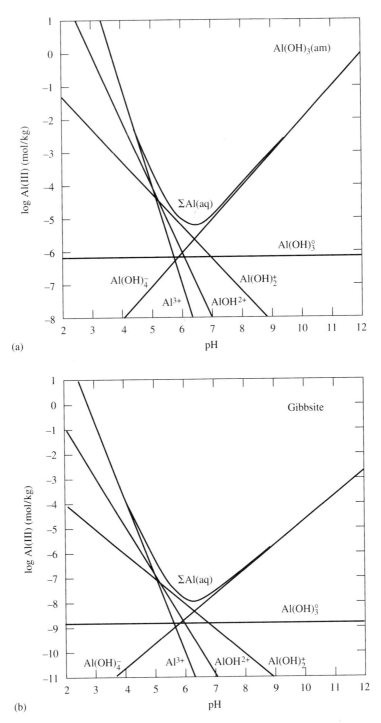

Figure 7.8 Solubility of **(a)** amorphous $Al(OH)_3$, $K_{sp} = 10^{-31.2}$, and **(b)** gibbsite $[Al(OH)_3]$, $K_{sp} = 10^{-33.9}$, as a function of pH at 25°C. Also shown are lines indicating the solubility contributions of Al^{3+} and individual Al-hydroxy complexes.

The equilibrium constant expression

$$*K_0 = \frac{[Al^{3+}]\,[H_4SiO_4^\circ]}{[H^+]^3} \tag{7.36}$$

can be solved for $[Al^{3+}]$ to give

$$[Al^{3+}] = \frac{*K_0[H^+]^3}{[H_4SiO_4^\circ]} \tag{7.37}$$

In the equilibrium constant, the superscript asterisk indicates that the dissolution reaction is written in terms of protons. The subscript following K (zero in this case) is the number of hydroxyl ions associated with Al^{3+} in the reaction. According to Nordstrom et al. (1990), $*K_0 = 10^{3.72}$. Reactions similar to (7.35) can be written in which kaolinite is dissolved to form each of the aluminum hydroxy complexes listed in mass-balance equation (7.22) for total aluminum. These reactions can be generated by successively adding the cumulative reactions for the Al-hydroxy complexes—Eqs. (7.23) to (7.26)—to Eq. (7.35). For example, adding Eqs. (7.35) and (7.23) and their log K_{eq} values for 25°C, we obtain

	log K_{eq}	
$\frac{1}{2}Al_2Si_2O_5(OH)_4 + 3H^+ = Al^{3+} + H_4SiO_4^\circ + \frac{1}{2}H_2O$	3.72	(7.35)
$Al^{3+} + H_2O = AlOH^{2+} + H^+$	−5.00	(7.23)
$\frac{1}{2}Al_2Si_2O_5(OH)_4 + 2H^+ + \frac{1}{2}H_2O = AlOH^{2+} + H_4SiO_4^\circ$	−1.28	(7.38)

The equilibrium constant expression for reaction (7.38) is:

$$*K_1 = *K_0 \, *\beta_1 = \frac{[Al(OH)^{2+}]\,[H_4SiO_4^\circ]}{[H^+]^2} = 10^{-1.28} \tag{7.39}$$

In general we find $*K_n = *K_0 \times *\beta_n$. The remaining reactions and their equilibrium constant expressions are:

$$\frac{1}{2}Al_2Si_2O_5(OH)_4 + H^+ + \frac{3}{2}H_2O = Al(OH)_2^+ + H_4SiO_4^\circ \tag{7.40}$$

$$*K_2 = *K_0 \, *\beta_2 = \frac{[Al(OH)_2^+]\,[H_4SiO_4^\circ]}{[H^+]} = 10^{-6.38} \tag{7.41}$$

$$\frac{1}{2}Al_2Si_2O_5(OH)_4 + \frac{5}{2}H_2O = Al(OH)_3^\circ + H_4SiO_4^\circ \tag{7.42}$$

$$*K_3 = *K_0 *\beta_3 = [Al(OH)_3^\circ][H_4SiO_4^\circ] = 10^{-13.18} \tag{7.43}$$

$$\frac{1}{2}Al_2Si_2O_5(OH)_4 + \frac{7}{2}H_2O = Al(OH)_4^- + H_4SiO_4^\circ + H^+ \tag{7.44}$$

$$*K_4 = *K_0 *\beta_4 = [Al(OH)_4^-][H_4SiO_4^\circ][H^+] = 10^{-18.98} \tag{7.45}$$

These $*\beta_n$ expressions are next solved for their respective aluminum species. Resultant expressions are then substituted into Eq. (7.22) to give an equation for $\Sigma Al(aq)$ in terms of H^+, the constants and dissolved silica. Factoring out the $[H_4SiO_4^\circ]$ term, the general result is

$$\Sigma Al(aq) = \frac{1}{[H_4SiO_4^\circ]}\left(*K_0[H^+]^3 + *K_1[H^+]^2 + *K_2[H^+] + *K_3 + \frac{*K_4}{[H^+]}\right) \tag{7.46}$$

Now if we assume a fixed concentration of $H_4SiO_4^\circ$ and introduce $*K_n$ values into Eq. (7.46), the solubility of kaolinite can be computed and shown on a plot of dissolved $\Sigma Al(aq)$ versus pH

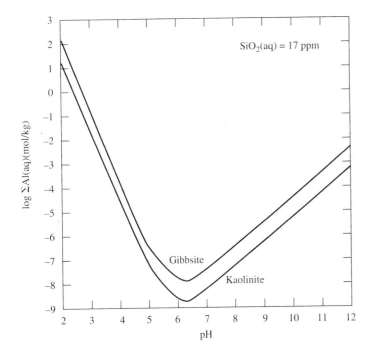

Figure 7.9 Solubilities of gibbsite [Al(OH)$_3$] and kaolinite [Al$_2$Si$_2$O$_5$(OH)$_4$] as ΣAl(aq) as a function of pH at 25°C. The solubility of kaolinite has been computed assuming SiO$_2$(aq) = 17 ppm, the average value for groundwater.

(Fig. 7.9). Because kaolinite solubility is in terms of the Al-OH complexes, the solubility curve for kaolinite at fixed H$_4$SiO$_4^\circ$ has the same shape as the solubility curves for gibbsite or Al(OH)$_3$(am). As suggested by Fig. 7.9, kaolinite is generally at least an order of magnitude less soluble than the aluminum oxyhydroxides at typical silica concentrations in natural waters. This justifies our assumption of Al immobility in the weathering example with kaolinite as the weathering product.

Although kaolinite is thermodynamically stable in most soils, it is usually not the first aluminosilicate mineral to precipitate from supersaturated soil solutions. Steefel and van Cappellen (1990) point out that in soil profiles an irreversible sequence of precipitation may begin with relatively soluble allophane ([Al(OH)$_3$]$_{(1-x)}$[SiO$_2$]$_x$), followed by less soluble halloysite (Al$_2$Si$_2$O$_5$(OH)$_4 \cdot$ 2H$_2$O), and finally by kaolinite. The halloysite and kaolinite grow at the expense of their more soluble precursor phases, in a process that apparently takes thousands of years. This sequence of replacement by progressively more stable phases is known as the Ostwald step rule (Steefel and van Cappellen 1990; see Chap. 9). The allophane and halloysite are typically found deep in the soil, where weathering of parent materials is most active, with kaolinite most abundant near the surface. Allophane and halloysite may dominate the soil in climates that have dry seasons, but be absent in climates with high rainfall and no dry periods, where instead kaolinite is the chief clay, sometimes accompanied by gibbsite (cf. Cozzarelli et al. 1987).

Allophane is an X-ray amorphous phase that may be viewed as a complete solid solution between Al(OH)$_3$ and SiO$_2$. Paces (1978) writes allophane dissolution as

$$[\text{Al(OH)}_3]_{(1-x)}[\text{SiO}_2]_x + (3 - 3x)\text{H}^+ = (1 - x)\text{Al}^{3+} + x\text{H}_4\text{SiO}_4^\circ + (3 - 5x)\text{H}_2\text{O} \qquad (7.47)$$

Based on 152 laboratory and groundwater analyses between pH 1.8 and 9.2, he proposes that x and K_{sp} are both functions of pH, with $x = 1.24 - 0.135$ pH and log $K_{sp} = -5.89 + 1.59$ pH. In his review of the saturation state of the 152 waters, Paces (1978) found that kaolinite was supersaturated by $10^{3.0}$ to $10^{4.8}$ times, whereas allophane and halloysite were both generally close to equilibrium.

The structure of halloysite is equivalent to that of kaolinite, but has a layer of water molecules between each pair of silica and alumina layers (see Chap. 9; Deer et al. 1992). Writing halloysite dissolution in the same form as in Eq. (7.35) for kaolinite, its solubility product is $K_{eq} = 10^{5.64}$ (Hem et al. 1973; Steefel and van Cappellen 1990), versus $K_{eq} = 10^{3.72}$ for kaolinite. In other words, halloysite is about 80 times more soluble than kaolinite. If plotted in Fig. 7.9, also assuming $SiO_2(aq) = 17$ ppm, the solubility curve for halloysite is parallel to but 1.9 log units above the curve for kaolinite at any pH.

Unless soils and their parent materials are capable of neutralizing acid precipitation (see Chap. 8) pH values may drop to 4 or less in soil moisture and local surface-waters and groundwaters. Spodosols are among the most vulnerable soils to such acidification (Schecher and Driscoll 1987). Under acid conditions dissolved Al concentrations may be as high as 8 to 10 mg/L from the leaching of aluminous minerals such as kaolinite and gibbsite (cf. Bottcher et al. 1985). The dissolved Al then buffers the pH at acid values because of reactions that include

$$Al(OH)_3 + 3H^+ = Al^{3+} + 3H_2O \tag{7.48}$$

Detailed geochemical controls on the mobility of aluminum in a soil were evaluated in an elegant field study by Cozzarelli et al. (1987). The authors used lysimeters to sample fast-flow waters (macropore; subject to gravity) and slow-flow waters (micropore; subject to capillary forces) in a highly permeable inceptisol in the Blue Ridge Mountains, Virginia. They also sampled an adjacent stream. Their soils were 40 to 120 cm deep and capped with a 5-cm-thick O horizon with 72% organic matter. Soil and stream samples had pH 3.81 to 7.04 and total aluminum ($\Sigma Al[aq]$) values from 6 to 1070 μg/L. By chemical analysis the authors speciated the $\Sigma Al(aq)$ into acid soluble, organic monomeric and inorganic monomeric fractions, and computer modeled the water analyses with WATEQF (Plummer et al. 1984). The study showed that the deep soil was a sink for the Al dissolved by acid waters in shallow soils. Aluminum mobilized by organic acids in the O and A horizons was precipitated in micropores deeper in the soil, probably as kaolinite and perhaps as gibbsite. Fast-moving macropore waters remained undersaturated with Al phases, except in the deeper C horizon where kaolinite may have equlibrated. Most of the Al dissolved in soil waters was present as Al complexes. A large percentage occurred as strong organic complexes, with most of the inorganic Al found as fluoride complexes.

Example 7.1

Tabulated here is the hypothetical analysis of dissolved species (in μM) in micropore water from the C horizon of a soil similar to that just described with pH = 5.65 and $T = 10°C$.

Species	Ca	Mg	K	$\Sigma Al(aq)$	SO_4	F	SiO_2
Concentration	15	40	80	6.4	105	5	160

Enter all the data except for $\Sigma Al(aq)$ in MINTEQA2. Make kaolinite an infinite solid, but change its K_{sp} to the value assigned in this chapter from Nordstrom et al. (1990).

1. What percentages of $\Sigma Al(aq)$ are present as Al^{3+} and Al-OH complexes, and as fluoride complexes?

2. If the modeling results accurately represent conditions in the soil and the tabulated $\Sigma Al(aq)$ value is correct, what percentage of Al may be present as other species (i.e., as organic complexes)?

3. What is the saturation state of the water with respect to quartz, $Al(OH)_3$(am), gibbsite, and halloysite?

The modeling shows that Al(aq) $= 0.20$ μM, of which $Al^{3+} = 1.1\%$, Al^{3+} plus the Al-OH complexes $= 26\%$, and Al-F complexes $= 74\%$, and thus dominate the inorganic Al speciation. The percentage of Al that may be present as organic (?) complexes is: $[(6.4 - 0.2)/6.4] \times 100 = 97\%$. (This is probably unrealistically high.) The saturation indices requested are: quartz, 0.44; $Al(OH)_3$(am), -3.2; gibbsite -1.4; and halloysite -1.7.

The aluminum oxyhydroxides, kaolinite, and halloysite dissolve to form cationic aluminum species at low pH and the anionic species $(Al(OH)_4^-)$ at high pH. The same amphoteric behavior is also true of the Fe(III) oxyhydroxides, although the latter are much less soluble, in general, under oxidizing conditions. We next compute the solubilities of the ferric oxyhydroxides as a function of pH. The approach is identical to that described above for the Al-oxyhydroxides.

Given below are cumulative formation constants for the ferric-hydroxyl complexes at zero ionic strength and 25°C from Macalady et al. (1990).

Reaction	$-\log {}^*\beta_n$
$Fe^{3+} + H_2O = Fe(OH)^{2+} + H^+$	2.19
$Fe^{3+} + 2H_2O = Fe(OH)_2^+ + 2H^+$	5.67
$Fe^{3+} + 3H_2O = Fe(OH)_3^\circ + 3H^+$	12.56
$Fe^{3+} + 4H_2O = Fe(OH)_4^- + 4H^+$	21.6

Starting with a mass-balance equation for total iron, and using an approach identical to that used to derive Eq. (7.32) for the Al-oxyhydroxides, we obtain Eq. (7.49), which describes the solubility of the ferric oxyhydroxides as a function of pH,

$$\Sigma Fe(III)(aq) = \frac{K_{sp}}{(K_w)^3}\left([H^+]^3 + {}^*\beta_1[H^+] + {}^*\beta_2[H^+] + {}^*\beta_3 + \frac{{}^*\beta_4}{[H^+]}\right) \tag{7.49}$$

The pK_{sp} ($-\log K_{sp}$) values for amorphous $Fe(OH)_3$ (ferrihydrite) and the mineral goethite (α-FeOOH) are 37.1 and 44.2, respectively. Substituting these values and the above cumulative formation constants for the Fe(III)-hydroxy complexes into Eq. (7.49), we may compute the pH-dependent solubilities of the solids at 25°C. The results are shown in Figs. 7.10 and 7.11. Figure 7.11 also shows the fields of dominance of the several Fe-OH complexes as a function of pH. Clearly the ferric oxyhydroxides are less soluble than the aluminum oxyhydroxides in general, which explains their persistence in the weathered soil. These figures also demonstrate why concentrations of dissolved Al and Fe in surface waters are generally below 1 ppm, except in acid mine waters. For iron, this statement applies only to oxidized environments. Under reducing conditions such as can develop in wetlands or water-logged soils, ferric iron is reduced to much more soluble ferrous iron and so may be leached from associated soils or sediments (cf. Chap. 11).

In previous discussion, calculations were simplified by assuming that the Fe(III) and Al oxyhydroxides and clays are pure phases as defined in the text. For the oxyhydroxides, effects of the full range of differences in their crystallinity, degrees of hydration, and particle size were implicitly accounted for by comparing the solubilities of their most amorphous and crystalline end members. However, as a further complication, the oxyhydroxides and clays often occur as solid solutions in soils and sediments. For example, on a mole-fraction basis, goethite may contain up to 33% Al,

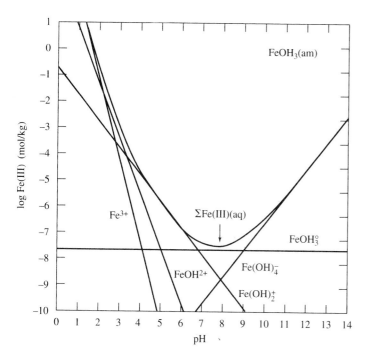

Figure 7.10 Solubility of amorphous $Fe(OH)_3$, $K_{sp} = 10^{-37.1}$, as a function of pH at 25°C. Also shown are lines indicating the solubility contributions of Fe^{3+} and individual Fe-hydroxy complexes.

hematite up to 25% Al, and kaolinite as much as 3% Fe^{3+} (Tardy and Nahon 1985). Assuming the applicability of ideal solid-solution concepts, Tardy and Nahon (1985) and Trolard and Tardy (1987) have estimated the solubilities and relative stabilities of such solid solutions, correcting for particle-size effects, and changes in the activity of water.

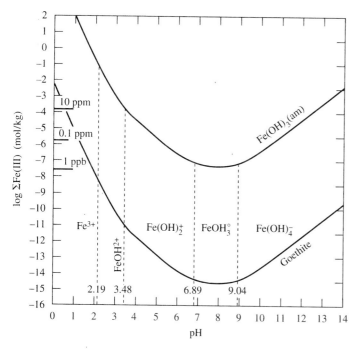

Figure 7.11 Solubility of amorphous $Fe(OH)_3$, $K_{sp} = 10^{-37.1}$ (top curve), and goethite [α-FeOOH], $K_{sp} = 10^{-44.2}$ (bottom curve), as a function of pH at 25°C. Also shown are the fields of dominance of Fe^{3+} and of individual Fe^{3+}-hydroxy complexes that contribute to the total solubilities. Vertical dashed lines, which define the fields of species dominance, are drawn where pH = $-\log {}^*K_n$, for $n = 0$ to 3. ${}^*K_n$ are stepwise formation constants for the complexation reactions written in the form $Fe(OH)_n^{3-n} + H_2O = Fe(OH)_{n+1}^{2-n} + H^+$.

7.8 CHEMICAL WEATHERING MODELS

For a variety of practical reasons, great effort has been expended in recent years to understand controls on and to predict rates of chemical weathering in soils. A principle reason has been a need to assess the effects of acid precipitation on the chemistry of soils and thus on the health of affected plants and trees. Soil acidification from acid precipitation has been a serious problem in industrialized areas where soils are thin or absent and bedrock and resultant soils lack carbonate or reactive silicate minerals (cf. Likens et al. 1977; Berner and Berner 1996). Soil acidification in such areas has caused the acidification of adjacent streams and even underlying groundwaters (cf. Bottcher et al. 1985; Hansen and Postma 1995).

Three general approaches have been taken to evaluate silicate weathering rates (Apello and Postma 1993). These have included: (1) historical rates; (2) mass-balance calculations; and (3) experimental laboratory rates. Historical rates are computed from changes in soil composition from the top down to the unweathered parent material, assuming the age of the soil is known. For example, if we could assume that the C-horizon composition in Table 7.2 represents most of that soil, and it was formed in the last 10^5 years, and that none of the soil has been removed by physical erosion (generally a very poor assumption), we can compute a rate of chemical weathering. An inert, relatively insoluble tracer found in the parent material, such as quartz, zircon (ZrO_2), or rutile (TiO_2) is needed for the calculation (cf. Bain et al. 1990; Nesbitt and Wilson 1992). An example calculation of weathering rate using historical data is given in Table 7.7 from Sverdrup and Warfvinge (1988), which indicates an overall weathering rate of 0.16 keq/ha/y, or 1.6×10^{-2} eq/m^2/y. The historic method often leads to rates that appear too low in comparison to the results of other methods (Bain et al. 1991).

In the mass-balance approach an attempt is made to determine field weathering rates from element flux calculations, usually focusing on the plant-nutrient important base cations (Ca^{2+}, Mg^{2+}, K^+, and Na^+) or on silica (cf. Velbel 1985). In some studies balances are also computed for Al (Swoboda-Colberg and Drever 1992) and/or the the major anions and nitrogen species (Likens et al. 1977; Mast et al. 1990). A general mass-balance equation for the base cations (BC) might be

$$[\text{BC released by weathering}] = [\text{BC removed in runoff and groundwater}] - [\text{BC from atmosphere}]$$
$$\pm [\text{change in exchange pool (BC adsorbed on soil clays)}]$$
$$\pm [\text{BC change in biomass}]$$

TABLE 7.7   Weathering rates, R_w (keq/ha/y), calculated from compositional differences between the top and bottom of a soil near Gardsjon, Sweden

Mineral	R_w	Bottom of profile composition (wt %)	Ions released
Microcline (K-feldspar)	0.03	16	K^+
Plagioclase feldspar	0.04	16	Na^+, Ca^{2+}
Hornblende	0.03	1.0	Mg^{2+}, Ca^{2+}
Epidote	0.04	2.0	Ca^{2+}
Biotite	0.02	1.0	Mg^{2+}, K^+
Sum	0.16		

Note: R_w is the weathering rate expressed as release of Ca^{2+}, Mg^{2+}, Na^+ and K^+ in kiloequivalents per hectare (1 ha = 10^4 m^2) per year.

Source: After H. Sverdrup and P. Warfvinge. Weathering of primary silicate minerals in the natural soil environment in relation to a chemical weathering model. *Water, Air and Soil Pollution* 38: 387–408. © 1988 by Kluwer Academic Publ. Used by permission.

Such studies have generally been limited to small watersheds, where the parent material and soils are of one type and preferably one mineralogy. Mass-balance studies have also mostly been restricted to watersheds where a chloride mass-balance calculation showed that the total yearly chloride inputs in precipitation equaled their outputs in the stream draining the watershed (cf. Velbel 1985). This permitted the neglect of BC removal (or addition) by groundwater. Usually, a watershed in which steady-state conditions existed has been sought such that significant changes in BC in the exchange pool or in biomass were not likely (cf. Cronan 1985; Mast et al. 1990). With these simplifications the mass-balance equation is

[BC released by weathering] = [BC removed in runoff] − [BC from atmosphere]

To solve this equation, species concentrations and volumes of precipitation and stream runoff are measured to obtain the annual flux of species into and out of the watershed. The relative and absolute amounts and chemical compositions of minerals being weathered must also be determined. Then, based upon close examination of the soil and personal experience, the researcher writes a series of hypothetical reactions that could describe the weathering processes. Such an exercise is detailed in Table 7.8 from a study by Katz et al. (1985). The authors list average precipitation (input) and stream-water runoff (output) chemistry, and their difference, corrected for the addition of de-icing salts. This difference is the amount of each species either consumed (only H^+ is consumed) or released from the watershed. The balanced weathering reactions indicate the number of moles of each mineral that must be dissolved to contribute a mole of each species in the runoff and thus provide the stoichiometries of the mass-balance equations. The suite of simultaneous equations is solved by hand or with a computer model such as NETPATH (Plummer et al. 1991) to obtain the amounts of each mineral weathered, with the results given at the bottom of the table. Katz et al. (1985) find that 703 mol/ha/y (7.0×10^{-2} mol/m^2/y) of minerals are weathered, releasing 2200 eq/ha/y of solutes (excluding silica), which is equivalent to 22×10^{-2} eq/m^2/y released. This is a relatively high weathering rate, probably because it is dominated by rapidly weathered minerals such as calcite, chlorite, and actinolite. An interesting conclusion of this study is that although calcite may be much less than 1% of the dominantly silicate bedrock in a watershed, it plays a major role in neutralizing soil acidity and contributing solutes to runoff. This is because the calcite solution rate is 10^6 to 10^9 times faster than that of most silicates (Table 7.10). Similarly, Mast et al. (1990) reported that only 0.005 to 0.4% calcite in a silicate-rock watershed in Colorado produced nearly 40% of the total cations in runoff.

Weathering rates obtained in this and other mass-balance studies of watersheds lie in the surprisingly narrow range between 0.02 and 0.2 eq/m^2/y (0.2 to 2.0 keq/ha/y) (Table 7.9). Laboratory mineral dissolution rates measured at a given pH can also be expressed in moles Si or moles cation/m^2/y (cf. Table 7.10). However, the area is that of the mineral surface, not of the watershed. To apply the lab results we must know the mineral area exposed to weathering per catchment area. Schnoor (1990) estimated 10^5 m^2 of mineral grains available to weathering per m^2 of catchment. Such estimates are approximate, at best, since they depend on soil type and thickness and on such soil properties as the grain size and the reactive and wetted surface areas of the minerals. Mineral surfaces are in continuous contact with well-mixed solutions in the laboratory studies, whereas the minerals in soil macropores and micropores have variable contact times with water. For these and other reasons, field rates computed from mass-balance calculations have often been roughly 10^1 to 10^3 times slower than predicted from laboratory rates, or even from the rates measured on small field plots (cf. Velbel 1985; Swoboda-Colberg and Drever 1992; 1993).

Warfinge and Sverdrup (1992) and Sverdrup and Warfinge (1995) have developed a comprehensive computer model (PROFILE) to predict the chemistry of soil water, groundwater recharge, and a watershed stream (see also Sverdrup and Warfinge 1993; Jonsson et al. 1995; and Alveteg et al.

TABLE 7.8 Weathering reactions among the main minerals in greenstone, H^+ from atmospheric deposition, and soil-generated carbonic acid

Constituent	H^+	Ca^{2+}	Mg^{2+}	Na^+	K^+	Cl^-	HCO_3^-	$H_4SiO_4^\circ$
Output	0.5	510	653	499	37.5	558	1280	933
Input	812	38	10	80	23.4	89	0	0
Output − Input	−812	472	643	419	14.1	469	1280	933
De-icing salt .75 $CaCl_2$ .25 NaCl	0	176	0	103	14.1	469	0	0
Consumed or released in watershed	−812	296	643	316	0	0	1280	933

Reactions

1. $NaAlSi_3O_8 + H^+ + \frac{9}{2}H_2O \rightleftarrows Na^+ + \frac{1}{2}Al_2Si_2O_5(OH)_4 + 2H_4SiO_4^\circ$
2. $CaCO_3 + H^+ \rightleftarrows Ca^{2+} + HCO_3^-$
3. $Mg_5Al_2Si_3O_{10}(OH)_8 + 10H^+ \rightleftarrows 5Mg^{2+} + H_4SiO_4^\circ + Al_2Si_2O_5(OH)_4 + 5H_2O$
4. $Ca_2(Mg_3Fe_2)Si_8O_{22}(OH)_2 + 14H^+ + 8H_2O \rightleftarrows 2Ca^{2+} + 3Mg^{2+} + 2Fe^{2+} + 8H_4SiO_4^\circ$
5. $H_2CO_3^\circ = H^+ + HCO_3^-$

Mass-Balance Equations

1. $[Na^+]$ = albite
2. $[Ca^{2+}]$ = calcite + 2 actinolite
3. $[Mg^{2+}]$ = 3 actinolite + 5 chlorite
4. $[H_4SiO_4^\circ]$ = 2 albite + chlorite + 8 actinolite
5. $[HCO_3^-]$ = calcite + $[H_2CO_3^\circ]$
6. $[H^+]$ = albite + calcite + 10 chlorite + 14 actinolite − $[H_2CO_3^\circ]$

Minerals consumed		Dissolved constituents produced		H^+ and $H_2CO_3^\circ$ reacted	
albite	316	Na^+	316	HCO_3^-	1280
calcite	249	Ca^{2+}	296	H^+	812
chlorite	115	Mg^{2+}	643	$H_2CO_3^\circ$	1031
actinolite	23.3	$H_4SiO_4^\circ$	933		

Note: Values in moles/hectare. Output is Hauver Branch stream water in 1982. Input is average precipitation composition in 1982.

Source: From B. G. Katz et al. Geochemical mass-balance relationships for selected ions in precipitation and stream water, Catoctin Mountains, Maryland. *Am. J. Sci.* 285: 931–62. © 1985 by and reprinted by permission of *American Journal of Science.*

1995). According to the authors, PROFILE predicts field weathering rates within ±20% of rates obtained by the historic or mass-balance methods. Required inputs to the model are average annual precipitation amounts, their average chemistry, and detailed soil characterization data. Also required are temperature, soil CO_2 pressure, and the mineralogy, particle-size distribution, and soil-moisture saturation of individual soil layers or horizons. Program output includes the soil pH and overall silicate weathering rate and the chemistry of the stream that may be compared to its measured chemistry.

The general sequence of the calculations in PROFILE is given in Fig. 7.12. The acid neutralization capacity (ANC) of soil moisture (the alkalinity minus the acidity) and the soil pH are com-

TABLE 7.9   Rates of cation export per square meter of catchment for some watersheds underlain by silicate rocks, based on mass-balance calculations

Geographic area	Equivalents of cations/m²/y
Hubbard Brook, NH (USA)[†]	20×10^{-2}
Swiss Alps[‡]	4–10×10^{-2}
Trnavka River Basin (CZ)[§]	2×10^{-2}
Old Forge, NY (USA)[‖]	15×10^{-2}
Hauver Branch, MD (USA)[#]	22×10^{-2}
Coweeta Watershed, NC (USA)[††]	4.6×10^{-2}
Adirondack Mountains, NY (USA)[‡‡]	5–6×10^{-2}
Lockvale Watershed, CO (USA)[§§]	3.9×10^{-2}
Three watersheds near Gardsjon (SW)[‖‖]	4–10.5×10^{-2}

Source: [†]Likens et al. (1977); [‡]Zobrist and Stumm (1979); [§]Paces (1983); [‖]Cronan (1985); [#]Katz et al. (1985); [††]Velbel (1985); [‡‡]April et al. (1986); [§§]Mast et al. (1990); [‖‖]Sverdrup (1990).

puted from a mass-balance equation that considers the effect on ANC and pH of the rate of nitrogen and sulfur species uptake, rate of acid deposition, weathering rate, and rate of H^+ release from the soil caused by its exchange with base cations adsorbing on the clays. The silicate weathering rate is computed assuming steady-state conditions, using a data base of rate constants for 12 important rock-forming minerals, corrected for temperature. These rate constants have been derived through the application of transition-state theory to published laboratory mineral dissolution rates. The model indicates that weathering rates are strongly temperature-dependent and increase by about 30% for a 4° temperature increase.

TABLE 7.10   Laboratory weathering (dissolution) rates of calcite, corundum, and some silicate minerals measured at pH = 4

Mineral	mol Si/m²/y
Calcite[†]	1.6×10^3 (mol Ca^{2+})
Plagioclase feldspars[‡]	
Bytownite (An_{76})	1.9
Andesine (An_{46})	0.25
Oligoclase (An_{13})	3.8×10^{-2}
Olivine[§]	2×10^{-2}
Enstatite[‖]	1.3×10^{-3}
Biotite[#]	1.2×10^{-3}
Albite (An_0)[††]	7.9×10^{-5}
Muscovite[‡‡]	4.5×10^{-5}
Quartz[§§]	6.3×10^{-6}
Kaolinite[‖‖]	5.0×10^{-6}
Corundum (α-Al_2O_3)[‖]	3.2×10^{-6} (mol Al^{3+})

Source: [†]Plummer et al. (1979); [‡]Oxburgh et al. (1994); [§]Wogelius and Walther (1991); [‖]Brady and Walther (1989); [#]Grandstaff (1986); [††]Chou and Wollast (1985); Hellman (1994); the rate for K-feldspar roughly equals that of albite (Helgeson et al. 1984); [‡‡]Wieland et al. (1988); [§§]Berger et al. (1994); [‖‖]Carroll and Walther (1990).

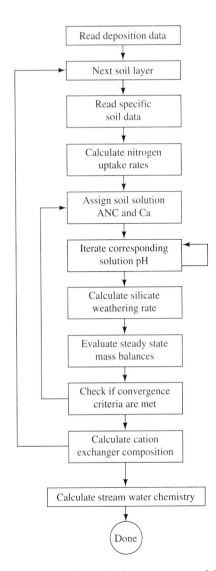

Figure 7.12 The order of calculations in the PROFILE model of P. Warfinge and H. Sverdrup. From Calculating critical loads of acid deposition with PROFILE—a steady-state soil chemistry model. *Water, air and soil pollution* 63: 119–143. © 1992 by Kluwer Academic Publ. Used by permission.

The equation for the total weathering rate summed for all minerals in all soil horizons is

$$R_W = \sum_{j=1}^{\text{horizons}} \sum_{j=1}^{\text{minerals}} r_j \cdot A_W \cdot x_j \cdot \theta \cdot z \tag{7.50}$$

where R_W = total weathering rate of all minerals in the soil (keq/m$_{ss}^2$/s), r_j = rate of dissolution of mineral j (keq/m^2/s), A_W = total exposed surface area of the minerals in a volume of soil (m^2/m^3), x_j = fraction of A_W exposed by mineral j, the soil-moisture saturation θ = (m^3/m$_{\text{soil}}^3$), and z = soil-layer thickness (m). In m$_{ss}^2$ the subscript denotes soil-surface area. The product $A_W \cdot x_j \cdot \theta$ in Eq. (7.50) describes the fraction of the total mineral area occupied by the wetted surface of mineral j. This makes the point that only wetted surfaces can be weathered. Assuming that the degree of surface wetting is proportional to the soil moisture saturation, θ is obtained from

$$\theta = \frac{\rho_{\text{solids}} \cdot \Theta}{\rho_{\text{solids}} + 1000 \cdot \Theta - \rho_{\text{bulk}}} \tag{7.51}$$

where $\rho_{solids} = 2700$ kg/m³, the density of soil particles, $\rho_{bulk} =$ bulk soil density (kg/m³), and Θ is the soil-water content (m³/m³$_{soil}$). A_W is derived from an estimate of the fractions of soil in different size ranges with the equation

$$A_W = 8.0 \cdot x_{clay} + 2.2 \cdot x_{silt} + 0.3 \cdot x_{sand} \tag{7.52}$$

The size fractions must sum to unity,

$$x_{clay} + x_{silt} + x_{sand} + x_{coarse} = 1 \tag{7.53}$$

By definition particle sizes of the fractions are $x_{clay} = <2$ μm, $x_{silt} = 2$ to 60 μm, $x_{sand} = 60$ to 250 μm, and $x_{coarse} = >250$ μm.

The form of the rate equation for the weathering of individual minerals j and the release rate of base cations is instructive,

$$r = k_{H^+} \cdot \frac{[H^+]^{n_H}}{f_H} + \frac{k_{H_2O}}{f_{H_2O}} + k_{CO_2} \cdot P_{CO_2}^{n_{CO_2}} + k_R \cdot \frac{[R^-]^{n_R}}{f_R} \tag{7.54}$$

The first term in this equation describes the effect of pH on the rate of weathering. The second and third terms define the effects of hydrolysis (H_2O) and of CO_2 or carbonic acid on the rate, while the final term describes the contribution of organic acids to the weathering rate. In Eq. 7.54, $k_{H^+} =$ rate constant for reaction with H^+ (m/s), $k_{H_2O} =$ rate constant for reaction with H_2O (keq/m²/s), $k_{CO_2} =$ rate constant for reaction with CO_2 (keq/atm/m²/s), and $k_R =$ rate constant for reaction with DOC (m/s). Exponential n values (n_H, etc.) are the reaction orders of individual rate terms. Brackets enclose dissolved species concentrations (kmol/m³ = M), $P_{CO_2} =$ partial pressure of CO_2 in the soil (atm), $R^- =$ organic anion concentration, and f_H, f_{H_2O}, f_R and are rate-reduction factors. They reflect the fact that concentrations of dissolved Al^{3+} and base metals produced by weathering can cause significant back reaction and so reduce mineral-weathering rates.

Example 7.2

A 1-m-thick soil contains, on average, 25% K-feldspar, 40% plagioclase feldspar, 5% vermiculite clay, and 30% quartz. Soil moisture pH = 5.0, $T = 8$°C, and $P_{CO_2} = 10^{-2}$ bar. Volumetric soil-water content $\Theta = 0.2$ m³/m³ (20%), $\rho_{bulk} = 1500$ kg/m³, and the total exposed surface area of minerals $A_W = 1.5 \times 10^6$ m²/m³. Assuming a single soil horizon, calculate the rate of weathering of the soil. Neglect back reactions and the effect of organic matter on the rate. Rate constants and reaction orders for dissolution of the three minerals from Sverdrup and Warfvinge (1993) are listed below along with mineral weight fractions (x_j). Notice that quartz is inert and so has been ignored.

Mineral	x_j	pk_H	n_H	pk_{H_2O}	pk_{CO_2}	n_{CO_2}
K-feldspar	0.25	16.0	0.5	17.2	17.2	0.6
Plagioclase feldspar	0.40	15.9	0.5	17.0	15.8	0.6
Vermiculite (clay)	0.05	14.8	0.6	17.6	16.5	0.5

Substituting the pH and rate-related data into Eq. (7.54) we solve for the weathering rate of each mineral and find: r(K-feldspar) $= 7.0 \times 10^{-18}$ keq/m²/s, r(plagioclase) $= 1.9 \times 10^{-17}$ keq/m²/s, and r(vermiculite) $= 7.3 \times 10^{-18}$ keq/m²/s. With the bulk density and moisture content, Eq. (7.51) gives a soil moisture saturation of $\theta = 0.39$. The total weathering rate, R_W, the sum

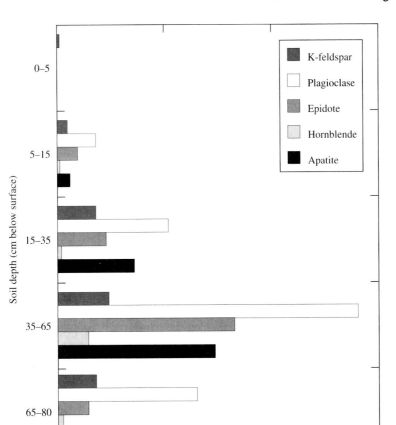

Figure 7.13 Contributions of different minerals to the total weathering rate as a function
of soil depth in four soil layers (horizons) at the Gardsjon site. Reprinted from *Applied Geochemistry*, 8, H. Sverdrup and P. Warfvinge, Calculating field weathering rates with a mechanistic geochemical model PROFILE, 273–83, with permission from Elsevier Science Ltd.,
The Boulevard, Langford Lane, Kidlington OX5 1GB, U.K.

of the rates for the three minerals, may be obtained with Eq. (7.50), which gives $R_W = 5.7 \times 10^{-12}$ keq/m^2/s, or 0.18 eq/m^2/y.

PROFILE computes weathering rates of individual minerals in each soil layer. Such a result
is shown in Fig. 7.13 for the Gardsjon site in Sweden (Sverdrup and Warfvinge 1993). The soil is
73% quartz, with decreasing amounts of K-feldspar (18%), plagioclase (12%), hornblende (0.5%),
epidote (0.5%), and apatite (0.35%). The order of weathering rates is, however, quite different, with
plagioclase > epidote > apatite > K-feldspar. The example makes the point that small amounts of
minerals such as epidote, apatite, hornblende, and plagioclase feldspar tend to dominate weathering.
When present, trace calcite (not considered in the PROFILE model) is an even more important con-
tributor to weathering.

Figure 7.14 compares weathering rates computed with PROFILE for 15 international sites that have been thoroughly characterized, with rates for those same sites obtained using historic or mass-balance methods (Sverdrup and Warfvinge 1993). Agreement among the methods is generally excellent and is best when the DOC correction term in the PROFILE weathering rate equation (Eq. [7.54]) is neglected, suggesting the need to improve on that aspect of the model.

STUDY QUESTIONS

1. Know how and why the rate and character of chemical weathering is dependent on temperature, soil organic/biological activity, and moisture infiltration rate.

2. Compared to parent rocks, the mineral products of chemical weathering are oxidized and much more hydrated, and are lower in cations relative to the concentrations of Al and Si. Explain this statement.

3. The major dissolved species in natural waters are derived from the chemical weathering of rocks. Which species are so derived, and from what minerals?

4. Only alumino-silicate minerals can weather to produce clays. Explain.

5. Weathering is accelerated because most of the minerals being weathered were formed at elevated temperatures and pressures and are thermodynamically unstable under ambient conditions. Defend this statement with an example that considers the composition of a typical natural water and the theoretical solubility of a silicate mineral being weathered.

6. In order to reconstruct the process of chemical weathering, it is necessary to select a substance in the source rock and resultant soil that can be considered relatively inert to weathering, with which concentrations of other soil constitutents can be compared. Al_2O_3 and TiO_2 have been used for such purposes. Once an inert "tracer" has been selected, how is such an analysis of weathering performed?

7. Discuss the origin of the horizons in a soil and explain how the thickness, processes, and compositions of those horizons can vary with parent rock, drainage, time, and climate.

8. Define mature and immature soils, aridisol, entisol, oxisol, and ultisol.

9. Understand how soils can cause filtration, sorption, and oxidation of organic and inorganic contaminants and under what conditions they may fail to remove such contaminants.

10. Quartz and amorphous silica have the lowest and the highest solubilities among the several polymorphs of silica. Because of its slow dissolution and precipitation kinetics, quartz is practically inert in low-temperature soils and sediments. Explain both statements and discuss some of their environmental implications.

11. How are the solubilities of quartz and amorphous silica affected by temperature? How and why do their solubilities depend on the pH?

12. Explain why silica concentrations in most surface- and groundwaters are remarkably constant.

13. Know how to calculate the solubility of a mineral such as gibbsite, kaolinite, or $Fe(OH)_3$ (ferrihydrite) as a function of pH, both by hand and with a geochemical code such as MINTEQA2.

14. Explain controls on the mobilization of Al and Fe in the O and A horizons and their precipitation in the B horizon of a spodosol.

15. How are the precipitation of allophane and halloysite often related to the formation of kaolinite in a soil and why?

16. How can differences in the thermodynamic stabilities of amorphous versus crystalline forms of SiO_2, $Al(OH)_3$ and $Fe(OH)_3$, affect their occurrence in soils and sediments?

17. The rate of chemical weathering has been evaluated using historic, mass-balance, and theoretical-empirical mineral dissolution rate approaches, often with different results. What are these approaches and why may their results differ?

18. In the PROFILE model of Sverdrup and Warfinge (1992), what is the significance of each term in the rate equation for mineral weathering?

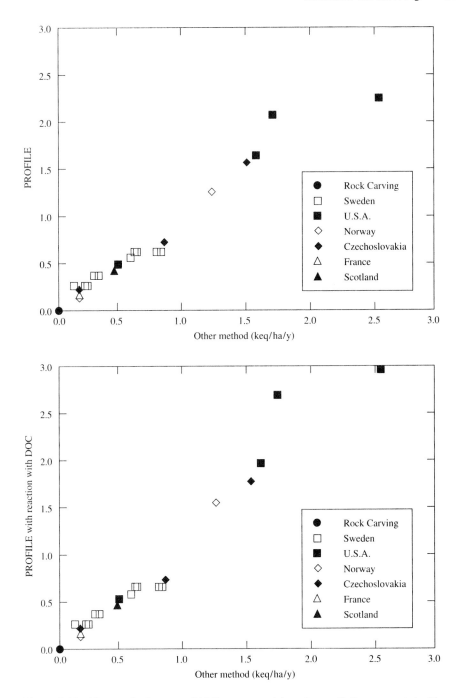

Figure 7.14 The weathering rate of 15 European and American soil sites computed with the PROFILE model with and without correction for the effect of DOC, compared to the weathering rate estimated using historic and mass-balance methods. Reprinted from *Applied Geochemistry*, 8, H. Sverdrup and P. Warfvinge, Calculating field weathering rates with a mechanistic geochemical model PROFILE, 273–83, with permission from Elsevier Science Ltd., The Boulevard, Langford Lane, Kidlington OX5 1GB, U.K.

PROBLEMS

1. This question relates to the weathering example in Table 7.1.
   (a) The plagioclase feldspar in the granite is about 73% albite and 27% anorthite. Assume that the anorthite fraction is completely destroyed by chemical weathering. Further, assuming that quartz is not lost or produced by the weathering, compute the percentages lost due to weathering of the K-feldspar, plagioclase feldspar, biotite, and hornblende.
   (b) List the minerals from part (a) in their order of increasing weathering rate {−log rate (mol/m²/s)}, as given in Fig. 7.1 and Table 7.1. Also list their percentages lost due to chemical weathering as computed in part (a). In this exercise, assume that chlorite is equivalent to biotite, and that hornblende is equivalent to tremolite. Discuss the significance of the table.

2. It is sometimes assumed that the minimum solubilities of amphoteric metal oxyhydroxide minerals roughly equal their solubilities written solely in terms of the neutral metal hydroxide complex. Using this approach and assuming 25°C, calculate the minimum solubilities of gibbsite and of kaolinite (at 17 mg/L $SiO_2[aq]$) in terms of total dissolved aluminum, and goethite in terms of total dissolved ferric iron. Compare these results to the minimum solubilities of these minerals for the same conditions but determined when all the metal hydroxy complexes are considered.

3. (a) Kaolinite is often present in soils and sediments affected by acid mine drainage, in which it is the least soluble of the clay minerals. It is probably often the chief source of the dissolved aluminum in acid mine drainage. The solubility of kaolinite as total dissolved aluminum ($\Sigma Al(aq)$) can be plotted versus pH, if one fixes the dissolved silica concentration. (See the text and Fig. 7.9.) In the presence of high sulfate concentrations in acid mine drainage, Al-sulfate complexes are formed. These can substantially increase the solubility of kaolinite. Given the following stability constant data for 25°C and assuming the total dissolved silica activity as $H_4SiO_4^\circ$ is fixed at $10^{-3.55}$ m (17 mg/L as $SiO_2(aq)$), calculate and plot the solubility of kaolinite in the presence of a total dissolved sulfate concentration of $10^{-2.00}$ mol/L, from pH 2 to 12, ignoring activity coefficients or other complexes. In your calculation do not forget that total sulfate has its own mass-balance equation. Compare your plot to the solubility plot of kaolinite in pure water and discuss reasons for any differences.

Reaction	$\log K_{eq}$
$Al^{3+} + H_2O = AlOH^{2+} + H^+$	−5.00
$Al^{3+} + 2H_2O = Al(OH)_2^+ + 2H^+$	−10.1
$Al^{3+} + 3H_2O = Al(OH)_3^\circ + 3H^+$	−16.9
$Al^{3+} + 4H_2O = Al(OH)_4^- + 4H^+$	−22.70
$H^+ + SO_4^{2-} = HSO_4^-$	1.99
$Al^{3+} + SO_4^{2-} = AlSO_4^+$	3.02
$Al^{3+} + 2SO_4^{2-} = Al(SO_4)_2^-$	4.92
$\frac{1}{2}Al_2Si_2O_5(OH)_4 + 3H^+ = Al^{3+} + H_4SiO_4^\circ + \frac{1}{2}H_2O$	3.72
kaolinite	

   (b) Use MINTEQA2 to perform the same calculation of kaolinite solubility as $\Sigma Al(aq)$, at integer pH's from 2 to 12, and compare your results to those obtained in part (a). Hints: Kaolinite may be defined as an infinite solid and the ionic strength set to zero in the computer model to make the results comparable to those you have hand-calculated.

8

General Controls on
Natural Water Chemistry

8.1 THE HYDROLOGIC CYCLE, RESIDENCE TIME, AND WATER-ROCK RATIO

The geochemical behavior and chemical and isotopic properties of natural waters are related to their location in the hydrosphere, that is, as precipitation, stream flow, soil water, groundwater, ocean water, etc. A goal of this section is to make the reader aware of the unique and different chemical and isotopic properties of each of these waters and explain why their chemistries differ.

The hydrologic cycle can be thought of as involving the movement of water liquid and vapor between reservoirs, including the ocean, atmosphere, lakes and streams, soils and groundwater. The water has a *residence time* (t_R) in each of these reservoirs, which is defined as

$$t_R = \frac{\text{amount of water in reservoir (g)}}{\text{flux into reservoir (g/time)}} \tag{8.1}$$

Water amounts or fluxes can be measured in volumes or weights. In terms of weights, t_R equals the weight of the reservoir divided by the weight rate of flux into (or out of) the reservoir. For the *ocean*, which contains 94% of Earth's water

$$t_R = \frac{13700 \times 10^{20} \text{ g}}{\underset{\substack{\text{rivers} + \text{ground} \\ \text{water discharge}}}{0.36 \times 10^{20} \text{ g/y}} + \underset{\substack{\text{precipitation onto} \\ \text{ocean surface}}}{3.5 \times 10^{20} \text{ g/y}}} = 3550 \text{ years} \tag{8.2}$$

For the *atmosphere*,

$$t_R = \frac{0.13 \times 10^{20} \text{ g}}{\underset{\substack{\text{evaporation} \\ \text{oceans}}}{3.8 \times 10^{20} \text{ g/y}} + \underset{\substack{\text{evaporation lakes} \\ \text{and rivers}}}{0.63 \times 10^{20} \text{ g/y}}} = 11 \text{ days} \tag{8.3}$$

(see Freeze and Cherry 1979). If we can assume the ice caps hold 2% of the world's water and continental waters the 4% remaining, and of that 4% groundwater is 95%, surface waters 3.5%, and soil moisture 1.5%, and if we use the above figure of 0.36×10^{20} g/y as a conservative estimate of the groundwater discharge rate (because it includes rivers), then this suggests a groundwater reservoir of 550×10^{20} g. Mason and Moore (1982) suggest 840×10^{20} g.

Assuming 600×10^{20} g for the groundwater reservoir, we compute

$$t_R = \frac{600 \times 10^{20} \text{ g}}{0.36 \times 10^{20} \text{ g/y}} = 1700 \text{ y} \tag{8.4}$$

as a mean residence time of the world's groundwaters. Age dating and tracing of groundwaters show their actual ages range from a few days to 10,000 years and more.

Other approximate reservoir residence times include: 10 years for lakes and reservoirs, 2 weeks for rivers, and 10 to thousands of years for ice caps and glaciers.

An important point to consider is that for a given amount and density of air or water pollution, if the pollution can only be remediated by dilution (mixing) and dispersion, *the time it takes for nature to clean up stable chemical pollutants will be directly proportional to reservoir residence time.* You might think about this idea as it applies to air pollution and surface- versus groundwater pollution.

8.2 WATER IN THE HYDROSPHERE

We will begin this section with some important definitions needed for the following discussion. (See Figs. 8.1 and 8.2.) For more detailed explanation of these and related concepts, not all of which will be defined below, the reader is referred to Freeze and Cherry (1979), Fetter (1988), and Domenico and Schwartz (1990).

1. *Evaporation*   The vaporization of water from free water surfaces and soil surfaces.
2. *Transpiration*   The loss of water from the surfaces of plants.
3. *Evapotranspiration*   The total water loss from evaporation and transpiration.
4. *Ephemeral streams*   Streams that flow intermittently.
5. *Perennial streams*   Streams that flow all year.
6. *Base flow*   The flow of a stream that is entirely due to groundwater discharging into the stream.
7. *Infiltration*   The flow of water downward from the land surface into and through the upper soil layers.
8. *Vadose or unsaturated zone water*   Water in the zone of aeration that is held by adsorption or capillary forces, except when net infiltration is occurring.
9. *Water table*   The surface of an unconfined aquifer at which the pore water pressure is atmospheric. Also, the level to which water will rise in a well installed a few feet or meters into the saturated zone.
10. *Groundwater*   Subsurface-water that occurs beneath the water table in fully-saturated soils and formations.
11. *Aquifer*   A formation that is saturated and sufficiently permeable to transmit economic quantities of water to wells and springs.
12. *Aquitard*   A formation that has too low a permeability or hydraulic conductivity to yield economic quantities of water to a well.

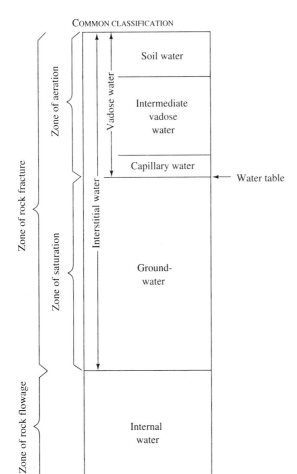

Figure 8.1 Classification of sub-surface-waters. Modified after S. N. Davis and R. J. M. DeWiest. *Hydrogeology.* Copyright © 1966. Used by permission. (See also Freeze and Cherry 1979.)

13. *Unconfined aquifer* (water-table aquifer)   An aquifer that is not separated from the surface by an aquitard or confining bed.

14. *Confined aquifer* (artesian aquifer)   An aquifer that is confined between two aquitards.

15. *Artesian well*   A well in a confined aquifer in which the water rises above the top of the aquifer.

16. *Perched aquifer*   A locally saturated zone or lense in the unsaturated zone that may overlay a low-permeability unit.

If one is interested in groundwater recharge (infiltration) at a particular site, it is instructive to consider a water budget at that site, which can be expressed as an equation. Thus,

$$I \quad = \quad P \quad \pm \quad R \quad - \quad ET \quad - \quad MR \qquad (8.5)$$

$$\begin{array}{ccccc} \text{ground-} & \text{precipi-} & \text{runoff} & \text{evapo-} & \text{moisture} \\ \text{water} & \text{tation} & & \text{transpi-} & \text{retention} \\ \text{recharge} & & & \text{ration} & \end{array}$$

Notice that runoff can add to or subtract from infiltration depending upon the topography. Runoff moves toward low-lying areas or wetlands ($+R$) and away from hill slopes ($-R$), for exam-

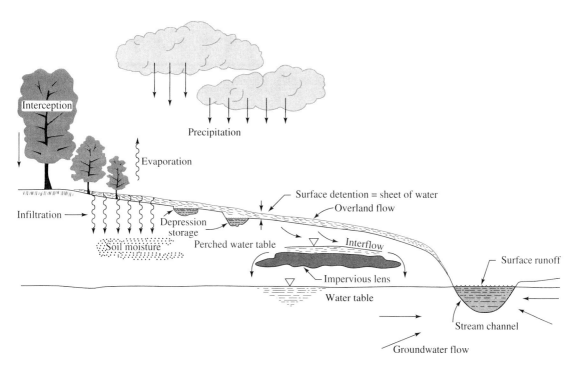

Figure 8.2   Simple hydrologic cycle, showing the different kinds of water in the cycle. From S. N. Davis and R. J. M. DeWiest. *Hydrogeology.* Copyright © 1966. Used by permission.

ple. Moisture retention is the moisture held in soils and sediments by capillary forces. Each water type in this equation has its own chemical quality.

Figure 8.1 defines the different types of subsurface-waters. The zone above the water table is called the zone of aeration, vadose zone, or unsaturated zone. Moving downward in this zone we encounter first soil water or soil moisture, then intermediate vadose water, and then capillary water. Capillary water rises into rock and sediment pores above the water table because of capillary forces. The capillary zone may be as much as 30 ft (9 m) thick in clayey sediments. To the extent the unsaturated zone can truly be called the zone of aeration, then air and atmospheric oxygen are present and conditions are oxidizing.

Water above the water table exists at pressures below atmospheric pressure, whereas water below the water table is at higher pressures than atmospheric pressure. Groundwater (also called phreatic water) is in the zone of saturation. Groundwater can yield water to wells or springs. Groundwater is chiefly meteoric water (water from precipitation), generally of increasing age with depth. At greater depths, which can range from a few hundred feet in crystalline rocks such as granite to tens of thousands of feet in sedimentary basins, water in the rock occurs only in unconnected pores, or in the crystal structures of minerals such as clays and micas. Such water is also called structural or internal water.

Some of these definitions and concepts are shown schematically in Fig. 8.2, which describes a simple hydrologic cycle. The figure introduces the important concept of a *perched water table,* which may result from ponding of infiltration on a clay layer or "impervious lens." Also shown is *depression storage*, where runoff accumulates in surface depressions or trenches, such as those associated with land disposal of solid and liquid wastes. Until recently surface depressions that received liquid wastes were often called *evaporation ponds*, although they usually leaked, causing increased infiltration and groundwater pollution.

8.3 CONTROLS ON THE COMPOSITION OF SUBSURFACE-WATERS

The composition of subsurface waters is a complex function of many variables, including:

1. The composition of groundwater recharge, which is influenced by the chemistry of precipitation, leaching of salts accumulated by evapotranspiration (ET) between infiltration events, organic activity in the soil, and weathering of soil materials.
2. The petrologic and mineralogic composition of subsurface rocks. Among common rock-forming minerals, halite (NaCl, $K_{sp} = 37.6$) is the most soluble, and gypsum (CaSO$_4$ · 2H$_2$O, $K_{sp} = 10^{-4.58}$) is the next most soluble. Carbonate rocks such as limestones (chiefly calcite, CaCO$_3$, $K_{sp} = 10^{-8.48}$) are next most soluble. When a rock contains merely 1% of any of these minerals, their dissolution will define the water chemistry. On the other hand, silicate and aluminosilicate rocks and minerals are generally either less soluble or their rates of dissolution are slower than solution rates of the carbonates, gypsum, or halite. The silicates, therefore, contribute less dissolved species to the water than do these more soluble minerals, when they are present.
3. The hydrogeologic properties of rocks have a strong influence on the extent of water/rock reaction. High groundwater-flow velocities usually imply groundwaters that are relatively low in dissolved solids because of short rock-contact times and high water/rock ratios, and conversely.

The hydraulic conductivities and porosities of common rocks are compared in Table 8.1. Clay has the highest porosity, but is among the least permeable of materials. Its low permeability is why clay is used to line the bottom of waste ponds, for example. Basalt and limestone may have low total porosities, but because groundwater flow in basalt and limestone may occur in large fractures and also in cavernous zones in limestone, these rocks often have high permeabilities.

Darcy's law states: $v = K \times (dh/dl)$, where v is termed the specific discharge or the Darcy velocity or flux (Fetter 1988; Heath 1989). The linear or average groundwater velocity is $\bar{v} = v/n$, where n is the connected, fractional porosity of the media. K is the hydraulic conductivity, and dh/dl is the hydraulic gradient or the slope of the water table (h/l = height/length). The average groundwater velocity, $\bar{v}$, then is proportional to K and dh/dl. Groundwater residence time, t_R, is thus inversely proportional to $\bar{v}$ and K and to dh/dl.

High groundwater velocities are typical in coarse-grained sediments such as sands and gravels, cavernous limestones, and highly fractured or weathered near-surface igneous and metamorphic rocks. Such conditions are usually found in shallow water-table aquifers, not in deep, confined aquifers. High velocities imply a thermodynamically open system, short residence times, and a continuous supply of fresh recharge. Chemical weathering is then very active, and few secondary minerals reach saturation at such high water/rock ratios, unless they are very insoluble. Examples of the latter are kaolinite and oxyhydroxides of aluminum and ferric iron.

Low groundwater velocities are most common in highly impermeable, low K rocks, and deep systems with near-zero hydraulic gradients. Example rocks are unfractured metamorphic and ig-

TABLE 8.1   The water-bearing properties of common rocks, listed in order of decreasing hydraulic conductivity and decreasing porosity

Highest hydraulic conductivity	$\log K$ (m/s)	Highest porosity	Porosity (%)
Well-sorted gravel	0 to −2	Soft clay	40 to 70
Clean sand, cavernous limestone	−2 to −6	Silt, tuff	35 to 50
Permeable basalt	−2 to −7	Well-sorted sand	25 to 50
Well-sorted sands	−3 to −5	Poorly-sorted sand & gravel	25 to 40
Fractured igneous & metamorphic rocks	−4 to −8.5	Gravel	25 to 40
Silty sands, fine sand	−5 to −7		
Silt, loess	−5 to −9	Sandstone	5 to 30
Sandy silts, clayey sands	−6 to −8		
Limestone & dolomite	−6 to −9.5	Fractured basalt	5 to 50
Sandstone	−6 to −10	Cavernous limestone & dolomite	0 to 20
Glacial till	−6 to −12	Fractured crystalline rock	0 to 10
Unweathered marine clay	−9 to −12.5		
Shale	−9 to −13	Shale	0 to 10
Dense crystalline igneous & metamorphic rocks	−10 to −14	Dense crystalline igneous & metamorphic rocks	0 to 5
Lowest hydraulic conductivity		Lowest porosity	

Source: Based on Freeze and Cherry (1979) and Fetter (1988).

neous rocks, clays and shales, and deep-basin sediments, generally. Such systems tend to be confined or artesian and thermodynamically closed (see Chap. 1) at a scale that may exceed cubic kilometers. The groundwater chemistry is rock-dominated because of low water/rock ratios and long residence times. Thus, many minerals in the rock are in chemical equilibrium with the groundwater. The rock tends to be unweathered, and secondary minerals are common in fractures and rock matrix. The groundwaters involved are usually relatively saline. In the extreme, such conditions are typical of diagenesis (the lithification of deeply buried sediments due to chemical and physical processes including compaction but not metamorphism).

The extent of surface weathering of crystalline rocks or of sedimentary rocks such as shales or carbonates, and thus rock permeability (and yield to wells), decreases rapidly with depth. Also, rock weathering is deeper under valley bottoms than on ridges or hill slopes. This reflects the fact that the weathering, which is facilitated by joints, fractures, and faults, tends to create valley bottoms in the first place. Valley bottoms continue to concentrate runoff (R is then a positive term in the infiltration equation) and so remain the locus of deeper development of secondary rock porosity and permeability and thus of enhanced groundwater storing and transmitting capacity.

In sedimentary rocks such as shales, deep burial and attendant compaction leads to a reduction in rock porosity and permeability with depth. In general, groundwater systems in sedimentary rocks become more artesian (confined) with depth, and the water becomes more saline, anaerobic and isolated from fresh recharge. The different kinds of conditions that can produce artesian groundwater

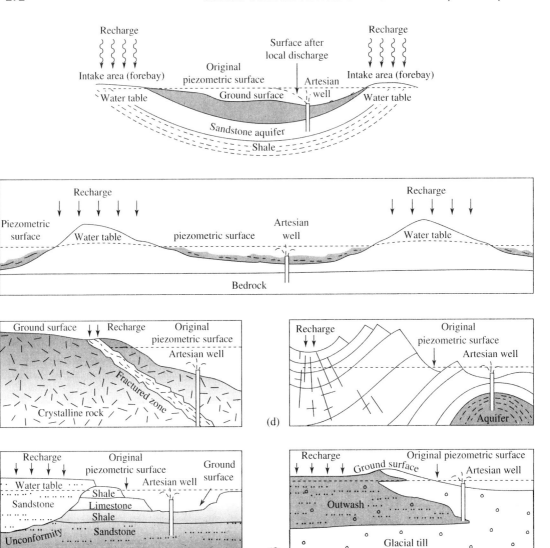

Figure 8.3 Different geologic and topographic conditions that can result in artesian groundwaters. Artesian conditions require that there be a confining "bed" and elevated heads elsewhere in a groundwater recharge zone. **(a)** Shows elements of a "classical" artesian system within a syncline. Artesian wells are shown in **(b)** stabilized sand dunes, **(c)** crystalline rock, **(d)** folded and fractured sedimentary rocks, **(e)** flat-lying sedimentary rocks, and **(f)** glacial sediments. From S. N. Davis and R. J. M. DeWiest. *Hydrogeology.* Copyright © 1966. Used by permission.

systems are shown schematically in Fig. 8.3. Major chemical differences between artesian and water-table groundwaters, generally, are summarized in Table 8.2.

The chemical compositions of a shallow water-table and a deeper and semiconfined groundwater in carbonate rocks near State College, PA, are contrasted in Fig. 8.4 and Table 8.3. The figure shows the variation of specific conductance of Rock Spring, which is the shallow groundwater ($t_R =$

TABLE 8.2   Contrast between some chemical and isotopic properties of artesian (confined) and water-table (unconfined) groundwaters

Artesian	Water table
Nearly constant head with time	Heads may vary with season & recharge events
O_2 usually below detection	O_2 near saturation
pH usually near 7 or higher	pH often below 7 and as low as 5 in recharge zones
No NO_3^{-}[†]	NO_3^{-} may be present
Fe(II) may be present	No Fe(II)
May contain H_2S, CH_4, and/or H_2[†]	No H_2S, CH_4, or H_2
Nearly constant composition and T	Variable composition and T
Many minerals at saturation (long residence times)	Few minerals at saturation (short residence times)
Relatively high TDS[‡]	Relatively low TDS
Constant or nearly constant stable isotope compositions	Deuterium (^2H)- and oxygen-stable isotopy may vary spatially and temporally, suggesting local sources and seasons (temperatures) for groundwater recharge
Water old enough to date with tritium (^3H; $t_{1/2} = $ 12.4 y) or ^{14}C ($t_{1/2} = 5570$ y) or longer-lived radionuclides such as ^{36}Cl ($t_{1/2} = 3.1 \times 10^5$ y)	Water may be as young as a few days, or old enough to date with tritium, but generally younger than artesian waters in the same area

[†]Due to microbial activity under anaerobic conditions

[‡]TDS: total-dissolved-solids residue is the weight of solids obtained on evaporation of a liter of the sample.

a few days). In Table 8.3 Rock Spring chemistry is compared to that of Thompson Spring, the deeper, semiconfined groundwater (t_R = months to years). Differences in these groundwaters are evident from the coefficients of variation (V) of some of their chemical and physical properties, where $V = (100 \times SD)$/mean. As expected, the discharge and chemistry of Rock Spring is more highly variable with time than that of Thompson Spring. Rock Spring is also more dilute, generally, than Thompson Spring, except for a few species, including Na^+, SO_4^{2-}, Cl^-, and NO_3^-. These latter species were probably introduced into Rock Spring during recharge events in part by washout of pollutant salts from the soil.

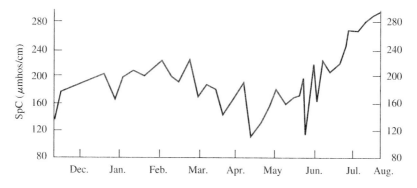

Figure 8.4   Specific conductance of Rock Spring near State College, Pennsylvania, from December 1971 to August 1972. The spring responded rapidly to snow melt (January 1972) and storm runoff (April, May, and June 1972) as evidenced by specific conductance minima during those months. From Jacobson and Langmuir (1974).

TABLE 8.3 Coefficients of variation (V) of some chemical and physical properties of two spring waters near State College, Pennsylvania, from January to August 1972

Parameter	Thompson Spring	Rock Spring
Ca^{2+}	6.4	26
Mg^{2+}	2.0	28
Na^+	**24**	**26**
K^+	10	16
HCO_3^-	2.7	29
SO_4^{2-}	**18**	**13**
Cl^-	**29**	18
NO_3^-	**20**	**32**
SpC	6.2	23
pH	0.6	1.7
Discharge	26	175

Note: $V = (100 \times SD)/mean$, where SD is the standard deviation of the measurements of a given property.

Boldface concentrations of species are thought to have been influenced by the washout of pollutant salts.

Source: From Jacobson and Langmuir (1974a).

8.4 PRECIPITATION CHEMISTRY AND ACID RAIN

8.4.1 The pH of Rain Due to Atmospheric Carbon Dioxide

Before we can know that precipitation has been subject to pollution, we must know what its "natural" composition is. This will depend on a variety of natural contributions and effects that are discussed in this chapter. Let us first examine the question: What is the pH of "natural" rain? To answer this question we will follow a method that applies to calculations of equilibria in natural waters, generally. The first question we ask is *what information is given or assumed?* In this case we will assume 25°C, pure water, and an atmospheric CO_2 pressure of $P_{CO_2} = 10^{-3.5}$ bar. The second question to ask is *what chemical equilibria apply to the problem?* In this case the equilibria are

$$K_{CO_2} = \frac{[H_2CO_3^\circ]}{P_{CO_2}} = 10^{-1.47}, \qquad K_1 = \frac{[H^+][HCO_3^-]}{[H_2CO_3^\circ]} = 10^{-6.35} \qquad (8.6)$$

We can combine these equilibria expressions, eliminating $H_2CO_3^\circ$, and solving for $[H^+]$, to obtain:

$$[H^+] = \frac{K_{CO_2} \times K_1 \times P_{CO_2}}{[HCO_3^-]} \qquad (8.7)$$

Third is the *charge-balance equation*, which describes the fact that the total of the equivalents (eq) or milliequivalents (meq) of cations in a given volume or weight of water must equal the same sum for the anions. The charge-balance equation in this example is

$$mH^+ = mHCO_3^- + 2mCO_3^{2-} + mOH^- \qquad (8.8)$$

In general problems in solution chemistry there is also a fourth kind of equation called a *mass-balance equation*. Mass-balance expressions relate the total concentration of a species reported in the chemical analysis to concentrations of the several forms of that species in solution. For example, the total calcium (ΣCa(aq)) reported in the chemical analyses equals the sum of all species of calcium in the water. For seawater, the calcium mass balance is ΣCa(aq) $= Ca^{2+} + CaSO_4^\circ + CaHCO_3^+ + CaCO_3^\circ$. The expression for carbonate alkalinity (C_B), which equals

$$C_B = HCO_3^- + 2CO_3^{2-} + OH^- - H^+ \tag{8.9}$$

is also a mass-balance equation.

It turns out, however, that we will not need this or any other mass-balance equation to solve our rain pH problem. For rain, we know the pH is generally below about 6, so that mOH^- and mCO_3^{2-} are negligible relative to mH^+ and $mHCO_3^-$. As a rule of thumb, any species whose concentration is less than 1% (two orders of magnitude) of another species' concentration present is negligible in a charge-balance or mass-balance calculation. (How do we know these two species are negligible at this pH?) With this simplification, and ignoring activity coefficients in such a dilute water, we find expression (8.8) reduces to simply $mH^+ = mHCO_3^-$.

Substituting H^+ for HCO_3^- in expression (8.7) and introducing the equilibrium constant values and partial pressure of CO_2 in the resultant equation, we find

$$[H^+]^2 = K_{CO_2} P_{CO_2} K_1 = 10^{-1.47} \times 10^{-3.5} \times 10^{-6.35} = 10^{-11.32} \tag{8.10}$$

or pH = 5.7. Activity coefficients have been ignored in this calculation, so it is not justified to express the final result to more than one decimal place.

One can argue that the pH of rain due solely to carbonic acid has not been "natural" since the industrial revolution and the onset of major burning of fossil fuels. Other causes of CO_2 increase during this period have included the burning of biomass and deforestation. Analyses of gases trapped in ice cores indicate that in 1850 the atmosphere contained about 265 ppm CO_2 by volume (ppmv) or $10^{-3.58}$ bar CO_2 pressure (Berner and Berner 1987). Shown in Fig. 8.5 is the CO_2 content of the atmosphere measured at Mauna Loa, Hawaii, since 1958, at which time the CO_2 pressure had risen to 315 ppmv or $10^{-3.50}$ bar. The plot shows that CO_2 pressures have been increasing at a rate of 0.4%/y since about 1975. Based on this rate, the CO_2 content of the atmosphere in 1996 just exceeds 360 ppmv, or $10^{-3.44}$ bars. The annual CO_2 oscillation of about 6 ppmv in the plot is caused by seasonal changes in biological activity.

The pH of present-day rain is usually below 5.7, mostly because of the presence SO_2 gas, which leads to sulfuric acid, and of smaller amounts of HNO_3 and HCl, all chiefly from the burning of fossil fuels. Also, even in pristine locations, natural organic acids from plants are volatilized and dissolved in water droplets contributing to rain acidity (Klusman 1996).

8.4.2 The General Composition of Precipitation

Following discussion is derived in large part from Berner and Berner (1987, 1996) who offer an in-depth summary and analysis of rainfall chemistry. Rainfall is typically a slightly acid, dilute solution, which generally contains a few parts per million dissolved solids, except where affected by air pollutants. The solute content of rain is derived from the dissolution of atmospheric gases and solid particles called aerosols. Aerosols tend to settle out naturally at sizes in excess of about 20 μm. They are mostly water-soluble salts in the marine atmosphere and over coastal areas. According to Berner and Berner (1987) aerosols, in general, are comprised of about 27% wind-blown sea salts, 17% soluble salts of sulfate, nitrate, and ammonia formed from natural and anthropogenic gaseous emissions, and about 41% soil and rock dust and volcanic debris. These amounts can be highly variable both spatially and temporally (Klusman 1996). Aerosols also include a few percent each of debris from forest fires and slash-and-burn agriculture, anthropogenic direct emissions, and hydrocarbons from fossil fuels and released as plant exudates.

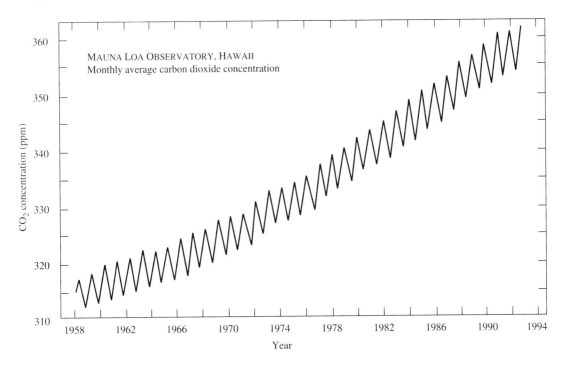

Figure 8.5   Mean monthly concentrations of atmospheric CO_2 at Mauna Loa, Hawaii. The yearly oscillation is explained mainly by the annual cycle of photosynthesis and respiration of plants in the Northern Hemisphere. From E. K. Berner and R. A. Berner. *Global environment: Water, air, and geochemical cycles.* Copyright © 1996. Used by permission of Prentice Hall, Inc., Upper Saddle River, NJ.

Table 8.4 details sources of the individual ions in rainwater according to Berner and Berner (1996). The composition of precipitation is summarized in Table 8.5, which shows that global rain (which falls mostly over the oceans) is on average a sodium chloride solution. Precipitation over the continents also contains important amounts of Ca^{2+}, NH_4^+, H^+, SO_4^{2-}, and NO_3^-, with Ca^{2+} typically exceeding Na^+. If the pH is above about 4.5, rain may also contain HCO_3^-. The total dissolved solids (TDS) of precipitation ranges from 4 to about 17 mg/L, with values in excess of about 10 mg/L affected by air pollution.

The maps in Fig. 8.6(a) and Fig. 8.6(b) show average concentrations of Na^+ and Cl^- in rain over the contiguous United States for 1994 (NADP/NTN, 1996).[†] The high concentrations along coastlines reflect the contribution of wind-blown salt spray, particularly within 100 to 300 km of the ocean.

Globally, about half of all atmospheric sulfate is derived from combustion of fossil fuels and half from natural sources (Berner and Berner 1987). It has been estimated that anthropogenic sources are responsible for 90% of the total atmospheric sulfur deposition in eastern North America, which occurs as dry deposition of SO_2 gas and sulfate particles or dissolved in rain. The highest amounts of sulfate and nitrate in U.S. rain, which are found in the northeast [Fig. 8.7(a) and Fig. 8.7(b)], are

---

[†]Since 1986, the National Atmospheric Deposition Program (NADP/NTN) has collected weekly precipitation samples from about 200 monitoring sites nationwide in the United States. Sites are away from major point sources and large urban centers to provide a more regional/national perspective on precipitation chemistry.

TABLE 8.4   Sources of individual ions in rainwater

	Origin		
Ion	Marine input	Terrestrial input	Pollutive input
Na^+	Sea salt	Soil dust	Biomass burning
Mg^{2+}	Sea salt	Soil dust	Biomass burning
K^+	Sea salt	Biogenic aerosols Soil dust	Biomass burning Fertilizer
Ca^{2+}	Sea salt	Soil dust	Cement manufacture Fuel burning Biomass burning
H^+	Gas reaction	Gas reaction	Fuel burning
Cl^-	Sea salt	—	Industrial HCl

SO_4^{2-}	Sea salt
	DMS from biological
	   decay
		DMS, H_2S etc., from
		   biological decay
		Volcanoes
		Soil dust
			Fossil fuel burning
			Biomass burning

NO_3^-	N_2 plus lightning	NO_2 from biological decay N_2 plus lightning	Auto emissions Fossil fuels Biomass burning Fertilizer

NH_4^+	NH_3 from biological
	   activity
		NH_3 from bacterial
		   decay
			NH_3 fertilizers
			Human, animal waste
			   decomposition
			   (Combustion)

PO_4^{3-}	Biogenic aerosols adsorbed on sea salt	Soil dust	Biomass burning Fertilizer
HCO_3^-	CO_2 in air	CO_2 in air Soil dust	—
SiO_2, Al, Fe	—	Soil dust	Land clearing

Source: From E. K. Berner and R. A. Berner. *Global environment: Water, air, and geochemical cycles.* Copyright © 1996. Used by permission of Prentice Hall, Inc., Upper Saddle River, NJ.

mostly from heavy industrial activity in Michigan, Indiana, Ohio, and western Pennsylvania. Somewhat elevated NO_3^- levels in the midwest may be derived from fertilizer use. Figure 8.8 shows that SO_2 emissions in the United States have been declining since about 1970, with NO_3^- declining slightly since about 1980. In 1979, roughly 56% of the nitrogen oxides originated from combustion for heating and electrical power and 40% from fuel burning for transportation. The U.S. SO_2 decline reflects the institution of controls on emissions from the burning of fossil fuels for heat and electrical power (the source of 78% of SO_2 emissions in 1975), controls on emissions from industrial smelting and refining (the source of 20% of SO_2 emissions in 1975) (Berner and Berner 1987), and on a shift from the burning of high-sulfur coal mined in the eastern United States to low-sulfur coal mined in the Rocky Mountain states.

In Europe SO_4^{2-} and NO_3^- increased rapidly after 1950, but have stabilized between 1972 and 1986 and may now be declining. Also, European sulfate and nitrate declines probably reflect emission controls (Berner and Berner 1996). In spite of a leveling or decline in SO_2 emissions in Europe and the United States, global SO_2 emmisions continue to climb (Fig. 8.8). This reflects the increased burning of fossil fuels and coal, in particular, for electrical power and heat generation by developing nations such as China.

TABLE 8.5 The chemical composition of global atmospheric precipitation

Constituent	Global mean estimate[†]	Continental rain[‡]	Marine & coastal rain[‡]	Northern Europe[§]	Northeastern United States[§]	European USSR[§]
Na$^+$ (μM)	86	9–40	40–200	13–85	5	39–96
K$^+$ (μM)	8	3–8	5–15	4–7	2	10–20
Mg^{2+} (μM)	12	2–20	16–60	5–16	2	8–12
Ca^{2+} (μM)	2	3–80	5–40	16–33	4	10–52
NH$_4^+$ (μM)	—	6–30	0.6–3	6–48	12	22–61
H$^+$ (μM)	2	1–100 (pH = 6–4)	1–10 (pH = 6–5)	16–38	74	0.2–79
HCO$_3^-$ (μM)	2	0–?	?	21	0	30–79
Cl$^-$ (μM)	107	6–60	30–300	11–98	7	23–107
NO$_3^-$ (μM)	—	6–21	2–8	5–36	12	6–16
SO$_4^{2-}$ (μM)	6	10–30	10–30	10–63	30	38–94
Total mean (mg)	7.2			10.2	4.0	16.6

Source: [†]Garrels and Mackenzie (1971). [‡]Berner and Berner (1996). [§]Lerman (1979).

8.4.3 Acid Rain

Rain more acid than about pH 5.7 results from reactions of the acid gases SO_2, NO_2, NO, and, to a lesser extent, HCl. The S and NO_x species are derived chiefly from the combustion of fossil fuels. Important reactions forming sulfuric acid are

$$SO_2 + 2OH \rightarrow H_2SO_4 \rightarrow 2H^+ + SO_4^{2-} \tag{8.11}$$

and

$$SO_2 + H_2O_2 \rightarrow H_2SO_4 \rightarrow 2H^+ + SO_4^{2-} \tag{8.12}$$

where OH is a free radical. Reaction (8.11), which takes place in the gaseous atmosphere and involves neutral OH radicals, produces a sulfuric acid aerosol. Reaction (8.12) can proceed in liquid cloud droplets (Berner and Berner 1996). Both reactions involve the oxidation of sulfur dioxide. Overall, the oxidation of $SO_2(g)$ can be written (Manahan 1994)

$$SO_2 + \tfrac{1}{2}O_2 + H_2O \rightarrow 2H^+ + SO_4^{2-} \tag{8.13}$$

Nitric acid in rain is formed by the reaction

$$NO_2 + OH \rightarrow HNO_3 \rightarrow H^+ + NO_3^- \tag{8.14}$$

where OH is a free radical. Overall NO_2 oxidation is described by

$$2NO_2 + \tfrac{1}{2}O_2 + H_2O \rightarrow 2H^+ + 2NO_3^- \tag{8.15}$$

(Manahan 1994).

Figure 8.9 shows the pH of rainfall in the eastern United States and adjacent Canada in 1955 to 1956 and 1972 to 1973, and for the contiguous states and Canada in 1980. A map of average precipitation pH for the contiguous United States in 1995 is shown in Fig. 8.10(a). The maps indicate that from 1955 to 1980 the lowest rainfall pH values declined from about 4.4 to 4.1. During the same period, the region having pH values below 5 expanded from the heavily industrialized northeast to

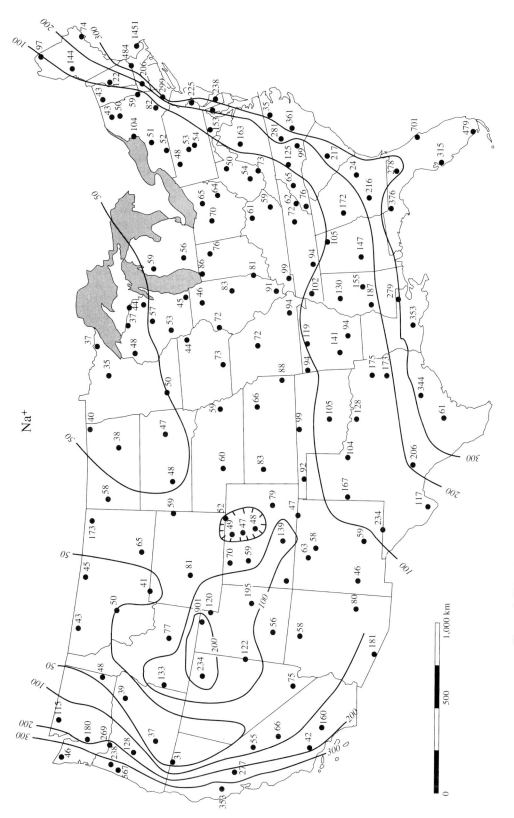

Figure 8.6(a) Generalized isoconcentration contours for Na$^+$ (μg/L) in atmospheric precipitation of the contiguous United States in 1995, based on the plotted average Na$^+$ concentrations reported for stations in the National Atmospheric Deposition Program/National Trends Network (1996).

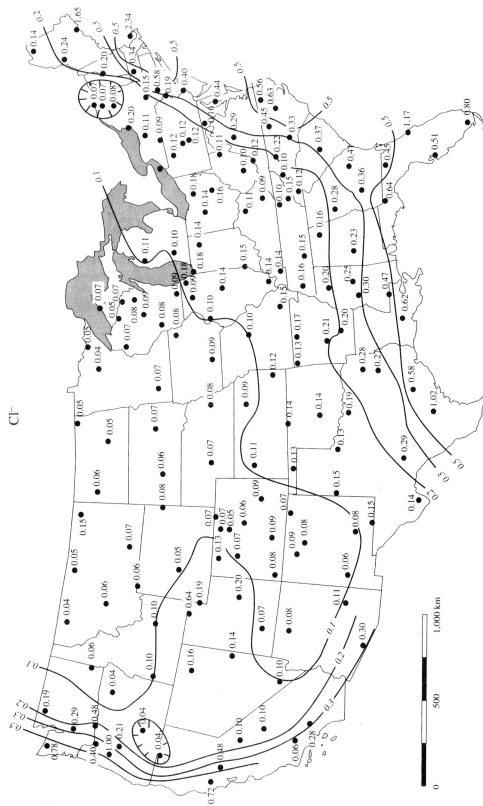

Figure 8.6(b) Generalized isoconcentration contours for Cl⁻ (mg/L) in atmospheric precipitation of the contiguous United States in 1995, based on the plotted average Cl⁻ concentrations reported for stations in the National Atmospheric Deposition Program/National Trends Network (1996).

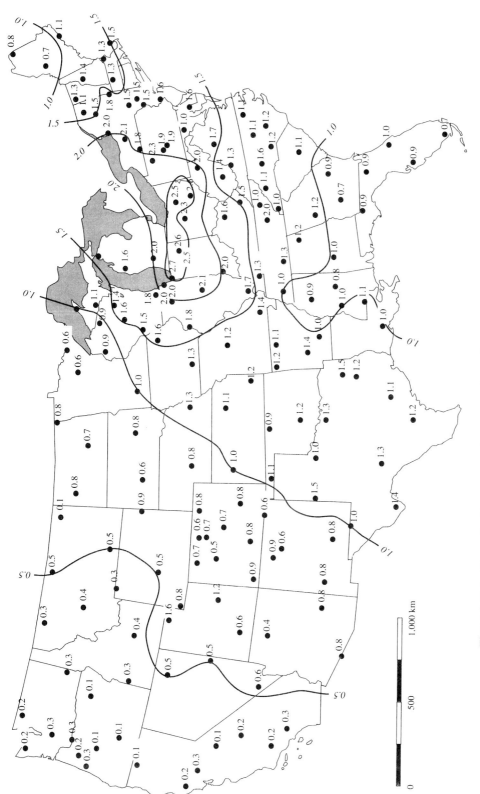

SO_4^{2-}

Figure 8.7(a) Generalized isoconcentration contours for SO_4^{2-} (mg/L) in atmospheric precipitation of the contiguous United States in 1995, based on the plotted average SO_4^{2-} concentrations reported for stations in the National Atmospheric Deposition Program/National Trends Network (1996).

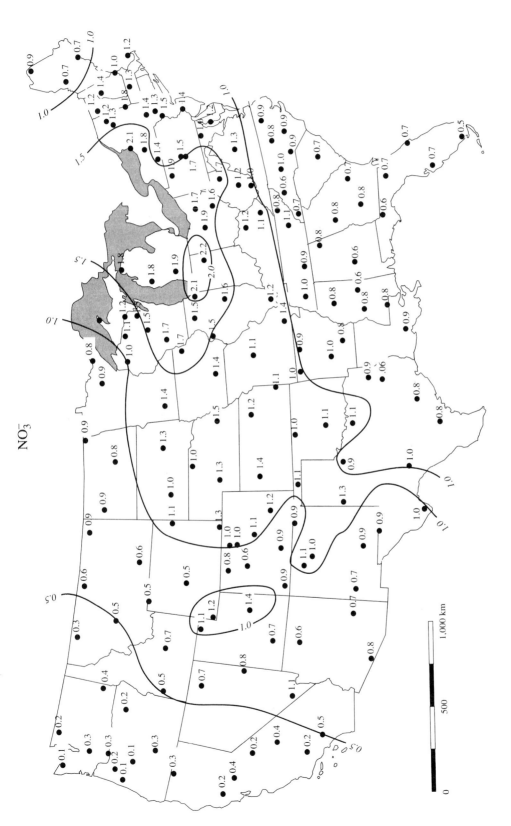

Figure 8.7(b) Generalized isoconcentration contours for NO_3^- (mg/L) in atmospheric precipitation of the contiguous United States in 1995, based on the plotted average NO_3^- concentrations reported for stations in the National Atmospheric Deposition Program/National Trends Network (1996).

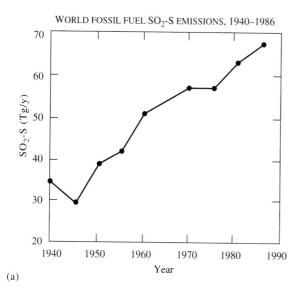

(a)

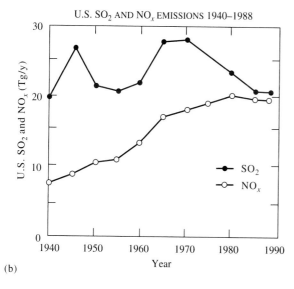

(b)

Figure 8.8   The rise in world SO_2 emissions from 1940 to 1986, and of U.S. SO_2 and NO_x emissions from 1940 to 1988. (1 Tg = 10^6 metric tons = 10^{12} g.) From E. K. Berner and R. A. Berner. *Global environment: Water, air, and geochemical cycles.* Copyright © 1996. Used by permission of Prentice Hall, Inc., Upper Saddle River, NJ.

encompass the entire eastern half of the country, north to James Bay, and across Labrador, Canada. This apparently reflected increased industrialization during that period, particularly of the southeastern states. The wider dispersal of acid rain in the eastern and northeastern states was apparently furthered by the installation of high stacks by industry and utilities. Such stacks reduce local pollution, but aggravate regional pollution problems downwind.

   The 1980 and 1995 maps show pH values above 5.7 at sites in 5 western states and western Canada. These high values reflect the absence of nearby industrial activity, but also reflect neutralization of the acidity by wind-blown carbonates. Calcium carbonate and sulfate salts accumulate in the topsoil of arid western prairies and can be transported in windblown dust for distances up to 300 to 650 km (Berner and Berner 1996). The highest resulting Ca^{2+} concentrations in precipitation

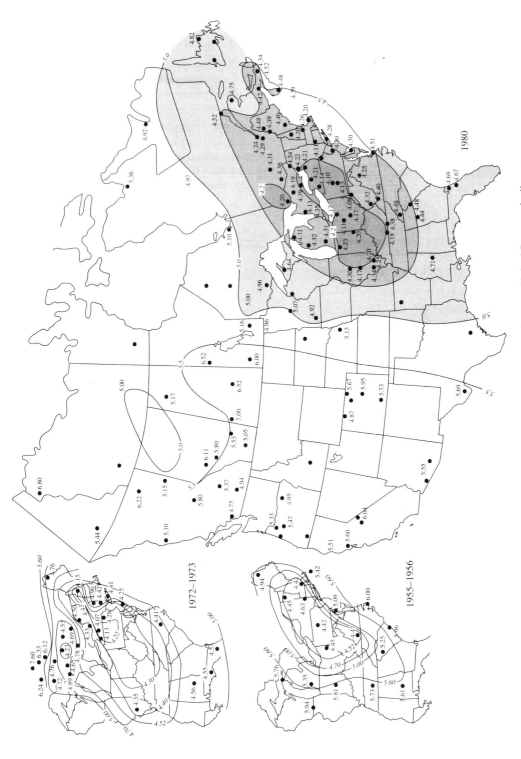

Figure 8.9 Average pH of annual precipitation in the northeastern United States and adjacent Canada in 1955 to 1956 and 1972 to 1973, and for the contiguous United States and Canada in 1980. From E. K. Berner and R. A. Berner. *The global water cycle, geochemistry and environment.* Copyright © 1987. Used by permission of Prentice Hall, Inc., Upper Saddle River, NJ.

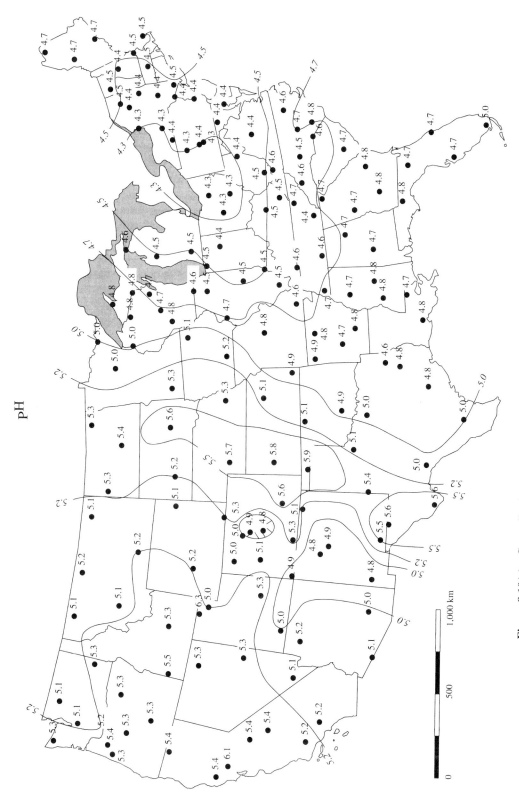

Figure 8.10(a) Generalized contours of the average lab-measured pH of precipitation of the contiguous United States in 1995, based on the plotted average reported for stations in the National Atmospheric Deposition Program/National Trends Network (1996).

(>0.3 mg/L) in Fig. 8.10(b), roughly coincide with the highest pH values (>5.2) in Fig. 8.10(a). Generally less important, the acidity of rain is also neutralized by $NH_3(g)$ released from fertilizers and biological sources, through the reaction $NH_3(g) + H^+ = NH_4^+$ (cf. Berner and Berner 1996).

Initial rainfall from a given event is the most acid, because of rain washout of atmospheric acid gases and aerosols. This effect produces initial rain pH's below 3 in central Pennsylvania, for example. In cities such as Los Angeles and London, when fog has been present and the air stagnant, pH values between 2 and 3 have been observed, due in part to evaporative concentration, with serious consequences for people with respiratory problems.

Acid rain has damaged carbonate stone buildings and monuments, particularly in urban areas (cf. Baedecker et al. 1990), and can kill fish and vegetation. Acid rain is not a serious problem in the east-central United States where carbonate-bearing rocks and soils are common, and can neutralize the rain. However, in New York, New England, and Quebec, Canada, where soils have developed on aluminosilicate rocks, they may be incapable of neutralizing acid rain pH's because of their high quartz content and/or slow rates of weathering (cf. Baker et al. 1993). The result can be acid soils, streams, and lakes, with stressed and dead plants, trees, and fish (cf. Ember 1981; Schindler 1988; Smith 1991; Berner and Berner 1996; Likens et al. 1996), as well as acid groundwaters (Hansen and Postma 1995).

Example 8.1

Manahan (1991) proposes a global average $SO_2(g)$ concentration of 0.02 ppmv away from the effects of "gross pollution" and derived chiefly from anthropogenic sources, combined with smaller amounts from photochemical and volcanic sources. Maximal sulfate concentrations were about 35 μM in rain over the northeastern United States in 1979. Given that K(Henry) = 1.18 mol/L bar for the dissolution of $SO_2(g)$ in water at 25°C (see Chap. 1), and assuming that all of the atmospheric $SO_2(g)$ is ultimately converted into sulfate acid and that $SO_2(aq)$ equals $H_2SO(aq)$:

a) Back-calculate the $SO_2(g)$ ppmv concentration that would result in a sulfate concentration of 35 μM in rain. Compare your answer to the atmospheric sulfur dioxide concentrations suggested above and by Manahan, and discuss any differences.

b) What rain pH should result from the sulfate concentration of 35 μM? Neglect the effect of atmospheric CO_2 on the rain pH in your calculation. A 1980 map of rain pH for the northeastern states (Fig. 8.9) shows values as low as pH 4.1 coinciding with the high sulfate rain values. How does this pH compare with your calculated value?

c) If you had considered the effect of $P_{CO_2} = 10^{-3.5}$ bar in your rain pH calculation in (b), what would the computed pH have been? Was the assumption that you could ignore CO_2 in this case valid?

d) The highest rain pH's measured in North America (about pH = 6.5) reflect the dissolution of calcite ($CaCO_3$) dust in arid midcontinent areas. Assuming that the rain is initially pure and its initial pH controlled by normal atmospheric CO_2 pressure, how many milligrams of calcite dust (GFW = 100) must dissolve in a liter of rain to produce a rain pH of 6.5? Ignore ion activity coefficients and ion pairs.

Solution

a) The overall reaction of $SO_2(g)$ to produce SO_4^{2-} is

$$SO_2(g) + H_2O + \tfrac{1}{2}O_2 \rightleftarrows 2H^+ + SO_4^{2-} \tag{8.16}$$

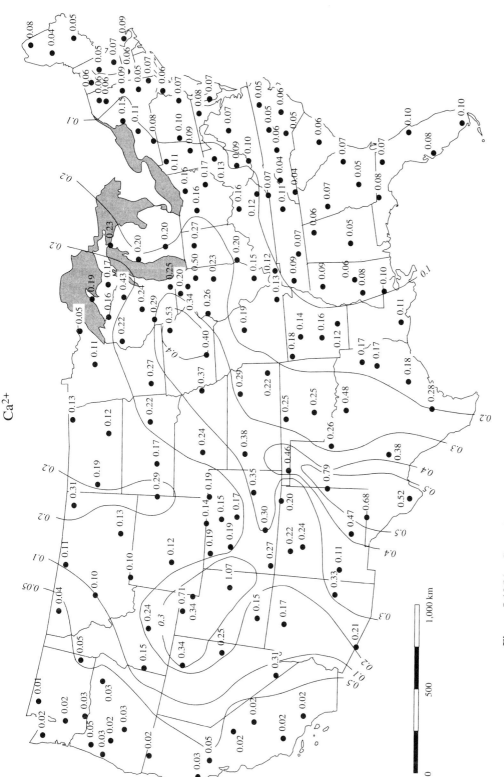

Figure 8.10(b) Generalized isoconcentration contours for Ca^{2+} (mg/L) in atmospheric precipitation of the contiguous United States in 1995, based on the plotted average Ca^{2+} concentrations reported for stations in the National Atmospheric Deposition Program/National Trends Network (1996). High mid-continent Ca^{2+} values in precipitation are chiefly from windblow carbonate and sulfate salts.

The reaction of SO_2 to produce $H_2SO_3^\circ$ is

$$SO_2(g) + H_2O \rightleftarrows H_2SO_3^\circ \tag{8.17}$$

The equilibrium expression is

$$K_{SO_2} = 1.18 = \frac{[H_2SO_3^\circ]}{P_{SO_2}} \qquad \text{Henry's law} \tag{8.18}$$

If all the $H_2SO_3^\circ$ [$= SO_2(g)$] is converted to SO_4^{2-}, then the initial $H_2SO_3^\circ = 35 \ \mu M = 3.5 \times 10^{-5}$ M. Now, via the K_{SO_2} expression,

$$P_{SO_2} = \frac{[H_2SO_3^\circ]}{K_{SO_2}} = \frac{3.5 \times 10^{-5}}{1.18} = 2.97 \times 10^{-5} \text{ bar} \tag{8.19}$$

or 29.7×10^{-6} bar, or 29.7 ppmv. This is much higher (about 1500 times higher) than the background values (≤ 0.02 ppmv) suggested by Klusman and Manahan and clearly reflects major air pollution.

b) Because 35 μM of SO_4^{2-} is produced, and 2 μM of H^+ are formed for every 1 μM of SO_4^{2-}, according to the overall reaction, we must form 70 μM H^+, or 7×10^{-5} M $= 10^{-4.15}$ M H^+, or pH $= 4.15$.

c) To consider the added effect of P_{CO_2} on this pH, we must look back at the equations used in Chap. 6 to calculate the pH of "natural" rain in the presence of $P_{CO_2} = 10^{-3.5}$ bar and see which, if any, are changed once SO_2 is added. Clearly, only the charge-balance equation is affected, and becomes

$$H^+ = HCO_3^- + 2SO_4^{2-} \tag{8.20}$$

From Chap. 5 we know, in general, that

$$[H^+] = \frac{K_{CO_2} K_1 P_{CO_2}}{[HCO_3^-]} \tag{8.21}$$

which, at 25°C in contact with the air equals

$$[H^+] = \frac{10^{-1.47} \times 10^{-6.35} \times 10^{-3.5}}{[HCO_3^-]} = \frac{10^{-11.32}}{[HCO_3^-]} \tag{8.22}$$

Eliminating HCO_3^- in the charge-balance equation by substitution, given that the SO_4^{2-} concentration is 3.5×10^{-5} M, and ignoring γ's, we obtain

$$[H^+] = \frac{10^{-11.32}}{[H^+]} + 7.0 \times 10^{-5} \tag{8.23}$$

Reformatting gives

$$[H^+]^2 - 7.0 \times 10^{-5} [H^+] - 4.79 \times 10^{-12} = 0 \tag{8.24}$$

which can be solved by completing the squares:

$$([H^+] - 3.50 \times 10^{-5})^2 = [3.51 \times 10^{-5}]^2 \tag{8.25}$$

$$[H^+] = 7.01 \times 10^{-5} = 10^{-4.15} \text{ M}$$

pH $= 4.15$. The CO_2 system has a negligible effect. Rain pH is controlled by the SO_2 system.

d) The water is free of SO_2 and in equilibrium with $P_{CO_2} = 10^{-3.5}$ bar. The same general equation between $[H^+]$ and $[HCO_3^-]$ applies:

$$[H^+] = \frac{K_{CO_2} K_1 P_{CO_2}}{[HCO_3^-]} = \frac{10^{-11.32}}{[HCO_3^-]} \qquad (8.26)$$

But we are told pH = 6.5, so $[H^+] = 3.16 \times 10^{-7}$ M, and

$$[HCO_3^-] = \frac{10^{-11.32}}{10^{-6.5}} = 10^{-4.82} = 1.51 \times 10^{-5} \text{ M} \qquad (8.27)$$

The charge-balance equation is

$$H^+ + 2Ca^{2+} = HCO_3^- \qquad (8.28)$$

so
$$2Ca^{2+} = 1.51 \times 10^{-5} - 0.03 \times 10^{-5} \qquad (8.29)$$

$$[Ca^{2+}] = 7.39 \times 10^{-6} \text{ M}$$

or the same number of moles of $CaCO_3$, which weigh $(100 \text{ g/mol}) \times (7.36 \times 10^{-6} \text{ M})/(10^3 \text{ mg/g}) = 0.74$ mg/L of rain.

8.4.4 Trace Elements in Rain

Most trace elements deposited by precipitation are in particulate form, associated with the wash-out of windblown dust or aerosols. However, some trace elements are borne in the atmosphere as true gases; that is, they have sufficient volatility to exist as a vapor in the atmosphere. A measure of this tendency is the boiling point of an element's oxide. The atmospheric enrichment factor for several elements is plotted against the boiling point of that element's oxide in Fig. 8.11. The enrichment factor is defined as [X atm/Al atm]/[X rocks/Al rocks], where [X atm/Al atm} is the ratio of the weight concentration of the element to that of aluminum in the atmosphere. [X rocks/Al rocks] is the corresponding ratio observed in average crustal rocks. The plot assumes that aluminum, because of its high oxide boiling point, has no volatility in the atmosphere. Element concentrations are then compared to that of aluminum in the atmosphere and in average crustal rocks. The plot suggests a high gaseous mobility for Cd, Se, As, and Hg, and a moderately high gaseous mobility for Sb, Zn, Pb, Ni, and V. As a hard base (see Chap. 3), oxygen forms strong bonds with hard acid cations (Al, Sc, Ti, Cr, Ba, and Mn), which explains the low volatility of their oxides. The oxides of soft or borderline soft acids (As, Cd, Hg, Ni, Pb, Sb, Se, Zn) are less stable, thus more volatile.

Gaseous metals (which may form aerosols) in the atmosphere are derived from perhaps six sources. These include: (1) volcanism; (2) release from biological activity (Hg, As, Se); (3) sea surf fractionation during production of atmospheric sea salt particles; (4) burning of fossil fuels; (5) smelting of metal ores (Nriagu 1996); and (6) incinerator combustion of urban wastes (cf. Fernandez et al. 1992). It is well known, for example that Pb is concentrated along roadsides because of the combustion of leaded gasolines and that As, Cd, and Pb concentrations are high in soils downwind from sulfide ore-smeltering operations.

Boutron et al. (1991) studied the change in atmospheric emissions of Cd, Pb, and Zn before and after 1971 by analyzing Greenland ice cores. They found that Pb values were 200 times higher in 1971 than before the industrial revolution, with 99% of the lead from anthropogenic sources. Since 1971 and the introduction of unleaded gasoline, atmospheric lead values have declined by 7.5 times in the United States. The same authors also reported that Cd and Zn concentrations in the ice have

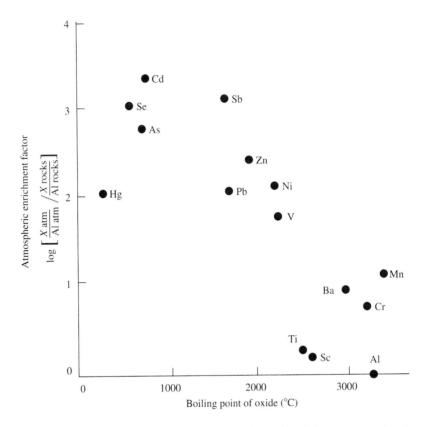

Figure 8.11 Plot of the atmospheric enrichment factor (AEF) for some metals where AEF = log [(X atm/Al atm)/(X rocks/Al rocks)], versus the boiling points of their oxides, where the latter is taken as a measure of metal volatility. Adapted from F. T. Mackenzie and R. Wollast. In *The sea.* Copyright © 1977 by John Wiley & Sons. Reprinted by permission of John Wiley & Sons, Inc.

declined by about 2.5 times since 1971 due to controls on industrial emissions in the northern hemisphere. In contrast, analysis of the Hg content of dated sediment cores from remote lakes in Minnesota and Wisconsin (Swain et al. 1992) shows that the rate of atmospheric Hg deposition has increased from 3.7 μg/m^2 in 1850 to 12.5 μg/m^2 in 1990, or by 3.4 times.

8.5 SOIL-MOISTURE CHEMISTRY

In high-rainfall climates the chemistry of soil water may resemble that of the local precipitation. Such conditions exist where evapotranspiration (ET) losses are minimal because of low temperatures and/or high humidity, or where the soils are quartz-sand–rich and deficient in weatherable minerals.

Table 8.6 contrasts soil moisture chemistry in a variety of soils and climates. All of the tabulated locations, except for the Sierra Nevada soil, are near the ocean, so that one would expect their precipitation to be dominated by Na and Cl. If ET alone controls infiltrating soil-water chemistry, this same dominance by Na and Cl should persist in the soil water. Such is the case for moisture from the Kauaian soil, which is also in an area of very high rainfall, consistent with its low TDS content of about 30 mg/L.

TABLE 8.6 The chemical composition of some soil waters

Locality	SiO_2	Ca	Mg	Na	K	HCO_3	SO_4	Cl	pH	TDS	$\dfrac{Na}{(Na+Ca)}$
Mor humus layer, English Lake District	—	1.1	0.9	9.2	5.8	—	17.5	11.5	4.2	(45)	0.89
Mull humus layer, English Lake District	—	10.0	1.6	8.7	4.4	8.1	9.6	7.6	6.2	(50)	0.47
Soil on Wissahickon schist, MD, USA	5.45	4.15	3.29	5.45	4.86	—	—	—	5.45	(35)	0.57
Soil on Woodstock granite, near Baltimore, MD, USA	23	5.8	15	4.9	0.9	80	11.6	0.8	8.4	142	0.46
Soil on granite, Sierra Nevada Mtns., CA, USA	23	7.0	0.3	3.9	1.0	34	0.0	0.4	6.9	70	0.36
Soil on basalt, Kauai, HI, USA	1.2	0.9	1.9	6.3	0.6	6	4.0	9.1	5.1	30	0.88

Note: Except for pH, concentrations are in mg/L. Values in parentheses are estimates. The Na/(Na+Ca) ratio is in terms of species weights. Woodstock granite soil water also contains Al = 0.1 mg/L, and Fe = 0.00 mg/L. Sierra Nevada granite soil water has Al = 0.00 mg/L and Fe = 0.12 mg/L. Kauai soil water has Al = 0.00 mg/L, and Fe = 0.18 mg/L. All soil waters are from the A horizon except the Woodstock granite soil, which is from the deep C horizon.

Source: Modified from R. M. Garrels and F. T. Mackenzie. *Evolution of sedimentary rocks.* Copyright 1971. Used by permission.

On an equivalent concentration basis, Na and Cl also dominate in the Mor and Mull humus waters and Wissahickon schist soil water. The Mor and Mull humus soil waters have TDS values of 45 and 50 mg/L respectively, reflecting a concentration increase of perhaps 10 times from the TDS of precipitation, probably chiefly because of ET losses. However, in waters in Mull humus, Woodstock granite, and Sierra Nevada soils, the significant concentrations of Ca^{2+} and HCO_3^- indicate that chemical weathering has contributed important ion concentrations to the soil water. This is consistent with the high TDS of the Woodstock granite (142 mg/L), and Sierra Nevada (70 mg/L) soil waters. Another important point is that the Woodstock granite soil is a deep soil (from the C horizon; see the discussion of soils in Chap. 7) so that its moisture has had a longer residence time in which to weather soil minerals, whereas the other soil waters are from near-surface A horizons. The extent of weathering is also evident from soil-moisture pH, which is highest, as expected, ranging from 6.2 to 8.4 in the soils containing important concentrations of Ca^{2+} and/or Mg^{2+} and HCO_3^-.

A detailed look at the evolution of soil-moisture chemistry was reported by Sears (1976). In his study Sears assumed the average composition of precipitation shown in Table 8.7. Table 8.7 also lists analyses of the soil moisture he collected from suction lysimeters at 1- and 3-m depths in respective B- and C-horizon soils formed by the weathering of underlying sandy dolomite. The 1-m sample is chiefly a Na^+-NO_3^- water, with the nitrate probably from fertilizer. The TDS is about 70 mg/L at 1 m and has increased to 500 mg/L at the 3-m depth. In order to explain changes occurring between the 1- and 3-m depth, it is useful to select a solute we can assume to be practically unreactive in the soil. The best common species for this purpose is probably Cl, with which we can then compare other species' concentrations. Relative increases from 1- to 3-m depth are shown in the third column. Increases compared to chloride are given in the fourth column.

The most obvious changes with depth are the creation of a dominantly Ca^{2+}-HCO_3^- water from the chemical weathering of residual dolomite fragments in the deeper soil. Absolute increases in species such as Na^+, SO_4^{2-}, NO_3^-, and Cl^-, which are unlikely to derive from rock weathering, prob-

TABLE 8.7 Composition of average precipitation and soil moisture in soils on weathered sandy dolomite near State College, Pennsylvania

Species/ parameter	Average precipitation	Soil 1 m deep	Soil 3 m deep	Increase 1 m to 3 m	Net change relative to Cl^- 1 m to 3 m
Ca^{2+}	1.2	3.3	108	33 times	2.9 times
Mg^{2+}	0.4	0.8	7.3	9 times	0.8 times
Na^+	0.5	6.9	21	3.0 times	0.3 times
K^+	1.0	1.8	5	2.7 times	0.2 times
HCO_3^-	0.0	1.9	144	76 times	6.6 times
SO_4^{2-}	3.6	2.9	17	5.9 times	0.5 times
NO_3^-	1.4	24	53	2.2 times	0.2 times
Cl^-	1.0	4.8	55	11.5 times	—
$SiO_2(aq)$	0.0	45	95	2.1 times	0.2 times
TDS (sum)	9.1	70	500	5.6 times	0.5 times
pH	4.5	5.9	6.5	(H 0.74 times lower)	0.02 times
P_{CO_2} (bar)	0.00032	0.003	0.04	13 times	1.2 times

Note: The sample from 1 m deep was collected May 1, 1974, the sample from 3 m deep on October 12, 1974. Concentrations are in mg/L unless otherwise noted.

Source: Data from Sears (1976).

ably result from their concentration by evapotranspiration. Relative decreases in Mg^{2+}, Na^+, and K^+ are probably due to their uptake by ion exchange and clay formation. Relative declines in SO_4^{2-} and especially NO_3^-, which are soluble and unlikely to react inorganically, probably reflect the biological uptake of these species as nutrients. The absolute increase in silica results from chemical weathering of unstable silicate minerals, whereas its relative decline may reflect silica adsorption by clays. The 13-fold increase in CO_2 with depth is consistent with the downward migration of relatively heavy CO_2 gas produced chiefly in the shallow A horizon, and its upward diffusive loss from the l-m soil to the atmosphere (cf. Wood 1985).

An important conclusion one can draw from the Sears study and other, similar studies is that in areas with well-developed soils, the chemistry of underlying shallow groundwaters is often chiefly determined by processes taking place in the soil and less by processes operating in the saturated zone below. In this case, groundwaters from the underlying dolomite are similar to the water from the soil 3 m deep. Similarly, the chemical composition of effluent streams (streams fed chiefly by groundwater) is also quickly determined by interactions with adjacent soils (see Section 7.8).

8.6 GENERAL CHEMISTRY OF SURFACE- AND GROUNDWATERS

Before we consider controls on the chemistry of surface and groundwaters in some detail, we will examine the general composition of such waters. Listed in Table 8.8 are the median concentrations of major, minor, and trace constituents in surface- and groundwaters (Turekian 1977), where "major," "minor," and "trace" concentrations are above 1 mg/L, between l mg/L and 1 μg/L, and less than 1 μg/L, respectively.

It is important to note that water analyses rarely include the concentrations of dissolved gases or carbonic acid ($H_2CO_3^\circ$). To be complete, the tabulated average concentrations in Table 8.10 should list the carbonic acid concentration, which, for typical surface- and groundwaters at near-neutral pH's, equals several to tens of milligrams per liter. The following concentrations of carbonic acid are present at 25°C, depending on the CO_2 pressure

$-\log P_{CO_2}$ (bar)	$H_2CO_3^\circ$ (mg/L)
3.5	0.6
2.5	6
2.0	18
1.0	185

Dissolved oxygen and nitrogen concentrations in air-equilibrated surface water at 25°C are also significant and equal about 8 and 14 mg/L, respectively, which makes both species major constituents.

Figures 8.12 and 8.13 are frequency distribution plots of species concentrations in potable water (chiefly surface-waters) and groundwaters, respectively. The important reference point is where the curves cross the 50% line, which is the median concentration of the species being considered. The first important observation one can make from these figures and Table 8.8 is that groundwaters contain more dissolved solids than do surface-waters, generally. This chiefly reflects their longer contact times with the geomedia, and their isolation from dilution by fresh precipitation. A second and related observation is that sulfate is a more important species in most groundwaters than in surface-waters. A third observation is that the concentrations of metals such as Fe and Mn, which are most mobile in their (2+) reduced forms, are higher in groundwaters than surface-waters. This is consistent with the fact that groundwaters tend to be more reducing (anaerobic) than surface-waters.

TABLE 8.8   Median values of pH and TDS and of major (>1 mg/L) constituents in surface waters and groundwaters

Major Constituents

Constituent or parameter	Surface-water (mg/L)	Groundwater (mg/L)
HCO_3^-	58	200
Ca^{2+}	15	50
Cl^-	7.8	20
K^+	2.3	3
Mg^{2+}	4.1	7
Na^+	6.3	30
SO_4^{2-}	3.7	30
$SiO_2(aq)$	14	16
pH	—	7.4
TDS	120	350

Trace and Minor Elements

Element	Median content (μg/L)	Element	Median content (μg/L)
Al	10	La	0.2
Ag	0.3	Li	3
As	2	Mn	15
Au	0.002	Mo	1.5
B	10	Nb	1
Ba	20	Ni	1.5
Be	5	P	20
Bi	0.005	Pb	3
Br	20	Rb	1
Cd	0.03	Sb	2
Co	0.1	Se	0.4
Cr	1	Sn	0.1
Cs	0.02	Sr	400
Cu	3	Th	0.1
F	100	Ti	3
Fe	100	W	0.03
Hg	0.07	U	0.5
I	7	V	2
		Zn	20

Note: Median values of minor elements (<1 mg/L and >1 μg/L), and trace elements (<1 μg/L) in both surface-waters and groundwaters.

Source: Values after Turekian (1977).

Figure 8.12 shows that on a weight basis, the order of abundance of major species in the average surface-water is: Ca > Na > Mg > K, and HCO_3 > Cl > SO_4 (ignoring silica). Similarly, for the average groundwater (Fig. 8.13) the order of abundance by weight is: Ca > Na > Mg > K, and HCO_3 > SO_4 > Cl (again, ignoring silica). In other words, the *prevalent chemical character* of most fresh surface-waters and groundwater is calcium bicarbonate. This reflects the processes involved in chemical weathering of silicates and the common occurrence of calcium carbonate (Chap. 6).

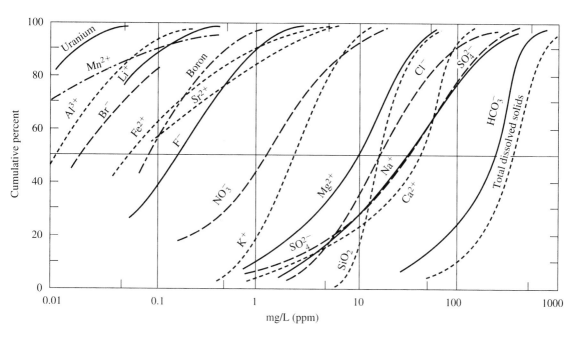

Figure 8.12 Cumulative percentages showing the frequency distribution of various constituents in potable (chiefly surface-) waters. From S. N. Davis and R. J. M. DeWiest. *Hydrogeology.* Copyright © 1966 by John Wiley & Sons. Reprinted by permission of John Wiley & Sons, Inc.

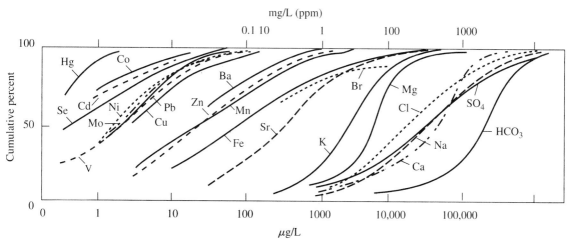

Figure 8.13 Cumulative percentages of some major and trace elements in groundwaters. Number of analyses: 13,000 to 18,000 for major elements; 750 to 8,000 for trace elements. Figure plotted by A. Rose using data in U.S. Geological Survey data bank. From A. W. Rose, H. E. Hawkes, and J. S. Webb. *Geochemistry in mineral exploration,* 2d ed. Copyright 1979. Used by permission.

Why is such concentration information important? One reason is that we need to know what the *natural* or *background concentrations* of species are, before we can decide what has been added as a consequence of land-use and waste-disposal practices. The frequency plots show that in some natural waters there are relatively high concentrations of such species as Fe, F, Mn, Zn, Pb, Cu, Ni, V, and Mo, for example. It is important for us to know how and when such elevated concentrations occur naturally, and not simply from waste disposal. The most obvious source of elevated concentrations of these elements is the weathering of ore deposits or leaching of their mine wastes. Second, under acid conditions the solubilities and mobilities of many trace species are greatly increased. Third, evaporation can also greatly concentrate levels of trace species. And fourth, reducing conditions in groundwaters tend to dissolve Fe(III) and Mn(III,IV) oxyhydroxides by reducing these ions to more soluble Fe(II) and Mn(II). This releases to the groundwater their adsorbed and coprecipitated content of metals such as Cu, Zn, Pb, Ni, Co, and Cd, for example.

8.7 CONTROLS ON THE CHEMICAL COMPOSITION OF RIVERS

Gibbs (1970) proposed an elegant and simple explanation for the general chemistry of streams and rivers.[†] He argued that surface-water chemistry was determined by: (1) rainfall (rain dominated), (2) rock-weathering reactions (rock dominated), or (3) evaporation-crystallization; or by combinations of these influences. Figures 8.14 and 8.15 summarize Gibbs's ideas in plots of TDS versus the $[Na^+/(Na^+ + Ca^{2+})]$ ratio and TDS versus relative mole fractions of $Ca(HCO_3)_2$ versus NaCl in the water. Because the composition of world rainfall (unpolluted) is chiefly determined by the NaCl of sea salt, rainfall plots in the lower right-hand corner of each figure at low TDS, as do streams whose chemistry is dominated by rainfall such as the Negro River, a tributary of the Amazon River.

Weathering of continental rocks increases their TDS and concentrations of calcium and bicarbonate relative to sodium and chloride. The composition of streams so affected plot to the left of both diagrams and include the Columbia, Mississippi, Yukon, and Thames rivers. Evaportranspiration from arid climate drainage basins and streams such as the Colorado, Pecos, and Jordan rivers, which receive soil runoff and irrigation return waters, further increase the Na^+ and TDS content of streams. Concomitant precipitation of $CaCO_3$ further shifts the prevalent chemical character of such streams back toward NaCl and the chemistry of seawater.

8.8. COMPARISON OF MEAN RIVER WATER AND SEAWATER

Shown in Table 8.9 are chemical analyses of mean river water and seawater, along with residence times of the species and a comparison of relative concentrations in mean river water and the ocean. As in the discussion of the hydrologic cycle, the residence time of species in seawater, t_R, equals the amount of that species in the reservoir (the ocean), divided by its rate of input or output (which must be equal at steady state). The input is in streams and groundwaters discharging into the ocean. The output is through adsorption and/or precipitation in and on solids that end up in marine sediments.

Chloride and sodium, those species most inert (least reactive) in seawater, have the longest residence times and are most concentrated in seawater relative to their amounts in rivers. Iron and alu-

---

[†]Gibbs's classification scheme has proven controversial. See Berner and Berner (1996) for further discussion.

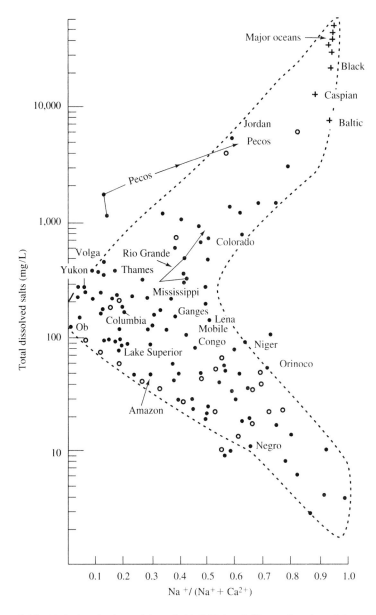

Figure 8.14 Variations in the weight ratio $Na^+/(Na^+ + Ca^{2+})$ as a function of the total dissolved solids (TDS) content of the world's surface-waters. From R. J. Gibbs. Mechanisms
controlling world water chemistry. *Science,* 170:1088–90. Copyright 1970 American Association for the Advancement of Science. Reprinted by permission.

minum are rapidly precipitated upon the mixing of rivers and ocean water because of the high pH of
the ocean (about 8.1) and its high salinity (TDS = 34,500 mg/L). Low silica values in ocean water
chiefly reflect silica removal from the water through biological activity.

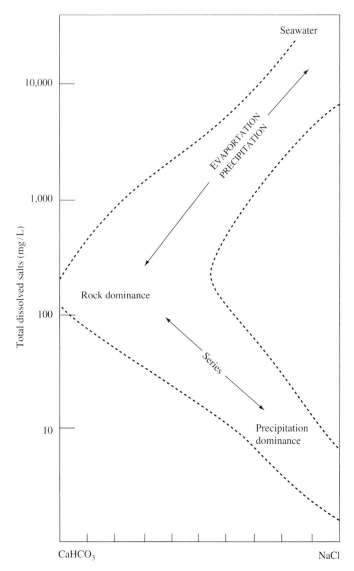

Figure 8.15   Schematic description of processes controlling the chemistry of world surface-waters. From R. J. Gibbs. Mechanisms controlling world water chemistry. *Science,* 170:1088–90. Used by permission.

8.9 WATER QUALITY VERSUS QUANTITY

8.9.1 Mixing of Waters

As emphasized early in this chapter, water quality and quantity are inherently interdependent. Mixing among surface-waters, groundwaters and wastewaters is a common occurrence, often with important consequences to water quality. Geochemical programs such as PHREEQC (Parkhurst 1995) can predict the consequences of mixing two waters of known composition in assumed proportions, including changes in the mixture as it attains chemical equilibrium. Often we have a different but related mixing problem, which is to reconstruct the amounts of two or more waters that could have created another water by mixing (cf. Hem 1985). We may have chemical analyses of the several wa-

TABLE 8.9   Chemical analyses of mean river water and seawater, along with residence times of the species and a comparison of relative concentrations in mean river water and the ocean

Species	Mean river water[†] (mg/L)	Seawater[‡] (mg/L)	Residence time in seawater (my)	Concentration in seawater relative to river water
Ca	15	410	1.2	27 times
Mg	4.1	1,350	15	330 times
Na	6.3	10,500	190	1670 times
K	2.3	390	8	170 times
HCO_3	60	142	—	2.4 times
Cl	7.8	19,000	300	2,400 times
SO_4	11	2,700	22	245 times
SiO_2	13.1	6.4	0.016	0.49 times
Fe	0.67	0.003	(0.003?)	0.004 times
Al	0.07	0.001	(0.003?)	0.014 times
TDS	120	34,500	—	288 times

Source: [†]Livingstone (1963). [‡]Hem (1985).

ters, but otherwise have no way to estimate their volumes. The mixing of two waters can be described by the mass-balance equation

$$C_m V_m = C_1 V_1 + C_2 V_2 \tag{8.30}$$

where subscripts m, 1, and 2 denote concentrations (C) and volumes (V) of the mixture and of mixed solutions 1 and 2, respectively. Given that $V_m = V_1 + V_2$, V_m may be eliminated from Eq. (8.30) to give

$$R_V = \frac{V_1}{V_2} = \frac{(C_2 - C_m)}{(C_m - C_1)} \tag{8.31}$$

R_V and the volume ratios involved in the mixing can be determined from the chemical analysis of a single species in all three waters. The species we choose must be conserved in the mixing process. In other words it must not be removed from or added from another source during mixing. The obvious choices are chloride or bromide ions, which are generally inert and unreactive, although other species known to be unreactive during a specific mixing event can also be used. Because each conserved species should give the same answer, it is important to use two or more in each mixing calculation and compare the results.

Example 8.2

A phenol manufacturing company injected its chemical wastewaters into formation brines in a sandstone (A) at a depth of about 5600 ft. In 1989, after about 20 years, a monitoring well in a sandstone (B), about 1400 ft above the injection horizon, was found contaminated by the injected wastes. Given the average composition of the injectate, and of the original (1968) formation brines, we are asked to estimate the proportions of injectate, and the 1968 groundwaters that could have been mixed to create the contaminated 1989 waters. Concentration data are given in Table 8.10.

Geochemical computer modeling and the abundance of limestone indicates saturation of the groundwater with respect to anhydrite and calcite. Calcium, sulfate alkalinity, and TDS

TABLE 8.10 Composition of brine in sandstones A and B in 1968 before deep-well injection, and in 1989 after both have been affected by the injection

Species	Sandstone A 1968	Sandstone A 1989	Sandstone B 1968	Sandstone B 1989	Average injectate
Na	58,300	14,000	54,000	$45,600 \pm 1340$	$8,770 \pm 2590$
Mg	7,080	424	7,610	$5,490 \pm 830$	10 (est.)
Ca	50,600	2,850	39,800	38,500	25 ± 11
Cl	200,000	21,200	176,000	$165,000 \pm 7,100$	$1,630 \pm 930$
Br	2,160	200	1,950	1,630	5 (est.)
SO$_4$	140	2,490	74	355 ± 35	$8,790 \pm 5,100$
TDS	316,000	48,400	278,000	$279,000 \pm 12,000$	$21,500 \pm 5,900$
Acetone	—	800	—	141 ± 98	818 ± 641
Acetophenone	—	35	—	6.3 ± 0.4	344 ± 337
Analine	—	3.1	—	—	123 ± 126
Formic acid	—	1335	—	175 ± 8	$1,430 \pm 1,190$
Phenol	—	940	—	460 ± 85	$5,100 \pm 2,100$

Note: Also given is average injectate composition with most samples obtained after 1981. Where duplicate samples were available, uncertainties equal one SD from the mean. Concentrations are in mg/L.

therefore cannot be assumed conserved in mixing. Among the remaining species, the least reactive are Na, Cl, and Br, which are thus emphasized in the mixing calculations. The results summarized in Table 8.11, suggest that 1989 sandstone A waters were created by mixing about 9.2 parts of injectate with 1 part of initial (1968) groundwater. The acetophenone, analine, and phenol have low R_V values consistent with their substantial removal from the groundwater system. Acetone and formic acid may not have been attenuated by the mixing (see Table 8.11 footnotes).

Sodium and bromide data suggest that 1989 sandstone B water might be a mixture of 0.22 parts injectate and 1 part uncontaminated sandstone B water. However, inconsistency of the important chloride data with this inference supports a more complex origin for the 1989 water. As a closing caution, the large uncertainties in the average injectate composition in Table 8.10, argue against overconfidence in the conclusions drawn from our mixing calculations.

8.9.2 Stream Flow Versus Quality

Changes in the flow rate or discharge of a stream or groundwater are usually accompanied by changes in its water quality. Groundwater inflow or baseflow may dominate the chemical quality of a stream during extended dry periods, whereas overland flow or interflow (see Fig. 8.2) can have an important effect on stream quality following storm events. Because groundwater is generally more concentrated in solutes than overland flow or interflow, increases in stream flow are usually accompanied by decreases in solute concentrations. A positive correlation between stream discharge and concentration may be found, however, when the substances involved are chiefly adsorbed or otherwise associated with stream sediment load, which generally increases with discharge. Such a positive correlation has been observed for phosphorus [Fig. 8.16(a)], and can be expected for most heavy metals in alkaline pH streams where they are insoluble, but may be strongly adsorbed on suspended sediments.

TABLE 8.11   Volume ratios (R_V) asssuming mixing of injectate with uncontaminated 1968 waters from sandstones A and B to produce the contaminated 1989 waters and predicted concentrations of the contaminated 1989 waters in mg/L, assuming average R_V values

Species	A(1968) + Injectate = A(1989)			B(1968) + Injectate = B(1989)		
	R_V	1989 concentration	1989 concentration if $R_V = 9.2$[†]	R_V	1989 concentration	1989 concentration if $R_V = 0.22$[‡]
Na	8.5	14,000	13,600	0.23	45,600	45,900
Mg	16	424	703	0.39	5,490	6,240
Cl	9.1	21,200	21,100	0.07	165,000	145,000
Br	10	200	217	0.20	1,630	1,600
TDS	9.9	48,400	50,300	undefined[§]	279,000	232,000
Acetone	undefined[‖]	800	738	0.32	141	104
Aceto-phenone	0.11	35	310	0.02	6.3	62
Analine	0.03	3.1	111			
Formic acid	undefined[#]	1,335	1,288	0.21	175	258
Phenol	0.23	940	4,600	0.10	460	919

[†]$R_V = 9.2$ is the average of tabulated values for Na, Cl, and Br.

[‡]$R_V = 0.22$ is the average of tabulated values for Na and Br.

[§]TDS values lead to an impossible negative R_V for sandstone B mixing.

[‖]$R_V = 44$, an impossible value for sandstone A mixing. If average injectate acetone is instead 887 mg/L (well within the uncertainty), then $R_V = 9.2$.

[#]$R_V = 14$, a very unlikely value for sandstone A mixing. If average injectate formic acid is instead, 1480 mg/L (well within the uncertainty), then $R_V = 9.2$.

If overland flow and interflow from storms only dilute the concentrations of solutes in streams, then such concentrations can be simply determined as a function of discharge (cf. Hem 1985). First, rewrite Eq. (8.30) in terms of discharge, Q (measured in ft^3/s (cfs); or m^3/s; 1 cfs = 2.83×10^{-2} m^3/s; 1 m^3/s = 10^3 L^3/s).

$$C_m Q_m = C_1 Q_1 + C_2 Q_2 \tag{8.32}$$

Now assume the lowest flow in a stream is baseflow of composition C_1 and discharge Q_1. If Q_2 represents the flow due to overland flow and interflow, and $C_2 = 0$, and since $Q_m = Q_1 + Q_2$, we obtain

$$C_m = \frac{C_1 Q_1}{Q_1 + Q_2} = \frac{C_1 Q_1}{Q_m} \tag{8.33}$$

the equation of a hyperbola. If we take the logarithm this expression becomes

$$\log C_m = \log C_1 Q_1 - \log Q_m \tag{8.34}$$

Assuming constant baseflow discharge and composition, this is the equation of a straight line with a slope of -1 and an intercept of $\log C_1 Q_1$.

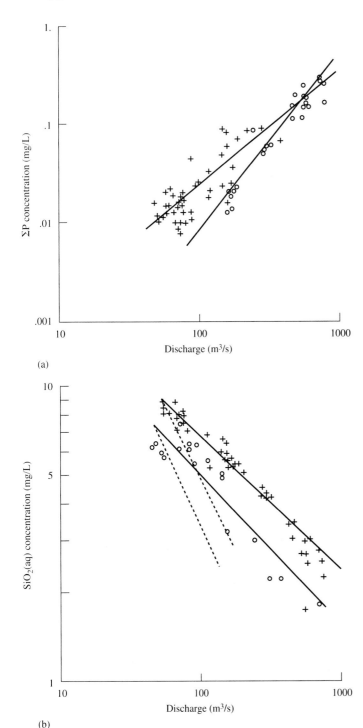

(a)

(b)

Figure 8.16   (a) Relationship between discharge and phosphorus content (chiefly on suspended sediment) Squamish River, British Columbia. The positive correlation with ΣP is consistent with the dependence of the stream's suspended sediment load on its discharge. **(b)** Relationship between discharge and dissolved silica concentration. Theoretical dilution curves (the dashed lines) have been added to the original figure. Regression lines in both plots are fit to data for the period December through June (circles) and July through November (crosses). From P. Kleiber and W. E. Erlebach. Limitations of single water samples in representing mean water quality. III. Effect of variability in concentration measurements on estimates of nutrient loadings in the Squamish River. B. C. Tech. Bull. 103. Copyright 1977 by Fisheries & Enviro. Canada. Reprinted by permission of the Minister of Supply and Services Canada, 1996.

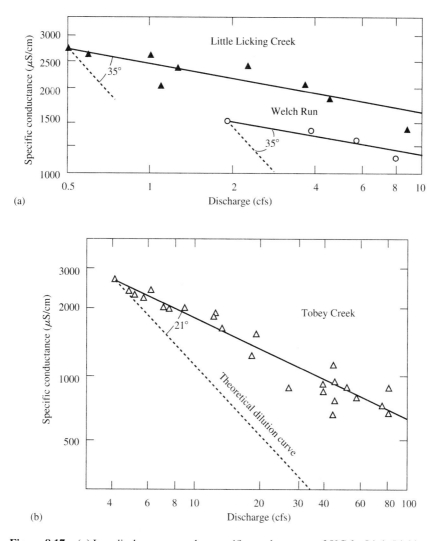

Figure 8.17   **(a)** Log discharge versus log specific conductance at 25°C for Little Licking Creek and Welch Run. **(b)** Log discharge versus log specific conductance at 25°C for Tobey Creek. Dashed lines show theoretical dilution of base-flow concentrations. From M.A. Gang and D. Langmuir. Controls on heavy metals in surface and ground waters affected by coal mine drainage. *Proc. 5th Symp. Coal Mine Drainage Research.* Copyright 1974 by Natl. Coal Assoc. Used by permission.

Stream-discharge quality trends are usually displayed in log-log plots (cf. Gang and Langmuir 1974; Lerman 1979; Levinson 1980; Hem 1985). Such plots are often linear or nearly so. Figure 8.16(b) shows dissolved silica versus discharge for a river in British Columbia (Kleiber and Erlebach 1977). Theoretical dilution curves based on Eq. (8.33) have been added to the figure. The difference between a dilution curve and data regression line is a measure of the amount of additional dissolved silica contributed to the stream by increased flow.

Shown in Fig. 8.17 are discharge versus specific conductance plots for three acid streams (pH 2.8 to 6.6) in an area of bitumnous coal strip-mining in northwestern Pennsylvania (Gang and

Langmuir 1974). The small (21°) angle between the data regression line and dilution curve for Tobey Creek is consistent with the fact that only 14% of the Tobey Creek basin has been strip-mined. Little Licking Creek and Welch Run, both with a 35° separation angle, have been 26% and 47% strip-mined. Soluble sulfate salts accumulated in the waste rock in strip-mined areas are dissolved and flushed into the streams during storms, contributing to their dissolved load. The effect is roughly proportional to the amount of strip-mining in each watershed. Some of the increased load with discharge must also come from increased groundwater flow to the streams after storms.

Heavy metals behave similarly in the three streams as shown in Fig. 8.18. Detailed interpretation is complicated by differences in basin geology and in the behavior of iron, manganese, and zinc. For example, the relatively low angles for iron in Tobey Creek and Little Licking Creek reflect not only dilution, but also Fe precipitation at the higher pH values found with higher flows (Gang and Langmuir 1974).

When strongly correlated and reproducible, concentration versus discharge plots can be used to predict stream flow from stream chemistry and conversely. The plots also suggest that the heavy metal load of a given stream may be predictable from its specific conductance or discharge. When identified, such relationships can be useful for the monitoring and management of surface-water quality.

8.10 THE IMPORTANCE OF DEFINING BACKGROUND WATER QUALITY

For reasons unrelated to human activity, many surface- and groundwaters contain natural (background) concentrations of one or more substances that exceed the U.S. Environmental Protection Agency (EPA) drinking water standards (see Table 8.12). Changes in land use, including urbanization, mining, and the disposal of liquid and solid wastes, will usually increase the number and severity of such violations. When a regulatory decision is made to restore affected waters to their pristine state, it is obviously unrealistic to assign clean-up goals that are below natural background levels. Runnells et al. (1992) point out, for example, that many streams, springs, and deeper groundwaters in mineralized areas unaffected by mining contain highly elevated background metal concentrations. We must, therefore, answer the question, what is the natural background quality?

How can we determine background? The preferred approach is to take samples from the stream or geologic formation of interest somewhere nearby, but where the water quality has presumably not been altered by human activity (cf. Banks et al. 1995; Lahermo et al. 1995). Unfortunately, in populated areas it is often difficult to locate a portion of a stream or groundwater aquifer that remains pristine. A second approach, then, is to assume that background water quality is the same as measured in similar streams and groundwater systems (in similar geological formations) in the same general area. Lacking this information, White et al. (1963) have usefully summarized typical compositions of groundwaters from different rock types. As perhaps a better approach to identifying background water quality, Klusman (1996) suggests the construction of cumulative probability plots of all the data for an area of interest (cf. Haan 1977; Levinson et al. 1987). Such plots may allow the classification of samples into uncontaminated and contaminated groups, with an estimate of the median and standard deviation for each.

The identification of background concentrations in surface and groundwaters is also of interest to those involved in mineral exploration (cf. Goleva 1968; Rose et al. 1979). The discovery of anomalously high concentrations of metals and related species in groundwater has been assumed a good prospecting tool for nearby ore deposits (cf. Langmuir and Chatham 1980; Wanty et al. 1987).

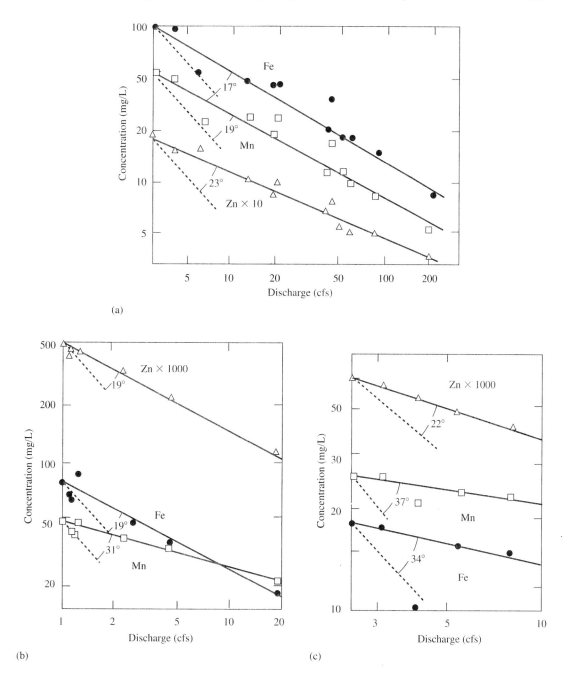

Figure 8.18   Dissolved Fe, Mn, and Zn versus discharge for **(a)** Tobey Creek, **(b)** Little Licking Creek, and **(c)** Welch Run. Dashed lines show theoretical dilution of base-flow concentrations. From M.A. Gang and D. Langmuir. Controls on heavy metals in surface and ground waters affected by coal mine drainage. *Proc. 5th Symp. Coal Mine Drainage Research.* Copyright 1974 by Natl. Coal Assoc. Used by permission.

TABLE 8.12 National interim (1993) drinking water standards for inorganic species in community water supplies as set by the U.S. Environmental Protection Agency (EPA)

Contaminant	MCL (mg/L)	Source	Contaminant	SMCL (mg/L)	Source
Antimony	0.006	1	Chloride	250	2
Arsenic	0.05	2	Color	15 units	2
Barium	1.0	2	Copper	1.0	2
Beryllium	0.004	1	Corrosivity	noncorrosive	2
Cadmium	0.010	2	Foaming agents (MBAS)[†]	0.5	2
Chromium	0.05	2	Hydrogen sulfide	not detectable	2
Cyanide	0.2	1	Iron	0.3	2
Fluoride	2.2	2	Manganese	0.05	2
Lead	0.005	2	Odor[‡]	3	2
Mercury	0.002	2	Sulfate[§]	250	2
Nickel	0.1	1	TDS residue[‖]	500	1
Nitrate (as N)	10	2	Zinc	5	2
Selenium	0.01	2			
Silver	0.05	2			
Thallium	0.002	1			

Note: Maximum contaminant levels (MCLs) are enforceable standards for substances that may constitute a health hazard at higher concentrations. Secondary maximum contaminant levels (SCMCLs) are set for aesthetic reasons, to avoid tastes, odors, and staining of plumbing fixtures.

[†]MBAS are methylene-blue active substances, a measure of detergents.

[‡]Units for odor are threshold numbers.

[§]The SMCL of 250 mg/L sulfate has been removed in the 1993 EPA interim standards (Hanson 1992).

[‖]Total dissolved solids residue is the weight of solids obtained on evaporation of a liter of the sample.

Source: (1) Hanson (1992); (2) Fetter (1988).

In order to define background and anomalous groundwater concentrations, Barnes and Langmuir (1978) examined 27,000 chemical analyses of groundwaters from the United States, available in the U.S. Geological Survey WATSTORE data base. The analyses of individual species in five rock types were plotted in histograms. Examples of two extreme types of frequency distributions are shown in Fig. 8.19. The distribution of Cu in shale groundwaters is almost lognormal in Fig. 8.19(a), whereas that for Cd in sandstone groundwaters in Fig. 8.19(b) is bimodal. The peaks at lowest concentrations in each figure lump all concentrations below the detection limit. If the "real" aqueous concentrations for these samples were known, a smooth tail for the low concentration samples would probably result.

    If one can assume that a species' concentrations have a lognormal distribution, as in Fig. 8.19(a), the distribution would appear as in Fig. 8.19(c). The plot shows that for such a lognormal distribution, the value of the mean plus two standard deviations equals the 97.7 percentile. Barnes and Langmuir (1978) concluded that the 2.3% of samples with greater concentrations included most significant anomalies and only a small proportion of the background population. Such an analysis is, however, not simply applied to the distributions shown in Figs. 8.19(a) and 8.19(b). First, it is not possible to calculate an accurate mean and standard deviation when a large number of analyses are below detection. Second, the Cd distribution is bimodal, not lognormal. Regardless, as a first approximation the observed 97.7 percentile level was chosen to separate background and anomalous concentrations of the 32 elements, bicarbonate, sulfate, and specific conductance in groundwater. The results are listed in Table 8.13. Background concentrations will generally be less than these threshold values. Exceedence of these values can be viewed as evidence of nearby mineralization or groundwater contamination. The same frequency distribution data for groundwater

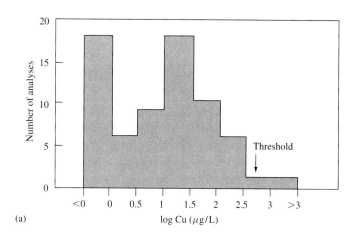

(a)

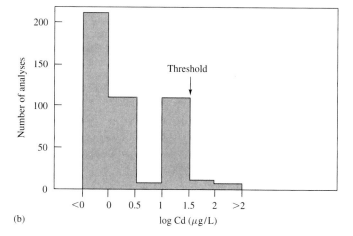

(b)

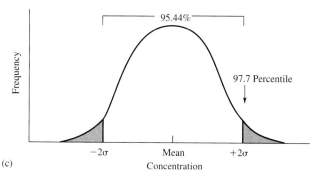

(c)

Figure 8.19 Frequency distributions: **(a)** of copper in shale groundwaters; and **(b)** of cadmium in sandstone groundwaters. Peaks at lowest concentrations are the total of samples below detection. The threshold concentration is two standard deviations above the mean. **(c)** Frequency distribution plot for a normally distributed sample population. 97.7% of samples are below the threshold concentration at two standard deviations above the mean.

species was also used to construct Fig. 8.13. In arid climates, concentration of the salts in groundwater recharge by evapotranspiration may make it difficult to apply this approach.

STUDY QUESTIONS

1. Discuss differences and reasons for those differences in the general chemical composition of precipitation, soil moisture, surface-waters and shallow (water-table) and deep (artesian) groundwaters.

TABLE 8.13 Generalized threshold values as log molality[†] of aqueous components in groundwaters associated with specific rock groups

	Carbonate		Sandstone[‡]		Shale[§]		Igneous-I[‖]		Igneous-II[#]		Cumulative[††]	
HCO_3	−1.99	(588)	−1.71	(1,009)	−1.68	(115)	−2.54	(30)	−2.13	(78)	−1.75	(1,820)
N	−2.74	(194)	−2.17	(95)	—		—		—		−2.62	(290)
P	−5.23	(43)	−4.29	(251)	−5.06	(60)	−4.98	(29)	−5.67	(21)	−4.97	(404)
Ca	−0.59	(592)	−2.05	(1,016)	−1.83	(116)	−1.99	(30)	−2.40	(78)	−1.54	(1,832)
Mg	−1.19	(592)	−2.34	(1,008)	−1.98	(116)	−2.14	(30)	−2.39	(78)	−1.53	(1,824)
Na	−0.35	(493)	−1.42	(933)	−1.20	(114)	−1.63	(30)	−2.28	(78)	−1.22	(1,648)
K	−1.92	(489)	−3.19	(915)	−3.18	(114)	−3.18	(30)	−3.40	(78)	−2.88	(1,626)
Cl	−0.52	(730)	−1.62	(1,241)	−0.88	(127)	−1.37	(30)	−2.22	(78)	−0.55	(2,206)
SO_4	−1.64	(600)	−1.89	(990)	−1.64	(121)	−2.17	(30)	−2.49	(78)	−1.82	(1,819)
As	−6.58	(351)	−6.58	(549)	−6.87	(92)	−7.59	(29)	−6.00	(40)	−6.57	(1,061)
Ba	−5.11	(192)	−5.13	(300)	−5.11	(55)	−6.13	(26)	−5.87	(13)	−5.12	(586)
Bi	−7.56	(6)	−7.28	(53)	−6.87	(7)	—		—		−7.24	(66)
Cd	−7.03	(322)	−6.58	(445)	−7.04	(61)	−8.06	(27)	−7.76	(13)	−6.76	(868)
Cr	−6.44	(274)	−6.70	(301)	−6.70	(46)	−6.70	(26)	−6.21	(35)	−6.22	(682)
Co	−7.08	(171)	−6.75	(252)	−6.83	(50)	−7.78	(27)	−7.46	(13)	−6.86	(513)
Cu	−5.66	(384)	−5.85	(604)	−5.17	(69)	−5.01	(27)	−5.25	(43)	−5.64	(1,127)
Fe	−3.73	(438)	−3.73	(813)	−3.91	(114)	−4.11	(30)	−4.46	(23)	−3.73	(1,418)
Pb	−6.72	(395)	−6.96	(624)	−6.66	(72)	−6.62	(27)	−6.74	(43)	−6.76	(1,161)
Mn	−4.52	(409)	−5.09	(699)	−4.80	(97)	−5.13	(29)	−5.01	(24)	−4.95	(1,258)
Mo	−6.31	(72)	−6.47	(174)	−6.51	(40)	—		—		−6.47	(288)
Ni	−6.78	(183)	−6.54	(222)	−6.34	(53)	−5.85	(27)	−6.06	(29)	−6.18	(514)
Ag	−7.02	(136)	−7.02	(308)	−7.05	(24)	−9.99	(26)	−9.99	(13)	−7.02	(507)
Sr	−3.43	(297)	−4.02	(379)	−4.00	(23)	—		−5.11	(23)	−3.67	(723)
V	−5.36	(91)	−6.14	(206)	−6.67	(36)	—		—		−5.93	(335)
Zn	−4.65	(397)	−4.11	(621)	−4.31	(72)	−4.63	(27)	−4.10	(43)	−4.29	(1,160)
Sb	−8.11	(4)	−9.10	(3)	—		—		—		−8.11	(7)
Sn	−6.34	(6)	−6.05	(55)	−6.63	(7)	—		—		−6.05	(68)
Se	−6.87	(93)	−6.77	(410)	−6.61	(87)	−7.29	(27)	−7.43	(13)	−6.82	(630)
Ti	−4.68	(6)	−6.39	(56)	−6.66	(7)	—		—		−6.11	(69)
U	−7.76	(3)	−7.89	(7)	—		—		—		−7.76	(10)
F	−3.78	(520)	−3.66	(935)	−3.63	(111)	−4.53	(30)	−2.86	(77)	−3.58	(1,673)
I	−3.19	(33)	−4.00	(33)	—		—		—		−3.39	(66)
Br	−2.87	(18)	−3.39	(35)	−4.00	(16)	—		−5.92	(3)	−2.88	(72)
Hg	−8.30	(253)	−8.30	(229)	−8.34	(53)	−8.59	(26)	−8.59	(41)	−8.30	(602)
SpC[‡‡]	28,561	(710)	4,010	(1,220)	9,541	(115)	5,429	(30)	1,931	(75)	8,209	(2,150)

Note: Numbers in parentheses represent number of analyses.

[†]log $m = −5$ and −8 are roughly a ppm and ppb, respectively.

[‡]Includes quartzites, arkoses, greywackes, and conglomerates.

[§]Includes shales, clays, siltstones, and slates.

[‖]Includes felsic to intermediate igneous and meta-igneous rocks.

[#]Includes mafic and ultramafic igneous and meta-igneous rocks.

[††]Cumulative for all rock types.

[‡‡]Specific conductance in μS/cm, 25°C.

Source: Barnes and Langmuir (1978).

2. The long-term persistence of pollutants in water and the difficulty of cleaning them up are related both to the residence time of waters in the hydrosphere and to the residence time of substances dissolved in those waters. Explain this statement.

3. Define the terms in the equation: $I = P \pm R - ET - MR$, and explain its applicability to groundwater pollution problems created by the disposal of liquid or solid wastes at the land surface.

4. The composition of soil moisture is influenced by processes and inputs listed below. Explain how each of these can contribute to soil-water chemistry:
   (a) the chemistry of precipitation;
   (b) leaching of salts accumulated by evapotranspiration (ET);
   (c) organic activity in the soil; and
   (d) the weathering of soil materials.

5. Chloride (or bromide) is often used to determine the extent to which precipitation and soil-moisture solute concentrations have been influenced by processes other than ET. Explain.

6. Contrast the chemistry of water-table and artesian groundwaters in terms of:
   (a) the applicability of equilibrium versus kinetic concepts;
   (b) rock versus water domination of the chemistry;
   (c) thermodynamically open versus closed systems;
   (d) the redox condition and salinity of the water;
   (e) the variability of water composition and temperature with time.

7. List the major species dissolved in natural waters, including the dissolved gases, in their approximate order of decreasing concentrations.

8. Toxic metals can occur at unusually high concentrations in natural waters from certain sources, and under certain conditions. Explain with examples.

9. Write the overall reactions that define the pH of background rain and acid rain. What causes acid rain? Discuss the ability of lakes, soils, and different geological materials to neutralize it.

10. Natural waters can be described as having chemical compositions that reflect domination by rocks, precipitation, or evaporation. Explain with examples.

11. The composition of seawater relative to that of rivers is related to the residence times and reactivities of the constituents in seawater. Explain with reference to Na^+ and Fe(III).

12. Given a surface- or groundwater mixture, be able to compute relative volumes or mixing proportions of the several waters involved when their chemical compositions are known.

13. Stream composition is a function of stream discharge. What is a theoretical dilution curve? How is stream chemistry related to discharge:
   (a) for soluble species, and
   (b) for species associated with suspended stream sediments?

14. Why is it important to establish local background concentrations before initiating a water pollution remediation program? How might one determine the background concentration of cadmium in a groundwater?

PROBLEMS

1. The CO_2 content of the atmosphere has increased from 270 ppmv (ppm by volume) in 1800 to 350 ppmv in 1991, due to burning of fossil fuels. The following question is intended to relate this CO_2 increase to its possible effect on the carbonate content of the oceans and on calcite solubility in the oceans. In this problem, assume that all activity coefficients equal unity, concentrations are in mol/kg, and $T = 25°C$. Remember, the definition of total carbonate (C_T) in water is

$$C_T \text{ (mol/kg)} = H_2CO_3^\circ + HCO_3^- + CO_3^{2-}$$

In seawater having a salinity of 35 g/kg or ppt (TDS = 35,000 ppm), the following concentration constants apply (Stumm and Morgan 1981).

$$K_{CO_2} = 10^{-1.55} = \frac{(H_2CO_3^\circ)}{P_{CO_2} \text{ (bars)}}$$

$$K_1 = 10^{-6.00} = \frac{(H^+)(HCO_3^-)}{(H_2CO_3^\circ)}$$

$$K_2 = 10^{-9.11} = \frac{(H^+)(CO_3^{2-})}{(HCO_3^-)}$$

(a) Compute C_T (molal) in the oceans in 1880 and in 1991, assuming the ocean pH has been constant at 8.1. (A constant pH may or may not be a good assumption.)

(b) Given the calcite dissolution reaction,

$$CaCO_3(\text{calcite}) + CO_2(g) + H_2O \rightleftarrows Ca^{2+} + 2HCO_3^-$$

and assuming that the solubility of calcite is proportional to the Ca^{2+} concentration (see Chap. 6), compute the *relative or fractional* increase in the equilibrium solubility of calcite in the oceans for the conditions just described, between 1800 and 1991.

2. You sample shallow infiltrating waters from suction lysimeters installed in holes drilled by a truck-mounted auger in a humid climate soil at depths of 1 m and 3 m. The following mg/L concentrations are obtained at the two depths.

Species	C_1 (1 m)	C_2 (3 m)
Ca^{2+}	3.3	108
Na^+	6.9	21
Cl-	4.8	55
HCO_3^-	1.9	144

(a) Compare the concentrations at 1- and 3-m depths. How much of the mg/L change in these species' concentrations from a 1- to 3-m depth is due to evapotranspiration (ET) alone? Hint: You must assume one species is an inert tracer.

(b) What is the mg/L change in these species' concentrations over this depth, which results from processes other than ET?

(c) There are fragments of dolomite $[CaMg(CO_3)_2]$ rock in the soil. How many mg/L of the HCO_3^- at a 3-m depth could have come from weathering of the dolomite?

(d) Write and balance the reaction for weathering of the dolomite by carbonic acid.

(e) In light of parts (c) and (d), has Ca^{2+} increased or decreased relative to its amounts produced by ET concentration and by weathering of dolomite? Can you suggest what has happened to Ca^{2+} and calculate how much Ca^{2+} has been involved?

(f) Na^+ has been involved in a process with depth in the soil. How many mg/L of Na^+ have been so involved, and what might the process be?

3. This question relates to the chemical analysis of mean river water from Livingstone (1963), which is given in Table 8.9.

(a) Compute the charge balance of that water in meq/L. (Be careful of which species are truly dissolved and ionized, and which are not.)

(b) If the charge balance is off, add sufficient Na^+ or Cl^+ (as needed) to make it balance and compute the molal ionic strength.

(c) Using the Debye-Hückel equation, and assuming 25°C and atmospheric CO_2 pressure ($10^{-3.5}$ bar), calculate the pH of the "average" river.

(d) Assuming K_{sp} (calcite) $= 10^{-8.42}$, ignoring complexes, but using activity coefficients, calculate the saturation index of calcite for the "average" river Ca^{2+} and HCO_3^- concentrations. At what pH would the river be at saturation with respect to calcite?

4. An important consequence of acid rain and dry deposition of SO_2 and HNO_3 is the rapid weathering and destruction of exposed carbonate stone buildings and monuments (cf. Baedecker et al. 1990). Lipfert (1989) has proposed a theoretical damage function equation to describe the weathering of carbonate stone between pH 3 and 5, which is

$$\text{Stone loss } (\mu m/m/y) = [18.8 + 0.016\,(H^+) + 0.18(V_{dS}\,SO_2 + V_{dN}\,HNO_3)]/R$$

In this equation stone loss is in μm of stone per meter of rain per year, (H^+) is in $\mu mol/L$, V_{dS} and V_{dN} are impact velocities of SO_2 and HNO_3 on the stone surface in cm/s. Atmospheric concentrations of SO_2 and HNO_3 are in $\mu g/m^3$, and R is meters of rain per year.
   (a) Explain the significance of each term in the equation.
   (b) What is the stone loss for 1 year in which we assume: 1 m of rain, rain pH $= 4.2$, $V_{dS} = 0.3$ cm/s, and $SO_2 = 40\ \mu g/m^3$? Ignore the HNO_3 term.
   (c) For this example, which is typical of a location in the humid eastern U.S., what percentage of stone loss is due to acid rain and dry deposition of SO_2?

5. When oil wells pump petroleum from deep formations, saline groundwaters associated with the oil may contaminate shallow fresh groundwaters or discharge into nearby streams. For damage assessment purposes, we may wish to determine the contribution of a brine-contaminated tributary to the chemistry and flow of a major stream. However, the tributary's flow may be too low to measure by stream gaging. In this problem we will calculate the flow of a small tributary from its chemistry using mixing equations and the measured flow of a larger stream.

   The specific conductance (SpC) of Fox tributary is 4510 μS/cm. The flow of Caddo Creek measured by current meter is 6.63 cfs (ft³/s). Above its confluence with Fox tributary the specific conductance of Caddo Creek is 2010 μS/cm, whereas 500 ft below this confluence at a distance sufficient to assure complete mixing of both waters, the SpC of Caddo Creek is 2130 μS/cm. Compute the flow of Fox tributary.

6. Plot all the precipitation, soil-, surface-, and groundwater analyses tabulated in this chapter on a plot of log TDS (mg/L) versus the $Na^+/(Na^+ + Ca^{2+})$ weight ratio as was done by Gibbs (1970) in Fig. 8.14. Explain and discuss the appearance of your plot.

9

The Geochemistry
of Clay Minerals

9.1 INTRODUCTION

Clay minerals are practically ubiquitous in modern soils and sediments. They are also the primary component phases in shales, which constitute the surface or nearest-surface bedrock on most of the continents. The chemistry of natural waters is strongly affected by the presence of clays. This, in part, results from the small size of clay mineral particles, which are often less than one micrometer ($<10^{-4}$ cm) in diameter, placing them in the colloidal size range. Because of their small particle size, clays have large surface area to mass ratios. Their reactions with natural waters are, therefore, relatively rapid, although they may not attain true thermodynamic equilibrium in low-temperature environments. Small particle sizes and a unique crystal chemistry also give the clays a high surface-charge density to mass ratio. Important reactions involving clays include: (1) a tendency of clay minerals to react and attempt to equilibrate in water/sediment systems; (2), adsorption by clays, especially of cations, because of clays' relatively high negative (usually) surface charge; and (3) colloidal behavior and colloidal transport of clays, particularly by streams and shallow, turbulent groundwaters.

A general understanding of the crystal chemistry and atomic structure of the important clay minerals helps to explain their occurrence and behavior in environmental systems. In this chapter we will first consider the crystal chemistry of the common clays and of the micas, which are structurally related to the clays. Next we will discuss the natural occurrence of the clays and their interaction with natural waters in near-surface environments. This will include an evaluation of the role of clays in controlling water chemistry, whether by a tendency toward mineral equilibration or through ion exchange. A more in-depth discussion of clay and mica geochemistry than that offered here is given by Drever (1988). The ion exchange behavior of clays is discussed in some detail in Chap. 10.

9.2 CRYSTAL CHEMISTRY OF IMPORTANT CLAY MINERALS

The clays and micas are compositionally complex hydrous aluminum silicates. Structurally they are called *phyllosilicates*, from the Greek word *phyllon*, or leaf. The phyllosilicates are continuous sheet

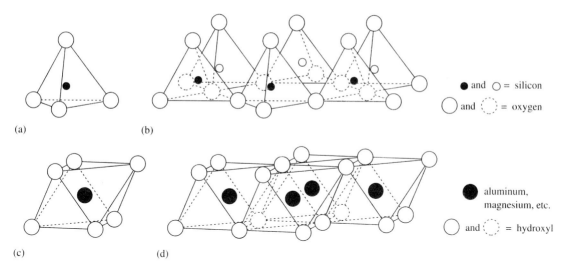

Figure 9.1   Top diagram is of a single silica tetrahedron **(a)** and sheet structure of silica tetrahedra arranged in a hexagonal network **(b)**. Lower diagram shows a single octahedral unit **(c)** and a sheet structure of octahedral units **(d)**. After *Clay mineralogy* by R. E. Grim. Copyright © 1968 by McGraw-Hill, Inc. Used with permission of The McGraw-Hill Book Companies.

structures, comprising a stacking of layers of cations coordinated with O^{2-} and/or OH^- groups. The layers may be separated by adsorbed or relatively fixed cations, or by water molecules. The layers that form clays are described as *octahedral* and *tetrahedral* (Fig. 9.1). The octahedral layers are called *dioctahedral* if they contain two Al^{3+} in sixfold coordination with O^{2-} or OH^-, in which case one of three cation sites remain vacant. Such a layer is also termed a *gibbsite layer*, for it has the composition of gibbsite, $Al(OH)_3$.

If the octahedral layers are made up of three Mg^{2+} coordinated with six O^{2-} or OH^-, then all the cation sites are occupied by Mg^{2+}. The layers are then termed *trioctahedral*, or *brucite layers*, because they are compositionally equivalent to the mineral brucite, $Mg(OH)_2$.

Tetrahedral layers are composed of cations fourfold coordinated with O^{2-} or OH^-. The dominant cation in such layers is usually Si^{4+}, but may also be Al^{3+}. The atomic arrangements of octahedral and tetrahedral layers are shown schematically in Fig. 9.1.

9.2.1 The Kaolinite Group and Other Two-Layer Phyllosilicates

The important clay mineral, kaolinite, is constructed of a gibbsite octahedral layer and a Si^{4+} tetrahedral layer, with O^{2-} at the apexes of the silica tetrahedra replacing OH^- groups of the octahedral layer, and shared between the two layers [Fig. 9.2(a)]. The two-layered unit in kaolinite is about 7 Å thick.[†] Kaolinite is a relatively pure mineral, with a formula always close to stoichiometric, and has the composition

$$Al_2Si_2O_5(OH)_4$$

[†]The angstrom (Å), which equals 10^{-8} cm, is the common unit used to describe C-axis dimensions of the clay minerals (the vertical axis in Figs. 9.1 and 9.2).

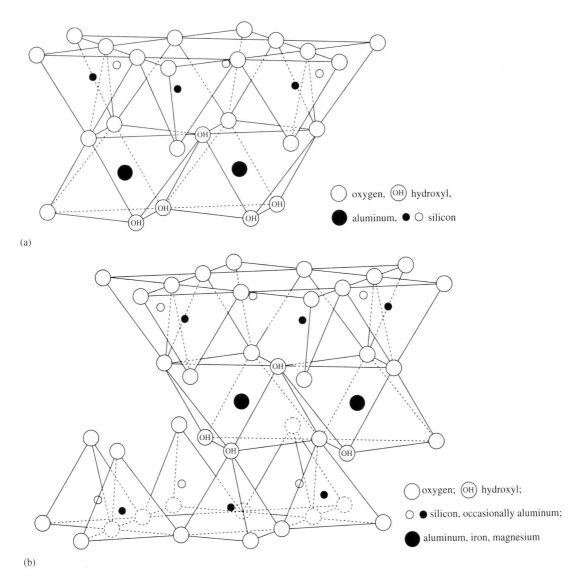

(a)

(b)

Figure 9.2 **(a)** A T:O or two-layer structure, which is composed of a tetrahedral and an octahedral layer bonded through shared oxygen and hydroxyl groups. A T:O structure is common to the kaolinite and serpentine mineral groups. **(b)** A T:O:T or three-layer structure. A T:O:T structure is found in the micas, and in the smectite, vermiculite, and chlorite group minerals. From *Clay mineralogy* by R. E. Grim. Copyright © 1968 by McGraw-Hill, Inc. Used with permission of The McGraw-Hill Book Companies.

There is no excess or deficiency of charge at the surface of kaolinite crystal plates because there is no excess or deficiency of charge within the crystal structure.

Because of their alternating tetrahedral (T) and octahedral (O) layers, the two-layer phyllosilicates are said to have T:O structures. Listed in Table 9.1 are other T:O minerals, including the kaolinite polymorphs nacrite, dickite, and halloysite, and the trioctahedral serpentine minerals lizardite, antigorite, and chrysotile, in which brucite layers alternate with layers of silica tetrahedra.

TABLE 9.1   Structures and idealized or typical compositions of some common layer silicates

---

I DIOCTAHEDRAL

(gibbsite-type layers)

Two-layer structures

Kaolinite Group: $Al_2Si_2O_5(OH)_4$
  Kaolinite, nacrite, dickite, & halloysite (hydrated)

Three-layer structures

Pyrophyllite: $Al_2Si_4O_{10}(OH)_2$
Smectite Group: $\{Na,Ca_{0.5}\}_{0.7}(Mg,Fe,Al)_4(Al,Si)_8O_{20}(OH)_4$[†]
  Montmorillonite: $\{Na,Ca_{0.5}\}_{0.7}(Mg_{0.7}Al_{3.3})Si_8O_{20}(OH)_4$[†]
    Wyoming type: $\{Ca_{0.5}\}(Mg_{0.5}Fe^{3+}_{0.5}Al_3)(Al_{0.5}Si_{7.5})O_{20}(OH)_4$[‡]
    Cheto WY type: $\{Ca_{0.5}\}(MgAl_3)Si_8O_{20}(OH)_4$[‡]
  Beidellite: $\{Na,Ca_{0.5}\}_{0.7}Al_4(Al_{0.7}Si_{7.3})O_{20}(OH)_4$[†]
    $\{Ca_{0.5}\}Al_4(Al,Si_7)O_{20}(OH)_4$[‡]
  Nontronite: $\{Na,Ca_{0.5}\}_{0.7}(Fe^{3+})_4(Al_{0.7}Si_{7.3})O_{20}(OH)_4$[†]
    $\{Ca_{0.5}\}(Fe^{3+}_3Al)(Al,Si_7)O_{20}(OH)_4$[‡]
Muscovite: $KAl_2(AlSi_3O_{10})(OH)_2$
Illite Group: $(K_{1.5-1.0})Al_4(Al_{1.5-1.0}Si_{6.5-7.0})O_{20}(OH)_4$[†]
  Illite (ideal average): $(K_{1.5})Al_4(Al_{1.5}Si_{6.5})O_{20}(OH)_4$[†]
  Illite (typical composition): $(K_{1.5})(Mg_{0.5}Fe^{3+}_{0.5}Al_3)(Al,Si_7)O_{20}(OH)_4$[‡]
  Ferric illite: $K(Fe^{3+}_3Al)(Al,Si_7)O_{20}(OH)_4$[‡]

II TRIOCTAHEDRAL

(brucite-type layers)

Two-layer structures

Serpentine Group: $Mg_3[Si_2O_5](OH)_4$
  Lizardite, antigorite, & chrysotile (fibrous)

Three-layer structures

Talc: $Mg_3Si_4O_{10}(OH)_2$
Vermiculite Group: $\{Mg,Ca\}_{0.6-0.9}(Mg,Fe^{3+},Al)_{6.0}(Al,Si)_8O_{20}(OH)_4$[†]
    $\{K,Mg_{0.5}\}(Mg_{4.0}Fe^{2+}_{1.5}Fe^{3+}_{0.5})(Al_{1.5}Si_{6.5})O_{20}(OH)_4$[‡]
Smectite Group: $\{Na,Ca_{0.5}\}_{0.7}(Mg,Fe,Al)_6(Al,Si)_8O_{20}(OH)_4$[†]
  Saponite: $\{Na,Ca_{0.5}\}_{0.8}Mg_6(Al_{0.8}Si_{7.2})O_{20}(OH)_4$[†]
    $\{Ca_{0.4}\}(Mg_{4.0}Fe^{2+}_{1.6}Fe^{3+}_{0.3}Al_{0.1})(Al_{1.2}Si_{6.8})O_{20}(OH)_4$[‡]
  Hectorite: $\{Na,Ca_{0.5}\}_{0.7}(Li_{0.7}Mg_{5.3})(Si)_8O_{20}(OH)_4$[†]
Phlogopite: $KMg_3(AlSi_3O_{10})(OH)_2$
Biotite: $K(Mg,Fe)_3(AlSi_3O_{10})(OH)_2$
Chlorite Group: $(Mg,Fe^{2+},Fe^{3+},Mn,Al)_{12}(Al,Si)_8O_{20}(OH)_{16}$[†]
  Sedimentary: $Mg_{5.0}Fe^{2+}_{4.5}Fe^{3+}_{0.5}Al_{2.0}(Al_{2.5}Si_{5.5})O_{20}(OH)_{16}$[‡]

Note: Cations listed first in curved brackets for the smectites and vermiculites (Na, Ca, K, and Mg) are present as exchangeable interlayer ions. All the smectites and vermiculites (and thus interlayer illite-smectites) have important amounts of interlayer water, the amount of which depends upon the clay and the nature of interlayer cations (cf. Brindley and Brown 1980). As is customary, these waters are left out of the mineral formulae.

Source: [†]Deer et al. (1992)
      [‡]Slaughter (1992)

9.2.2 Three-Layer Phyllosilicates

The three-layer phyllosilicates include talc and pyrophyllite, illite and the smectite group clays, various mixed-layer clays, vermiculite, and the micas (e.g., muscovite, phlogopite, and biotite). We will limit ourselves to a discussion of the more environmentally important of these minerals, which include the micas, the smectites and illites, interlayered (mixed-layer) smectite-illites and vermiculite.

The smectite group [see Table 9.1 and Fig. 9.3(a)] includes any clay whose interlayer repeat distance (thickness of individual T:O:T layers plus interlayer spacing) expands to 17 Å on treatment with ethylene glycol (cf. Drever 1988). This is indicative of a structure in which the number of interlayer cations is smaller than 0.65 (Greenland and Hayes 1978), and is usually about 0.2 to 0.5 per formula unit $O_{10}(OH)_2$ (Drever 1988). The interlayer cations are adsorbed, and are necessary to balance the unsatisfied net charge (usually negative) of the clay crystal lattice caused by structural substitutions or vacancies in the octahedral and/or tetrahedral layers. For example Mg^{2+} or another divalent cation may substitute for Al^{3+} in the octahedral layer, or Al^{3+} or Fe^{3+} may replace Si^{4+} in the tetrahedral layer.

Deer et al. (1992) give the following general formulas for dioctahedral and trioctahedral smectite (neglecting interlayer water):

Dioctahedral $\{Na,Ca_{0.5}\}_{0.7}(Al,Mg,Fe)_4(Si,Al)_8O_{20}(OH)_4$

Trioctahedral $\{Na,Ca_{0.5}\}_{0.7}(Mg,Fe,Al)_6(Si,Al)_8O_{20}(OH)_4$

In these formulas Na and Ca are the exchangeable ions. Smectites normally occur in extremely small crystals with diameters of less than 1 μm. This makes them too small to identify with the optical microscope. Identification is instead usually accomplished by X-ray diffraction techniques.

Water is normally present in the interlayer spaces of smectites. When Mg^{2+} or Ca^{2+} occur in the interlayer space, about two water layers separate the clay layers. This creates a basal spacing of about 14 Å, as determined by X-ray diffraction. On the other hand, when Na^+ is the interlayer cation large amounts of water can enter the interlayer space and swelling of the clay ensues. This decreases the permeability of smectite-rich soils and can make such soils susceptible to plastic flow under pressure. The swelling and plastic flow of smectites can cause the rupture of overlying floors, basements, parking lots, and highways. Smectite clays, whether pure or in mixtures with sand, have very low water permeabilities, which can be further reduced by their swelling behavior after emplacement. Mechanical compaction of such mixtures makes them even less water permeable. For these reasons such materials are commonly used to line waste-disposal ponds and impoundments to prevent waste infiltration and groundwater pollution. Bentonite clays, which are formed by the weathering of volcanic glasses, are composed of the common dioctahedral smectite-group minerals montmorillonite and/or beidellite (cf. Deer et al. 1992). Compacted bentonite-crushed rock mixtures are being considered by many countries as backfill to surround and isolate high-level nuclear wastes in a geological repository (cf. OECD 1993; Murakami and Ewing 1995).

Among the smectites, montmorillonite and beidellite form a series with the general formula

$$X^+_{x+y} Mg_yAl_{4-y}[Al_xSi_{8-x}O_{20}](OH)_4$$

in which X is the number of interlayer, adsorbed, monovalent or equivalent divalent cations. For montmorillonite $y > x$, and for beidellite $y < x$. Idealized formulas for montmorillonite and beidellite are given below.

Montmorillonite $X_{0.7}(Mg_{0.7}Al_{3.3})Si_8O_{20}(OH)_4$

Beidellite $X_{0.7}(Al_4)(Si_{7.3}Al_{0.7})O_{20}(OH)_4$

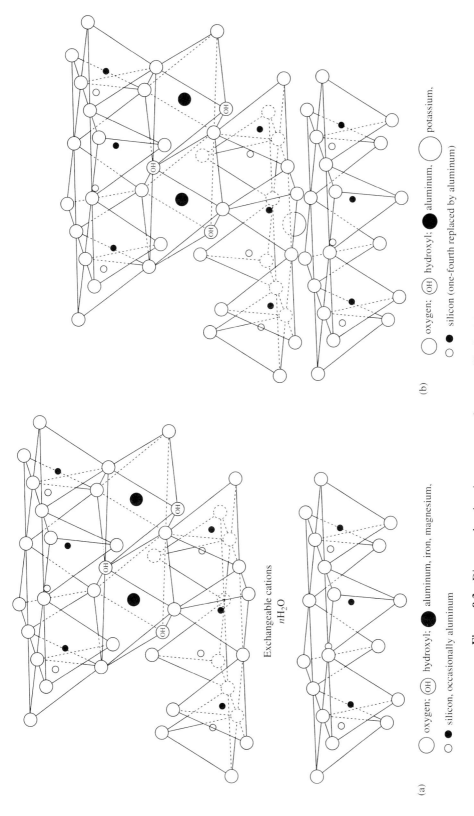

Figure 9.3 Diagram showing the structures of montmorillonite **(a)** and muscovite **(b)**. After *Clay mineralogy* by R. E. Grim. Copyright © 1968 by McGraw-Hill, Inc. Used with permission of The McGraw-Hill Book Companies.

(a)

○ oxygen; ⊙ hydroxyl; ● aluminum, iron, magnesium,
○ silicon, occasionally aluminum

Exchangeable cations
nH_2O

(b)

○ oxygen; ⊙ hydroxyl; ● aluminum. ○ potassium.
● silicon (one-fourth replaced by aluminum)

In these formulas, X denotes a monovalent cation such as Na^+ and/or $\frac{1}{2}Ca^{2+}$, occupying interlayer exchange sites. The formulas show that beidellite is magnesium-free and montmorillonite magnesium-rich. Other smectite minerals include nontronite, hectorite, saponite, and sauconite (Table 9.1). All the smectites (and vermiculites) have swelling properties.

The general formula of the clay vermiculite is

$$\{Mg,Ca\}_{0.6-0.9}(Mg,Fe^{3+},Al)_{6.0}[(Al,Si)_8O_{20}](OH)_4$$

where Mg and Ca are the adsorbed interlayer cations. Both vermiculites and smectites have about 0.7 moles of interlayer cations, however, because the interlayer cations in vermiculite are divalent rather than monovalent, as in the smectites; the vermiculites have the highest cation exchange capacities among the clays.

Vermiculite clay can form as a result of the weathering and partial degradation of biotite or phlogopite micas (cf. Drever 1988) or of other iron- and magnesium-rich aluminosilicates (Deer et al. 1992). Vermiculite is structurally similar to smectite. However, the interlayer charge between T:O:T units is greater in vermiculite than in smectite (about 0.5 to 0.7 per $O_{10}(OH)_2$ structural unit versus that of 0.2 to 0.5 in the smectites) so that vermiculites do not swell as much as smectites when treated with ethylene glycol. The interlayer charge in vermiculite results chiefly from substitution of Al^{3+} for Si^{4+} in the tetrahedral layer. Magnesium is the chief cation in the octahedral layer. Macroscopic vermiculites are dominantly trioctahedral; however, fine-grained soil vermiculites may be more dioctahedral in structure. Drever (1988) notes than when Mg^{2+} is the interlayer cation, two water layers occupy the interlayer space, and the basal spacing is about 14 Å. The structural formula of a frequently studied vermiculite called *Llano vermiculite* is

$$\{Mg_{0.93}\}(Mg_{5.62}Fe^{3+}_{0.13}Al_{0.21})(Al_{2.21}Si_{5.79})O_{20}(OH)_4$$

(Brown et al. 1978) where the interlayer Mg^{2+} is given first in the formula.

In the mica group of T:O:T minerals an Al^{3+} ion substitutes for about one of every four Si^{4+} ions in the tetrahedral layer. The resultant excess negative charge is compensated for by interlayer K^+ ions that are held electrostatically and fairly strongly between adjacent tetrahedral layers. The micas are, therefore, relatively hard and elastic, and the potassium ions not readily exchangeable. The micas include muscovite (dioctahedral), phlogopite (trioctahedral), and biotite, which is similar to phlogopite but with Fe^{2+} substitution for Mg^{2+} in some octahedral sites. The micas have a repeat basal spacing of about 10 Å. Muscovite and biotite are common in igneous and metamorphic rocks. Ideal compositions these micas are

Muscovite: $KAl_2(AlSi_3O_{10})(OH)_2$

Phlogopite: $KMg_3(AlSi_3O_{10})(OH)_2$

Biotite: $K(Mg,Fe)_3(AlSi_3O_{10})(OH)_2$

In biotite up to approximately 75% of the octahedral Mg in phlogopite can be replaced by Fe^{2+} (Deer et al. 1992).

Because they are the dominant mineral in shales, illites, and illite-smectites (see below) are the most abundant of all the clays. Illites are defined as micalike materials less than 2 μm in size, which, like the micas, have a basal spacing of 10 Å (Drever 1988). Most illites are dioctahedral and structurally similar to muscovite, although some are trioctahedral like biotite. Illites contain less K^+ and Al^{3+} and more Si^{4+} than muscovite. They also usually contain some Mg^{2+} and Fe. The irregularity of occurrence of interlayer K^+ makes bonding between the layers weaker than in muscovite. Illitic clays are usually mixed-layer clays, with the illite making up perhaps 80% of the clay layers and smectite the rest. Robinson and Haas (1983) give as the composition of a unit cell of a natural illite

$$K_{1.50}(Mg_{0.50}Al_{3.50})(Al_{1.0}Si_{7.0})O_{20}(OH)_4$$

Deer et al. (1992) suggest the following general illite formula

$$K_{1.5-1.0}Al_4(Al_{1.5-1.0}Si_{6.5-7.0})O_{20}(OH)_4$$

The Marblehead illite used in hydrothermal laboratory experiments by Aja et al (1991a, 1991b) has the structural formula

$$(K_{0.79}Na_{0.02})(Al_{1.43}Fe^{3+}_{0.11}Mg_{0.37}Ti_{0.08})(Si_{3.55}Al_{0.45})O_{10}(OH)_2$$

All three of these illites contain smectite layers of montmorillonite or biedellite composition (Deer et al. 1992) and are, therefore, termed *mixed-layer* illite-smectite (I/S) clays. As a shorthand to identify the composition of a given I/S, the structural formula of a half-unit cell of the clay is sometimes written $K_n/O_{10}(OH)_2$, where n is the number of moles of K in the clay. By definition, end-member illite (I) contains K that equals or exceeds $K_{0.8}$ to $K_{0.9}/O_{10}(OH)_2$, whereas end-member smectite (S) has K roughly equal to or less than $K_{0.4}/O_{10}(OH)_2$ (cf. Aja et al. 1991a). The illite-smectites have been considered: (a) a single solid solution; (b) two separate solid solutions (Ransom and Helgeson 1993); or (c) mixtures of two or more defined compositions of illite and smectite (cf. Garrels 1984; Wilson and Nadeau 1985). Based on illite solution equilibria measurements up to 250°C, Aja et al. (1991a, 1991b) and Aja and Rosenberg (1992) have suggested that the illite-smectites have four distinct compositions: $K_{0.29}/O_{10}(OH)_2$ (S), $K_{0.50}/O_{10}(OH)_2$ (IS), $K_{0.69}/O_{10}(OH)_2$ (ISII), and $K_{0.85}/O_{10}(OH)_2$ (I), with the illite-rich forms more stable at elevated temperatures.

As mixed-layer I/S clays become smectite-rich, their ion exchange and swelling properties approach those of the pure smectites. Other three-layer clays, including the chlorites and vermiculites, also commonly occur in soils in mixed-layer form (e.g., mixed-layer chlorite-smectite [-vermiculite]) (cf. Wilson and Nadeau 1985; Drever 1988).

9.3 THE OCCURRENCE OF COMMON CLAY MINERALS

Generalized in Fig. 9.4 from Brady (1974) are the conditions, including extent of weathering, that lead to the formation of clay minerals and secondary metal oxyhydroxides in soils.

Vermiculite may form from the alteration of biotite or phlogopite, or of other Fe- and Mg-rich aluminosilicates, such as chlorites and hornblende in igneous or metamorphic rocks.

Smectite clays such as montmorillonite are favored by relatively poor drainage, alkaline pH's, high concentrations of silica and divalent cations, and low potassium concentrations. They tend to form early in the weathering of unstable Fe-, Mg-, and Ca-rich minerals in igneous or metamorphic rocks.

Illite clays may result from the weathering of micas and feldspars. Their formation in soils and sediments is favored by high K^+ and moderate silica concentrations. When smectites or mixed-layer smectite/illite clays are buried in deep sedimentary basins, they are gradually transformed into more stable illites by a combination of time and temperature (diagenesis) (cf. Velde and Vasseur 1992; Huang et al. 1993; Cuadros and Linares 1996). The reaction involved might be

$$2Na_{0.4}(Al_{1.47}Mg_{0.29}Fe_{0.18})Si_4O_{10}(OH)_2(\text{smectite}) + 0.85K^+ + 1.07H^+$$

$$\rightarrow 1.065K_{0.80}(Al_{1.98}Mg_{0.02})(Si_{3.22}Al_{0.78})O_{10}(OH)_2(\text{illite}) + 4.6\ SiO_2(aq)$$

$$+ 0.36Fe(OH)_3(s) + 0.56Mg^{2+} + 0.8Na^+ + 0.9H_2O$$

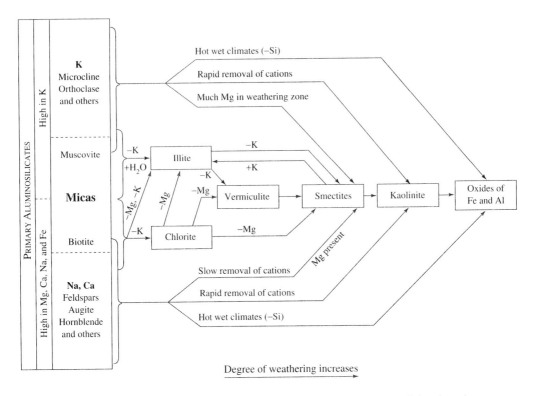

Figure 9.4 Weathering products of primary minerals and sequence of forming clays. Modified from N. C. Brady. *The nature and prospecting of soils*, 8th ed. Copyright © 1974. Used by permission of Macmillan Publ. Co., New York.

The rate of the smectite → illite reaction is thus directly proportional to K^+ and H^+, but is retarded by Mg^{2+} and by dissolved silica and Na^+. In deepening sedimentary basins, the extent of the reaction at any depth also depends on the local thermal gradient (temperature) and the sediment burial rate (reaction time). Because smectites of small particle size are the least stable, they alter to illite at lower temperatures than do coarser-grained smectites (Fig. 9.5).

Example 9.1

Huang et al. (1993) derive a rate equation for the smectite to illite reaction at low Mg^{2+} and low to moderate Na^+ concentrations, which is

$$-\frac{dS}{dt} = A \exp\left(\frac{-E_a}{RT}\right)(K^+)\, S^2$$

where S = fraction of smectite layers in the I/S, t = time in seconds, A = frequency factor = 8.08×10^4 M/s, E_a = activation energy = 28 kcal/mol, R = gas constant = 1.987 cal/mol K, T = degrees K, and (K^+) the molar concentration in the fluid. A 100% smectite clay is buried at 3800-m depth and 125°C in contact with groundwater that contains 200 mg/L K^+. How long will it take to convert 80% of the smectite to illite?

The integral of the rate equation is:

$$S = S_0/(1 + S_0(K^+)A \exp(-E_a/RT)t)$$

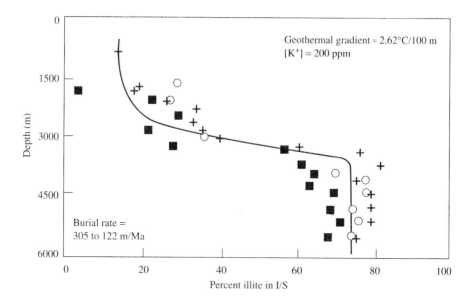

Figure 9.5 Comparison of modeled and observed smectite/illite conversion found in CRWU Gulf Coast Well 6 shale (Hower et al. 1976). Kinetic model predicts the conversion using the estimated temperature and age at each depth. Line is the modeled result. Symbols represent average particle sizes of the initial clay: $+$ = <0.1 μm, $\bigcirc$ = 0.1 to 0.5 μm, $\blacksquare$ = 0.5 to 2 μm, 1 Ma = 10^6 y. From W-L. Huang, J. M. Longo and D. R. Pevear in *Clays & Clay Minerals*, copyright 1993. Used by permission.

(Huang et al. 1993), where S_o is the initial fraction of smectite layers. Solving for t after substitution gives $t = 1.41 \times 10^{12}$ s or about 45,000 y.

Because kaolinite lacks structural cations other than relatively insoluble aluminum and silicon, it is the clay most stable under acid-weathering conditions. As such, kaolinite is common in humid-climate, well-leached acid soils, and stream sediments where there is good drainage. Kaolinite often forms from weathering of the potassium feldspar (microcline and orthoclase) and muscovite mica in rocks such as granite. For extreme conditions of high rainfall and good drainage, kaolinite is destroyed through loss of its silica, becoming unstable relative to gibbsite [$Al(OH)_3$].

Increased rainfall and infiltration lowers both the pH of soil moisture (Fig. 9.6) and cation and silica concentrations. Shown as a function of precipitation in Fig. 9.7 is the relative abundance of these clays and of gibbsite formed in residual soils over some igneous rocks in California. Dark igneous rocks, such as basalt, are high in Fe, Ca, and Mg aluminosilicates. Light-colored igneous rocks, including granites, are high in K and Na, and are more Al-rich. Within the residual soils formed in arid climates on both rock types, smectite is the most abundant clay. It can be assumed that soil moisture is alkaline and high in cation and silica concentrations in low-rainfall, poorly leached soils and becomes acid and more dilute as rainfall amounts increase. Because it contains essential K$^+$, illite generally forms from the weathering of potassic minerals such as K-feldspar and muscovite. Illite may also simply be inherited from the weathering of underlying shales. The occurrence of vermiculite often reflects the partial degradation of T:O:T micas such as biotite and phlogopite. Rainfall amounts from about 50 to 80 inches (130 to 200 cm) tend to leach away metal cations and silica and so favor the destruction of smectite, illite, and vermiculite. The residual soils may then contain kaolinite and halloysite, smaller amounts of vermiculite, and increasing percentages of gibbsite.

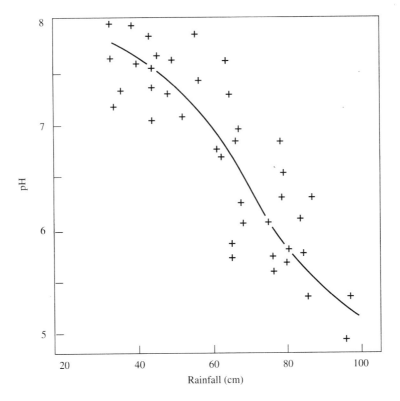

Figure 9.6   The effect of rainfall on soil pH. The line has no statistical significance. After *Factors of soil formation* by H. Jenny. Copyright © 1941. Used with permission.

The occurrence of specific clays in soils depends both on the presence of particular source rocks and minerals and on the intensity of leaching, which reflects climate, drainage, and soil permeability. The long-term persistence of smectites, vermiculite, or illite is possible only in poorly leached soils, where relatively high pH's and high silica and cation concentrations can be maintained by rock weathering and evapotranspiration. In humid climate soils, infiltrating soil moisture will tend to follow pathways of enhanced permeability (macropores) around plant roots and away from clay-rich layers or lenses. Based on the oxygen- and deuterium-stable isotopy of precipitation and soil moisture in some Pennsylvania soils, Sears and Langmuir (1982) concluded that precipitation infiltrated to a soil depth of 9 m within 3 months after a storm event. However, this rapidly infiltrating macropore water, which bypasses most of the soil, was only about 1% of total soil moisture. The micropore soil water is probably undiluted by fresh recharge for many years, even in humid climates (cf. Cozzarelli et al. 1987). Such conditions can lead to high soil moisture pH's, and high silica and cation concentrations, assuring the persistence of illite, smectite, and vermiculite clays.

9.4 APPLICABILITY OF EQUILIBRIUM CONCEPTS

Whether clay minerals, in particular mixed-layer clays, can attain thermodynamic equilibrium in low-temperature soils and groundwater systems, has long been a controversial issue (cf. May et al. 1986; Aja and Rosenberg 1992). In theory, time must eventually favor formation of the thermody-

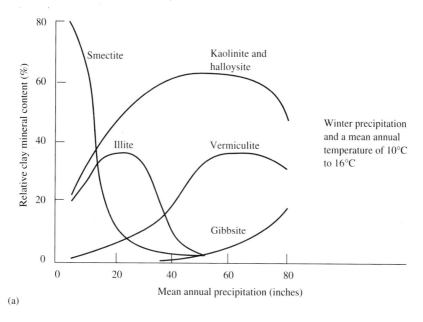

(a)

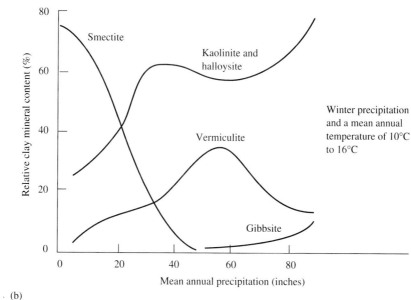

(b)

Figure 9.7   Clay mineral composition of the surface layers of residual soils formed on
(a) quartz- and feldspar-rich rocks, and **(b)** Fe- and Mg-rich igneous rocks in California.
After I Barshad. The effect of variation in precipitation on the nature of clay mineral formation in soils from acid and basic igneous rocks, *Proc. Int. Clay Conf.,* © 1966 by the Geological Survey of Norway. Used by permission.

namically most stable clay or clays in a soil exposed to a particular climate and water composition.
However, at low temperatures reversible equilibrium involving clay minerals and contacting waters
has been difficult to demonstrate. Kaolinite has been shown to dissolve and precipitate reversibly

and thus attain thermodynamic equilibrium at 25°C (May et al. 1986; Nagy et al. 1991). Allophane and halloysite (a metastable polymorph of kaolinite) are also capable of reversible equilibration at 25°C (cf. Steefel and Van Cappellen 1990). However, May et al. (1986) have questioned whether it is possible for complex aluminosilicates of variable composition, such as the illites and smectites, to reversibly equilibrate. Aja and Rosenberg (1992) disagree and argue that solubility experiments have proven the attainment of equilibrium by the I/S clays and chlorite at low temperatures (see also Aja et al. 1991a, 1991b). Although the equilibrium may be metastable, it still describes a condition that can persist and control the chemistry of soil moisture and groundwater.

The formation and survival of unstable or metastable micas and clays in sediments and soils at low temperatures reflects kinetic as well as thermodynamic factors. First, the rates of reactions involving solid-aqueous and especially solid-solid transformations in dilute solutions are very slow at low temperatures (most natural waters are "dilute"). The slow kinetics of clay transformations reflects small differences in free energy between stable and metastable clays. Also, the occurrence of specific clays is related to the chemistry and crystal structure of source minerals. Thus, illite often results from the weathering of muscovite, and vermiculite results from the weathering of biotite (cf. Drever 1988), consistent with the similar chemistries and structures of these pairs of T:O:T minerals.

The empirical observation that unstable or metastable minerals form first in the weathering process, followed by progressively more stable minerals, is explained by *the Gay-Lussac-Ostwald (GLO)* or *Ostwald step rule* (cf. Sposito 1989, 1994; Steefel and Van Cappellen 1990). Steefel and Van Cappellen define the GLO step rule as follows:

> The (soil-water) system preferentially forms the phase with the fastest precipitation rate under the prevailing conditions. The nucleation of a more soluble phase (such as an illite, smectite, or vermiculite) is kinetically favored over that of a less soluble analogue (such as kaolinite) because the more soluble phase has the lower mineral-solution interfacial energy, (and thus the lower nucleation energy). Hence, when the supersaturation of the solution is sufficiently elevated, the metastable mineral may have the higher rate of precipitation. (Simply stated, the solid that reduces oversaturation the most rapidly will form first.) This explains the frequently observed formation of metastable aluminosilicates in weathering profiles. As the solution composition evolves, however, the relative rates of precipitation and dissolution will change, gradually resulting in the formation of more stable secondary mineral assemblages.

9.5 CLAY MINERAL EQUILIBRIA AND PHASE DIAGRAMS

Although many minerals never reach true thermodynamic equilibrium in a soil or sediment, their stability constants provide insights as to their general behavior and occurrence. Such behavior is conveniently examined through the use of phase diagrams. Before considering such diagrams in detail, let us examine the assumptions inherent in their construction, which may limit the applicability of the phase diagrams to natural systems.

1. *Mineral/aqueous-solution phase diagrams assume chemical equilibrium can be attained among all phases shown.* In other words, reaction rates among the minerals and aqueous species are assumed fast enough so that equilibrium is reached within the time frame being considered. This assumption is especially questionable in low-temperature, surface and near-surface environments such as streams and high permeability soils, given the short residence times of water in those systems, seasonal and storm-related changes in their moisture content and temperature, and the slow rates of mineral/mineral and mineral/solution reactions.

2. *The phases plotted are assumed to be pure and fixed in composition, and to correspond to the phases being considered in the natural system of interest.* This assumption is often tenuous, particularly when the minerals involved are complex and highly variable solid solutions such as the micas and clays, which usually contain several more chemical components than can be meaningfully depicted in a two-dimensional diagram, and which may not have been chemically analyzed.

3. *Accurate and meaningful thermodynamic data are available for all the solids and aqueous species being considered.* This assumption may be questionable given the highly complex and variable compositions, thus stabilities, of the clays and micas in particular. Naturally occurring clays and micas rarely have the same composition as the specific clays and micas for which experimentally determined thermodynamic data have been obtained.

4. *In many such diagrams, aluminum is assumed insoluble and conserved within reactant and product solid phases.* We have already noted that for acid and alkaline conditions, aluminum becomes soluble below about pH 5 and above pH 9, for which conditions it will not be conserved in the solids (see Chap. 7).

In spite of these caveats, we will find that phase diagrams involving clays, micas, and related phases provide us with useful insights regarding such processes as weathering and mineral alteration and formation in soils and sedimentary rocks.

The best known of such phase diagrams is probably the simplest—the $\log([K^+]/[H^+])$ versus $\log[H_4SiO_4^\circ]$ diagram given in Fig. 9.8. If we assume that aluminum is conserved within reactant and product solid phases (assumption number 4 above), the phases that can be shown on such a diagram are listed below, with their ideal compositions.

Mineral/phase	Ideal formula
quartz, amorphous silica	$SiO_2(c)$, $SiO_2(am)$
gibbsite	$Al(OH)_3$
kaolinite	$Al_2Si_2O_5(OH)_4$
muscovite (K-mica)	$KAl_3Si_3O_{10}(OH)_2$
K-feldspar	$KAlSi_3O_8$

Boundaries on the diagram are defined by equilibrium conditions between the above phases. Accordingly we write the various reactions between these phases and their equilibrium constant expressions. For simplicity, mineral names are used instead of chemical formulas. The equilibrium constant values which are from Hess (1966), do not differ much from such values in recent use by others (cf. Drever 1988). These K_{eq} values are based in large part on the observed natural behavior of the minerals (cf. Feth et al. 1964; Garrels and Christ 1965; Helgeson 1969).

The reaction between K-feldspar and kaolinite may be written

$$\text{K-feldspar} + H^+ + \tfrac{9}{2}H_2O = \tfrac{1}{2}\text{kaolinite} + K^+ + 2H_4SiO_4^\circ \tag{9.1a}$$

for which,

$$K_{eq} = \frac{[K^+]}{[H^+]}[H_4SiO_4^\circ]^2 = 10^{-0.9} \tag{9.1b}$$

Taking logarithms and transposing, we obtain

$$\log = \frac{[K^+]}{[H^+]} = -0.9 - 2 \log [H_4SiO_4^\circ] \tag{9.1c}$$

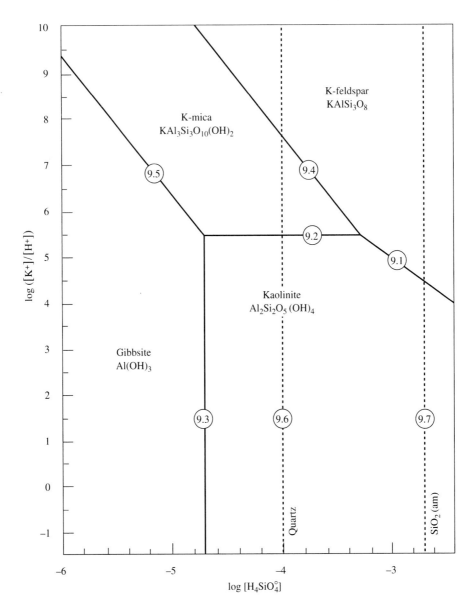

Figure 9.8   Log([K⁺]/[H⁺]) versus log[H₄SiO₄°] diagram at 25°C and 1 bar pressure, showing the stability fields of gibbsite, K-mica, K-feldspar, and kaolinite. Circled numbers are the same numbers that have been assigned to the equilibrium reactions between these phases in the text.

This is the equation of a straight line with an intercept of -0.9 and slope of -2 on a plot of $\log([K^+]/[H^+])$ versus $\log[H_4SiO_4^\circ]$ (Fig. 9.8). Note that the intercept is determined by the equilibrium constant, whereas the slope is defined solely by the reaction stoichiometry and is independent of relative or absolute mineral stabilities.

The reaction between K-mica and kaolinite is

$$\text{K-mica} + H^+ + \tfrac{3}{2}H_2O = \tfrac{3}{2}\text{kaolinite} + K^+ \tag{9.2a}$$

for which

$$K_{eq} = \frac{[K^+]}{[H^+]} = 10^{5.5} \tag{9.2b}$$

and taking logarithms

$$\log = \frac{[K^+]}{[H^+]} = 5.5 \tag{9.2c}$$

Equilibrium between gibbsite and kaolinite may be written

$$\text{gibbsite} + H_4SiO_4^\circ = \tfrac{1}{2}\text{kaolinite} + \tfrac{5}{2}H_2O \tag{9.3a}$$

for which $K_{eq} = 1/[H_4SiO_4^\circ] = 10^{4.7}$. Transposing and taking logarithms gives

$$\log[H_4SiO_4^\circ] = -4.7 \tag{9.3b}$$

The K-feldspar/K-mica reaction is

$$\tfrac{3}{2}\text{K-feldspar} + H^+ + 6H_2O = \tfrac{1}{2}\text{K-mica} + K^+ + 3H_4SiO_4^\circ \tag{9.4a}$$

with $K_{eq} = 10^{-4.1}$. The log-linear equation for the reaction boundary is

$$\log = \frac{[K^+]}{[H^+]} = -4.1 - 3 \log [H_4SiO_4^\circ] \tag{9.4b}$$

The K-mica/gibbsite reaction is

$$\text{K-mica} + H^+ + 9H_2O = 3\text{gibbsite} + K^+ + 3H_4SiO_4^\circ \tag{9.5a}$$

Given that $K_{eq} = 10^{-8.6}$, the log-linear equation that defines the K-mica/gibbsite boundary is given by

$$\log = \frac{[K^+]}{[H^+]} = -8.6 - 3 \log[H_4SiO_4^\circ] \tag{9.5b}$$

The dissolution reactions for quartz and amorphous silica are both written

$$SiO_2(s) + 2H_2O = H_4SiO_4^\circ \tag{9.6a, 9.7a}$$

with equilibrium constants of $K_{eq}(\text{quartz}) = 10^{-4.0}$, and $K_{eq}[SiO_2(\text{am})] = 10^{-2.7}$, from which we obtain for quartz

$$\log [H_4SiO_4^\circ] = -4.0 \tag{9.6b}$$

and for amorphous silica

$$\log [H_4SiO_4^\circ] = -2.7 \tag{9.7b}$$

The diagram derived using the K_{eq} values given above is shown in Fig. 9.8. To construct it we have assumed that the solids shown are of known and fixed composition and thermodynamic stability and that identical characteristics apply to the same minerals in the natural environments being modeled. The Gibbs free energy of kaolinite, for example, can range from −900 to −908.1 kcal/mol, although that of well-crystallized material is probably near −907.7 kcal/mol (Bassett et al. 1979; Devidal et al. 1996). Figure 9.8 has been constructed using equilibrium constants that reflect *differences* in free energies, not their absolute values. Little is known of the stabilities of the other minerals in Fig. 9.8 when they are in equilibrium with kaolinite of a given stability at 25°C. The *behavior* of kaolinite in environments where it weathers to gibbsite is at least consistent with the K_{eq} for reaction (9.2a) above.

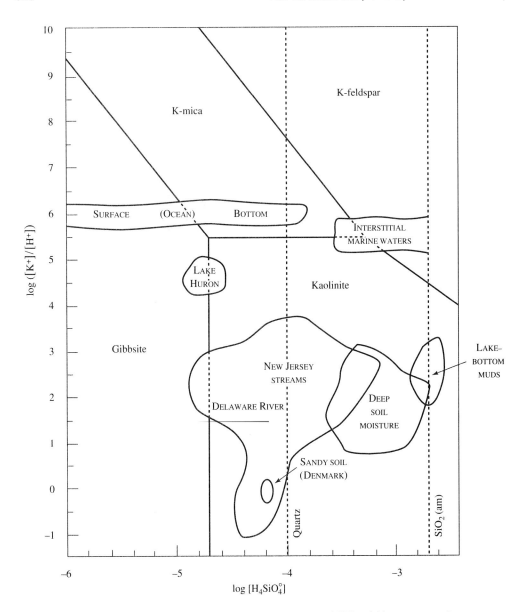

Figure 9.9 Log([K$^+$]/[H$^+$]) versus log[H$_4$SiO$_4^\circ$] diagram at 25°C and 1 bar pressure, show-ing the stability fields of gibbsite, K-mica, K-feldspar, and kaolinite. The compositions of some natural waters have been plotted on the diagram.

We will next orient ourselves by calculating where certain water chemistries plot on a log([K$^+$]/[H$^+$]) versus log[H$_4$SiO$_4^\circ$] diagram. Seawater, for example, has K$^+$ = 10$^{-2.00}$ mol/L (391 mg/L) and pH = 8.15. Accordingly, log([K$^+$]/[H$^+$]) = 6.15. Silica in seawater ranges from about 0.01 (surface) to 7 mg/L (bottom water) as SiO$_2$ (GFW = 60.085) or 10$^{-6.8}$ to 10$^{-3.9}$ mol/L as H$_4$SiO$_4^\circ$. These compositional ranges have been plotted in Fig. 9.9 along with the chemistry of interstitial wa-ters in marine muds (cf. Lafon and Mackenzie 1974). Surface seawater is low in silica because of its scavenging from the water by planktonic (floating) organisms such as the diatoms, which use silica in their skeletons. As these organisms die their skeletons sink and begin dissolving; thus silica concentrations increase toward the ocean bottom. High silica concentrations in interstitial marine muds result from the dissolution of buried diatoms and other siliceous organisms.

Generally, freshwaters have lower K^+ concentrations (for example 10 mg/L or $10^{-3.6}$ mol/L), and lower pH values (typically pH 5 to 8) than seawater. In other words, their $\log([K^+]/[H^+])$ values will range from roughly 1.4 to 4.4. Also, freshwaters usually have higher silica concentrations than seawater (about $10^{-3.6}$ mol/L as SiO_2 or $H_4SiO_4^\circ$ for the average surface or groundwater; see Chap. 7). Such compositions generally place these waters within the kaolinite stability field. This is consistent with the common occurrence of kaolinite in humid-climate soils and sediments.

Shown in Fig. 9.9 are water-composition ranges for some humid-climate streams (in New Jersey), a dilute, freshwater lake (Lake Huron) and lake-bottom muds from the Great Lakes (Sutherland 1970), and deep-soil moisture from Pennsylvania (Sears 1976; Sears and Langmuir 1982). Lake Huron and the Delaware River are dilute, humid-climate waters. They both plot near the kaolinite-gibbsite boundary. Their composition can be described as *water dominated*. In other words, their chemistries are controlled chiefly by dilution with fresh rainfall and runoff, not by reactions with geological materials. In a study of acid rain (water-dominated) control of soil moisture and groundwater chemistry of a sandy aquifer in Denmark, Hansen and Postma (1995) found that pore waters were close to equilibrium with gibbsite and supersaturated with kaolinite (Fig. 9.9). Precipitation pH = 4.34 at the site, and $\log([K^+]/[H^+]) = -0.95$.

Rock-dominated waters are those in relatively closed systems, out of contact with fresh, diluting recharge. Such waters can approach equilibrium with respect to contacting silicates and aluminosilicates, and so tend to have relatively high pH's (low H^+ concentrations) and high K^+ and silica concentrations. Such waters (plotted in Fig. 9.9) include interstitial waters in marine muds and lake-bottom muds and some C-horizon soil moisture in micropores. That some soil moisture has high pH and solute contents may also reflect the concentrating effect of evapotranspiration. The high dissolved-silica levels in Great Lakes bottom mud may have resulted from the dissolution of buried siliceous diatoms.

Water compositions that plot within the stability field of a mineral in Figs. 9.8 or 9.9, may not be in equilibrium with that mineral.[†] For example, global mean rainfall (Table 8.5, Chap. 8) has $K^+ =$ 8 μM and $H^+ =$ 2 μM, so that $\log([K^+]/[H^+]) = 0.6$. The dissolved silica content of mean rainfall is probably much less than 10^{-5} mol/L, in general. Thus, rainfall plots within the gibbsite field. The H^+ content of mean rainfall is equivalent to a pH of 5.7 and a gibbsite solubility as $\Sigma Al(aq)$ of about 10^{-7} mol/L (see Fig. 7.8, Chap. 7). Information on the dissolved Al content of rain is generally lacking, although it seems likely that normal rain is undersaturated with respect to gibbsite. Particularly for precipitation-dominated dilute systems, it is, therefore, useful to imagine a third axis labeled $-\log [Al^{3+}]$ in Figs. 9.8 and 9.9, in order to plot the chemical evolution of waters that are initially undersaturated with respect to the minerals in both figures.

Many researchers have employed $\log([M^{n+}]/[H^+]^n)$ versus $\log[H_4SiO_4^\circ]$ and similar diagrams to explain and interpret natural water systems, where M denotes a cation and n its charge (cf. Helgeson et al. 1969; Lippmann 1979; Faust and Aly 1981; Bowers et al. 1984; Drever 1988; Nesbitt and Wilson 1992; Anderson and Crerar 1993; Nordstrom and Munoz 1994; Cramer and Smellie 1994). Of necessity, simplified clay compositions have been assumed in order to show them on such diagrams. Assumptions and limitations comparable to those that were inherent in the construction and use of Figs. 9.8 and 9.9 apply to all such diagrams.

[†]Obtaining positive proof that a water is in equilibrium with gibbsite or with any other mineral phase in which field it plots involves two steps. First, a saturation-state calculation based on a complete analysis of the water should indicate that the mineral occurs under equilibrium conditions. Second, examination of the mineral should reveal evidence that its crystals are growing.

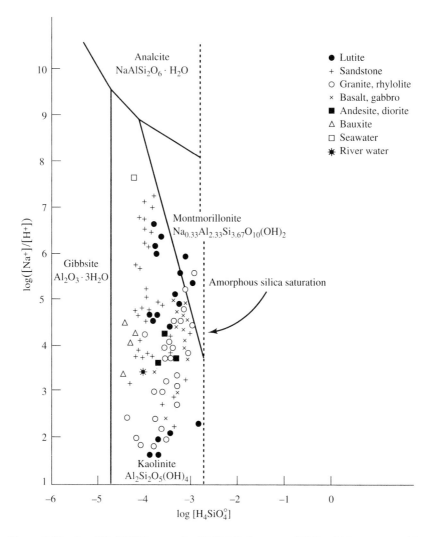

Figure 9.10 Log([Na$^+$]/[H$^+$]) versus log[H$_4$SiO$_4^\circ$] diagram at 25°C and 1 bar pressure. The figure shows a stability field for an idealized sodic montmorillonite. Plotted on the diagram are analyses of groundwaters from various rock types. A lutite is a shale or mudstone that probably contains illite and kaolinite, with smaller amounts of smectite clays such as montmorillonite. Sandstones include feldspars as well as quartz. Note that most of the water analyses fall in the kaolinite field. After O. P. Bricker and R. M. Garrels, Mineralogic factors in natural water equilibria. In *Principles and applications of natural water chemistry*, ed. S. Faust and J. V. Hunter. Copyright © 1965. Reprinted by permission.

An example of such a plot is a log([Na$^+$]/[H$^+$]) versus log[H$_4$SiO$_4^\circ$] diagram (Bricker and Garrels 1965), in Fig. 9.10. In order to show montmorillonite in such a figure, its composition has obviously been idealized. Also given is the stability field of the zeolite mineral analcite. As in Fig. 9.9, most water analyses plot in the kaolinite field. The montmorillonite/ kaolinite (mont/kaol) equilibrium line is an upper bound for most of the analyses. The *slope* of this boundary in Fig. 9.10 is defined by the stoichiometry of the mont/kaol reaction. This is determined by the formula that Bricker and Garrels (1965) chose for the montmorillonite. The *position* of the boundary line defines K_{eq} for the reaction and may be used to calculate ΔG_f° for the montmorillonite, if ΔG_f° for kaolinite is known.

Example 9.2

Given the montmorillonite/kaolinite (mont/kaol) boundary drawn in Fig. 9.10, compute K_{eq} for the reaction between the clays. Then with the ΔG_f° data here, calculate ΔG_f° for the montmorillonite.

Species or mineral	ΔG_f°(kcal/mol)	Source
Na$^+$	−62.60	Wagman et al. 1982
H$_2$O	−56.69	Wagman et al. 1982
H$_4$SiO$_4^\circ$	−312.58	Langmuir 1978
Al$_2$Si$_2$O$_5$(OH)$_4$ (kaolinite)	−908.1	Robie et al. 1978

We first write a balanced reaction

$$6Na_{0.33}Al_{2.33}Si_{3.67}O_{10}(OH)_2 + 23H_2O + 2H^+ = 7Al_2Si_2O_5(OH)_4 + 8H_4SiO_4^\circ + 2Na^+$$

Extension of the mont/kaol boundary line to the log [H$_4$SiO$_4^\circ$] axis, where [Na$^+$]/[H$^+$] = 10.0 and [H$_4$SiO$_4^\circ$] = $10^{-2.0}$ indicates

$$K_{eq} = \left(\frac{[Na^+]}{[H^+]}\right)^2 [H_4SiO_4^\circ]^8 = 10^{-14.0}$$

Therefore, $\Delta G_r^\circ = -19.1$ kcal/mol, and with free energies from the table ΔG_f°(mont) = −1283.0 kcal/mol.

Based on an evaluation of the literature, Garrels (1984) proposed that I/S clays were not a solid solution, but could be considered mixtures in various proportions of a single illite [K$_{0.8}$Al$_{1.9}$(Al$_{0.5}$Si$_{3.5}$)O$_{10}$(OH)$_2$] and single montmorillonite [K$_{0.3}$Al$_{1.9}$Si$_4$O$_{10}$(OH)$_2$]. He deduced the stabilities of these hypothetical clays relative to that of kaolinite by plotting analyses of waters that had been in contact with the clays on a log ([K$^+$]/[H$^+$]) versus log [H$_4$SiO$_4^\circ$] diagram (Fig. 9.11). The relatively tight trend of these analyses was assumed to represent equilibrium between kaolinite and montmorillonite and between montmorillonite and illite. Slopes of the equilibrium boundaries were defined by the mineral compositions and reaction stoichiometries. Mineral stabilities could then be obtained as descibed above in Example 9.2. Free energies used in Garrels's calculation and derived from it are given here.

Species or mineral	ΔG_f° (kcal/mol)	Source
K$^+$	−67.52	Robie et al. 1978
H$_2$O	−56.68	Robie et al. 1978
H$_4$SiO$_4^\circ$	−312.6	Robie et al. 1978
K-feldspar [KAlSi$_3$O$_8$]	−894.4	Robie et al. 1978
Muscovite [KAl$_3$Si$_3$O$_{10}$(OH)$_2$]	−1338.6	Robie et al. 1978
Kaolinite [Al$_2$Si$_2$O$_5$(OH)$_4$]	−908.1	Robie et al. 1978
Pyrophyllite [Al$_2$Si$_4$O$_{10}$(OH)$_2$]	−1259.4	Garrels 1984
Gibbsite [Al(OH)$_3$]	−277.0[†]	Garrels 1984
Illite [K$_{0.8}$Al$_{1.9}$(Al$_{0.5}$Si$_{3.5}$)O$_{10}$(OH)$_2$]	−1307.8	Garrels 1984
Montmorillonite [K$_{0.3}$Al$_{1.9}$Si$_4$O$_{10}$(OH)$_2$]	−1267.5	Garrels 1984

[†]Apparently computed assuming [H$_4$SiO$_4^\circ$] = $10^{-4.5}$ M (1.9 mg/L) at the kaolinite/gibbsite boundary.

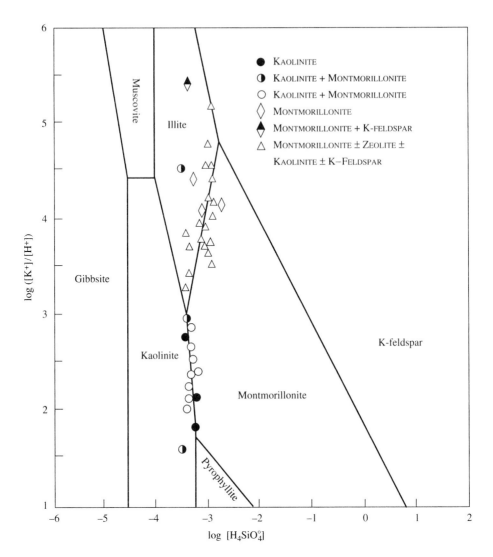

Figure 9.11   Log($[K^+]/[H^+]$) versus log[$H_4SiO_4^\circ$] diagram at 25°C with plotted chemical analyses of waters in contact with clays, as reported by Aagaard and Helgeson (1983). Phase boundaries are drawn consistent with the analyses and with the illite and montmorillonite compositions proposed in the text. Agreement of the data and boundaries suggest equilibrium between the phases and support the idea that illite and montmorillonite behave as two discrete phases. After R. M. Garrels in *Clays & Clay Minerals*, 32: 161–66. Copyright 1984.

Using the I/S compositional shorthand introduced earlier, Garrels's illite and montmorillonite may be written $K_{0.8}/O_{10}(OH)_2$, and $K_{0.3}/O_{10}(OH)_2$, respectively.

Based on their hydrothermal experiments, Aja et al. (1991a, 1991b) and Aja and Rosenberg (1992) have concluded that the I/S clays are not two-phase mixtures, but have several defined compositions. They have inferred these compositions from the solution chemistry and the slopes of reaction boundaries. For example, their data in Fig. 9.12, suggest the most stable I/S solids are $K_{0.48}/O_{10}(OH)_2$ and $K_{0.69}/O_{10}(OH)_2$ at 25°C, and $K_{0.31}/O_{10}(OH)_2$ and $K_{0.85}/O_{10}(OH)_2$ at 125°C. The

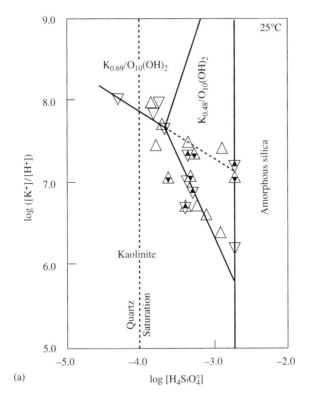

(a)

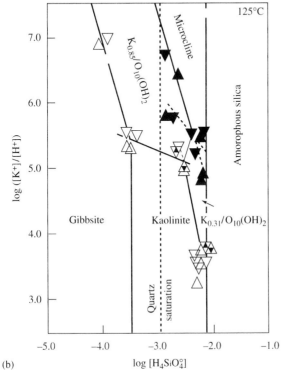

(b)

Figure 9.12   Isothermal and isobaric $\log([K^+]/[H^+])$ versus $\log[H_4SiO_4^\circ]$ diagrams for 25°C **(a)** and 125°C **(b)** for Marblehead illite in the presence of kaolinite-gibbsite, kaolinite, or of microcline plus quartz. Open (2.0 M KCl solutions) and partially open (0.2 M KCl solutions) triangles represent excess kaolinite; solid triangles represent excess microcline. Top and bottom apices of triangles indicate direction of approach to equilibrium. Reprinted from *Geochim. et Cosmochim. Acta 55*, U. Aja, P. E. Rosenberg, and J. A. Kittrick, Illite equilibria in solutions: I. Phase relationships in the system K_2O-Al_2O_3-SiO_2-H_2O between 25 and 250°C, 1353–64, © 1994, with permission from Elsevier Science Ltd, The Boulevard, Langford Lane, Kidlington OX5 1GB, U.K.

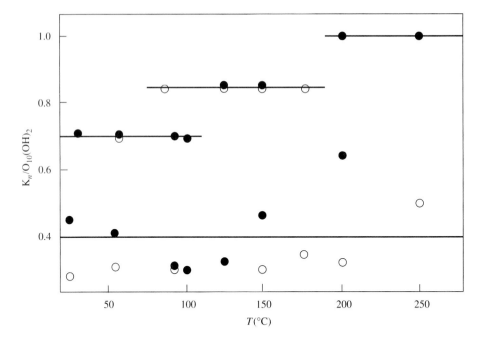

Figure 9.13 Temperature dependence of the compositions of illitic solubility-controlling phases in the presence of kaolinite and quartz betweeen 25°C and 250°C. From S. U. Aja and P. E. Rosenburg in *Clays & Clay Minerals* 40(3):292–99. Copyright 1992. Used by permission.

temperature dependence of I/S compositions is summarized in Fig. 9.13. Although the 25°C results appear inconsistent with those of Garrels (1984), the higher temperature results are not.

So far we have employed $\log([M^{n+}]/[H^+]^n)$ versus $\log[H_4SiO_4^o]$ diagrams to relate the stabilities of individual minerals to the chemistry of natural waters. The same diagrams may be used to follow the chemical evolution of waters as they weather fresh silicate minerals. The approach is also based on the assumption that the water can equilibrate with all the minerals shown. (We should remember when examining the results of this exercise, that this may be a doubtful assumption.) In the following exercise we consider the changes in water chemistry and mineralogy that might occur in a humid climate, as fresh recharge percolates downward in a residual soil formed on a rock such as granite. The granite contains abundant K-feldspar subject to weathering. As shown in Fig. 9.14, which repeats the $\log([K^+]/[H^+])$ versus $\log[H_4SiO_4^o]$ diagram introduced in Figs. 9.8 and 9.9 (cf. Drever 1988; Parkhurst 1995),[†] fresh rainfall plots behind point A on the figure along a hypothetical third axis, such as $-\log[Al^{3+}]$ or $-\log\Sigma Al(aq)$, because it is unsaturated with respect to gibbsite. Weathering proceeds according to a reaction such as

$$KAlSi_3O_8 + 8H_2O \rightarrow K^+ + Al(OH)_4^- + 3H_4SiO_4^o$$

K-feldspar

[†]Steinman et al. (1994) note that the reaction paths for this problem are, in fact, curves, not straight lines.

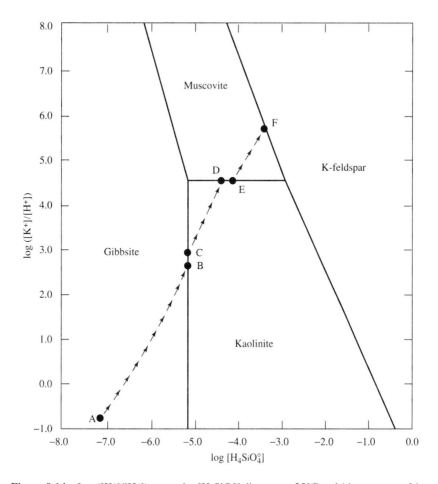

Figure 9.14   Log($[K^+]/[H^+]$) versus log$[H_4SiO_4^\circ]$ diagram at 25°C and 1 bar pressure. Line A→B→C→D→E→F describes the paths of solution composition as K-feldspar reacts with water. After Parkhurst et al. (1990).

until saturation with gibbsite is reached at point A, at which point the weathering reaction that takes place along line A → B is

$$(A \rightarrow B) \qquad KAlSi_3O_8 + H^+ + 7H_2O \rightarrow Al(OH)_3 + K^+ + 3H_4SiO_4^\circ$$

$$\text{K-feldspar} \qquad\qquad\qquad \text{gibbsite}$$

Continued dissolution of K-feldspar as the water moves downward through the soil tends to raise the pH and concentrations of both dissolved potassium and silica in soil waters. At point B saturation with kaolinite is reached, and the weathering reaction along line B → C is

$$(B \rightarrow C) \qquad 2KAlSi_3O_8 + 4Al(OH)_3 + 2H^+ \rightarrow 3Al_2Si_2O_5(OH)_4 + 2K^+ + H_2O$$

$$\text{K-feldspar} \quad \text{gibbsite} \qquad\qquad \text{kaolinite}$$

As long as gibbsite is still present, soil-water composition continues along line B → C at equilibrium with both gibbsite and kaolinite

$$2Al(OH)_3 + H_4SiO_4^\circ = \tfrac{1}{2}Al_2Si_2O_5(OH)_4 + \tfrac{5}{2}H_2O$$

$$\quad\; \text{gibbsite} \qquad\qquad\qquad \text{kaolinite}$$

at a fixed silica concentration (see Eq. [9.3], above). Once all the gibbsite has been consumed (at point C), weathering corresponds to line C → D and the reaction

(C → D) $2KAlSi_3O_8 + 2H^+ + 9H_2O \rightarrow Al_2Si_2O_5(OH)_4 + 2K^+ + 4H_4SiO_4^\circ$

$$\qquad\qquad \text{K-feldspar} \qquad\qquad\qquad\quad \text{kaolinite}$$

Starting at point D and along segment D → E, K-feldspar reacts with kaolinite to form muscovite (K-mica)

(D → E) $KAlSi_3O_8 + Al_2Si_2O_5(OH)_4 + 3H_2O \rightarrow KAl_3Si_3O_{10}(OH)_2 + 2H_4SiO_4^\circ$

$$\qquad\qquad \text{K-feldspar} \qquad \text{kaolinite} \qquad\qquad\qquad \text{K-mica}$$

Along D → E, kaolinite and K-mica will also be in equilibrium with each other, as described by the reaction

$$KAl_3Si_3O_{10}(OH)_2 + H^+ + \tfrac{3}{2}H_2O = \tfrac{3}{2}Al_2Si_2O_5(OH)_4 + K^+$$

$$\qquad\quad \text{K-mica} \qquad\qquad\qquad\qquad\qquad \text{kaolinite}$$

which corresponds to a fixed $[K^+]/[H^+]$ ratio (Eq. [9.2], above). Once all kaolinite has been consumed in the formation of K-mica at E, soil-solution composition follows segment E → F, which is defined by the reaction

(E → F) $\tfrac{3}{2}KAlSi_3O_8 + H^+ + 6H_2O \rightarrow \tfrac{1}{2}KAl_3Si_3O_{10}(OH)_2 + K^+ + 3H_4SiO_4^\circ$

$$\qquad\qquad \text{K-feldspar} \qquad\qquad\qquad\quad \text{K-mica}$$

Finally, equilibrium between K-mica and K-feldspar, as defined by the same reaction, is attained at F. At this point or position in the soil, the weathering reaction obviously stops.

One might imagine a soil having this sequence of phases (gibbsite, kaolinite, K-mica, and K-feldspar) appearing in order from the surface down to bedrock. However, in a real soil illite or mixed-layer illite-smectite are usual soil phases rather than K-mica.

The extremes of conditions described in Fig. 9.14 range from undersaturation of the soil water with respect to gibbsite (described by a hypothetical third [aluminum] axis "behind" A) to saturation of the water with respect to K-feldspar at F. These extremes correspond to weathering the feldspar in a high-rainfall, water-dominated, open soil system in which all weathering products are destroyed, to a rock-dominated, low-rainfall, relatively closed, deep soil system in which K-feldspar persists fresh and unweathered. Depending on climatic and drainage conditions, a real soil profile could, theoretically, stop its development anywhere between these extremes. Steinmann et al. (1994) have also shown that the actual weathering pathways followed in Fig. 9.14 are strongly dependent on the initial pH of the solution.

In their study of the Canadian uranium deposit at Cigar Lake, Cramer and Smellie (1994) have plotted data for K^+, Na^+, Ca^{2+}, and Mg^{2+}, in site waters on $\log([M^n]/[H^+]^n)$ versus $\log[H_4SiO_4^\circ]$ diagrams. In Fig. 9.15, the illite phase field is contoured to show the stabilities of different illite fractions in I/S. The plot describes the evolution of water chemistry from atmospheric precipitation and surface-waters (lakes and streams) to infiltrating soil water and groundwater above, and then in contact with, the orebody. In the soil, kaolinite and illite (the dominant clay), quartz, and feldspars are

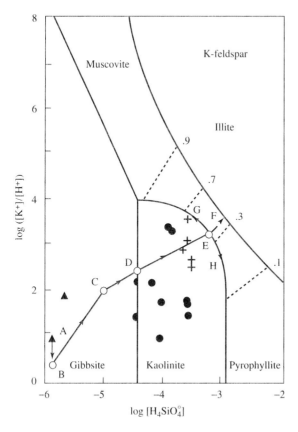

Figure 9.15 Log([K⁺]/[H⁺]) versus log[H₄SiO₄⁰] diagram at 25°C and 1 bar pressure for waters from the area of the Canadian Cigar Lake uranium deposit, with the composition of precipitation (triangles), surface waters (dots), and overburden groundwaters (crosses) plotted. Line segments A through E, G, and H, describe chemical changes from precipitation to soil-zone moisture (groundwater recharge) and to ground-water in the underlying sandy over-burden. From B to E the reaction is driven by dissolution chiefly of kaolin-ite and illite. From J. J. Cramer and J. A. Smellie, eds. *Final report of the AECL/SKB Cigar Lake analog study.* Copyright 1994 Whiteshell Labora-tories. Used by permission.

dissolved. Evolution of the water proceeds with gibbsite precipitation, followed by precipitation of kaolinite. Finally, in the overburden the groundwater is in equilibrium with kaolinite and illite.

The schematic $\log([M^{n+}]/[H^+]^n)$ versus $\log[H_4SiO_4^0]$ diagram in Fig. 9.16, summarizes the general appearance and message of Figs. 9.8 to 9.11. Primary silicate minerals, including the feldspars and micas, are stable only at high pH and high silica and cation concentrations, such as those that occur in rock-dominated systems. Smectite, illite, and vermiculite clays are stable at somewhat lower (but still alkaline) pH's and lower (but still elevated) silica and cation levels. Such conditions are found in soils that are incompletely leached. Kaolinite and finally gibbsite are the stable phases under precipitation-dominated, highly-leached and weathered conditions where the waters are relatively acid and dilute. Even when kaolinite is thermodynamically the most stable clay, muscovite may first weather to form illite and biotite to form vermiculite (cf. Likens et al. 1977). This is because biotite and vermiculite have similar T:O:T structures to the parent micas and so have lower nucleation energies than kaolinite with its T:O structure.

9.6 THE THERMODYNAMIC STABILITY OF COMPLEX CLAY MINERALS

The thermodynamic properties of clay minerals have been obtained from calorimetric measurements (cf. Robie et al. 1978), from laboratory solubility and phase equilibria measurements (Aja and

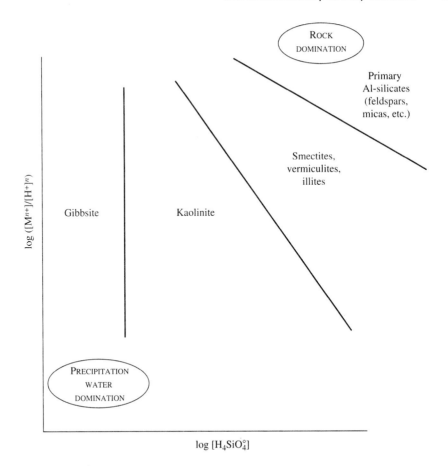

Figure 9.16 Schematic $\log([M^{n+}]/[H^+]^n)$ versus $\log[H_4SiO_4^\circ]$ diagram showing the general phase relations among primary silicates such as K-feldspar, the clays smectite, vermiculite, illite, and kaolinite, and the mineral gibbsite.

Rosenberg 1992), and from the chemistry of natural waters in contact with the clays (Hess 1966; Garrels 1984). For the silicates in general and clay minerals in particular, the most accurate way to obtain their ΔG_f° values is usually through solubility measurements. However, the wide range of possible compositions of nonstoichiometric clay minerals makes it impractical for us to measure the thermodynamic properties of all of them in the laboratory. Fortunately, in recent years methodologies for estimating S°, ΔG_f°, and ΔH_f° for such minerals have been greatly refined. Resultant estimates, which usually require a structural chemical analysis, are probably of adequate accuracy for most geochemical modeling purposes ($\pm 0.2\%$ in ΔG_f° at 25°C) (cf. Vieillard and Tardy 1988 [ΔH_f°]; Holland 1989 [S°]; Chermak and Rimstidt 1989 [ΔG_f°, ΔH_f°], 1990 [$\Delta G_f^\circ(T)$]; Varadachari et al. 1994 [ΔG_f°]).

So far we have used phase diagrams to visualize clay mineral stabilities and phase relations involving the clays in natural waters. Given the complex chemistries of mixed-layer clays in particular, geochemical computer codes offer a more rigorous way to evaluate their stabilities. The thermodynamic data bases of most of these codes list stability constants for a variety of clay minerals which, except for kaolinite, are usually of nonideal composition. Most of these stability constants have been obtained from solubility measurements and are of mixed reliability. It is appropriate to ask if they can help us to understand the behavior of similar clay minerals (e.g. illites or smectites) that may have different or unknown compositions in the water-rock systems we are studying. The answer to this question is probably a qualified yes.

Example 9.3

Tabulated here is the composition of average river water (Livingstone 1963). Assume pH = 7.0 and 25°C.

Species	mg/L	Species	mg/L
Ca^{2+}	15	Cl^-	7.8
Mg^{2+}	4.1	SO_4^{2-}	11
Na^+	6.3	$H_4SiO_4^\circ$	21
K^+	2.3	$\Sigma Fe(III)$	0.67
HCO_3^-	60	ΣAl	0.07

Given the insolubility of Fe(III)- and Al-oxyhydroxides and clay minerals at pH 7, the Fe and Al are probably present as suspended solids.

Assuming that goethite and kaolinite are *finite solids*, enter the chemical analysis into MINTEQA2, except for the Fe and Al concentrations. Compare computed $\Sigma Fe(III)(aq)$ and $\Sigma Al(aq)$ concentrations at saturation with goethite and kaolinite to the values in the table. How much Fe and Al may be in suspension? Next assume that the water is in equilibrium with goethite and montmorillonite (make them finite solids). Now what is the computed $\Sigma Al(aq)$ concentration?

If goethite and kaolinite solubility limited Al and Fe, their concetrations would be $\Sigma Fe(III)(aq) = 7.99 \times 10^{-13}$ M (4.46×10^{-8} mg/L) and $\Sigma Al(aq) = 7.14 \times 10^{-10}$ M (1.93×10^{-5} mg/L). Essentially all of both metals must, therefore, be in suspension. Montmorillonite is slightly more soluble, so that at montmorillonite saturation $\Sigma Al(aq) = 1.32 \times 10^{-9}$ M (3.56×10^{-5} mg/L).

At present, there are serious limitations to the use of geochemical codes to study clay mineral solution-equilibria. These include the sparse and often dubious clay mineral stabilities given in program data bases and the fact that water analyses rarely include reliable data for both dissolved Si and Al. Also, dissolved Al is usually below detection above pH 5 to 6. When reported at higher pH's, aluminum is mostly present in suspended form, as suggested by Example 9.3. For such reasons, many researchers still prefer using graphic methods to depict the stabilities and behavior of clay minerals.

STUDY QUESTIONS

1. Know the basic structures of the major clay and mica mineral groups and the definitions of octahedral, dioctahedral, gibbsite, brucite, and tetrahedral layers.

2. What is the origin of the ion exchange behavior of smectite and vermiculite clays?

3. Be able to calculate the ion exchange capacity of a smectite or vermiculite clay (meq/100 g clay) from the chemical formula of the clay.

4. Know the general conditions of occurrence and stability of the major clay mineral groups, including the roles played by parent mineralogy, temperature, precipitation rate, soil drainage, and soil maturity in clay occurrence.

5. What is the Gay-Lussac-Ostwald (GLO) step rule and how does it help explain the formation and persistence of metastable clay minerals in soils?

6. Know how to construct a log $([M^{n+}]/[H^+]^n$ versus log $[H_4SiO_4^o]$ diagram, starting with the appropriate Gibbs free energy data or equilibrium constants.

7. Be able to plot natural water compositions on a log $([M^{n+}]/[H^+]^n$ versus log $[H_4SiO_4^o]$ diagram and discuss their significance.

8. What assumptions are made if one uses a log $([M^{n+}]/[H^+]^n$ versus log $[H_4SiO_4^o]$ diagram to predict and explain the occurrences of minerals that have stability fields shown on the figure?

9. Know how to follow hypothetical changes in the chemical composition of waters that are weathering minerals such as the feldspars under water-dominated or rock-dominated conditions, using a log $([M^{n+}]/[H^+]^n$ versus log $[H_4SiO_4^o]$ diagram.

10. How can the stabilities of coexisting clay minerals be estimated from the composition of soil moisture or groundwater?

11. Contrast the applicability and use of phase diagrams and geochemical codes for purposes of evaluating clay mineral equilibria in natural waters.

PROBLEMS

1. **(a)** On the following figure plot the compositions:
   - **(i)** seawater, given that $mNa^+ = 0.460$, pH = 8.15, and $SiO_2(aq) = 0.01$ mg/L (surface ocean water) to 7 mg/L (ocean bottom);
   - **(ii)** rainwater, with $Na^+ = 86$ μmol/L, pH = 4.4, and $\Sigma Al(aq) = 5$ μmol/L, and $SiO_2(aq) = 0.3$ mg/L;
   - **(iii)** a stream, assuming $Na = 10^{-3.00}$ M, pH = 6.5, and $SiO_2(aq) = 14$ mg/L.

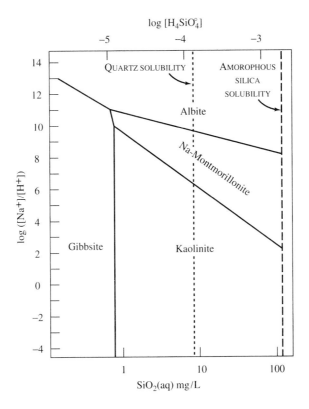

(b) What conditions do the lines that separate the mineral stability fields represent?

(c) Can you confidently argue that when a composition plots in a mineral stability field on the diagram that water is in equilibrium with the mineral? Explain your answer with reference to the rainwater analysis.

(d) A deep soil water has a pH of 7.5 and Na concentration of 2×10^{-3} M. Roughly how high must the $SiO_2(aq)$ concentration be to make Na-montmorillonite thermodynamically stable in the soil?

(e) Comment on the use of the diagram to describe the weathering of the feldspar in wet, high-rainfall conditions versus in an arid, organic-free soil.

2. In their study of the chemical weathering of basalts, Nesbitt and Wilson (1992) used a plot of $\log [(Ca^{2+})/(H^+)^2]$ versus $\log [SiO_2(aq)]$. An important boundary on their plot is that for calcite saturation (labeled C_{sat}), which represents an upper limit for calcium concentrations in weathering environments. Their plot is shown here. The data points are for rainwater (black squares) and for soil and groundwaters from basaltic terranes.

(a) Calculate and show the position of the $\log [(Ca^{2+})/(H^+)^2]$ boundary for calcite on this plot for CO_2 pressures of $10^{-3.5}$ bar (atmospheric) and $10^{-1.0}$ bar (a roughly maximum soil and groundwater value).

(b) Comment on the significance of this boundary relative to the occurrence of other phases plotted in the figure.

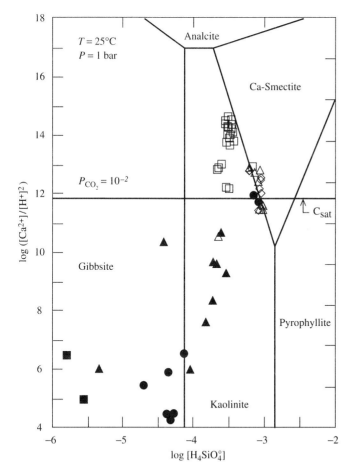

H. W. Nesbitt and R. E. Wilson, 1992. Recent chemical weathering of basalts. From *American Journal of Science,* 292:740–77. Reprinted by permission of *American Journal of Science.*

3. Sears (1976) observed that kaolinite and illite coexist in deep Pennsylvania soils under conditions such that the soil moisture is practically isolated from dilution by fresh recharge. His chemical analyses of soil moisture are plotted on the $\log([K^+]/[H^+])$ versus $\log[H_4SiO_4^\circ]$ diagram for two different sampling sites. Assuming that the *highest* silica and K^+ concentrations are those most likely to approach equilibrium with both

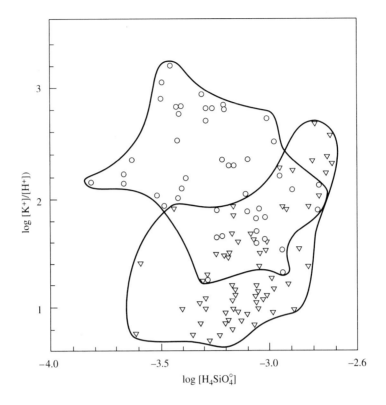

illite and kaolinite, estimate an approximate stability for illite relative to that of kaolinite. Use the generalized illite compositions suggested by Hess (1966) $[KAl_5Si_7O_{20}(OH)_4]$ and by Deer et al. (1992) $[K_{1.5}Al_{5.5}Si_{6.5}O_{20}(OH)_4]$, and compare the results.

4. This groundwater was collected from a well 167 ft. deep in granitic gneiss in Baltimore County, MD (Hem 1985). Assume Eh = 0.0 V and $T = 12°C$.

Species	mg/L	Species	mg/L
Na^+	6.8	F^-	0.1
K^+	4.2	HCO_3^-	121
Mg^{2+}	1.9	NO_3^-	0.2
Ca^{2+}	28	SO_4^{2-}	1.4
Fe^{2+}	2.7	Al^{3+}	0.2
Mn^{2+}	0.22	$SiO_2(aq)$	31
Cl^-	1.0		

Note: pH 6.9

Input this data into SOLMINEQ.88 (Kharaka et al. 1988), which has a relatively large thermodynamic data base for clay minerals. Assume the groundwater contains no dissolved organic ligands.

(a) What is the saturation state of the water with respect to the clays chamosite (7 Å) (a chlorite), illite, and kaolinite?

(b) Rerun the program, precipitating Al^{3+} as kaolinite to attain equilibrium with kaolinite. Compare the $\Sigma Al(aq)$ concentration at kaolinite saturation to its chemical analytical value reported in the table. Compare the saturation indices of the clays considered in part (a) to their indices computed assuming equilibrium of the water with respect to kaolinite.

10

Adsorption-Desorption Reactions

10.1 PROPERTIES OF SORBENT MATERIALS

10.1.1 Particle Size and Surface Area

Particles less than about 1 μm (10^{-4} cm) in diameter have a significant percentage of their atoms at particle surfaces. At such sizes and smaller (in the colloidal size range), particles have important surface properties, whereas larger particles generally do not. Such surface properties have at least three significant consequences:

1. They cause an important increase in the solubilities of small particles (an increase in a solid's ΔG_f° value) (cf. Langmuir and Whittemore 1971; Stumm and Morgan 1996).
2. Colloidal-sized particles can remain in stable suspension and be transported by natural waters, including by groundwaters (cf. McCarthy and Degueldre 1993).
3. Such particles usually have a significant unsatisfied surface charge that is related to their colloidal behavior, but also makes them potential sorbents for dissolved species in water.

Shown in Fig. 10.1 are the size ranges of some common natural materials that occur in colloidal and larger particle sizes in water. Colloids are stable in solution when the solution chemistry does not cause particle agglomeration and settling. Under such conditions colloidal particles remain in suspension because of mutual charge repulsion and Brownian motion (cf. Stumm and Morgan 1996) and can be transported by natural waters moving at any velocity. Larger particles tend to settle out gravitationally, except in fast-moving streams and groundwaters, particularly when their flow is turbulent.

The size ranges of some filter types used to remove particulate materials from natural waters are also shown in Fig. 10.1. The standard 0.45 μm-membrane filter clearly has pores too large to filter out many colloidal-sized materials, which can include metal oxyhydroxides, clays, and viruses. Removal of colloidal-sized particles by filtration often takes place, however, when soil water or

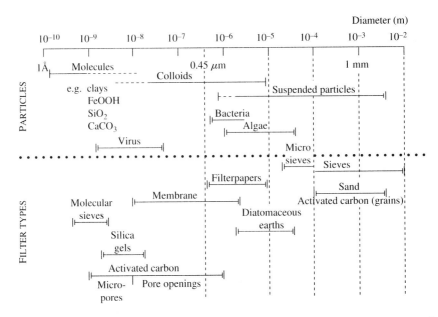

Figure 10.1   The sizes of some natural particulate materials found in surface-water and groundwaters, and the pore sizes of filters used to remove them. After W. Stumm and J. J. Morgan. Copyright © 1981 by John Wiley & Sons. Reprinted by permission of John Wiley & Sons, Inc.

groundwater moves through fine-grained rock matrix or clay-rich soils or sediments. Davis and Kent (1990) (see also Beven and Germann 1982) suggest that the pores in rock and soil be termed micropores if less than 2 nm ($<2 \times 10^{-7}$ cm) in diameter. Mesopores range in size from 2 to 50 nm (2×10^{-7} to 5×10^{-6} cm), with macropores larger than 50 nm. Micropores and mesopores can filter out most of the colloidal materials carried by soil water and groundwater.

10.1.2 Surface Charge and Surface-Site Density

The importance of the surface properties (including sorptive properties) of a given weight of material exposed to solution, increases in proportion to the surface area of that material, and to its surface charge (or site) density or number of charged sites per unit area or weight.

The following example demonstrates how remarkably large the total area of a substance becomes when a given amount of it occurs in smaller and smaller particles. The example substance is the mineral hematite (α-Fe$_2$O$_3$). Table 10.1 shows the profound increase in surface area that results

TABLE 10.1   The total surface area of a 1 cm³ volume of the mineral hematite (α-Fe$_2$O$_3$), density = 5.26 g/cm³, GFW = 160 g/mol, when it is subdivided into smaller and smaller cubes

Hematite cube diameter	Number of cubes	Area (m²/g)	Mol % on surface
1 cm	1	0.00011	0.00012
10^{-4} cm (1 μm)	10^{12}	1.1	0.12
10^{-6} cm (0.01 μm)	10^{18}	110	12

Note: All atoms within 2 Å of the hematite surface are assumed to be on the surface.

TABLE 10.2   Some measured or estimated surface areas (S_A) and maximum surface-site densities (N_S) (chiefly negative) for geological materials

Mineral/phase	S_A (m²/g)	Source	N_S (sites/area or wt)	Source
α-FeOOH (goethite)	45 to 169	†	2.6 to 16.8/nm²	‡
			18/nm²;	§
			1.35×10^{-3} mol/g	
α-Fe$_2$O$_3$ (hematite)	1.8 (natural)	§	5 to 22/nm²	‡
	3.1 (synthetic)			
Fe(OH)$_3 \cdot n$H$_2$O (ferrihydrite)	250, 306	§	20/nm²	‡
	600	‖	0.1 to 0.9 mol per	§
			mol of Fe	
MnO$_2$ (synthethic and natural	290 (fresh)	†	18/nm²	†
birnessite)	143 (aged)		2/nm²	‡
	180 (natural)			
SiO$_2$ (quartz)	0.14	‖	4.2 to 11.4/nm²	#
SiO$_2$ gel [SiO$_2$(am)]	53, 292	‡	4.5 to 12/nm²	‡
	170, 180	††	5/nm²	#
α-Al(OH)$_3$ (gibbsite)	120	‖	2 to 12/nm²	‡
γ-Al(OH)$_3$ (bayerite)	156	#	6 to 9/nm²	‡
TiO$_2$ (rutile)	5 to 19.8	††	5.8/nm²	††
			12.2/nm²	#
kaolinite	10 to 38	‡	1.3 to 3.4/nm²	‡
	12	‖	1.2 to 6.0/nm²	‡‡, §§
illite	65 to 100	‖	0.4 to 5.6/nm²	‡‡, §§
montmorillonite (Na form)	600 to 800 (esp.	‡, ‖	0.4 to 1.6/nm²	‡‡, §§
	interlayer)			
organic substances in soils;	260 to 1300	§§	2.31/nm² assumed	§§
humic materials			1 to 5×10^{-3} mol/g	
bulk composite geological	600	‡	range 1 to 7/nm²	‡
materials (except smectites);			mean value of	
assumes Fe$_2$O$_3 \cdot$ H$_2$O with an			2.31/nm²; 3.84	
area of 600 m²/g and			μmol sites/m²	
0.205 mol sites/mol Fe				

Source: †Catts 1982. ‡Davis and Kent 1990. §Hsi 1981. ‖Schwarzenbach et al. 1993. #James and Parks 1982. ††Kent et al. 1986. ‡‡Computed from information given by Schwarzenbach et al. 1993. §§Estimated from CEC values in Table 10.4.
Sposito (1989) suggests -9×10^{-3} to $+1 \times 10^{-3}$ mol/g.

from subdividing 1 cm³ of hematite (0.033 mol; 1 g/0.19 cm³) into smaller and smaller particle sizes. It is interesting to note that 100 m² = 10^{-4} km² = 0.0247 acres. Thus, 40 cm³ (210 g) of hematite, if present in 0.01 μm-diameter cubes, has a total surface area of about 1 acre.

Many minerals occur routinely in even smaller particle sizes and therefore have remarkably large surface areas, thus high surface reactivities for their weights. Given in Table 10.2 are surface areas for some geological materials for comparison with the hematite illustration in Table 10.1. Also tabulated are ranges or example values of maximum surface-site densities for the same materials. The sorptive abilities of minerals are proportional to their surface areas and surface-site densities, which can range widely, reflecting differences in how they are measured (cf. James and Parks 1982; Kent et al. 1986, Davis and Kent 1990).

In calculations of adsorption from natural waters, it is convenient to measure the concentration of sorbing surface sites (Γ_{SOH}) in units of moles of monovalent sites exposed to a liter of solution. Using the approach taken in MINTEQA2, Γ_{SOH} equals

$$\Gamma_{SOH}(\text{mol sites/L}) = \frac{N_S(\text{sites/m}^2) \times S_A(\text{m}^2/\text{g}) \times C_S(\text{g/L})}{N_A(\text{sites/mol sites})} \tag{10.1}$$

where N_S, S_A, C_S, and N_A are, respectively, the surface-site density, surface area per weight of sorbent, weight of sorbent in contact with a liter of solution, and Avogadro's number (6.022×10^{23} sites/mol of sites). PRODEFA2, the input file for MINTEQA2, requests values for Γ_{SOH}, S_A, and C_S, and computes a corresponding value for N_S. The following conversions are useful among the units listed in Table 10.2: $N_S(\text{sites/nm}^2) = N_S(\mu\text{mol sites/m}^2) \times 0.6022$; $N_S(\text{sites/m}^2) = 10^{18} \times N_S(\text{sites/nm}^2)$; and $N_S(\text{mol sites/g}) = N_S(\text{sites/nm}^2) \times 1.661 \times 10^{-6} \times S_A(\text{m}^2/\text{g})$. Thus, for goethite in Table 10.2, it can be shown that the listed N_S values of 18 sites/nm^2 and 1.35×10^{-3} mol sites/g, correspond to a surface area (S_A) of 45 m^2/g.

Soil scientists have historically reported the site density (generally termed the exchange capacity) on a per weight basis, in units of meq/100 g of sorbent. A sorbent phase with a net negative surface charge exhibits cation exchange capacity (CEC), whereas one with a net positive surface charge has anion exchange capacity (AEC). For a negatively charged surface, the site density and cation exchange capacity are related through the expression $N_S(\text{sites/nm}^2) \times S_A(\text{m}^2/\text{g}) \times 0.1661 = $ CEC(meq/100 g).

Surface charge may be *permanent* and independent of solution composition, or *variable*, changing with solution composition. The CEC of the smectite and vermiculite clays and the zeolites is largely permanent and independent of solution chemistry. In contrast, the metal oxyhydroxides and kaolinite clay have a surface charge that is strongly pH-dependent; they are net positive at low pH values and net negative at higher pH values (cf. James and Parks 1982; Davis and Kent 1990). Reasons for such behavior are presented in following sections.

For the clays there are three important causes of surface charge (generally negative). These include:

1. *Isomorphous substitution* in the crystal lattice; for example in the clays, Al^{3+} replaces Si^{4+} in the tetrahedral layer and Mg^{2+} replaces Al^{3+} in the octahedral layer. This results in an excess of O^{2-} bonds and is important for illites; it is the chief cause of negative surface charge for the smectites and vermiculites.

2. *Lattice imperfections or defects,* such as a deficit in octahedral Al^{3+} or interlayer K^+, which leads to a net negative surface charge. This is important for smectites, and, to a lesser extent, illites.

   The permanent charge of clay minerals is due to lattice imperfections or defects, plus isomorphous substitutions. Sposito (1989) suggests that the permanent negative charge of illites, smectites, and vermiculites in mol sites/kg, ranges from 1.9 to 2.8, 0.7 to 1.7, and 1.6 to 2.5, respectively.

3. *Broken or unsatisfied bonds* at crystal plate corners and edges lead to the ionization of surface groups, usually resulting in a net negative surface charge due to exposed O^{2-} and OH^-. This mechanism of charge development affects all clays, as can be seen in Fig. 10.2, which is a plot of CEC versus pH for kaolinite, illite, and bentonite (a smectite) clay. All the CEC versus pH curves in Fig. 10.2 have a slight negative slope, showing that metal adsorption decreases with decreasing pH. This is because the increasing H^+ ion concentrations compete more effectively with fixed metal cation concentrations for adsorption sites on the clay. Broken bonds

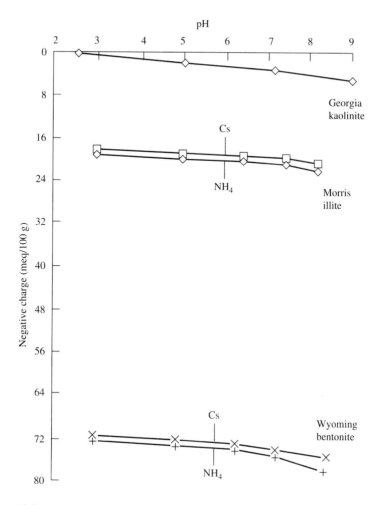

Figure 10.2   Negative surface charge (CEC) of some clays between pH 2 and 9 determined using 1.0 M CsCl or 1.0 M NH₄Cl. Modified after D. J. Greenland and C. J. B. Mott, 1978, Surfaces of Soil Particles, in D. J. Greenland and M. H. B. Hayes, eds., *The chemistry of soil constituents*. Copyright © 1978 by John Wiley & Sons, Ltd. Reprinted by permission of John Wiley & Sons, Ltd.

are the chief source of surface charge for kaolinite, but are proportionately less important for clays such as the illites and smectites, which have a much higher total surface charge than kaolinite because of permanent charge effects.

As clay particles decrease in size, the relative importance of broken bonds increases. This effect is illustrated in Fig. 10.3, which plots clay CEC versus particle size. As expected, the size effect is most important for kaolinite and decreasingly important for illite and the smectite clays nontronite and saponite. Thus, in the size range from 1.0 to 0.1 μm, the CEC of the Georgia kaolinite increases from about 3.4 to 7.3 meq/100 g. The plot shows that this increase of about 4 meq/100 g is practically negligible for the smectites, which have CEC values of 60 meq/100 g or more.

The zeolites are a group of hydrated aluminosilicate minerals with open framework structures of $(Si,Al)O_4$ tetrahedra, charge-balanced mostly by Ca^{2+}, Na^+, or K^+ (cf. Mumpton 1977). A general

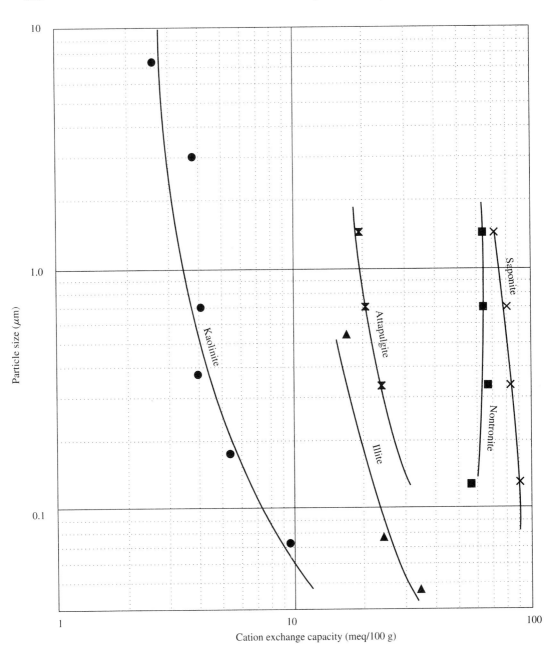

Figure 10.3 Negative surface charge (CEC) of some clays as a function of particle size. Values are from Grim (1968). See also Bodek et al. (1988).

formula for the zeolites is $(Na_2,K_2,Ca,Ba)[(Al,Si)O_2]_n \cdot nH_2O$ (Deer et al. 1992). The cations and water molecules occupy open cavities in the zeolite structure. The isomorphous substitution of tetrahedral Al^{3+} for Si^{4+} creates a negative charge within the cavities. Monovalent and divalent cations in the cavities are weakly held and readily exchangeable. Because cation exchange takes place within the zeolite crystal lattice, it is practically independent of particle size. Since it results from isomorphous substitutions, the CEC of zeolites is permanent and nearly pH-independent.

The surface charge of oxides, hydroxides, phosphates, and carbonates is produced chiefly by ionization of surface groups, or surface chemical reactions. Resultant adsorption sites on oxides and hydroxides are sometimes written as SOH_n^{n-1}, where S denotes the major structural cation at the surface and $n = 0$ to 2. Alternatively, the major structural cation (e.g., Fe, Si, Al) is written in lieu of S. In either case the cation may be preceded by a triple equal sign. For example, the pH-dependent surface charge of a silicate might reflect the presence of surface species written symbolically as $\equiv SiOH_2^+$, $\equiv SiOH$, and $\equiv SiO^-$ (see Fig. 10.4). Similarly on a ferric oxide surface, the surface complexes contributing to surface charge with increasing pH might be described as $FeOH_2^+$, $FeOH$, and FeO^-. For both examples, the surface species are positively charged at low pH and deprotonate as pH increases to form neutral and negatively charged species at intermediate and higher pH's, consistent with the amphoteric behavior of the oxide or hydroxide. The surface charge of such materials is thus strongly pH-dependent.

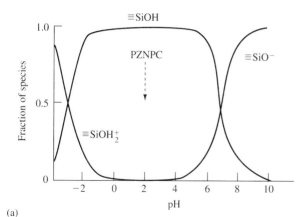

(a)

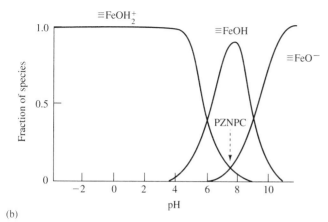

(b)

Figure 10.4   Schematic distribution of charged surface species (denoted by the triple dashes) on **(a)** silica gel and **(b)** ferrihydrite as a function of pH, showing for both the predominance of positive, neutral, and negatively charged surface species with increasing pH. The pH of the PZNPC is found where the net surface charge is zero (i.e., $[\equiv SiOH_2^+] = [\equiv SiO^-]$ and $[\equiv FeOH_2^+] = [\equiv FeO^-]$). Based on Healy (1974).

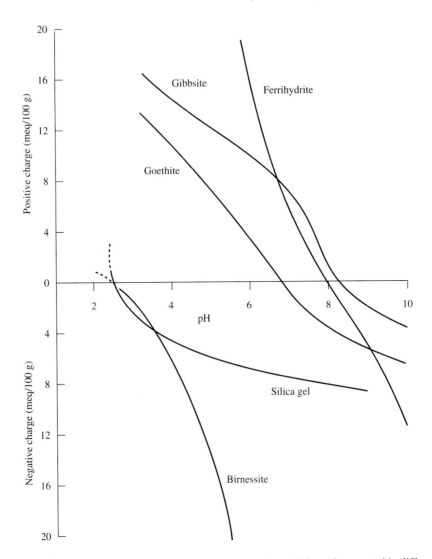

Figure 10.5 Surface charge of some oxyhydroxides from pH 2 to 10 measured in different electrolyte solutions shown in parentheses. Ferrihydrite $[Fe(OH)_3 \cdot nH_2O]$ (0.001 M $NaNO_3$) from Hsi (1981); gibbsite $[Al(OH)_3]$ and silica gel $[SiO_2 \cdot nH_2O]$ (1.0 M CsCl) based on Greenland and Mott (1978); goethite $[\alpha\text{-}FeOOH]$ (0.005 M CsCl) based on Greenland and Mott (1979) (see also Hsi [1981]); birnessite $[\sigma\text{-}MnO_2]$ (0.001 M $NaNO_3$) from Catts (1982).

Figure 10.5 shows the strong surface charge variation with pH of several common oxyhydroxides of Al, Fe(III), Mn(IV), and Si. As shown in Fig. 10.2, the surface charge of kaolinite also exhibits a marked pH dependence, behaving like that of the oxides and hydroxides in Fig. 10.5.

The pH at which the mineral surface charge changes sign is called the *zero point of charge (ZPC)* or *point of zero charge (PZC)*. If a surface changes its net surface charge from positive to negative solely because of the adsorption of H^+ or OH^- ions, the pH of charge reversal is called the *PZNPC* or *point of zero net proton charge*—sometimes also called the *PPZC* or *pristine point of zero charge* or *isoelectric point (IEP)* (cf. Davis and Kent 1990; Lewis-Russ 1991). Some PZNPC val-

TABLE 10.3 The pH of the point of zero net proton charge (PZNPC) of some minerals and organic substances

Solid	pH_{PZNPC}	Solid	pH_{PZNPC}
SiO_2 (quartz)	1 to 3 (2.91)	MnO_x (general)	1.5 to 7.3
SiO_2 (amorphous)	3.5 (3.9)	α-MnO_2 (cryptomelane)	4.5
Na-feldspar	6.8 (5.2)	β-MnO_2 (pyrolusite)	4.6 to 7.3 (4.8)
K-feldspar	(6.1)	CuO (tenorite)	9.5 (8.6)
Montmorillonite	$\leq$ 2 to 3	MgO (periclase)	12.4 (12.24)
Kaolinite	$\leq$ 2 to 4.6 (4.66)	α-$Al(OH)_3$ (gibbsite)	10.0 (9.84)
Muscovite	(6.6)	$CaCO_3$ (calcite)	8.5, 10.8
Mg-silicates	9 to 12	$Ca_5(PO_4)_3F$ (fluorapatite)	4 to 6
α-Fe_2O_3 (natural hematite)	4.2 to 6.9	$Ca_5(PO_4)_3(OH)$ (hydroxyapatite)	$\leq$ 8.5
$Fe(OH)_3$ (amorphous)	8.5 to 8.8	$FePO_4 \cdot 2H_2O$ (strengite)	2.8
α-FeOOH (goethite)	5.9 to 6.7	$AlPO_4 \cdot 2H_2O$ (variscite)	4
Mn(II) manganite	1.8	Algae	2
σ-MnO_2 (birnessite)	1.5 to 2.8	Sewage effluent (bacteria, etc.)	2

Source: Values are from Parks (1965), Sverjensky (1994), and Stumm and Morgan (1996). Values estimated by Sverjensky (1994) are given in parentheses.

ues are given in Table 10.3. The PZNPC is measured in an *indifferent electrolyte,* that is in a solution where the cations and anions of the major salt tend not to be adsorbed by the solid. The PZNPC is a property of the solid, independent of solution composition (Lewis-Russ 1991; Sverjensky 1994). Parks (1965) and more recently Sverjensky (1994), have shown how to estimate PZNPC values from sorbent properties, including the sorbent's dielectric constant, Pauling electrostatic bond strength, and cation-hydroxyl bond length.

10.1.3 Cation Exchange Capacity of Some Natural Materials

Table 10.4 lists the cation exchange capacities (CEC's) and their pH dependences for some important sorbing minerals and other phases. Dependencies characterized as absent, negligible, or slight indicate that the sorbent obtains its CEC chiefly because of interior lattice charge imbalances. Strong dependencies indicate that the surface charge results from reactions at the sorbent-solution interface.

TABLE 10.4 Cation exchange capacities (CEC's) of some substances measured at pH 7 and their pH dependences

Substance	CEC(meq/100 g)	pH dependence
Kaolinite	3 to 15	strong
Glauconite (green sand)	11 to 20	slight
Illite and chlorite	10 to 40	slight
Smectite-montmorillonite	80 to 150	absent or negligible
Vermiculite	100 to 150	negligible
Zeolites	100 to 400	negligible
Organics in soils, humic materials	100 to 500	strong
Mn(IV) and Fe(III) oxyhydroxides	100 to 740	strong
Synthetic cation exchange resins	290 to 1020	slight

Source: Grim (1968); Brady (1974); Mumpton (1977); Bodek et al. (1988); Lide (1995).

TABLE 10.5 Measured cation exchange capacities, in situ soil pH values and percentages of major exchangeable ions of some soils at pH = 7

Soil	pH	CEC	Equivalent Percent of CEC					
			H	K	Na	Ca	Mg	Mn
Morrison sandy loam subsoil (5 ft depth)	~5	3.4	50	3	3	3	24	18
Lanna soil	4.6	17.3	34	2	1	48	16	—
California soils	7	20.3	—	6	3	66	26	—
Holland soils	7	38.3	—	2	6	79	13	—
Merced clay loam	10	18.9	0	5	95	0	0	—

Note: Dashes indicate no available data.

Source: From Grim (1968), Wiklander (1964), and Apgar and Langmuir (1971).

Given in Table 10.5 are measured, total CEC values for some soils, along with the equivalent percents of different adsorbed cations making up the exchangeable CEC and corresponding pH's. The CEC values reported in the soils literature have usually been measured at pH 7. Figures 10.2 and 10.5 show that CEC values for the oxides and kaolinite are a strong function of pH, so that a CEC measured at pH 7 for these phases is in serious error if assumed for other pH's. For the bentonite (a smectite clay) and the illite, surface charge is relatively independent of pH, so that the CEC measured at pH 7 has more general value.

Note the relative concentrations of H^+ ion and other cations as percents of the soil CEC values in Table 10.5. Important ions adsorbed on soils, but rarely measured, may include NH_4^+, and on waterlogged soils may also include Fe^{2+} and Mn^{2+}. The data in Table 10.5 show that below about pH 5, protons occupy a large fraction of clay surface sites. Below pH 3 to 4, protons occupy all the sites and tend to destroy clay structures. This, of course, makes it difficult to study the surface and sorptive properties of clays in acid solutions. The proton competes effectively with other cations for exchange sites, even when its concentration is 10 to 100 times less than that of the cations.

The total CEC values of soils reflect the variable contributions of organic matter, clay minerals, and ferric and Mn oxyhydroxides. The high CEC of natural organic matter (humic substances), which is typically between 150 to 300 meq/100, derives from its carboxyl (COOH) groups, which deprotonate at the pH of most natural waters (Thurman 1985). The high CEC values for waterlogged soils (up to 15 meq/100 g for New Jersey soils, according to Toth and Ott 1970) probably result from their high organic contents. The CEC of soils is generally highest in the A and B horizons, where humic substances and clays are most abundant (Thurman 1985). Apgar and Langmuir (1971) measured CEC values of only a few meq/100 g in C-horizon soils in central Pennsylvania. In these soils organic matter was absent, and clays such as kaolinite and illite made up about 20 to 40% of the soil.

According to Toth and Ott (1970), about 80% of the CEC of modern river and estuarine sediments is due to fresh organic matter. These authors found that sediments from the Delaware River, Chesapeake Bay, and Barnegat Bay were 2 to 24% organic matter and had a mean CEC of 57 meq/100 g. Of this mean CEC, organic sorbents contributed 41 meq/100 g, and inorganic sorbents 16 meq/100 g.

In soils and in bottom sediments associated with streams and other surface-waters, the CEC is highest for fine-grained materials, and decreases rapidly with increasing particle sizes. Bodek et al. (1988) report the following CEC values in meq/100 g for different soil-size ranges: sand, 2 to 3.5;

TABLE 10.6 Cation exchange capacity and time for 50% exchange of K^+ onto Mattole River sediments of increasing size from clay to fine gravel

Size fraction	Total CEC (meq/100 g)	CEC organic fraction (meq/100 g)	Time for 50% exchange
Clay ($<2\ \mu$m)	53 to 38	47 to 34	2 to 3 s
Silt (2 to 62 μm)	18 to 6	16 to 6	3 to 5 s
Sand (0.06 to 2 mm)	8 to 7	7.5 to 6	8 to 3,000 s
Fine gravel (2 to 9.5 mm)	8 to 6.4	7.4 to 6.4	10,000 s (3 h)

Source: Data from Malcolm and Kennedy (1970).

sandy loam, 2.3 to 17.1; loam, 7.5 to 15.9; silt-loam, 9.4 to 26.3; and clay and clay-loam, 4 to 57.5. For Mattole River sediments, Kennedy and Malcolm (1977) observed that clay-sized materials ($<2\ \mu$m) had a CEC of 38 to 53 meq/100 g. The CEC of fine gravel (2 to 9.5 mm) ranged from 6.4 to 8 meq/100 g.

In a survey of U.S. stream sediments, Kennedy (1965) concluded that the makeup and properties of the stream sediments essentially equaled that of local soils. In the eastern states (50 to 150 cm precipitation), dominant clays in the $<4\ \mu$m (0.004 mm) fraction were illite, kaolinite, vermiculite, and interlayered clays, with a CEC of 14 to 28 meq/100 g. In central and west-central states (25 to 100 cm precipitation) Kennedy found dominant smectite, vermiculite, mixed-layer illite, kaolinite, quartz, and feldspar in the $<4\ \mu$m fraction, with a CEC range of 25 to 65 meq/100 g. In California and Oregon, because of the wide range of wet and dry conditions (<25 to >200 cm precipitation), clays were highly variable, and had a range of CEC's from 18 to 65 meq/100 g for the $<4\ \mu$m fraction.

The rate of attainment of equilibrium in exchange reactions is fastest when clay-sized materials are involved and sorbate ions can directly contact surface sites. In larger-sized materials, including rock fragments and in rock matrix, however, the attainment of adsorption equilibrium is limited by the much slower rate of diffusion of sorbing species into and out of rock pores (cf. Jenne 1995). This point is illustrated by the measurements of Malcolm and Kennedy (1970), who studied the rate of exchange of K^+ ion by different size fractions of Mattole River sediments using a potassium-specific ion electrode. Their results are summarized in Table 10.6. Complete exchange took 1 to 2 days in the fine gravel. Such effects are also important in fractured rock where most groundwater flow and solute transport occurs in fractures, but diffusion of sorbing solutes from the fractures into adjacent rock matrix takes place (cf. Skagius and Neretnieks 1988; Schwarzenbach et al. 1993).

10.2 SORPTION ISOTHERMS AND THE DISTRIBUTION COEFFICIENT

10.2.1 The Freundlich Adsorption Isotherm and the Distribution Coefficient K_d

It is generally observed that as the concentration of a species in solution increases, the amount of it sorbed (attached in some way) to contacting solid (sorbent) surfaces also increases. A plot that describes the amount of a species sorbed (the sorbate) as a function of its concentration in solution, measured at constant temperature, is called a sorption isotherm. A simple sorption isotherm is shown

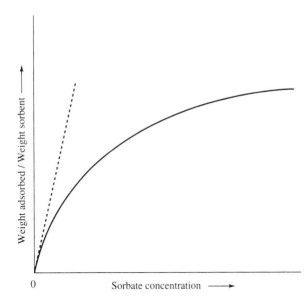

Figure 10.6   The appearance of a typical adsorption isotherm. The maximum percentage of the species adsorbed is found at the lowest sorbate concentration. The dashed line is the tangent to the isotherm at the origin. Its slope equals the distribution coefficient K_d.

in Fig. 10.6. The plot indicates that the highest fraction of the sorbate species sorbed is observed at the *lowest* sorbate concentrations, corresponding to the steepest part of the isotherm plot. Such behavior is typical of all dissolved species. Stated differently, the lower the concentration of a dissolved substance in a natural water, the greater the fraction of it that will be sorbed on solids in, for example, a stream bed or a soil. This behavior is typical of trace organic and inorganic substances at $\mu g/L$ concentrations or lower.

The simple isotherm plot also suggests that at high sorbate concentrations, a smaller and smaller fraction of the sorbate species is sorbed. At high enough concentrations, the substance has occupied all available sorption sites and is no longer sorbed at all. This has been termed the "breakthrough" concentration of the sorbate and can be of considerable environmental concern if the sorbate is a toxic substance, for soil materials no longer have the capacity to retard or prevent its migration.

For some inorganic sorbates, the dissolved sorbate concentration may reach a maximal value at and above which a mineral is precipitated. The adsorption isotherm ends at this saturation concentration, which is indicated on an isotherm plot by a vertical upward line (Fig. 10.7).

The simplest mathematical models to describe sorption from solution are the so-called isotherm equations. The simplest of these is the Freundlich isotherm equation, which may be written:

$$\frac{x}{m} = KC^n \tag{10.2}$$

where x/m is the weight of sorbate divided by the weight of sorbent (usually in $\mu g/g$ or mg/g), K is a constant, C the aqueous concentration (usually in mg/ml or g/L), and n a constant. Some users prefer to write the Freundlich isotherm equation as $x/m = KC^{1/n}$. Values of n in Eq. (10.2) range from about 0.6 to 3.3, but usually lie between 0.9 and 1.4 (Lyman et al. 1982). For pesticides, n in Eq. (10.2) averages about 1.15. When $n = 1$ in expression (10.2), $K = (x/m)/C = K_d$, the so-called distribution coefficient, which is the slope of the isotherm at the origin, or $K_d =$ (wt adsorbed/wt sorbent)/(solute conc.), with units usually in ml/g or L/kg.

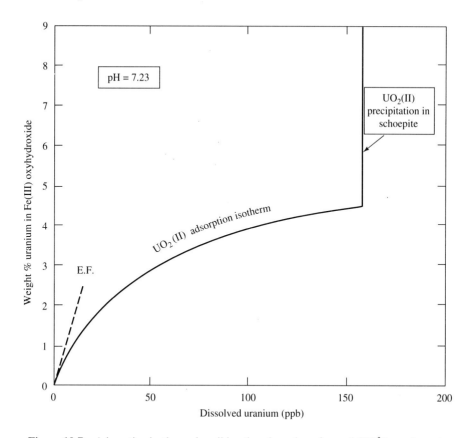

Figure 10.7   Adsorption isotherm describing the adsorption of uranyl (UO_2^{2+}) species onto suspended amorphous ferric hydroxide at pH 7.23 and 25°C. The vertical line denoting saturation with respect to schoepite [$UO_2(OH)_2 \cdot H_2O$] has been computed from the pH and dissolved uranyl concentration. The enrichment factor, E.F., equals K_d. Reprinted from *Geochim. et Cosmochim. Acta*, Vol. 53(6), D. Langmuir, Uranium solution-mineral equilibria at low temperatures with applications to sedimentary ore deposits, pp. 547–569, © 1978, with permission from Elsevier Science Ltd., The Boulevard, Langford Lane, Kidlington OX5 1GB, U.K.

   To decide if the experimental sorption behavior of a substance obeys the Freundlich isotherm or K_d model, the approach generally used is to first linearize the isotherm by taking the logarithm of expression (10.2), which yields

$$\log \frac{x}{m} = \log K + n \log C \tag{10.3}$$

The Freundlich isotherm is assumed obeyed if the experimental sorption data plotted in the form of Eq. (10.3), yields a straight line. The intercept of the line then equals $\log K$ and the slope is n. The Freundlich isotherm and K_d approaches assume an infinite supply of unreacted sorption sites. Values of $K_d = 0$ for a nonreactive sorbate, and may exceed 100 for a reactive one.

   An increase in the pH of natural systems tends to favor removal of most trace elements (especially cations) from solution. Their removal is usually by adsorption and/or by precipitation in solids that become insoluble with increasing pH. The adsorbing solids already may be present in a soil or sediment or may be solids that have precipitated, such as metal carbonates and Al, Fe(III), and Mn(IV) oxyhydroxides. The trace elements may precipitate in their own pure solids, but most often are coprecipitated as trace species in the major element solids. The K_d approach has been used to describe the removal of trace elements from solution by all of these mechanisms combined. Such a lumped process approach is strongly discouraged in that resultant K_d values only describe the behavior of the system in which they were measured. Assuming a single K_d value for predicting the removal of a dissolved species from streams, soils, and groundwaters, usually leads to serious error for species that form strong complexes, precipitate in insoluble solids or are subject to oxidation or reduction. Lumped-process K_d values (apparent K_d's) for such species can vary nonlinearly by many orders of magnitude between pH 2 and 6. (See Fig. 10.8.)

When K_d values dominantly reflect a sorption process, it has been found that log K_d often varies linearly with pH for at least several pH units. For example, measurements between pH 3.1 and 4.6 reported by van der Weijden and van Leeuwen (1985) lead to log K_d (ml/g) = $-1.073 + 0.8218$ pH ($r = 0.993$) for uranyl adsorption by peat. Andersson et al. (1982) have reported linear log K_d versus pH plots for Cs^+ and Sr^{2+} adsorption by a number of igneous rocks and minerals between pH 5 to 11.

When a contaminant species of interest is already present in the soil or groundwater, it is highly recommended that the water and soil be sampled and analyzed for the contaminant to determine an *in situ* K_d value. This is, after all, the K_d we are interested in. Its value often differs substantially from a K_d measured in the laboratory. For example, Fruchter et al. (1985) report respective laboratory K_d's of 3560, 400, and 19 ml/g for ^{60}Co, ^{106}Ru, and ^{125}Sb. Respective *in situ* K_d's for these radionuclides equaled 9, 6, and 4, indicating that, in fact, they were relatively mobile in the groundwater. The authors point out that the laboratory K_d's were measured in noncomplexing electrolyte solutions, whereas in the field the radionuclides were complexed, probably by organic substances, making them less likely to be adsorbed.

The best use of K_d is probably to model the sorption of trace molecular organic species such as pesticides, or other weakly sorbed nonionized organic species. The adsorption of dissolved molecular organic substances by soils and sediments has been found to be proportional to the amount of solid organic matter present and relatively independent of the weight of associated inorganic materials. This observation, which is particularly true in soils that contain 0.1 to 20% organic carbon (cf. Lyman et al. 1982; Thurman 1985; Schwarzenbach et al. 1993), has led to the concept of the adsorption or distribution coefficient for organic carbon or K_{oc} defined as

$$K_{oc} = \frac{\mu \text{g adsorbed/g organic C}}{\mu \text{g/ml in solution}} \tag{10.4}$$

The weight ratio of total organic matter to organic carbon ranges from about 2.0 in soil fulvic acids to 2.7 in humin (Schwarzenbach et al. 1993). Assuming average organic matter is approximately CH_2O, the weight ratio is 2.5. The distribution coefficient for organic substances sorbed by total organic matter (K_{om}) is then related to K_{oc} by $K_{oc} = 2.5 \times K_{om}$. Values of K_{oc} range from 1 to 10^7 ml/g of organic carbon. Example K_{oc} values (in ml/g) for some common organic contaminants, including pesticides, are: benzene, 83; atrazine, 160; lindane, 1080; malathion, 1800; parathion, 1.06×10^4; DDT, 2.43×10^5; and mirex, 2.4×10^7 (Lyman et al. 1982).

For organic sorbates, the distribution coefficient K_d is related to K_{oc} through the expression

$$K_d = K_{oc} \times f_{oc} \tag{10.5}$$

where f_{oc} is the weight fraction of organic carbon in the sediment. Frequently K_{oc} values have not been measured. K_{oc} is, however, related to other important properties of organic substances, includ-

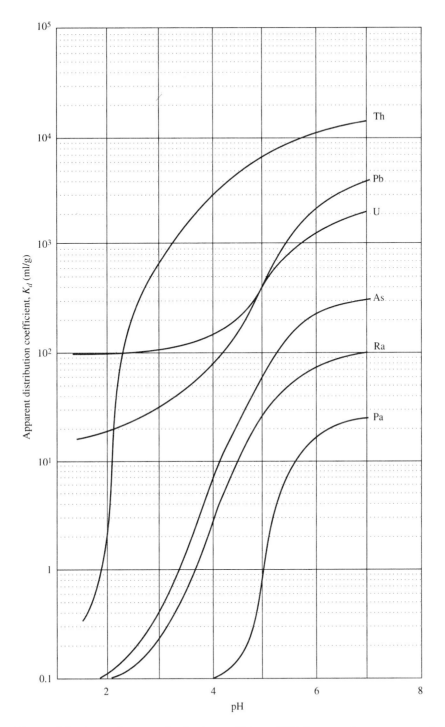

Figure 10.8 Lumped-process distribution coefficients (K_d's) for some elements measured during laboratory neutralization of acid solutions with calcium carbonate (Taylor 1979). These K_d values represent the combined effects of element removals by adsorption, coprecipitation, and/or precipitation onto and in freshly precipitated Fe(III), Al, and Mn(IV) oxyhydroxides and hydroxysulfates. As such they are system specific and cannot be assumed to apply to other systems under different conditions or having different compositions.

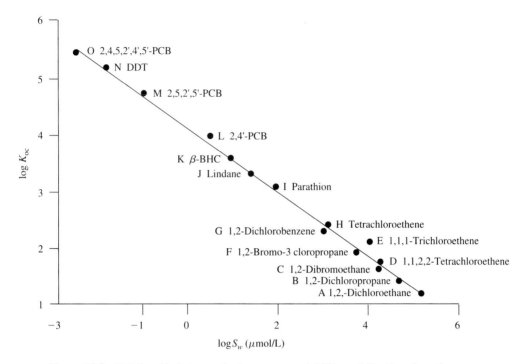

Figure 10.9 Relationship between S_w, the aqueous solubility, and K_{oc}, the adsorption or distribution coefficient for some molecular organic substances on organic carbon in soil. From C. T. Chiou et al. A physical concept of soil-water equilibria for nonionic organic compounds, *Science* 206:831–832. Copyright © 1979 by *Science*. Used by permission.

ing their solubility (S_w), octanol-water partition coefficient (K_{ow}), and bioconcentration factor, (BCF).

The aqueous solubilities of 15 chlorinated hydrocarbons in μmol/L are plotted against their log K_{oc} values in Fig. 10.9, from Chiou et al. (1979) (see also Chiou et al. 1983). The regression line through the data has the equation

$$\log K_{oc} = 4.040(\pm 0.038) - 0.557(\pm 0.012) \log S_w \ (\mu M) \tag{10.6}$$

with $r^2 = 0.99$. Solubilities of the same substances identified by letter in Fig. 10.9 are listed below in ppm (mg/L) with their respective K_{oc} values according to Chiou et al. (1979).

	S_w (ppm)	K_{oc}		S_w (ppm)	K_{oc}		S_w (ppm)	K_{oc}
A	8450	19	F	1230	75	K	5	2900
B	3570	27	G	148	180	L	0.64	8000
C	3529	36	H	200	210	M	0.027	4700
D	3230	46	I	24	1160	N	0.004	140000
E	1360	104	J	7.8	1730	O	0.000095	220000

Based on a regression of the solubility data for 106 substances that were mostly pesticides, Lyman et al. (1982) reported

$$\log K_{oc} = 3.64 - 0.55 \log S_w \ (mg/L) \qquad (r^2 = 0.71) \tag{10.7}$$

(See also Lyman et al. 1982, Fig. 4-1.)

About 95% of natural dissolved organic carbon (humic and fulvic acid) is ionic (Thurman 1985). Such substances have been described as hydrophylic ("water-loving"). Their high solubilities correspond to a low tendency to be sorbed by organic matter and thus to K_{oc} values near zero. In contrast, highly sorbed organic substances, such as the pesticide DDT, are hydrophobic ("water-hating") and insoluble in water.

The value of K_{ow}, which describes the relative solubility of an organic substance in octanol versus its solubility in water, is defined as

$$K_{ow}(\text{unitless}) = \frac{\text{concentration in the octanol phase}}{\text{concentration in water}} \tag{10.8}$$

The solubility of organic substances in human fatty tissues (in lipids) has been found to be roughly proportional to K_{ow} values. The bioconcentration factor (BCF) defined as

$$\frac{\text{BCF}}{(\text{unitless})} = \frac{\text{concentration of substance at equilibrium in the organism (wet wt)}}{\text{mean concentration of substance in water}} \tag{10.9}$$

is thus generally proportional to K_{ow}. In this expression, the concentrations are often measured in $\mu g/g$.

Shown in Figs. 10.10 and 10.11 are plots of K_{ow} and BCF versus aqueous solubility for a number of organic substances from Chiou et al. (1977). The figures show the strong negative correlation

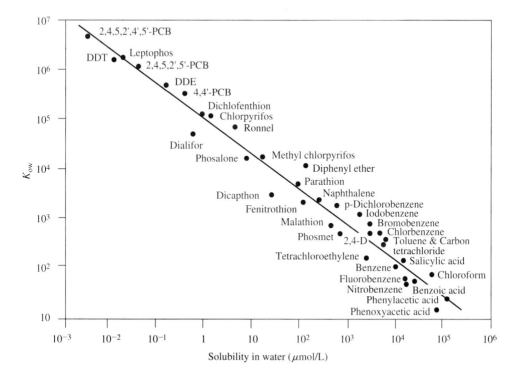

Figure 10.10   Aqueous solubility (S_w) versus octanol-water partition coefficient (K_{ow}) for a variety of organic solutes. Note that highly soluble compounds are poorly sorbed and vice versa. Reprinted with permission from Chiou et al. *Envir. Sci. & Technol.* 11:475–477. Copyright 1977 American Chemical Society.

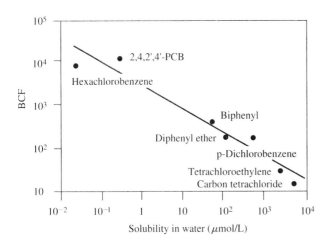

Figure 10.11   Aqueous solubility (S_w) versus bioconcentration factor (BCF) in rainbow trout. Reprinted with permission from Chiou et al., *Envir. Sci. & Technol.* 11:475–477. Copyright 1977 American Chemical Society.

between solubility and both K_{ow} and the bioconcentration factor, consistent with the strong positive correlations between K_{ow} and K_{oc}, and BCF and K_{oc}.

Numerous equations have been derived from empirical data relating K_{oc}(ml/g) to K_{ow} or BCF. For example, based on measurements involving 45 pesticides, Lyman et al. (1982) suggest

$$\log K_{oc} = 1.377 + 0.544 \log K_{ow} \qquad (r^2 = 0.74) \tag{10.10}$$

From a study of 22 substances, chiefly pesticides, the same authors report

$$\log K_{oc} = 1.886 + 0.681 \log \text{BCF} \qquad (r^2 = 0.83) \tag{10.11}$$

As an example calculation, we will estimate K_{oc}, K_d and x/m for hexachlorobenzene (C_6Cl_6) in a soil that contains 2% organic carbon and 0.001 mg/L of dissolved C_6Cl_6, given that the aqueous solubility, S_w, equals 0.0204 μM or 0.0058 mg/L (Schwarzenbach et al. 1993). For simplicity we will assume $n = 1$ in the Freundlich isotherm equation and thus $K_d = (x/m)/C$. Substituting into Eq. (10.6) gives $K_{oc} = 1.7 \times 10^5$ ml/g. Similarly, Eq. (10.7) leads to $K_{oc} = 7.4 \times 10^4$ ml/g. These results may be compared to a measured K_{oc} value of 3.2×10^5 ml/g (Schwarzenbach et al. 1993). Given that $K_d = K_{oc} \times f_{oc}$, our two estimates of K_{oc} correspond to respective K_d values of 3400 and 1500 ml/g. Next, since $x/m = K_d$ (ml/g) $\times$ C (mg/L), and mg/L = μg/ml, we obtain $x/m = 3.4$ and 1.5 μg/g for respective K_d values of 3400 and 1500 ml/g.

A number of variables largely ignored in the above discussion can complicate the measurement and application of organic substance adsorption or partitioning concepts to natural systems. This is particularly true for organic substances that can occur in both ionized and molecular forms. For example, Lyman et al. (1982) suggest that the adsorption of pesticides is best modeled with a nonlinear isotherm and an average value of $n = 1.15$ in the Freudlich isotherm equation. Variables that should be considered in a rigorous analysis of organic adsorption include the pH as a measure of the degree of dissociation of organic acids, the ionic strength (activity coefficient effects), and the surface area and surface charge of sorbents, including minerals in the soil. These effects and others are discussed in detail by Schwarzenbach et al. (1993).

10.2.2 The Langmuir Adsorption Isotherm

The Langmuir adsorption isotherm equation may be written two ways:

$$\frac{x}{m} = \frac{aC}{1 + bC} = \frac{bCN_{max}}{1 + bC} \tag{10.12}$$

where a and b are constants, C the aqueous concentration of the sorbate, and $a = bN_{max}$. N_{max} is the maximum possible sorption by the solid, usually assumed to represent monolayer surface coverage and the value of x/m where the isotherm curve flattens. Other assumptions inherent in the Langmuir isotherm are given by Adamson (1990). (See also Rubin and Mercer 1981.) The existence of a value for N_{max} indicates the presence of a finite supply of sorption sites.

The Langmuir isotherm is usually linearized by inversion and so used to test whether it is obeyed by experimental data. The inversion is often done incorrectly, producing an induced correlation in C, thus

$$\frac{C}{x/m} = \frac{1}{bN_{max}} + \frac{C}{N_{max}} \tag{10.13}$$

A linearization that avoids this criticism is

$$\frac{1}{x/m} = \frac{b}{a} + \frac{1}{aC} = \frac{1}{bCN_{max}} + \frac{1}{N_{max}} \tag{10.14}$$

When C is very small, the isotherm reduces to the Freundlich isotherm with $x/m = aC = KC$, $n = 1$, and $K = K_d$. When C is very large, $x/m = a/b = N_{max}$, which is equivalent to the Freundlich isotherm with $n = 0$.

Competition between adsorbing species for the same site on a sorbent has been modeled with the so-called competitive Langmuir model. In this model an individual isotherm equation with its own constants is written for each species sorbed by a given sorbent (see Table 10.7).

10.2.3 General Discussion of the Adsorption Isotherm Models

Some examples of simple and complex Freundlich and Langmuir isotherm plots are shown in Fig. 10.12 from Domenico and Schwartz (1990).

Experiments to determine constants for the isotherms are usually performed in laboratory batch tests, that is in vessels that contain a known total amount of sorbent solid mixed with a known total amount of a potentially sorbing aqueous species. The approach is often to systematically vary sorbent and/or sorbate concentrations and pH, for example, in a series of centrifuge tubes. After centrifugation the amount sorbed is then deterimined by its difference from the total sorbate added (cf. Catts and Langmuir 1986).

The value of K_d may also be obtained from the results of column test experiments using the expression

$$K_d = \left(\frac{C_i - C_f}{C_f}\right)\frac{V}{M} \tag{10.15}$$

where C_i and C_f are initial and final concentrations of the sorbate, V is the volume of solution passed through the column, and M is the mass of solids in the column. K_d values for radioactive substances can be determined with the same equation, where instead of initial and final concentrations, initial and final solution radioactivities are employed. Properly designed column experiments generally give results more representative of the *in situ* behavior of soils and sediments than do batch tests. Unfortunately, values of K_d obtained from batch and column studies are often in disagreement. Pavik et al. (1992) have suggested corrections to batch experimental results to reduce or eliminate such differences.

TABLE 10.7   The adsorption isotherm models as defined and used in MINTEQA2

Surface reaction	Equilibrium constant expression
Activity K_d Adsorption Model	
$SOH + M = SOH \cdot M$	$K_d^{act} = \dfrac{(SOH \cdot M)}{\gamma_m(M)}$ $(SOH \cdot M) = K_d^{act}\,\gamma_m(M)$
Activity Freundlich Adsorption Model	
$SOH + (n)M = SOH \cdot M$	$K_f^{act} = \dfrac{(SOH \cdot M)}{\gamma_m^n(M)^n}$ $(SOH \cdot M) = K_d^{act}\,\gamma_m^n(M)^n$
Activity Langmuir Adsorption Model	
$SOH + M = SOH \cdot M$	$K_L^{act} = \dfrac{(SOH \cdot M)}{\gamma_m(M)(SOH)}$ Assuming: $(SOH)_T = (SOH \cdot M) + (SOH)$ $(SOH \cdot M) = \dfrac{K_L^{act}(SOH)_T\,\gamma_m(M)}{1 + K_L^{act}\,\gamma_m(M)}$
Competitive Langmuir Adsorption Model	
$SOH + M_1 = SOH \cdot M_1$ $SOH + M_2 = SOH \cdot M_2$ $\vdots$ $SOH + M_n = SOH \cdot M_n$	Individual expressions for K_{Ln}^{act} equivalent to the single term expression for K_L^{act} $(SOH)_T = (SOH) + \displaystyle\sum_1^n (SOH \cdot M_n)$

Source: Allison et al. (1991).

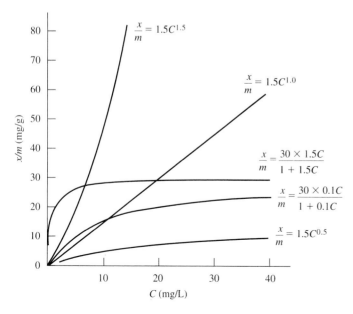

Figure 10.12   Examples of Langmuir and Freundlich isotherms. From P. A. Domenico and F. W. Schwartz, *Physical and chemical hydrogeology.* Copyright © 1990 by John Wiley & Sons. Reprinted by permission of John Wiley & Sons, Inc.

A measure of the limited applicability of the isotherm models to complex natural systems is suggested by the number of adjustable parameters that each can consider.

Model	Number of adjustable parameters
K_d	2
Freundlich	3
Langmuir[†]	3

[†]The one-term, noncompetitive isotherm equation.

Except for the competitive Langmuir isotherm (Table 10.7), the simple isotherm models cannot consider competition between multiple sorbates for a single sorption site. They are most applicable to physical adsorption of molecular species where the sorbate is held by much weaker hydrogen bonds or residual bonds. Adsorption of ions and complexed species is generally best studied with models that can consider more independent variables. To accurately predict adsorption of ions that compete for adsorption sites and also form important complexes, may require a model that considers as many as a dozen or more variables.

The fallacy of assigning single K_d values to the sorption of metal cations is evident from Figs. 10.13 and 10.14, which show Pb adsorption by kaolinite and Cd by montmorillonite. Lead adsorption has been measured as a function of pH. Clearly, a sorption isotherm equation is needed to

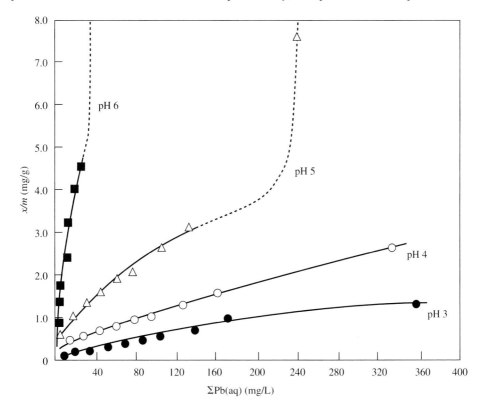

Figure 10.13   Adsorption of Pb from DuPage landfill leachate by kaolinite at 25°C, as a function of pH. Dashed vertical lines show the Pb concentration at saturation with Pb-hydroxy-carbonate solid. Reprinted with permission from R. A. Griffin and N. F. Shimp. *Envir. Sci. Technol.* 10(13):1256–1261. Copyright 1976 American Chemical Society.

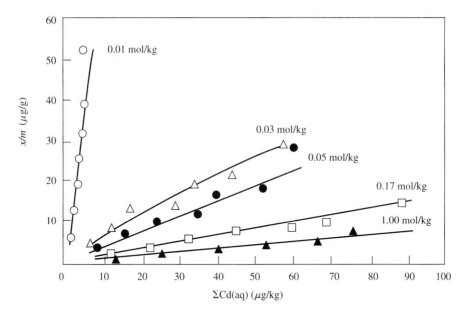

Figure 10.14   Influence of Cd-chloride complexing on cadmium adsorption by montmorillonite. Molal values in the figure are the concentrations of NaCl present for each isotherm curve. Modified after Garcia-Miragaya and Page (1976), *Soil Sci. Soc. Am. J.,* 40(5):658–63. Used by permission of Soil Science Society of America.

describe sorption at each pH. The increase in sorption with pH may reflect stronger sorption of the complex $PbOH^+$ than of the free Pb^{2+} ion. The complex becomes more important as pH increases. ($K = 10^{-6.3}$ for $Pb^{2+} + H_2O = PbOH^+ + H^+$.) This plot also shows that at a certain total lead concentration, precipitation of a solid lead hydroxy-carbonate phase [hydrocerrusite; $Pb_3(CO_3)_2(OH)_2$)] occurs, and a further Pb concentration increase cannot take place.

    The Cd sorption plot in Fig. 10.14 shows that a different isotherm is needed to describe Cd sorption at each NaCl concentration. This reflects weak adsorption of Cd-Cl complexes relative to free Cd^{2+} and increasing competition of Na^+ for Cd^{2+} sorption sites on the clay as the NaCl concentration increases. Obviously, such effects cannot be accounted for using an isotherm or K_d model.

10.2.4 The Adsorption Isotherm Models as Defined in MINTEQA2

As presented above, and in most of the published literature, the isotherm model equations are expressed in terms of total sorbate concentrations. In MINTEQA2, however, aqueous species activities are used rather than their concentrations. This is preferable in that reactions are written in terms of activities. Also, so that the concentrations of sorption sites, sorbed species, and dissolved species are computationally equivalent, the former are entered in MINTEQA2 as moles of sites per liter of solution. Activity coefficients of sorption sites and sorbed species are generally taken to be equal to unity by most authors and in MINTEQA2.

    Given in Table 10.7 are the surface reactions and corresponding "activity" adsorption isotherm model equations used in MINTEQA2 as presented by Allison et al. (1991). In these expressions SOH and SOH · M represent unoccupied surface sites and surface sites occupied by species M. Because the K_d and Freundlich isotherm models assume an infinite number of available sorption sites, the concentration of such sites may be assumed equal to unity, or (SOH) = 1. The Langmuir isotherm assumes a finite total number of sorption sites, designated as SOH_T in MINTEQA2, where SOH_T (mol/L) = SOH + SOH · M. SOH_T may be obtained from the expression: SOH_T (mol/L) = $N_S S_A C_S / N_A$, where N_S is the surface-site density (sites/m^2), S_A the specific surface area of the solid sorbent (m^2/g), C_S the weight of sorbent in suspension or in contact with a liter of solution (g/L), and N_A is Avogadro's number = 6.022×10^{23} sites/mol sites.

10.3 ION-EXCHANGE TYPE MODELS AND CONCEPTS

10.3.1 Simple Ion Exchange

Adsorption of a dissolved ionic species is always part of an exchange reaction that involves a competing ionic species. The desorbing species creates the vacant site to be occupied by the adsorbing one. Soil scientists use a variety of ion exchange models (e.g., the Gaines-Thomas, Gapon, Vanselow, and Rothmund-Kornfeld models) in which different conventions are used to write the concentrations of dissolved and adsorbed species. These models are adequately described and compared elsewhere (cf. Bolt 1979; Sposito 1981, 1989; Appelo and Postma 1993).

It is likely that the adsorption sites on most solids are monovalent (Jenne 1995). Nevertheless, to avoid having to write fractional moles of divalent cations, sorption sites are assumed divalent in the following discussion. Thus, in the above example, where Na$^+$ competes with Cd^{2+} for sorption sites on the montmorillonite, the competitive reaction is written

$$2Na^+ + CdX = Cd^{2+} + Na_2X \tag{10.16}$$

and X^{2-} denotes a divalent negative surface site on the clay. In general terms, for simple binary exchange the reaction is

$$aA^z + B_bX = bB^y + A_aX \tag{10.17}$$

where A and B are ions, a and b are mole numbers, z and y are valences, and X is again the exchanger phase. The equilibrium or exchange constant is

$$K_{ex} = \frac{[B^y]^b [A_aX]}{[A^z]^a [B_bX]} \tag{10.18}$$

where the brackets denote activities. If $a = b$, then $z = y$, and we have "homovalent" exchange, such as Ca^{2+} for Mg^{2+}, or:

$$K_{ex} = \frac{[Mg^{2+}] [CaX]}{[Ca^{2+}] [MgX]} \tag{10.19}$$

If $a = 2b$, example reactions are Eq. (10.16) and

$$2Na^+ + CaX = Ca^{2+} + Na_2X \tag{10.20}$$

for which:

$$K_{ex} = \frac{[Ca^{2+}] [Na_2X]}{[Na^+]^2 [CaX]} \tag{10.21}$$

How can one deal with the activities of species in the equilibrium expression for ion exchange? For the aqueous ions we can often use the Debye-Hückel or Davies equation to obtain ion activity coefficients, as is done in MINTEQA2. However, for the solids, determining activities is very difficult. One approach is to define a *"rational" activity coefficient* of the sorbing site. For example,

$$[CaX] = \lambda_{CaX} N_{CaX} \tag{10.22}$$

where N is the mole fraction of X^{2-} sites occupied by Ca^{2+}, and $N_{NaX} + N_{CaX} = 1$. Historically, soil scientists and geochemists have often ignored the aqueous and solid mole fraction activity coefficients and expressed the exchange reaction in terms of easily measureable aqueous concentrations and solid component mole fractions. The concentration quotient is then defined as equal to a distribution or selectivity coefficient, D, which for exchange reaction (10.20) equals

$$D = \frac{(mCa^{2+})(N_{Na_2X})}{(mNa^+)^2(N_{CaX})} \tag{10.23}$$

The value of D may be relatively constant for exchange reactions involving major concentrations of ions and sorbents such as smectite clays or zeolites (cf. Langmuir 1981). However, D for exchange reactions between major and trace constituents can change by orders of magnitude as their concentrations change.

10.3.2 The Power Exchange Function

It has been found that for binary exchange, a mass action expression in which the ratio of sorbed cation mole fractions is raised to the nth power ($n \neq 1$), can fit a wider range of empirical sorption data than when $n = 1$. Thus, for the homovalent exchange reaction,

$$A + BX = B + AX \tag{10.24}$$

the general exchange equilibrium expression is

$$K_{ex} = \frac{[B]}{[A]}\left(\frac{AX}{BX}\right)^n \tag{10.25}$$

where [A] and [B] are ion activities in solution, and (AX) and (BX) are sorbed equivalent fractions. This has been called a power exchange function (Langmuir 1981). Taking logs and transposing gives

$$\log\frac{[B]}{[A]} = \log K_{ex} + n\log\left(\frac{BX}{AX}\right) \tag{10.26}$$

A linear plot of $\log[B]/[A]$ versus $\log(BX/AX)$ has a slope of n and intercept of $\log K_{ex}$. Combined literature review and laboratory study shows that the exchange behavior of H^+, Na^+, K^+, Ca^{2+}, Mg^{2+}, Cd^{2+}, Co^{2+}, Ni^{2+}, Pb^{2+}, UO_2^{2+}, and Zn^{2+}, and their hydroxy complexes on a variety of adsorbents (smectites, illites, ferric oxyhydroxides, zeolites, carbonates, soils, and humic materials), can be accurately described for a wide range of competing sorbate concentrations and ratios using from one to three power exchange expressions (cf. Frizado 1979; Langmuir 1981; Chapelle and Knobel 1983; Zachara et al. 1988).

In the simplest case, $K_{ex} = n = 1$ in Eq. (10.25). For such conditions the reaction is called *Donnan exchange*. Donnan exchange behavior has several implications:

1. the sorbent has a constant surface charge or CEC;
2. the exchange reaction is controlled by valence differences of the ions only, not size differences. Sorption occurs between sorbate ions and a surface of opposite sign and the bonding is strictly electrostatic;

3. the sorbate ions are not differently complexed; and
4. the activity coefficients of both dissolved and sorbed ions have equal ratios and those ratios are constant and equal unity.

Donnan exchange best applies to the sorption of major dissolved cations ($>10^{-4}$ to 10^{-3} mol/kg) such as Na^+, Ca^{2+}, and Mg^{2+} by minerals of practically constant surface charge, such as the smectite and vermiculite clays and zeolitic minerals.

When an empirical fit of exchange data can be accomplished with a power exchange function in which $n = 1$, but $K_{ex} \neq 1$, the reaction can be termed *simple ion exchange*. That a sorption reaction obeys simple ion exchange implies that

1. the sorbate mineral has a constant surface charge or CEC;
2. sorption is controlled by the valences and sizes of the ions, and the bonding is strictly electrostatic. Sorption takes place between sorbate ions and a surface of opposite sign;
3. there is minimal differential complexing of sorbate ions; and
4. the activity coefficients of both dissolved and sorbed ions have constant, but not equal, ratios in both cases.

Simple ion exchange describes the competitive adsorption onto clays of most metal cations present in solution at concentrations from about 10^{-4} to 10^{-2} mol/kg. It has most often been used to describe the sorption of alkaline earth and alkali metal cations onto clays. In the case of minerals having pH-dependent surface charge (e.g., kaolinite, metal oxyhydroxides) simple ion exchange or the power-exchange function (see below) may also fit the adsorption data measured in systems at constant pH.

The general *power-exchange function* applies when $n \neq 1$ and $K_{ex} \neq 1$. The exchange reaction that corresponds to this function obviously lacks charge balance. For example, the Na^+-Fe^{2+} exchange reaction with a power-exchange equilibrium expression written

$$K_{ex} = \frac{[Fe^{2+}]}{[Na^+]^2} \left(\frac{Na_2X}{FeX} \right)^{0.83} \tag{10.27}$$

(Langmuir 1981) suggests the nonstoichiometric reaction

$$2Na^+ + 0.83FeX = Fe^{2+} + 0.83Na_2X \tag{10.28}$$

Such unbalanced stoichiometries imply that the sorption capacity for Na^+ is different from that for Fe^{2+} and/or that other cations are being exchanged (cf. Anderson et al. 1973; Benjamin and Leckie 1981). The power-exchange approach can model such behavior, but obviously does not help us to understand it.

The value of n includes within it information on the ratio of the rational activity coefficients of the sorbed ions. Thus, for homovalent exchange, as in reaction (10.24), it can be shown that $\lambda_{AX} = (AX)^{n-1}$ and $\lambda_{BX} = (BX)^{n-1}$ (Langmuir 1981). Adsorption of heavy metals at aqueous concentrations between about 10^{-7} and 10^{-3} mol/kg fits power-exchange functions with $n = 0.8$ to 2.0. Values of n generally increase with decreasing concentration of the trace ion. In other words, as the trace metal concentration drops relative to that of a competing major ion, adsorption of the trace species is increasingly favored relative to competing major species.

Values of K_{ex} and n for some exchange reactions on ferric oxide, montmorillonite clays, and soils are given in Table 10.8. Example plots of Eq. (10.26) for Cd^{2+} and Pb^{2+} exchange onto calcium montmorillonite are shown in Fig. 10.15. The plots show that as the concentrations of Cd^{2+} and Pb^{2+} decrease below those of Ca^{2+}, K_{ex} values drop below unity. This indicates that the montmorillonite

TABLE 10.8   Some K_{ex} and n values for studies involving the binary exchange of ions including H^+, Na^+, Mg^{2+}, Ca^{2+}, and Sr^{2+}, and divalent Cd, Co, Cu, Fe, Pb, or Zn onto Na- and Ca-montmorillonite clays, ferric oxide, and some soils

A^{a+}: Concentration range	B^{b+}: Concentration range	K_{ex}	n	pH	Adsorbent
Na^+: 2 to 17×10^{-4} M	Cd^{2+}: 9 to 180×10^{-6} M	0.28 ± 0.03	1.00 ± 0.10	5.5	Na-mont-
	2 to 7.5×10^{-4} M	0.32 ± 0.03	0.80 ± 0.08	to	morillonite[†]
	Co^{2+}: 9 to 180×10^{-6} M	0.25 ± 0.03	1.00 ± 0.10	6.0	
	2 to 7.5×10^{-4} M	0.32 ± 0.03	0.80 ± 0.08		
	Cu^{2+}: 9 to 180×10^{-6} M	0.25 ± 0.03	0.95 ± 0.10		
	2 to 7.5×10^{-4} M	0.32 ± 0.03	0.80 ± 0.08		
	Zn^{2+}: 9 to 180×10^{-6} M	0.27 ± 0.03	1.00 ± 0.10		
	2 to 7.5×10^{-4} M	0.32 ± 0.03	0.80 ± 0.08		
Ca^{2+}: 8 to 10×10^{-5} M	Cd^{2+}: 1 to 6.5×10^{-4} M	0.91 ± 0.05	2.00 ± 0.10	4.8	Ca-mont-
1 to 8.7×10^{-4} M		0.98 ± 0.05	1.01 ± 0.05	to	morillonite[‡]
Ca^{2+}: 8 to 10×10^{-5} M	Pb^{2+}: 3 to 9.2×10^{-4} M	0.50 ± 0.05	1.30 ± 0.06	6.5	
1 to 8.7×10^{-4} M		0.61 ± 0.10	0.95 ± 0.05		
Na^+: 2 to 20×10^{-3} M	Fe^{2+}: 2 to 20×10^{-3} M	1.05 ± 0.09	0.83 ± 0.04	3.3	Na-mont-morillonite[§]
Ca^+: 2 to 25×10^{-3} M	Zn^{2+}: 3 to 35×10^{-5} M	0.42 ± 0.06	0.96 ± 0.05	6.6	Ca-mont-
	3.5 to 9×10^{-4} M	1.53 ± 0.08	1.40 ± 0.05		morillonite[‖]
H^+: 1×10^{-5} M	Pb^{2+}: 1 to 60×10^{-4} M	3.0×10^7	1.93 ± 0.07	5.0	ferric oxide[#]
Ca^{2+}: 7.9 to 9×10^{-4} M	Cd^{2+}: 2 to 200×10^{-6} M	0.56 ± 0.05	1.16 ± 0.06	4.1	Ca-mont-
	2 to 2.4×10^{-4} M	0.89 ± 0.06	2.00 ± 0.11	to	morillonite[††]
	Pb^{2+}: 4 to 890×10^{-7} M	0.25 ± 0.05	1.25 ± 0.06	8.2	
Na^+: 6 to 57×10^{-3} M	Ca^{2+}: 1 to 18×10^{-3} M	1.12 ± 0.22	1.00 ± 0.05		silty clay[‡‡]
		2.85 ± 0.09	1.03 ± 0.03	?	clay loam[‡‡]
		2.37 ± 0.15	1.05 ± 0.06		sandy loam[‡‡]
Ca^{2+}: 1 to 20×10^{-4} M	Sr^{2+}: 5 to 2000×10^{-6} M	1.10 ± 0.10	1.00 ± 0.10	4.0	mont-
Mg^{2+}: 1 to 20×10^{-4} M	Sr^{2+}: 5 to 2000×10^{-6} M	1.40 ± 0.15	1.00 ± 0.10	to	morillonite[§§]
Na^+: 1 to 5×10^{-2} M	Sr^{2+}: 5 to 2000×10^{-6} M	1.20 ± 0.10	1.00 ± 0.10	5.0	

Source: From Ozsvath (1979). Data from: [†]Maes et al. (1975); [‡]Bittel and Miller (1974); [§]Singhal et al. (1975); [‖]DiGiacomo (1976); [#]Gadde and Laitinen (1975); [††]Ozsvath (1979); [‡‡]Levy and Hillel (1968); and [§§]Kown and Ewing (1969).

preferentially adsorbs Cd^{2+} and Pb^{2+} over Ca^{2+} at low lead and cadmium concentrations. The plots also show that slopes of the functions steepen (n values increase) and K_{ex} values decline at lower lead and cadmium concentrations. This corresponds to an even greater increase in the selectivity of the clay for the heavy metals as their amounts approach trace levels.

In summary, Donnan exchange, simple ion exchange, and more general power-exchange function behavior can accurately model binary exchange of many cationic species at trace and major concentration levels. Donnan exchange and simple ion exchange model simple electrostatic behavior between sorbate ions and oppositely charged surface sites. These models do not adequately describe most trace element adsorption behavior that can occur on surfaces of changing surface charge and against the net surface charge. Such adsorption can be modeled using the more general power-exchange function. However, that function provides no mechanistic explanation for the sorption process.

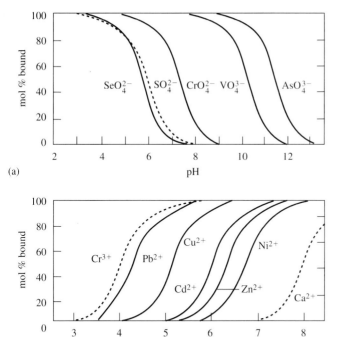

(a)

(b)

Figure 10.17   Adsorption of various metal cations and oxyanions, (each at a trace concentration of 5×10^{-7} M), by ferrihydrite ($\Sigma Fe(III) = 10^{-3}$ M) as a function of pH at an ionic strength of 0.1 mol/kg. There are 2×10^{-4} M of reactive sites on the oxyhydroxide. The dashed curves have been calculated. The plots are after W. Stumm, *Chemistry of the solid-water interface*, Copyright © 1992 by John Wiley & Sons, Inc. Reprinted by permission of John Wiley & Sons, Inc.

Complex adsorption models based on double-layer theory, which take a mechanistic and atomic-scale approach to adsorption, can be used to model and predict these observations. Such models have been called *electrostatic adsorption models* or *surface complexation models*. They can consider simultaneously such important system properties as changes in pH, aqueous complex formation and solution ionic strength (solution speciation), and the acid-base and complexing properties of one or more sites on several sorbing surfaces simultaneously.

A number of electrostatic adsorption models have been developed, particularly since the late 1970s (cf. James and Healy 1972; Davis et al. 1978; Davis and Leckie 1978; Davis and Kent 1990). Westall and Hohl (1980) (see also Dzombak and Morel 1987) have shown that any of five electrostatic models can fit the same set of experimental data equally well. In this section we will focus on use of the three electrostatic adsorption models available in MINTEQA2, which were the three compared by Dzombak and Morel (1987) and were among the five considered by Westall and Hohl (1980). These are the constant capacitance (CC), diffuse-layer (DL), and triple-layer (TL) models. Our goal is limited to providing the reader with a general understanding of these models and their use. We will also discuss approaches to estimating model parameters when such parameters have not been measured and note some limitations to model use in solving environmental problems. Many authors, including Brown and Allison (1987) and Allison et al. (1991), have done an excellent job of covering much the same material. Readers interested in learning how to parameterize the electrostatic adsorption models, starting with laboratory experimental measurements, should study one of the many papers and books on this subject (e.g., Davis et al. 1978; Huang 1981; James and Parks 1982; Dzombak and Morel 1987, 1990; Stumm 1987, 1992).

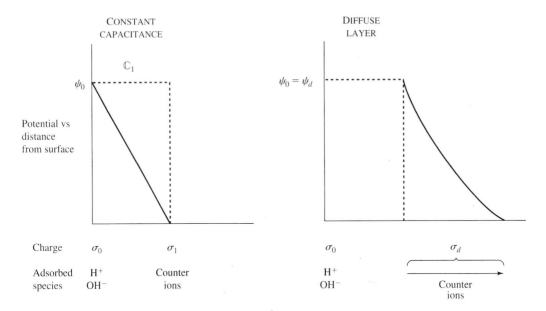

Figure 10.18 Schematic plot of surface species and charge (σ) and potential (ψ) relationships versus distance from the surface (at the zero plane) used in the constant capacitance (CC) and the diffuse-layer (DL) models. The capacitance, $\mathbb{C}_1$ is held constant in the CC model. The potential is the same at the zero and d planes in the diffuse-layer model ($\psi_0 = \psi_d$). Reprinted from *Adv. Colloid Interface Sci.* 12, J. C. Westall and H. Hohl, A comparison of electrostatic models for the oxide/solution interface, pp. 265–294, Copyright 1980 with kind permission of Elsevier Science-NL, Sara Burgerhartstraat 25, 1055 KV Amsterdam, The Netherlands.

10.4.2 General Assumptions and Attributes of Three Models

The surface complexation (SC) or electrostatic models differ conceptually and mathematically from the simpler isotherm and ion-exchange models in several important ways. As used in MINTEQA2, all the models consider solution speciation and aqueous ion activities. In addition, the SC models employ electrical double-layer (EDL) theory. EDL theory assumes that the + or − surface charge of a sorbent in contact with solution generates an electrostatic potential that declines rapidly away from the sorbent surface. (See Figs. 10.18 and 10.19.) EDL theory further assumes that an excess of *counterions* (ions of opposite charge to the fixed charge of the surface) are present near the surface in the EDL. Away from the surface, the double layer usually (but not always) contains a deficit of ions of the same charge as the surface charge (*coions*), while at a greater distance from the surface, the bulk solution is charge balanced in cations and anions. The SC models thus consider the adsorption process in a relatively more atomistic and mechanistic way than do the simpler models. This provides the SC models with a greater capability than the simple models have of predicting adsorption for conditions beyond those used to determine model parameters.

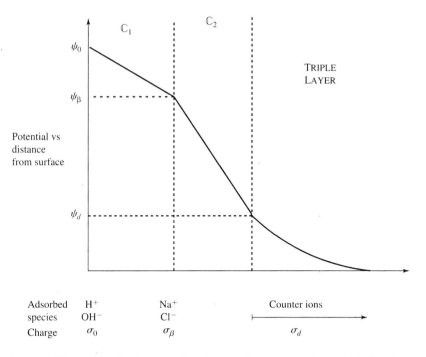

Figure 10.19   Schematic plot of surface species, charge (σ), and potential (ψ) relationships versus distance from the surface, used in the triple-layer model. Integral capacitances $\mathbb{C}_1$ and $\mathbb{C}_2$ are assigned to regions between the 0 and β and the β and d planes, respectively. Electrolyte ions are adsorbed at the β plane. Reprinted from *Adv. Colloid Interface Sci.* 12, J. C. Westhall and H. Hohl, A comparison of electrostatic models for the oxide/solution interface, pp. 265–294, copyright 1980 with kind permission of Elsevier Science-NL, Sara Burgerhartstraat 25, 1055 KV Amsterdam, The Netherlands.

Some common assumptions inherent in the SC models include:

1. The sorbing surface is composed of specific functional groups that react with sorbing solutes to form surface complexes (inner sphere or true complexes, or ion pairs) in a manner analogous to the formation of aqueous complexes in the bulk solution.
2. Surface complexation and ionization reactions can be described with mass-action equations, corrected for electrostatic effects using EDL theory.
3. Surface charge (σ) and electrical potential (ψ) are consequences of chemical reactions involving the surface functional groups.
4. The apparent binding constants determined for the mass-law adsorption equations are empirical parameters related to thermodynamic constants (so-called intrinsic constants) via activity coefficients of the surface species.

Allison et al. (1991) state that "the activity difference between ions near the surface and those far away is the result of electrical work in moving the ions across the potential gradient between the charged surface and the bulk solution." The activity change of an ion moved from the surface to the bulk solution is described by EDL theory with an exponential Boltzmann expression

$$(X_s^z) = (X^z)[e^{-\psi F/RT}]^z \tag{10.29}$$

in which z is the charge of ion X, (X_s^z) is the activity of ion X of charge z near the surface, (X^z) is the corresponding activity of X in the bulk solution unaffected by the surface, $e^{-\psi F/RT}$ is the Boltzmann factor (unitless), ψ is the potential in volts at the plane of adsorption, and F (96,480 C/mol), R (8.314 J/mol K), and T are the Faraday constant, ideal gas constant, and absolute temperature, respectively.

The SC models differ among themselves in how they conceptualize the structure of the double layer and describe changes in surface potential and surface charge from the surface of the sorbent phase to the bulk solution (Figs. 10.18 and 10.19). The models make different assumptions regarding the positions of adsorbed species, which are considered *specifically adsorbed* if located at the sorbent surface. Hydrogen and hydroxyl ions are assumed specifically adsorbed in all three of the models. In the constant capacitance (CC) and diffuse-layer (DL) models, all adsorbed species are considered specifically adsorbed at the zero plane, and the adsorption of individual electrolyte ions is ignored. The triple-layer (TL) model, however, can assign adsorbed species to either a zero plane or more distant β plane and considers the adsorption of electrolyte ions.

Metal cations in the solid sorbent surface are symbolized by S in MINTEQA2. At mineral surfaces exhibiting amphoteric behavior, that behavior is often attributed to the successive dominance of surface SOH_2^+, SOH, and SO^- species with increasing pH (Fig. 10.4). In MINTEQA2, surface protonation and deprotonation reactions that are basic to the three SC models are written in the form

$$SOH + H_s^+ = SOH_2^+ \tag{10.30}$$

and
$$SOH - H_s^+ = SO^- \tag{10.31}$$

where H_s^+ represents a hydrogen ion at the surface. Corresponding mass-action expressions and their intrinsic constants are written

$$\frac{1}{K_{a1}^{int}} = \frac{(SOH_2^+)}{(SOH)(H_s^+)} \tag{10.32}$$

$$K_{a2}^{int} = \frac{(SO^-)(H_s^+)}{(SOH)} \tag{10.33}$$

The value of H_s^+, which cannot be directly measured, is assumed related to the bulk H^+ activity through a Boltzmann factor

$$(H_s^+) = (H^+) \exp(-\psi_o F/RT) \tag{10.34}$$

This expression can be substituted for H_s^+ in the mass-action quotients to give

$$\frac{1}{K_{a1}^{int}} = \frac{(SOH_2^+)}{(SOH)(H^+)\exp(-\psi_o F/RT)} \tag{10.35}$$

$$K_{a2}^{int} = \frac{(SO^-)(H^+)\exp(-\psi_o F/RT)}{(SOH)} \tag{10.36}$$

Boltzmann factor terms can be considered equivalent to activity coefficients for the sorption process (cf. Morel and Herring 1993). All three models assume protons are adsorbed at the surface of the sorbent and ψ_o is the potential at that surface, with the potential of the bulk solution assumed equal to zero.

For an oxide surface on which H^+ and OH^- are the only specifically adsorbed ions, at the pH of the point of zero net proton charge or pH_{PZNPC} (see Table 10.3) the net surface potential and net surface charge are both zero ($\psi_o = \sigma_o = 0$) and $(SOH_2^+) = (SO^-)$. Thus at the pH_{PZNPC} Eqs. (10.35) and (10.36) can be combined to give

$$pH_{PZNPC} = 0.5(pK_{a1}^{int} + pK_{a2}^{int}) \tag{10.37}$$

Accordingly, values of pK_{a1}^{int} and pK_{a2}^{int} should be equal pK units above and below the pH_{PZNPC}.

The intrinsic constant expressions written with Boltzmann function terms in Eqs. (10.35) and (10.36), correspond to hypothetical protonation and deprotonation reactions written

$$SOH + H^+ + \exp(-\psi_o F/RT) = SOH_2^+ \tag{10.38}$$

$$SOH - H^+ - \exp(-\psi_o F/RT) = SO^- \tag{10.39}$$

In MINTEQA2, which uses this approach, the Boltzmann terms are considered separate "electrostatic" components in model calculations.

A similar approach is taken with adsorbed metal cations and anions. Thus, for cation M^{2+} the reaction may be written:

$$SO^- + M^{2+} = SO \cdot M^+ \tag{10.40}$$

Assuming M^{2+} adsorption at the zero plane, then $(M_s^{2+}) = (M^{2+})\exp(-\psi_o F/RT)^2$. Unmeasurable (M_s^{2+}) can now be eliminated by substitution and the intrinsic constant expression becomes

$$K_{M^{2+}}^{int} = \frac{(SO \cdot M^+)}{(SO^-)\,(M^{2+})\,\exp(-\psi_o F/RT)^2} \tag{10.41}$$

The sorption of M^{2+} is often written, instead, as a proton exchange

$$SOH + M_s^{2+} - H_s^+ = SO \cdot M^+ \tag{10.42}$$

After substitution of Boltzmann factors, the intrinsic constant expression is

$$*K_{M^{2+}}^{int} = \frac{(SO \cdot M^+)\,(H^+)\,\exp(-\psi_o F/RT)}{(SOH)\,(M^{2+})\,\exp(-\psi_o F/RT)^2} = \frac{(SO \cdot M^+)\,(H^+)}{(SOH)\,(M^{2+})\,\exp(-\psi_o F/RT)} \tag{10.43}$$

for which the equivalent MINTEQA2 reaction is

$$SOH + M^{2+} - H^+ + \exp(-\psi_o F/RT) = SO \cdot M^+ \tag{10.44}$$

As before, the cation has been assumed adsorbed at the zero plane. Examination of the above intrinsic constant expressions shows that

$$*K_{M^{2+}}^{int} = K_{a2}^{int} \times K_{M^{2+}}^{int} \tag{10.45}$$

The same general approach can be used to describe adsorption of chloride ion, which may be written

$$SOH_2^+ + Cl_s^- = SOH_2 \cdot Cl \tag{10.46}$$

for which

$$K_{Cl^-}^{int} = \frac{(SOH_2 \cdot Cl)}{(SOH_2^+)\,(Cl_s^-)} \tag{10.47}$$

If we assume Cl^- adsorption into the β plane in the TL model, then $(Cl_s^-) = (Cl^-)\exp(+\psi_\beta F/RT)$. If described, instead, as a proton exchange, chloride adsorption and its intrinsic constant expression are

$$SOH + H_s^+ + Cl_s^- = SOH_2 \cdot Cl \tag{10.48}$$

$$\frac{1}{*K_{Cl^-}^{int}} = \frac{(SOH_2 \cdot Cl)}{(SOH)(H^+)(Cl^-)\exp([\psi_\beta - \psi_o]\,F/RT)} \tag{10.49}$$

This corresponds to

$$*K_{Cl^-}^{int} = \frac{K_{a1}^{int}}{K_{Cl^-}^{int}} \tag{10.50}$$

Intrinsic constant expressions so derived are of the same general form in all three models. However, because the models conceive different structures for the EDL and different species locations within the EDL, values of the intrinsic constants and capacitances assumed in each model will generally not be equal (cf. Schindler and Stumm 1987).

MINTEQA2 is capable of considering simultaneously five adsorbing surfaces, each having two types of binding sites. In the modeling of adsorption by oxides and clays it has sometimes been found necessary to assume two types of amphoteric binding sites to optimally fit empirical adsorption data. These have been labeled SOH and TOH sites and weak (S^wOH) and strong (S^sOH) sites. Such an approach has been used by Dzombak and Morel (1990) in the DL model as embodied in MINTEQA2 and in TL modeling of adsorption by clays (cf. James and Parks 1982; Mahoney and Langmuir 1991).

Solving problems that involve the electrostatic adsorption models requires the solution of a number of simultaneous equations. These include intrinsic constant adsorption expressions (mass-action equations), such as just described, and mass- and material-balance and charge-balance equations that account for total surface sites and sorbate species associated with the surface and the solution (cf. Davis et al. 1978; Dzombak and Morel 1987; Allison et al. 1991). These equations are considered in more detail in discussions of the three models.

10.4.3 Diffuse-Layer and Constant Capacitance Models

The origin and history of the constant capacitance (CC) and diffuse-layer (DL) models have been well documented by Davis and Kent (1990). Depicted schematically for the two models in Fig. 10.18 are the locations of adsorbed ions and assumed charge-potential relationships within the double layer. The DL model assumes that potentials measured at the zero plane and diffuse-layer plane are equal, or $\psi_o = \psi_d$. In both models, overall solution charge balance dictates that the charge due to the surface and its adsorbed species (σ_o), plus that contributed by ions in the diffuse layer (σ_d), must equal zero, or $\sigma_o + \sigma_d = 0$.

Figure 10.18 shows H^+ and OH^- specifically adsorbed onto surface sites in the zero plane in both models. Other specifically adsorbed species are also assumed to occupy the zero plane. With protonated and deprotonated surface sites symbolized as SOH_2^+, SOH and SO^-, a specifically adsorbed cation M^{2+} and anion A^{2-} may be represented as SOM^+, and SOH_2A^-. The surface charge, which of course neglects uncharged SOH, then equals

$$\sigma_o\left(\frac{C}{m^2}\right) = \frac{F}{S_A C_S}[(SOH_2^+) + (SOM^+) - (SO^-) - (SOH_2A^-)] \tag{10.51}$$

where the units of charge are coulombs (C) per square meter, F (the Faraday) is 96,480 C/mol, S_A is sorbent surface area in m^2/g, C_S is the weight of sorbent in contact with a liter of solution in g/L, and concentrations of the adsorbed species are given in mol/L. Surface-charge density Γ_0 in mol/L simply equals the bracketed term in this equation.

In the DL model, assuming Gouy-Chapman theory for a symmetrical electrolyte of charge z and 25°C, the charge density, σ, at some distance away from the surface, is given by

$$\sigma \left(\frac{C}{m^2}\right) = 0.1174 \, I^{1/2} \sinh\left(\frac{z\psi F}{2RT}\right) \tag{10.52}$$

which, for low surface potentials ($\psi < 25$ mV) reduces to $\sigma = 2.5 \, I^{1/2} \, \psi$, where I is the ionic strength in mol/L (cf. Dzombak and Morel 1990). The charge density in units of mol/L may be obtained from the relationship $\Gamma_0(\text{mol sites/L}) = [\sigma_0(C/m^2) \times S_A(m^2/g) \times C_S(g/L)]/F(C/mol)$.

Dzombak and Morel (1990) have published an exhaustive treatment of the DL model, including a compilation of measured or estimated intrinsic constants for the adsorption of 20 divalent cations and 14 anions by hydrous ferric oxide (HFO). Their results are summarized in Table 10.9. MINTEQA2 includes an intrinsic constant data base for the DL model from Dzombak and Morel (1990) in the file *feo-dlm.dbs*. This file lists K^{int} values for the adsorption by HFO of 9 cations (H^+, divalent Zn, Cd, Cu, Be, Ni, Pb, Ca, and Ba) and 10 anions, including $H_nSO_4^{n-2}$ ($n = 0, 1$), $H_nPO_4^{n-3}$ ($n = 0$ to 2), $H_2AsO_3^\circ$, $H_nAsO_4^{n-3}$ ($n = 0$ to 2), and $H_2BO_3^\circ$. The file lists K^{int} values for the simultaneous adsorption of most of the cations by high- and low-energy HFO surface sites (cf. Loux et al. 1989; Dzombak and Morel 1990). As data input to the DL model, MINTEQA2 asks the user to enter K^{int} values and values of Γ_0, S_A, and C_S. For the solution of DL problems in MINTEQA2 that involve adsorption by HFO of any of the species found in the *feo-dlm.dbs* file, that file can simply be appended to the problem input file. Further discussion and parameterization of the DL model, including adsorption by oxides of Al, Si, and Ti, is offered by Hohl et al. (1980), James and Parks (1982), and Stumm (1992, 1993).

Example 10.1

This problem involves the diffuse-layer (DL) model. We will solve the problem by hand and compare our results to the answer obtained with MINTEQA2.

A solution at 25°C and pH = 6 contains a 10^{-3} molar suspension of hydrous ferric oxide (HFO). The ionic strength due to a 1:1 electrolyte is fixed at 0.01 M. Assume that the only specifically adsorbed species (potential determining ions) in the solution are H^+ and OH^-. The HFO surface area, $S_A = 600$ m^2/g, and its gram formula weight (as FeOOH) is 89 g/mol. There are 0.2 moles of active (strong) sorption sites per mole of Fe in the HFO. Using the DL model, calculate the surface speciation of the HFO and the surface potential and surface charge. Dzombak and Morel (1990) give $K_{a1}^{int} = 10^{-7.29}$ and $K_{a2}^{int} = 10^{-8.93}$ for the HFO surface protonation-deprotonation reactions. The same authors tabulate values of the exponential Boltzmann term, which are a function of pH and ionic strength, listing $\exp(-\psi F/RT) = 10^{-1.85}$, for pH = 6 and $I = 0.01$ M.

The information given leads to $C_S = 89$ g/mol $\times 10^{-3}$ M $= 8.9 \times 10^{-2}$ g/L, and $\Gamma = 0.2$ mol active sites/mol Fe $\times 10^{-3}$ M FeOOH $= 2 \times 10^{-4}$ mol active sites/L. The species and variables to consider are: aqueous H^+ (given) and OH^-; surface species $FeOH_2^+$, $FeOH$, and FeO^-; surface potential ψ (fixed because pH and I are known), and surface charge, σ. The mole balance for the surface corresponds to

$$\Gamma = 2 \times 10^{-4} \text{ (M)} = FeOH_2^+ + FeOH + FeO^- \tag{10.53}$$

TABLE 10.9 Diffuse-layer model surface complexation (intrinsic) constants for adsorption of species by hydrous ferric oxide (HFO) at 25°C

Ion	$\log K_1^{int}$	$\log K_2^{int}$	$\log K_3^{int}$	Note	Ion	$\log K_1^{int}$	$\log K_2^{int}$	$\log K_3^{int}$	Note
H^+	7.29			†	Cr^{3+}	2.06			‖
Ba^{2+}	5.46	−8.93		‡	SO_4^{2-}		7.78	0.79	#
Sr^{2+}	5.01	−6.58	−17.60	‡	SO_3^{2-}		(11.6)	(4.3)	#
Ca^{2+}	4.97	−5.85		‡	$S_2O_3^{2-}$		(7.5)	0.49	#
Mg^{2+}		(−4.6)		‡	SeO_4^{2-}		7.73	0.80	#
Be^{2+}	(5.7)	(3.3)		‡	SeO_3^{2-}		12.69	5.17	#
Ag^+	−0.72	(−5.3)		§	PO_4^{3-}	31.29	25.39	17.72	††
Mn^{2+}	(−0.4)	(−3.5)		‡	AsO_4^{3-}	29.31	23.51	10.58 (K_4^{int})	††, ‖‖
Co^{2+}	−0.46	−3.01		‡	$H_3AsO_3^\circ$	5.41			‡‡
Ni^{2+}	0.37	(−2.5)		‡	$H_3BO_3^\circ$	0.62			‡‡
Cd^{2+}	0.47	−2.90		‡	$SiO_3^{2-}=H_2SiO_4^{2-}$	(15.9)	(8.3)		#
Zn^{2+}	0.99	−1.99		‡	VO_4^{3-}			13.57 (K_4^{int})	††, ‖‖
Cu^{2+}	2.89	(0.6)		‡	CrO_4^{2-}		10.82	(3.9)	#
Pb^{2+}	4.65	(0.3)		‡	WO_4^{2-}		(9.2)	(2.1)	#
Hg^{2+}	7.76	6.45		‡	MoO_4^{2-}		(9.5)	(2.4)	#
Sn^{2+}	(8.0)	(5.9)		‡	$SbO(OH)_4^-$		(8.4)	(1.3)	§§
Pd^{2+}	(9.6)	(7.7)		‡	CNO^-		(8.9)	(1.8)	§§
UO_2^{2+}	(5.2)	(2.8)		‡	CNS^-		(7.0)	(0.1)	§§
PuO_2^{2+}	(5.4)	(3.0)		‡	CN^-		(13.0)	(5.7)	§§
NpO_2^{2+}	(5.9)	(3.6)		‡	F^-		(8.7)	(1.6)	§§

Note: Values are consistent with the following assumptions: HFO has the formula FeOOH (88.85 g/mol), $S_A = 600$ m²/g, 0.005 mol/mol Fe of high energy, or strong sites (Fe^sOH sites), and 0.2 mol/mol Fe of low energy or weak sites (Fe^wOH sites). Parenthetic values are estimates.

† $1/K_{a1}^{int}$: $FeOH + H^+ = FeOH_2^+$
K_{a2}^{int}: $FeOH - H^+ = FeO^-$

‡ K_1^{int}: $Fe^sOH + M^{2+} = Fe^sOHM^{2+}$
K_2^{int}: $Fe^wOH + M^{2+} - H^+ = Fe^wOM^+$
K_3^{int}: $Fe^wOH + M^{2+} + H_2O - 2H^+ = FeOHOM$

§ K_1^{int}: $Fe^sOH + Ag^+ - H^+ = Fe^sOAg$
K_2^{int}: $Fe^wOH + Ag^+ - H^+ = Fe^wOAg$

‖ K_1^{int}: $Fe^sOH + Cr^{3+} + H_2O - H^+ = Fe^sOOHCr^+$

\# K_1^{int}: $Fe^sOH + A^{2-} - H_2O + H^+ = Fe^sA^-$
K_3^{int}: $Fe^sOH + A^{2-} = FeOA^{2-}$

†† K_1^{int}: $Fe^sOH + A^{3-} - H_2O + 3H^+ = FeH_2A$
K_2^{int}: $Fe^sOH + A^{3-} - H_2O + 2H^+ = FeHA^-$
K_3^{int}: $Fe^sOH + A^{3-} - H_2O + H^+ = FeA^{2-}$
K_4^{int}: $Fe^sOH + A^{3-} = FeOHA^{3-}$

‡‡ K_1^{int}: $FeOH + H_3AsO_3^\circ - H_2O = FeH_2AsO_3$
K_1^{int}: $FeOH + H_3BO_3^\circ - H_2O = FeH_2BO_3$

§§ K_2^{int}: $FeOH + A^- - H_2O + H^+ = FeA^-$
K_3^{int}: $FeOH + A^- = FeOA^-$

‖‖ These are K_4^{int} values for AsO_4^{3-} and VO_4^{3-}. K_3^{int} values are not available.

Source: Reported by Dzombak and Morel (1990).

Intrinsic constant equations for the surface are

$$K_{a1}^{int} = \frac{(FeOH)\,(H^+)\,\exp(-\psi F/RT)}{(FeOH_2^+)} = 10^{-7.29} \tag{10.54}$$

$$K_{a2}^{int} = \frac{(FeO^-)\,(H^+)\,\exp(-\psi F/RT)}{(FeOH)} = 10^{-8.93} \tag{10.55}$$

First, we solve for ψ, given that $\exp(-\psi F/RT) = 10^{-1.85}$. With $F = 96,480$ C/mol, $R = 8.314$ J/mol K, and $T = 298.15$ K (1 volt-coulomb = 1 joule). The result is $\psi = 0.109$ V. At 25°C, the Gouy-Chapman relationship is

$$\sigma(C/m^2) = 0.1174\ I^{1/2} \sinh(z\psi \times 19.46) \tag{10.56}$$

With $I = 0.01$ M and $z = 1$, we find $\sigma = 0.0483$ C/m². Substituting $10^{-1.85} = \exp(-\psi F/RT)$ and pH = 6 into Eqs. (10.54) and (10.55), results in $(FeOH_2^+) = 2.75$ (FeOH), and $(FeO^-) = 8.32 \times 10^{-2}$ (FeOH). These expressions can be solved simultaneously with Eq. (10.53), to obtain the surface concentrations in mol/L: $(FeOH) = 1.47 \times 10^{-4}$, $(FeOH_2^+) = 4.06 \times 10^{-5}$, and $(FeO^-) = 1.22 \times 10^{-5}$.

If the problem is input into MINTEQA2, the solution is $\psi = 0.110$ V, $\sigma = 0.0492$ C/m², $(FeOH) = 1.48 \times 10^{-4}$, $(FeOH_2^+) = 3.98 \times 10^{-5}$, and $(FeO^-) = 1.26 \times 10^{-5}$. The slight difference in results reflects the fact that in our hand calculation we assumed $\gamma_{H^+} = 1.00$, whereas MINTEQA2 assigns $\gamma_{H^+} = 0.901$ at $I = 0.01$ M.

Example 10.2

With the diffuse-layer model and the sweep option in MINTEQA2, calculate the adsorption of zinc onto HFO at 0.5-pH-unit increments between pH 4 and 8.5 and determine corresponding surface speciation. Comment on the relationship of surface speciation to the pH_{PZNPC} and on the behavior of surface charge and surface potential around the pH_{PZNPC}. Finally, compute and plot K_d for zinc adsorption from pH 4 to 8.5. Assume the same system conditions as in the previous problem, but with a total added zinc concentration of 10^{-6} M. System conditions include: $I = 0.01$ M, $S_A = 600$ m²/g, $C_S = 8.9 \times 10^{-2}$ g/L, and $\Gamma = 2 \times 10^{-4}$ mol active sites/L.

Results of the calculations are summarized in Table 10.10 and Fig. 10.20, which show that the zinc is almost entirely adsorbed between about pH 4.5 and 6.5. The adsorption edge

TABLE 10.10. Adsorption of Zn^{2+} by HFO computed in Example 10.2 using the sweep option in MINTEQA2 with the diffuse-layer model

pH	ψ volts	σ C/m²	$(FeOH_2^+)$ M	$(FeOH)$ M	(FeO^-) M	$(FeOZn^+)$ M	% Zn adsorbed
4.0	0.184	0.212	−3.93	−4.10	−5.91	−8.43	0.4
4.5	0.169	0.158	−4.04	−3.97	−5.54	−7.56	2.8
5.0	0.152	0.112	−4.14	−3.90	−5.26	−6.76	17.5
5.5	0.132	0.0763	−4.30	−3.85	−5.05	−6.21	61.4
6.0	0.110	0.0496	−4.41	−3.83	−4.90	−6.03	92.5
6.5	0.0865	0.0305	−4.49	−3.82	−4.79	−6.00	99.0
7.0	0.0609	0.0174	−4.56	−3.82	−4.72	−6.00	99.9
7.5	0.0341	0.0084	−4.60	−3.82	−4.67	−6.00	100.0
8.0	0.00066	0.0015	−4.64	−3.82	−4.63	−6.00	100.0
8.5	−0.02100	−0.0049	−4.67	−3.82	−4.60	−6.00	100.0

Note: Listed are the surface potential (ψ) and surface charge (σ), and molar concentrations of surface species.

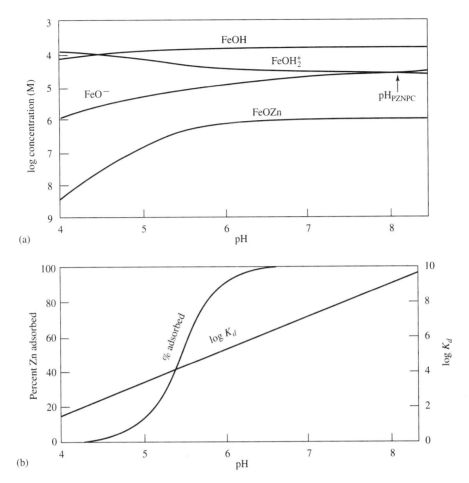

Figure 10.20   **(a)** Concentrations of surface species on HFO computed with the DL model as a function of pH, for $\Sigma Zn = 10^{-6}$ M and $I = 0.01$ M, $C_S = 0.089$ g/L, $S_A = 600$ m^2/g, and $\Gamma = 2 \times 10^{-4}$ mol active sites/L. Note that $FeOH_2^+ = FeO^-$ at pH_{PZNPC}. **(b)** Plot of percent Zn adsorbed and log K_d for Zn adsorption for the system described in (a).

(pH of 50% adsorbed) is near pH 5.4. Figure 10.20(a) indicates that sorbed zinc as FeOZn$^+$ increases rapidly above pH 4. The table and Fig. 10.20(a) show that FeO$^-$ and FeOH$_2^+$ become equal near pH 8.1, the pH of the PZNPC, with FeO$^-$ > FeOH$_2^+$ above that pH. (Remember that $pH_{PZNPC} = 0.5[pK_{a1}^{int} + pK_{a2}^{int}]$). Note that above pH 4.4, most surface sites are neutral FeOH sites. The table shows that both surface charge and potential are positive below the pH_{PZNPC}, equal zero at that pH, and turn negative above it.

The distribution coefficient, K_d, can be obtained through the equation

$$K_d = \left(\frac{Zn \text{ sorbed (M)}}{Zn \text{ dissolved (M)}} \right) \left(\frac{10^3 \text{ (mg/g)}}{(8.9 \times 10^{-2} \text{ (mg/ml) sorbent)}} \right) \qquad (10.57)$$

Values of K_d computed from the MINTEQA2 sweep output are plotted in Fig. 10.20(b). As often observed for the adsorption of weakly complexed metal cations, the plot of log K_d versus pH closely approximates a straight line. The equation of the line is log $K_d = -5.48 + 1.77$ pH.

The constant capacitance (CC) model (Fig. 10.18) can be viewed as a special case of the DL model that applies to conditions of low surface potential (<25 mV) and high and constant ionic strength. For such conditions the surface charge versus potential relationship in Eq. (10.52) reduces to

$$\sigma_o \left(\frac{C}{m^2} \right) \approx \mathbb{C} \psi_o \qquad (10.58)$$

where $\mathbb{C}$ is a constant capacitance value in farads/m^2 (1 farad = 1 coulomb/volt) and ψ_o is in volts. MINTEQA2 suggests 1.4 farads/m^2 for $\mathbb{C}$, which seems atypically high. To optimize a CC model fit of adsorption data measured at different ionic strengths, it is necessary to adjust values of the capacitance (cf. Dzombak and Morel 1987). A capacitance of about 1.0 F/m^2 has been proposed for TiO$_2$, γ-Al$_2$O$_3$, and α-FeOOH in 0.1 M ionic strength solutions, with $\mathbb{C}_1 = 0.72$ F/m^2 ($I = 0.01$ M) and 0.53 F/m^2 ($I = 0.001$ M) suggested for TiO$_2$ (Westall and Hohl 1980; Goldberg and Sposito 1984). Graphic analysis of charge-potential plots for Fe- and Al-oxide-dominated soils shown by Greenland and Hayes (1978) suggests the data can be fit assuming $\mathbb{C}_1 = 0.4$ to 0.5 F/m^2 ($I = 1$ M), and 0.3 F/m^2 ($I = 0.01$ M).

As data input to the CC model, MINTEQA2 asks the user to enter K^{int} values and values of Γ_o, S_A, C_S, and $\mathbb{C}_1$. Theoretically, intrinsic constants in the CC model apply only to a fixed or narrow range of electrolyte compositions and ionic strengths. All adsorbed species, including H$^+$ and OH$^-$, are assumed specifically adsorbed and thus found at the zero plane. The model has been applied to the adsorption of transition metal cations, including Cd^{2+}, Pb^{2+}, and borate and phosphate, by hydrous oxides of Al, Fe, and Si (cf. Davis and Kent 1990; Brady 1992). Brady has also considered the effect of temperature on adsorption by silica. The CC model has been used to describe adsorption data at low ionic strengths (cf. Goldberg and Sposito 1984; Schindler and Stumm 1987; Goldberg and Glaubig 1988) for which conditions it is not equivalent to the DL model.

A collection of adsorption parameters and intrinsic constants used in the CC model for $I = 0.1$ M is given in Table 10.11. Schindler and Stumm (1987) report intrinsic constants for additional metal cations and ligands, chiefly measured at higher ionic strengths. For internal consistency it is preferable to use the adsorption parameters from single sources.

10.4.4 Triple-Layer Model

The triple-layer (TL) model as used in MINTEQA2 (Fig. 10.19) was developed by Davis et al. (1978) and Davis and Leckie (1978, 1980). The model is more versatile than the DL and CC models in that it allows for the observation that adsorption of some species involves strong chemical bonding, while others experience relatively weak electrostatic attraction to surfaces. There is recent evidence that strongly adsorbed species, such as divalent transition-metal ions, and selenite, arsenate, and phosphate, are located in the zero plane at the sorbent surface. Weakly adsorbed alkali and alkaline earth cations and halides, nitrate, and carbonate are positioned in the β plane, farther from the surface (Davis and Kent 1990). The surface bonding of strongly adsorbed species is assumed comparable to the bonding experienced by aqueous cations and ligands in inner-sphere solution complexes. Species adsorbed in the β plane, separated as they are from the surface by waters of hydration and by adsorbed species in the zero plane, are surface bonded via long-range, weak coulombic forces. Such bonding is equivalent to that experienced by ions that form aqueous ion pairs or outer-sphere complexes. As a guide to the positions of adsorbed species in the double layer, it has been suggested that when adsorption is found independent of ionic strength (as is the adsorption of Pb^{2+} and Cd^{2+} by α-FeOOH, for example) sorbate species occupy the zero plane. Conversely, ionic strength-dependent adsorption indicates that a sorbate ion is located in the β plane (cf. Smith and Jenne 1991).

TABLE 10.11 Constant capacitance model parameters for adsorption by SiO_2(am), $Al(OH)_3$(am), γ-Al_2O_3, hydrous ferric oxide [HFO or $Fe(OH)_3$(am)], and goethite (α-FeOOH) in 0.1 M nitrate or perchlorate solutions

	SiO_2 (am)	$Al(OH)_3$ (am)	γ-$Al(OH)_3$	$Fe(OH)_3$ (am)	α-FeOOH	Source
S_A (m²/g)	105	41		193		†
					45	§
N_S (mol sites/m²)	8.3×10^{-6}	1.3×10^{-5}		4.4×10^{-5}		†
					3×10^{-5}	§
$\mathbb{C}_1$(F/m²)	1.25	1.06		1.25		†
					1.06	§
log K^{int} values						
$(H^+)_{a1}$		6.8		5.0		†
		7.4		6.6		‡
					7.31	§
$(H^+)_{a2}$	-7.5	-11.0		-9.4		†
	-6.8	-9.24		-9.1		‡
					-8.80	§
Ag^+		-4.86		-4.5		†
Ba^{2+}			-6.6			‡
Ca^{2+}			-6.1			‡
Cd^{2+}	-7.35	-2.82		-3.14		†
Cu^{2+}			-2.1			‡
Mg^{2+}			-5.4			‡
Pb^{2+}			-2.2			‡
Zn^{2+}	-4.17	-0.76		-2.68		†
PO_4^{3-}		5.0		7.27	20.53	§
HPO_4^{2-}		14.0			14.34	§
$H_2PO_4^-$					12.73	§
SeO_3^{2-}		13.6		12.9		†
$H_3SiO_4^-$				4.15		†
$H_2SiO_4^{2-}$				-3.85		†

Note: Intrinsic constant expressions for cations (M^z) and anions (A^z) are written in the form: $SOH + M^z = SOM^{z+1} + H^+$; and $SOH + A^z + H^+ = SA^{z+1} + H_2O$.

Silicic acid species adsorption reactions are written: $SOH + H_4SiO_4^\circ = SH_3SiO_4 + H_2O$; and $SOH + H_4SiO_4^\circ = SH_2SiO_4^- + H^+ + H_2O$.

Source: †Anderson and Benjamin (1990a, 1990b). ‡Schindler and Stumm (1987). §Goldberg and Sposito (1984).

MINTEQA2 assumes that adsorbed H^+ and OH^- occupy the zero plane, but allows the user to position other adsorbed species in either the zero or beta plane in the TL model, consistent with their probable behavior as just described. However, most of the extensive published literature and reported K^{int} values for the TL model have been derived assuming that only H^+ and OH^- occupy the zero plane with all other adsorbed species positioned in the β plane (cf. Kent et al. 1986; Smith and Jenne 1988, 1991). We are, therefore, obliged to make this same assumption in our MINTEQA2 TL-modeling calculations to maintain internal consistency.

An example of the positions of adsorbed species and general structure of the double layer in the TL model is shown schematically in Fig. 10.19. Assuming that $n = 2$, the net surface charge in the zero plane equals

$$\sigma_0 = \frac{F}{S_A C_S} [(SOH_2^+) + (SOH_2A^-) - (SO^-) - (SOM^+)] \tag{10.59}$$

The charge-balance calculation for the zero plane ignores species A^{2-} and M^{2+} in the beta plane and considers only the zero plane species to which these ions are adsorbed, plus unfilled adsorption sites (SO^- sites) and sites occupied by specifically adsorbed protons (SOH_2^+ sites). With A^{2-} and M^{2+} assigned to the beta plane, the net charge of that plane is given by

$$\sigma_\beta = \frac{F}{S_A C_S} [2(SOM^+) - 2(SOH_2A^-)] \tag{10.60}$$

Net charge is written in terms of the free ions themselves, since they occur as such in the beta plane. Overall charge balance requires

$$\sigma_0 + \sigma_\beta + \sigma_d = 0 \tag{10.61}$$

The triple-layer model calls for capacitances $\mathbb{C}_1$ and $\mathbb{C}_2$, corresponding to zones between the zero and beta planes and beta and d planes, respectively. The capacitances are related to the net charge and potentials of those planes through the expressions

$$\sigma_0 = \mathbb{C}_1(\psi_0 - \psi_\beta) \tag{10.62}$$

$$\sigma_\beta = \mathbb{C}_1(\psi_\beta - \psi_0) + \mathbb{C}_2(\psi_\beta - \psi_d) \tag{10.63}$$

$$\sigma_d = \mathbb{C}_2(\psi_d - \psi_\beta) \tag{10.64}$$

Usually $\mathbb{C}_2$ is assumed constant at 0.2 F/m^2, with $\mathbb{C}_1$ and K^{int} values adjusted to optimize the fit of the TL model to empirical adsorption data. Published values of $\mathbb{C}_1$ for hydrous ferric oxide (HFO), for example have ranged from 0.90 to 1.4 F/m^2 (Hsi and Langmuir 1985), with $\mathbb{C}_1 = 2.4$ F/m^2 proposed for δ-MnO$_2$ (Catts and Langmuir 1986). Riese (1982) suggested $\mathbb{C}_1 = 1.3$ F/m^2 for quartz (α-SiO$_2$) and $\mathbb{C}_1 = 2.4$ F/m^2 for kaolinite. On the other hand, Silva and Yee (1981) used $\mathbb{C}_1 = 0.9$ F/m^2 for quartz.

Kent et al. (1986) have tabulated reported TL intrinsic constants and $\mathbb{C}_1$ values for a variety of metal oxyhydroxides of Al, Si, Fe, and Ti. More recently, Smith and Jenne (1988, 1991) reevaluated published TL modeling of adsorption by ferric oxyhydroxide solids and by δ-MnO$_2$. Their analysis led to a set of intrinsic constants based on measurement and estimation that have been reproduced in Tables 10.12 and 10.13. The intrinsic constants in the tables were derived independent of values for $\mathbb{C}_1$ chosen by others. Few studies have applied TL modeling to adsorption by clays, although James and Parks (1982) and Mahoney and Langmuir (1991) TL-modeled alkali metal and alkaline earth adsorption by clays, including beidellite, illite, kaolinite, and montmorillonite.

Riese (1982) measured and TL-modeled the adsorption of Ra and Th by quartz and kaolinite in the presence of Ca-Ra competition and thorium-sulfate complexing. Some of his results are shown in Figs. 10.21, 10.22, and 10.23, which are drawn in terms of log dissolved concentration versus pH. This plotting approach provides much more information at the low concentrations typical of natural waters than does a plot of percent adsorbed.

Accurate modeling of his Th adsorption data as a function of pH required Riese (1982) to assume that the metal-hydroxy complexes were strongly adsorbed. Other researchers have often found this same assumption necessary when TL modeling the adsorption of strongly OH-complexed cations. Tables 10.12 and 10.13 list K^{int} values for a number of metal-hydroxy complexes.

TABLE 10.12. Triple-layer model intrinsic constants for adsorption by both amorphous hydrous ferric oxide (HFO) and goethite (α-FeOOH)

Species	pK^{int}	Species	pK^{int}	Species	pK^{int}
[†]Ag^+	5.2	[†]Fe^{2+}	(5.1)	[‡,§]SbO_2^-	(14.1)
[†]$AgOH^\circ$	12.3	[†]$FeOH^+$	(11.1)	[‡,§]SbO_3^-	(9.6)
[‡]$HAsO_4^{2-}$	26.1	[†]Hg^{2+}	(1.0)	[‡]$S_2O_3^{2-}$	9.9
[‡]$H_2AsO_4^-$	31.8	[†]$HgOH^+$	(3.1)	[‡]SO_4^{2-}	10.5
[‡,§]AsO_2^-	(12.9)	[†]K^+	9.3	[‡]HSO_4^-	16.2
[†]Ba^{2+}	(7.8)	[†]Mg^{2+}	6.8	[‡]SeO_4^{2-}	10.5
[†]$BaOH^+$	(16.3)	[†]$MgOH^+$	15.6	[‡]$HSeO_4^-$	15.9
[†]Ca^{2+}	6.6	[†]Mn^{2+}	(5.4)	[‡]SeO_3^{2-}	12.3
[†]$CaOH^+$	15.9	[†]$MnOH^+$	(12.1)	[‡]$HSeO_3^-$	19.4
[†]Cd^{2+}	5.0	[†]Na^+	9.3	[†]Tl^+	(8.1)
[†]$CdOH^+$	11.3	[‡]NO_3^-	7.5	[†]$Tl(OH)^\circ$	(16.4)
[‡]Cl^-	6.2	[†]NpO_2^+	4.1	[†]UO_2OH^+	6.9
[†]Co^{2+}	5.0	[†]Pb^{2+}	4.0	[†]$(UO_2)_3(OH)_5^+$	13.9
[†]$CoOH^+$	11.8	[†]$PbOH^+$	7.5	[∥]$UO_2(CO_3)_2^{2-}$	−29.0
[‡]CrO_4^{2-}	10.6	[†]$PuOH^{3+}$	−1.1	[∥]$UO_2(CO_3)_3^{4-}$	−42.0
[‡]$HCrO_4^-$	18.2	[†]$Pu(OH)_2^{2+}$	3.4	[†]Zn^{2+}	5.0
[†]Cu^{2+}	4.3	[†]$Pu(OH)_3^+$	7.3	[†]$ZnOH^+$	10.6
[†]$CuOH^+$	8.8	[†]$Pu(OH)_4^\circ$	13.4		

Notes: Estimates are shown parenthetically. Values are consistent with a surface-site density of 18 sites/nm^2 (3.0×10^{-5} mol sites/m^2), log $K_{a1}^{int} = -5.0 \pm 0.5$, log $K_{a2}^{int} = -10.9 \pm 0.5$, $C_1 = 1.25$ F/m^2, and $C_2 = 0.2$ F/m^2.
[†]$SOH + M^{z+} + nH_2O - (n + 1)H^+ = SOM(OH)_n^{z-n}$
[‡]$SOH_2H_nA^{n-z} - A^{z-} - (n + 1)H^+ = SOH$
[§]Equivalent to As(OH)$_4^-$, Sb(OH)$_4^-$, or Sb(OH)$_6^-$
[∥]$SOH_2UO_2(CO_3)_n^{2-2n} - UO_2^{2+} - nCO_3^{2-} - H^+ = SOH$
Source: Smith and Jenne (1988, 1991).

For a good approximation the adsorption of aqueous Th-sulfate complexes by quartz or kaolinite could be ignored in Riese's model-fitting of the adsorption data. Figure 10.24 shows that the distribution coefficient for Th(IV) adsorption is a strong function of both pH and total sulfate. Sulfate complexing clearly inhibits Th adsorption. Others have also found that nonhydroxyl metal com-

TABLE 10.13 Triple-layer model intrinsic constants for adsorption by birnessite (δ-MnO$_2$)

Species	p*K^{int}	Species	p*K^{int}	Species	p*K^{int}
Ag^+	(3.4)	Cu^{2+}	0.1	$MnOH^+$	(10.6)
$AgOH^\circ$	(13.0)	$CuOH^+$	7.5	$Mn(OH)_2^\circ$	(19.9)
Ba^{2+}	(4.5)	$Cu(OH)_2^\circ$	13.4	Na^+	3.5
Ca^{2+}	5.3	Fe^{2+}	(1.8)	Pb^{2+}	−1.8
Cd^{2+}	(2.0)	$FeOH^+$	(9.4)	$PbOH^+$	6.5
$CdOH^+$	(10.1)	$Fe(OH)_2^\circ$	(18.3)	Tl^+	(4.8)
$Cd(OH)_2^\circ$	(18.3)	Hg^{2+}	(−2.3)	$TlOH^\circ$	(15.3)
Co^{2+}	(1.6)	$HgOH^+$	(0.4)	Zn^{2+}	1.5
$CoOH^+$	(9.3)	Mg^{2+}	5.9	$ZnOH^+$	8.8
$Co(OH)_2^\circ$	(16.6)	Mn^{2+}	(2.1)	$Zn(OH)_2^\circ$	15.0

Note: Estimates are shown parenthetically. Values are consistent with log $K_{a1}^{int} = -6.2$, log $K_{a2}^{int} = 1.6$, $C_1 = 2.4$ F/m^2, and $C_2 = 0.2$ F/m^2. Adsorption reactions are written in the form: $SOH + M^{z+} + nH_2O - (n + 1)H^+ = SOM(OH)^{nz-n}$.
Source: Reported by Smith and Jenne (1988, 1991).

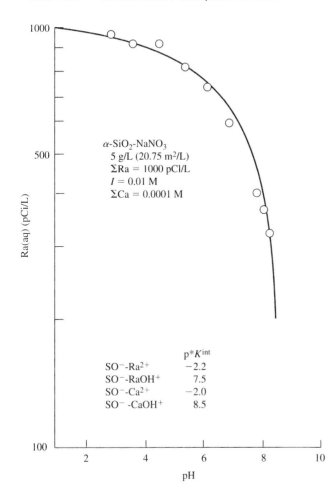

	p*K^{int}
SO$^-$-Ra^{2+}	-2.2
SO$^-$-RaOH$^+$	7.5
SO$^-$-Ca^{2+}	-2.0
SO$^-$ -CaOH$^+$	8.5

Figure 10.21   Adsorption of radium by α-SiO$_2$(quartz) at 25°C in 0.01 M NaNO$_3$ solution. Circles are experimental data. Solid curve is TL model calculated. Other experimental conditions and parameters include: $S_A = 4.15$ m^2/g, $N_S = 4.5$ sites/nm^2, p$K_{a1}^{int} = -0.95$, p$K_{a2}^{int} = 6.95$, p*$K_{Na}^{int} = 6.6$, $\mathbb{C}_1 = 1.3$ F/m^2 and $\mathbb{C}_2 = 0.2$ F/m^2. From A. C. Riese, © 1982. Used by permission.

plexes, such as those formed with carbonate, sulfate, and fluoride, may be poorly adsorbed by clays and oxyhydroxides. However, carbonate species themselves are strongly adsorbed by goethite, and so compete with the adsorption of other anions, such as chromate (Van Geen et al. 1994). Carbonate, as one would expect, is also strongly adsorbed by carbonate minerals (Zachara et al. 1993).

At moderate to high ionic strengths, electrolyte ions occupy most sorbent surface sites (cf. Mahoney and Langmuir 1991). This is not specifically quantified in the CC or DL models, but is considered in the TL model, which assumes that electrolyte ions are found in the β plane. Figure 10.25 is a distribution plot showing the percent of surface sites occupied by electrolyte ions (Na$^+$ and NO$_3^-$) and U(VI) species on goethite. At ΣU(VI) = 10^{-5} M and NaNO$_3$ = 0.1 M, uranyl (UO$_2^{2+}$) species are largely adsorbed below the zero point of charge and occupy less than 1% of surface sites that are chiefly SOH and SOH$_2^+$-NO$_3^-$ at this ionic strength.

10.4.5 Comparison of the Models

A list of the input parameters used in the three electrostatic adsorption models and required by MINTEQA2 is given in Table 10.14. For measurements over a range of ionic strengths, the CC model requires one or more values for $\mathbb{C}_1$ (cf. Dzombak and Morel 1987). The DL model is both simpler

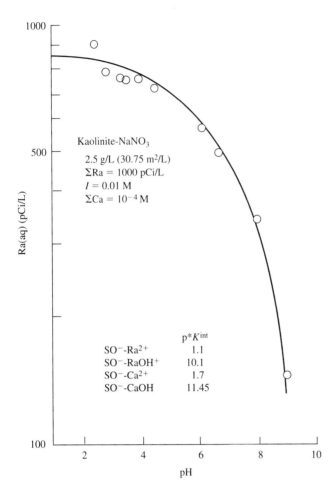

Figure 10.22 Adsorption of radium onto kaolinite at 25°C in 0.01 M $NaNO_3$ solution. Circles are experimental data. Solid curve is TL model calculated. Other experimental conditions and parameters include: $S_A = 12.3$ m²/g, $N_S = 6.0$ sites/nm², pK_{a1}^{int}(SOH sites) = 1.75, pK_{a2}^{int}(SOH sites) = 6.25, pK_{a2}^{int}(TOH sites) = 7.5, pK_{Na}^{int}(SOH sites) = 3.5, pK_{Na}^{int}(TOH sites) = 7.5, $C_1 = 2.4$ F/m², and $C_2 = 0.2$ F/m². From A. C. Riese, © 1982. Used by permission.

and more versatile in that it corrects adsorption for ionic strength using Gouy-Chapman theory (Stumm and Morgan 1996) and does not demand an input capacitance. Neither model explicitly considers the adsorption of electrolyte ions that compete for adsorption sites, particularly at moderate to high ionic strengths. The TL model does account for the adsorption of electrolyte ions, but then requires a second capacitance value (C_2) and intrinsic constants for the electrolyte species.

As noted before, all three models can fit adsorption data equally well. Of the three models, the DL model demands the least number of input parameters. However, the assumption of weak and strong bonding sites in the DL model increases its complexity to approach that of the TL model (cf. Meseure and Fish 1992a, 1992b). The TL model has the most parameter requirements, but exhibits the greatest versatility. It is capable of modeling complex systems over a range of pH's and ionic strengths and in which sorbate species, including electrolyte ions, compete for surface sites and form aqueous complexes of differing sorptive tendencies. Accurate modeling then requires the determination of intrinsic constants for the complexes, as well as the free ions.

Probably the TL model offers the most accurate description of the sorption process on an atomic scale. However, most published model parameters have been derived assuming that only H^+ and OH^- occupy the zero plane and that all other sorbate species are found in the β plane (cf. Smith

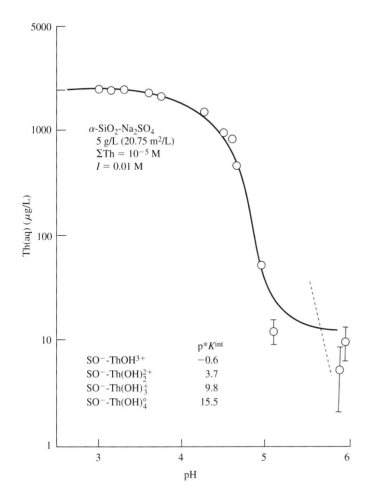

Figure 10.23 Adsorption of Th(IV) by quartz at 25°C in 0.01 M Na_2SO_4 solution (Riese 1982). Circles are experimental data. The solid curve is TL model-calculated assuming that Th-sulfate complexes are not adsorbed. The dashed line is the computed solubility of ThO_2. $*K_{Th}^{int}$ reactions are written in the form: $SOH + Th_s^{4+} + nH_2O =$ $SOTh(OH)_n + (n + 1)H_s^+$. Experimental conditions and parameters not given in the figure are listed in the caption to Fig. 10.21. From A. C. Riese, ©1982. Used by permission.

In figure:

α-SiO_2-Na_2SO_4
5 g/L (20.75 m^2/L)
$\Sigma Th = 10^{-5}$ M
$I = 0.01$ M

	p$*K^{int}$
SO^--$ThOH^{3+}$	-0.6
SO^--$Th(OH)_2^{2+}$	3.7
SO^--$Th(OH)_3^{+}$	9.8
SO^--$Th(OH)_4^{o}$	15.5

and Jenne 1991). Recent evidence shows that other strongly adsorbed species also occupy the zero plane. When published K^{int} values for the TL model are inconsistent with this evidence, adsorption-modeling results must be treated with caution, particularly when the modeling has been applied to conditions that differ substantially from those used for model parameterization.

10.4.6 Estimation of Intrinsic Adsorption Constants

Cations that form strong inner-sphere aqueous complexes are often strongly and specifically adsorbed when the same ligand is involved in solution and at the sorbent surface. Conversely, species forming weak aquocomplexes or ion pairs tend to be weakly adsorbed. These observations have formed the basis for most estimates of intrinsic adsorption constants (cf. Dzombak and Morel 1990; Smith and Jenne 1991). The extensive literature of solution complexation constants (cf. Smith and Martell 1976; Baes and Mesmer 1976, 1981) has facilitated such estimation methods. As an example, log K^{int} values for cation adsorption by specific metal oxyhydroxides plotted against the first hydrolysis constants of the same cations are often strongly correlated. Such plots, called linear free

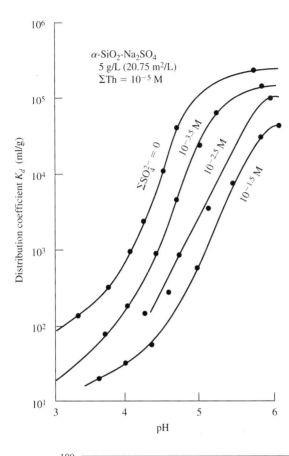

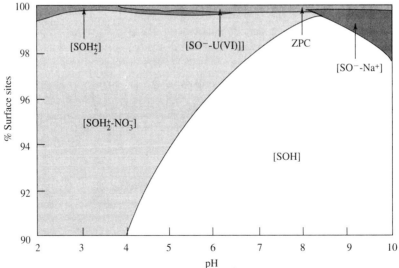

Figure 10.24 Adsorption of Th(IV) by quartz as a function of total sulfate for the same conditions as given in Fig. 10.23, plotted as distribution coefficients. Note that sulfate complexing of Th(IV) inhibits its adsorption, particularly at low pH's, increasing the pH of the adsorption edge. From A. C. Riese, ©1982. Used by permission.

Figure 10.25 Distribution plot of percent surface species on goethite at 25°C in 0.1 M $NaNO_3$ solutions for $\Sigma U(VI) = 10^{-5}$ M computed with the TL model. $C_S = 1$ g/L, $S_A = 45$ m^2/g (Hsi 1981; Hsi and Langmuir 1985).

TABLE 10.14 Parameters in the constant capacitance (CC), diffuse-layer (DL), and triple-layer (TL) models, as required inputs to MINTEQA2

Model	C_S g/L	S_A m²/g	Γ mol/L	$\mathbb{C}_1$ F/m²	$\mathbb{C}_2$ F/m²	K_{a1}^{int} K_{a2}^{int}	K_M^{int} K_L^{int}	K_{el}^{int}
CC	✓	✓	✓	✓		✓	✓	
DL	✓	✓	✓			✓	✓	
TL	✓	✓	✓	✓	✓	✓	✓	✓

Note: MINTEQA2 suggests $\mathbb{C}_1 = 1.4$ F/m² for use in the CC model. Most workers assume $\mathbb{C}_2 = 0.2$ F/m² in the TL model. K_M^{int} and K_L^{int} are intrinsic constants for strongly adsorbed cations and ligands, respectively. K_{el}^{int} denotes intrinsic constants for electrolyte cations and anions. Intrinsic constants for the adsorption of complexes are also commonly used in the TL model.

energy (LFER) plots, have been used to correlate and predict intrinsic constants for all three electrostatic adsorption models. (See Figs. 10.26, 10.27, and 10.28; Schindler et al. 1976; Hingston 1981.)

Most of the recent adsorption literature has emphasized the importance of the acid-base properties of oxide surfaces when explaining or estimating their sorption behavior. However, Sverjensky (1993) has shown that log K^{int} values for the adsorption of a specific cation by multiple mineral sorbents are a simple linear function of $1/\epsilon$, where ϵ is the dielectric constant of each mineral. He has used this approach to estimate CC and TL model K^{int} values for the adsorption of up to 18 cations on 7 oxide and silicate mineral surfaces.

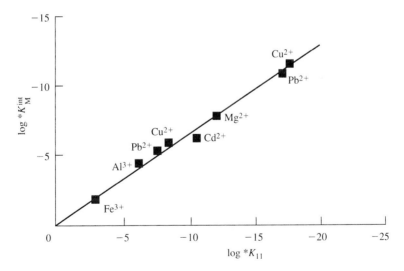

Figure 10.26   Correlation plot for some metal cations, of their first hydrolysis constants (*K_{11}) versus intrinsic surface complex constants (*K_M^{int}) for their adsorption by SiO_2(am) assuming the constant capacitance model. The equation of the solid line is log *$K_M^{int} = 0.09$ + 0.62 log *K_{11}. Hydrolysis and adsorption reactions are written *K_{11}; $M^{z+} + H_2O = H^+ +$ MOH^{z-1}, and *K_M^{int}; $M^{z+} + SOH = H_s^+ + SOM^{z-1}$. After P. W. Schindler and W. Stumm, The surface chemistry of oxides, hydroxides, and oxide minerals, In *Aquatic Surface Chemistry,* W. Stumm, ed. Copyright 1987 by John Wiley & Sons, Inc. Used by permission of John Wiley & Sons, Inc.

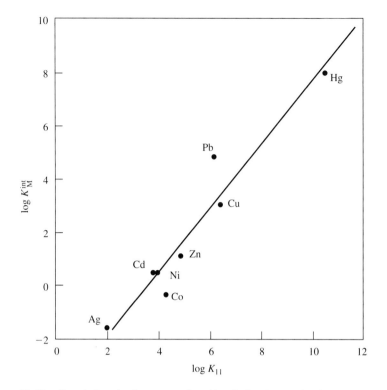

Figure 10.27   Correlation plot for some soft and borderline hard-soft acid cations, of their first hydrolysis constants (K_{11}) versus intrinsic surface complex constants (K_M^{int}) for their adsorption by hydrous ferric oxide (HFO), assuming the diffuse-layer model. Hydrolysis and adsorption reactions are written: K_{11}; $M^{z+} + OH^- = MOH^{z-1}$, and K_M^{int}; $M_s^z + SOH^- = SOHM^{z-1}$. After D. A. Dzombak and F. M. M. Morel, *Surface complexation modeling. Hydrous ferric oxide.* Copyright © 1990 by John Wiley & Sons, Inc. Reprinted by permission of John Wiley & Sons, Inc.

A number of observers have reported on the adsorption affinity of different metal cations for clays (cf. Farrah et al. 1980) and for hydrous oxides, organic matter, soils, and sediments (cf. Kinniburgh and Jackson 1982; Loux et al. 1989). There is evidence that the adsorption tendency of divalent transition elements by hydrous oxides often follows the Irving-Williams order (see Chap. 3). This suggests that the sequence of decreasing adsorption tendency is

$$Cu > Zn \geq Ni > Co > Fe > Mn > Ca$$

The relative percent adsorption of Cu, Zn, Ni, and Ca by ferric oxyhydroxide, shown in Fig. 10.17(b), obeys the Irving-Williams order. The plot suggests an overall adsorption affinity of

$$Pb > Cu > Cd > Zn > Ni > Ca$$

One can expect that K_M^{int} values in any of the electrostatic adsorption models would decrease, in the same order, for adsorption of these metal cations by other hydrous oxides, as well as by many other phases.

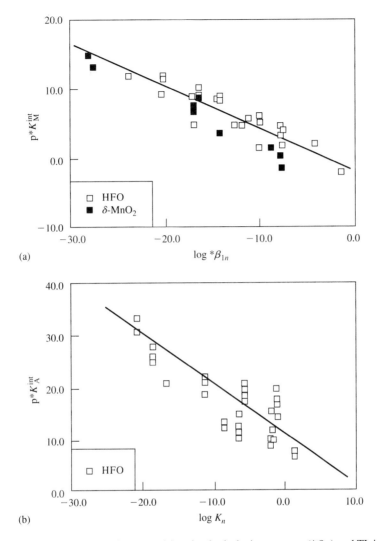

Figure 10.28   Correlations between: **(a)** cation hydrolysis constants ($*\beta_{1n}$) and TL intrinsic adsorption constants for cations ($*K_M^{int}$) adsorbed by HFO (goethite and amorphous HFO) or by δ-MnO$_2$; and **(b)** acid dissociation constants (K_n) and TL intrinsic adsorption constants for anions ($*K_A^{int}$) adsorbed by HFO. Reactions are written as follows: Cation hydrolysis, $M^z + nH_2O = M(OH)_n^{z-n} + nH^+$ ($n = 1, 2$); Acid dissociation, K_n, $H_nA^{z-n} = nH^+ + A^z$ ($n = 1$–3); Cation adsorption, $M_s^z + SOH + nH_2O = nH_s^+ + SOM(OH)_n^{z-1}$; Anion adsorption, $SOH_2A^{z-1} = nH_s^+ + A^z + SOH$. The lines drawn have no statistical significance. From Smith and Jenne (1988).

10.4.7 Application of the Electrostatic Adsorption Models to Natural Systems

Most adsorption modeling has focused on the laboratory behavior of single sorbent phases. In contrast, natural soils, sediments, and geologic formations typically contain multiple sorbents. For example, important sorbents in shallow soils and aquatic sediments often include organic matter plus clays. In deep soils important sorbents may be Fe- and Mn-oxyhydroxides and clays. Altered and weathered volcanic rocks may contain zeolites, smectite clays, and Fe- and Mn-oxyhydroxides. In fractured crystalline rocks the most important sorbents in fractures are likely to be Fe- and Mn-oxyhydroxides and clays.

MINTEQA2 is capable of considering simultaneous adsorption by one or two sorption sites on up to five sorbents. The model then assumes that the adsorptive behavior of multiple sorbents and sites is additive and proportional to the relative and absolute amounts of each sorbent phase in a mixture. This approach has been termed the linear adsorptivity model or LAM (cf. Honeyman 1984). Several recent studies have tested the LAM, chiefly in batch tests using the CC model (cf. Anderson and Benjamin 1990a, 1990b, 1990c). The simple LAM has failed when mixed, relatively amorphous sorbents of Al, Si, and Fe suspended in batch tests agglomerated and interfered with each other's surface sites. For example, positively charged $Fe(OH)_3$(am) and negatively charged SiO_2(am) particles tended to agglomerate and coagulate (Anderson and Benjamin 1990a). Similarly, amorphous $Al(OH)_3$ tended to coat and interfere with adsorption by SiO_2(am) in their mixtures (Meng and Letterman 1993). For such reasons and under such conditions, the LAM usually overestimates adsorption. These studies demonstrate the problem of applying the LAM to mixed, relatively amorphous sorbents suspended in some surface and shallow groundwaters, where interaction of sorbent particles is likely.

The LAM must also be used with caution in soils or sediments that contain significant amounts of organic matter. Davis (1984) (see also Schlautman and Morgan 1994) found that dissolved organic carbon (DOC) coated metal oxide surfaces in such systems. Thus, Cu^{2+}, which forms strong metal-organic complexes, was strongly adsorbed to the DOC coatings and not to exposed oxide sorption sites. The oxide adsorption of Cd^{2+}, however, which forms weak complexes with DOC, was unaffected by the presence of the DOC.

Siegel et al. (1992) tested the LAM using mixtures of goethite and montmorillonite clay, with quartz as an inert matrix. They modeled adsorption by goethite with the TL model and by montmorillonite using an ion-exchange approach. Adsorption of Ni, Pb, and Sr on the mixtures obeyed the LAM, probably in part because the sorbents were well-defined, crystalline mineral phases. One can tentatively reason that where suspended sorbent phases are relatively crystalline, or occur chiefly in the soil or rock itself, immobile and separate, the LAM approach should work.

Potential pitfalls to avoid and approaches to take in the application of adsorption models to complex soils and sediments have been described by Bolt and van Riemsdijk (1987), Davis and Kent (1990), and Warren and Zimmerman (1994). Based in part on these authors, if our goal is to predict or model adsorption in a soil, for example, questions we might ask include the following:

1. What is the chemical composition and range of compositions of soil waters and what concentrations of sorbate species of interest are or might be present? Does computer modeling of the water chemistry indicate that a potential sorbate species is at or above saturation with respect to a possible mineral? Sorption of that species cannot be meaningfully examined, except at concentrations below mineral saturation.

2. When multiple sorbent phases are present, which are most likely to interact with the sorbate species of interest?

3. What are the absolute and relative abundances of important sorbent solids and what fraction of their surface areas are exposed to flowing water? Any adsorption model we select that assumes a finite number of sorption sites, requires, as input, the area of a sorbing phase exposed to a given volume of water [e.g., C_S(g/L) $\times S_A$(m^2/g)] and a surface site density [N_S(sites/m^2)] for that phase. Can we measure or estimate these values? Such measurements and estimates are extremely difficult for metal adsorption by modern stream sediments, which may be mix-

tures of organic matter, and Fe- and Mn-oxyhydroxides and clays (cf. Warren and Zimmerman 1994).

Optical microscopic study of rock thin sections using dye tracers is a way to determine mineral-water contact (physical surface) areas if the water flows in rock matrix or fine fractures. The surface areas of particulate materials can be computed from particle size and geometry (cf. Sverdrup and Warfvinge 1993) or measured by BET gas adsorption methods. White and Peterson (1990) point out, however, that measured or computed surface areas of geological materials generally exceed their reactive surface areas. The reactive surface area (as defined by S_A) is what we need to model sorption or reaction rates in porous media.

4. When multiple sorbents are present, can we simplify the problem and limit our analysis to adsorption by a single sorbent because of its greater abundance or higher sorptive capacity for a group of sorbate species? The most successful modeling and prediction of adsorption in natural systems has been in such situations. For example, HFO is often the dominant sorbent in acid mine surface- and groundwaters. The diffuse double-layer model has been highly successful modeling metal and ligand adsorption by HFO in such waters (cf. Loux et al. 1990; Smith 1991; Smith et al. 1991; Stollenwerk 1994). Also, pesticide adsorption is chiefly by humic organic materials and may be independent of other sorbents present. As a first approximation in these cases, secondary sorbents can be ignored.

5. What is the simplest adsorption model that can adequately define sorption in the system of interest for purposes of our study? The simpler the model, the less information is needed to parameterize it. The distribution coefficient model requires only entry of the mass of sorbent in contact with a volume of water and a value for K_d. Pesticide adsorption can often be modeled adequately using a simple K_d (K_{oc}) approach (cf. Lyman et al. 1982). For smectite and vermiculite clays and zeolites that have dominantly pH-independent surface charge, ion-exchange or power-exchange models may accurately reproduce adsorption of the alkaline earths and alkali metals. If the system of interest experiences a wide range of pH and solution concentrations, and adsorption is of multivalent species by metal oxyhydroxides, then an electrostatic model may be most appropriate.

Loux et al. (1989) used MINTEQA2 to predict the adsorption and precipitation behavior of eight metals in an oxidized, sandy aquifer as a function of pH. Aquifer material contained 0.34% carbon, 0.077% Fe, and 0.0089% Mn. With the DL model, and assuming that the only important sorbent was HFO, they concluded that adsorption adequately described the behavior of Pb, Zn, and Ni. Cd behavior was better described assuming its precipitation as $CdCO_3$ (otavite). Changes in Cu, Ba, Be, and Tl could not be simply explained; the change in Cu probably reflected the role of organic matter which was ignored in sorption modeling. The mixed results obtained in this study suggest the limits to our present capability of predicting the simultaneous behavior of multiple sorbing species and multiple sorbents in complex natural systems.

10.5 ADSORPTION MODELS AND CONTAMINANT TRANSPORT MODELING

A variety of adsorption models, from K_d to the electrostatic adsorption models in MINTEQA2, have been coupled with hydrologic transport models. Available coupled codes and their attributes have been described and compared, in some detail, by Mangold and Tsang (1991) (see also Lichtner et al. 1996) and will not be considered here.

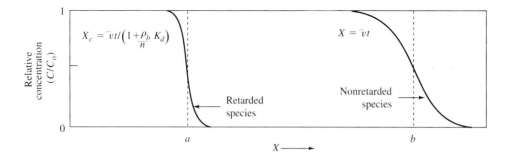

Figure 10.29 Advance of an adsorbing (retarded) species and nonadsorbing (nonretarded) species through a column of porous materials. Relative concentration (C/C_o) is the ratio of the concentration at time t to the input concentration at $t = 0$. Distances a and b traveled in time t are measured for $C/C_o = 0.5$. After R. A. Freeze and J. A. Cherry, *Groundwater.* Copyright © 1979. Used by permission of Prentice-Hall, Inc., Upper Saddle River, NJ.

The most commonly used adsorption model in contaminant transport calculations is the distribution coefficient, K_d model. In large part this reflects the simplicity of including a K_d value in transport calculations (cf. Freeze and Cherry 1979; Domenico and Schwartz 1990; Stumm 1992). Nevertheless, such applications should be limited to conditions where K_d values can be expected to remain near constant during transport (cf. Reardon 1981). Alternatively, if a K_d can be confidently shown to be a maximal or minimal possible value, such calculations can provide bounding or conservative information on contaminant transport. The bounding minimum K_d approach has become standard in the modeling radionuclide transport, for example (cf. Meijer 1992).

System assumptions that should be valid for such applications include: fluid flow in the porous media is isotropic and adsorption is fast, reversible, and linear (cf. Freeze and Cherry 1979). Given these constraints, the comparative transport of a conserved (nonadsorbed) tracer, such as Br^-, and an adsorbed or retarded species, such as Am^{3+}, can be described as shown in Fig. 10.29. A comparison of migration distances of the two species after time t, is made at concentrations where C(measured)$/C_o$(initial) $= 0.5$ for the conserved and adsorbed species. The migration distance X of the conserved species after time t is a measure of the average groundwater velocity ($\bar{v}$), or $X = \bar{v}t$. Similarly, the migration distance of the adsorbed species (X_c) is related to its velocity of movement ($\bar{v}_c$) by $X_c = \bar{v}_c t$. The retardation factor (R_d) for the adsorbed species is then given by

$$R_d = \frac{\bar{v}}{v_c} = \frac{X}{X_c} = 1 + \left(\frac{\rho_b}{n}\right) K_d \qquad (10.65)$$

where ρ_b is the bulk density of the geomedia, and n is the porosity given as a volume fraction. The average density of silicate rocks is 2.65 g/cm³, and the fractional porosity of unconsolidated granular deposits lies in the range $n = 0.2$ to 0.4 (Freeze and Cherry 1979). These values correspond to bulk mass densities of 1.6 to 2.1 g/cm³ and ρ_b/n between 4 and 10 g/cm³. With this information we may compute for $K_d = 0$, $R_d = 1$, and there is no retardation. For $K_d = 1$, $R_d = 5$ to 11, and a contaminant moves 9 to 20% as far as the groundwater. If $K_d = 10$, $R_d = 41$ to 101, and the contaminant moves only 1 to 2.4% of the distance traveled by the groundwater.

Example 10.3

The K_d for Ra^{2+} in groundwater systems is generally in the range of 25 to 250 cm³/g. Assuming $K_d = 100$ cm³/g, how much faster does groundwater move than Ra^{2+}? With this K_d we find

$R_d = \bar{v}/\bar{v}_c = 1 + (4 \text{ to } 10) \times 100 = 400 \text{ to } 1000$. The groundwater moves 400 to 1000 times faster than the radium.

STUDY QUESTIONS

1. Discuss the significance of the fact that substances occurring in small particles have an important percentage of their atoms or molecules at or near particle surfaces, as well as having very large molar surface areas.

2. The sorptive capability of solids for charged species is proportional to their surface area per weight and the density of charged sites on their surfaces (their surface-site density or exchange capacity). Explain this statement as it applies to the sorptive capacity of natural hematite versus that of ferrihydrate.

3. The cation exchange capacity of clays results from lattice imperfections or defects, isomorphous substitutions, and/or broken bonds on clay particle surfaces. Explain how the CEC's of kaolinite, the smectites, and illite, and their variation with pH, reflect these sources of their surface charge.

4. A few minerals can be assumed to have a constant surface charge, practically independent of pH, whereas most have a strongly pH-dependent surface charge that results from ionization of surface groups or surface chemical reactions. Explain this statement and illustrate its application to several example minerals.

5. Cation adsorption on natural materials is a function of pH, absolute and relative amounts of organic matter, clays, metal oxyhydroxides, and the particle-size distribution of sorbent phases. Explain how this statement applies to arid versus humid climate soils and to estuarine muds.

6. The rate of sorption may be diffusion rate limited. Explain how this statement may apply to sorption from groundwater by the rock matrix adjacent to a fracture and from a stream by a mixture of silt and gravel in a stream bed.

7. Draw a typical sorption isotherm; show on it and discuss the conditions for which the K_d concept applies, for which the breakthrough concentration is defined, and for which mineral saturation is attained.

8. As concentrations of trace species decrease, their tendency to be sorbed rather than dissolved increases. Explain how this observation is consistent with the appearance of the typical adsorption isotherm plot.

9. Write the Freundlich and Langmuir isotherm equations and explain how one determines their applicability. Discuss important applications and limitations of the isotherm equations, particularly in their use to describe the adsorption of multivalent cations and organic contaminants, such as pesticides, and the effects of ionic strength, pH, competitive adsorption, and complexation.

10. The Freundlich isotherm and the distribution coefficient (K_d) adsorption models assume an infinite number of sorption sites are available, whereas the Langmuir isotherm and ion-exchange models assume a limited or maximum number of sorption sites. Write sorption reactions that correspond to each of these models and explain the above statements in terms of those reactions.

11. For calculations of sorption, MINTEQA2 requires that the concentration of sorption sites be available in units of mol/L. Explain how this information can be obtained from the surface-site density (N_s, sites/m^2), specific surface area of the sorbing solid (S_A, m^2/g), concentration of solid in contact with a liter of soil water (C_s, g/L), and Avogadro's number (N_A).

12. Define K_{oc} and explain its applicability. How is it related to K_d? How and why is K_{oc} related to K_{ow}, to the solubility of organic substances (S), and to their bioconcentration factor (BCF)?

13. What is the retardation factor, R_d, used in discussions of contaminant transport in porous media and how is it related to K_d?

14. What is the power-exchange function and how is it related to ion exchange and Donnan exchange? Give examples of the applicability of each of these approaches to competitive cation adsorption. How are activity coefficients dealt with in ion-exchange reactions?

15. MINTEQA2 includes three electrostatic adsorption models. What are they? Compare their treatment of:
    (a) the distribution of charge and potential from the surface toward the bulk solution; and
    (b) the positions occupied by adsorbed H$^+$ and OH$^-$ and other ions at and near the surface.

16. Explain how the electrostatic models relate concentrations of ions that are sorbed to concentrations of the same ions in the bulk solution.

17. Be able to write a balanced sorption reaction in which a cation or anion is adsorbed onto a protonated or deprotonated surface site, using the format required for input into MINTEQA2. What is an intrinsic constant?

18. MINTEQA2 treats electrostatic terms (Boltzmann function terms) as if they were separate components. Explain.

19. The tendency of cations to be adsorbed by oxide and hydroxide minerals with increasing pH, is usually proportional to their tendency to form hydroxyl complexes in solution. Explain this statement. How would this general rule help you to decide under what different pH conditions equal concentrations of Cr^{3+}, Ca^{2+}, and Na^+ would tend to be adsorbed by the same solid?

20. The adsorption of cations increases with increasing pH, whereas the adsorption of anions generally decreases as the pH goes up. Explain this statement with a schematic figure and examples.

21. Specific adsorption is an important adsorption mechanism for trace substances in water, but not for major ionic species. Define specific adsorption and explain this statement with examples.

22. Given a soil containing a few percent of one or more sorbing phases, and with the pore-water volume fraction and soil-water composition known, describe in detail what you would need to know to compute the adsorption of a contaminant cation or anion by the soil using:
    (a) the K_d approach;
    (b) an ion-exchange model with H^+ or OH^- as competing ions;
    (c) the triple-layer model.

23. Discuss limitations to the potential accuracy and applicability of modeling results obtained in the previous problem.

PROBLEMS

1. A pesticide is retarded in a groundwater system relative to the groundwater flow and has been found to have a laboratory K_d in similar materials of 12 ml/g. If the groundwater moves 100 m in 1 year, how far does the pesticide move?

2. Sorption of uranyl (UO_2^{2+}) species (the oxidized form of dissolved uranium) by $Fe(OH)_3$ was measured in the laboratory at a constant pH of 7.23 and 25°C. In the experiment, 51.3 mg of ferrihydrite, $Fe(OH)_3$ (weight as Fe) was suspended in 50 ml of a 0.01 M KCl solution. The unpublished experimental data given here are from van der Weijden et al. (1976). Speciation calculations using MINTEQA2 show that the dominant form of dissolved uranyl (~97%) in these experiments is the $UO_2(OH)_2^0$ complex.

ΣU added (mg/L)	ΣU added (M) × 10^{-5}	$\Sigma U(aq)$ at equilibrium ($\mu g/L$)	$\Sigma U(aq)$ at equilibrium (M) × 10^{-8}	U sorbed (mg/L)	U sorbed (M) × 10^{-5}
1.00	0.42	0.6	0.252	0.9994	0.42
2.00	0.84	0.9	0.378	1.999	0.84
4.00	1.68	2.5	1.05	3.998	1.68
6.00	2.52	5.0	2.10	5.995	2.52
10.00	4.20	10.6	4.45	9.989	4.20
12.00	5.04	12.8	5.38	11.99	5.04
14.00	5.88	15.4	6.47	13.99	5.88
16.00	6.72	19.5	8.19	15.98	6.71
20.00	8.40	27.0	11.3	19.97	8.39

Note the data indicate that in excess of 99.8% of the total uranyl is sorbed at all total uranyl concentrations.

(a) Test to see if the data obey a Freundlich isotherm, solving for the constants in the isotherm equation. Use the approach suggested in MINTEQA2, where dissolved and sorbed concentrations are given in mol/L, but ignore activity coefficients. Is the Freundlich isotherm an appropriate model to be using in this case? Explain.

(b) Test to see if the data obey a Langmuir isotherm, solving for the constants in the isotherm equation. Use the approach suggested in MINTEQA2, where dissolved and sorbed concentrations are given in mol/L, but ignore activity coefficients. Is the Langmuir isotherm an appropriate model to be using in this case? Explain.

(c) What are some limitations of modeling the adsorption of uranyl using the isotherm equations?

3. O'Connor and Renn (1964) measured Zn^{2+} adsorption on suspended stream sediments at a constant total zinc concentration of 200 $\mu g/L$, and 200 mg/L of dissolved solids (an ionic strength of about 0.004 M), using 100 mg/L of sediment from the Potomac River at Point of Rocks, MD. Their measured data are tabulated here.

Zn^{2+}(aq) ($\mu g/L$)	pH	Zn^{2+}(aq) ($\mu g/L$)	pH
185	5.60	100	7.67
160	6.65	88	7.75
140	7.20	55	8.00
120	7.50	40	8.15

(a) Assuming that the river is in equilibrium with atmospheric CO_2 pressure, input the data into MINTEQA2 and calculate the saturation state of the water with respect to any minerals that might precipitate and complicate interpretation of the sorption data.

(b) Calculate K_d (ml/g) for sorption of zinc by the sediment, ignoring complexing and ion activity coefficients, and plot log K_d as a function of pH. Discuss the significance of your plot.

4. A vermiculite clay has the structural formula $\{K,Mg_{0.5}\}(Mg_{4.0}Fe^{2+}_{1.5}Fe^{3+}_{0.5})(Al_{1.5}Si_{6.5})O_{20}(OH)_4$. Calculate its cation-exchange capacity in meq/100 g and its surface-site density in mol/g. Assuming the surface area of the clay is 30 m^2/g, what is its surface-site density in nm^2 and $\mu mol/m^2$?

5. Consider the exchange reaction on a clay:

$$Ca^{2+} + Na_2X = 2Na^+ + CaX$$

Assuming simple ion-exchange behavior, we can write

$$K_{ex} = \frac{[Na^+]^2 \, (CaX)}{[Ca^{2+}] \, (Na_2X)} = 1.00$$

where the adsorbed cation concentrations are given in mole fractions.

(a) If Ca^{2+} occurs in solution at $10^{-3.00}$ M, at what dissolved Na^+ concentration will equal moles of sodium and calcium be adsorbed on the clay?

(b) If the dissolved concentrations of Na^+ and Ca^{2+} are diluted by 10 times, what happens to the concentrations of these ions adsorbed by the clay?

(c) Comment on the significance of the above calculations as they shed light on the relative behavior of monovalent and divalent ions in ion-exchange reactions in natural waters.

6. A clay-loam soil has a CEC of 20 meq/100 g. All exchange sites on the soil are initially occupied by Na^+. We place 10 g of the soil in a 1-liter solution that contains 20 mg/L of Na^+ and 40 mg/L of Ca. Experimental measurements of the soil suspended in batch solution have shown the following ion-exchange equilibrium is obeyed

$$\frac{(CaX_2)}{(NaX)^2} = 2.85 \frac{[Ca^{2+}]}{[Na^+]^2}$$

where X^{-1} represents soil exchange sites, CaX_2 and NaX are the molar concentrations of the ions adsorbed on the soil, and the bracketed terms are aqueous molar concentrations of the ions. Assuming a fixed pH of 7, ignoring ion-activity coefficients and assuming that the only cations adsorbed are Ca^{2+} and Na^+, calculate their final concentrations in solution and sorbed to the soil.

7. Cadmium-contaminated wastewaters are being released to a pond underlain by a silty sandstone. The silt fraction of the sandstone contains Ca-montmorillonite clay that has a CEC of 100 meq/100 g. The montmorillonite constitutes 12% by weight of the overall rock. The rest is quartz. Assume the total rock porosity of 33% is also the effective porosity. As an approximation, assume, as well, that the overall rock, the quartz, and the montmorillonite all have a density of 2.65 g/cm³.

    Near pH 7, the exchange reaction onto the montmorillonite,

$$Ca^{2+} + CdX = Cd^{2+} + CaX$$

obeys the power-exchange function

$$K_{ex} = \frac{[Cd^{2+}]}{[Ca^{2+}]}\left(\frac{CaX}{CdX}\right)^{1.16} = 0.56$$

where CaX and CdX are the mole fractions of the ions adsorbed by the clay, when other competing ions are absent. Assume that Cd-Ca exchange by the quartz obeys the same power-exchange function.
    (a) Calculate how many equivalents of clay-exchange sites will be in contact with 1 liter of pore water in the rock. The quartz in the sandstone has a CEC of 1.0 meq/100 g. What are the equivalents of quartz-exchange sites and the total of clay- and quartz-exchange sites in contact with a liter of pore water in the rock?
    (b) Assuming that Ca^{2+} and Cd^{2+} are the only ions initially present in the groundwater, and that background concentrations of these ions are $[Ca^{2+}] = 10^{-3.00}$ M and $[Cd^{2+}] = 10^{-7.00}$ M (assume activities equal concentrations), compute (CaX/CdX), the molar ratio of sorption sites in the rock occupied by $[Ca^{2+}]$ to those occupied by $[Cd^{2+}]$, and the concentrations of CaX and CdX.
    (c) A slug of cadmium-rich wastewater is discharged into the pond. Assuming that this wastewater contains 112 mg/L Cd^{2+} and 40 mg/L Ca^{2+} and that it instantaneously displaces preexisting ground water in the rock, calculate the Cd^{2+} and Ca^{2+} concentrations that will be found in a liter of the groundwater after it has reequilibrated with the silty sandstone and the corresponding molar concentrations of the ions sorbed by the rock.
    (d) Comment on the ability of the groundwater system to minimize groundwater contamination by the cadmium.
    (e) If all exchange sites on the rock were ultimately occupied by Cd^{2+}, would the groundwater system be easy or difficult to clean up? Explain.

8. A clay liner dried of its water content is 30% montmorillonite clay with a CEC of 120 meq/100 g. The porosity of the clay liner is 40%. Assume the dry density of the clay and other minerals in the liner is 2.65 g/cm³. At pH 6, the exchange reaction onto the clay involving Pb^{2+} and Ca^{2+} may be written

$$Ca^{2+} + PbX = Pb^{2+} + CaX$$

The sorption equilibrium obeys a power-exchange function

$$K_{ex} = \frac{[Pb^{2+}]}{[Ca^{2+}]}\left(\frac{CaX}{PbX}\right)^{1.25} = 0.25$$

where CaX and PbX are the mole fractions of the ions adsorbed by the clay when other ions are absent.
    (a) How many mole equivalents of clay-exchange sites will be in contact with 1 liter of pore water in the clay liner?
    (b) Assuming that Ca^{2+} and Pb^{2+} are the only ions initially present in the water within the clay liner and that their background concentrations are $[Ca^{2+}] = 10^{-3.00}$ M and $[Pb^{2+}] = 10^{-8.00}$ M (assume activities

equal concentrations, ignore complexes), compute (CaX/PbX) the molar ratio of sorption sites on the clay in the liner occupied by Ca versus Pb and calculate CaX and PbX in mol/L.

(c) If a slug of lead-rich wastewater is now pumped into the pond, calculate the final concentrations of lead and calcium in the water and sorbed on the clay. Assume that the wastewater contains 103.6 mg/L Pb and 40 mg/L Ca and that it instantaneously displaces all water within the clay liner.

9. The following problem involves calculations using the constant capacitance model in MINTEQA2. A sandy soil with an average mineral density of 2.63 g/cm^3 has a porosity of 20%. Soil grains have a 1.0%-by-weight coating of goethite (FeOOH). The goethite can be assumed to have a surface area of 45 m^2/g and a surface-site density of 1.35×10^{-3} mol/g. Assume the soil is water-saturated, the soil solution contains 0.1 M NaCl and 0.3 mg/L phosphorus as PO$_4$, and its pH is 6.00. The capacitance of the goethite can be assumed equal to 1.06 F/m^2. The following adsorption reactions and intrinsic constants for the constant capacitance model, where S denotes the surface, can be obtained or derived from Goldberg and Sposito (1984).

Surface reaction	Intrinsic constant
$SOH + H^+ = SOH_2^+$	$\log [1/K_{a1}^{int}] = 7.31 \pm 1.11$
$SOH - H^+ = SO^-$	$\log K_{a2}^{int} = -8.80 \pm 0.80$
$SOH + 3H^+ + PO_4^{3-} - H_2O = SH_2PO_4$	$\log K_1^{int} = 32.28 \pm 1.4$
$SOH + 2H^+ + PO_4^{3-} - H_2O = SHPO_4^-$	$\log K_2^{int} = 26.69 \pm 1.4$
$SOH + H^+ + PO_4^{3-} - H_2O = SPO_4^{2-}$	$\log K_3^{int} = 20.53 \pm 1.4$

Program the information given and compute and tabulate the speciation and concentrations of phosphate in the soil water and sorbed on the soil. What percentage of the total phosphate is adsorbed?

10. This problem is designed to provide insight into the fundamentals of the diffuse-layer (DL) model. We will solve the problem by hand and compare our results to the answer obtained by MINTEQA2.

A solution at 25°C and pH = 6 contains a 10^{-3} molar suspension of hydrous ferric oxide (HFO). The ionic strength due to a monovalent electrolyte is fixed and equals 0.01 M. Assume that the only specifically adsorbed species (potential determining ions) in the solution are H$^+$ and OH$^-$. The HFO surface area (S_A) is 600 m^2/g, and its gram formula weight (as FeOOH) is 89 g/mol. Assume that there are 0.2 moles of active sorption sites per mole of Fe in the HFO. Using the DL model, calculate the surface speciation of the HFO and the surface potential and surface charge. According to Dzombak and Morel (1990), the intrinsic constants for surface protonation-deprotonation reactions are $K_{a1}^{int} = 10^{-7.29}$ and $K_{a2}^{int} = 10^{-8.93}$. Values of the exponential Boltzmann term, which are a function of pH and ionic strength, have been tabulated by Dzombak and Morel (1990). For pH = 6 and $I = 0.01$ M, they give $\exp(-\psi F/RT) = 10^{-1.85}$.

11. With the diffuse layer model and the sweep option in MINTEQA2, calculate the adsorption of zinc onto HFO at 0.5-pH-unit increments between pH 4 and 8.5 and determine corresponding surface speciation. Comment on the relationship of surface speciation to the pH$_{PZNPC}$ and on the behavior of surface charge and surface potential around the pH$_{PZNPC}$. Assume the same system conditions as in the previous problem, but with a total added zinc concentration of 10^{-6} M. System conditions include: $I = 0.01$ M, $S_A = 600$ m^2/g, $C_S = 8.9 \times 10^{-2}$ g/L, and $\Gamma = 2 \times 10^{-4}$ mol active sites/L.

12. In this problem we use the diffuse-layer model to compute the adsorption of orthophosphate species by goethite between pH 3 and 10. Intrinsic constants for the adsorption reactions are available in the MINTEQA2 file, *feo-dlm.dbs*. Similar calculations have been discussed and performed by Hohl et al. (1980).

(a) Assume 0.6 g/L of goethite in suspension, with a surface-site density of 1.35×10^{-3} moles of sites per gram. The system contains $\Sigma Na = 3.0 \times 10^{-3}$ M, and $\Sigma PO_4 = 1.0 \times 10^{-3}$ M. Define a single adsorption site type in PRODEFA2, and in response to the query, "Do you want to attach an auxiliary data base of adsorption reactions?" answer, "yes," and name the *feo-dlm.dbs* file. In Edit Level III you will see that the complete file of adsorption reactions and intrinsic constants for the double-layer model from *feo-dlm.dbs* are listed. You can delete all reactions not of interest, or simply ignore their presence. MINRUN will ignore all reactions that involve species for which there is no concentration data entered in PRODEFA2. Note that there are intrinsic constants given for both high-energy sites (SO1 sites) and low-energy sites (SO2 sites) in *feo-dlm.dbs*. We will assume all sorption is by the SO1 sites only.

Use the sweep option to calculate the mole percent of dissolved and adsorbed phosphate species at integer pH values between pH 3 and 10 and tabulate the results.

(b) Perform the same calculation as in (a), except assume 6.0 g/L of goethite are present. Compare your results in parts (a) and (b) and discuss the difference. Comment on the relevance of your results to the mobility of phosphate in soils.

13. In this problem you are to compare the predicted adsorption of phosphate by goethite as computed using the constant capacitance and diffuse-layer models. Assume the same general system conditions as given in Problem 12. In other words, there are 0.6 g/L of goethite in suspension. The surface-site density of goethite equals 1.35×10^{-3} mol sites/g, and its surface area is 45 m^2/g. Total phosphate is 1.0×10^{-3} M and ΣNa = 3.0×10^{-3} M.

The surface capacitance for the constant capacitance model is 1.06 F/m^2. The intrinsic constants to be used in the constant capacitance model calculation are tabulated in Problem 9. Intrinsic constants for the double-layer model are listed in the MINTEQ file *feo-dlm.dbs*. You can either enter them individually in PRODEFA2, or simply attach that file to your problem. Assign all phosphate adsorption to the high-energy SO1 sites.

(a) Using the sweep option in MINTEQA2, calculate the percent of phosphate adsorbed between pH 3 and 10 as determined by both models. Plot and compare the model results and discuss.

(b) Compute the distribution coefficient of phosphate, K_d, in ml/g from the output results of part (a), and plot and compare log K_d versus pH as derived from the constant capacitance and diffuse-layer model calculations.

14. This problem involves use of the triple-layer adsorption model in MINTEQA2 to calculate the effect of pH on the adsorption of Pb^{2+} by the mineral goethite (α-FeOOH) in a soil.

The soil has an average mineral density of 2.65 g/cm^3 and a porosity of 30%. The dry soil contains 5.0% by weight of goethite as a coating on quartz sand grains. The goethite has a surface area of 45 m^2/g, and surface-site density of 18 sites/nm^2. The inner-layer capacitance of the goethite ($\mathbb{C}_1$) can be assumed equal to 1.4 F/m^2. The outer-layer capacitance, $\mathbb{C}_2$ is given as 0.20 F/m^2.

The following adsorption reactions and intrinsic constants are available for the triple-layer model, where FeO$^-$, FeOH, and FeOH$_2^+$ are surface sites.

Surface reaction	Intrinsic constant
FeOH + H$^+$ = FeOH$_2^+$	log $[1/K_{a1}^{int}] = 4.9 \pm 0.5^\dagger$
FeOH $-$ H$^+$ = FeO$^-$	log $K_{a2}^{int} = -10.9 \pm 0.5^\dagger$
FeOH + Na$^+$ $-$ H$^+$ = FeO-Na	log $K_{Na}^{int} = -9.1^\dagger$
FeOH + Cl$^-$ + H$^+$ = FeOH$_2$-Cl	log $K_{Cl}^{int} = -6.6^\dagger$
FeOH + Pb^{2+} $-$ H$^+$ = FeOPb$^+$	log $K_{Pb}^{int} = 0.38^\ddagger$

Source: †Smith and Jenne 1988. ‡Smith 1986

A total lead concentration of $10^{-5.00}$ M is introduced into the soil solution. Assume the soil is water saturated and that the soil solution also contains 0.01 M NaCl. Assume that Na$^+$ and Cl$^-$ are sorbed at the β plane (denoted by the hyphens), whereas H$^+$ and Pb^{2+} are sorbed at the zero plane. Given the above information, with the sweep option in MINTEQA2, compute lead sorption by goethite in the soil at increments of 0.5 pH units between pH 5 and 9. The following tasks can be accomplished using the output files from the sweep computer runs. Plot and discuss the significance of the following.

(a) Plot both percent Pb adsorbed and the total dissolved Pb versus pH from pH 5 to 9.

(b) Plot the mole percent distribution of dissolved and adsorbed lead species versus pH.

(c) Plot the $-$log concentration of surface species from pH 5 to 9.

(d) Calculate and plot K_d, the distribution coefficient for lead adsorption versus pH.

(e) Examine the saturation indices of lead minerals from pH 5 to 9 and tabulate the *SI* value of the most nearly saturated mineral. Discuss the significance of the lead mineral saturation versus adsorption results.

15. An acid-tailings solution from the mineral processing of uranium ore (pH = 1.2) is to be discharged into holding ponds in a sandstone bedrock that contains a few percent each of calcium carbonate and clay minerals. At issue environmentally is what will happen to the concentrations of toxic substances in the acid wastewater as it percolates into the underlying bedrock. An approach taken to answer this question involved neutralizing the tailings solution with lime [Ca(OH)$_2$] to pH values of 2, 4, and 6 in contact with the atmosphere. The residual solutions were analyzed to identify changes in concentrations resulting from the neutralization. The acid tailings had an Eh of 0.78 volts. Chemical and isotopic analyses of the tailings and neutralized solutions are given in the table below. Concentrations are in mg/L unless otherwise noted. Analyses were performed in duplicate. The uncertainties are as reported by the chemical analytical laboratory. Concentrations of Mg shown in parentheses have been estimated from the charge balance. Barium concentrations were below detection (<0.1 mg/L) at all pH's.

The concentration trends of the major species excluding Mg are plotted here. Computer modeling of the chemical analyses (assuming Fe is present as ferric iron) suggests that the drop in sulfate is caused by gypsum precipitation and that decreases in the concentrations of Al, Si, and Fe may result from supersaturation and consequent precipitation of phases, including allophane, alunite, basaluminite, Al(OH)$_3$(am), SiO$_2$(am), ferrihydrite, jarosite, and jurbanite.

The table also shows reductions in concentrations of As, Cd, Cu, Ni, Pb, and Zn with increasing pH. Among these species, computer modeling indicates that Pb is at saturation with anglesite (PbSO$_4$) at pH 1.2 and 2.0, but not at pH 4.0 and 6.0. The modeling shows that the other five elements are undersaturated

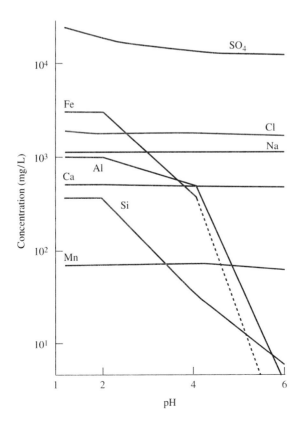

pH	1.2	2.0	4.0	6.0
Al	$1,020 \pm 10$	$1,009 \pm 15$	458 ± 1.5	4 ± 0
As	0.60 ± 0.03	0.56 ± 0.01	0.089 ± 0.003	0.02
B	1.7 ± 0.10	1.65 ± 0.07	0.65 ± 0.07	0.20 ± 0.0
Ca	576 ± 4	569 ± 2	493 ± 5.0	482 ± 0.0
Cd	1.03 ± 0.15	1.1 ± 0.0	0.4 ± 0.00	0.10 ± 0.01
Co	1.13 ± 0.15	1.2 ± 0.0	0.8 ± 0.0	0.4 ± 0.00
Cr	1.9 ± 0.0	1.9 ± 0.0	0.45 ± 0.07	0.35 ± 0.07
Cu	3.17 ± 0.12	3.1 ± 0.0	2.15 ± 0.07	<0.1
Fe	$2,430 \pm 20$	$2,330 \pm 28$	389 ± 1.5	<0.1
Mg	$(2,800)$	$(2,700)$	$2,300 \pm 41$	$2,835 \pm 21$
Mn	75.1 ± 0.6	76.75 ± 0.78	75.5 ± 0.40	65.2 ± 0.10
Mo	1.33 ± 0.58	1.0 ± 0.6	<1.0	<1.0
Na	$1,100 \pm 20$	$1,100 \pm 14$	$1,065 \pm 7.0$	$1,090 \pm 14$
Ni	1.43 ± 0.10	1.4 ± 0.0	1.2 ± 0.15	0.7 ± 0.0
Pb	4.3 ± 0.6	5.0 ± 2.0	1.50 ± 0.71	<0.1
Si	442 ± 10	408 ± 1.4	40.5 ± 0.7	6.0 ± 0.0
Sr	16.6 ± 1.0	12.8 ± 0.1	8.1 ± 0.0	6.6 ± 0.0
Zn	5.7 ± 0.35	5.75 ± 0.21	4.85 ± 0.21	1.0 ± 0.0
NO_3	31 ± 0	31 ± 0	40 ± 30	56 ± 10
SO_4	$24,400 \pm 75$	$20,600 \pm 75$	$14,690 \pm 440$	$12,340 \pm 220$
Cl	$1,630 \pm 32$	$1,584 \pm 40$	$1,670 \pm 0.0$	$1,584 \pm 40$
Se	4.0 ± 0.0	4.0 ± 0.0	2.0 ± 0.0	2.0 ± 0.0
NH_4	35.7 ± 2.3	38.5 ± 2.1	36.5 ± 5.0	30.0 ± 0.0
^{210}Pb (pCi/L)	$24,224 \pm 23$	$19,365 \pm 360$	$2,625 \pm 28$	<50
^{238}U (pCi/L)	$6,565 \pm 400$	$6,575 \pm 394$	$4,315 \pm 172$	<30
^{230}Th (pCi/L)	$149,302 \pm 2370$	$138,616 \pm 896$	$7,240 \pm 1401$	<400
^{226}Ra (pCi/L)	$3,334 \pm 330$	$2,923 \pm 250$	$1,784 \pm 177$	<150

Note: Concentrations are in mg/L unless otherwise indicated.

with respect to their pure solids at all pH's. Their declines in concentration with increasing pH must, there-fore, reflect adsorption or coprecipitation. Based on the precipitation of major species, the most likely sor-bents and coprecipitants in decreasing order may be solids of Fe, Al, and Si.

(a) As a first approximation, assume that the reduction in concentrations of As, Cd, Cu, Ni, Pb, and Zn ob-served during neutralization of the pH 1.2 tailings solution is caused by their adsorption onto ferrihy-drite or hydrous ferric oxide (HFO). Note that the amount of HFO sorbent increases with increasing pH. With this assumption, and using the double-layer adsorption model in MINTEQA2 (intrinsic con-stants are given in the file *feo-dlm.dbs*), compute the predicted aqueous concentrations of these species at pH 2, 4, and 6, and plot your predicted values against the measured concentrations. Discuss possi-ble reasons for any differences between measured and predicted values. Make the following assump-tions in the adsorption modeling, as suggested by Dzombak and Morel (1990): (1) sorbent HFO gram formula weight 89 g/mol Fe (this corresponds with a goethite or FeOOH formula for the HFO); (2) 0.005 high-energy sorption sites per mole of Fe and 0.200 low-energy adsorption sites per mole of Fe in the HFO; and (3) a specific surface area of 600 m²/g sorbent.

(b) Double the concentration of both low- and high-energy sorption sites over their number present at pH 6.0 in part (a), and compare and discuss your results.

(c) The chemical behavior of Pb^{2+} and ^{210}Pb should be the same. Can adsorption of ^{210}Pb by HFO explain reductions in the ^{210}Pb activity with increasing pH?

(d) The file *feo-dlm.dbs* includes intrinsic constants for Ba^{2+}. Assuming these constants are valid for $^{226}Ra^{2+}$ (a reasonable assumption), predict radium adsorption by HFO and compare it to the measured values. Radium is not in the MINTEQA2 database. However, given the similarity between Ba^{2+} and Ra^{2+} and the fact that Ba^{2+} concentrations were below detection in the tailings solution, you can input and fol-low the Ra^{2+} concentration as Ba^{2+}.

11

Oxidation-Reduction Concepts

11.1 REDOX THEORY AND MEASUREMENT

Many elements can occur in nature in more than one oxidation state. (See Table 11.1 showing the oxidation states of 45 elements.) The major elements with such "redox" behavior (defined here as occurring in solution typically at concentrations roughly above 1 mg/L) include H, O, C, S, N, and Fe. In a few subsurface (reducing) systems, Mn is also a major redox element. Important redox-sensitive elements that generally occur at lower, "minor" concentrations, except in some metal waste-waters, mine-tailings waters, or in subsurface-waters near their ore deposits include U, Cr, As, Mo, V, Se, Sb, W, Cu, Au, Ag, and Hg. The redox state of an element can be of considerable interest, because it often determines the chemical and biological behavior, including toxicity of that element as well as its mobility in the environment.

As a general rule, most reactions that involve electrons also involve protons. Oxidation usually releases protons or acidity. This is a basic cause of acid-mine drainage. Conversely, reduction usually consumes protons, and the pH rises.

11.1.1 General Redox Reaction

The general *half-reaction* written as a *reduction reaction* in conformity with the Stockholm or IUPAC (International Union of Pure and Applied Chemistry) convention, with electrons on the left, is

$$a\mathrm{A} + b\mathrm{B} + ne^- = c\mathrm{C} + d\mathrm{D} \qquad (11.1)$$

$$\underset{\text{oxidized state}}{} \quad \underset{\text{reduced state}}{}$$

where uppercase and lowercase letters denote the species involved and their stoichiometric coefficients, and n is the number of electrons (e^-). The theoretical voltage corresponding to this general half-reaction is given by

$$\mathrm{Eh(volts)} = E^\circ + \frac{RT}{nF} \ln \frac{(\mathrm{A})^a(\mathrm{B})^b}{(\mathrm{C})^c(\mathrm{D})^d} \qquad (11.2)$$

TABLE 11.1 Oxidation states of some important elements as they occur in natural waters and mineral systems

Element	Symbol	Number of protons (atomic number)	Oxidation states
Aluminum	Al	13	3+
Antimony	Sb	51	3+, 5+
Arsenic	As	33	3+, 5+, (0)
Barium	Ba	56	2+
Beryllium	Be	4	2+
Bismuth	Bi	83	3+, (0)
Boron	B	5	3+
Bromine	Br	35	1−, 0
Cadmium	Cd	48	2+
Calcium	Ca	20	2+
Carbon	C	6	4+, (0), 4−, 2−
Chlorine	Cl	17	1−
Chromium	Cr	24	6+, 3+
Cobalt	Co	27	2+, (3+)
Copper	Cu	29	2+, 1+, (0)
Fluorine	F	9	1−, 0
Gold	Au	79	3+, 1+, (0)
Hydrogen	H	1	1+, 0
Iron	Fe	26	3+, 2+
Iodine	I	53	5+, 0, 1−
Lead	Pb	82	2+, (4+), (0)
Lithium	Li	3	1+
Magnesium	Mg	12	2+
Manganese	Mn	25	2+, (3+), (4+)
Mercury	Hg	80	2+, 1+, (0)
Nickel	Ni	28	2+, (3+)
Nitrogen	N	7	5+, 3+, 0, 3−
Oxygen	O	8	2−, 0
Phosphorus	P	15	5+
Platinum	Pt	78	4+, 2+
Potassium	K	19	1+
Radium	Ra	88	2+
Selenium	Se	34	6+, 4+, (0), 2−
Silicon	Si	14	4+
Silver	Ag	47	1+, (0)
Sodium	Na	11	1+
Strontium	Sr	38	2+
Sulfur	S	16	6+, 4+, 0, (1−), 2−
Thorium	Th	90	4+
Tin	Sn	50	4+
Titanium	Ti	22	4+
Tungsten	W	74	6+
Uranium	U	92	6+, 4+
Vanadium	V	23	5+, 4+, 3+
Zinc	Zn	30	2+

Note: Values in parentheses are found in mineral systems only.

where $R = 0.001987$ kcal/mol deg; T is absolute temperature; and F, the Faraday constant, is 23.061 kcal/volt g eq. The term $(2.303RT/F)$ is called the *Nernst factor* and equals 0.05916 volts at 25°C (note: 2.303 log $X = \ln X$). $E°$, the standard potential of the half-reaction in volts, is related to $\Delta G_r°$ in kcal/mol by

$$E°(\text{volts}) = \frac{-\Delta G_r°}{nF} \tag{11.3}$$

There is a logical reason for writing the Eh reaction as given above so that all the terms are positive, and the *oxidized* species always appears in the numerator on the right. This conforms to the rationale that, as the concentrations of oxidized species increase, the system becomes more oxidizing, that is, the Eh increases.

All redox reactions involve an *oxidizing agent* (a substance that accepts or takes electrons) and a *reducing agent* (a substance that gives or donates electrons). For example, to go forward, the *redox couple reaction*

$$Fe^{3+} + e^- = Fe^{2+} \tag{11.4}$$

must involve a reducing agent (a second couple), such as

$$4H^+ + O_2(g) + 4e^- = 2H_2O \tag{11.5}$$

which is equivalent to

$$H^+ + \tfrac{1}{4}O_2(g) + e^- = \tfrac{1}{2}H_2O \tag{11.6}$$

Subtracting to eliminate electrons gives the *overall redox reaction,*

$$Fe^{3+} + \tfrac{1}{2}H_2O = Fe^{2+} + H^+ + \tfrac{1}{4}O_2(g) \tag{11.7}$$

$$\begin{array}{cccc} & (O^{2-}) & & (O°) \\ \text{oxid.} & \text{red.} & \text{red.} & \text{oxid.} \\ \text{agent} & \text{agent} & \text{agent} & \text{agent} \end{array}$$

Major redox couples tend to drive minor couples, such as

$$Fe^{3+} + e^- = Fe^{2+} \qquad (\text{major: } \Sigma mFe > 10^{-4} \text{ to } 10^{-5}) \tag{11.8}$$

$$\tfrac{1}{2}Cu^{2+} + e^- = \tfrac{1}{2}Cu° \qquad (\text{minor: } \Sigma mCu < 10^{-5}) \tag{11.9}$$

The overall redox reaction is

$$Fe^{2+} + \tfrac{1}{2}Cu^{2+} = Fe^{3+} + \tfrac{1}{2}Cu°(\text{native Cu}) \tag{11.10}$$

$$\begin{array}{cccc} \text{red.} & \text{oxid.} & \text{oxid.} & \text{red.} \\ \text{agent} & \text{agent} & \text{agent} & \text{agent} \end{array}$$

for which, from Gibbs free-energy data, we obtain

$$K_{eq} = 10^{-7.27} = \frac{[Fe^{3+}]}{[Fe^{2+}][Cu^{2+}]^{1/2}} \tag{11.11}$$

The relative concentrations of ferrous and ferric iron usually determine which direction reaction (11.10) will go.

Much of the recent literature has abandoned the use of Eh in favor of pE, which is the negative common logarithm of the electron concentration, that is, $pE = -\log_{10}(e^-)$ (cf. Stumm and Morgan 1981; Drever 1988). pE is related to Eh by the expression

$$pE = \frac{[nF]\,Eh}{2.303RT} = \frac{Eh(\text{volts})}{0.05916} \tag{11.12}$$

TABLE 11.2   Facts and assumptions involved in the definition of pH compared with corresponding concepts in the definition of pE (and Eh) according to Thorstenson (1984)

pH	Redox potential
H^+(aq) concentrations > 0	e^- concentrations ≈ 0
Electrodes respond to H^+(aq) from solutes and *solvent.*	Electrodes respond to electron transfer from *solutes.*
$H_2(g) = 2H^+(aq) + 2e^-$(Pt) approaches reversibility.	Red(aq) = Ox(aq) + e^-(Pt) is generally irreversible.
$10^{-15} < mH^+(aq) < 10$	$m[e^-](aq) \approx 0; \ 10^{-65} < a[e^-](aq) < 10^{-35}$
In pure water, pH ~ 7; $mH^+(aq) \approx 10^{-7}$.	In pure water, pE ~5.9; pE(aq) ~55.
H^+(aq) is a weak oxidizing agent; it can be reduced to $H_2(g)$ in aqueous solution.	e^-(aq), once formed, is a stronger reducing agent than metallic Na.

in general and at 25°C. Using this approach avoids the issue of whether or not Eh is measureable and, if measured, is thermodynamically meaningful. pE equations and diagrams are strictly theoretical. In any case, pE-pH and Eh-pH equations and diagrams have a similar appearance and both can be used for the same purposes—to describe and visualize theoretical relationships among redox-sensitive elements. We will limit ourselves to the Eh-pH approach in the following discussion.

There are profound differences between pE (or Eh) and pH, which have been inventoried in Table 11.2 (Thorstenson 1984). For example, protons exist in water as such, or, for example as H_3O^+ (hydronium ion), whereas electrons do not exist as free species in water. Also, most reactions involving protons are reversible, whereas those involving electrons usually are not (see also Hostettler 1984; Stumm and Morgan 1985; Scott and Morgan 1990).

Writing a balanced half-cell reaction can be a complex undertaking and justifies a brief review. Assuming our goal is to write such a reduction reaction, the steps, in order, are:

1. Write the species with the oxidized form of the element of interest on the left side of the equation and its reduced form on the right side.
2. Balance the elements except for H and O on both sides.
3. Balance the number of oxygen atoms by adding H_2O.
4. Balance the number of hydrogen atoms by adding H^+.
5. Achieve electroneutrality by adding electrons to the left side.

We will test this procedure by balancing the reduction reaction for sulfate to bisulfide. First, write $SO_4^{2-} = HS^-$, which is already balanced in sulfur atoms. Now, balancing the oxygen and then the hydrogen, we have $SO_4^{2-} + 9H^+ = HS^- + 4H_2O$. Adding eight electrons to the left side balances the charge and we have

$$SO_4^{2-} + 9H^+ + 8e^- = HS^- + 4H_2O \tag{11.13}$$

If our intent is to write an overall redox reaction, this procedure is followed first to obtain each half-reaction. The stoichiometry of one reaction is then adjusted so that the electrons are cancelled out by subtraction (cf. Appelo and Postma 1993).

11.1.2 The Standard Hydrogen Electrode

The standard hydrogen electrode is the ultimate reference for Eh (and pH) measurements. This electrode is formed by bubbling hydrogen gas at 1 bar pressure over a platinum electrode in a 1 N HCl solution (cf. Bates 1964). The electrode reaction is that of the H_2 gas-H^+ ion couple

$$H^+ + e^- = \tfrac{1}{2}H_2(g) \tag{11.14}$$

for which

$$Eh = E° + \frac{RT}{F} \ln \frac{[H^+]}{[P_{H_2}]^{1/2}} \tag{11.15}$$

Since $E° = -\Delta G_r°/nF$ and $\Delta G_r° = 0$, then $E° = 0$. Substituting for the Nernst factor gives

$$Eh = -0.0592 \log [P_{H_2}]^{1/2} - 0.0592 \text{ pH} \tag{11.16}$$

or

$$Eh = -0.0296 \log P_{H_2} - 0.0592 \text{ pH} \tag{11.17}$$

Shown in Fig. 11.1 is a hydrogen electrode used in an apparatus to measure the standard potential of the redox couple

$$Cu^{2+} + 2e^- = Cu° \tag{11.18}$$

which is $E° = 0.340$ volts at 25°C. In the left-hand cell, pure $H_2(g)$ is bubbled over a Pt electrode ($P_{H_2} = 1$) in a strong acid solution where $[H^+] = 1$. The corresponding redox half-cell reaction is the reverse of reaction (11.14). The oxidation reaction produces protons and electrons. In the right-hand cell, the electrons are discharged from a Pt electrode reacting with cupric ion to produce metallic copper. The salt bridge, which contains a concentrated KCl solution, completes the circuit. The overall redox reaction is the sum of the two half-cell reactions or couples

$$Cu^{2+} + H_2(g) = 2H^+ + Cu° \tag{11.19}$$

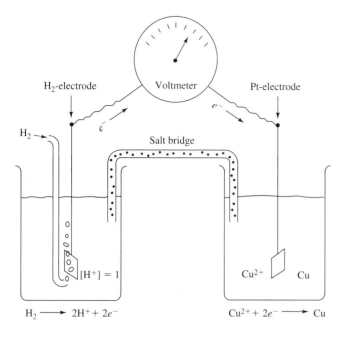

H₂-electrode   Voltmeter   Pt-electrode

H₂

Salt bridge

$[H^+] = 1$ Cu^{2+} Cu

H₂ ⟶ 2H⁺ + 2e⁻   Cu²⁺ + 2e⁻ ⟶ Cu

Figure 11.1   A standard hydrogen electrode (SHE) being used as the reference electrode in a circuit designed to measure the potential of the Cu^{2+}/Cu metal couple.

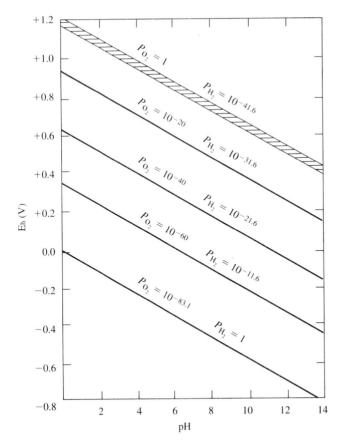

Figure 11.2 The stability field of water as a function of Eh and pH at 25°C and 1 bar pressure. Contours showing partial pressures of hydrogen and oxygen at intermediate Eh values have been computed with Eqs. (11.17) and (11.23). The crosshatched area is the locus of Eh values computed assuming the reaction $4H^+ + O_2(g) + 4e^- = 2H_2O$ is at thermodynamic equilibrium and dissolved oxygen is at or above a detection limit of 5 μg/L.

The measured Eh, which depends on the concentration of Cu^{2+} in the right-hand cell, is described by

$$Eh(\text{volts}) = E° + \frac{0.0592}{2} \log [Cu^{2+}] = 0.340 + 0.0296 \log [Cu^{2+}] \qquad (11.20)$$

It is generally both impractical and cumbersome to use a hydrogen electrode for routine Eh measurement. Instead, such measurements are usually performed with a platinum or glassy carbon indicator electrode and a calomel ($Hg_2Cl_2/Hg°$) or silver-silver chloride (Ag/AgCl) reference electrode of known potential (cf. Langmuir 1971a; Macalady et al. 1990).[†] This measurement is then corrected to Eh, the potential with reference to the standard hydrogen electrode, or $Eh = E_{\text{measured}} - E_{\text{ref. electrode}}$.

11.1.3 The Eh-pH Stability Field of Water

The theoretical stability field of water in Eh and pH terms (see Fig. 11.2) bounds all theoretical redox reactions taking place in water. The upper boundary of the water stability field is defined by Eh and pH values for which liquid water is in equilibrium with $O_2(g)$ at 1 bar pressure. The lower boundary is defined by Eh and pH values for which liquid water is in equilibrium with $H_2(g)$ at 1 bar pressure.

[†]The calomel electrode reaction $Hg_2Cl_2(s) + 2e^- = 2Hg(l) + 2Cl^-$ has $E° = 0.2682$ V at 25°C. The Ag-AgCl electrode reaction is $AgCl(s) + e^- = Ag(s) + Cl^-$, with $E° = 0.2223$ V at 25°C.

The equation of the lower boundary can be derived from Eq. (11.17) for the hydrogen electrode. At the lower stability field of water, $P_{H_2}(g) = 1$ bar, and this equation reduces to

$$Eh(volts) = -0.0592 \text{ pH} \qquad (11.21)$$

The $P_{O_2}(g)$ pressure at the lower stability limit can be computed from the reaction

$$2H_2O = 2H_2(g) + O_2(g) \qquad (11.22)$$

for which we find

$$K_{eq} = 10^{-83.1} = (P_{H_2})^2(P_{O_2}) \qquad (11.23)$$

If $P_{H_2} = 1$ bar, then $P_{O_2} = 10^{-83.1}$ bar. For $P_{O_2} = 1$ bar at the upper-stability limit, this equation indicates $P_{H_2} = 10^{-41.6}$ bar.

The Eh-pH boundary for the upper stability field can be computed from the reaction

$$4H^+ + O_2(g) + 4e^- = 2H_2O \qquad (11.24)$$

for which
$$Eh = E° + \frac{0.0592}{4} \log P_{O_2} [H^+]^4 \qquad (11.25)$$

and the free-energy data yield $E° = 1.23$ volts. Simplifying and expanding terms gives

$$Eh = 1.23 + 0.0148 \log P_{O_2} - 0.0592 \text{ pH} \qquad (11.26)$$

At the upper boundary where $P_{O_2}(g) = 1$ bar, this equation reduces to

$$Eh(volts) = 1.23 - 0.0592 \text{ pH} \qquad (11.27)$$

Under atmospheric conditions where $P_{O_2} = 0.2$ bar, this equation is practically unchanged and Eh = $1.22 - 0.0592$ pH. The chemical analytical detection limit for dissolved oxygen is from 0 to 10 μg/L (Macalady et al. 1990). Assuming it is 5 μg/L (1.56×10^{-7} M) with $K_{Henry} = 1.26 \times 10^{-3}$ M/bar, we find $P_{O_2} = 1.24 \times 10^{-4}$ bar and Eh = $1.17 - 0.0592$ pH. This equation, which plots as the lower boundary of the crosshatched area in Fig. 11.2, defines the lower limit of Eh values one would expect in the presence of measureable oxygen, assuming the Eh computed from thermodynamic data equals the measured Eh.

11.1.4 Measured versus Theoretical Redox Potentials

The water stability boundaries and the locus of measured Eh and pH measurements in natural waters, as reported by Baas-Becking et al. (1960), are shown in Fig. 11.3 (see also Fig. 11.4). It has been observed that frequently the Eh values measured with a Pt electrode differ significantly from values computed from Gibbs free energies or standard potentials and solution concentrations. When they exist, there are two important reasons for such differences. These include: (1) misbehavior of the Pt or other indicator electrode; (2) the irreversibility or slow kinetics of most redox couple reactions and resultant disequilibrium between and among different redox couples in the same water; and (3) the common existence of mixed potentials in natural waters (see below).

Whitfield (1974) noted that most of the measured Eh data (Fig. 11.3) were consistent with reactions involving the Pt electrode itself. For example, highest Eh values clustered near boundaries for the Pt°/PtO and PtO/PtO$_2$ couples. For the Pt°/PtO couple

$$Eh = 0.88 - 0.059 \text{ pH} \qquad (11.28)$$

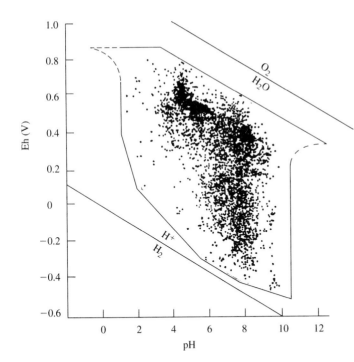

Figure 11.3   Locus of measured Eh values. After L. G. M. Baas-Becking et al., Limits of the natural environment in terms of pH and oxidation-reduction potentials, *J. Geol.* 68:243–84. Copyright © 1960 by The University of Chicago Press. Used by permission.

suggesting that the Pt electrode behaves as a pH electrode in oxygenated waters. Whitfield (1974) further noted that in the presence of hydrogen sulfide, PtS forms on the electrode surface, again complicating interpretation of measurements at low Eh values. Careful and frequent cleaning of Eh electrodes of Pt and Au can sometimes avoid measurement problems caused by surface reactions (cf. Langmuir 1971a). Alternatively, Macalady et al. (1990) used a glassy, carbon Eh electrode, which is less vulnerable to surface poisoning than a Pt or Au electrode.

The equations for the oxidation and reduction of water given above are theoretical and define conditions at chemical equilibrium. Chemical equilibrium among most redox reactions in natural waters is unusual because of their sluggish kinetics. If redox equilibrium were attained for reaction (11.24) in natural waters, the Eh and pH of waters containing measureable oxygen would lie within the crosshatched area in Figs 11.2 and 11.4. However, the Eh measured in such waters lies nearly 0.2 volts lower on the diagram. This partly reflects the extremely slow kinetics of water oxidation at 25°C, such that the exchange of electrons that should take place with oxidation of water cannot be measured with an Eh electrode. Sato and Mooney (1960) proposed that the highest Eh values in Fig. 11.3, which were measured in the presence of atmospheric oxygen, correspond instead to the reaction

$$O_2(g) + 2H^+ + 2e^- = H_2O_2 \tag{11.29}$$

for which $E° = 0.682$ V and

$$\text{Eh (V)} = 0.682 + 0.0296 \log \frac{P_{O_2}}{[H_2O_2]} - 0.0592 \text{ pH} \tag{11.30}$$

at 25°C (see also Berner 1971). As shown in Fig. 11.5, Sato and Mooney's measured Eh values for oxidized mine waters are bracketed by $P_{O_2}/[H_2O_2]$ values between 1 and 10^6, consistent with the occurrence of hydrogen peroxide in natural waters (Berner 1971).

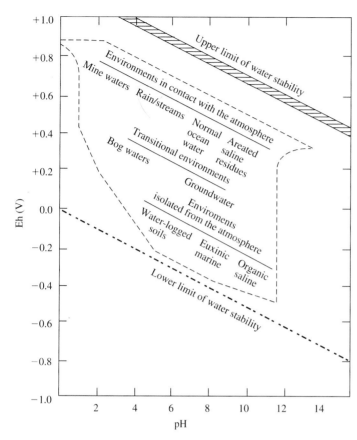

Figure 11.4 Approximate position of some natural environments in terms of Eh and pH. The dashed line represents the limits of measurements in natural environments, as reported by Baas-Becking et al. (1960) and shown in Fig. 11.3. The crosshatched area defines theoretical conditions under which waters are calculated to contain dissolved oxygen at or above a detection limit of 5 μg/L. Modified after R. M. Garrels and C. L. Christ (1965). Solutions, minerals and equilibria. Copyright © 1965 by Freeman, Cooper and Company. Used by permission.

Reversible, thermodynamically meaningful Eh measurements may be possible in waters that contain significant amounts of dissolved Fe, Mn, and sulfide sulfur (acid mine waters, Fe-rich groundwaters, and sulfide-rich sediments). Meaningful Eh measurements are often not possible when the dominant redox sensitive elements are C, N, O, H, and oxidized S, as is usually the case in surface waters and municipal wastewaters. Excellent agreement between computed Eh (Eh_c) and measured Eh (Eh_m) values for acid mine waters is evident from Fig. 11.6 (Nordstrom et al. 1979). Surprisingly good agreement was also found by Thorstenson as reported by Berner (1971) when Eh values computed from $N_2(aq)/NH_4^+$ concentrations were compared to the Eh computed from SO_4^{2-}/HS^- in Bermuda groundwaters. The agreement was less satisfactory when computed Eh values from concentrations of $N_2(aq)/NH_4^+$ and $HCO_3^-/CH_4(aq)$ couples were compared (Fig. 11.7).

Lindberg and Runnells (1984) constructed the comparison plot in Fig. 11.8 with Eh_m compared to Eh_c values computed from analyzed concentrations of redox couples of Fe, O, H, S, N, and C in groundwater. The generally poor agreement between Eh_m and Eh_c for couples involving O, H, N, C, and oxidized S is undoubtedly related to their nonelectroactive behavior. (A species in a redox couple that cannot be readily oxidized or reduced, that is, cannot exchange electrons at the surface of an Eh indicator electrode, is said to be nonelectroactive.) The poor agreement in Eh_m and Eh_c for Fe^{3+}/Fe^{2+}, $HS^-/S_{rhombic}$, and $Fe^{2+}/Fe(OH)_3$ is, however, suspect, in that these couples are electroactive and Eh_m would be expected to approach Eh_c. Patterson and Runnells (1992) suggest additional explanations for differences between Eh_m and Eh_c. The disparity may also reflect Eh measurement

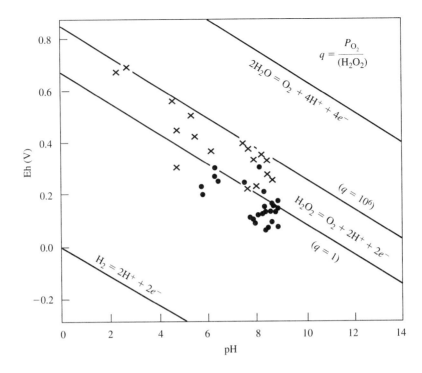

Figure 11.5 Measured Eh and pH values of some mine waters. $\times$ represents oxidized ore zone. $\bullet$ represents primary (unoxidized) ore zone. Boundaries for the hydrogen peroxide/oxygen reaction are plotted as a function of the P_{O_2}/H_2O_2 ratio. From M. Sato and H. M. Mooney. The electrochemical mechanism of sulfide self-potentials. *Geophysics* 25(1): 226–49. Copyright 1960 by Society of Exploration Geophysicists. Used by permission.

error. Reproducible, drift-free Eh measurements under field conditions are notoriously difficult (Langmuir 1971a).

Alternatively, the Eh_m values may be *mixed potentials* (cf. Stumm and Morgan 1981). The measured Eh is the potential recorded when there is zero current flow at the electrodes. As long as all redox couples contributing to the current flow are rapidly electroactive and sufficient in concentration, Eh_c for each couple will equal Eh_m for the system. If, however, one or more of the couples contributes a cathodic, but not an anodic current, or vice versa, or the current flows are vastly different, Eh_m, which is recorded at net zero current flow, will be a mixed potential and will not correspond to Eh_c for a single redox couple (Fig. 11.9). Mixed potentials can result when one species in a redox couple is unreactive or a species is present in too low a concentration to create a significant current flow at the measuring electrode. These are all possible explanations for the poor agreement between Eh_m and Eh_c reported by Lindberg and Runnells (1984). A detailed understanding of mixed potentials demands study of the electrode kinetics of individual couples. In such a study, Kempton et al. (1990) found that Fe(III)/Fe(II) was electroactive at the electrode surface, but that rates of the As(V)/As(III) and Se(VI)/Se(IV) couples were too slow to provide a Nernst Pt-electrode response. (See also Eary and Schramke 1990.)

Mixed potential problems have led some researchers to abandon or at least reexamine the idea of an overall system Eh for natural waters. Instead, measurements may be made of species concentrations in individual redox couples. Corresponding Eh_c values are then compared among different couples and compared to Eh_m (cf. Masscheleyn et al. 1991; Maest et al. 1992; Moncure et al. 1992).

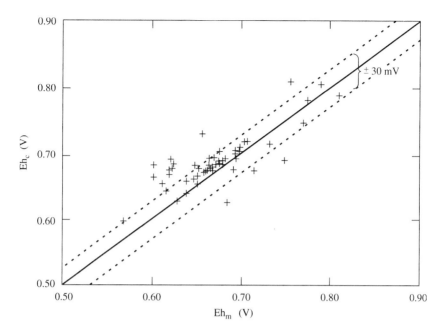

Figure 11.6   Measured Eh (Eh_m) versus computed Eh (Eh_c) values in some acid mine waters. Reprinted with permission from Nordstrom et al. A comparison of computerized chemical models for equilibrium calculations in aqueous systems. In *Chemical modeling in aqueous systems,* ed. E. A. Jenne. Am. Chem. Soc. Symp. Ser. 93, 857–892. Am. Chem. Soc. Copyright 1979 American Chemical Society.

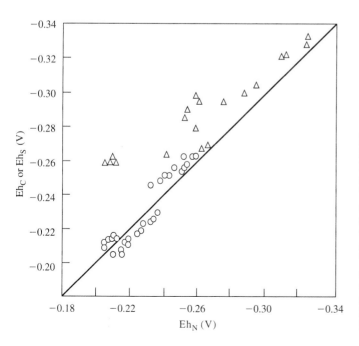

Figure 11.7   Eh values computed from measured concentrations of $N_2(aq)$ and NH_4^+ (Eh_N); SO_4^{2-} and HS^- (Eh_S); and HCO_3^- and $CH_4(aq)$ (Eh_C). The straight line is the locus of points corresponding to complete internal equilibrium: ○ Eh_S versus Eh_N; △ Eh_C versus Eh_N. Reprinted from *Geochim. et Cosmochim. Acta* 34(7), D. C. Thorstenson, Equilibrium distribution of small organic molecules in natural waters, 745–770, © 1970, with permission from Elsevier Science Ltd., The Boulevard, Langford Lane, Kidlington OX5 1GB, U.K.

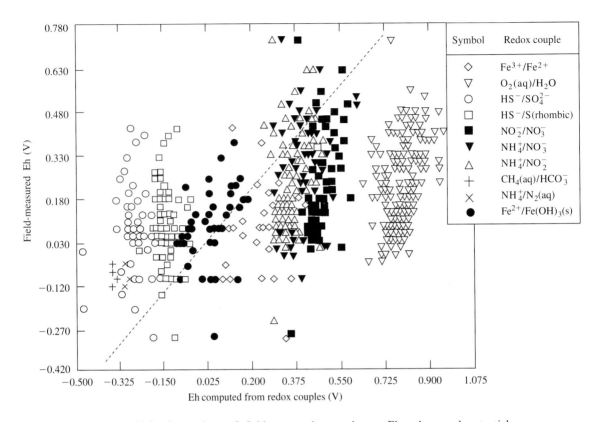

Figure 11.8 Comparison of field-measured groundwater Eh values and potentials computed from the concentrations of individual redox couples. From R. E. Lindberg and D. D. Runnells. Copyright 1984 by *Science*. Used by permission.

Recent years have seen the development and refinement of chemical analytical techniques for the determination of trace amounts of dissolved oxygen (DO). For example, Kent et al. (1994) reported DO measurements with a precision of $\pm 20\%$ below 30 μM and a detection limit of 0.03 μM. They found 1 μM DO in suboxic zone groundwaters down gradient from a sewage disposal site.

Lovley and Goodwin (1988), Chapelle and Lovley (1992), and Lovley et al. (1994), among others, have suggested that field measurements of dissolved H_2 in surface-water bottom sediments and in groundwater can be used to define a system redox state. This reflects the fact that bacteria employ H_2 in the reduction of nitrate, Mn(IV), Fe(III), sulfate, and carbon dioxide. Lovley and Goodwin (1988) have suggested the following H_2 concentrations are indicative: nitrate reduction, <0.05 nM; Fe(III) reduction, 0.2 nM; sulfate reduction, 1 to 1.5 nM; and methanogenesis, 7 to 10 nM.

When the overall oxidation state of a system is desired, unless a water is obviously anaerobic (e.g., it has an H_2S odor) one should first attempt to measure dissolved oxygen as an index of system redox state. Eh measurements are unlikely to be stable and thermodynamically meaningful in surface-waters, except in acid waters (where ferrous and ferric species are usually present). Eh measurements may be stable and meaningful in anaerobic sediments or groundwaters, when species of iron, sulfur, and manganese dominate the redox chemistry, but otherwise are of qualitative value only.

Stable Eh measurements are only possible in *well-poised* systems. Such systems, which are also described as having high *redox capacity*, tend to resist changes in Eh, just as systems with high

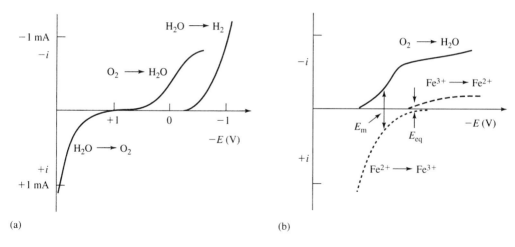

(a)   (b)

Figure 11.9   The measured electrode potential (E_m), is read by the meter at a net current flow of zero ($i = 0$). These schematic electrode polarization curves for oxygen-containing solutions show some reasons why Eh measurements may be ill-defined and/or not thermodynamically meaningful. The curves are generated by plotting the current flow measured as a variable voltage is applied at an inert electrode. Solution **(a)** is pure water and shows anodic (oxidizing) and cathodic (reducing) current flows for the couple $4H^+ + O_2 + 4e^- = 2H_2O$, but only a cathodic current flow for reduction of water to hydrogen gas. This results in a mixed potential. Further, because the net current in (a) is close to zero for a wide range of potentials, E_m is difficult to measure with accuracy. Solution **(b)** contains Fe^{2+} and perhaps one-tenth as much Fe^{3+}. The E_m, which also represents a mixed potential, is more positive than E_{eq}, the equilibrium potential contributed by the $Fe^{3+} + e^- = Fe^{2+}$ couple. From W. Stumm and J. J. Morgan, *Aquatic chemistry,* 2d ed. Copyright © 1981 by John Wiley & Sons, Inc. Used by permission of John Wiley & Sons, Inc.

pH buffer capacity resist changes in pH. The redox capacity ρ, at any Eh, is defined as the quantity of strong reductant in equivalents, which must be added to a liter of sample solution to lower the Eh by 1 volt (Nightingale 1958). That is,

$$\rho = \frac{dC_r}{d(\text{Eh})} \tag{11.31}$$

(Strictly speaking, σ is the oxidation capacity. The reduction capacity similarly defined is equal but opposite in sign to ρ). The units of ρ are, therefore, eq/L volt. In well-poised systems when oxidized and reduced species are electroactive at the electrode surface and both exceed roughly 10^{-5} to 10^{-3} M, Eh measurements are relatively easy and reproducible. Concentrations as low as 10^{-7} M can provide stable Eh readings in pure Fe(II)/Fe(OH)$_3$(s) and Fe(II)/Fe(III) systems (Macalady et al. 1990). On the other hand, in dilute and/or poorly poised waters, Eh measurements may drift without stabilizing for hours.

Chemical analysis for specific redox couples is especially useful in studies of groundwaters contaminated by waste disposal, where redox gradients are often steep, and the fate of redox-sensitive contaminants may be rate dependent or otherwise unpredictable. In such a study of sewage-contaminated groundwaters, Kent et al. (1994) found that reduction of O_2, Cr(VI), and Se(VI) did not occur, although it was thermodynamically favored. They concluded that the disequilbrium resulted from unfavorable microbial conditions and heterogeneity of groundwater flow.

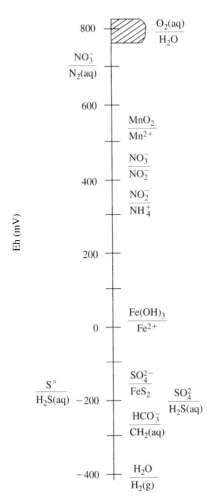

Figure 11.10 The theoretical Eh (mV) of some important oxidation-reduction couples at equal molar ion concentrations except as indicated below, at pH = 7 and 25°C. Cross-hatched area gives Eh's for $O_2(aq)/H_2O$, where $O_2(aq)$ ranges from 8.25 to 0.01 mg/L. Other conditions are: $NO_3^-/N_2(aq)$ at $N_2(aq) = 14$ mg/L (atmospheric $N_2 = 0.80$ bar), $NO_3^- = 62$ mg/L; MnO_2(pyrolusite)/ Mn^{2+} at $Mn^{2+} = 1$ mg/L; $Fe(OH)_3/Fe^{2+}$ at $Fe^{2+} = 1$ mg/L assuming K_{sp} for $Fe(OH)_3 = 10^{-38.5}$; SO_4^{2-}/FeS_2(pyrite) at $Fe^{2+} = 1$ mg/L and $SO_4^{2-} = 96$ mg/L; and S°(native sulfur)/$H_2S(aq)$ at $H_2S(aq) = 108$ mg/L ($10^{-1.5}$ mol/L). After D. Langmuir, Physical and chemical characteristics of carbonate water. In *Guide to the hydrology of carbonate rocks,* ed. P. E. Lamoreaux, B. M. Wilson, and B. A. Memeon. Copyright 1984 by UNESCO. Used by permission.

11.2 THE REDOX BEHAVIOR OF NATURAL SYSTEMS

11.2.1 Redox Reaction Sequences and Redox Ladders

At a given pH, the oxidized species of couples having more positive Eh values can theoretically oxidize the reduced species of couples having more negative Eh values, and vice versa. Figure 11.10 shows the Eh value at pH 7 and 25°C for many of the important redox couples in natural waters, for conditions specified in the figure. Excluding the water redox couples, the reduction reactions and Eh-pH equations that correspond to these couples at 25°C are given below in the order of decreasing Eh. The equations have been computed using free-energy data from Wagman et al. (1982), except for the free energy of pyrite from Robie et al. (1978) and $\Delta G_f^\circ = -31.0$ kcal/mol for $CH_2O(aq)$ as a proxy for organic matter from Latimer (1952).

$$NO_3^- + 6H^+ + 5e^- = \tfrac{1}{2}N_2(g) + 3H_2O \tag{11.32}$$

$$Eh = 1.24 + 0.0118 \log \frac{[NO_3^-]\,[H^+]^6}{(P_{N_2})^{1/2}} \tag{11.33}$$

Assuming $P_{N_2} = 0.8$ bar, and $NO_3^- = 10^{-3}$ M (62 mg/L), (11.33) reduces to

$$\text{Eh} = 1.21 - 0.071 \text{ pH} \tag{11.34}$$

$$MnO_2(\text{pyrolusite}) + 4H^+ + 2e^- = Mn^{2+} + 2H_2O \tag{11.35}$$

$$\text{Eh} = 1.23 + 0.0296 \log \frac{[H^+]^4}{[Mn^{2+}]} \tag{11.36}$$

With $Mn^{2+} = 1$ mg/L ($10^{-4.74}$ M), (11.36) reduces to

$$\text{Eh} = 1.37 - 0.118 \text{ pH} \tag{11.37}$$

$$NO_3^- + 2H^+ + 2e^- = NO_2^- + H_2O \tag{11.38}$$

$$\text{Eh} = 0.845 + 0.0296 \log \frac{[NO_3^-] [H^+]^2}{[NO_2^-]} \tag{11.39}$$

For equal molar nitrate and nitrite concentrations (the convention for solute-solute boundaries), this equation becomes

$$\text{Eh} = 0.845 - 0.0592 \text{ pH} \tag{11.40}$$

$$NO_2^- + 8H^+ + 6e^- = NH_4^+ + 2H_2O \tag{11.41}$$

$$\text{Eh} = 0.892 + 0.00986 \log \frac{[NO_3^-] [H^+]^8}{[NH_4^+]} \tag{11.42}$$

which, for equal nitrite and ammonium ion concentrations, gives

$$\text{Eh} = 0.892 - 0.0789 \text{ pH} \tag{11.43}$$

$$Fe(OH)_3(s) + 3H^+ + e^- = Fe^{2+} + 3H_2O \tag{11.44}$$

To express the Eh/pH equation for this couple, we need its $E°$ value, which in turn requires $\Delta G_f^°$ values for all reactants and products. All are known except $\Delta G_f^°$ for $Fe(OH)_3(s)$ which can vary widely. We will assume the solid is a relatively fresh, amorphous precipitate with $K_{sp} = [Fe^{3+}][OH^-]^3 = 10^{-38.5}$. From this information

$$\Delta G_r^° = -1.3642 \log(10^{-38.5}) = 52.52 \text{ kcal/mol} \tag{11.45}$$

for the reaction

$$Fe(OH)_3 = Fe^{3+} + 3(OH^-) \tag{11.46}$$

With $\Delta G_f^° = -1.1$ and -37.60 kcal/mol for Fe^{3+} and OH^-, respectively (from Wagman et al. 1982),

$$\Delta G_r^°(\text{kcal/mol}) = 52.52 = -1.1 - 112.8 - \Delta G_f^°[Fe(OH)_3]$$

and $\Delta G_f^° = -166.42$ kcal/mol. Combined with free-energy data for the other species in reaction (11.44), this gives $E° = 0.975$ V, and

$$\text{Eh} = 0.975 + 0.0592 \log \frac{[H^+]^3}{[Fe^{2+}]} \tag{11.47}$$

Assuming $Fe^{2+} = 1$ mg/L ($10^{-4.75}$ M), the final equation is

$$\text{Eh} = 1.26 - 0.178 \text{ pH} \tag{11.48}$$

$$2SO_4^{2-} + Fe^{2+} + 16H^+ + 14e^- = FeS_2(\text{pyrite}) + 8H_2O \tag{11.49}$$

$$\text{Eh} = 0.362 + 0.00423 \log [Fe^{2+}][SO_4^{2-}]^2[H^+]^{16} \tag{11.50}$$

For $SO_4^{2-} = 96$ mg/L (10^{-3} M) and $Fe^{2+} = 1$ mg/L ($10^{-4.75}$ M) this becomes

$$Eh = 0.317 - 0.0676 \text{ pH} \tag{11.51}$$

$$SO_4^{2-} + 10H^+ + 8e^- = H_2S(aq) + 4H_2O \tag{11.52}$$

$$Eh = 0.301 + 0.00740 \log \frac{[SO_4^{2-}][H^+]^{10}}{[H_2S]} \tag{11.53}$$

which, for equal concentrations of the sulfur species, leads to

$$Eh = 0.301 - 0.0740 \text{ pH} \tag{11.54}$$

$$S°(\text{native sulfur}) + 2H^+ + 2e^- = H_2S(aq) \tag{11.55}$$

$$Eh = 0.144 + 0.0296 \log \frac{[H^+]^2}{[H_2S]} \tag{11.56}$$

which, for $H_2S(aq) = 10^{-3}$ M (34 mg/L) becomes,

$$Eh = 0.233 - 0.0592 \text{ pH} \tag{11.57}$$

$$HCO_3^- + 9H^+ + 8e^- = CH_4(aq) + 3H_2O \tag{11.58}$$

$$Eh = 0.206 + 0.00740 \log \frac{[HCO_3^-][H^+]^9}{[CH_4]} \tag{11.59}$$

which, for $HCO_3^- = CH_4$, reduces to

$$Eh = 0.206 - 0.0666 \text{ pH} \tag{11.60}$$

$$HCO_3^- + 5H^+ + 4e^- = CH_2O(aq) + 2H_2O \tag{11.61}$$

$$Eh = 0.036 + 0.0148 \log \frac{[HCO_3^-][H^+]^5}{[CH_2O]} \tag{11.62}$$

and with $HCO_3^- = CH_2O(aq)$ the equation becomes

$$Eh = 0.036 - 0.0740 \text{ pH} \tag{11.63}$$

These redox reaction boundaries and their computed equilibrium Eh values for pH = 7.0 are summarized in Table 11.3. The Eh/pH equations generally have different slopes so that when plotted on an Eh-pH diagram some of the reaction boundaries cross above or below pH = 7. To illustrate the importance of this effect, Fig. 11.11 shows that for constant Eh and increasing pH, the solubilities of solids of Se, Tc, and U (radioelements important in nuclear waste disposal) may reverse in order, and also become more or less soluble than the corrosion products of possible waste-container metals Cu, Fe, and Ni.

Example 11.1

Nitrate is a common pollutant in aerated groundwaters, but is never found in groundwaters that contain significant amounts of dissolved ferrous iron. Show that in an aquifer at pH = 7.0, $10^{-4.0}$ M Fe^{2+} in equilibrium with $Fe(OH)_3(s)$ ($K_{sp} = 10^{-38.5}$) can quantitatively reduce nitrate to dissolved N_2. Assume 25°C and $\Sigma N = 10^{-3}$ M in the groundwater.

The Eh of the groundwater can be computed from the pH and Fe^{2+} concentration using Eq. (11.47), which leads to Eh = −0.031 V. The nitrate reduction reaction is written

$$NO_3^- + 6H^+ + 5e^- = \tfrac{1}{2}N_2(aq) + 3H_2O \tag{11.64}$$

TABLE 11.3   The standard potential, $E°$, and Eh at pH = 7.0 and 25°C of some redox couples, assuming thermodynamic equilibrium for conditions listed in the table

Reaction	$E°$ (volts)	Eh (volts) pH = 7.0	Assumptions
$4H^+ + O_2(g) + 4e^- = 2H_2O$	1.23	0.816	$P_{O_2} = 0.2$ bar
$NO_3^- + 6H^+ + 5e^- = \frac{1}{2}N_2(g) + 3H_2O$	1.24	0.713	$[NO_3^-] = 10^{-3}$ M $P_{N_2} = 0.8$ bar
$MnO_2(\text{pyrolusite}) + 4H^+ + 2e^- = Mn^{2+} + 2H_2O$	1.23	0.544	$[Mn^{2+}] = 10^{-4.74}$ M
$NO_3^- + 2H^+ + 2e^- = NO_2^- + H_2O$	0.845	0.431	$[NO_3^-] = [NO_2^-]$
$NO_2^- + 8H^+ + 6e^- = NH_4^+ + 2H_2O$	0.892	0.340	$[NO_3^-] = [NH_4^+]$
$Fe(OH)_3 + 3H^+ + e^- = Fe^{2+} + 3H_2O$	0.975	0.014	$[Fe^{2+}] = 10^{-4.75}$ M
$Fe^{2+} + 2SO_4^{2-} + 16H^+ + 14e^- = FeS_2(\text{pyrite}) + 8H_2O$	0.362	−0.156	$[Fe^{2+}] = 10^{-4.75}$ M $[SO_4^{2-}] = 10^{-3}$ M
$S°(\text{rhombic}) + 2H^+ + 2e^- = H_2S(aq)$	0.144	−0.181	$[H_2S] = 10^{-3}$ M
$SO_4^{2-} + 10H^+ + 8e^- = H_2S(aq) + 4H_2O$	0.301	−0.217	$[SO_4^{2-}] = [H_2S]$
$HCO_3^- + 9H^+ + 8e^- = CH_4(aq) + 3H_2O$	0.206	−0.260	$[HCO_3^-] = [CH_4]$
$H^+ + e^- = \frac{1}{2}H_2(g)$	0.0	−0.414	$P_{H_2} = 1.0$ bar
$HCO_3^- + 5H^+ + 4e^- = CH_2O(\text{organic matter}) + 2H_2O$	0.036	−0.482	$[HCO_3^-] = [CH_2O]$

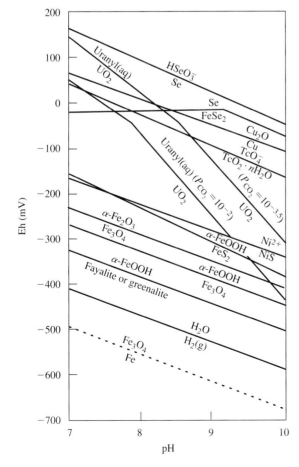

Figure 11.11   Some Eh-pH boundaries important to nuclear waste disposal at 25°C and 1 bar pressure and their conditions: $HSeO_3^-/Se$, $mHSeO_3^- = 10^{-7}$; $Se/FeSe_2$, $mFe^{2+} = 10^{-4}$ for pH < 8.91, α-FeOOH present for pH > 8.91; Uranyl(aq)/UO_2, $\Sigma mU(aq) = 10^{-8}$; $TcO_4^-/TcO_2 \cdot nH_2O$, $mTcO_4^- = 10^{-8}$; Ni^{2+}/NiS, $mNi^{2+} = 10^{-8}$, $\Sigma mS = 10^{-3}$; α-FeOOH/FeS_2, $\Sigma mS = 10^{-3}$; α-FeOOH/Fe_2SiO_4 or α-FeOOH/greenalite, $mH_4SiO_4° = 10^{-4}$. Minerals shown: native selenium (Se), cuprite (Cu_2O), copper metal (Cu), ferroselite ($FeSe_2$), uraninite (UO_2), goethite (α-FeOOH), hematite (α-Fe_2O_3), magnetite (Fe_3O_4), fayalite (Fe_2SiO_4), greenalite ($[Fe_3Si_2O_5(OH)_4]$, and iron metal (Fe). After Langmuir and Apted (1992). Used by permission.

With $\Delta G_f^\circ = 4.36$ kcal/mol for $N_2(aq)$, we find $E^\circ = 1.231$ V, and:

$$Eh = 0.734 + 0.0118 \log \frac{[NO_3^-]}{[N_2]^{1/2}} \tag{11.65}$$

from which (because $Eh = -0.031$ V)

$$\frac{[NO_3^-]}{[N_2]^{1/2}} = 10^{-64.6} \tag{11.66}$$

so that $[NO_3^-] \approx 0$ and $[N_2(aq)] = 5 \times 10^{-4}$ M.

11.2.2 General Controls on the Redox State of Natural Waters

In natural waters, atmospheric oxygen is the ultimate (major) oxidant and constitutes the ultimate oxidation reservoir. Organic matter is the ultimate (major) reductant. Most organic matter is unstable (or metastable) in water. Its stability plots below the $H_2O/H_2(g)$ boundary. Among the few exceptions are formic and acetic acids, some simple alcohols such as methanol, and methane, CH_4. When decaying organic matter is introduced into a system that has been isolated from atmospheric oxygen, aerobic decay takes place first, depleting all free O_2, after which, as the Eh drops, anaerobic conditions are established, with NO_3^-, NO_2^-, and then SO_4^{2-} reduced. If redox conditions below the H_2O/H_2 boundary can be attained, bacterial reduction of carbonate species to CH_4 (methanogenesis or methane fermentation) may occur (cf. Manahan 1994).

What determines the oxygen content of an environment in the hydrosphere? The answer is the rate of oxygen replenishment from the atmosphere, versus the rate of oxygen depletion or consumption by reductant organic matter, or reductant H_2S, NH_4^+, Fe^{2+}, Mn^{2+}, FeS, or FeS_2, etc. The movement of oxygen through soils is at a rate of about 1 ft/day, so that unsaturated soil moisture is usually oxidized, except in soil microenvironments rich in organic matter. The fact that atmospheric oxygen is relatively insoluble in water makes it easier to deplete the dissolved oxygen content of water. Air contains 286 mg/L of O_2 at 25°C (21% by volume). At saturation with atmospheric oxygen and 1 bar total pressure, water holds only 8.25 mg/L of O_2 (2.58×10^{-4} mol/L O_2) at 25°C. Assuming aerobic decay of organic matter, it takes only 3.1 mg/L of disssolved organic carbon or DOC (of C° oxidation state) to consume the 8.25 mg/L of DO (Table 11.4). This is not an unusually high

TABLE 11.4   Amounts of carbon species as C, NH_3 as N, H_2S as S, Mn, and Fe required to eliminate the oxygen content of a water at 25°C, assuming oxygen saturation of 8.25 mg/L ($O_2(aq) = 2.58 \times 10^{-4}$ M)

Reaction	Oxidation state	Reductant amount (mg/L)
$CH_4 + 2O_2 \rightarrow CO_2 + 2H_2O$	−4 (C)	1.6
$NH_3 + 2O_2 \rightarrow NO_3^- + H^+ + H_2O$	−3 (N)	1.8
$CH_2O + O_2 \rightarrow CO_2 + H_2O$	0 (C)	3.1
$H_2S + 2O_2 \rightarrow SO_4^{2-} + 2H^+$	−2 (S)	4.1
$HCOOH(aq) + \frac{1}{2}O_2 \rightarrow CO_2 + H_2O$	+2 (C)	6.2
$Mn^{2+} + \frac{1}{2}O_2 + H_2O \rightarrow MnO_2 + 2H^+$	+2 (Mn)	28.3
$Fe^{2+} + \frac{1}{4}O_2 + \frac{5}{2}H_2O \rightarrow Fe(OH)_3 + 2H^+$	+2 (Fe)	57.6

DOC content. In fact Leenheer et al. (1974) report a median DOC of 0.7 mg/L for U.S. groundwaters. As a rule of thumb, this author has found that a temperate climate groundwater with more than about 4 mg/L of dissolved organic carbon (DOC), total organic carbon (TOC), biochemical oxygen demand (BOD), or chemical oxygen demand (COD), will usually become anaerobic. This applies to the waters in many stream-bottom muds, organic-rich waste ponds, lake and reservoir bottoms, water-logged soils, and deeper groundwater systems.

The diffusion coefficient of O_2 is about 2.05×10^{-1} cm^2/s in air, and in water 10^{-5} cm^2/s, or 2×10^4 less (Lerman 1979). Insofar as the replenishment of locally depleted dissolved oxygen concentrations must occur by aqueous diffusion, the replenishment rate will thus be slow. The rate of replenishment of O_2 in soil gas is enhanced by the barometric pumping of the air caused by fluctuations in barometric pressure. The flow of streams is often turbulent, enhancing aeration of the stream and renewal of its oxygen content. In groundwaters or stratified, quiescent surface waters the flow is usually laminar and O_2 replenishment is by diffusion. For such conditions, small amounts of reductants such as DOC can deplete the water's oxygen content at a rate greatly exceeding its rate of replenishment. Only mixing with oxygenated waters, or the total lack of reductant minerals or organic matter, can maintain oxidizing conditions in groundwater.

Other important reductants dissolved in water-saturated soils and sediments include ammonia, hydrogen sulfide, Mn^{2+}, and Fe^{2+}. Table 11.4 shows that organic carbon (depending on its oxidation state) is generally a stronger reductant on a mole basis than any of these. Organic carbon is also usually more abundant than other potential reductants, particularly in modern aquatic sediments.

11.2.3 Berner's Redox Classification and Oxidative and Reductive Capacity

The relative constancy of pH in subaqueous sediments (usually 6 to 8, most often 6.5 to 7.5) and the difficulty of measuring a thermodynamically meaningful Eh, led Berner (1981b) to propose a simplified redox classification scheme. The scheme is broadly based on the presence or absence of dissolved oxygen or dissolved sulfide (Table 11.5). Measureable dissolved oxygen indicates an *oxic* environment, its absence an *anoxic* environment. More recently others have suggested that Berner's

TABLE 11.5 The redox classification of sedimentary environments showing bounding concentrations of dissolved oxygen (O_2) and total sulfide (as H_2S) in μM and index Fe and Mn minerals[†]

Environment	Characteristic phases
Oxic ($O_2 > 30\ \mu$M)	hematite, goethite, ferrihydrite, MnO_2-type phases, no organic matter
Suboxic ($O_2 \geq 1\ \mu$M and $< 30\ \mu$M)	hematite, goethite, ferrihydrite, MnO_2-type phases, minor organic matter
Anoxic ($O_2 < 1\ \mu$M)	
Sulfidic ($H_2S \geq 1\ \mu$M)	pyrite and marcasite, rhodocrosite, organic matter
Nonsulfidic ($H_2S < 1\ \mu$M)	
Postoxic	low-temperature Fe(II)-Fe(III) silicates, siderite, vivianite, rhodocrosite, no sulfide minerals, minor organic matter
Methanic	siderite, vivianite, rhodocrosite, earlier formed sulfide minerals, organic matter

[†]Hematite (α-Fe$_2$O$_3$), goethite (α-FeOOH), ferrihydrite (Fe(OH)$_3 \cdot n$H$_2$O), pyrite (FeS$_2$), marcasite (FeS$_2$), rhodocrosite (MnCO$_3$), siderite (FeCO$_3$), vivianite (Fe$_3$(PO$_4$)$_2 \cdot$ 8H$_2$O).

Source: Proposed by Berner (1981b) with minor modifications. The suboxic environment is described by Anderson et al. (1994). See also Kent et al. (1994).

oxic environments be subdivided into *oxic and suboxic* (cf. Anderson et al. 1994). With this revision, oxic environments contain more than 30 μM of DO with Mn^{2+} below detection. Organic matter is absent. Suboxic environments have a small amount of reactive organic matter and DO $\geq$ 1 μM but < 30 μM, with detectable Mn^{2+}, but with Fe^{2+} below detection.

Anoxic systems are divided into those with or without measureable sulfide, which Berner termed *sulfidic* and *nonsulfidic*. Nonsulfidic environments themselves are described as *postoxic* if too oxidized to permit sulfate reduction and *methanic* if strongly reduced with sulfate reduction and methane formation. Berner suggests that the presence or absence of specific iron and manganese minerals in Table 11.5 can be used to distinguish these different redox environments.

Scott and Morgan (1990) introduced the concept of *oxidative capacity* (OXC) and *reductive capacity* (RDC), which they defined as:

$$OXC = \Sigma n_i[Ox]_i - \Sigma n_i[Red]_i = -RDC \tag{11.67}$$

In this expression $[Ox]_i$ and $[Red]_i$ are the molar concentrations of individual oxidants and reductants, including both dissolved species and solid phases, and n_i is the number of equivalents of electrons that are transferred in the redox couple reaction (i.e., the stoichiometry of the electron). OXC is determined by titrating a system with a strong oxidizing agent. The corresponding *reductive capacity* (RDC) is measured by titrating with a strong reducing agent. OXC and RDC may also be computed from a water and sediment analysis. Scott and Morgan (1990) further define $OX_T = \Sigma n_i[Ox]_i$, and $RD_T = \Sigma n_i[Red]_i$, so that $OXC = OX_T - RD_T$. They suggest choosing an electron reference level (ERL) or pE on the redox ladder for any system of interest. Oxidized species in couples at Eh values above the ERL are capable of oxidizing the system, whereas reduced species at lower Eh values can be system reductants. When free oxygen is considered the only system oxidant, then H_2O is chosen at the ERL. If we wish to consider organic matter as the only reductant of interest, it is convenient to select HS^- as the ERL. Scott and Morgan relate the OXC concept to Berner's (1981b) redox classification scheme. Heron et al. (1994) propose a Ti^{3+}-EDTA extraction solution for measuring the OXC of sediments containing Fe- and Mn-oxyhydroxides. Heron and Christensen (1995) have applied the OXC concept to the Fe(II)/Fe(III) chemistry of an aquifer polluted by landfill leachate.

Shown in Fig. 11.12 is the computed reduction titration curve of a model groundwater, assuming the titrant is dissolved organic carbon (DOC). Note that the pH remains in a narrow range between 6.5 and 7.5 as the water drops in pE and Eh.

Example 11.2

How much organic carbon ($C°$ oxidation state) as DOC is needed to reduce the oxygenated water in Fig. 11.12 to just below the SO_4^{2-}/HS^- boundary? (The ERL is HS^-.) Electron stoichiometries in Table 11.3 indicate the following equation for OXC:

$$OXC = 4(O_2) + 5(NO_3^-) + 2(MnO_2) + Fe(OH)_3 + 8(SO_4^{2-}) - 4(CH_2O) = 0 \tag{11.68}$$

With mM concentrations listed in the figure and given that 8.0 mg/L DO is 0.25 mM of O_2, we compute $OXC(meq/L) = 2.6 - 4(CH_2O) = 0$, or $CH_2O = 0.65$ mM. This is equivalent to 7.8 mg/L DOC required to complete the reduction.

The units of OXC and RDC, as defined, are eq/L or meq/L. This makes OXC comparable to acidity and alkalinity, which are also given in eq/L or meq/L (Chap. 5). The pH *buffer capacity* measures the resistance of a system to pH change at a given pH, upon addition of a strong acid or base (Chap. 5). The concept of *redox capacity*, as proposed by Nightingale (1958) and defined above, is

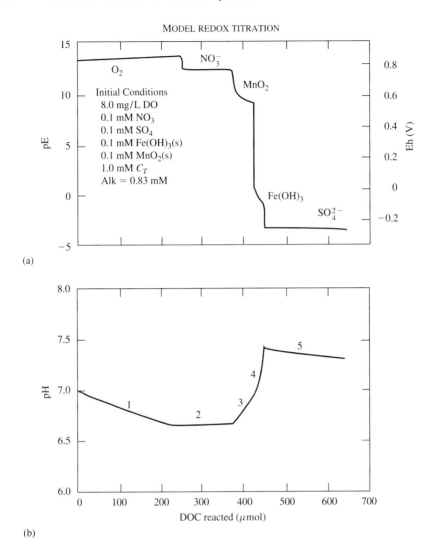

Figure 11.12   Redox titration curve of a model groundwater system of initial composition shown in (a), which also describes the computed response in pE and Eh as specific species are reduced during the titration. The computed pH change during the titration is shown in (b). Numbered segments correspond to sequential reduction: (1) $O_2(aq)$, (2) NO_3^-, (3) $MnO_2(s)$, (4) $Fe(OH)_3(s)$, and (5) SO_4^{2-}. From Scott and Morgan (1990). Reprinted with permission from M. J. Scott and J. J. Morgan. Energetics and conservative properties of redox systems. In *Chemical modeling of aqueous systems II*, ed. D. C. Melchior and R. L. Bassett. Am. Chem. Soc. Symp. Ser. 416, pp. 368–78. Copyright 1990 by the American Chemical Society.

completely analogous to the definition of buffer capacity. Unfortunately, Scott and Morgan's (1990) oxidative capacity concept is not. To avoid confusion, their *oxidative capacity* might conceivably be better termed *oxidative strength*.

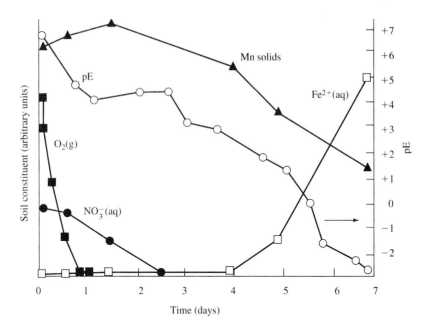

Figure 11.13 Relative changes in O_2, NO_3^-, Mn(IV) solids, Fe^{2+}(aq), and pE of a soil with time after flooding. From G. Sposito, *The chemistry of soils.* Copyright 1989 by Oxford University Press. Used by permission.

When a soil or sediment is flooded, limiting the availability of oxygen, its organic content will often drive reduction of the system with time "down" the redox ladder. Figure 11.13 shows that during the week following a soil's flooding, soil reduction progressed to below the $Fe(OH)_3(s)/Fe^{2+}$ boundary. Over longer periods rice paddies and other flooded soils often become even more anaerobic, undergoing sulfate reduction with consequent precipitation of metal sulfides and methanogenesis (cf. Van Breemen 1975; Bohn et al. 1985).

11.2.4 The Redox Interface

Because reductants are present in most water-saturated soils or sediments and oxygen is relatively unavailable, the Eh declines as the water moves into the subsurface. The decline may be from oxic to anoxic sulfidic or nonsulfidic levels (Table 11.5). The rate and extent of Eh decline with distance from the surface depends on the availability and reactivity of sediment organic matter and other reductants. In the sediments of flooded rice paddies, wetlands, estuaries, and shallow lakes, which may be especially rich in fresh organic matter, the *redox front or interface* (also termed a redox barrier or boundary by some), which is the zone of abruptly changing Eh values, may be only a few millimeters or centimeters thick.

Groundwater aquifers in older carbonate and sandstone rocks may have been leached of any reactive organic matter and of Fe(II) and sulfide minerals through time. In such systems, the paucity of reductants may create a redox interface that spans hundreds of meters to several kilometers. For example, Langmuir and Whittemore (1971) found that Eh_m gradually drops from 250 mV to -100 mV over a flow distance of about 6 km in the Potomac-Magothy-Raritan aquifer of coastal plain New

Jersey. Because of a paucity of reactive organic matter, redox conditions do not become sulfidic. Over the 6 km, nitrate drops from 3 to 5×10^{-3} to $<10^{-5}$ M, whereas ferrous iron increases from about 10^{-6} to 10^{-4} M (Langmuir 1969b).

The Chalk (limestone) aquifer lies within a synclinal basin near London, U.K. Groundwater in the Chalk is oxic where it is shallow and unconfined ($Eh_m = 330$ to 420 mV), but becomes anoxic where the aquifer deepens and is confined ($Eh_m \leq 160$ mV) (Edmunds et al. 1987). The process is later reversed and a second redox interface is present where groundwater flow returns to unconfined near-surface conditions. Edmunds et al. (1987) report the composition of redox-sensitive species and parameters in the groundwater along a north-south section through the aquifer. Their results, summarized in Fig. 11.14, show that once the aquifer becomes confined, DO values are depleted from about 9 mg/L to <1 mg/L about 8 km further downdip at the first redox interface. The most reduced species in the aquifer is Fe(II). The terminal electron-accepting process or TEAP (Lovley et al. 1994) is thus the $Fe^{2+}/Fe(OH)_3(s)$ couple. The plots show nitrate values dropping from 20 to 30 mg/L (3.2 to 4.8×10^{-4} M) to below detection across the first interface. Redox-sensitive species, unstable under oxic conditions, that become more abundant in the confined aquifer, include not only Fe^{2+}, but also Mn^{2+}, I^-, and NH_4^+.

Unlike the Chalk aquifer, the Lincolnshire Limestone aquifer in the U.K. contains lenses of limestone with 0.5% (by weight) organic matter, according to Edmunds (cf. Champ and Gulens 1979). The TEAP can thus be taken as the HCO_3^-/CH_2O couple, for example. Figure 11.15 shows trends of Eh_m, pH, O_2, and HS^- in the aquifer as it flows downdip from the outcrop area. DO values decline from about 8 mg/L, 4 km from the outcrop to below detection at the redox interface, 11 km from the outcrop. Eh_m values are near 400 mV in the recharge zone and drop to about -100 mV downdip. Sulfate reduction is favored beyond about 13 km downdip. Resultant sulfide can be expected to precipitate free Fe^{2+} as ferrous sulfide solids. In all of the examples of redox interfaces in aquifers, natural reductants and the unavailability of atmospheric oxygen under confined conditions causes the decline in Eh as groundwater moves across a redox interface.

Surface or subsurface disposal of organic-rich solid or liquid wastes in groundwater recharge zones can create a "reverse" redox interface (cf. Apgar and Langmuir 1971; Baedecker and Back 1979; Kent et al. 1994; Lovley et al. 1994). The wastes and their leachates are generally anoxic. As resultant leachate and waste plumes move away from a waste site, they are gradually oxidized and diluted by mixing with local groundwater and uncontaminated groundwater recharge. A redox interface is, therefore, created and the reduced species are oxidized as they move in the groundwater flow direction (Fig. 11.16). Such conditions were observed by Apgar and Langmuir (1971) in the unsaturated zone under a landfill in central Pennsylvania. Eight years after waste emplacement, the leading edge of the leachate plume was 17 m below the waste. Conditions farther away were oxic and uncontaminated. The most anoxic waters were found about 3.1 to 4.6 m under the waste, where $Eh_m \approx 80$ mV, BOD $\approx 10,000$ mg/L, Fe(II) ≈ 1000 mg/L and NH_4-N ≈ 40 mg/L. Sulfate reduction was not observed.

Kent et al. (1994) describe a plume of sewage-contaminated groundwater at the Otis Air Force Base in Massachusetts. Since waste disposal began in 1936, the plume has moved more than 4 km in the groundwater flow direction. The plume is suboxic near the disposal site and oxic 3 km down gradient. Pristine oxic waters have 250 μM (8 mg/L) DO and 80 μM SO_4^{2-} and negligible NO_3^-, NO_2^-, NH_4^+, Fe^{2+}, or Mn^{2+} concentrations. In contrast a suboxic groundwater has 430 μM SO_4^{2-}, 650 μM NO_3^-, 25 μM NO_2^-, 3 μM NH_4^+, 5 μM Mn^{2+}, and 1 μM DO. Elsewhere in the suboxic plume, concentrations are as high as Fe^{2+} 500 μM, Mn^{2+} 16 μM, NO_3^- >800 μM, and NH_4^+ 180 μM. The absence of sulfate reduction indicates that conditions were nonsulfidic in the plume.

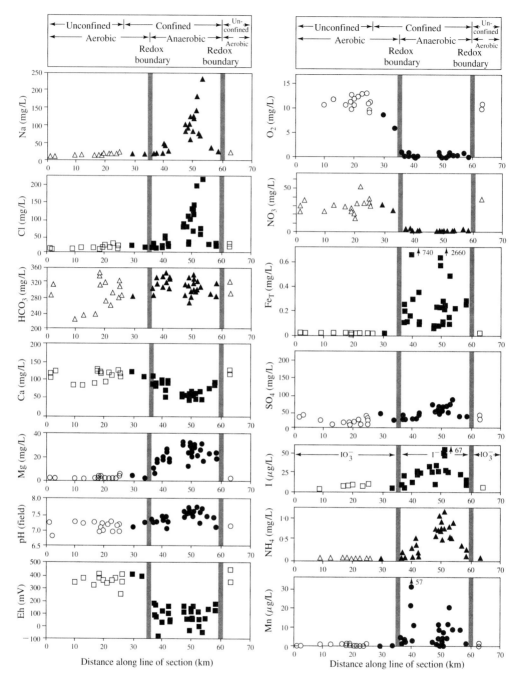

Figure 11.14 Changes in Eh and concentrations of major and redox-sensitive species in the direction of groundwater flow across two redox interfaces in the Chalk aquifer in the Berkshire syncline in the U.K. Distances measured from the outcrop area (north), where the groundwater is unconfined and aerobic, to confined, anaerobic deeper portions of the aquifer and back toward the surface where the aquifer is again unconfined and aerobic. Open symbols = unconfined groundwaters; solid symbols = confined groundwaters. From *Applied geochemistry,* 2, W. M. Edmunds, et al. Baseline geochemical conditions in the chalk aquifer, Berkshire, U.K.: A basis for groundwater quality management, pp. 251–274, with permission from Elsevier Science Ltd., The Boulevard, Langford Lane, Kidlington OX5 1GB, U.K.

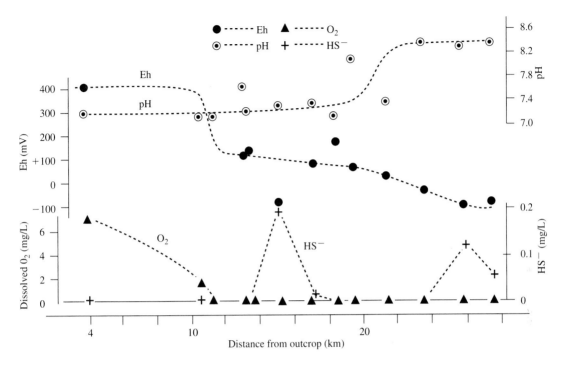

Figure 11.15 Changes in Eh, pH, dissolved oxygen and HS⁻ in the direction of groundwater flow in the Lincolnshire Limestone aquifer, U.K. Distances are measured from the outcrop area, which is oxidizing, to deeper, confined portions of the formation downdip, where the groundwater becomes anaerobic. After D. R. Champ and J. Gulens, Oxidation-reduction sequences in groundwater flow systems. *Can. J. Earth Sci.* 16:12–23. Copyright 1979 by NRC Research Press. Used by permission.

STUDY QUESTIONS

1. Which are the six or seven "major" elements that exhibit redox behavior in natural waters? In what form do they occur under oxidizing and under reduced conditions?

2. Write a balanced half reaction involving the species/solids of a redox-sensitive element and write the Eh equation that corresponds to that reaction. Define the terms in the Eh equation and know how to plot it on an Eh-pH diagram.

3. Define the Nernst factor, $E°$ and pE. For a given redox reaction how are they related to each other and to $\Delta G_r°$ and Eh?

4. Write the reactions that define the "upper" and "lower" boundaries of the Eh-pH stability field of liquid water?

5. Be able to draw an Eh-pH diagram showing schematically the stability field of water, and plot on it and label the Eh-pH conditions typical of some example surface-waters and groundwaters.

6. The Eh measured in the field with a platinum electrode and the theoretically computed Eh for the same water may be very different, particularly for surface-water systems where the dominant redox-sensitive elements are N, H, O, and C. Why?

7. In which natural water systems will the measured Eh most closely equal the Eh computed from solution concentrations and free-energy data?

8. When is a measurement of the dissolved oxygen concentration of a water a more meaningful assessment of its oxidation state than an Eh measurement?

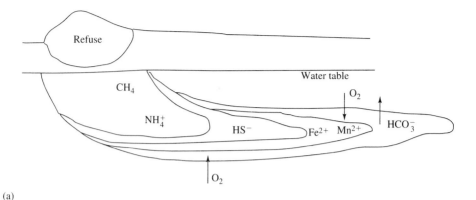

(a)

(b)

Figure 11.16   Schematic cross-sections of groundwater systems contaminated by organic-rich wastes. **(a)** Development of redox zones down gradient from a landfill in the ground-water flow direction (Baedecker and Back 1979). **(b)** Possible sequence of redox zones en-countered in the groundwater flow direction from a source of organic contamination. After D. R. Lovley, F. H. Chapelle, and J. C. Woodward, Use of dissolved H_2 concentrations to determine distribution of microbially catalyzed redox reaactions in anoxic groundwater. *Envir. Sci. & Technol.* 28(7):1205–10. Copyright 1994 by American Chemical Society.

9. Unlike pH and protons, an Eh measurement does not correspond to a measureable concentration of free electrons in the water. Discuss.

10. Define electroactive redox couples and mixed potentials and give examples.

11. What is the sequence of oxidation/reduction reactions of species of N, C, Mn, S, Fe, H, and O encountered in water at pH = 7, as it moves from an aerated environment to an organic-rich waterlogged one?

12. Groundwater is far more likely to be anaerobic than is soil moisture for reasons that relate to oxygen solubility and diffusion rates. Explain.

13. Why is it that a few ppm of DOC is often sufficient to make a subsurface-water anaerobic?

14. What is a redox interface and what determines its scale (inches or miles) in a stream- or lake-bottom mud versus a sandy aquifer?

15. Explain the statement that atmospheric oxygen is the ultimate (major) oxidant and constitutes the ultimate oxidation reservoir, whereas organic matter is the ultimate (major) reductant.

PROBLEMS

1. Groundwater pumped from an open, uncased borehole in western Australia has the following composition. Units are mg/L unless otherwise indicated.

Parameter or species	Value or concentration	Parameter or species	Value or concentration
$T(°C)$	25.5	Fe(II)(aq)	0.5
DO	1.3	Fe(III) (in suspension)	0.1
Eh(field) (V)	−0.195	Mn(II)(aq)	0.08
pH	7.15	SiO_2	20
Ca^{2+}	77	SO_4^{2-}	11
Mg^{2+}	42	NO_3^-	4
Na^+	661	Cl^-	812
K^+	6.4	HCO_3^-	491

(a) Study the analysis, including the DO and Eh values and concentrations of redox-sensitive dissolved species. Is the composition of this water internally consistent or inconsistent? Is it likely to be a mixture? Explain.

(b) If the measured DO controlled the Eh through the reaction

$$4H^+ + O_2(g) + 4e^- = 2H_2O$$

what would the Eh be?

(c) Input the chemical analysis including the pH and Eh values into MINTEQA2. Why is it incorrect to input the Fe(III) concentration given in the table? You will need to specify Fe^{2+}/Fe^{3+} and Mn^{2+}/Mn^{3+} redox couples with fixed activities in Edit Level II. Based on the mineral saturation results in the output, what reaction probably controls Fe(II) and Fe(III) concentrations and perhaps also the measured Eh?

(d) Calculate the oxidative capacity of the groundwater. If all redox-sensitive species in the water can come to thermodynamic equilibrium with each other, will the water be aerobic or anaerobic? Explain.

2. The drinking water standard for dissolved lead is 50 μg/L. Lead concentrations in drinking water often exceed this amount when the water has passed through lead pipes or been stored in lead tanks and its pH is below 7.0 to 7.5. High-lead problems caused by lead plumbing are most common in houses constructed prior to 1960 (cf. Raab et al. 1993).

(a) Assuming a potable water is flowing through lead pipe at 25°C, and that the water sits unmoving in the pipe and so can equilibrate with the Pb metal, calculate the Eh value at which the dissolved Pb just equals the drinking water standard. Assume 25°C, pH 7.0, $Na^+ = 75$ mg/L, and $HCO_3^- = 200$ mg/L. What is the speciation of the dissolved lead?

(b) It has been found that under oxidizing conditions hydrocerrusite $[Pb_3(OH)_2(CO_3)_2]$ may precipitate on the surface of lead plumbing, limiting maximum Pb concentrations in drinking water (cf. Hem and Durum 1973). Again using MINTEQA2, calculate the solubility of lead as $\Sigma Pb(aq)$ assuming Eh = +0.700 V (an aerated water), pH = 7.0, $Na^+ = 75$ mg/L, and $HCO_3^- = 200$ mg/L. Assume that hydrocerrusite solubility limits the maximum Pb concentration. There is good evidence that hydrocerrusite is less soluble than indicated in the MINTEQA2 file *thermo.dbs*. Make hydrocerrusite a finite solid and change its log K value as given in the database to +19.17.

(c) The solubility of hydrocerrusite is pH dependent. Accordingly, water departments near Edinburgh, Scotland, have added lime (CaO) to raise the pH of drinking water and so lower lead concentrations (Raab et al. 1993). Assuming the same solution conditions as given in part (b), that hydrocerrusite solubility limits Pb concentrations, and also assuming atmospheric CO_2 pressure (0.00032 bar), and $Cl^- = 105$ mg/L, how high would the pH have to be raised with lime to lower the total Pb concentration to its minimum possible value? Use the sweep option in MINTEQA2 to determine the solubility of hydrocerrusite at 0.5 pH unit increments from pH 6 to 9.5. How does the minimum Pb solubility compare to the drinking water standard? What is the Pb speciation of the water for these conditions?

3. Chromium concentrations are generally below detection in anaerobic groundwaters that contain ferrous iron. This may reflect reduction of soluble chromate (CrO_4^{2-}) to insoluble chromic ion (Cr(III)) by Fe^{2+} in a coupled redox reaction that can be written

$$CrO_4^{2-} + 3Fe^{2+} + 8H_2O = Cr(OH)_3 + 3Fe(OH)_3 + 4H^+$$

(See also Eary and Rai 1988.) The trace $Cr(OH)_3$ is coprecipitated with usually more abundant ferric oxyhdroxides. The equilibrium constant expression is

$$K_{eq} = 10^{12.8} = [H^+]^4/[CrO_4^{2-}] \, [Fe^{2+}]^3$$

Ignoring activity coefficients, assume that pH = 7.0 and $[Fe^{2+}] = 10^{-6}$ M. What is the equilibrium concentration of chromate?

4. Chao and Theobald (1976) measured the manganese oxide and minor metal contents of a wide range of soils and sediments that had Mn concentrations from about 300 to 90,000 ppm. Their data closely fit the linear equation

$$\log Mn \, (ppm) = -0.857 + 2.154 \log M \, (ppm)$$

where $M = Co + Cu + Ni + Pb + Zn$. The minor metals are apparently adsorbed on Mn oxides (see Chap. 10) or have been coprecipitated with the oxides or both. Minor metal scavenging by Mn oxides [or Fe(III) oxides] can prevent their concentrations from reaching toxic levels in some waters (cf. Hem et al. 1989).

According to Hem (1980) in oxidized waters with a pH high enough to precipitate MnO_2, the interaction of Co^{2+} and Mn^{2+} at a mixed-oxide surface may be generalized by the redox reaction

$$Mn^{2+} + O_2(aq) + 3Co^{2+} + 4H_2O = MnO_2(c) + Co_3O_4(c) + 8H^+$$

for which

$$K_{eq} = 10^{-17.81} = \frac{[H^+]^8}{[Mn^{2+}] \, [O_2] \, [Co^{2+}]^3}$$

A water in equilibrium with atmospheric oxygen, with $Na^+ = 23$ mg/L, $Cl^- = 35$ mg/L, and pH = 7.0, is in equilibrium with pyrolusite (MnO_2). Use MINTEQA2 to compute the dissolved Mn^{2+} concentration at equilibrium, and with the equation given above calculate Co^{2+} in equilibrium with a MnO_2-Co_3O_4 coprecipitate.

5. Recent work indicates that metallic iron emplaced in a permeable barrier or wall can degrade chlorinated solvents in groundwater by a coupled redox reaction in which the solvent is the oxidant and the reductant is Fe metal or product Fe^{2+} or H_2 (cf. Wilson 1995; Roush 1995). A similar Fe metal barrier has been used to reduce chromate concentrations to values below detection, and has been suggested as a means of cleaning up groundwaters high in nitrate (Wilson 1995).

(a) Write possible degradation reactions for chromate and nitrate assuming metallic Fe is the reductant.

(b) Can such degradation reactions be described using concepts of chemical equilibrium or chemical kinetics? Explain.

(c) If the iron metal barrier is emplaced in an oxidized groundwater flow system, what concerns might you have for its longterm effectiveness in removing CrO_4^{2-} and NO_3^-?

12

Iron and Sulfur Geochemistry

12.1 IRON GEOCHEMISTRY

12.1.1 Introduction

The primary sources of iron in the hydrosphere are the iron minerals in igneous and metamorphic rocks. Among the silicates and aluminosilicates these include olivine, the pyroxene and amphibole mineral groups, and the mica biotite (Deer et al. 1992). Pyrite (FeS_2) and magnetite (Fe_3O_4) are common minor minerals. Iron is largely mobilized and redistributed during the chemical weathering of igneous and metamorphic rocks. Mobilization is chiefly as dissolved Fe(II) under reducing conditions and as particulate Fe(III) oxyhydroxides in oxygenated environments. In sedimentary rocks and soils exposed to the atmosphere the iron is found chiefly as Fe(III) oxyhydroxides. Whereas in anaerobic systems, with decreasing Eh, the iron occurs as Fe(III) oxyhydroxides, the carbonate siderite, and the ferrous sulfides.

In oxidized surface waters and sediments, dissolved iron is mobile below about pH 3 to 4 as Fe^{3+} and Fe(III) inorganic complexes. Fe(III) is also mobile in many soils, and in surface and groundwaters as ferric-organic (humic-fulvic) complexes up to about pH 5 to 6 and as colloidal ferric oxyhydroxides between about pH 3 to 8. Under reducing conditions iron is soluble and mobile as Fe(II) below about pH 7 to 8, when it occurs, usually as uncomplexed Fe^{2+} ion. However, where sulfur is present and conditions are sufficiently anaerobic to cause sulfate reduction, Fe(II) precipitates almost quantitatively as sulfides. Discussion and explanation of these observations is given below. Thermodynamic data for iron aqueous species and solids at 25°C considered in this chapter are given in Table A12.1. Stability constants and ΔH_r° values computed from these data are considered more reliable than their values in the MINTEQA2 data base for the same species and solids.

12.1.2 Stability Constants of Aqueous Complexes

Cumulative formation constants of some iron aquocomplexes are listed in Tables 12.1 and 12.2. Examination of the tabulated data shows that Fe^{2+} generally forms weak complexes or ion pairs (except

TABLE 12.1 Cumulative formation constants for ferrous and ferric hydroxyl complexes written in the proton form (e.g., $nFe^{3+} + mH_2O = Fe_n(OH)_m^{3n-m} + mH^+$) at 25°C

Complex	$-\log {}^*K_{nm}$	Source	Complex	$-\log {}^*K_{nm}$	Source
$FeOH^+$	10.1	†	$Fe(OH)_3^\circ$	12.56	§
$Fe(OH)_2^\circ$	20.5	‡	$Fe(OH)_4^-$	21.6	§
$Fe(OH)_3^-$	29.4	‡	$Fe_2(OH)_2^{4+}$	2.95	‡
$FeOH^{2+}$	2.19	§	$Fe_3(OH)_4^{5+}$	6.3	‡
$Fe(OH)_2^+$	5.67	§			

Source: †Yatsimirskii and Vasilév (1966). See discussion in Chap. 3. ‡Baes and Mesmer (1976). §Macalady et al. (1990).

with bisulfide ion). In contrast, Fe^{3+} forms strong (chiefly inner-sphere) complexes with most ligands, and especially with OH^-, HPO_4^{2-} and F^-. Consistent with the stability constant data, Fe^{3+} is usually complexed, whereas Fe^{2+} occurs uncomplexed in most natural waters.

In our discussion of aqueous species of iron, it is appropriate to first consider the Fe(II) and Fe(III) hydroxyl complexes (Table 12.1). To judge their importance, we will construct plots to show the fractional distribution of these complexes as a function of pH. The computational approach is as presented in Chap. 3. Solute activity coefficients are ignored.

For ferrous iron, the mass-balance equation is

$$\Sigma Fe(II)(aq) = Fe^{2+} + FeOH^+ + Fe(OH)_2^\circ + Fe(OH)_3^- \tag{12.1}$$

We can substitute for and eliminate the complexes in this equation using the cumulative formation expressions for the complexes and values of their constants from Table 12.1. With $[Fe^{2+}]$ factored out, the resultant equation is

$$\Sigma Fe(II)(aq) = [Fe^{2+}]\left(1 + \frac{10^{-10.10}}{[H^+]} + \frac{10^{-20.51}}{[H^+]^2} + \frac{10^{-29.41}}{[H^+]^3}\right) \tag{12.2}$$

TABLE 12.2 Cumulative formation constants of Fe(II) and Fe(III) complexes at 25°C and zero ionic strength (unless otherwise indicated)

Complex	$\log \beta$	Source	Complex	$\log \beta$	Source
FeF^+	1.0	†	FeF_2^+	10.8	†
$FeCl^+$	0.14	†	FeF_3^+	14.0	†
$FeCO_3^\circ$	5.1	‡	$FeCl^{2+}$	1.48	†
$FeHCO_3^+$	2.0	†	$FeCl_2^+$	2.13	†
$FeSO_4^\circ$	2.25	†	$FeCl_3^\circ$	1.13	†
$FeHSO_4^+$	1.08	†	$FeHSO_4^{2+}$	2.48	†
$Fe(HS)_2^\circ$	8.94	§§	$FeSO_4^+$	4.04	†
$Fe(HS)_3^-$	10.97	§§	$Fe(SO_4)_2^-$	5.38	†
$FeH_2PO_4^+$	2.7	§	$FeH_2PO_4^{2+}$	4.17	‖
$FeHPO_4^\circ$	3.6	§	$FeHPO_4^+$	9.92	#
$FePO_4^-$	7.34	††	$FeH_3SiO_4^{2+}$	9.45	‡‡
FeF^{2+}	6.20	†			

Source: †Nordstrom et al. (1990). ‡Estimated from stability of the Fe(II) oxalate complex using method of Langmuir (1979). Bruno et al. (1992) report $\log K_{assoc} = 5.5 \pm 0.2$ for the complex, but neglecting $FeHCO_3^+$. §Smith and Martell (1976). ‖$\log K = 3.47$ at $I = 0.5$ mol/kg (Christensen et al. 1975), corrected to $I = 0$. #$\log K = 8.30$ at $I = 0.5$ mol/kg (Christensen et al. 1975) corrected to $I = 0$. ††Christensen et al. (1975). ‡‡Porter and Weber (1971) at $I = 0.1$ mol/kg corrected to $I = 0$. §§Naumov et al. (1974); $Fe(HS)_2^\circ$ at $I = 0.1$ mol/kg, $Fe(HS)_3^-$ at $I = 0.15$ mol/kg.

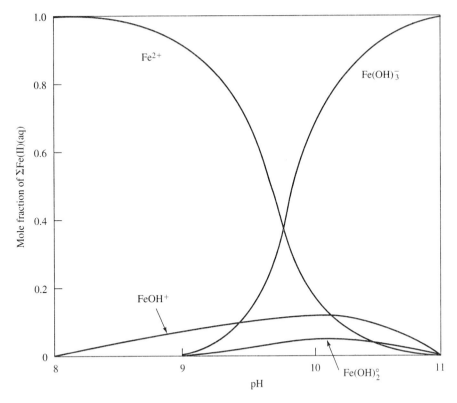

Figure 12.1   Mole fraction of total dissolved Fe(II) present as Fe^{2+} and Fe(II)-OH complexes as a function of pH in pure water at 25°C.

To determine the fractional importance of each complex we set $\Sigma Fe(II)(aq) = 1.0$. The value of $[Fe^{2+}]$ may now be obtained as a function of pH. Fractional concentrations of each complex at a given pH then equal $[Fe^{2+}]$ times the individual terms within the parentheses in Eq. (12.2). The results are tabulated below and plotted in Fig. 12.1.

	Mole Fraction at pH			
Species	8	9	10	11
Fe^{2+}	0.992	0.920	0.167	0.0
$FeOH^+$	0.008	0.073	0.133	0.002
$Fe(OH)_2^{\circ}$	0.0	0.003	0.051	0.008
$Fe(OH)_3^-$	0.0	0.004	0.646	0.990

The figure shows that only Fe^{2+} and $Fe(OH)_3^-$ are major species. Through substitution we find $[Fe^{2+}] = [Fe(OH)_3^-]$ at pH = 9.80, where the fraction of each equals 0.38.

Ignoring for the moment the two polynuclear complexes in Table 12.1, the mass-balance equation involving ferric hydroxyl species is

$$\Sigma Fe(III)(aq) = Fe^{3+} + FeOH^{2+} + FeOH_2^+ + Fe(OH)_3^{\circ} + Fe(OH)_4^- \qquad (12.3)$$

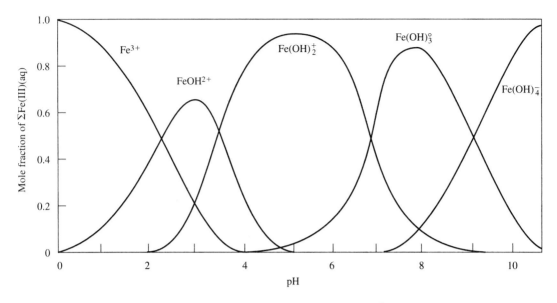

Figure 12.2 Mole fraction of total dissolved Fe(III) present as Fe^{3+} and Fe(III)-OH complexes as a function of pH in pure water at 25°C.

After substitution, as above, to eliminate the complexes in Eq. (12.3), using cumulative constant expressions and the constants from Table 12.1, this becomes

$$\Sigma Fe(III)(aq) = [Fe^{3+}]\left(1 + \frac{10^{-2.19}}{[H^+]} + \frac{10^{-5.67}}{[H^+]^2} + \frac{10^{-12.56}}{[H^+]^3} + \frac{10^{-21.6}}{[H^+]^4}\right) \tag{12.4}$$

We next set $\Sigma Fe(III)(aq) = 1.0$ and solve for $[Fe^{3+}]$ as a function of pH. With $[Fe^{3+}]$ now known, fractional amounts of the complexes are obtained by back substitution into the cumulative constant expressions. The results have been plotted in Fig. 12.2. These calculations show that the Fe(III)-OH complexes exceed 1% of $\Sigma Fe(III)(aq)$ above pH 0.2, and dominate when the pH $= p^*K_{11}$ exceeds 2.19. In contrast, the Fe(II)-OH complexes are less than about 1% of $\Sigma Fe(II)(aq)$ below about pH 8.

The polynuclear Fe(III)-OH complexes listed in Table 12.1 are less than 1% of $\Sigma Fe(III)(aq)$, unless iron concentrations exceed about 10^{-2} to 10^{-3} mol/kg, which is unlikely above pH $= 3$ (cf. Baes and Mesmer 1976). Formation of the dimer, $Fe_2(OH)_2^{4+}$, can be written $2FeOH^{2+} = Fe_2(OH)_2^{4+}$, for which, from the free energy data, we have

$$\frac{[Fe_2(OH)_2^{4+}]}{[FeOH^{2+}]^2} = \frac{1}{26.9} \tag{12.5}$$

This shows that the dimer increases rapidly in importance relative to the monomer as amounts of the monomer increase. Thus a twofold rise in $FeOH^{2+}$ corresponds to a fourfold increase in $Fe_2(OH)_2^{4+}$.

Cumulative constants of some ferrous and ferric complexes involving the same ligand are compared below in the order of increasing stability of the ferric ion complex. Strong (largely inner-sphere) complexes have been marked with an asterisk.

Ligand	$\log \beta_{11}$	
	Fe^{2+}	Fe^{3+}
Cl^-	0.14	1.48
SO_4^{2-}	2.25	4.04
$H_2PO_4^-$	2.7	4.17
F^-	1.0	6.20*
HPO_4^{2-}	3.6	9.92*
OH^-	3.9	11.81*
Salicylate $(C_6H_4(COO)O^{2-})$	7.4*	17.6*

Salicylate has been included as an example of a strong organic complexing ligand (cf. Sigg 1987). The importance of a given complex is determined by both its formation constant and the relative and absolute amounts of the metal cation and ligand in the water. These principles are illustrated by the examples below.

Example 12.1

Assuming that $\Sigma mFe(II)(aq) = 10^{-5.0}$ in seawater ($mCl^- = 0.56$), that the only important Fe(II) species in the water are Fe^{2+} and $FeCl^+$, and that activity coefficients can be ignored, compute the percent of Fe^{2+} and $FeCl^+$ present.

Substituting $mCl^- = 0.56$ into the cumulative constant expression we find

$$\frac{[FeCl^+]}{[Fe^{2+}]} = \frac{1}{1.29} \tag{12.6}$$

which indicates that $FeCl^+$ is 44% of $\Sigma Fe(II)(aq)$.

Example 12.2

Solve Example 12.1 more accurately with MINTEQA2. For the calculation assume that seawater is simply $mNa^+ = mCl^- = 0.56$, pH = 7.0, and $\Sigma Fe(II)(aq) = 10^{-5.0}$ mol/kg. The $FeCl^+$ complex (GFW 91.3) is not in the data base and must be introduced along with its stability constant. Use the Davies equation for ion activity coefficients. The result is $mFe^{2+} = 8.37 \times 10^{-6}$ (83.7%) and $mFeCl^+ = 1.63 \times 10^{-6}$ (16.2% of $\Sigma Fe[II](aq)$).

Example 12.3

FeF^{2+} is a strong complex. At a typical groundwater fluoride concentration of 0.2 mg/kg ($10^{-5.0}$ mol/kg), what percent of $\Sigma Fe(III)(aq)$ is found in the FeF^{2+} complex? As above, ignore ionic strength. Assume the only ferric species present are Fe^{3+} and the 1:1 complex.

Substitution into the cumulative constant expression leads to $[FeF^{2+}]/[Fe^{3+}] = 16/1$. In other words 94% of the iron is complexed.

12.1.3 Ferric Oxyhydroxides

Occurrence, thermodynamic stability, and solubility.   The conditions of formation and occurrence of the Fe(III) oxyhdroxides have been summarized by Langmuir and Whittemore (1971), Schwertmann and Taylor (1977), Bolt and Van Riemsdijk (1987), Macalady et al. (1990),

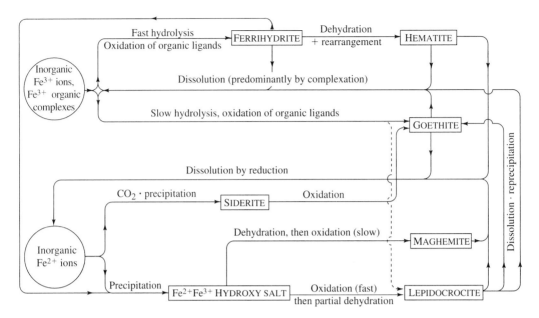

Figure 12.3 Possible pathways of Fe(III) oxide formation under near pedogenic conditions. Adapted from U. Schwertmann and R. M. Taylor, Iron oxides, in *Minerals in soil environments*, ed. J. B. Dixon and S. B. Weed. Copyright 1989 by Soil Science Society of America. Used by permission.

Combes et al. (1990), and Baltpurvins et al. (1996), among others (see Fig. 12.3 and Table 12.3). Thermodynamic stabilities of the oxyhydroxides as known are given in Table A12.1. The most abundant oxyhydroxide in sedimentary environments is goethite. In aquatic environments, which are dominantly sulfate rich when acid and bicarbonate rich at higher pH's, freshly precipitated oxyhydroxides are chiefly mixtures of amorphous material and goethite (Langmuir and Whittemore 1971). Thermodynamic data in Table A12.1 suggest that in water well-crystallized goethite is slightly more stable than hematite, with $\Delta G_r^\circ = -0.53$ kcal/mol at 25°C for the reaction

$$\alpha\text{-Fe}_2O_3 + H_2O(l) = 2\alpha\text{-FeOOH} \tag{12.7}$$

(Macalady et al. 1990). This corresponds to $[H_2O] = 0.41$ at equilibrium, indicating that in soils where the relative humidity is less than about 40%, hematite is the stable phase. The relative stabilities of hematite and goethite also depend on temperature (hematite is more stable above roughly 80°C), and especially depend on particle size (Langmuir 1971b). In desert soils and older sedimentary rocks hematite is more abundant than goethite, in part because it is thermodynamically favored by both increased temperature and dryness, but also because it is relatively unreactive once formed.

The partially amorphous Fe(III) oxyhydroxide common in soils and modern sediments is termed ferrihydrite or hydrous ferric oxide (HFO). When freshly precipitated, HFO has an apparent solubility product as pK_{sp} ($pK_{sp} = -\log[Fe^{3+}][OH]^3$) of about 37, which, because of partial crystallization, may rise to 38.5 to 39.5 after a short time. (Crystalline goethite and hematite pK_{sp} values are both near 44.) Crystallization of HFO takes years in waters low in iron, but may occur in a few hours or days in the presence of several mg/kg of dissolved iron. The growth rate of an X-ray or SEM-identifiable crystalline phase increases with increasing temperature and is roughly proportional to the aqueous Fe(II) and less so the aqueous Fe(III) present during crystallization (Macalady et al. 1990).

TABLE 12.3   Occurrence and stabilities of the important Fe(III) oxyhydroxides—$pK_{sp} = -\log ([Fe^{3+}][OH^-]^3)$

Name	Formula	pK_{sp}	Remarks
Ferrihydrite/ hydrous ferric oxide (HFO)	$Fe(OH)_3 \cdot nH_2O$	37 to 39	Initial precipitate from rapid hydrolysis (neutralization) of Fe(III) or oxidation of Fe(II). May crystallize to form goethite in cool regions and hematite in warm regions.
Goethite	α-FeOOH	$\leq 44.1 \pm 0.2$	Most abundant Fe(III) oxhydroxide in modern soils and sediments in cool climates. The thermodynamically most stable oxyhydroxide in water at low temperatures. Thought to form from HFO by dissolution-reprecipitation.
Hematite	α-Fe$_2$O$_3$	$\leq 43.9 \pm 0.2$	Results from the weathering and oxidation of Fe (esp. ferrous) silicate minerals or the aging of HFO. May form from HFO by direct solid-solid transformation. Common in warm climate soils, and the chief oxyhydroxide in red-bed sediments.
Lepidocrocite	γ-FeOOH	38.7 to ≥ 40.6	Minor constituent in humid and temperate climate noncalcareous soils, especially where alternating reducing and oxidizing conditions in seasonally waterlogged soils. Favored by rapid oxidation at low pH, low ΣFe, low T, and Fe(III)(aq) absent.
Maghemite	γ-Fe$_2$O$_3$	38.8 to ≥ 40.4	Same conditions as lepidocrocite, but formed in tropical and subtropical climates, especially in highly weathered soils.

Note: Well-crystallized lepidocrocite and maghemite are less stable than well-crystallized goethite and hematite, respectively.
Source: After Greenland and Mott (1978), Bolt and Van Riemsdijk (1987), Macalady et al. (1990), and Combes et al. (1990). pK_{sp} values based on Langmuir (1969a), free-energy data in Table A12.1 and Macalady et al. (1990).

The often important effect of particle size on solubility can be estimated through the equation

$$\log K_{sp} = \log K_{sp}(S = 0) + \gamma_s \left(\frac{S}{2.30RT} \right) \tag{12.8}$$

(Stumm and Morgan 1981) where S is the surface area of a mole of the solid, γ_s the interfacial energy, R the gas constant, and T the temperature in K. The particle-size correction term on the right in Eq. (12.8) can be symbolized as $\delta_s \log K_{sp}$. For very fine-grained goethite (GFW 88.85), assuming $S \approx 15,000$ m^2/mol (Table 10.2), $\gamma_s = 0.30$ cal/m^2 (Langmuir and Whittemore 1971) and with $R = 1.9872$ cal/mol K and $T = 298.15$ K, we find $\delta_s \log K_{sp} = 3.3$. Thus, assuming $\log K_{sp} = -44.2$ for well-crystallized goethite, we find $\log K_{sp} = -40.9$ for the fine-grained material.

When HFO is precipitated in acid solutions high in anions, such a chloride or sulfate, these anions often substitute for hydroxyl, forming metastable solids such as $Fe(OH)_{2.35}(Cl)_{0.65}$ and $Fe(OH)_{2.35}(SO_4)_{0.325}$, in which the Fe(III)/(OH) ratio is 1/2.35 (cf. Matijevic and Scheiner 1978; Fox 1988). (A related phase [$Fe(OH)_{2.7}(Cl)_{0.3}$] is listed in the MINTEQA2 data base.) After aging hours to days, hydroxyl tends to replace the other anions in these solids, and the Fe(III)/OH ratio approaches the theoretical 1/3 value of stoichiometric ferric oxyhydroxides (Macalady et al. 1990).

The solubility of the Fe(III) oxyhydroxides as a function of pH was considered in Chap. 7. As shown in Fig. 12.4, the oxyhydroxides are amphoteric and decrease in solubility by about 10^7-fold between amorphous material and well-crystallized goethite or hematite. Assuming a Fe detection

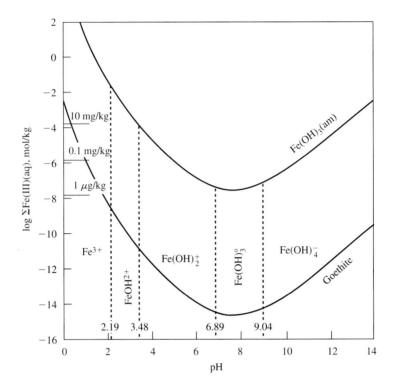

Figure 12.4 Solubility of amorphous $Fe(OH)_3$, $pK_{sp} = 37.1$ (top curve) and goethite [α-FeOOH], $pK_{sp} = 44.2$ (bottom curve) as a function of pH at 25°C. Also shown are the fields of dominance of Fe^{3+} and Fe^{3+}-OH complexes. Numbers along the abcissa are values of the stepwise constants, $p*K_n = pH$ for $n = 0$ to 3, for the complexation reactions written in the form $Fe(OH)_n^{3-n} + H_2O = Fe(OH)_{n+1}^{2-n} + H^+$.

limit of about 0.05 mg/kg ($\sim 10^{-6.0}$ mol/kg) and equilibrium with respect to the oxyhydroxides, Fe(III)(aq) concentrations will be at or well below detection limits when the pH is between 5 and 10.

Suspended/colloidal oxyhydroxides. Although dissolved Fe(III) is absent between pH 5 and 10, the Fe(III) oxyhydroxides occur in suspended form in surface- and groundwaters within this pH range. In colloidal sizes such materials can remain suspended indefinitely in fresh groundwaters. Langmuir (1969b) noted that 30 to 70% of up to 15 mg/L total iron in some confined New Jersey groundwaters was present as suspended Fe(III) oxyhydroxides in particle sizes mostly less than 5 μm. Similar materials and amounts are transported by streams (cf. Mayer 1982a; McKnight et al. 1988).

Suspensions of the colloidal-sized oxyhydroxides can be destabilized by increases in ionic strength (strong electrolyte concentrations) such as occur when a stream enters an estuary (Fig. 12.5). Small amounts of specifically adsorbed species at well-defined concentrations can also destabilize the suspended oxyhydroxides.

The pH at which the surface of the oxyhydroxides is uncharged in pure water is called the point of zero net proton charge (PZNPC) (see Chap. 10). For the ferric oxyhydroxides in pure water the PZNPC corresponds to the pH of minimum solubility, which is near pH 8 (Fig. 12.4). Below this pH, consistent with the charge of the predominant Fe(III)-OH aqueous complex, solid surfaces are OH^- deficient (relative to Fe(III)/OH = 1/3) and so have a net positive charge. Above pH 8 the surface has

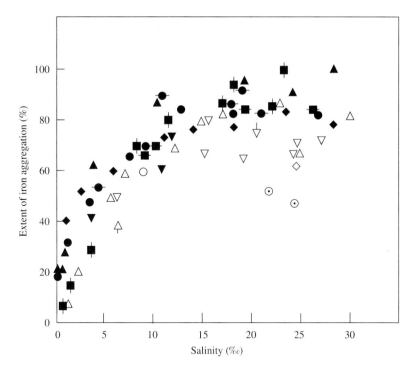

Figure 12.5   Extent of iron aggregation as a function of salinity in the Saco River estuary, calculated as the percent loss from a conservative mixing line connecting filtrable iron concentrations of the river water and seawater end members. Open symbols are samples with $T < 14°C$; solid symbols are samples with $T > 14°C$. Reprinted from *Geochim. et Cosmochim. Acta*, 46, L. M. Mayer, Aggregation of colloidal iron during estuarine mixing: Kinetics, mechanism and seasonality, 2527–35, © 1982, with permission from Elsevier Science Ltd., The Boulevard, Langford Lane, Kidlington OX5 1GB, U.K.

an OH^- excess and is net negative. At pH values far from the PZNPC, the surface-charge density of colloidal-sized oxyhydroxide particles is high and mutual particle repulsion stabilizes their suspensions. At the PZNPC, however, particles are not repelled and so tend to agglomerate or flocculate and settle out (cf. Liang and Morgan 1990).

The rate at which suspended oxyhydroxide particles coagulate may be described by the stability ratio, W, where

$$W = \exp\left(\frac{V_{max}}{\kappa T}\right) \tag{12.9}$$

V_{max} is the height of the free-energy barrier preventing particle coagulation, κ the Boltzmann constant, and T the temperature in kelvin (Everett 1988). Coagulation occurs most rapidly when $V_{max} = \kappa T$ and log $W = 0$. Compression of the diffuse layer by increased concentrations of strong electrolyte cations and anions can reduce particle repulsion and cause coagulation (cf. Stumm and Morgan 1996). An example of this effect is given in Fig. 12.6, which plots log W_{exp} (experimental) versus pH for hematite suspensions over a range of ionic strengths. The plot shows that rapid coagulation of hematite suspensions (log $W \approx 0$) is limited to about pH 8.5 when $I = 10^{-3}$ M, but expands to the pH range from 6 to 11 as the ionic strength is increased to 8.4×10^{-2} M.

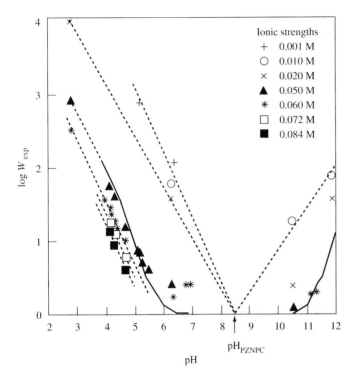

Figure 12.6 Experimentally derived stability ratio, W_{exp}, of a hematite suspension plotted as a function of pH for different ionic strengths. The pH of the PZNPC is indicated. Dashed lines are drawn through the experimental points as a guide. The solid line has been model-calculated. From *Aquatic Sciences* 52(1): 32–55, L. Liang and J. J. Morgan, Chemical aspects of iron oxide coagulation in water: Laboratory studies and implications for natural systems, Copyright 1990 by Birkhauser Verlag, Basel, Switzerland. Used by permission.

The concentration of an electrolyte or other species that causes the most rapid coagulation of a specific suspended solid is termed the critical coagulation concentration (ccc). At the ccc, $V_{max} = \kappa T$ and log $W = 0$. The effect of different concentrations of solutes on coagulation of 20 mg/L of suspended hematite at pH 6.5 is shown in Fig. 12.7. In this example the ccc is about 10^{-3} M for divalent electrolyte ions and near $10^{-1.4}$ M for Na^+. Higher concentrations of these weakly adsorbed species have little further effect. Much smaller amounts of specifically adsorbed humic and fulvic acids and phosphate can also destabilize the suspended hematite. The ccc for fulvic acid (FA) is about $10^{-6.2}$ M, equivalent to a FA concentration of 0.1 mg/L, an amount commonly found in surface- and groundwaters. At lower or higher FA concentrations, hematite particles are (+) and (−) charged, respectively, they repel each other, and the suspension is restabilized (Liang and Morgan 1990). Figure 12.7 shows that a shift in phosphate or humic acid (HA) concentrations away from their ccc values also restabilizes hematite in suspension.

As the pH of an oxyhydroxide suspension is varied, the ccc concentration of strongly adsorbed species also changes. For example, near pH = 3, 10 mg/L of goethite flocculates most effectively (log $W = 0$) for $\Sigma PO_4 = 10^{-4}$ M, whereas when ΣPO_4 is essentially zero (10^{-8} M), the optimum flocculation pH is about 8 near the pH of the PZNPC (Liang and Morgan 1990).

In seawater (pH $\approx$ 8.1) the reported iron concentration ranges from $10^{-6.0}$ to $10^{-8.75}$ M with a mean value of perhaps $10^{-7.4}$ M (cf. Holland 1978; Hem 1985). It is interesting to question whether this amount represents dissolved or suspended ferric iron. Examination of Fig. 12.4 shows that the dominant aqueous species in seawater at pH = 8.1 is $Fe(OH)_3^\circ$. (Fe(III) chloride complexing can be ignored.) Assuming that Fe(III) is controlled by the solubility of amorphous HFO ($pK_{sp} = 37.1$) as an upper limit, and with free energies from Table A12.1, we find for the reaction $Fe(OH)_3(am) = Fe(OH)_3^\circ$, that $K_{eq} = [Fe(OH)_3^\circ] = 10^{-7.7}$ M. This value is close to the mean iron concentration reported

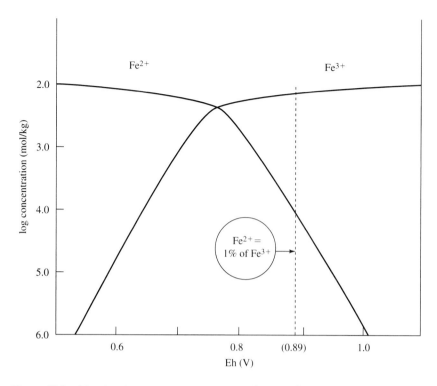

Figure 12.9   Plot showing the concentrations of Fe^{2+} and Fe^{3+} as a function of Eh at pH = 1.0. The plot indicates that Fe^{2+} equals 1% of Fe^{3+} at 0.89 V.

mol/kg), and pH = 3.0? Solve the problem with MINTEQA2 using the Debye-Hückel equation for ion activity coefficients.

Program output shows that $[Fe^{2+}] = 10^{-4.18}$ mol/kg, and $[Fe^{3+}] = 10^{-5.39}$ mol/kg. Substituting these values into Eq. 9 in Table A12.2, we compute Eh = 0.698 V. The output shows that 100% of Fe(II)(aq) is present as uncomplexed Fe^{2+}; however, only 9.4% of Fe(III)(aq) occurs as free Fe^{3+}. Ferric complexes as a percent of ΣFe(III)(aq) are: $FeCl^{2+}$ 1.7%, FeF_2^+ 3.8%, $FeOH_2^+$ 9.6%, FeF^{2+} 35.5%, and $FeOH^{2+}$ 39.7%. Complexing has enlarged the stability field of Fe(III)(aq) species, lowering the Fe^{3+}/Fe^{2+} boundary by 0.072 V.

We will next construct Eh-pH diagrams involving iron solids. The solid/aqueous species boundaries are drawn for a defined concentration of the aqueous species. One should select solids that have been found or are expected to occur in the environment under study. In surface or near-surface systems near a redox interface, as in waterlogged soils, or where the solids are rapidly precipitated in water treatment, one can expect them to be relatively amorphous and more soluble than their well-crystallized analogs. Assuming this is the case, we will consider $Fe(OH)_3(am)$ ($pK_{sp} = 37.1$), ferrosic hydroxide ($Fe_3(OH)_8$), and siderite ($FeCO_3$). The reactions and their Eh-pH equations are given in Table A12.2, Part III. Figure 12.10 shows the solid fields of $Fe(OH)_3(am)$ and $FeCO_3$, ignoring ferrosic hydroxide. In Fig. 12.11, $Fe_3(OH)_8$ is added. Both figures show the increased size of solid phase stability fields when solid/aqueous boundaries are drawn for $\Sigma Fe = 10^{-3}$ rather than 10^{-5} mol/kg. As a general rule, the larger the stability field of a solid between about pH 3 and 9 (the usual pH range of natural waters), the more common that solid should be in nature. Given this point,

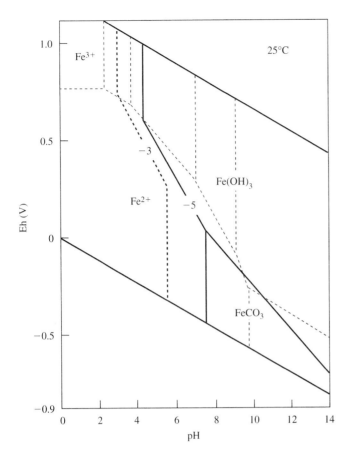

Figure 12.10 Eh-pH diagram for the system $Fe-O_2-CO_2-H_2O$ at 25°C ignoring ferrosic hydroxide $[Fe_3(OH)_8]$ and assuming $pK_{sp} = 37.1$ for amorphous $Fe(OH)_3$. Bicarbonate is fixed at $10^{-2.7}$ mol/kg. Aqueous/solid boundaries are drawn for total dissolved iron concentrations of 10^{-5} mol/kg (solid line) and 10^{-3} mol/kg (heavy dashed line). Lightly dashed lines show the positions of the aqueous/aqueous species boundaries identified in Fig. 12.8.

the large relative size of the ferrosic hydroxide field is somewhat surprising, given that it is infrequently reported, and may suggest that the tabulated free energy of the phase is too negative.

As we increase the thermodynamic stability (i.e., the pK_{sp}) of the ferric oxyhydroxide considered in our Eh-pH diagram, the size of its stability field increases. This is evident from Fig. 12.12, which shows the very large stability field of goethite ($pK_{sp} = 44.2$) relative to that of the amorphous phase ($pK_{sp} = 37.1$). The field of siderite practically disappears in equilibrium with goethite, suggesting that siderite and well-crystallized goethite (or hematite with a similar stability) should rarely be found together.

In iron-rich surface and groundwaters, whether pristine or contaminated, Fe(aq) concentrations are often buffered by equilibrium with ferric oxyhydroxides present in the soil or sediment or suspended in the water. This applies to acid mine waters as well as to groundwaters in sediments affected by organic contamination. Measured Eh (Eh_m) and pH values in a plume of landfill leachate in sandy soils under an 8-year-old Pennsylvania landfill are plotted in Fig. 12.13. The plume was sampled from suction lysimeters installed at depths from about 1 to 17 m under the landfill (Apgar and Langmuir 1971). BOD values generally exceeded 10 mg/kg with iron ranging from 0.1 to 885 mg/kg. The figure shows that Eh_m and pH are roughly consistent with the assumed equilibrium between the iron and amorphous Fe(III) oxyhydroxide ($pK_{sp} = 37.1$).

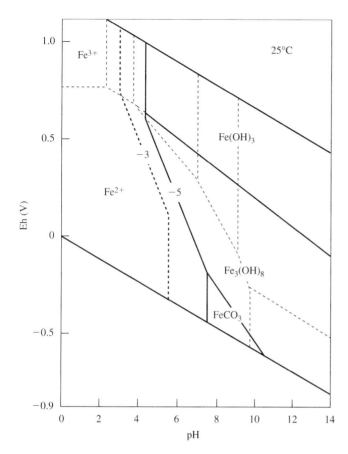

Figure 12.11   Eh-pH diagram for the system Fe-O_2-CO_2-H_2O at 25°C, including ferrosic hydroxide [$Fe_3(OH)_8$] and assuming $pK_{sp} = 37.1$ for amorphous $Fe(OH)_3$. Bicarbonate is fixed at $10^{-2.7}$ mol/kg. Aqueous/solid boundaries are drawn for total dissolved iron concentrations of 10^{-5} mol/kg (solid line) and 10^{-3} mol/kg (heavy dashed line). Lightly dashed lines show the positions of the aqueous/aqueous species boundaries identified in Fig. 12.8.

12.2 SULFUR GEOCHEMISTRY

12.2.1 Thermodynamic Data for Substances in the System S-O_2-H_2O

The stabilities of species in the system S-O_2-H_2O have been studied by numerous researchers. Much of this work has been summarized or critiqued by Garrels and Naeser (1958), Boulegue and Michard (1979), Morse et al. (1987), Schoonen and Barnes (1988), and Williamson and Rimstidt (1992). Thermodynamic data for some substances in the system S-O_2-H_2O are given in Table A12.3 in the chapter appendix. The many aqueous sulfur species that are either thermodynamically stable or are important metastably are shown in Fig. 12.14. Acid-base reactions among the sulfur species are generally rapid and reversible. The redox reactions, however, may be fast and reversible (e.g., H_2S/S_n^{2-}; see Boulegue and Michard 1979) or more often irreversible in the absence of bacterial activity (e.g., SO_4^{2-} reduction to H_2S).

12.2.2 Acid-Base Reactions

Dissociation constants and reaction enthalpy data for the stable and most important metastable sulfur species are summarized in Table 12.4. The log K values in Table 12.4 indicate the pH at which the acid and conjugate base have equal concentrations. Bisulfate (HSO_4^-) is a relatively strong acid,

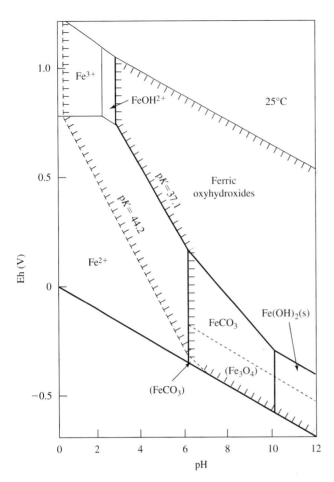

Figure 12.12 Eh-pH diagram for the system Fe-O$_2$-CO$_2$-H$_2$O, assuming total dissolved carbonate equals 10^{-3} mol/kg, and total dissolved iron is 10^{-3} mol/kg at aqueous/solid boundaries. Also shown is the position of aqueous/solid boundaries for amorphous Fe(OH)$_3$ with pK_{sp} = 37.1 and goethite with pK_{sp} = 44.2. The figure shows that magnetite (Fe$_3$O$_4$) and siderite (FeCO$_3$) are metastable in the presence of goethite. From *Ground Water* 13(4): 360–65, D. O. Whittemore and D. Langmuir, The solubility of ferric oxyhydroxides in natural waters. Copyright 1975 by Ground Water Publishing Company. Used by permission.

important in acid mine waters (see Chap. 5) where its concentration exceeds that of sulfate below about pH 2. Sulfate is a hard base and tends to form ion pairs or weak complexes with most monovalent and divalent cations (Chap. 3). Sulfite (SO$_3^{2-}$) and thiosulfate (S$_2$O$_3^{2-}$), on the other hand, are relatively soft ligands and thus form stronger complexes than sulfate with borderline or soft metal cations such as Cu^{2+}, Zn^{2+}, Pb^{2+}, Ag$^+$, Cd^{2+}, and Hg^{2+} (cf. Smith and Martell 1976).

The value of K_1(H$_2$S) has been measured by many researchers. Millero (Morse et al. 1987) proposes the temperature function

$$\text{p}K_1 = 32.55 + 1519.44/T - 15.672 \log T + 0.02722T \qquad (12.10)$$

based on Barbero et al. (1982), where T is in degrees K. This is considered accurate within ±0.05 units below 100°C and within ±0.47 units up to 300°C. At 25°C the function yields pK_1 = 6.99. Hersey et al. (1988) suggest instead the function

$$\text{p}K_1 = -98.080 + 5765.4/T + 34.6436 \log T \qquad (12.11)$$

for 0 to 300°C, which gives pK_1 = 6.98 at 25°C. These authors consider Eq. (12.11) more accurate than Eq. (12.10) at low temperatures. The two functions are largely in agreement at elevated tem-

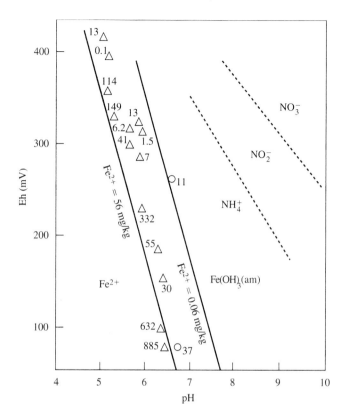

Figure 12.13 Eh-pH diagram showing total iron values (mg/kg) in unsaturated zone waters sampled at depths of 1 to 17 m under an 8-year-old municipal waste landfill emplaced in a valley bottom in central Pennsylvania. The solid lines are theoretical $Fe^{2+}/Fe(OH)_3$(am) boundaries computed assuming $Fe^{2+} = 56$ mg/kg and 0.06 mg/kg. The dashed lines are equal concentration boundaries among NO_3^-, NO_2^-, and NH_4^+. From *Ground Water* 9:76–96. D. O. M. A. Apgar and D. Langmuir, Ground water pollution potential of a landfill above the water table. Copyright 1971 by Ground Water Publishing Company. Used by permission.

peratures. As a soft base, bisulfide forms strong complexes with the borderline soft and soft metal cations. Stabilities of these complexes as a function of temperature are reported by Barnes (1979).

The second dissociation constant $K_2(H_2S)$ is poorly known. Reported constants range from $pK_2 = 12.44$ to 17.1 (cf. Morse et al. 1987). The most reliable value is probably that of Schoonen and Barnes (1988), who propose $pK_2 = 18.51 \pm 0.56$ at 20°C based on extrapolation of thermodynamic data for aqueous polysulfide species. (See also Williamson and Rimstidt 1992.)

Incomplete O_2 oxidation of H_2S at ΣS concentrations exceeding the solubility of sulfur ($\sim 5 \times 10^{-6}$ mol/kg) may lead to the precipitation of colloidal-sized elemental sulfur, which can then react with HS^- to form polysulfides (cf. Boulegue and Michard 1979; Morse et al. 1987). The successive reactions are

$$HS^- + \tfrac{1}{2}O_2 \rightarrow S° + OH^- \tag{12.12}$$

$$HS^- + (n-1)S° \rightarrow S_n^{2-} + H^+ \tag{12.13}$$

with $n = 1$ to 5 for the polysulfides. Giggenbach (1974) writes the formula of polysulfide ions as S_nS^{2-} to emphasize that n moles of the sulfur are zero valent. As the proportion of elemental sulfur to aqueous sulfide increases in solution, the chain length (value of n) of the resultant polysulfide increases. The polysulfides also form when a solution that contains both thiosulfate and sulfide species is acidified (Giggenbach 1974). Depending on the pH, polysulfide ions may react with H^+ to form HS_n^- or

Number of sulfur atoms

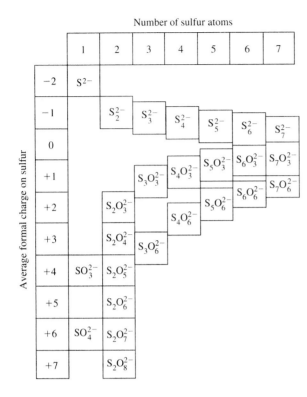

Figure 12.14 Diagram showing relationships among aqueous sulfur species that are either thermodynamically stable or have significant metastable persistence. Reprinted from *Geochim. et Cosmochim. Acta*, 56, M. A. Williamson and J. D. Rimstidt, Correlation between structure and thermodynamic properties of aqueous sulfur species, 3867–80, © 1992, with permission from Elsevier Science Ltd., The Boulevard, Langford Lane, Kidlington OX5 1GB, U.K.

H_2S_n species. Schoonen and Barnes (1988) suggest the following stepwise dissociation constants for the polysulfide acids:

Polysulfide	pK_1	pK_2
H_2S	7.05	(18.51)
H_2S_2	5.12	10.06
H_2S_3	4.32	7.86
H_2S_4	3.92	6.66
H_2S_5	3.58	6.02

TABLE 12.4 Hydrolysis constants for some aqueous species in the system $S-O_2-H_2O$ at 25°C and 1 bar pressure, computed from the data in Table A12.3

Species	Reaction	ΔH_r° kcal/mol	$\log K$
HSO_4^-	$HSO_4^- = H^+ + SO_4^{2-}$	−5.24	−1.99
HSO_3^-	$HSO_3^- = H^+ + SO_3^{2-}$	−2.72	−7.36
$HS_2O_3^-$	$HS_2O_3^- = H^+ + S_2O_3^{2-}$	−6.1	−1.75
H_2S	$H_2S = H^+ + HS^-$	5.3	−6.99
HS^-	$HS^- = H^+ + S^{2-}$	(12.2)	(−18.5)

Note: Values in parentheses are estimates or based on estimated values.

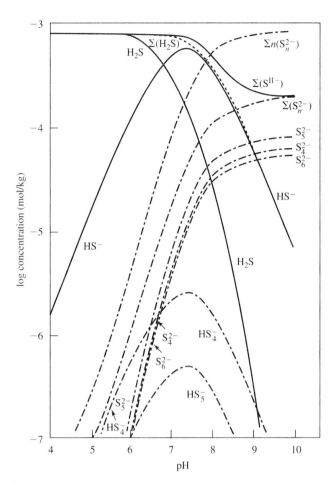

Figure 12.15   Distribution of sulfur species versus pH in the system H_2S-S(colloid)-H_2O-NaCl(0.7 M) for $\Sigma S(aq) = 10^{-3}$ mol/kg. Reprinted with permission from Am. Chem. Soc. Symp. Ser. 93, J. Boulegue and G. Michard, Sulfur speciations and redox processes in reducing environments. In *Chemical modeling in aqueous systems*. Copyright 1979 American Chemical Society.

The polysulfides are most stable in alkaline waters, and for $\Sigma S(aq) = 10^{-3}$ mol/kg they predominate over HS^- above pH 8 (Fig. 12.15). Polysulfide ions are soft ligands that can be expected to form strong complexes with soft or borderline soft metal cations; however, stability data for such complexes are lacking. Organically bonded polysulfides are important constituents in the pore waters of modern sediments (Boulegue et al. 1982).

12.2.3 Redox Reactions

The Eh-pH diagram for thermodynamically most stable sulfur species is shown in Fig. 12.16. The acid-base boundaries have been considered above. We will derive two of the redox boundaries. Probably the SO_4^{2-}/H_2S and SO_4^{2-}/HS^- boundaries are the most important. The SO_4^{2-}/H_2S redox reaction is

$$SO_4^{2-} + 9H^+ + 8e^- = HS^- + 4H_2O \tag{12.14}$$

for which, from the free-energy data in Table A12.3, we find $E° = 0.249$ V. The Eh-pH equation is

$$Eh(V) = 0.249 + \frac{0.05916}{8} \log \frac{[SO_4^{2-}] [H^+]^9}{[HS^-]} \tag{12.15}$$

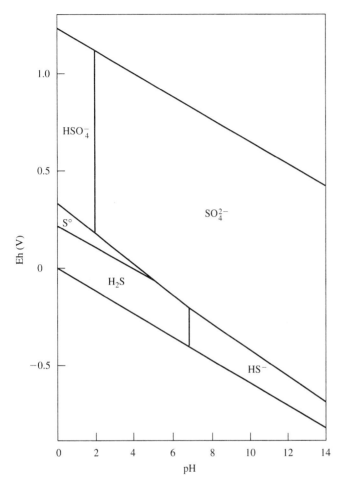

Figure 12.16 Eh-pH diagram for thermodynamically stable substances in the system S-O$_2$-H$_2$O at 25°C, showing the fields of predominance of the aqueous species and of elemental sulfur for ΣS(aq) = 10^{-3} mol/kg at aqueous/S° boundaries.

At the boundary [HS$^-$] = [SO$_4^{2-}$] and the expression simplifies to

$$Eh(V) = 0.249 - 0.0666 \, pH \tag{12.16}$$

The important SO$_4^{2-}$/S°(c) boundary corresponds to the reaction

$$SO_4^{2-} + 8H^+ + 6e^- = S° + 4H_2O \tag{12.17}$$

for which we find $E° = -0353$ V, and

$$Eh(V) = -0.353 + \frac{0.05916}{6} \log \, [SO_4^{2-}] \, [H^+]^8 \tag{12.18}$$

which becomes

$$Eh(V) = -0.353 + 0.00991 \log[SO_4^{2-}] - 0.0793 \, pH \tag{12.19}$$

The position of this boundary in Fig. 12.16 depends on the choice of total dissolved sulfur. The native sulfur field increases in size with increasing dissolved ΣS(aq).

Most surface-waters and groundwaters plot in the sulfate field, with acid mine waters close to or in the bisulfate field. Hydrogen sulfide and HS$^-$ are major species in organic-rich, anaerobic waterlogged soils and sediments. In fresh groundwaters hydrogen sulfide is usually less than 10^{-4} mol/kg (3 mg/kg). However, H_2S may exceed 3×10^{-3} mol/kg (100 mg/kg) in brines associated with petroleum (Hem 1985) and can approach 5×10^{-3} mol/kg (170 mg/kg) in the interstitual waters of marine sediments (Goldhaber and Kaplan 1974). A silver billet electrode can be coated with AgS and used as a sulfide electrode, responding to S^{2-} in natural waters and sediments. This approach was introduced by Berner (1963) and has been used by Boulegue (1978) and Boulegue and Michard (1979) to analyze for sulfide in some laboratory systems and natural waters.

Sulfate reduction.   All plants, animals, and bacteria metabolize sulfur in order to synthesize amino acids such as cysteine and methionine. The sulfur may be assimilated as sulfate or as organic molecules containing sulfate. The reduction of sulfate in biosynthesis is termed *assimilatory sulfate reduction* and can take place in aerobic or anaerobic environments (cf. Goldhaber and Kaplan 1974; Rheinheimer 1981; Cullimore 1991).

Inorganic sulfur species more oxidized than sulfide (sulfate, sulfite, and thiosulfate, for example) can act as electron acceptors in the oxidation of organic matter by bacteria. In the process the sulfur is reduced to sulfide. The reaction is described as *dissimilatory reduction*. The bacteria involved are of the genus *Desulfovibrio* or genus *Desulfotomaculum* and are heterotrophs, that is, they require organic matter for energy transfer. A wide range of organic substances can be used, including hydrocarbons, fatty acids, carbohydrates, and amino acids (Berner 1971). Lactic acid ($CH_3CHOHCOOH$) and pyruvic acid ($CH_3COCOOH$) and their salts are the preferred organic reactants, with acetate (CH_3COOH), a product that cannot be further metabolized by the bacteria. For example, with lactic acid as the electron donor, the reaction may be written

$$2CH_3CHOHCOOH + SO_4^{2-} \rightarrow CH_3COOH + H_2S + 2HCO_3^- \tag{12.20}$$

The sulfate-reducing bacteria *Desulfovibrio desulfuricans* prefers a pH between 6 and 8, but can function between pH 4.2 and 9.9 (Wallhauser and Puchelt 1966; Baas Becking et al. 1969; Karamenko 1969; Zehnder 1988). Sulfate-reducing bacteria can operate at temperatures as low as 0°C, and as high as 110°C, in deep-sea hydrothermal vent sediments (Jorgensen et al. 1992). At temperatures above 100 to 120°C sulfate reduction also proceeds at a measureable rate without bacterial participation.

Until recently it has been assumed that sulfate-reducing bacteria always required a strictly anaerobic environment. These environments are found in deep coastal-plain areas, oil-field brines, and in black (organic-rich), waterlogged soils and muds associated with rivers, lakes, and swamps. Sulfate reduction has also been observed in local microenvironments such as those created by the decay of a fish buried in otherwise oxidizing sediments (Berner 1971). Contrary to traditional belief, active sulfate reduction has also been observed in the presence of dissolved oxygen in the photosynthetic zone of microbial mats (Canfield and Des Marais 1991).

Oxidation of reduced sulfur species.   Oxidation of reduced sulfur species in the presence of oxygen can occur spontaneously, without bacterial mediation. Bacteria of the family *Thiobacteriaceae* are probably the most important bacteria involved in sulfur oxidation. Of these, bacteria of the genus *Thiobacillus* have been most studied (Goldhaber and Kaplan 1974; Cullimore 1991). The first product of sulfide oxidation abiotically or by *Thiobaccillus* is thought to be elemental sulfur according to

$$H_2S + \tfrac{1}{2}O_2 \rightarrow S° + H_2O \tag{12.21}$$

This reaction takes place in sulfide-rich flowing wells and springs in western Pennsylvania, where they discharge at the land surface, cloudy with a suspension of elemental sulfur.

As noted above, the polysulfides are formed by the reaction of elemental sulfur with bisulfide. Further oxidation of S° can produce sulfite. The sulfite may, in turn, be reduced to thiosulfate by reaction with S°,

$$S^\circ + O_2 + H_2O \rightarrow SO_3^{2-} + 2H^+ \tag{12.22}$$
$$SO_3^{2-} + S^\circ \rightarrow S_2O_3^{2-}$$

or be oxidized to sulfate,

$$SO_3^{2-} + \tfrac{1}{2}O_2 \rightarrow SO_4^{2-} \tag{12.23}$$

Oxidation of thiosulfate also produces small amounts of trithionate ($S_3O_6^{2-}$), tetrathionate ($S_4O_6^{2-}$), and pentathionate ($S_5O_6^{2-}$) (Goldhaber and Kaplan 1974). Summarized in Fig. 12.17 are possible oxidation and disproportionation pathways of reduced sulfur species leading toward sulfate that may be mediated by *Thiobacilli*. (Disproportionation pathways involve no electron transfer; see also O'Brien and Birkner 1977; Morse et al. 1987.) More recently, Jorgensen (1990) used radioactive ^{35}S to unravel the complex pathways of sulfide oxidation in sediments. He showed that thiosulfate disproportionation to sulfate and sulfide species

$$S_2O_3^{2-} + H_2O \rightarrow SO_4^{2-} + HS^- + H^+ \tag{12.24}$$

was a key reaction in anoxic sediments.

The polysulfides, thiosulfate and sulfite, which are generally metastable relative to sulfate, are most abundant near groundwater redox interfaces where they can complex with borderline or soft metal cations (cf. Barnes 1979; Daskalakis and Helz 1992). Many sedimentary and hydrothermal ore deposits are found at redox interfaces where the breakdown and formation of such complexes may have locally caused metal precipitation and remobilization (cf. Granger and Warren 1969; Boulegue and Michard 1979; Barnes 1979). Similar metal-sulfur species mobilization may also occur in the vicinity of some toxic metal-organic waste sites.

Less common among the metastable sulfur species are the polythionates ($S_nO_6^{2-}$ with $n = 4$ to 9). They are formed, for example, in volcanic lakes by the bubbling of H_2S gas through sulfurous acid (Takano 1987). The polythionates break down eventually into sulfate and elemental sulfur. The polysulfides and polythionates are themselves metastable relative to sulfite, thiosulfate, S°, and the sulfide species. In order to examine the redox-stability environments of these species, it is useful to construct an Eh-pH diagram ignoring sulfate, which is relatively inert. Such a plot is given in

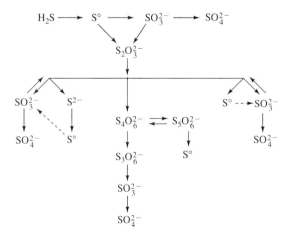

Figure 12.17 Possible oxidation pathways for reduced sulfur species to sulfate by *Thiobacilli*. From M. B. Goldhaber and I. R. Kaplan, The sulfur cycle, in *The sea*, Vol. 5, *Marine chemistry*, ed E. D. Goldberg. Copyright © 1974 by John Wiley & Sons, Inc. Reprinted with permission of John Wiley & Sons, Inc.

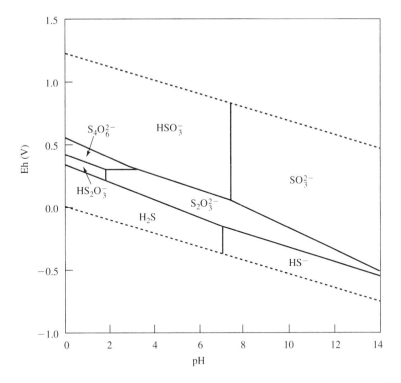

Figure 12.18   Eh-pH diagram for aqueous species in the system S-O$_2$-H$_2$O at 25°C neglecting sulfate, bisulfate, and elemental sulfur; showing significant stability fields for sulfite and bisulfite, thiosulfate and its conjugate acid, and tetrathionate. Reprinted from *Geochim. et Cosmochim. Acta*, 56, M. A. Williamson and J. D. Rimstidt, Correlation between structure and thermodynamic properties of aqueous sulfur species, 3867–80, © 1992, with permission from Elsevier Science Ltd., The Boulevard, Langford Lane, Kidlington OX5 1GB, U.K.

Fig. 12.18, which also ignores elemental sulfur. It shows that bisulfite and sulfite ions are the major metastable species in oxidized environments. Predominant in a narrow range of intermediate Eh values with increasing pH, are tetrathionate (S$_4$O$_6^{2-}$) and the thiosulfate species. Hydrogen sulfide and HS$^-$ ion have roughly the same stability range as shown in Fig. 12.16.

12.3 IRON-SULFUR REDOX CHEMISTRY

12.3.1 Occurrence and Solubility of Fe(II)-Sulfide Minerals

Pyrite (FeS$_2$) is by far the most abundant sulfide mineral, occurring in most types of geologic formations. Its less common polymorph, marcasite, usually forms in near-surface, low-temperature environments. At 25°C pyrite is more stable than marcasite by about −0.4 kcal/mol. The oxidative breakdown of these minerals as the result of exposure to aerobic conditions due to mining is the chief cause of acid mine waters.

　　　Pyrite and marcasite do not nucleate and precipitate directly from solution, but result from the successive sulfidation of a series of metastable Fe(II) sulfides. The experimental results of Schoonen and Barnes (1991a, 1991b) suggest that the initial amorphous FeS is precipitated by the reaction of dissolved Fe(II) with hydrogen sulfide

$$Fe^{2+} + H_2S(aq) \rightarrow FeS + 2H^+ \tag{12.25}$$

The subsequent sequence of sulfides leading to pyrite/marcasite is

$$FeS(am)(FeS_{0.90} \text{ to } FeS_{0.92}) \rightarrow \text{mackinawite } (FeS_{0.93} \text{ to } FeS_{0.96}) \tag{12.26}$$

$$\rightarrow \text{greigite } (Fe_3S_4) \rightarrow \text{pyrite/marcasite}(FeS_2)$$

The process involves greigite if conditions are slightly oxidizing, although greigite may be absent under highly anaerobic conditions. Sulfur present as sulfate, hydrogen sulfide, or bisulfide is probably little involved in the sulfidation process, as rates of sulfidation are very slow when these species provide the sulfur. If metastable sulfur species are available as reactants, however, the transformation to marcasite or pyrite is orders of magnitude faster. Overall FeS_2-forming reactions that involve polysulfides, thiosufate, and polythionate ions may be written

$$FeS + S_n^{2-} \rightarrow FeS_2 + S_{n-1}^{2-} \tag{12.27a}$$

$$FeS + S_2O_3^{2-} \rightarrow FeS_2 + SO_3^{2-} \tag{12.27b}$$

$$FeS + S_nO_6^{2-} \rightarrow FeS_2 + S_{n-1}O_6^{2-} \tag{12.27c}$$

Marcasite forms below about pH 5, when the neutral, undissociated polysulfide acids dominate, whereas pyrite is favored at pH values above 6, when polysulfide anions are the major species (Fig. 12.15). Schoonen and Barnes (1991a) offer detailed explanations for such complex behavior.

Because of the poorly known value for $K_2(H_2S)$, it has become traditional to write solubility product expressions for the Fe(II) sulfides in terms of bisulfide ion and elemental sulfur instead of S^{2-}. Assuming for simplicity that the amorphous sulfide and mackinawite are 1:1 solids, their dissolution expressions are:

$$FeS + H^+ = Fe^{2+} + HS^- \tag{12.28}$$

The dissolution of griegite and pyrite and marcasite may be written

$$\tfrac{1}{3}Fe_3S_4 + H^+ = Fe^{2+} + HS^- + \tfrac{1}{3}S^\circ \tag{12.29a}$$

and $$FeS_2 + H^+ = Fe^{2+} + HS^- + S^\circ \tag{12.29b}$$

All of these reactions have the same equilibrium constant expression

$$K_{eq} = \frac{[Fe^{2+}][HS^-]}{[H^+]} \tag{12.30}$$

which makes it easy to compare their solubilities. Free-energy data for the amorphous phase, mackinawite, and greigite in Table A12.1 have been recomputed from ΔG_f° data given by Berner (1971), revised to be consistent with the free energy of Fe^{2+} in the table. K_{eq} values based on these free energies are summarized here.

Sulfide	pK_{eq}
FeS(am)	2.96
FeS(mackinawite)	3.69
Fe$_3$S$_4$(griegite)	4.28
FeS$_2$(pyrite)	16.38
FeS$_2$(marcasite)	16.05

Morse et al. (1987) also recomputed these constants based largely on Berner's data. Their results are identical to those listed above, except for mackinawite, for which they give $pK_{eq} = 3.55$. Bagander and Carman (1994) measured the *in situ* solubility of amorphous FeS in near-shore bottom sediments of the Baltic Sea between 13 and 19°C. Corrected to zero ionic strength and 25°C their measurements lead to $pK_{eq} = 2.55$ for FeS(am) in reasonable agreement with the value given above.

Example 12.6

All of the Fe(II) sulfides have extremely low solubilities. Assuming pH = 8.0, and [HS$^-$] = $10^{-4.0}$ mol/kg, compute [Fe^{2+}] at equilibrium with the most soluble sulfide, FeS(am), given that $pK_{eq} = 2.96$. Substituting into Eq. (12.30) we obtain [Fe^{2+}] = 1.10×10^{-8} mol/kg, or about 0.6 μg/kg, well below a Fe(aq) detection limit of about 50 μg/kg.

FeS(am) may age and crystallize to form pyrite. Compute [Fe^{2+}] in equilibrium with pyrite for the same conditions considered above, but also assuming that elemental sulfur is present. Substituting into Eq. (12.30) leads to the incredibly low Fe^{2+} activity of 4.2×10^{-21} mol/kg at equilibrium with pyrite.

Sedimentary Fe(III) oxyhydroxides, of which goethite is the most common, are the source of iron for most precipitated Fe(II) sulfides. Pyzik and Sommer (1981) (see also Morse et al. 1987) have proposed a number of possible reaction mechanisms for the reduction of goethite to FeS. All involve initial reactant HS$^-$, with production of Fe^{2+}, and S°, polysulfide ions or thousulfate. These species then react to form the Fe(II) sulfides. Their initiating reactions are

$$2\alpha\text{-FeOOH} + 5\text{H}^+ + \text{HS}^- \rightarrow 2\text{Fe}^{2+} + \text{S}^\circ + 4\text{H}_2\text{O} \tag{12.31a}$$

$$3\alpha\text{-FeOOH} + 7\text{H}^+ + 2\text{HS}^- \rightarrow \tfrac{1}{2}\text{S}_4^{2-} + 3\text{Fe}^{2+} + 6\text{H}_2\text{O} \tag{12.31b}$$

$$8\alpha\text{-FeOOH} + 19\text{H}^+ + 5\text{HS}^- \rightarrow \text{S}_5^{2-} + 8\text{Fe}^{2+} + 16\text{H}_2\text{O} \tag{12.31c}$$

$$8\alpha\text{-FeOOH} + 16\text{H}^+ + 2\text{HS}^- \rightarrow \text{S}_2\text{O}_3^{2-} + 8\text{Fe}^{2+} + 13\text{H}_2\text{O} \tag{12.31d}$$

The overall reaction involving goethite may be written

$$2\alpha\text{-FeOOH} + 3\text{HS}^- + 3\text{H}^+ = 2\text{FeS} + \text{S}^\circ + 4\text{H}_2\text{O} \tag{12.32}$$

for which we compute from the free-energy data [H$^+$][HS$^-$] = $10^{-9.94}$. At pH = 8.0, [HS$^-$] = 1.15×10^{-2} mol/kg or 390 mg/kg as H$_2$S, an unusually high concentration. If instead the oxyhydroxide is relatively amorphous with a pK_{sp} of 38.5 and the reaction is

$$2\text{Fe(OH)}_3 + 3\text{HS}^- + 3\text{H}^+ = 2\text{FeS} + \text{S}^\circ + 6\text{H}_2\text{O} \tag{12.33}$$

we find [H$^+$][HS$^-$] = $10^{-13.67}$, and at pH = 8.0. [HS$^-$] = 2.14×10^{-6} mol/kg or 0.073 mg/kg as H$_2$S. These calculations indicate that high sulfide concentrations are required to convert crystalline goethite to FeS(am). In contrast, when amorphous HFO is the reactant, trace amounts of sulfide will transform it to FeS(am).

12.3.2 Eh-pH Relationships in the System Fe-O$_2$-CO$_2$-S-H$_2$O

An Eh-pH diagram for the Fe-O$_2$-CO$_2$-S-H$_2$O system at 25°C is given in Fig. 12.19. Comparison with Eh-pH diagrams for simpler systems in Figs. 12.10, 12.11, and 12.16 shows notable differences. Most obvious is the stability field for pyrite, which lies roughly over fields for aqueous H$_2$S and

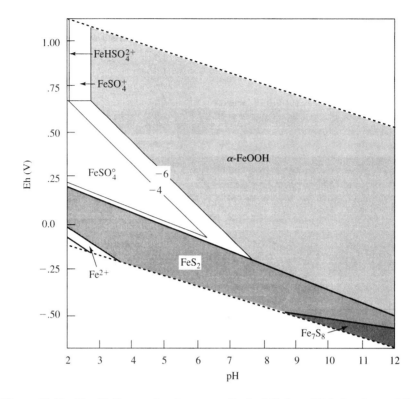

Figure 12.19 Eh-pH diagram for the system Fe-O$_2$-S-H$_2$O at 25°C showing stability fields of goethite (α-FeOOH), pyrite (FeS$_2$), and monoclinic pyrrhotite (Fe$_7$S$_8$ = Fe$_{0.877}$S) for ΣS(aq) = 10^{-2} mol/kg, and total carbonate 10^{-4} mol/kg. ΣFe(aq) = 10^{-6} and 10^{-4} mol/kg at aqueous/ solid boundaries. The diagram shows that aqueous iron occurs chiefly in sulfate complexes. From Barnes and Langmuir (1979).

HS$^-$. The lack of a stability field for siderite reflects the low carbonate and high sulfur concentrations chosen for the diagram, and the large sizes of both pyrite and goethite fields are consistent with the remarkable insolubilities of both minerals over a wide pH range.

In Example 12.5 it was shown how preferential hydroxyl and fluoride complexing of Fe^{3+} relative to Fe^{2+} enlarged the stability field of ferric iron toward lower Eh values at the expense of Fe^{2+}. Figure 12.19 shows the same effect caused by preferential sulfate complexing of Fe^{3+} over Fe^{2+}.

Environmentally, the most important boundary in Fig. 12.19 is the upper pyrite boundary with aqueous ferrous iron. The redox reaction is

$$Fe^{2+} + 2SO_4^{2-} + 16H^+ + 14e^- = FeS_2 + 8H_2O \qquad (12.34)$$

for which $E° = 0.355$ V, and the general redox equation is

$$Eh(V) = 0.355 + 0.00423 \log [Fe^{2+}][SO_4^{2-}]^2[H^+]^{16} \qquad (12.35)$$

To plot this equation we must fix the ΣS(aq) concentration and that of ΣFe(aq) at boundaries with Fe solids. We will assume ΣS(aq) = 10^{-2} mol/kg, and ΣFe(aq) = 10^{-4} mol/kg. With these substitutions the boundary equation is

$$Eh(V) = 0.321 - 0.676 \, pH \qquad (12.36)$$

For comparison purposes, it is instructive to derive the similar boundary for freshly precipitated, metastable FeS(am). The reaction is

$$Fe^{2+} + SO_4^{2-} + 8H^+ + 8e^- = FeS(am) + 4H_2O \qquad (12.37)$$

for which $E° = 0.271$ V, and the general Eh-pH equation is

$$Eh(V) = 0.271 + 0.00741 \log [Fe^{2+}][SO_4^{2-}][H^+]^8 \qquad (12.38)$$

With the same assumed $\Sigma S(aq)$ and $\Sigma Fe(aq)$ concentrations we have

$$Eh(V) = 0.227 - 0.0592 \text{ pH}$$

Substitution shows that at pH = 7 the $Fe^{2+}/FeS(am)$ boundary is 35 mV below that of Fe^{2+}/FeS_2 and within the pyrite field.

12.3.3 Acid Mine Waters

General controls on acid mine water production.   That pyrite or marcasite occur in coal, lignite, or metal sulfide deposits does not predetermine that mining of the rock will produce significant acid mine drainage (AMD). The amount and nature of the pyrite present is an important factor, however. Coal and lignite deposits that formed in ancient freshwater environments are generally low in sulfur. However, when these deposits were laid down in estuarine or marine environments they often contain several percent of Fe(II) sulfide sulfur. (The sulfate in seawater is 2700 mg/kg.) Most coals also contain organically-bound sulfur in amounts usually exceeding that present as FeS_2. However, the organic sulfur does not contribute to the acidity of mine waters (Casagrande et al. 1990). In the United States, coals in the Appalachian region (e.g., in Pennsylvania and West Virginia) are mostly of marine origin and high in sulfur, except in some shallow horizons where coal formation was lacustrine (associated with lakes). Coals in western states (e.g., New Mexico, Colorado, and Wyoming) are generally of freshwater origin, and so low in sulfur and of little concern as AMD producers (cf. McWhorter et al. 1974).

A second important factor is the particle size of the FeS_2. Caruccio et al. (1976) concluded that AMD production was a serious problem when exposed pyrite grains were about 0.25 μm in diameter in "framboidal" pyrite. This corresponds to a surface area of 4.8 m²/g. For crystal sizes in excess of 5 to 10 μm (5 μm ≡ 0.24 m²/g), AMD production was greatly diminished, and became unlikely when crystal sizes exceeded 400 μm. Particle sizes were smallest in marine coals and became larger in lacustrine coals (Caruccio and Ferm 1974). A third factor determining the risk of AMD is the proximity of carbonate rocks that can neutralize AMD acidity.

Usually base and precious metal sulfide deposits also have important amounts of pyrite. The pyrite in hydrothermal ore deposits is most often coarse grained and relatively unreactive. Mining and milling the rock to fine particle sizes for the purpose of metal extraction, vastly increases pyrite surface area and exposes the pyrite in waste-tailings piles to oxidation and weathering. Serious AMD releases can result.

Of basic importance when determining risk is the absolute and relative amounts of the exposed pyrite that occur above or below the water table. Above the water table, oxidation rates are usually catalyzed by bacteria and can be very fast in fine-grained waste-rock piles. Rates in water-saturated spoil materials or below the water table are generally much slower because of the unavailability of free oxygen.

Overall FeS$_2$ and Fe(II) oxidation reactions.   Oxidation of pyrite first occurs at its exposed surface, where pyritic sulfur with a formal charge of (−1) is oxidized to sulfate (+6) and the ferrous iron is released to solution. The stoichiometry of the oxidation half reaction (Eq. [12.34]) suggests that a mole of pyrite should produce 16 protons. However, only the complete oxidation reaction, which must include an oxidizing agent, can provide us with a valid reaction stoichiometry. Two possible overall redox reactions are

$$FeS_2 + \tfrac{7}{2}O_2 + H_2O \rightarrow Fe^{2+} + 2SO_4^{2-} + 2H^+ \tag{12.39a}$$

$$FeS_2 + 14Fe^{3+} + 8H_2O \rightarrow 15Fe^{2+} + 2SO_4^{2-} + 16H^+ \tag{12.39b}$$

Thus, if O_2 is the oxidizing agent, 2 moles of H^+ are produced by 1 mole of pyrite. In contrast, oxidation by Fe^{3+} releases 16 moles of protons for each mole of pyrite. Obviously, it is essential to know which reaction dominates in any attempt to predict the acid-generating capacity of a system.

The next step is the oxidation of ferrous to ferric iron. The rate of the Fe^{3+}/pyrite oxidation reaction is, in fact, limited by the availability of Fe^{3+}, which itself depends on the Fe(II) oxidation rate. Below about pH 3 the overall Fe(II) oxidation reaction is

$$Fe^{2+} + \tfrac{1}{4}O_2 + H^+ \rightarrow Fe^{3+} + \tfrac{1}{2}H_2O \tag{12.40}$$

The abiotic rate of this reaction is slow. The pH of streams and groundwaters that have moved away from source pyrite surfaces increases because of partial neutralization by carbonate and silicate minerals. Above roughly pH 3, Fe^{2+} oxidation is then described by the overall reaction

$$Fe^{2+} + \tfrac{1}{4}O_2 + \tfrac{5}{2}H_2O \rightarrow Fe(OH)_3 + 2H^+ \tag{12.41}$$

which has a rate that increases rapidly with increasing pH. (See kinetics Chap. 2.)

Abiotic oxidation kinetics of sulfur and Fe(II).   The abiotic oxidation rates of pyrite and ferrous iron have been studied extensively for more than 30 years (cf. Garrels and Thompson 1960; McKibben and Barnes 1986; Nicholson et al. 1988; Moses and Herman 1991). The work on pyrite-oxidation kinetics has been updated and critiqued by Williamson and Rimstidt (1994), while Wehrli (1990) has synthesized the published data on Fe^{2+} oxidation. (See also King et al. 1995.)

Based on statistical analysis of published FeS_2 oxidation-rate data and their own work, Williamson and Rimstidt (1994) derived three empirical rate law expressions, where $r = d(FeS_2)/dt$ is the rate of pyrite destruction in mol/m^2 s. From pH 2 to 10 when oxygen is the only oxidant (as perhaps above the water table in spoil materials or coarse-grained tailings) the rate law is

$$r = (10^{-8.10})\frac{(mO_2)^{0.5}}{(mH^+)^{0.11}} \tag{12.42}$$

Between pH 0.5 to 3.0, when Fe^{3+} is the oxidant and oxygen is absent (conditions that might be expected in an acid groundwater or in flooded-mine pools), the rate equation is

$$r = (10^{-8.58})\frac{(mFe^{3+})^{0.30}}{(mFe^{2+})^{0.47}\,(mH^+)^{0.32}} \tag{12.43}$$

In the same acid pH range, when $O_2(aq)$ is present along with Fe^{3+}, the rate is given by

$$r = (10^{-6.07})\frac{(mFe^{3+})^{0.93}}{(mFe^{2+})^{0.40}} \tag{12.44}$$

The latter conditions might be found near a fluctuating water table or within the capillary zone above the water table. The success of Williamson and Rimstidt's model fits to published experimental data is evident from Fig. 12.20. Regardless of the oxidant, the pyrite breakdown rate is seen to be practi-

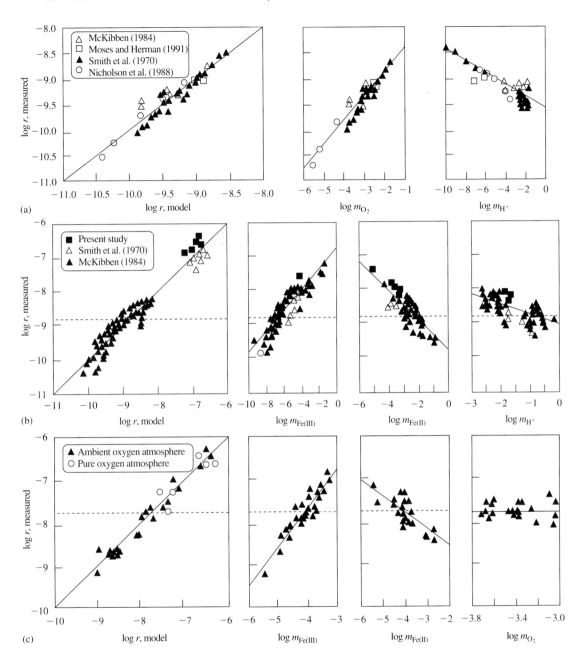

Figure 12.20   The rate of oxidation of pyrite ($r = d[\text{FeS}_2]/dt$ in mol/m^2 s) near 25°C and 1 bar pressure. Whole model and leverage plots for multiple linear regression analysis of published and measured rate data for the aqueous oxidation of pyrite: **(a)** Oxidation of pyrite by dissolved oxygen; **(b)** Oxidation of pyrite by ferric iron under an N$_2$ atmosphere; and **(c)** Oxidation of pyrite by ferric iron in the presence of dissolved oxygen. Reprinted from *Geochim. et Cosmochim. Acta*, 58, M. A. Williamson and J. D. Rimstidt, The kinetics and electrochemical rate-determining step of aqueous pyrite oxidation, 5443–54, © 1994, with permission from Elsevier Science Ltd., The Boulevard, Langford Lane, Kidlington OX5 1GB, U.K.

TABLE 12.5 Example calculations of the half-time ($t_{1/2}$) for oxidative destruction of pyrite at 25°C

Oxidants	Rate Equations and Their Applicability		
		pH range	Rate equation
O_2		2 to 10	(12.42)
Fe^{3+}		0.5 to 3.0	(12.43)
Fe^{3+} and O_2		0.5 to 3.0	(12.44)

Oxidants	Assumed Conditions and Predicted Half-times					
	pH	$\log mO_2(aq)$	$\log \Sigma mFe(aq)$	$\log mFe^{3+}$	$\log mFe^{2+}$	$t_{1/2}$
O_2	2.0	−7.0	—	—	—	780 y
		−3.6	—	—	—	16 y
Fe^{3+}	2.0	—	−2.0	−2.0	−4.0	4.4 d
		—	−2.0	−4.0	−2.0	150 d
Fe^{3+} and O_2	—	—	−2.0	−2.0	−4.0	2.1 d
		—	−2.0	−4.0	−2.0	2.6 y

Note: Computed using the empirical rate laws of Williamson and Rimstidt (1994), given an intial pyrite surface area of 0.05 m²/g. Other assumed conditions are given in the table.

cally independent of sulfate, chloride, or ionic strength variations. In spite of this excellent work, elementary reactions to describe the mechanisms of pyrite oxidation remain unknown.

It is instructive to compare pyrite oxidation half-times for some reasonable environmental conditions. Results of such a comparison are summarized in Table 12.5. Oxygen concentrations of $10^{-7.0}$ and $10^{-3.6}$ mol/kg used in the exercise are roughly the O_2(aq) detection limit (~5 μg/kg) and the air-saturated value (8 mg/kg). In order to integrate the rate equations, DO, total iron, and ferric and ferrous ion concentrations have been fixed. After substitution the rate equations all become pseudo zero-order with half-times given by $t_{1/2}(s) = 0.5A_o/k$, where k equals all terms on the right-hand side of Eqs. (12.42) to (12.44). (See Table 2.2.) The surface area of pyrite has been arbitrarily set equal to 0.05 m²/g, the value Williamson and Rimstidt (1994) measured and used in their own rate experiments. Results in Table 12.5 indicate that abiotic pyrite oxidation takes years when oxygen is the only oxidant. The rates are much faster and similar when ferric iron alone and ferric iron plus oxygen are present, with the relative rates of these two reactions depending on the Fe^{3+}/Fe^{2+} ratio. Thus oxygen accelerates the rate at high Fe^{3+}/Fe^{2+} ratios (high Eh), but slows it when this ratio is low (at low Eh).

Appearance of the Fe^{3+}/Fe^{2+} ratio in Eqs. (12.43) and (12.44) indicates that the rates are a function of the redox potential. Williamson and Rimstidt found that this ratio was thermodynamically consistent with solution Eh measurements. When both Fe^{3+} and DO are present, regression of their rate data leads to

$$r = 10^{-6.37(\pm 0.4)} \, Eh^{23.6(\pm 1.8)} \, pH^{4.38(\pm 1.2)} \tag{12.45}$$

(Williamson 1996). When DO is absent and Fe^{3+} is the only oxidant, the regression relationship is

$$r = 10^{-6.71(\pm 0.08)} \, Eh^{7.96(\pm 0.2)} \, pH^{1.06(\pm 0.1)} \tag{12.46}$$

(Williamson 1996), where r is in mol/m s, and Eh is in volts. The Eh-dependence of the rate suggests an electrochemical reaction mechanism rather than one involving site-specific adsorption of oxidants. That the log r versus log mO_2 plot in Fig 12.20 is linear for a wide range of O_2 concentrations shows that the rate is independent of the extent of pyrite surface coverage by O_2. This also argues for an electrochemical oxidative reaction, a reaction that is not surface-site specific (Williamson and Rimstidt 1994).

Recent work on Fe(II) oxidation has identified the elementary reactions involved (cf. King et al. 1995). Abiotic oxidation rates based on this recent work should differ little from rates computed from earlier empirical rate laws, particularly below about pH 6 (cf. Wehrli 1990; Stumm and Morgan 1996). The empirical rate data (see Fig. 2.9) indicate pH-independent rate behavior below about pH 3.5, and a rapid increase in the rate with pH above this value. The pH-independent rate is given by

$$\frac{d[\text{mFe(II)aq}]}{dt} = -k_1 [\text{mFe(II)aq}] P_{O_2} \tag{12.47}$$

where $k_1 = 10^{-3.2}$ bar/d. For $P_{O_2} = 0.21$ bar in air, $t_{1/2} = 15$ y. Abiotic oxidation of Fe(II)(aq) in acid waters is obviously very slow.

Above roughly pH 3.5 to 4, the empirical rate law for Fe(II)(aq) oxidation is

$$\frac{d[\text{mFe(II)aq}]}{dt} = -k_2 \frac{[\text{mFe(II)aq}] P_{O_2}}{[\text{H}^+]^2} \tag{12.48}$$

for which $k_2 = 1.2 \times 10^{-11}$ mol^2/d bar. At pH = 6.0 and $P_{O_2} = 0.21$ bar, the rate becomes pseudo first-order and $t_{1/2} = 7$ h ($t_{1/2} = 4.2$ min at pH = 7.0). The abiotic rate is thus greatly accelerated by an increase in pH.

The role of sulfur- and iron-oxidizing bacteria.   As already noted, the rates of FeS$_2$ and Fe(II) oxidation in environmental systems often differ substantially from the abiotic rates. Usually natural rates are much faster than laboratory abiotic rates. The reasons include inorganic catalysis and especially enzymatic oxidation by microorganisms. Oxidation of Fe(II), for example, is catalyzed by some clays and metals, including Al^{3+}, Fe^{3+}, Co^{2+}, Cu^{2+}, and Mn^{2+}, and also HPO_4^{2-} (Stumm and Morgan 1981).

Sulfur- and iron-oxidizing bacteria flourish at the oxidized side of redox interfaces. They are important especially because they catalyze and thus greatly accelerate reactions that are thermodynamically favored, but may be abiotically slow. The sulfur oxidizing Thiobacteria are all aerobic and autotrophs (they obtain carbon from carbonate species). Thiobacteria can oxidize the sulfur in sulfides, proteins, or elemental sulfur, producing sulfate and acidity. Among the individual strains are *Thiobacillus thiooxidans, Ferrobacillus ferrooxidans,* and *Thiobacillus ferrooxidans* (*T. ferrooxidans*). Some of these are filamentous and photosynthetic; others are simple bacteria. They can grow in and produce acidities as great as 10% H_2SO_4 (pH = 0) (cf. Manahan 1994). Karamenko (1969) noted that 200 times or more iron is dissolved from pyrite when these bacteria are present versus when they are not. Baas Becking et al. (1969) reported that when pyrite was placed in pure water of pH 6.2, after 24 hours without bacteria the pH was 3.1 and Eh = +650 mV. However, when Thiobacteria were present the pH dropped below 2 and the Eh rose to +860 mV. Because they require oxygen, Thiobacteria are most important to sulfur oxidation above the groundwater table, as in mine tailings and mine-spoil materials. They catalyze sulfur oxidation particularly where oxygen is the chief oxidant, so that abiotic rates would be very slow (Table 12.5).

All iron oxidizing bacteria are aerobic and are either *autotrophs* (get C from carbonate species), *heterotrophs* (need organic C), or *facultative* (get C from either source). Specific bacteria are: autotrophs, *Gallionella*; heterotrophs, *Sphaerotilus, Siderocapsa* and *L. Crassa*; facultatives, *Crenothrix, Polyspora* and some *Leptothrix*. Their favored pH range is 5 to 8, the same range in which inorganic rates are also fast and the oxidation product is precipitated HFO. Many of the bacteria are filamentous and can accumulate up to 500 times their cell weight in precipitated HFO. (They also oxidize Mn^{2+} to MnO_2.) The Fe(II)-oxidizing bacteria prefer Fe(II) > 0.1 mg/kg and operate with the lowest Fe(II) levels in moving water, such as in acid mine streams and pumping wells. Iron bacteria are common in stagnant water, mines, springs, quiet parts of streams, marshes and lagoons, reservoirs and water pipes, well screens and well casings. In other words, they are found wherever they can access Fe^{2+} at the oxidizing side of a redox interface. They are especially a problem in fouling of iron pipes in water supply systems and well screens. They can cause a loss of up to 90% in the productivity of a well (cf. Starkey 1945; Kleinmann et al. 1979; Hackett 1987). Under acid conditions the iron-oxidizing bacteria can speed up Fe(II) oxidation by up to 10^6 times (cf. Taylor et al. 1984a). Thus for the example calculation above, $t_{1/2}$ = 15 y for abiotic oxidation, but only 8 min when catalyzed by iron-oxidizing bacteria.

In a related study, Kim (1968) measured the rate of Fe(II) oxidation in some acid mine water that contained natural levels of iron-oxidizing bacteria in an experiment open to the atmosphere at pH 3.30. Some of her results are plotted in Fig. 12.21, which shows a reaction half-time of about 24 h, or an oxidation rate about 5500 times faster than the abiotic rate.

Sulfur and oxygen isotopes. Stable sulfur and oxygen isotopes can provide clues regarding the relative importances of O_2 versus Fe(III) oxidation of pyrite, and whether the oxidation is abiotic (sterile) or involves bacteria (Taylor et al. 1984a, 1984b). The isotopic data are reported in δ (per mil or ‰) units, where

$$\delta^{18}O \quad \text{or} \quad \delta^{34}S(\text{‰}) = \left(\frac{R_{\text{sample}} - R_{\text{std}}}{R_{\text{std}}}\right) \times 10^3 \tag{12.49}$$

and $R = {}^{18}O/{}^{16}O$ or ${}^{34}S/{}^{32}S$ of samples and standards (cf. Faure 1991). Reactions or processes often cause changes in $\delta^{18}O$ or $\delta^{34}S$ values, so that the heavier or the lighter isotope is favored in reaction products. Such a change in R is termed isotopic fractionation. Fractionation is minimized when reactions are fast. The nearly identical $\delta^{34}S$ content of sulfides and oxidation product SO_4^{2-} (i.e., the lack of fractionation) indicates that sulfide oxidation to sulfate is fast and that intermediate sulfur species (e.g. SO_3^{2-} and $S_2O_3^{2-}$) are transient and minor in amount.

Based on their experiments and field analyses, Taylor et al. (1984a, 1984b) proposed that the $\delta^{18}O$ composition of SO_4^{2-} could be used to determine dominant local mechanisms of pyrite oxida-

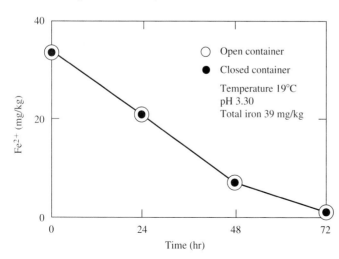

Figure 12.21 Rate of oxidation of Fe(II)(aq) in a natural sample of acid mine water that contained natural levels of iron-oxidizing bacteria. Presumably the water was initially saturated with atmospheric oxygen with DO = 9.5 mg/kg. Modified after G. A. Kim (1968). Copyright 1968 by the National Coal Association. Used by permission.

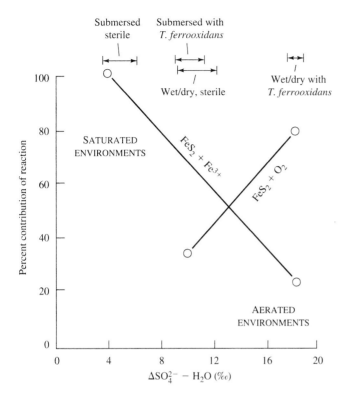

Figure 12.22   Percent contributions to overall pyrite oxidation, of FeS_2 oxidation by O_2 (Eq. [12.39a]), and by Fe^{3+} (Eq. [12.39b]), plotted against $[\Delta SO_4^{2-} - H_2O(‰)]$, the expected oxygen isotope fractionation between sulphate and water. End-member points (open circles) are plotted for extreme fractionations measured in the following conditions: anaerobic, sterile, and aerobic with *T. ferrooxidans* present. Bars denote the ranges of values measured in laboratory experiments. From *Nature* 308: 539–41, E. B. Taylor et al., Isotope composition of sulphate in acid mine drainage as measure of bacterial oxidation, copyright 1984 by Macmillan Magazines, Ltd. Used by permission.

tion. The stoichiometry of overall oxidation reactions in Eqs. (12.39a) and (12.39b) shows that when O_2 is the oxidant, 87.5% of the oxygen in product sulfate comes from the O_2, and 12.5% from the water. When Fe^{3+} is the oxidant, 100% of sulfate oxygen comes from the water. Atmospheric oxygen has an $\delta^{18}O$ of +23.8‰, whereas in surface water $\delta^{18}O$ values are generally negative.

It is instructive to calculate $\Delta SO_4^{2-} - H_2O(‰)$, which is the difference between the $\delta^{18}O$ values of SO_4^{2-} and H_2O in mine water (Fig. 12.22). In the absence of isotope fractionation and with Fe^{3+} as the oxidant, a zero difference would be expected. Taylor et al. (1984b) propose that Fe^{3+} dominates pyrite oxidation in submersed environments. For such conditions $\Delta SO_4^{2-} - H_2O(‰)$ values exceed zero due to isotope fractionation, ranging from about 4 to 6 in sterile systems to about 11 in the presence of *T. ferrooxidans*. Fractionation is caused by the greater reactivity of light ^{16}O than heavy ^{18}O, with ^{16}O thus favored in product sulfate. Bacterial activity (in this case *T. ferrooxidans*) causes more fractionation than takes place in the abiotic oxidation reaction. $\Delta SO_4^{2-} - H_2O(‰)$ values range up to about 18 when O_2 is the chief oxidant in the presence of bacteria. Such conditions are expected in aerated streams and in shallow unsaturated materials (Taylor et al. 1984a, 1984b).

Example 12.7

Assume no isotopic fractionation during pyrite oxidation in a periodically wet tailings pile, that oxidation is 20% by Fe^{3+} and 80% by O_2 (see Fig. 12.22), and that $\delta^{18}O = -10.9‰$ in H_2O and +23.0‰ in O_2. (a) Compute $\delta^{18}O$ for the sulfate. (b) Compute the difference $\Delta SO_4^{2-} - H_2O(‰)$. (c) How would isotopic fractionation by *T. ferrooxidans* change the values obtained in (a) and (b) and why?

Solution

(a) $\delta^{18}O$ for the sulfate may be determined from the stoichiometries of the ferric ion/pyrite and O_2/pyrite reactions noted above. The equation is

$$\delta^{18}O(SO_4^{2-}) = 0.2(-10.9) + 0.8 \times 0.125(-10.9) + 0.8 \times 0.875(+23) = 12.8\%_0$$

(b) The difference $\Delta SO_4^{2-} - H_2O(\%_0)$ equals $12.8 + 10.9 = 23.7\%_0$.

(c) Isotopic fractionation should enrich sulfate in ^{16}O. This reduces $\delta^{18}O$ for sulfate, which lowers $\Delta SO_4^{2-} - H_2O(\%_0)$ toward $18\%_0$, the value indicated in Fig. 12.22 from Taylor et al. (1984a) for the pyrite oxidation conditions specified in this problem.

The origin of deep acid mine waters.   Barnes et al. (1964) studied waters in three flooded coal-mine shafts in the north anthracite field of eastern Pennsylvania. Samples were obtained from depths up to 646 ft (197 m) below the water table. In the shaft pools Barnes et al. (1964) obtained the following compositional ranges: pH 3.4 to 5.4, Eh −0.10 to +0.57 V, ΣSO_4^{2-} 1260 to 6720 mg/kg, and $\Sigma Fe(II)(aq)$ 34 to 1463 mg/kg. Seven of their chemical analyses are reproduced in Table 12.6. Highest iron and sulfate concentrations were found at the greatest depths. Eh and pH values were less depth predictable. All of their Eh-pH data are listed in Table 12.6 and plotted in Fig. 12.23, which shows that several of the analyses lie at or below the Fe(II)(aq)/pyrite boundary, suggesting equilibrium with pyrite.

Sample	Mine	Sampling depth, in ft below LSD[‡]	Sample	Mine	Sampling depth, in ft below LSD
1	Loree 2	$\left\{\begin{array}{l}224\\223\end{array}\right.$	5	Loree 2	808
[†]2	Loree 2	$\left\{\begin{array}{l}292\\438\end{array}\right.$	6	Storrs 1	$\left\{\begin{array}{l}300\\363\\423\end{array}\right.$
[†]3	Loree 2	$\left\{\begin{array}{l}467\\579\end{array}\right.$	7	Storrs 1	550
4	Loree 2	751	8	South Wilkes-Barre 5	452

[†]All samples turbid. [‡]LSD denotes land surface datum.

Others plot along the Fe(OH)$_3$(am)/Fe(II)(aq) boundary, consistent with their turbid yellow appearance, presumably due to suspended HFO.

The high iron concentrations are difficult to explain. The stoichiometry of the O_2/FeS$_2$ oxidation reaction (Eq. 12.39a) indicates that 3.5 moles of O_2 are consumed to produce a mole of Fe^{2+}. Given the atmospheric O_2 solubility of 8.4 mg/kg at 25°C, for example, oxidation of pyrite by O_2 can produce a Fe^{2+} concentration of only 4.2 mg/kg. The low Eh values in the pools make it doubtful that measureable DO is present in any case. Based on the foregoing discussion of sulfur and oxygen isotopes and bacterial oxidation, pyrite oxidation in the pools is probably largely abiotic with ferric iron as the oxidant. Oxidation rates must be slow, given the fact that Fe(III) << Fe(II) at these Eh values.

If ferric iron is the oxidant, the thorny question is how to get oxidant Fe(III) deep in the relatively stagnant mine pools. One possibility is the flushing of particulate HFO materials into the pools by storm runoff, surface-waters, and actively flowing groundwaters, with the HFO solids then physically settling into the pools and dissolving to given Fe(III)(aq) at depth. Another explanation may relate to the development of an electrochemical cell involving the coal, pyrite, and groundwater

TABLE 12.6 Chemical analyses of waters from some flooded anthracite mine shafts

Depth (ft)	$T°C$	Eh (V)	pH	Color	Gas	Odor	Fe^{2+}	SO_4^{2-}	Mn^{2+}	Al^{3+}	Ca^{2+}	Mg^{2+}	SiO_2	Na^+	Cl^-
224L	14	+.322	3.41	clear	slightly effervescent	none	34	1260	12	7	180	148	14	—	—
292L	17	+.533	3.55	yellow turbid	very effervescent	none	478	3320	37	79	342	292	30	<25	—
467L	17	+.568	3.36	slightly yellow turbid	no gas	none	488	3650	44	86	363	326	33	<25	—
808L	17	-.103	3.92	clear	effervescent	H_2S odor	1463	6720	92	143	456	462	85	35	11
300St	16	-.030	5.41	clear	very effervescent	noxious H_2S?	107	1340	11	0	247	145	16	—	—
550St	16	+.078	3.95	clear	no gas	none	249	2090	20	24	301	193	25	<25	—
452So	19	-.012	4.20	clear	no gas	none	839	4530	38	20	386	278	49	305	10

Note: Sampling depths are in feet below the land surface. Letters after depths denote the different shafts: L is Loree no. 2; St is Storrs no. 1; and So is South Wilkes-Barre no. 5. Depths below the land surface of the surface of water in the shafts are: Loree no. 2, 162 ft; Storrs no. 1, 227 ft; and South Wilkes-Barre no. 5, 398 ft. Concentrations are total (not free-ion) values, and are in ppm or mg/kg. Potassium was below detection (<10 mg/kg) in all samples.

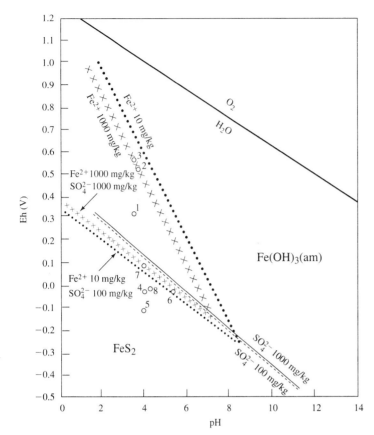

Figure 12.23   Measured Eh and pH values for waters sampled at different depths in three flooded anthracite mine shafts in eastern Pennsylvania (numbered open points). Also shown are theoretical equilibrium boundaries for $Fe^{2+}/Fe(OH)_3$(am) and Fe^{2+}/FeS_2(pyrite) reactions for Fe^{2+} = 10 and 1000 mg/kg, and SO_4^{2-} = 100 and 1000 mg/kg. From Barnes et al. (1964).

(cf. Rose et al. 1979). In this scenario the coal beds and associated pyrite are conductors that short-circuit Earth's normal electrochemical potential. The tops of the beds near the water table become enriched in electrons (negatively charged), and the deeper horizons are positively charged. Ground-water is the electrolyte completing the circuit. Such conditions would favor Fe^{3+} reduction to Fe^{2+} near the surface and, at depth, the reverse oxidation reaction $Fe^{2+} \rightarrow Fe^{3+} + e^-$.

Alternatively, Barnes et al. (1964) suggest that the reaction

$$FeS_2 + 8H_2O = Fe^{2+} + 2SO_4^{2-} + 2H^+ + 7H_2(g) \tag{12.50}$$

may be taking place deep in the mine pools. Free-energy data for this reaction at 25°C indicate

$$K_{eq} = 10^{-85.8} = [Fe^{2+}][SO_4^{2-}]^2[H^+]^2[P_{H_2}]^7 \tag{12.51}$$

To test the possible importance of reaction (12.50), chemical analyses from Table 12.6 for Loree no. 2 at 224 and 808 ft and for Storrs no. 1 at 550 ft were input into MINTEQA2. Program output included activities of the ions. Substituting their values into Eq. (12.51) we can solve for the equilibrium partial pressure of H_2. This may be compared with the apparent P_{H_2} value computed from the measured Eh and pH, via the redox couple $H^+ + e^- = \frac{1}{2}H_2(g)$. The results, which are summarized here, are consistent with the positions of the samples as plotted in Fig. 12.23.

Sample	Sample no. in Fig. 12.23	Sampling depth (ft)	Equilibrium $-\log P_{H_2}$ (bars)	Apparent $-\log P_{H_2}$ (bars)
Loree 2	1	224	10.1	17.5
Loree 2	5	808	10.3	4.3
Storrs 1	7	550	10.1	10.6

The Loree no. 2 (224 ft) sample, which has a computed hydrogen partial pressure $10^{7.4}$ bars under pyrite saturation, plots almost 200 mV above the pyrite boundary. Loree no. 2 (808 ft) lies well within the pyrite field, as it should, given its computed 10^6 bar excess in P_{H_2} over the equilibrium value. Storrs no. 1 (550 ft) plots on the Fe(II)/pyrite boundary consistent with nearly identical apparent and equilibrium hydrogen pressures.

This exercise suggests that reaction (12.50) may explain high iron and sulfate concentrations in the deep pools. Free hydrogen is very reactive and is used by bacteria to reduce Fe(III), sulfate, and carbon dioxide (cf. Lovley et al. 1994). If H_2 is continually removed from the system, pyrite should dissolve according to reaction (12.50). Information is lacking on the rate of this reaction. As is the case when Fe^{3+} is the oxidant (Eq. [12.39b]), all the oxygen in product sulfate in Eq. (12.50) comes from water. Thus, apart from isotope fractionation effects, the $\delta^{18}O$ content of both should be about the same, making oxidation via Eq. (12.50) isotopically indistinguishable from oxidation by ferric iron.

Oxidation and neutralization trends and secondary minerals.   Acid mine waters that are formed by pyrite oxidation in the saturated zone evolve very differently from those that develop in unsaturated materials above the water table. In their study of acid mine waters in bituminous coal mine areas of western Pennsylvania, Gang and Langmuir (1974) noted that the Fe(II) from pyrite oxidation remained largely unoxidized in the groundwaters (Table A12.4, samples T-1.1 through WR-W-3). The Eh and pH and Fe(II) content of the groundwaters was buffered by equilibrium with HFO solids (Fig. 12.24). Once the groundwater had discharged into streams, oxidation of the Fe(II) lowered stream pH values, making stream waters more corrosive to geological materials. At these low pH's, stream Eh and pH conditions were also buffered by equilibrium between the aqueous iron species and HFO (Fig. 12.24).

The pyrite in refuse piles, coal-storage piles, and mine tailings above the water table has usually been crushed, making it finer-grained and thus much more reactive than the FeS_2 in undisturbed rock in the saturated zone. Under such conditions, when Fe^{3+} and O_2 are readily available as oxidants, the breakdown of FeS_2 catalyzed by Thiobacteria is greatly accelerated. The result is often pH values from 0 to 2, and the accumulation of extraordinary concentrations of total iron and sulfate. For example, in column-leaching experiments involving eastern U.S. coals, Helz et al. (1987) reported that typical leachates had pH values below 2, and contained about 0.1 mol/kg total Fe (5600 mg/kg), and 0.2 mol/kg total SO_4 (about 20,000 mg/kg). (See also Anderson and Youngstrom 1976.) Similar waters develop in the pyritiferous tailings piles associated with precious and base-metal deposits. For example, Maley (1994) reported waters in sulfide tailings in Colorado with pH values from 1.3 to 2.8, up to 1300 mg/kg Cu and Mn, 18,500 mg/kg Fe, 1100 mg/kg Ca, and 152,000 mg/kg SO_4. When the coal or sulfide ore contains carbonates, pyrite oxidation is inhibited in tailings and leachates are less objectionable (cf. Table A12.4, well 201 analysis; and McWhorter et al. 1974).

High aluminum (and silica) concentrations in acid streams and in sulfidic-waste tailings and coal-storage piles derive from the weathering of aluminosilicate minerals. Kaolinite clay, for example, is common in the underclays beneath Pennsylvania coal beds (Gang and Langmuir 1974). Above

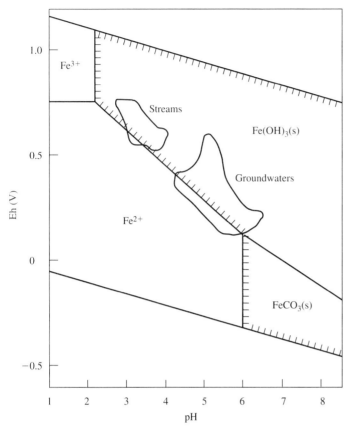

Figure 12.24   Eh-pH diagram showing the stability field of $Fe(OH)_3(s)$ assuming $pK_{sp} = 38.5$, and siderite, $FeCO_3$, at $10°C$. Mineral-solution boundaries are drawn for a dissolved ΣFe activity of $10^{-2.75}$ mol/kg (100 mg/kg as Fe), and a total alkalinity of $10^{-3.40}$ mol/kg (24 mg/kg as HCO_3^-). Field-measured Eh and pH values for 16 streams and 19 groundwaters from coal strip-mined areas of northwestern Pennsylvania are circled. From M.A. Gang and D. Langmuir (1974). Copyright 1974 by the National Coal Association. Used by permission.

roughly pH 4.6 maximum Al concentrations from the weathering of such phases are probably limited by the solubilities of microcrystalline gibbsite and/or amorphous $Al(OH)_3$ (Nordstrom and Ball 1986) (Fig. 12.25). Dissolved Al increases further as the pH drops below pH $\approx$ 4.6. This increase, often parallel to an increase in sulfate, suggests that the Al is unreactive. However, under acid conditions Al concentrations may sometimes be limited by the solubilities of sulfate minerals such as alunite. For example, alunite dissolution via the reaction

$$KAl_3(SO_4)_2(OH)_6 + 6H^+ = K^+ + 3Al^{3+} + 2SO_4^{2-} \tag{12.52}$$

has $K_{eq} = 10^{-1.6}$. Assuming pH = 3.0 and reasonable acid mine water values of $[K^+] = 10^{-3}$ mol/kg and $[SO_4^{2-}] = 10^{-2}$ mol/kg, we obtain $[Al^{3+}] = 10^{-4.2}$ mol/kg at equilibrium, a value similar to those plotted in Fig. 12.25 for pH = 3.

A host of such secondary solids approach saturation in the groundwaters, streams, and leachates associated with the oxidation and weathering of sulfide minerals. Some of these phases are listed in Table 12.7. Alunite and jarosite form a solid solution with Fe^{3+} for Al^{3+} substitution. Written in the general form: $AB_3(XO_4)_2(OH)_6$, the A position can be occupied by large cations such as H_3O^+, K^+, Na^+, Ag^+, Rb^+, Tl^+, NH_4^+, $\frac{1}{2}Ca^{2+}$, $\frac{1}{2}Sr^{2+}$, $\frac{1}{2}Ba^{2+}$, $\frac{1}{2}Pb^{2+}$, or $\frac{1}{2}Cu^{2+}$. The B position holds eightfold coordinated Fe^{3+} or Al^{3+}, Cu^{2+} or Zn^{2+}. In the anion (XO_4), X may be S^{6+}, P^{6+}, As^{5+}, or Si^{4+} (cf. Scott 1987; Alpers et al. 1992). Pure metal-OH or hydrated sulfates are also formed by monovalent Ag and divalent Cu, Mn, Zn, Co, Pb, and UO_2 (Palache et al. 1951). Solubility products of some of the single metal jarosites are: Ag-jarosite, 99.1; $Cu_{0.5}$-jarosite, 98.3; and $Pb_{0.5}$-jarosite, 97.6.

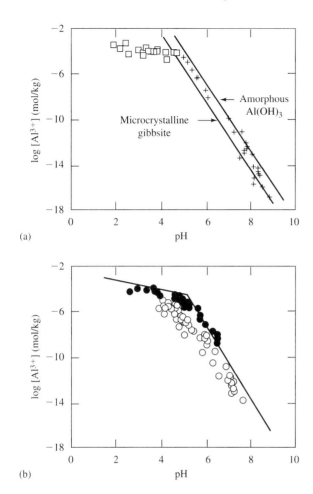

Figure 12.25 **(a)** Plot of log $[Al^{3+}]$ versus pH for 64 samples from a drainage basin affected by acid mine waters from the Leviathan mine, California-Nevada. Points shown as open squares have pH <4.6, plus symbols are those with pH >4.9. The solid lines are theoretical solubilities of amorphous $Al(OH)_3$ and microcrystalline gibbsite. **(b)** Plot similar to (a) for acid mine drainage from Appalachia (solid circles) and Adirondack lake waters affected by acid precipitation (open circles). From *Science* 232:54–56, D. K. Nordstrom and J. W. Ball, The geochemical behavior of aluminum in acidified surface waters. Copyright 1986 by Science-AAAS. Used by permission.

The insolubility of such phases and of their solid solutions makes them potential sinks for heavy metals in acid sulfate systems. This is important, in part, because such metals are poorly adsorbed by phases such as hydrous ferric oxyhydroxide (HFO) under acid conditions (see Chap. 10) and in fact the HFO itself becomes soluble below pH 2 to 3 (Fig. 12.4).

Alunite-jarosite minerals (and the other secondary sulfates) form, in part, because of evaporative concentration of pore and capillary waters in pyritic materials, and also at depth in saturated tailings (Dubrovsky et al. 1985). They are found in soils beneath acid sulfate evaporation ponds (Peterson et al. 1983), and are also precipitated directly from acid mine waters (cf. Filipek et al. 1987; Alpers et al. 1989). The alunite-jarosite mineral group also occurs in the weathered zones on top of metal sulfide deposits (Scott 1987) and in the sediments of acid hypersaline lakes (Alpers et al. 1992).

The aluminum sulfates (e.g., alunite, alunogen, basaluminite) do not significantly buffer acid pH values (see Chap. 5). However, the ferric and ferrous sulfates (e.g., coquimbite, the jarosites, melanterite, and szomolnokite) are strong acid buffers that can keep pH values at or below 3 until they are dissolved. The upper pH expected when jarosite is present probably reflects its equilibrium with practically ubiquitous HFO in such systems and the reaction

$$KFe_3(SO_4)_2(OH)_6 + 3H_2O = K^+ + 3Fe(OH)_3(ppt) + 2SO_4^{2-} + 3H^+ \qquad (12.53)$$

TABLE 12.7 Solubility products of some secondary aluminum oxyhydroxides and sulfate minerals observed resulting from the interaction of acid mine waters with geological materials

Mineral or solid phase	Formula	$-\log K_{sp}$	Source
Allophane	$[Al(OH)_3]_{1-x}[SiO_2]_x$	$5.89 - 1.59$ pH	†
Amorphous $Al(OH)_3$	$Al(OH)_3$(am)	31.2	‡
Alunite	$KAl_3(SO_4)_2(OH)_6$(am)	83.4?	§
	$KAl_3(SO_4)_2(OH)_6$(c)	85.6	
Alunogen	$Al_2(SO_4)_3 \cdot 17H_2O$	7.0	§
Anglesite	$PbSO_4$	7.76	‖
Anhydrite	$CaSO_4$	4.36	#
Aphthitalite	$NaK_3(SO_4)_2$	3.80	††
Barite	$BaSO_4$	9.97	‡‡
Basaluminite	$Al_4SO_4(OH)_{10} \cdot 5H_2O$(am)	116	§
	$Al_4SO_4(OH)_{10} \cdot 5H_2O$(c)	117.7	
Celestite	$SrSO_4$	6.62	‡‡
Coquimbite	$Fe_2(SO_4)_3 \cdot 9H_2O$	3.58	§§
Gibbsite	$Al(OH)_3$	33.9	‡
Gypsum	$CaSO_4 \cdot 2H_2O$	4.59	‖‖
Jarosites:			##
Hydronium jarosite	$(H_3O^+)Fe_3(SO_4)_2(OH)_6$	75.4	
Natrojarosite	$NaFe_3(SO_4)_2(OH)_6$	89.3	
"Jarosite"	$KFe_3(SO_4)_2(OH)_6$	93.2 (94.6 to 98.8)	
Jurbanite	$AlSO_4OH \cdot 5H_2O$	17.8	†††
Kieserite	$MgSO_4 \cdot H_2O$	0.12	††
Melanterite	$FeSO_4 \cdot 7H_2O$	2.21	‡‡‡
Syngenite	$K_2Ca(SO_4)_2 \cdot H_2O$	7.45	††
Szomolnokite	$FeSO_4 \cdot H_2O$	0.91	‡‡‡

Sources: †Paces (1978). ‡Nordstrom et al. (1990). §Nordstrom (1982); MINTEQA2 (Allison et al. 1991) lists $pK_{sp} = 117.3$ for basaluminite. ‖Paige et al. (1992). #Langmuir and Melchior (1985); MINTEQA2 gives pK_{sp} (anhydrite) = 4.637 in serious error. ††Harvie et al. (1984). ‡‡Langmuir and Melchior (1985). §§Computed from ΔG_f° data given by Naumov et al. (1974), and adapted in MINTEQA2. ‖‖Langmuir and Melchior (1985); MINTEQA2 lists pK_{sp}(gypsum) = 4.848 in serious error. ##Alpers et al. (1989) use these single values. The jarosites are an extensive series of solid solutions. MINTEQA2 lists $pK_{sp} = 82.1$ for hydronium jarosite. The range of parenthetic pK_{sp}'s is based on geochemical modeling of waters with which jarosites, usually of unknown composition, are thought to have been in equilibrium, but are assumed to be pure K-jarosite. See also Baron and Palmer (1996). †††Nordstrom (1982); MINTEQA2 lists $pK_{sp} = 17.23$ for $AlOHSO_4$. ‡‡‡Reardon and Beckie (1987); MINTEQA2 lists pK_{sp}(melanterite) = 2.47.

for which if pK_{sp}(jarosite) = 93.0, and pK_{sp}(HFO) = 39.0, we find

$$K_{eq} = [K^+][SO_4^{2-}]^2[H^+]^3 = 10^{-19.5} \tag{12.54}$$

Assuming $[K^+] = 10^{-4}$ mol/kg, and $[SO_4^{2-}] = 10^{-2}$ mol/kg, then pH = 3.8 at equilibrium. The stability field of jarosite and the jarosite/HFO boundary are shown on the Eh-pH diagram in Fig. 12.26. Apparent Eh-pH buffering of waters in saturated mine tailings by equilibrium with respect to HFO and jarosite is suggested by the analyses plotted in Fig. 12.27.

Filipeck et al. (1987) computed the saturation state of the Al-sulfates in waters from West Squaw Creek drainage, California. As shown in Fig. 12.28, Al concentrations appear limited by sat-

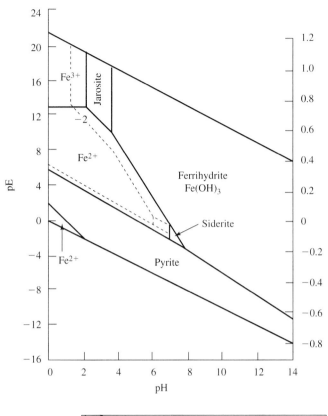

Figure 12.26   pE-pH/Eh-pH diagram for the system Fe-K-S-CO_2-H_2O-O_2 at 25°C and 1 bar pressure, assuming $\Sigma Fe(aq) = 10^{-4}$ mol/kg at solid/solution boundaries, $\Sigma K(aq) = 10^{-4}$ mol/kg, and $\Sigma S(aq) = 10^{-2}$ mol/kg, and $P_{CO_2} = 10^{-2}$ bar. Ferrihydrite [$Fe(OH)_3$)] is assumed to have $K_{sp} = 10^{-39}$, siderite [$FeCO_3$] $K_{sp} = 10^{-10.5}$, and jarosite [$KFe_3(SO_4)_2(OH)_6$] $K_{sp} = 10^{-93}$. Dashed line is solids/solution boundary for $\Sigma Fe(aq) = 10^{-2}$ mol/kg. Figure is modified after Nordstrom and Munoz (1985).

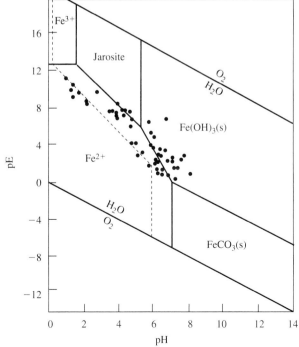

Figure 12.27   pE-pH plot for iron species and solid phases showing measured pE and pH values of groundwaters from Nordic Main, Nordic West Arm, and Lacnor tailings (Canada). The pK_{sp} for $Fe(OH)_3(s)$ is 38.5. The unbroken Fe(aq)/solid boundary is drawn for [Fe(aq)] = 10^{-4} mol/kg, the dashed boundary for [Fe(aq)] = 10^{-2} mol/kg. [K^+] = 10^{-3} mol/kg. From *Canadian Geotech. J.* 22(1):110–28, N. M. Dubrovsky et al., Geochemical evolution of inactive pyritic tailings in the Elliot Lake Uranium District: 1 The groundwater zone. Copyright 1985 by NRC Research Press. Used by permission.

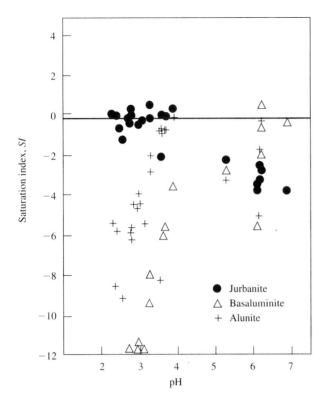

Figure 12.28 Plot of saturation indices for jurbanite, basaluminite and alunite as a function of pH for water samples from West Squaw Creek drainage, West Shasta Mining District, California. Reprinted with permission from *Envir. Sci. & Technol.* 21(4): 388–96, L. H. Filipek, D. K. Nordstrom, and W. H. Ficklin, Interaction of acid mine drainage with waters and sediments of West Squaw Creek in the West Shasta Mining District, California. Copyright 1987 American Chemical Society.

uration with respect to jurbanite, basaluminite, and alunite. Above pH 5 to 6 many of the same waters were supersaturated with respect to gibbsite and kaolinite. Under these conditions the Al-sulfates are metastable and should ultimately alter to phases such as gibbsite and kaolinite (see Figs. 12.29 and 12.30).

The mineral phases most generally found to be at or near saturation in acid mine waters are the ferric oxyhydroxides and gypsum in high-sulfate waters. The Ca^{2+} for gypsum saturation is released by the acid weathering of associated limestone and silicate rocks such as the plagioclase feldspars. Several high-sulfate waters in Tables 12.6 and A12.4 are saturated with respect to gypsum. Gypsum limits maximal calcium and sulfate concentrations in acid mine surface-water and groundwaters as shown by Fig. 12.31 (see also Helz et al. 1987). That gypsum precipitation has taken place as sulfate increased is suggested by the anomalously low molar Ca/Mg ratios of several waters in Table A12.4. (Usually mCa $\geq$ mMg in natural waters.) Because $pK_a = 1.99$ for $HSO_4^- = H^+ + SO_4^{2-}$, sulfate concentrations increase rapidly above pH 2 in acid mine waters at the expense of bisulfate. Thus, even at constant $\Sigma SO_4(aq)$, sulfate minerals can rapidly approach saturation as pH increases.

Example 12.8

Input the analysis of acid spring water T-1.1 from Table A12.4 into MINTEQA2 (ignore species <5 mg/kg), to compute the saturation state of the water with respect to possible secondary minerals. Correct pK_{sp}(gypsum) in the data base to 4.59 for 25°C (Table 12.7). Computer output shows that, among the sulfates, jurbanite and the Na and H jarosites are slightly supersaturated. Gypsum and anhydrite are slightly undersaturated ($SI_{gyps} = -0.15$).

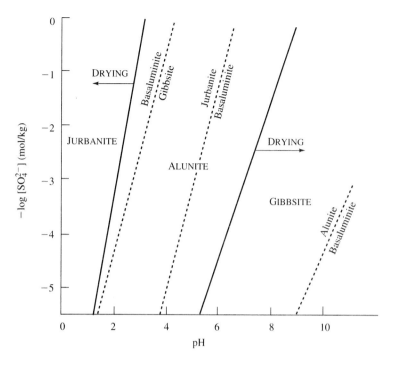

Figure 12.29   Stability fields of phases in the system K^+-Al^{3+}-SO_4^{2-}-H_2O at 25°C and 1 bar pressure, as a function of pH and sulfate activity for $K^+ = 10^{-2}$ mol/kg. The following pK_{sp} values are assumed: 33.96 for gibbsite, 88.4 for alunite, 17.8 for jurbanite, and 116 for basaluminite. Dashed lines denote metastable equilibria. The arrows labeled "Drying" show that the stability field of alunite increases in size with decreasing water activity (drying) at the expense of the jurbanite and gibbsite fields.

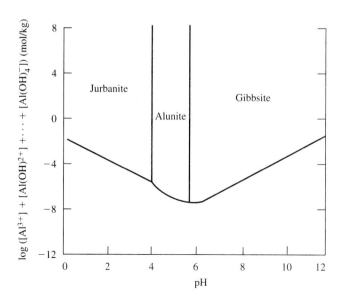

Figure 12.30   Stabililty fields and solubilities of jurbanite, alunite and gibbsite at 25°C, in terms of the sum of Al species activities, assuming $[K^+] = 10^{-4}$ mol/kg and $[SO_4^{2-}] = 10^{-2}$ mol/kg. At lower sulfate concentrations the gibbsite/alunite boundary moves to lower pH's. Reprinted from *Geochim. et Cosmochim. Acta*, 46, D. K. Nordstrom, The effect of sulfate on aluminum concentrations in natural waters: Some stability relations in the system Al_2O_3-SO_3-H_2O at 298 K, 681–92, © 1982, with permission from Elsevier Science Ltd., The Boulevard, Langford Lane, Kidlington OX5 1GB, U.K.

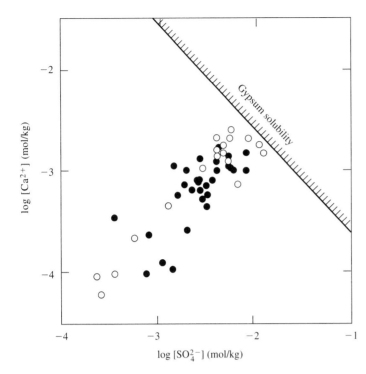

Figure 12.31 Plot of log [Ca^{2+}] versus log [SO$_4^{2-}$] for 16 streamwaters and 19 groundwaters from coal strip-mined areas of northwestern Pennsylvania, showing maximum activities are limited by the solubility of gypsum at 10°C. Solid circles denote stream waters, open circles groundwaters. From M. A. Gang and D. Langmuir (1974). Copyright 1974 by the National Coal Association. Used by permission.

Now perform the calculation with the same water analysis, but assume pH = 1.0. MINTEQA2 output shows that gypsum is now undersaturated with $SI_{\text{gyps}} = -0.55$. SI values for the other sulfates are even more negative.

Depending on the mix and history of surface runoff and groundwater inflow feeding the streams that discharge from acid mine country, solute concentrations may increase or decrease with discharge. Where the fraction of a watershed that has been mined is small, overland flow may dilute increased streamflows. However, when storms cause a major increase in acid groundwater flows into streams or wash out important amounts of acid sulfate salts accumulated in spoils and tailings piles, streams may be degraded by storm events. Gang and Langmuir (1974) found that several acid streams in Pennsylvania were diluted by storms, but that overall stream solute and metal loads were increased, probably chiefly due to the washout of sulfate salts (see Figs. 8.17 and 8.18). The acid discharge of a coal mine in Ohio increased in acidity with discharge, probably because of the washout of accumulated salts from mine workings by rising groundwater levels in response to storms (Shumate and Smith 1968).

Reliable models have been developed for predicting the formation of acid mine drainage and its fate in laboratory column and batch tests (cf. Sullivan et al. 1986; Felmy et al. 1987; Davis and Runnells 1987). Comprehensive modeling attempts have been less successful when applied to natural systems because of their complexity and a lack of subsurface information, particulary on abandoned mines. Among the more successful efforts at comprehensive mathematical modeling are the studies of Shumate and Smith (1968) and Morth et al. (1972). Their model considers a complex suite of geologic, hydrologic, and geochemical variables, including pyrite oxidation kinetics.

STUDY QUESTIONS

1. Discuss the conditions of formation, particle size, crystallinity, moisture content, [Fe(III)/OH] solid ratio, and thermodynamic stability of Fe(III) oxyhydroxide solids such as ferrihydrite and goethite. What happens to the solution pH, stoichiometry, and solubility of such solids when they are precipitated and aged under acid versus under alkaline conditions?

2. Given Gibbs free-energy data, be able to draw and explain a species distribution diagram for Fe(II) and Fe(III)-OH complexes as a function of pH.

3. Be able to construct an Eh(or pE)-pH diagram for a redox-sensitive element in the O_2-H_2O-CO_2 system, where the element forms insoluble solids and aqueous acid-base species, such as oxyanions and metal-OH complexes.

4. Sketch an Eh-pH plot for the S-H_2O-O_2 system, and comment on the relevance of kinetic versus equilibrium concepts and of microorganisms in facilitating the acid-base and redox reactions described by boundaries on the diagram.

5. Comment on the conditions of formation and occurrence of thermodynamically metastable versus stable aqueous sulfur species and Fe(II)-S solids.

6. Discuss the origins, occurrences, and stabilities of the Fe(II) sulfides.

7. Draw a schematic Eh-pH diagram for the Fe-K-S-CO_2-H_2O-O_2 system at 25°C as it might apply to the production of acid mine drainage. Label stability fields of the minerals and aqueous species and show on the diagram how acid mine drainage in seepages and streams evolves from the oxidation of pyrite or marcasite exposed in mine wastes or tailings.

8. What are the two key pyrite-marcasite oxidation reactions that cause the weathering of FeS_2? Compare and discuss the meaning of the acidity produced by the overall oxidation reactions to that of the corresponding oxidation half-cell reactions.

9. Comment on the rate laws that describe pyrite and dissolved Fe(II) oxidation as a function of pH. Compare the inorganic rates to the reaction rates observed when microorganisms are present. What role do bacteria play in the oxidation/reduction reactions of Fe and S and in the production of acid mine waters?

10. What are possible explanations for acid, high Fe(II), and sulfate waters in deep mine pools?

11. What insights do $\delta^{18}O$ measurements provide regarding the origins of acid mine waters?

12. Acid mine waters are produced when pyrite and/or marcasite are important minerals in coal-mine wastes, or in Zn, Pb, and Cu metal sulfide ore wastes. The Zn, Pb, and Cu sulfides themselves do not produce important amounts of acidity. Why not? Hint: Write the weathering/oxidation reactions for the Zn, Pb, and Cu sulfides.

13. How and why do changes occur in the concentrations of acidity and of dissolved metals and other species in streams draining acid mine drainage country as a function of stream discharge?

14. If a Fe(III)-rich acid mine water with minor Fe(II) (pH = 3) flows into a confined formation comprising minerals such as calcite and feldspars, but without reductants, its Eh will drop as its pH rises. Why?

PROBLEMS

1. Maximum concentrations of Fe(III) in natural waters are usually limited by the solubility of the Fe(III) oxyhydroxides. The Fe(III) concentration in turn limits maximum concentrations of dissolved phosphate through precipitation of the mineral strengite ($FePO_4 \cdot 2H_2O$). With MINTEQA2 and the sweep option, calculate total concentrations of Fe and P at equilibrium with both ferrihydrite and strengite (both present as finite solids) from pH 2 to 9 at even pH increments. Assume also that $Na^+ = Cl^- = 0.01$ mol/L. Compare the concentrations of Fe(III) and PO_4 as a function of pH.

2. If oxidized and reduced forms of an element are present in a water and a complexing ligand is introduced, the redox form that participates in the strongest complex or complexes with the introduced ligand will have

the size of its Eh stability field increased at the expense of the field of the weakly complexing form. The effect is magnified if the species that is most strongly complexed also has the smaller ion activity coefficient. The following problem illustrates this principle using the Fe^{3+}/Fe^{2+} redox pair. In general for this pair at 25°C

$$Eh(\text{volts}) = 0.771 + 0.0592 \log \frac{[Fe^{3+}]}{[Fe^{2+}]}$$

where brackets enclose activities of free ions. Note that in the absence of complexing of either ion, at equal concentrations and at zero ionic strength, $Eh = E° = 0.771$ V.

    (a) Given that pH = 3.00, and $\Sigma Fe(III)(aq) = \Sigma Fe(II)(aq) = 10.^{-3.00}$ mol/L, use MINTEQA2 to calculate the corresponding Eh. You will have to assign the redox pair Fe^{2+}/Fe^{3+} a fixed activity in PRODEFA2. Explain how aqueous iron speciation and the activity coefficients of Fe^{3+} and Fe^{2+} influence the computed Eh value.

    (b) Now add $\Sigma SO_4(aq) = 10^{-2.00}$ mol/L to the solution in (a). What is the Eh? Tabulate the activities of the major species. What are the mole percent concentrations of $\Sigma Fe(III)(aq)$ and $\Sigma Fe(II)(aq)$ present as complexes? Compare the speciation to what you computed in part (a). Explain the differences in Eh and in the speciation.

3. The measured Eh of natural waters (particularly of surface-waters) often represents a mixed potential and thus has little or no thermodynamic meaning. For this reason many researchers have chosen to chemically analyze natural waters for their concentrations of individual redox pairs, rather than computing questionable redox pair concentrations from the measured Eh via the Nernst equation. The analysis of leachate from Well 201 in Table A12.4 includes measured Eh and pH values and concentrations of Fe(III)(aq), Fe(II)(aq), As(III)(aq), and As(V)(aq). Separately compute the Eh that corresponds to the directly measured concentrations of each redox pair and compare these values to the measured Eh. Solve the problem using MINTEQA2 and comment on the results.

4. **(a)** Given the composition of the leachate from Coal A in Table A12.4, and assuming its temperature is 25°C, calculate the half-time for oxidation of pyrite in the coal in contact with the leachate. Assume only abiotic oxidation, a constant pyrite surface area of 0.05 m²/g, and that both O_2 and Fe^{3+} act as oxidants. Hint: you will need to use MINTEQA2 to compute the Fe^{2+} concentration in order to solve this problem. Assume a fixed Fe^{2+}/Fe^{3+} activity ratio in PRODEFA2.

    (b) Are the assumptions being used in this problem realistic? What is likely to happen to the relative and absolute concentrations of aqueous Fe species and to the oxidation rate of FeS_2 within the coal pile as time passes? Explain.

    (c) Read the paper by Elberling and Nicholson (1996) and discuss the probable importance of oxygen diffusion and moisture content to the rate of pyrite oxidation in coal piles and metal mine tailings piles.

5. Helz et al. (1987) found that the 40-day leachate from Coal A in Table A12.4 was at saturation with respect to gypsum, melanterite, and goethite. Input the analysis into MINTEQA2, adjusting K_{sp} values for sulfates as necessary (Table 12.7), and compute *SI* values for these minerals. Are other sulfates near saturation in the leachate? Based upon your calculations, if the minerals reported by Helz et al. (1987) are not at saturation, discuss possible reasons.

6. Most current gold mining in the western U.S. involves the open-pit excavation of FeS_2-bearing gold ore, which is then crushed and heap-leached with a high-pH cyanide solution to extract the gold. Surface- and groundwater pollution have been caused by escape of the gold cyanide leaching solutions (cf. King 1995), and result from the acidities generated by oxidation of FeS_2 exposed in heap-leach piles and waste-rock piles. After mine closure the open pit gradually fills with inflowing groundwaters and runoff from pit walls. The lake is diluted by rainfall but concentrated by evaporation. Given such complexities, predicting the long-term chemistry of pit lakes is a difficult challenge (cf. Miller et al. 1996).

    The following problem is a simplified calculation to predict the chemistry of a pit lake. We will consider the mixing of runoff from pit walls with inflowing groundwater in the pit, but will neglect the effects of rainfall and evaporation. There are two important questions: (1) will the lake be acid or alkaline; and (2) will it contain toxic concentrations of metals and species such as arsenic that would harm wildlife? The predicted average compositions of pit wall runoff (based on laboratory tests) and groundwater inflow (based on groundwater chemical analyses) are given below. Concentrations are in mg/L unless otherwise indicated.

Species or parameter	Pit wall runoff	Groundwater inflow	Species or parameter	Pit wall runoff	Groundwater inflow
pH	3.50	7.37	Mg	28.5	31.5
Eh(mV)	1000	750	Mn(II)(aq)	1.6	0.6
Al	9.5	0.1	PO_4	0.02	0.35
As	0.010	0.515	K	3.1	7.4
Ca	20.8	69	SiO_2	4.6	18
Cl	0.7	35.5	Na	2.0	138
F	0.3	0.3	SO_4	255	530
Fe(III)(aq)	1.9	0.9	Zn	1.71	0.12

(a) Hydrogeologic studies indicate that 9.4 gpm of pit wall runoff will mix with 10 gpm of groundwater inflow in the pit lake. Using a geochemical code such as SOLMINEQ.88 (Kharaha et al. 1988) or PHREEQC (Parkhurst 1995), determine the composition of the mixture. What is its pH and alkalinity?

(b) Mixing creates a pit lake that is supersaturated with respect to numerous solids, some of which may precipitate. Input the composition of the mixture from (a) into MINTEQA2. Assume the following six possible solids could precipitate (their ID numbers are given in parentheses): $Al_4(OH)_{10}SO_4$ (6003001), ferrihydrite (2028100), gibbsite (2003003), strengite (7028100), pyrolusite (2047000), and hydroxyapatite (7015003). Reset the log K_{eq} value for ferrihydrite in PRODEFA2 to -3.0 (equivalent to $pK_{sp} = 39.0$). Which of the possible solids will precipitate? What is the composition of the pit lake after they have precipitated?

(c) Precipitated phases can beneficially adsorb toxic species such as As from the lake. With the diffuse-layer model in MINTEQA2, compute adsorption by the ferrihydrite (HFO) precipitate. In the modeling assume a HFO surface area of 600 m^2/g, and two adsorption sites; site 1 with 0.005 mol sites/mol Fe, and site 2 with 0.2 mol sites/mol Fe (Dzombak and Morel 1990). Use the diffuse-layer database file *feo-dlm.dbs*. Comment on the importance of adsorption to pit-lake chemistry.

(d) Triple the amount of HFO used in part (c); recompute the adsorption of arsenic from the pit lake and discuss.

7. (a) Given the thermodynamic data in these tables, construct an Eh-pH diagram for the $Mn-O_2-H_2O-CO_2$ system for $P_{CO_2} = 10^{-2.00}$ bar, and with solid/aqueous species boundaries at $Mn^{2+} = 10^{-5.00}$ mol/kg. Consider the following solids for your diagram: hausmanite, rhodocrosite, todórokite, bixbyite, manganite, birnessite, and pyrolusite. Hints: Ignore complexing of Mn^{2+}. In order to decide which minerals are thermodynamically stable, it is useful to first assign to each of them an *average* valence for Mn. Then write reactions between minerals of the same valence to decide which one is thermodynamically most stable. Next, write reactions between the most stable minerals of *proximal valence* for Mn, and plot their reaction boundaries on the Eh-pH diagram.

Aqueous species or solid phase	ΔG_f° (kcal/mol)	Source	Aqueous species or solid phase	ΔG_f° (kcal/mol)	Source
Mn^{2+}	-54.5	†	$Mn^{II}(Mn^{IV})_3O_7$ (todorokite)	-415.1	§
γ-MnOOH (manganite)	-133.3	‡	$MnCO_3$ (rhodocrosite)	-195.2	†
δ-MnO_2 (birnessite)	-108.3	†			
β-MnO_2 (pyrolusite)	-111.2	†	$H_2O(l)$	-56.687	‖
			OH^-	-37.604	‖
MnO_4^-	-106.9	†	$CO_2(g)$	-94.254	‖
MnO_4^{2-}	-119.7	†	$H_2CO_3^\circ$	-148.94	†
α-Mn_2O_3 (bixbyite)	-210.6	†	HCO_3^-	-140.26	†
Mn_3O_4 (hausmanite)	-306.7	†	CO_3^{2-}	-126.17	†

Sources: †Wagman et al. (1982); ‡Bricker (1965); §Based on composition of soil moisture in C horizon in apparent equilibrium with todorokite; ‖CODATA (1976).

Some of these minerals are thermodynamically unstable relative to others and will not have stability fields on the diagram if only the most stable solids are considered. Which are they? Explain. Construct the diagram between pH 2 and 10, showing the full stability field of water in that pH range.

(b) Based on the Mn Eh-pH diagram and similar diagrams for iron, explain how Mn and Fe can be separated from each other in a soil- or groundwater. Why is it more difficult to remove dissolved Mn^{2+} from drinking water than dissolved Fe^{2+} (for which reason in part, their drinking water standards are 0.05 and 0.3 mg/L, respectively)?

CHAPTER 12 APPENDIX

The tables in this section present thermodynamic data and equilibria for aqueous species and solids of iron and sulfur and equations for constructing Eh-pH and Eh-concentration diagrams, and chemical analyses of some acid mine waters and tailings waters.

TABLE A12.1 Thermodynamic data for iron aqueous species and solids at 25°C and 1 bar pressure

Mineral or aqueous species	ΔH_f° (kcal/mol)	ΔG_f° (kcal/mol)	S° (cal/mol K)	Source
α-Fe(c)	0	0	6.52	1
Fe^{2+}	−21.3	−18.85	−32.9	2
$FeOH^+$	−76.42	−61.76	−18.1	3
$Fe(OH)_2^\circ$	−129.3	−104.25	2.6	4
$Fe(OH)_3^-$		−148.8		5
$FeCl^+$	−53.5	−49.5	4.2	6
$Fe_{0.947}O(c)$	−63.64	−58.59	13.76	1
$FeSO_4^\circ$	−237.1	−199.8	−13.	7
$FePO_4^-$	−313.8	−272.4	−9.4	8
$FeHPO_4^\circ$		−284.1	(−13)	9,10
$FeH_2PO_4^+$		−292.7	(−25)	9,10
$Fe_3(PO_4)_2(H_2O)_8$ vivianite		−992.6	(132)	11
FeS(am)		−20.0		12
FeS(c) troilite	−24.13	−24.22	14.42	1
FeS(c) mackinawite		−21.0		12
Fe_3S_4(c) griegite		−65.4		12
FeS_2(c) marcasite	−40.5	−37.86	12.88	1
FeS_2(c) pyrite	−41.0	−38.3	12.65	1
$Fe_{0.877}S$ monoclinic pyrrhotite	−25.2	−25.6	14.5	13
$FeHCO_3^+$		−161.84		14
$FeCO_3^\circ$		−151.98		15
$FeCO_3$(c) siderite	−176.96	−159.48	22.82	16
FeAs(c)	−5.50	−10.54	25.12	17
$FeAs_2$(c) loellingite	−11.10	−15.94	33.02	17
FeAsS(c) arsenopyrite	−26.71	−30.04	27.24	17
FeSbS(c) gudmundite	−24.41	−24.43	25.16	17
$FeSb_2S_4$(c) berthierite	−66.06	−67.97	58.77	17
$FeSe_2$(c) ferroselite			20.76	1
Fe_2SiO_4(c) fayalite	−353.58	−329.68	35.45	1
$Fe_3Si_2O_5(OH)_4$(c) greenalite	(−783.0)	−710.0	(66.7)	18
Fe^{3+}	−11.6	−1.10	−75.5	2
$FeOH^{2+}$	−69.5	−54.80	−34	2,4
$Fe(OH)_2^+$		−106.74		19
$Fe(OH)_3^\circ$		−154.03		19
$Fe(OH)_4^-$		−198.38	(−1.1)	19
$Fe_2(OH)_2^{4+}$	−146.3	−111.55	−86	4
$Fe_3(OH)_4^{5+}$	−293.76	−221.46	−141	4
FeF^{2+}	−94.05	−76.90	−43	20
FeF_2^+	−168.4	−150.49	−21	20
FeF_3°	−247.65	−222.2	−6.1	20
$FeCl^{2+}$	−45.93	−34.50	−36	7
$FeCl_2^+$		−66.77	(−18)	7
$FeSO_4^+$	−223.0	−184.56	−32.4	7
$Fe(SO_4)_2^-$		−364.34	(-9)	7
$FeHPO_4^+$	−316.66	−275.0	−25	21
$FeH_2PO_4^{2+}$	(−324.5)	−277.0	(−45)	22,23
$FePO_4 \cdot 2H_2O$ strengite	−451.30	−397.45	40.9	1
$FeH_3SiO_4^{2+}$	(−363.7)	−313.18	(−45)	23,24
$Fe(OH)_3$(am)		−164.52		25
$Fe(OH)_3$(s)		−166.43		25
α-FeOOH goethite	−134.27	−117.36	14.43	26

TABLE A12.1 *(continued)*

Mineral or aqueous species	ΔH_f° (kcal/mol)	ΔG_f° (kcal/mol)	S° (cal/mol K)	Source
γ-FeOOH lepidocrocite		−112.61		27
α-Fe$_2$O$_3$ hematite	−197.09	−177.51	20.89	1
γ-Fe$_2$O$_3$ maghemite		(−168)		28
Fe$_3$O$_4$ magnetite	−266.67	−242.01	34.93	1
Fe$_3$(OH)$_8$ ferrosic hydroxide		−459.22		29
KFe$_3$(SO$_4$)$_2$(OH)$_6$(c) jarosite		−791.1		30
Fe$_4^{II}$Fe$_2^{III}$(OH)$_{12}$SO$_4 \cdot$ 3H$_2$O		−1046.8		31

Note: Values in parentheses are estimates. The thermodynamic data in this table have been computed using auxiliary thermodynamic data for related substances as published by Wagman et al. (1982), which often differ from such data given by Cox et al. (1989).

Source: [1]Robie et al. (1978).

[2]Wagman et al. (1982).

[3]K from Yatsimirskii and Vasil'ev (1960). ΔH_r° from Baes and Mesmer (1976).

[4]K and ΔH_r° from Baes and Mesmer (1976).

[5]K from Baes and Mesmer (1976).

[6]Based on K data from 25 to 350°C in Barnes (1979).

[7]K and ΔH_r° from Smith and Martell (1976).

[8]Christensen et al. (1975).

[9]K from Smith and Martell (1976).

[10]S° assumed equal for the complexes: FeHPO$_4^\circ$ = FeSO$_4^\circ$ and FeH$_2$PO$_4^+$ = FeHPO$_4^+$.

[11]Al-Borno and Thomson (1994) obtain pK_{sp} = 25.76 at 25°C. Their solubility measurements from 5 to 90°C lead to the tabulated free energy, ΔH_f° = −1193.1 kcal/mol, and an impossible negative entropy for the solid. S° has instead been estimated using Latimer's method (Naumov et al. 1974). Nriagu (1972) obtained pK_{sp} = 36 for vivianite.

[12]ΔG_f° recomputed from free energies given by Berner (1971) consistent with the tabulated ΔG_f° for Fe^{2+}.

[13]Bezman and Smolyarova (1977).

[14]Based on K given by Nordstrom et al. (1990).

[15]K_{assoc} = 10$^{5.1 \pm 0.2}$ estimated by the author using the oxalate comparison method (Langmuir 1979). Bruno et al. (1992) suggest K_{assoc} = 10$^{5.5 \pm 0.2}$ using a different solution model.

[16]S° from Robie et al. (1984). pK_{sp} = 10.68 ± 0.02 (30°C) recomputed from solubility data of Smith (1918), which with ΔH_r° = −6.0 ± 0.1 kcal/mol and the van't Hoff equation, gives pK_{sp} = 10.60 at 25°C.

[17]Barton and Skinner (1979).

[18]K_{sp} from Ball et al. (1980). S° estimated using Latimer's method (Naumov et al. 1974).

[19]Macalady et al. (1990).

[20]K_{assoc} from Roberson and Barnes (1978). ΔH_r° from Smith and Martell (1976).

[21]K_{assoc} = 10$^{8.30}$ for I = 0.5 (Smith and Martell 1976) corrected to K_{assoc} = 10$^{9.92}$ for I = 0. This may be compared to K_{assoc} = 10$^{9.75}$ at I = 0 reported by Sillen and Martell (1964). ΔH_r° from Christensen et al. (1975).

[22]K_{assoc} = 10$^{4.17}$ estimated by this author.

[23]S° for complex estimated from plot of empirically obtained monatomic entropies of Fe aquospecies versus the charge of those species on a monatomic Fe basis (cf. Langmuir and Herman 1980).

[24]K measured at I = 0.1 (Porter and Weber 1971) corrected to I = 0.

[25]ΔG_f° values computed for freshly precipitated amorphous material (pK_{sp} = 37.1), and for briefly aged material (pK_{sp} = 38.5) (cf. Macalady et al. 1990).

[26]S° from Robie et al. (1978). ΔG_f° based on ΔG_r° = 0.53 kcal/mol for the reaction: 2α-FeOOH = α-Fe$_2$O$_3$ + H$_2$O (Macalady et al., 1990).

[27]Based on pK_{sp} = 40.6 for γ-FeOOH + H$_2$O = Fe^{3+} + 3OH$^-$ (Schuylenborgh 1973).

[28]ΔG_f° estimated from the assumption that ΔG_r° = 0.53 kcal/mol for the reaction: 2γ-FeOOH = γ-Fe$_2$O$_3$ + H$_2$O (see note 26).

[29]Lindsay (1979).

[30]Baron and Palmer (1996).

[31]Bruun Hansen et al. (1994).

TABLE A12.2 Equilibria and equations for constructing Eh-pH and Eh-concentration diagrams at 25°C and 1 bar pressure for the systems Fe-O_2-H_2O and Fe-O_2-CO_2-H_2O in Figs. 12.8 to 12.12

Part I. The pH of Hydrolysis Reaction Boundaries, Computed for Equal Activities of the Two Fe Species

Reaction equation no.	Reaction pair or boundary	Reaction	$-\log K =$ pH of boundary
1	$Fe^{3+}/FeOH^{2+}$	$Fe^{3+} + H_2O = FeOH^{2+} + H^+$	2.19
2	$FeOH^{2+}/Fe(OH)_2^+$	$FeOH^{2+} + H_2O = Fe(OH)_2^+ + H^+$	3.48
3	$Fe(OH)_2^+/Fe(OH)_3^\circ$	$Fe(OH)_2^+ + H_2O = Fe(OH)_3^\circ + H^+$	6.89
4	$Fe(OH)_3^\circ/Fe(OH)_4^-$	$Fe(OH)_3^\circ + H_2O = Fe(OH)_4^- + H^+$	9.04
(5)	$Fe^{2+}/FeOH^+$	$Fe^{2+} + H_2O = FeOH^+ + H^+$	10.10
(6)	$FeOH^+/Fe(OH)_2^\circ$	$FeOH^+ + H_2O = Fe(OH)_2^\circ + H^+$	10.41
(7)	$Fe(OH)_2^\circ/Fe(OH)_3^-$	$Fe(OH)_2^\circ + H_2O = Fe(OH)_3^- + H^+$	8.90
8	$Fe^{2+}/Fe(OH)_3^-$	$\frac{1}{3}Fe^{2+} + H_2O = \frac{1}{3}Fe(OH)_3^- + H^+$	9.80

Part II. Redox Reactions among Aqueous Species; Boundaries Are Computed for Equal Activities of the Two Species

Reaction equation no.	Reaction pair or boundary	Reaction	E° (V)	Eh/pH or other equilibrium expression
9	Fe^{3+}/Fe^{2+}	$Fe^{3+} + e^- = Fe^{2+}$	0.770	$Eh = 0.770 + 0.0592 \log \dfrac{[Fe^{3+}]}{[Fe^{2+}]}$ $Eh = 0.770$ V
10	$FeOH^{2+}/Fe^{2+}$	$FeOH^{2+} + H^+ + e^- = Fe^{2+} + H_2O$	0.899	$Eh = 0.889 + 0.0592 \log \dfrac{[FeOH^{2+}]}{[Fe^{2+}]} - 0.0592$ pH $Eh = 0.899 - 0.0592$ pH
11	$Fe(OH)_2^+/Fe^{2+}$	$Fe(OH)_2^+ + 2H^+ + e^- = Fe^{2+} + 2H_2O$	1.105	$Eh = 1.105 + 0.0592 \log \dfrac{[Fe(OH)_2^+]}{[Fe^{2+}]} - 0.118$ pH $Eh = 1.105 - 0.118$ pH
12	$Fe(OH)_3^\circ/Fe^{2+}$	$Fe(OH)_3^\circ + 3H^+ + e^- = Fe^{2+} + 3H_2O$	1.513	$Eh = 1.513 + 0.0592 \log \dfrac{[Fe(OH)_3^\circ]}{[Fe^{2+}]} - 0.177$ pH $Eh = 1.513 - 0.117$ pH
13	$Fe(OH)_4^-/Fe^{2+}$	$Fe(OH)_4^- + 4H^+ + e^- = Fe^{2+} + 4H_2O$	2.048	$Eh = 2.048 + 0.0592 \log \dfrac{[Fe(OH)_4^-]}{[Fe^{2+}]} - 0.237$ pH $Eh = 2.048 - 0.237$ pH
14	$Fe(OH)_4^-/Fe(OH)_3^-$	$Fe(OH)_4^- + H^+ + e^- = Fe(OH)_3^- + H_2O$	0.308	$Eh = 0.308 + 0.0592 \log \dfrac{[Fe(OH)_4^-]}{[Fe(OH)_3^-]} - 0.0592$ pH $Eh = 0.308 - 0.0592$ pH

Part III. Reactions Involving Solids and Aqueous Species

Reaction equation no.	Reaction pair or boundary	Reaction	$E°$(V) or log K	Eh/pH or other equilibrium expression
15	$Fe(OH)_3(am)/Fe(OH)_2^+$	$Fe(OH)_3(am) + H^+$ $= Fe(OH)_2^+ + H_2O$	log $K = -0.80$	pH = 4.20
16	$Fe(OH)_3(am)/Fe^{2+}$	$Fe(OH)_3(am) + 3H^+ + e^-$ $= Fe^{2+} + 3H_2O$	$E° = 1.065$	Eh = 1.065 − 0.0592 log[Fe^{2+}] − 0.177 pH Eh = 1.361 − 0.177 pH
17	$Fe(OH)_3(am)/FeCO_3(c)$	$Fe(OH)_3(am) + HCO_3^- + 2H^+ + e^-$ $= FeCO_3(c) + 3H_2O$	$E° = 1.078$	Eh = 1.078 + 0.0592 log[H^+]^2[HCO_3^-] Eh = 0.918 − 0.118 pH
18	$FeCO_3(c)/Fe^{2+}$	$FeCO_3(c) + H^+ = Fe^{2+} + HCO_3^-$	log $K = -0.22$	pH = 7.48
19	$Fe(OH)_3(am)/Fe_3(OH)_8(s)$	$3Fe(OH)_3(am) + H^+ + e^-$ $= Fe_3(OH)_8(s) + H_2O$	$E° = 0.969$	Eh = 0.969 − 0.0592 pH
20	$Fe_3(OH)_8(s)/Fe^{2+}$	$Fe_3(OH)_8(s) + 8H^+ + 2e^-$ $= 3Fe^{2+} + 8H_2O$	$E° = 1.102$	Eh = 1.102 + 0296 log $\dfrac{[H^+]^8}{[Fe^{2+}]^3}$ Eh = 1.368 − 0.237 pH
21	$Fe_3(OH)_8(s)/FeCO_3(c)$	$Fe_3(OH)_8(s) + 3HCO_3^- + 5H^+ + 2e^-$ $= 3FeCO_3(c) + 8H_2O$	$E° = 1.126$	Eh = 1.126 + 0.0296 log [HCO_3^-]^3[H^+]^5 Eh = 0.886 − 0.148 pH

Note: Reaction equation numbers in parentheses indicate reactions involving minor species that do not appear in the figures. Boundaries in Part III are computed for a Fe(aq) activity of 10^{-5} mol/kg (0.56 mg/kg). Their position is also shown as a dashed line for 10^{-3} mol/kg Fe(aq) (56 mg/kg). $K_{sp} = 10^{-37.1}$ is assumed for Fe(OH)$_3$(am). [HCO_3^-] = $10^{-2.7}$ mol/kg (122 mg/kg).

TABLE A12.3 Thermodynamic data for some substances in the system S-O_2-H_2O at 25°C and 1 bar pressure

Substance	ΔH_f° (kcal/mol)	ΔG_f° (kcal/mol)	S° (cal/mol K)	Source
S(c) rhombic	0	0	7.6	†
$H_2S(g)$	−4.93	−8.02	49.16	†
$H_2S(aq)$	−9.49	−6.65	28.9	†
HS^-	−4.2	2.89	15.0	†
S^{2-}	(7.96)	(28.1)	(−28.7)	‡
H_2S_2	−12.5	−1.40	39.7	§
HS_2^-	−5.86	5.40	23.8	§
S_2^{2-}	7.20	19.75	5.51	§
H_2S_3	−9.54	1.10	47.7	§
HS_3^-	−3.88	6.90	31.80	§
S_3^{2-}	6.16	17.68	17.04	§
H_2S_4	−8.35	2.20	55.60	§
HS_4^-	−3.28	7.50	39.70	§
S_4^{2-}	5.50	16.62	27.12	§
H_2S_5	−6.86	2.70	63.60	§
HS_5^-	−2.19	7.50	47.70	§
S_5^{2-}	5.46	15.69	37.04	§
S_6^{2-}	5.66	16.01	43.70	§
SO_4^{2-}	−217.40	−177.95	4.50	‖
HSO_4^-	−212.16	−180.67	31.2	#
HSO_3^-	−149.67	−126.20	33.6	‡
SO_3^{2-}	−152.39	−116.16	−9.2	‡
$HS_2O_3^-$	−151.63	−132.36	55.3	‡
$S_2O_3^{2-}$	−157.70	−129.97	26.9	‡
$S_4O_6^{2-}$	−295.48	−252.27	63.7	‡

Note: Values in parentheses are estimates or have been computed from estimated values.

Source: †Wagman et al. (1982). ‡Williamson and Rimstidt (1992). ΔG_f° for S^{2-} based on log K estimated by Schoonen and Barnes (1988). Entropies have been computed from the tabulated enthalpy and free-energy data. For $S_4O_6^{2-}$ Wagman et al. (1982) give $\Delta H_f^\circ = -292.59$ kcal/mol, $\Delta G_f^\circ = -248.66$ kcal/mol, and $S^\circ = 61.5$ cal/mol K. §Data from Murowchick and Barnes (1986). Williamson and Rimstidt (1992) have systematized and modeled free-energy and enthalpy data for the polysulfides. ‖CODATA (1977). #Values adjusted relative to values for SO_4^{2-} assuming properties of the reaction $HSO_4^- = H^+ + SO_4^{2-}$ unchanged from those properties computed using values given by Wagman et al. (1982).

TABLE A12.4 Chemical analyses of some acid mine waters, tailings waters, and a wastewater (tailings solution)

								Bluewater Mill	Coal A	Well 201
	ST-3	ST-6	T-1.1	T10-20	WR-W-3	IMJ-5				
Sample type	Stream	Stream	Spring	Flowing well	Flowing well	Stream	Tailings solution	Leachate	Saturated tailings
$T(°C)$	16.6	16.4	21.0	9.0	10.0	25	25	25	13.6
Ca	91	153	357	48	191	265	576	377	561
Mg	49	151	985	121	100	710	633	29	184
Na	17	6.8	6.8	7	54	48	1100	1.5	248
K	4.1	4.2	1.1	10	7.7	0.014	90	1.3	26.8
SiO_2	8.5	13	70	8.6	9.5	153	442	—	—
HCO_3	0	0	0	10	12	0	0	—	53
Acidity as H^+	23	1.9	51	2.7	4.6	—	—	—	0.2
SO_4	619	1220	6230	1050	1130	29,500	24,400	22,900	2210
Cl	31	19	17	55	23	—	1630	—	322
SpC (μS/cm^2)	1190	2290	5900	2300	1450	—	—	3903	—
pH	3.65	2.90	2.99	5.99	6.50	1.66	1.2	1.76	7.60
Eh(V)	+.570	+.754	+.428	+.094	+.172	—	+.78	+.716	+.234
DO	5.0	9.4	8.9	0.6	0	—	—	—	—
Fe(III)	4.1	5.2	5.5	0.10	0.10	7100	Σ2430	Σ5860	0.08
Fe(II)	110	0.31	135	390	120	<1	0.34	—	—
Mn	5.3	51	281	10	21	13.9	75	4700	1.2
Al	2.8	55	201	0	0	1075	1020	46	0.027

TABLE A12.4 (continued)

				Sample			Bluewater Mill	Coal A	Well 201
	ST-3	ST-6	T-1.1	T10-20	WR-W-3	IMJ-5			
Zn	0.11	1.8	11	0.03	0.003	1,450	5.7	6200	0.0039
Co	0.19	1.5	4.8	0.16	0.32	—	1.13	—	0.006
Ni	0.16	1.4	7.5	0.11	0.15	1.8	1.43	—	0.008
Cu	.012	.070	.270	.0061	.0067	174	3.17	5300	0.005
Cr	.003	.0082	.120	.0048	.0040	—	1.9	—	0.005
Cd	.0019	.0021	.013	.0003	.001	12.1	1.03	280	0.003
Ag	.0003	.0002	.0023	.0006	.0005	—	—	—	—
Pb	.003	.004	.004	<.001	.001	—	4.3	—	0.010

Note: Values are in mg/kg except as indicated. All samples except the Bluewater Mill sample acquired their initial acidities and solutes because of pyrite oxidation. The Bluewater Mill tailings solution is derived from sulfuric acid leaching of uraniferous sandstone ore. Although listed as Fe(III), iron concentrations in the Bluewater Mill solution and Coal A leachate are total iron values. Except for Bluewater Mill and Coal A laboratory studies, Eh and pH values were measured in the field. All samples were filtered through $0.45 \mu m$ filters prior to acidification, except for Well 201.

Data sources: ST-3 through WR-W-3 are waters from northeastern Pennsylvanian anthracite coal regions (Gang and Langmuir 1974).

IMJ-5 an acid mine streamwater from Iron Mountain, California, stored for 11 years, and found to be at equilibrium with a jarosite of composition $K_{.77}Na_{.03}(H_3O)_{.20}Fe_3(SO_4)_2(OH)_6$ (Alpers et al. 1989).

The Bluewater Mill sample is a uranium mill tailings solution from near Grants, New Mexico, analyzed in 1980. The solution also contains the following mg/kg concentrations: As, 0.6; Mo, 1.33; Sr, 16.6; NO_3, 31; NH_4, 35.7; and Se, 4.0. Radioisotopes in pCi/L are ^{210}Pb, 24,224; ^{238}U, 6565; ^{230}Th, 149,302; and ^{226}Ra, 3334. Data from Langmuir and Nordstrom (1995).

Coal A is an Appalachian (eastern U.S.) coal with 3.38% S and a specific surface of 23 cm^2/g (especially 0.5 to 5 mm sizes). Chemical analysis is the mean of samples from three identical replicate runs collected after 40 leaching days. The sample also has 50 mg/kg Be (Helz et al. 1987).

Well 201 is taken from about 50 ft deep, in water-saturated tailings at Midvale, Utah, limed for pH control during disposal. The tailings contain FeS_2, minor ZnS, PbS, and $CuFeS_2$, with gangue carbonates. The analysis also includes in mg/kg: $\Sigma As(III)$, 1.03; $\Sigma As(V)$, 0.040; Sb, 0.028; Se, 0.010; V, 0.004; and F, 4.6. (Sharon Steel/Midvale Project, Final Report, U.S. EPA Contract No. 68-W9-0021, 1990).

13

Actinides and Their Daughter and Fission Products

Radioactive elements, or radionuclides, are found throughout the environment. Many occur naturally (e.g., U, Th, Ra, Rn), while others are mostly or entirely manufactured (e.g., Tc, Pu, Np, Am). In some occurrences they may be employed as valuable geochemical tools, in others they constitute a health hazard. Beneficial uses include the dating of natural waters and rocks (Pearson et al. 1991) and assessing if a geochemical system has exhibited open- or closed-system behavior over geologic time. Hazardous or potentially hazardous occurences are in waste materials from uranium mining and milling and nuclear power generation and from radionuclides employed in research and medical radiation therapy. In countries that have used nuclear power for electrical power generation, scientists and engineers are attempting to evaluate the long-term risk of disposing of nuclear waste in geological repositories.

This chapter first introduces the fundamentals of radioactivity and its environmental significance. The following sections focus on the geochemistry of uranium and uranium ore deposits as the basis of the nuclear fuel cycle.[†] Later sections consider nuclear power and the geochemistry of important radionuclides in nuclear wastes, with emphasis on the actinide elements[‡] and some of their fission products which make nuclear wastes a potential problem for future generations because of their very long half-lives.

[†]The nuclear fuel cycle describes the sequence of activities beginning with the mining of uranium ore, to the fabrication of nuclear fuel and its use in power plants, to reprocessing of the spent fuel for reuse, or its disposal in a geologic repository (cf. Berlin and Stanton 1989).

[‡]The actinides are radioactive elements of atomic number 89 or greater.

13.1 RADIOACTIVITY[†]

13.1.1 Stable and Unstable Nuclei

The bulk of the mass of all atoms comprises protons, neutrons, and electrons. The standard description of an isotope of element E is of the form ^aE_z, where a is the number of particles in the nucleus or protons plus neutrons, and z, the atomic number, is the number of protons or positive charge of the nucleus. In an electrically neutral atom, z is also the number of electrons. With the exception of $^{209}Bi_{83}$, all elements with atomic numbers greater than 82 (Pb) consist of only radioactive isotopes. Elements with atomic numbers less than or equal to 92 (U) are naturally occurring, and so there are many naturally radioactive heavy elements, in addition to numerous lighter unstable nuclei, scattered throughout the periodic table (Durrance 1986). Nuclear instability arises when the coulombic repulsion of the protons in the nucleus exceeds the stabilizing influence of neutrons and other short-range attractive forces (Ivanovich 1992). This chapter focuses on the radioactive decay series of U and Th and on the long-lived elements produced in nuclear reactors.

13.1.2 Modes of Radioactive Decay

There are three principle modes of radioactive decay, denoted by α, β, and γ. Alpha particles are identical to the nuclei of helium atoms, with a mass of 4 atomic mass units (amu; 1 amu = 1.661 × 10^{-27} kg) and a charge of +2. The radioactive parent with initial nomenclature ^aP_z ends up as $^{a-4}D_{z-2}$ after undergoing a single α-decay. Following α-decay, the α-particle rapidly accumulates two electrons from its environment, thus balancing its charge and becoming a helium atom ($^4He_2$). Beta particles are identical to electrons (although they are emitted from the nucleus), with a mass of 5.49 × 10^{-4} amu and a charge of −1. Gamma rays are electromagnetic radiation with wavelengths somewhat shorter than those of X-rays. A fourth mode of decay, termed K-capture, occurs when an electron in an orbital shell is absorbed into the nucleus. Some radioisotopes are known to decay by more than one mechanism. For example, ^{214}Bi can decay by β-emission (99.96% probability) or α-decay (0.04% probability). In this case, the two modes of decay effectively compete with one another. In many cases, a radioisotope will emit more than one type of radiation simultaneously. For instance, many isotopes in the ^{238}U decay series (see below for a description of decay series), decay by emitting α- or β-particles, with a simultaneous emission of γ-radiation (Ivanovich 1992).

Radioactive decay is irreversible and spontaneous, releasing energy. The amount and type of energy depends on the parent element and mode of decay. All radioactive decay results in the release of heat. In fact radioactive decay of long-lived ^{238}U, ^{235}U, ^{232}Th, and ^{40}K is the primary source of heat in Earth's core and mantle (Mason and Moore 1982). The energy released during radioactive decay, and the possible ionizing nature of the radiation, constitute the primary concerns to human health that are associated with radioactivity.

13.1.3 Units of Radioactivity and Decay Laws

All units used to quantify radioactive decay are defined in terms of the number of decays per unit of time. The most fundamental expression of radioactivity is the number of decays per second (1 decay

[†]Most of the text of Section 13.1 was written by Richard B. Wanty of the U.S. Geological Survey, Mail Stop 973, Denver Federal Center, Denver, CO.

per second equals 1 becquerel, Bq). The standard SI unit of radioactivity is the becquerel per cubic meter. A curie, defined as 3.7×10^{10} decays per second, equals the number of α-particles emitted by 1 gram of ^{226}Ra in 1 second. A curie represents an enormous number of radioactive decays. In natural systems, picocuries (1 pCi $= 10^{-12}$ curie) are generally used to describe the radioactivity of rocks and water samples. A picocurie equals 0.037 Bq.

For a parent element decaying to produce a single daughter element, the abundance of the parent remaining can be described using first-order kinetics:

$$N = N_o \, e^{-kt} \tag{13.1}$$

(Patel 1991), where N_o and N are the number of atoms of the parent at $t = 0$ and after time t respectively, and k is the decay constant (see Chap. 2). The decay rate per unit time, Nk, decreases with time (assuming no production of the parent radionuclide). The decay constant is an intrinsic property of a given radioisotope, and is, therefore, unaffected by environmental variables such as time, temperature, pressure, or element abundance.

From the expression for radioactive decay, one can derive the half-life ($t_{1/2}$) of a given radionuclide. The half-life is the time it takes for half of the atoms to decay, in other words, the time at which $N/N_o = 0.5$. Thus, the half-life equals

$$t_{1/2} = \frac{-\ln(0.5)}{k} \tag{13.2}$$

This expression shows that $t_{1/2}$ is inversely related to k. Thus radionuclides with short half-lives have relatively larger values of k. After two half-lives, one-fourth of N_o remains, and in general, after n half-lives, the fraction remaining is 2^{-n}, assuming that no new atoms of the parent are introduced to the system. After seven half-lives, more than 99% of the parent atoms have decayed.

13.1.4 Natural Thorium, Uranium, and Plutonium

Naturally radioactive elements occur throughout the periodic table. Many of these elements (e.g., ^{40}K, ^{232}Th, ^{238}U, and ^{235}U) are found in common silicate minerals. Typical abundances of U (chiefly ^{238}U) and Th (chiefly ^{232}Th) in Earth-surface materials are given in Table 13.1. In general, U and Th are enriched in silica-rich igneous rocks (Fig. 13.1). Thus, granites usually have several times more U and Th than do basalts. Similarly, some pegmatites contain extremely high U and Th concentrations. Sedimentary rocks derived from igneous rocks will, in general, exhibit U and Th concentrations typical of the parent material. However, significant departures may occur in sedimentary environments due to water-rock interactions. For example, black shales may contain up to 50 μg/g U or more, because of secondary enrichment of U in the strongly reducing geochemical environment found in black shales.

Plutonium present in the earth at its time of formation has long since decayed because of its relatively short half-life ($t_{1/2} = 24{,}360$ y for ^{239}Pu). Most Pu in the environment is derived from nuclear-weapons testing or from nuclear wastes (cf. Hanson 1980; Kathren 1984). However, small amounts of natural ^{239}Pu are produced through neutron capture by ^{238}U (see Eq [13.13]). Analyses of ^{239}Pu in a number of uranium ore deposits have shown it to be near secular equilibrium with ^{238}U (see Section 13.1.6), with a weighted average Pu/U atomic ratio of $(3.1 \pm 0.4) \times 10^{12}$, which nearly equals $(3.0 \pm 0.5) \times 10^{12}$, the ratio at secular equilibrium (Curtis et al. 1992, 1994).

TABLE 13.1 Typical natural abundances of uranium and thorium in Earth's crust

Material	U (μg/g)	Th (μg/g)
Earth's crust (average continental)	2.7	9.6
Granites (average)	4.4	16
Basalt	0.8	2.7
Shale	3.8	12
Phyllosilicates (biotite, muscovite)	20	25
K-feldspar	1.5	5.0
Zircon	2500	2000

Water	U (μg/L)	Th (μg/L)
Seawater	3.3	0.0015
Chemically oxidizing groundwater	0.1 to 100	<1
Chemically reducing groundwater	<0.1	

Source: From R. B. Wanty and D. K. Nordstrom. Natural radionuclides. In *Regional ground-water quality*, ed. W. M. Alley. Copyright © 1995 by Van Nostrand Reinhold. Used by permission.

13.1.5 Radioactive Decay Series

In nature, the radioisotopes ^{238}U, ^{235}U, and ^{232}Th each begin a cascade of daughter products referred to as a radioactive decay series. The ^{238}U decay series is shown schematically in Fig. 13.2. All three decay series are given in Table 13.2. In these series, the number of atoms of any radioisotope in the series at a given time equals its initial abundance, plus the amount produced by the decay of parents, minus the amount lost in its own time-dependent decay. In most cases of geologic interest, the initial amount of a daughter has undergone complete decay over geologically significant time spans.

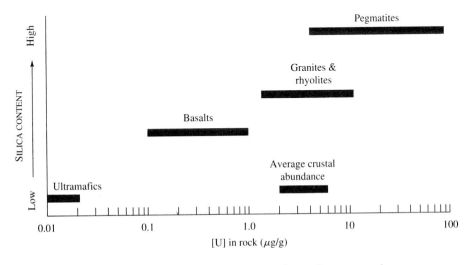

Figure 13.1   Plot showing that the uranium content of some important rock-types generally increases with their increasing silica content.

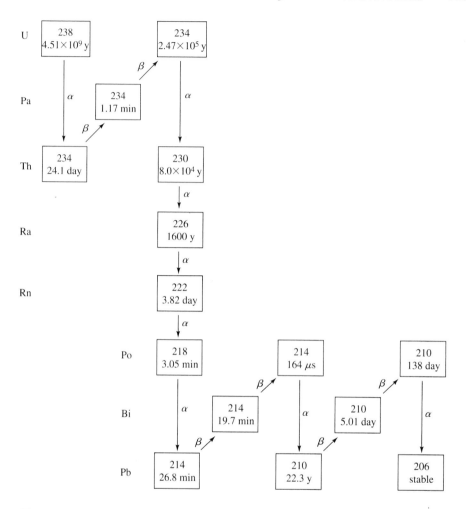

Figure 13.2 Simplified schematic representation of the ^{238}U decay series.

This simplification arises from the fact that, in most cases, the ultimate parents (^{238}U, ^{235}U, and ^{232}Th) have half-lives many orders of magnitude greater than that of any of the daughters.

In the ^{232}Th decay series there is a significant branch in the series at ^{212}Bi, which undergoes either α- or β-decay. The half-lives of the two decay modes of ^{212}Bi both are given as 1.15×10^{-4} y. This is the cumulative half-life of both types of decay. The other two decay series have minor branches (less than 1% branch in all but one case), and are shown in simplified form in Table 13.2.

The rate law given in Eq. (13.1) for simple one-step radioactive decay does not hold for decay series. Decay series can be represented schematically as in Fig. 13.2 and in the form of a series of chemical reactions as

$$P \rightarrow D_1 \rightarrow D_2 \rightarrow D_3 \rightarrow \cdots \rightarrow D_s \tag{13.3}$$

where the ultimate parent of the series, P, decays to form a succession of radioactive daughters, D_i, until a stable daughter, D_s, is formed. The abundance of the parent P can be described by the relatively simple expression given above, but the abundance of the succession of daughters is given by a series of equations first derived by Bateman (1910). Although hand calculation of parent and

TABLE 13.2   The decay series for ^{238}U, ^{235}U, and ^{232}Th

^{238}U			^{235}U			^{232}Th		
Element	Decay mode	$t_{1/2}$†	Element	Decay mode	$t_{1/2}$	Element	Decay mode	$t_{1/2}$
^{238}U	α	4.51×10^9	^{235}U	α	7.1×10^8	^{232}Th	α	1.41×10^{10}
^{234}Th	β	6.60×10^{-2}	^{231}Th	β	2.91×10^{-3}	^{228}Ra	β	5.77
^{234}Pa	β	2.23×10^{-6}	^{231}Pa	α	3.25×10^4	^{228}Ac	β	6.99×10^{-4}
^{234}U	α	2.47×10^5	^{227}Ac	β	21.6	^{228}Th	α	1.91
^{230}Th	α	8.00×10^4	^{227}Th	α	5.07×10^{-2}	^{224}Ra	α	9.97×10^{-3}
^{226}Ra	α	1.60×10^3	^{223}Ra	α	3.13×10^{-2}	^{220}Rn	α	1.74×10^{-6}
^{222}Rn	α	1.05×10^{-2}	^{219}Rn	α	1.27×10^{-7}	^{216}Po	α	4.75×10^{-9}
^{218}Po	α	5.80×10^{-6}	^{215}Po	α	5.64×10^{-11}	^{212}Pb	β	1.21×10^{-3}
^{214}Pb	β	5.10×10^{-5}	^{211}Pb	β	6.86×10^{-5}	^{212}Bi‡	β	1.15×10^{-4}
^{214}Bi	β	3.75×10^{-5}	^{211}Bi	α	4.09×10^{-6}	^{212}Po	α	9.63×10^{-15}
^{214}Po	α	5.20×10^{-12}	^{207}Tl	β	9.09×10^{-6}	^{208}Pb	stable	—
^{210}Pb	β	22.3	^{207}Pb	stable	—	^{212}Bi	α	1.15×10^{-4}
^{210}Bi	β	1.37×10^{-2}				^{208}Tl	β	5.89×10^{-6}
^{210}Po	α	0.378				^{208}Pb	stable	—
^{206}Pb	stable	—						

†Half-lives are given in years.
‡Indicates a branch in the decay series. 63.4% of the ^{212}Bi decays by β, the remainder by α.

daughter abundance as a function of time is very tedious, simple computer codes can be written to perform this calculation. The Bateman equations are described in greater detail in numerous texts devoted to nuclear physics and chemistry (cf. Harvey 1962; Kathren 1984; Ivanovich 1992).

Example 13.1

A field technician, armed with water-sampling equipment, a portable gas sampler, and an α-counting device, wishes to analyze a water for ^{222}Rn, but not ^{220}Rn. Radon gas for analysis is obtained by first filling a bottle with a known volume of the water sample, with a known volume of head space remaining. The capped bottle is then shaken to degass Rn into the head-space gas, which is extracted with a syringe and injected into the counting chamber. What is the best procedure for determining ^{222}Rn, but not ^{220}Rn in the water?

From Table 13.2, the half life of ^{220}Rn is 1.74×10^{-6} y, or about 55 s. The half-life of ^{222}Rn is 3.82 days, or 3.3×10^5 s. The ^{220}Rn will have mostly decayed after 7 half-lives (less than 1% of the original remains), or in just under 7 minutes. If the technician draws the sample into the syringe, then waits 7 to 10 minutes before injecting it into the counter, the ^{220}Rn will be essentially gone, whereas ^{222}Rn will not yet have undergone significant decay.

13.1.6 Radioactive Equilibrium and Steady State

As radioactive decay progresses in a decay series, steady-state conditions may develop between some or all of the daughters. These conditions are sometimes referred to as radioactive equilibria (or nonequilibria). However, radioactive decay reactions are irreversible and so cannot reach chemical equilibrium. They are, therefore, more properly referred to as occurring under steady-state conditions. Wanty and Nordstrom (1993) discuss this distinction.

For a parent-daughter pair of a radioactive series in a closed system the relationship between parent and daughter is

$$N_D = \frac{N_{o,P}(k_P)\,(e^{-tk_P} - e^{-tk_D})}{(k_D - k_P)} \tag{13.4}$$

(Atkins 1978), where the subscripts P and D refer to parent and daughter and $N_{o,P}$ is the original abundance of the parent element. If $k_D \gg k_P$, Eq. (13.4) reduces to

$$N_D k_D = N_{o,P}\, k_P \tag{13.5}$$

In other words, the decay rates of parent and daughter are equal, and their concentrations are equal, when expressed in terms of Bq/m^3 or other units of radioactivity. This condition is known as secular equilibrium. Attainment of secular equilibrium between a parent and daughter can occur only when $k_D \gg k_P$. In a closed system, secular equilibrium among all the daughters in a decay series may also be attained if the decay constant of the initial parent is much less than that of any of its daughters. Such is the case for ^{238}U, ^{235}U, and ^{232}Th. If secular equilibrium is attained in such a series, then each daughter decays at the rate at which it is produced, that is, its concentration is constant in time as long as the system remains closed in a thermodynamic sense (cf. Nordstrom and Muñoz 1994). If $k_P < k_D$ (but not much less), then a condition known as transient equilibrium may be attained. If $k_P > k_D$, equilibrium is never attained, and the supply of parent is exhausted as daughter is produced. At secular equilibrium, the relative abundances of parent and daughter, expressed in terms of their masses, equals the ratio of their half-lives.

Whether or not the various forms of radioactive equilibria are attained in a system depends on several factors, for example: (1) the open or closed nature of the system; (2) the passage of sufficient time for the buildup of daughters (referred to as ingrowth); (3) the relative values of k for each parent-daughter pair in the series; (4) the relative geochemical mobility of each radioisotope; and (5) the definition of the system (e.g., water plus rock, water only). At secular equilibrium, the abundance of each daughter in a series is kept constant by the constant decay of its parent, thus the daughter is said to be supported by its parent. For example, consider the ^{238}U decay series which includes among its daughters ^{226}Ra and ^{222}Rn. A common observation in numerous studies (Tanner 1964; Wanty et al. 1992) is that ^{222}Rn concentrations in groundwater are rarely supported by concentrations of its direct parent ^{226}Ra or its ultimate parent ^{238}U in the same groundwater. However, if the definition of the system is expanded to include the groundwater plus aquifer materials, many systems approach secular equilibrium (Wanty et al. 1992).

The attainment of secular equilibrium in a radioactive series is a strong indicator of a closed system, although it is possible, but unlikely, that there are equal fluxes of each daughter into and out of an open system. Radioactive equilibrium has many applications in geochemistry, including radioactive dating techniques (Faure 1991), radiometric surveys, and determination of groundwater-flow properties (Folger 1995). A more detailed description of radioactive equilibria, with specific examples, can be found in Wanty and Nordstrom (1993).

13.1.7 Alpha Recoil and Radon Emanation

The α-particle is the most massive of all radioactive decay products (not including, of course, daughter products and fission products). As an α-decay occurs, a daughter element is produced along with the α-particle. This process occurs with conservation of momentum. Therefore, the newly produced daughter element recoils in the opposite direction from that in which the α-particle was emitted. This recoil is similar to that of a gun (and the person holding it) as a bullet is fired. The distance of the

α-recoil depends on the physical surroundings. Recoil distances for ^{222}Rn as it is produced by ^{226}Ra decay, are on the order of angstroms (1 Å = 10^{-4} μm) in most solids (cf. Fleischer 1988; Semkow 1990). Greater recoil distances are observed in less-crystalline solids. Recoil distances in water and air are significantly longer than in solids.

When a decay occurs in the environment, the daughter element often has vastly different chemical properties from its parent. If the daughter is a gas, it may partition into more mobile phases, such as air or water. For example, in the decay series beginning with ^{238}U, ^{222}Rn is produced by ^{226}Ra decay. Radon is a noble gas, and is expected to be in the gaseous state under normal Earth-surface conditions. However, if ^{222}Rn is produced in a rock, it may not escape to an adjacent pore space and thus may not become mobile in the environment. An example of immobile ^{222}Rn might be that produced in a zircon ($ZrSiO_4$) crystal, which originally had larger concentrations of U. In this case, the U and its daughters are unable to escape from the crystal lattice. In contrast, consider ^{222}Rn produced by the decay of U adsorbed on HFO that is coating a grain of biotite. In this case, the radioactive elements are at the solid surface and any ^{222}Rn produced can easily enter adjacent pore space. Emanation is defined as that fraction of a gaseous radiogenic element produced in a rock that enters the pore space. It is also referred to as emanating power. Emanation coefficients must be between 0 and 1 and are usually on the order of 0.2 to 0.5 (Flügge and Zimens 1939). Thus, most of the ^{222}Rn produced in a rock remains there.

Radon as a groundwater tracer.   The natural accumulation of Rn (actually ^{222}Rn) in groundwater depends on variables that include: (1) the groundwater flow rate; (2) the rock/water ratio (porosity); (3) the flux (emanation) of Rn from the rock to the groundwater; (4) the distribution of radon parents in the rock (i.e., within mineral grains or on mineral surfaces); and (5) the nature of the flow system (i.e., fractured versus porous media). Groundwater flow may limit the accumulation of Rn because at faster flow rates less Rn is transferred to a parcel of groundwater before it moves farther down a flow path. However, if the rock emanates Rn uniformly over a sufficient distance to produce a steady-state Rn concentration, such dilution may not occur. If Rn emanation is not spatially homogeneous, then faster flow rates will create lower average Rn concentrations, but the Rn may be spread over a greater area.

Porosity affects groundwater Rn concentrations because, other things being equal, greater porosities lead to increased dilution of dissolved Rn. Greater emanation rates for Rn will obviously lead to greater dissolved Rn concentrations. Distribution of the parent radionuclides of Rn, especially ^{226}Ra and ^{238}U, affects dissolved Rn concentrations because the Rn is more or less likely to enter the groundwater, depending on the distribution of the parent. The nature of the flow system is important because of the degree of contact of groundwater with the bulk rock and the effects of flow on other variables. Several studies have attempted to quantify the many factors that affect Rn concentrations in groundwater. The effort demands that the water/rock system be characterized in great detail, using a combination of geologic, hydrologic, geochemical, and geophysical methods (Folger 1995; Folger et al. 1996).

Radon and human health.   Approximately 100,000 to 150,000 deaths from lung cancer are reported annually in the United States. Of these, 80 to 85% are caused by cigarette smoking, but the remainder have no known direct cause. According to the Environmental Protection Agency (EPA 1986) some of these deaths may result from lifetime exposures to the natural ^{222}Rn that accumulates in houses. At high levels, Rn is a suspected carcinogen because it and two rapidly produced daughter radionuclides decay by α-emission. If Rn is inhaled and adheres to the lung tissue, these α-particles may damage lung cells and lead to cancer. Indoor Rn has become an important health issue over the past two decades, although the exact mechanism of carcinogenesis remains unknown. The lifetime

Rn dose of nonsmoking individuals with lung cancer is difficult to estimate because people move from house to house and place to place over their lifetimes; thus levels of exposure that may be considered dangerous are not well established. For these and other reasons, indoor Rn remains a contentious issue in scientific and regulatory circles (cf. Hopke 1987; Nazaroff and Nero 1988).

Because ^{238}U and ^{235}U occur in all rocks, daughters in these decay series (Table 13.2) also occur in all rocks and soils. The decay of ^{226}Ra in the ground produces ^{222}Rn, which may enter the pore-filling medium in proportion to the emanation coefficient. In the unsaturated zone, the Rn enters soil gas, soil moisture, and capillary water. In the saturated zone, Rn enters the groundwater. The ^{222}Rn concentration in groundwater or soil gas is proportional to the abundance of ^{226}Ra in the rock or soil. The ^{222}Rn accumulates in houses, either because of the flux of soil gas into the lowest level of the house or because of ^{222}Rn present in a potable water supply, which degasses into the indoor air as the water is used (cf. Gesell and Prichard 1980; Hess et al. 1987; Folger et al. 1994). Usually the Rn contributed to indoor air from soil gas exceeds that from degassing of a water supply, but the latter sometimes leads to short-lived, extremely high Rn levels in indoor air. (See Chap. 2, Problem 5.)

Example 13.2

A soil with 5 pCi/g of ^{226}Ra, has an emanation coefficient of 0.20, an average porosity of 0.3, and a solid density of 2.7 g/cm^3. Therefore. the bulk density is about 1.9 g/cm^3. Calculate the average ^{222}Rn concentration in the soil gas, assuming secular equilibrium.

If secular equilibrium is attained, then 1 gram of soil, along with its entrained air, also contains 5 pCi of ^{222}Rn, 1.0 pCi (20%) of which escapes to the soil gas. With a porosity of 0.3, 1 gram of soil contains 0.16 cm^3 of air. The ^{222}Rn concentration in soil gas is thus 6.3 pCi/cm^3, or 6300 pCi/L of soil gas. Compared to the U.S. EPA's recommended maximum level of 4 pCi/L of ^{222}Rn in indoor air, this is a huge ^{222}Rn concentration!

13.1.8 Measuring Radioactivity and the Mass of Radionuclides

Natural radionuclide concentrations are usually reported in units of radioactivity (e.g., Bq/g or pCi/g). However, to study their geochemical reactions, it is necessary to know radionuclide concentrations in mass-based units, such as moles per liter.

Properly, radionuclide concentrations are reported in units consistent with the method used for their determination. A radioanalytical technique such as liquid scintillation or alpha spectrometry yields a result expressible in terms of radioactivity. A mass-based analytical technique, such as laser phosphorimetric determination of U(aq) yields mass-based concentrations such as micromoles per liter. Detection limits using radioactivity measuring methods are often many orders of magnitude lower than those possible with mass-based methods (cf. Krieger and Whittaker 1980). For example, 100 pCi/L of ^{222}Rn, a concentration easily determined by liquid scintillation, equals 2.93×10^{-18} M, or approximately 1.76×10^6 atoms/L. At standard temperature and pressure, this amount of radon has a minuscule partial pressure of 6.6×10^{-17} bar, assuming ideal gas behavior.

If a conversion is to be made between mass-based and radioactivity-based concentrations (or vice versa), the specific activity (SpA) is used. Specific activity is the mass per unit radioactivity of an individual radioisotope and may be calculated with one of the following equations:

$$\text{SpA (Bq/g)} = 1.324 \times 10^{16}/(m_{amu} \cdot t_{1/2}) \qquad (13.6)$$

$$\text{SpA (pCi/g)} = 3.578 \times 10^{17}/(m_{amu} \cdot t_{1/2}) \qquad (13.7)$$

TABLE 13.3 Specific activities for radionuclides in the ^{238}U decay series

Radionuclide	SpA (pCi/g)	Radionuclide	SpA (pCi/g)
^{238}U	3.33×10^5	^{218}Po $(\alpha, 99.98\%)$	2.83×10^{20}
^{234}Th	2.32×10^{16}	^{214}Pb	3.28×10^{19}
^{234}Pa	1.99×10^{18}	^{214}Bi $(\beta, 97\%)$	4.46×10^{19}
^{234}U	6.19×10^9	^{214}Po	3.22×10^{26}
^{230}Th	1.94×10^{10}	^{210}Pb $(\beta, > 99.99\%)$	8.11×10^{13}
^{226}Ra	9.89×10^{11}	^{210}Bi $(\beta, 99\%)$	1.24×10^{17}
^{222}Rn	1.53×10^{17}	^{210}Po	4.50×10^{15}

where m_{amu} is the atomic mass (in amu), and $t_{1/2}$ is the half-life (in years). Specific activities for radionuclides in the ^{238}U decay series are given in Table 13.3. When using SpA to convert between radioactivity and mass concentrations, to avoid serious errors, it is critical to understand what analytical methods have been used and the nature of sample collection and handling techniques (cf. Welch et al. 1995).

13.2. AQUEOUS GEOCHEMISTRY OF URANIUM

13.2.1 Introduction

As the most abundant actinide element, U averages 1.2 to 1.3 μg/g in sedimentary rocks, ranges from 2.2 to 15 μg/g in granites, and from 20 to 120 μg/g in phosphate rocks (Langmuir 1978; Eisenbud 1987; see also Table 13.1). Uranium occurs in 4+, 5+, and 6+ oxidation states, which are usually written U(IV), U(V) and U(VI). Most important in nature are the uranous [U(IV)] and uranyl [U(VI)] oxidation states.

Seawater contains 2 to 3.7 μg/L U (Kathren 1984). Uranium concentrations are usually between 0.1 and 7 μg/L in U.S. and Russian streams (Rogers and Adams 1970; see also Titayeva 1994), but may exceed 20 μg/L in streams that receive irrigation return flows in the arid southwestern U.S. because of evaporative concentration (Zielinski et al. 1995). Groundwaters in granite have some of the highest U concentrations, although they rarely exceed 20 μg/L (Gascoyne 1989). The maximum acceptable U concentration in Canadian drinking waters (Canadian Drinking Water Quality Guidelines) is 100 μg/L. The proposed U.S. Environmental Protection Agency drinking water standard for U is 20 μg/L (EPA 1991), indicating that U concentrations in natural waters are usually not high enough to constitute a health risk. However, in uraniferous areas of the U.S., U ranges from 1 to 10 μg/L in streams and 1 to 120 μg/L in groundwaters. In uranium mines the U in groundwater is typically 15 to 400 μg/L (Fix 1956). The highest U concentrations are probably found in leachates from the mill tailings produced by U mining and milling, which often contain 10 to 20 mg/L U (Langmuir and Nordstrom 1995; see also Kathren 1984).

Of concern in many countries is the proposed disposal of spent fuel (largely UO_2) from nuclear power plants in a geological respository. Leaching of the spent fuel by groundwater could release to the environment the U and associated radionuclides, including ^{90}Sr, ^{99}Tc, ^{125}I, and ^{137}Cs, and long-lived radioisotopes of Am, Np, and Pu. Because of their importance in nuclear waste disposal, the geochemistry of I, Tc, Am, Np, and Pu are considered briefly later in this chapter.

Uranous ion (U^{4+}) and its aqueous complexes predominate in groundwaters of low Eh. U(IV) is the major oxidation state in the most common uranium ore minerals uraninite [UO_2(c)]—pitchblende is roughly UO_2(am)—and coffinite ($USiO_4$). The U(IV) concentrations in groundwater at low

Eh are usually less than 10^{-8} M because of the extremely low solubilities of these solids. In the U(V) oxidation state, uranium occurs as the UO_2^+ ion which forms relatively weak complexes (Grenthe et al. 1992). This species is only found at intermediate oxidation potentials and low pH's and is unstable relative to U(IV) and U(VI). In oxidized surface- and groundwater-uranium is transported as highly soluble uranyl ion (UO_2^{2+}) and its complexes, the most important of which are the carbonate complexes. The thermodynamic properties of these minerals and aqueous species must be known if we are to understand the reactions that may control U concentrations in natural waters.

13.2.2 Selected Thermodynamic Data

Publication of the Nuclear Energy Agency's (NEA) data base for uranium (Grenthe et al. 1992) has been a major contribution to uranium geochemistry. The authors have sought to update earlier uranium data bases published by Langmuir (1978), Lemire and Tremaine (1980), and Hemingway (1982), among others. However, there is strong evidence that the thermodynamic data for several geochemically important species are seriously in error in Grenthe et al. (1992). The likelihood of some of these errors has recently been acknowledged by the same authors (Grenthe et al. 1995). Presented in the appendix to this chapter in Tables A13.1, A13.2, and A13.3 are thermodynamic data for uranium aqueous species and solids of geochemical interest, and auxiliary thermodynamic data. The U data are based largely on Grenthe et al. (1992), but with several important corrections and additions that are explained in table footnotes.

Among aqueous species, the most important corrections are for stabilities of the complexes $UO_2(OH)_2^\circ$ and $U(OH)_4^\circ$, which are apparently less stable than proposed by Grenthe et al. (1992) by about 2.4 and 10.6 kcal/mol, respectively. At near neutral pH's, stabilities of these complexes define the minimal respective solubilities of U(VI) and U(IV) minerals in groundwater. These errors have important implications to nuclear waste disposal, where the solubilities of U(IV) and U(VI) minerals are being used to define maximum possible uranium concentrations that might be released from a geological repository for nuclear waste (cf. McKinley and Savage 1994).

The version of MINTEQA2 available from the Environmental Protection Agency in Athens, Georgia, has a uranium thermodynamic data base from Langmuir (1978). More recently, D. R. Turner[†] (Turner et al. 1993) has added to MINTEQA2 the extensive data base for radionuclides from the EQ3/6 data base of Wolery (1992b). In this data base uranium entries are taken in part from 1989 and 1990 drafts of Grenthe et al. (1992), as well as from the final draft. These data are not fully consistent. Also included are data for minerals suggested by Langmuir (1978) and estimated by Hemingway (1982), among other sources (Turner 1993).[‡] The Turner version is used to solve most problems in this text, but with K_{eq} values revised when so noted to agree with the values in the chapter tables.

13.2.3 Aqueous Speciation and Solution-Mineral Equilibria

Above about pH 5, U(VI) generally occurs as aquocomplexes in natural waters. The relative importances of the U(VI) hydroxyl complexes are shown in Fig. 13.3 for a typical groundwater U con-

---

[†]Dr. David R. Turner, Center for Nuclear Waste Regulatory Analyses, 6220 Culebra Rd., P.O. Drawer 28510, San Antonio, TX 78228. In a 1996 revision of the MINTEQA2 data base, Turner has added the thermodynamic data for Am from Silva et al. (1995).

[‡]Literature sources of individual K_{eq} and ΔH_r° values for reactions involving the radioactive elements are listed in the files *species.dat* and *radrefs.dat* in Turner's version of MINTEQA2.

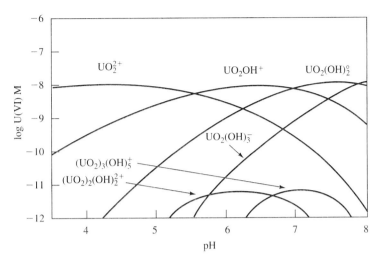

Figure 13.3   Distribution of U(VI) species at 25°C and $I = 0.1$ M for $\Sigma U(VI) = 10^{-8}$ M. $P_{CO_2} = 0$ bar. Reprinted from *Geochim. et Cosmochim. Acta*, 58, T. O. Waite, J. A. Davis, T. E. Payne, G. A. Waychunas, and N. Xu, Uranium (VI) adsorption to ferrihydrate: Application of a surface complexation model, pp. 5465–78, © 1994, with permission from Elsevier Science Ltd., The Boulevard, Langford Lane, Kidlington OX5 1GB, U.K.

centration of about 10^{-8} M. At this concentration stepwise monomeric species are seen to dominate at all pH values. However, model calculations show that at pH = 7 with $\Sigma U(VI) = 10^{-7}$ M, the 3:5 polynuclear complex equals 6% of total U(VI). At even higher U(VI) levels the 3:5, 3:7, and 2:2 polynuclear species become the major hydroxyl complexes.

Uranyl ion forms strong carbonate complexes in most natural waters. Their importance as a function of pH at atmospheric CO_2 pressure ($10^{-3.5}$ bar) and for a typical groundwater CO_2 pressure ($10^{-2.0}$ bar) is shown in Fig. 13.4, which indicates that these complexes largely replace the U(VI)-hydroxyl complexes above pH 6 to 7. The carbonate complexes are extremely important because they greatly increase the solubility of uranium minerals, facilitate U(IV) oxidation, and also limit the extent of uranium adsorption in oxidized waters, thus increasing uranium mobility. The mononuclear carbonate complexes predominate at typical groundwater CO_2 pressures. Other important U(VI) complexes are formed with fluoride, phosphate, and sulfate ligands, for example. The effect of the carbonate complexes on mineral solubilities is evident from Figs. 13.5 and 13.6, which show schoepite and carnotite solubilities as a function of pH and CO_2 pressure.

Uranium U(VI) minerals are most often products of the oxidation and weathering of nearby primary U(IV) ore minerals such as uraninite [$UO_2(c)$] and coffinite [$USiO_4(c)$] (cf. Pearcy et al. 1994). They also form by evaporative concentration of dissolved U(VI), particulary under arid conditions. Schoepite (β-$UO_3 \cdot 2H_2O$) is fairly soluble and, therefore, is a rare mineral, whereas carnotite [$K_2(UO_2)_2(VO_4)_2$] and tyuyamunite [$Ca(UO_2)_2(VO_4)_2$], which have lower solubilities (particularly above pH 5) are the chief oxidized ore minerals of uranium. The plots in Figs. 13.5 and 13.6 indicate that uranyl minerals are least soluble in low-CO_2 waters, and, therefore, are most likely to precipitate from such waters. This is consistent with the occurrence of carnotite and tyuyamunite in oxidized arid environments with poor soil development (Chap. 7), such as in the calcrete deposits in Western Australia (cf. Mann 1974; Dall'Aglio et al. 1974), and in the sandstone-hosted uranium deposits of the arid southwestern United States (cf. Hostetler and Garrels 1962; Nash et al. 1981). The

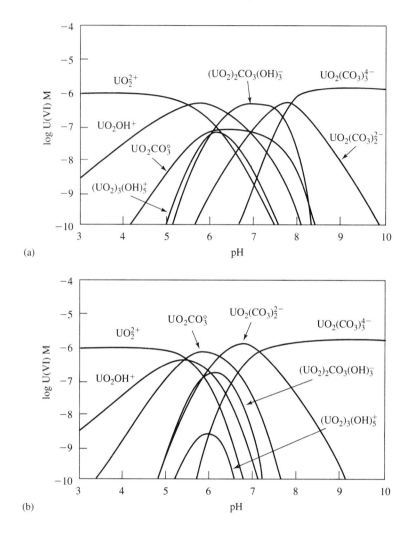

Figure 13.4   Distribution of U(VI) species at 25°C and $I = 0.1$ M for ΣU(VI) = 10^{-6} M, for **(a)** $P_{CO_2} = 10^{-3.5}$ bar, and **(b)** $P_{CO_2} = 10^{-2.0}$ bar. Reprinted from *Geochim. et Cosmochim. Acta,* 58, T. O. Waite, J. A. Davis, T. E. Payne, G. A. Waychunas, and N. Xu, Uranium (VI) adsorption to ferrihydrate: Application of a surface complexation model, pp. 5465–78, © 1994, with permission from Elsevier Science Ltd., The Boulevard, Langford Lane, Kidlington OX5 1GB, U.K.

autunite (cation-U(VI) phosphate) minerals are slightly more soluble (compare K_{sp} values of autunite and carnotite in Table A13.3) and are, therefore, less common than the vanadates. They have been described in the weathered oxidized zone of the Alligator River uranium deposit in Australia, for example (ANSTO 1992).

Example 13.3

Yucca Mountain, Nevada, is being considered as the site for deep geological disposal of U.S. high-level nuclear wastes. Any release of uranium (or other radionuclides) from the waste to

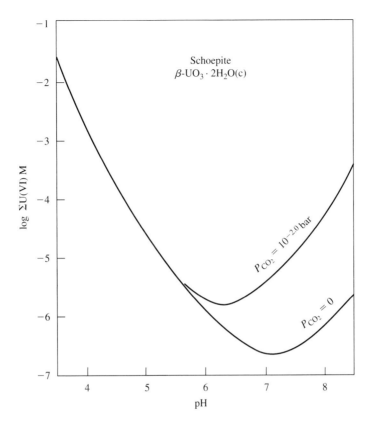

Figure 13.5   Solubility of schoepite at 25°C and 1 bar total pressure as a function of pH for $P_{CO_2} = 0$ and $10^{-2.0}$ bar.

the accessible environment would be transported by groundwater beneath the site. The composition of such groundwater sampled from well J-13 is given here (Ogard and Kerrisk 1984).

Species or parameter	mM	Species or parameter	mM
Ca	0.29	F	0.11
Mg	0.072	Cl	0.18
Na	1.96	SO_4	0.19
K	0.136	NO_3	0.16
Li	0.009	HCO_3	2.34
Fe	0.0008	PO_4	0.00125
Mn	0.00002	pH	7.0
Al	0.00010	DO	0.18
SiO_2	1.07	Eh	700 mV

(a) Assuming the groundwater has 1.0×10^{-8} M dissolved U(VI), input the analysis in MINTEQA2 and determine the speciation of uranium. Before inputing the data, as necessary, change the stability constants of the U(VI) hydroxide and carbonate complexes in MINTEQA2 to their values given in the tables in this chapter. (Instructions for making such changes and additions are given in the *database.txt* file of MINTEQA2.)

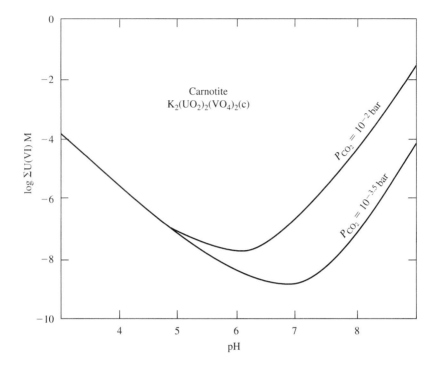

Figure 13.6   Solubility of carnotite at 25°C and 1 bar total pressure as a function of pH, with $K^+ = 10^{-3}$ M, $\Sigma V = 10^{-6}$ M, and $P_{CO_2} = 10^{-3.5}$ and $10^{-2.0}$ bar.

(b) Tabulate the molar percentages of the most important complexes from the output. Also list the molar concentrations of the single most important hydroxyl, fluoride, and phosphate complexes.

    The results are given below.

Species	M	Percent of U(VI)
$UO_2CO_3^\circ$	7.86×10^{-10}	7.9
$UO_2(CO_3)_2^{2-}$	8.31×10^{-9}	83.1
$UO_2(CO_3)_3^{4-}$	7.83×10^{-10}	7.8
UO_2F^+	7.41×10^{-13}	0.007
$UO_2(OH)_2^\circ$	6.11×10^{-12}	0.06
$UO_2PO_4^-$	8.21×10^{-11}	0.8

Example 13.4

Dall'Aglio et al. (1974) placed carnotite-bearing rock in contact with distilled water in a well-mixed laboratory vessel for 1 month. The general composition of the final solution is given here. Final pH was 7.1.

Species	mM	Species	mM
Ca	0.055	U(VI)	8.82×10^{-6}
Mg	0.038	SO_4	0.14
Na	0.71	Cl	0.56
K	0.31	SiO_2	1.3
HCO_3	0.31	VO_4	8.05×10^{-4}

(a) Enter the stability constants of carnotite and tyuyamunite from Table A13.3 into the MINTEQA2 data base. (See instructions in Example 13.3.)

(b) Input the water analysis assuming 25°C and determine the uranium speciation and the saturation state of the water with respect to carnotite and tyuyamunite.

The modeling output shows that U(VI) is entirely complexed, as: $UO_2CO_3^\circ$ 6.4%; $UO_2(CO_3)_2^{2-}$ 78.3%, $(UO_2)_2CO_3(OH)_3^-$ 5.5%; and $UO_2(OH)_3^-$ 1.6%. Written as solubility products, K_{sp}(carnotite) $= 10^{-56.3}$ and K_{sp}(tyuyamunite) $= 10^{-53.3}$ (Table A13.3). The saturation indices (SI) of carnotite and tyuyamunite both exceed -0.3 or are within 0.5% of the log K_{sp} values. Given the uncertainties in the thermodynamic data, both minerals can be considered at saturation in the water.

Many researchers have attempted to measure the solubility of amorphous to crystalline UO_2 (uraninite) as a function of pH. Some of this work is summarized in Table 13.4 and Fig. 13.7. Solubility measurements have been complicated by the fact that the UO_2 solids were often of different and poorly known crystallinity and particle size. Further, oxygen (and possible CO_2) contamination invalidated the results of most early measurements. With oxygen contamination, the measured solubility becomes that of a mixed oxidation-state oxide or a U(VI) solid such as schoepite. Oxygen contamination apparently invalidates the results of Gayer and Leider (1957) and Bruno et al. (1987) who obtained solubilities roughly equal to that of $UO_3 \cdot H_2O$ (as reported by Gayer and Leider 1955) or to the solubility of schoepite as shown in Fig. 13.5. Measurements by Rai et al. (1990), Torrero et al. (1991), and Yajima et al. (1995) (see also Parks and Pohl 1988) indicate that the solubility of amorphous to more crystalline UO_2 is independent of pH above about pH 4 to 4.5. This indicates that the dissolution reaction is

$$UO_2(s) + 2H_2O = U(OH)_4^\circ \tag{13.8}$$

and that the species $U(OH)_5^-$, which would lead to a solubility increase at high pH, can be neglected. These and other authors (cf. Grenthe et al. 1992) have also concluded that the measured solubility of UO_2 can be accurately described considering only U^{4+} and the hydroxyl complexes UOH^{3+} and $U(OH)_4^\circ$. The complexes $U(OH)_2^{2+}$, $U(OH)_3^+$ and $U(OH)_5^-$ can, therefore, be ignored and removed from thermodynamic data bases.[†] Geochemical modeling shows that in most low-Eh groundwaters the species $U(OH)_4^\circ$ predominates over other U(IV) complexes and, therefore, defines the minimum solubility of UO_2.

The solubility measurements of Rai et al. (1990) have been used to derive the stabilities of UO_2(am) and $U(OH)_4^\circ$ given in Tables A13.1 and A13.3, as explained in the footnotes to Table 13.4. Based on the solubility reaction

$$UO_2(am) + 2H_2O = U^{4+} + 4OH^- \tag{13.9}$$

[†]The $U(OH)_n^{4-n}$ complexes are effectively "removed" from MINTEQA2 by entering log $K = -30.0$ in *thermo.dbs* for their formation reactions, which are written $U^{4+} + nH_2O = U(OH)_n^{4-n} + nH^+$.

TABLE 13.4   Solubility products ($pK_{sp} = -\log K_{sp}$) of UO_2, based on the reaction $UO_2 + 2H_2O = U^{4+} + 4OH^-$, for UO_2 solids from amorphous [$UO_2(am)$] to uraninite [$UO_2(c)$], and their molal solubilities as $U(OH)_4^\circ$ at 25°C and 1 bar total pressure

pK_{sp}	UO_2 crystallinity	Solubility as $-\log[U(OH)_4^\circ]$	Source
51.9 (u)[†]	amorphous (XRD)	8.0	Rai et al. (1990)
52 ± 0.4 (s)	amorphous	—	Stepanov and Galkin (1960)
52.6 (s, u)[‡]	microcrystalline (XRD)	8.7	Yajima et al. (1995)
(51.6)[§]	$pK_{sp}(am) = pK_{sp}(c) - 9.4$	—	Langmuir (1978)
(53.0)[‖]	$pK_{sp}(am) = pK_{sp}(c) - 8.0$	—	Rai et al. (1987)
53.0 (u)[‖]	amorphous (XRD)	—	Rai et al. (1996)
53.4 (u, s)[#]	more crystalline (XRD)	9.47	Parks and Pohl (1988)
56.1 (u)[††]	amorphous (XRD) (probably = schoepite)	4.45	Bruno et al. (1987)
61.0 (computed)[‡‡]	well-crystallized	17.1 (computed)	Grenthe et al. (1992)

Note: Experimental solubility measurements run from undersaturation to supersaturation and are denoted by u or s, respectively. Parentheses enclose estimated values. XRD indicates that crystallinity was determined by X-ray diffraction methods. [†]$K = 10^{3.5 \pm 0.8}$ for $UO_2(am) + 3H^+ = UOH^{3+} + H_2O$ (Rai et al. 1990). With this value and $\Delta G_f^\circ(UOH^{3+}) = -182.25$ kcal/mol, we obtain $\Delta G_f^\circ = -234.15$ kcal/mol for $UO_2(am)$, and $pK_{sp} = 51.9$. Rai et al. (1990) report $pK_{sp} = 52.0 \pm 0.8$. Their measured $UO_2(am)$ solubility of $10^{-8.0}$ M as $U(OH)_4^\circ$ leads to $\Delta G_f^\circ[U(OH)_4^\circ] = -336.6$ kcal/mol. In serious disagreement Grenthe et al. (1992) propose $\Delta G_f^\circ[U(OH)_4^\circ] = -347.2$ kcal/mol. [‡]Working with X-ray microcrystalline UO_2, Yajima et al. (1995) measured a solubility of $10^{-8.7}$ M as $U(OH)_4^\circ$, which, with ΔG_f° data from Tables A13.1 and A13.2, and from table note †, results in $pK_{sp} = 52.6$. [§]The difference in solubilities of M^{4+} oxides and hydroxides of Hf, Th, Ti, and Zr, written as MO_2 average 9.4 pK_{sp} units according to Langmuir (1978), or 8.0 pK_{sp} units according to Rai et al. (1987). Assuming $pK_{sp} = 61.0$ for $UO_2(c)$ (this table), this suggests that for $UO_2(am)$, $pK_{sp} = 51.6$ or 53.0, respectively. [‖]Rai et al. (1996) measured the solubility of $UO_2(am)$ in up to 6.0 molal NaCl and 3.0 molal $MgCl_2$ solutions, extrapolating their results to $I = 0$ with the Pitzer ion-interaction model. Given the many assumptions and corrections involved, their proposed pK_{sp} of 53.0 is probably less accurately known than earlier values measured at lower ionic strength. [#]Parks and Pohl (1988) measured the solubility of a more crystalline UO_2 between 100 and 300°C from pH 1 to 10, and obtained $10^{-9.47}$ M as $U(OH)_4^\circ$. The solubility was independent of temperature, which suggests $pK_{sp} = 53.4$ at 25°C. [††]Bruno et al. (1987) reported a $UO_2(am)$ solubility of $10^{-4.45}$ M as $U(OH)_4^\circ$. This roughly corresponds to the solubility of schoepite [β-$UO_3 \cdot 2H_2O(c)$] and suggests that the experiments were contaminated with oxygen and CO_2. [‡‡]The ΔG_f° value for uraninite [$UO_2(c)$] computed by Grenthe et al. (1992) from calorimetric measurements of its enthalpy and entropy, leads to $pK_{sp} = 61.0$ for $UO_2(c)$.

the results of Rai et al. (1990) lead to $K_{sp}[UO_2(am)] = 10^{-51.9}$. This is in excellent agreement with MINTEQA2-modeled analyses of 11 groundwaters in contact with U(IV) ore deposits. The three most saturated of these groundwaters have an average $[U^{4+}][OH^-]^4$ product of $10^{-51.63 \pm 0.02}$.

Coffinite ($USiO_4$) is the dominant uranium ore mineral in many sandstone-type deposits in the Colorado Plateau region of the southwestern United States (cf. Goldhaber et al. 1987), where it occurs with quartz, fine-grained uraninite, and organic material (Nord 1977). It is also common in other world deposits (cf. Pacquet et al. 1987; Pearcy et al. 1994). Langmuir (1978) estimated coffinite stability from that of $UO_2(c)$, and the high silica concentrations present in waters associated with coffinite and uraninite, ~$10^{-3.0}$ M $SiO_2(aq)$, via the reaction

$$UO_2(c) + H_4SiO_4^\circ = USiO_4(c) + 2H_2O \tag{13.10}$$

This approach leads to $\Delta G_f^\circ(coffinite) = -449.9$ kcal/mol. In good agreement, based on the high-temperature decomposition of coffinite to $UO_2(c)$ and silica glass, Hemingway (1982) estimated $\Delta G_f^\circ(coffinite) = -450.76$ kcal/mol, which is the value listed in Table A13.1. Because these estimates are based on the stability of $UO_2(c)$, they probably refer to the stability of $USiO_4(c)$. A more direct

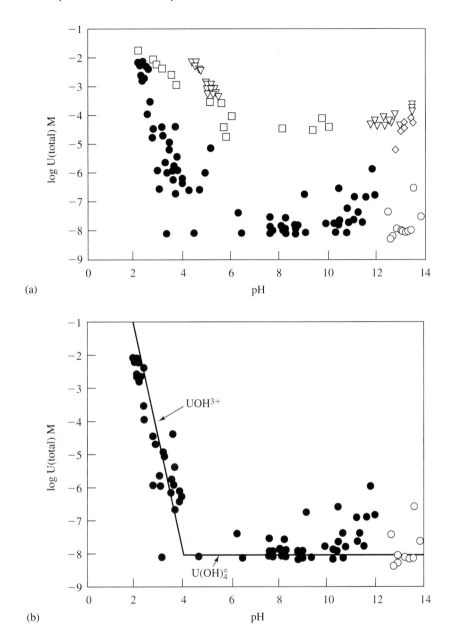

(a)

(b)

Figure 13.7 (a) The measured solubility of $UO_2 \cdot xH_2O(am)$ [$U(OH)_4(am)$] at 25°C as a function of pH as reported by Gayer and Leider (1957) ($\Diamond$); Ryan and Rai (1983) ($\bigcirc$); Bruno et al. (1987) ($\square$); and Rai et al. (1990) ($\bullet$). The solubility of $UO_3 \cdot H_2O$ from Gayer and Leider (1955) ($\triangledown$) is also shown for comparison. (b) The measured solubility of $U(OH)_4(am)$ as a function of pH according to Ryan and Rai (1983) ($\bigcirc$) and Rai et al. (1990) ($\bullet$). The solid lines are best fit lines for the reactions $U(OH)_4(am) + 3H^+ = UOH^{3+} + 3H_2O$ and $U(OH)_4(am) = U(OH)_4^\circ$. Reprinted with permission from *Inorg. Chem.* 29:260–64, D. Rai et al., Uranium (IV) hydrolysis constants and solubility product of $UO_2 \cdot xH_2O(am)$. Copyright © 1990 American Chemical Society.

approach to coffinite stability is to compute it using the chemical analyses of low Eh groundwaters from coffinite-bearing ore zones. This leads to $K = 10^{0.50 \pm 0.03}$ for the reaction

$$USiO_4(s) + 4H^+ = U^{4+} + H_4SiO_4^\circ \tag{13.11}$$

which corresponds to the average highest coffinite solubility in three groundwaters from the Cigar Lake and Palmottu uranium deposits (see Table A13.4). These same waters are at saturation with respect to $UO_2(am)$, suggesting that the associated coffinite phase may also be amorphous. This calculation is the basis for $\Delta G_f^\circ[USiO_4(am)]$ listed in Table A13.1. The difference in ΔG_f° values for amorphous and crystalline $USiO_4$ in Table A13.1, indicates a solubility (pK_{sp}) difference of $10^{8.9}$ times, which seems reasonable compared to $pK_{sp} = 10^{9.1}$ times between $UO_2(am)$ and $UO_2(c)$ (Table A13.3).

An Eh-pH diagram for the system $U-O_2-H_2O$ at 25°C and a typical groundwater uranium concentration of $\Sigma U(aq) = 10^{-8}$ M is given in Fig. 13.8. The plot shows the stability fields of the dominant aqueous species and the large size of the stability field of uraninite $[UO_2(c)]$. If instead the stability field of $UO_2(am)$ is plotted, it almost exactly overlaps the field of $U(OH)_4^\circ$. Of particular

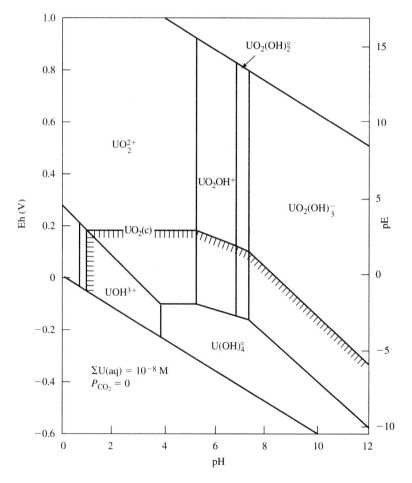

Figure 13.8 Eh-pH diagram for aqueous species in the $U-O_2-H_2O$ system in pure water at 25°C and 1 bar total pressure for $\Sigma U = 10^{-8}$ M. The $UO_2(c)$ solid/solution boundary for $\Sigma U = 10^{-8}$ M is stippled.

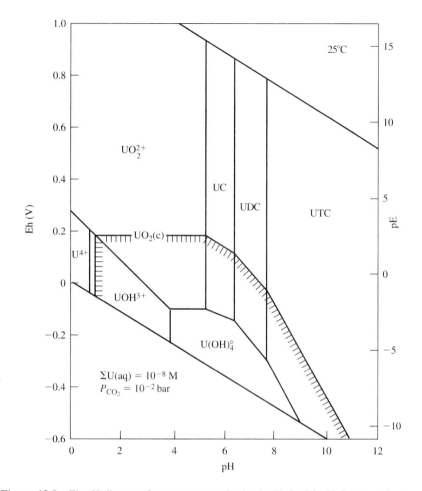

Figure 13.9 Eh-pH diagram for aqueous species in the U-O$_2$-CO$_2$-H$_2$O system in pure water at 25°C and 1 bar total pressure for ΣU = 10^{-8} M and P_{CO_2} = 10$^{-2.0}$ bar. UC, UDC, and UTC denote the aqueous complexes UO$_2$CO$_3^\circ$, UO$_2$(CO$_3$)$_2^{2-}$, and UO$_2$(CO$_3$)$_3^{4-}$, respectively. The position of the UO$_2$(c) solid/solution boundary for ΣU = 10^{-8} M is stippled.

interest, the diagram shows the predominance of the uranyl-hydroxy complexes at low Eh values in the presence of uraninite, with U(OH)$_4^\circ$ only important in waters where the Eh is less than about −100 to −200 mV.

At a typical groundwater CO$_2$ pressure of 10^{-2} bar, the highly stable uranyl carbonate complexes predominate above about pH 5 (Fig. 13.9). Comparison of Figs. 13.8 and 13.9 indicates that these complexes are stable relative to U(OH)$_4^\circ$ under highly reducing conditions. Accordingly, above pH 5, the oxidation of U(IV)(aq) and dissolution of UO$_2$(s) can occur at lower Eh values when high carbonate concentrations are present. (The U(IV)-carbonate complexes are unstable relative to U(OH)$_4^\circ$ under these conditions and so do not stabilize U(IV)(aq).)

Most natural uraninites or pitchblendes (and probably coffinites) are partially oxidized, with compositions between UO$_{2.00}$ and UO$_{2.67}$ (U$_3$O$_8$) (cf. Giblin 1987; Ahonen et al. 1993; Sunder et al. 1994). The stabilities of oxide phases of mixed oxidation state (between IV and VI) are shown in

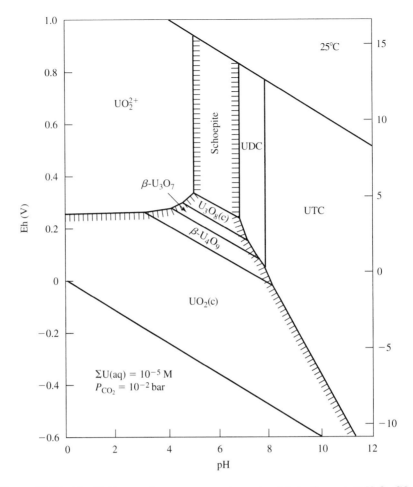

Figure 13.10 Eh-pH diagram for aqueous species and solids in the system U-O_2-CO_2-H_2O at 25°C and 1 bar total pressure. Solid/aqueous boundaries (stippled) are drawn for $\Sigma U = 10^{-5}$ M. UDC and UTC are $UO_2(CO_3)_2^{2-}$ and $UO_2(CO_3)_3^{4-}$, respectively.

Fig. 13.10, along with stability fields of stoichiometric uraninite and schoepite. The plot indicates that the intermediate oxides have a small stability range in Eh-pH space, however their stability fields occur under conditions commonly encountered in groundwater. Ahonen et al. (1993) have suggested, in fact, that their measured Eh and pH values in three drill holes in the Finnish Palmottu uranium deposit may be in equilibrium with $UO_{2.33}$ (U_3O_7) (Fig. 13.11). (See also Cramer and Smellie 1994, regarding the Cigar Lake deposit.)

The Eh-pH diagram in Fig. 13.12 shows the large stability field of carnotite under oxidizing conditions (the field of tyuyamunite is similar), consistent with its common occurrence in the weathered zone of U(IV) uranium deposits. The figure, which is from Langmuir (1978), is little changed by revisions in the thermodynamic data base in Table A13.1.

In recent years numerous complete analyses of groundwaters have become available that include Eh and pH values and uranium analyses. Some examples of these are given in Table A13.4.

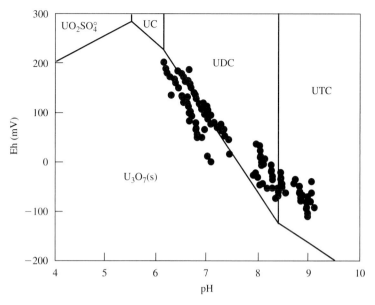

Figure 13.11   Eh-pH diagram showing measured Eh and pH values in groundwaters in the Palmottu uranium deposit, Finland. The diagram is drawn for C_T (total carbonate) = 2 × 10^{-3} M and 10^{-2} M SO_4^{2-}, with $\Sigma U = 10^{-8}$ M at the U_3O_7(s) boundary. This solid (equivalent to $UO_{2.33}$(s)) is considered a possible control on dissolved U concentrations. UC, UDC and UTC are defined in Fig. 13.9. Modified from Ahonen et al., *Uranium mineral-groundwater equilibration at the Palmottu natural analogue study site,* Finland. Mat. Res. Soc. Symp. Proc. 294:497–504. Copyright 1993 by Materials Research Society. Used by permission.

Such sampling and analyses have had several purposes. One goal has been to characterize geologic systems analogous to those that might be considered for nuclear waste disposal (analyses 3 and 10) (see also Pearson et al. 1989; Pearson and Scholtis 1992). Because the spent nuclear fuel requiring disposal is chiefly UO_2, numerous studies of groundwaters associated with uraninite deposits have been studied as analogs for a potential geological repository (analyses 1, 2, 6 to 9) (cf. Pearcy et al. 1994). Groundwater analyses have also been used to prospect for uranium deposits (analyses 4 and 5) (see also Langmuir and Chatham 1980; Giblin 1987), and as part of the effort to solution-mine (*in situ* leach) such deposits to extract uranium (analysis 11) (cf. Deutsch et al. 1985).

An interesting conclusion from studies of low Eh groundwaters from crystalline rocks has been that most appear at saturation with some form of UO_2 and perhaps $USiO_4$ (and ThO_2) (cf. Paces 1969; Krupka 1983; Langmuir 1987; Gascoyne 1989; Pearson and Scholtis 1992). For example, MINTEQA2 modeling of analysis 3 (Äspö, Sweden) and analysis 10 (Stripa, Sweden), which contain 1.9×10^{-9} and 4.4×10^{-10} M U(IV)(aq), gives the following respective saturation indices: for UO_2(am) −0.9 and −1.4; for $USiO_4$(am) −1.4 and −0.7. In other words, the groundwaters are saturated with respect to phases that are slightly less soluble than the amorphous forms, but much more soluble than crystalline UO_2 or $USiO_4$. The implication is that the UO_2 of spent nuclear fuel might be close to equilibrium in such groundwaters and so not tend to dissolve. This would beneficially limit the release to groundwater and the accessible environment of more radioactive and hazardous elements such as Pu, Np, and Am, in the spent fuel.

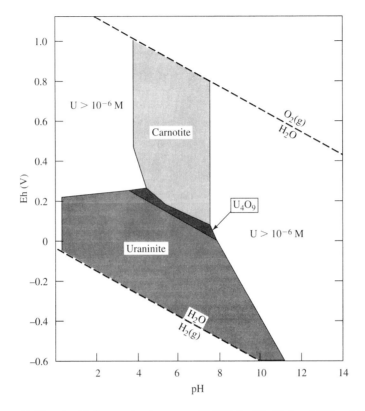

Figure 13.12   Eh-pH diagram for the system K-U-V-O_2-CO_2-H_2O at 25°C and 1 bar total pressure, for $K^+ = 10^{-3}$ M, $\Sigma V = 10^{-6}$ M (0.1 mg/L as VO_4), and $P_{CO_2} = 10^{-2}$ bar, showing aqueous fields where $\Sigma U > 10^{-6}$ M. Solid/solution boundaries are drawn for $\Sigma U = 10^{-6}$ M (0.24 mg/L). Reprinted from *Geochim. et Cosmochim. Acta*, 42(6), D. Langmuir, Uranium solution-mineral equilibria at low temperatures with applications to sedimentary ore deposits, pp. 547–69, © 1978, with permission from Elsevier Science Ltd., The Boulevard, Langford Lane, Kidlington OX5 1GB, U.K.

Example 13.5

Input analysis no. 2 (Table A13.4) from Cigar Lake into MINTEQA2, and determine (a) total dissolved concentrations of U(IV) and U(VI); (b) the speciation of U(IV) and U(VI); and (c) *SI* values for UO_2(am) and $USiO_4$(am).

Total concentrations are $\Sigma U(IV) = 2.38 \times 10^{-8}$ M, and $\Sigma U(VI) = 9.02 \times 10^{-9}$ M. The U(IV) is 100% as $U(OH)_4^\circ$. U(VI) occurs as 14.6% $UO_2CO_3^\circ$, 77.5% $UO_2(CO_3)_2^{2-}$, 3.5% $UO_2(CO_3)_3^{4-}$, 1.2% $UO_2(OH)_2^\circ$, and 1.9% $UO_2(OH)_3^-$. The *SI* values are 0.22 for UO_2(am) and 0.07 for $USiO_4$(am), probably indicating that the groundwater is at saturation with both phases within uncertainties in the chemical analyses and thermodynamic data.

13.2.4 Adsorption-Desorption Reactions and Models

Above it was shown that U(aq) concentrations may be at saturation with respect to U minerals in some low-Eh groundwaters in granites and in waters associated with primary and secondary U mineral deposits. At more usual, lower concentrations, dissolved U and other trace elements, including other actinides, will always partition themselves between the water and the surfaces of contacting solids in soils, sediments, and rocks. In fact, in typical soils and groundwater systems (pH >5) more than 99% of individual trace elements will be associated with solid surfaces and less than 1% dissolved. Equilibration of trace U (or other actinide) species in subsurface waters is thus usually with sorption sites. To understand U mobility we must, therefore, also understand its adsorption behavior.

The pH range of minimum solubility of the uranyl minerals is also the pH range of maximal U(VI) sorption on most important natural sorbents, including organic matter (van der Weijden and Van Leeuwen 1985), Fe(III) oxyhydroxides, Mn and Ti oxyhydroxides, zeolites and clays (cf. Langmuir 1978; Turner 1995). In terms of approximate K_d values, where K_d (ml/g) = (wt adsorbed/wt sorbent)/(solute concentration) (see Chap. 10), maximum K_d values for U(VI) sorption by the above phases are: $Ti(OH)_4$(am), 8×10^4 to 10^6; HFO, 1.1×10^6 to 2.7×10^6; peat, 10^4 to 10^6; fine-grained natural goethite (α-FeOOH), 4×10^3; phosphorites, 15; montmorillonite, 6; and kaolinite, 2 (Langmuir 1978).

Because of their common occurrence in soils and sediments and strong sorptive behavior toward U(VI), the Fe(III) oxyhydroxides are generally the most important potential sorbents for U, with organic matter (peat, for example) second in importance. Shown in Fig. 13.13 are typical U(VI) concentration versus pH plots for U(VI) adsorption by HFO and goethite. The similarity between these plots and the pH-solubility curves for schoepite (Fig. 13.5) and carnotite (Fig. 13.6) is readily apparent. Once U(VI) has been adsorbed, it may be reduced to U(IV) in uraninite or coffinite by mobile reductants such as H_2S, CH_4, or Fe^{2+} (cf. Goldhaber et al. 1987) or by the sorbent itself if the latter is organic matter (cf. Schmidt-Collerus 1967; Meunier et al. 1987). If reduction does not follow adsorption, uranyl can be desorbed by an increase in alkalinity at constant pH, or by raising the pH. Such changes increase the extent of uranyl carbonate complexing (see Fig. 13.4), and because the carbonate complexes are poorly adsorbed, cause the desorption and remobilization of uranyl species.

As just noted, U(VI) adsorption behavior depends on its aqueous speciation, which is a function of pH and redox conditions, and of the absolute and relative concentrations of ligands that form U(VI) complexes, for the complexes have different tendencies to be adsorbed by different sorbents. Given the plethora of independent variables involved, the modeling of U(VI) adsorption is best accomplished using an electrostatic adsorption model, such as the constant capacitance (CC), diffuse-layer (DL), or triple-layer (TL) model (Chap. 10). Hsi and Langmuir (1985) modeled U(VI) adsorption by HFO, goethite, and hematite with the TL model (Davis et al. 1978). They proposed that U(VI) was adsorbed chiefly as its UO_2OH^+ and $(UO_2)_3(OH)_5^+$ aquocomplexes between about pH 3 to 8, and assumed that the adsorbed surface complexes had the same stoichiometry as the aqueous complexes. Using a different thermodynamic data base for U(VI) aqueous species with the incorrect ΔG_f° value for $UO_2(OH)_2^\circ$ from Grenthe et al. (1992) (see Table A13.1), Turner (1995) has successfully modeled the data of Hsi and Langmuir (1985) for U(VI) adsorption by HFO and goethite, assuming adsorption is of any one of the three complexes, $UO_2(OH)_2^\circ$, $UO_2(OH)_3^-$, or $UO_2(OH)_4^{2-}$. Turner (1995) also noted that the U(VI) adsorption data could be just as accurately modeled with the CC or DL models as with the TL model.

In an effort to understand adsorption mechanisms, Waite et al. (1994) (see also Chisholm-Brause and Morris 1992) examined the character of U(VI) adsorption sites on the HFO surface with uranium EXAFS spectroscopy. They concluded that a single inner-sphere, mononuclear, bidentate complex, $(\equiv FeO_2)UO_2$, could explain their low pH-adsorption results and that U(VI) desorption at alkaline pH's could be modeled assuming a $(\equiv FeO_2)UO_2CO_3^{2-}$ surface species. Waite et al. (1994) used the DL model in their study and assumed the existence of both weak and strong adsorption sites (see Fig. 13.14 and Chap. 10).

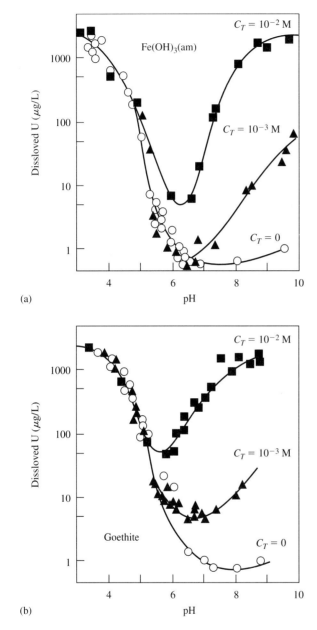

(a)

(b)

Figure 13.13   The effect of changes in total carbonate (C_T) on the adsorption of U(VI) onto 1 g/L suspensions of **(a)** Fe(OH)$_3$(am) or HFO and **(b)** α-FeOOH (goethite) in 0.1 M NaNO$_3$ solutions at 25°C, as a function of pH for ΣU = 10^{-5} M. Reprinted from *Geochim. et Cosmochim. Acta,* 49(11), C. K. D. Hsi and D. Langmuir, Adsorption of uranyl onto ferric oxyhydroxides: Application of the surface complexation site binding model, pp. 2423–32, © 1985, with permission from Elsevier Science Ltd., The Boulevard, Langford Lane, Kidlington OX5 1GB, U.K.

As shown in Fig. 13.14, most researchers display their adsorption results in plots of percent adsorbed versus pH. The selection of K^{int} values for trace element adsorption is then based on the fit of model-derived adsorption curves to the data plotted as percent adsorbed. This is a poor procedure because the important adsorption data below 1% or above 99% adsorbed cannot be distinguised and, therefore, may not be accurately modeled. If this approach had been used to present the adsorption data given in Fig. 13.13, none of the important data for U(VI)(aq) below about 24 μg/L [<1% of ΣU(VI)(aq)] could be distinguished from zero percent adsorbed.

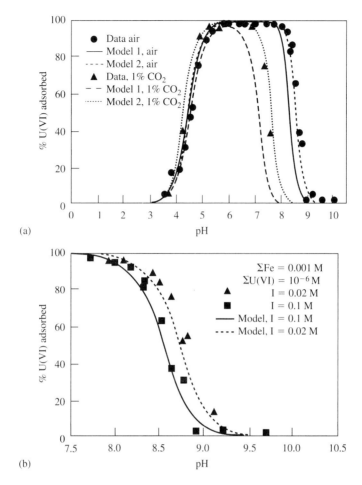

Figure 13.14 Diffuse layer (DL) modeling of U(VI) adsorption by HFO at 10^{-3} M as Fe, as a function of pH, for $\Sigma U(VI) = 10^{-6}$ M. Model 1 assumes one U(VI) species is adsorbed at two surface sites. Model 2 assumes two species are adsorbed at two surface sites. (**a**) Model fits to adsorption data at $I = 0.1$ M, and CO_2 pressures of 0.01 bar (1%) and 0.00032 bar (air) between pH 3 to 10. (**b**) Effect of different ionic strengths ($I = 0.02$ M and 0.1 M) on model 2 adsorption fits for 0.00032 bar CO_2 pressure between pH = 7.5 to 10. Reprinted from *Geochim. et Cosmochim. Acta,* 58, T. O. Waite, J. A. Davis, T. E. Payne, G. A. Waychunas, and N. Xu, Uranium (VI) adsorption to ferrihydrate: Application of a surface complexation model, pp. 5465–78, © 1994, with permission from Elsevier Science Ltd., The Boulevard, Langford Lane, Kidlington OX5 1GB, U.K.

Example 13.6

Assuming that uranyl carbonate complexes are not adsorbed, using the triple layer (TL) model compute the adsorption of uranyl species by hydrous ferric oxide (HFO) at pH = 7 in the absence of CO_2, and for $P_{CO_2} = 10^{-2.0}$ bar, given that total U(VI) = 10^{-5} M. Assume the solution electrolyte is 0.1 M $NaNO_3$, and that 1 g/L of HFO is in contact with the solution, with a surface area of 700 m²/g, corresponding to 0.021 mol sites/L. Also, $\mathbb{C}_1 = 1.25$ F/m², $\mathbb{C}_2 = 0.2$ F/m².

The following adsorption reactions and intrinsic constants must be entered in PRODEFA2 (see Table 10.12).

Expression	$\log K^{int}$
$SOH + H^+ = SOH_2^+$	+5.0
$SOH - H^+ = SO^-$	−10.9
$SOH + Na^+ - H^+ = SONa$	−9.3
$SOH + H^+ + NO_3^- = SOH_2NO_3$	+7.0
$SOH + UO_2^{2+} + H_2O - 2H^+ = SOUO_2OH$	−6.9
$SOH + 3UO_2^{2+} + 5H_2O - 6H^+ = SO(UO_2)_3(OH)_5$	−13.9

Modeling output shows that U(VI)(aq) = 1.39×10^{-10} M in CO_2-free water and 1.18×10^{-7} M when $P_{CO_2} = 10^{-2.0}$ bar. In other words, complexing of uranyl by carbonate reduces its adsorption and leads to a U(VI)(aq) concentration 850 times greater when $P_{CO_2} = 10^{-2}$ bar than when no CO_2 is present.

13.3 URANIUM ORE DEPOSITS

13.3.1 Origin of Low-Temperature Uranium Deposits

Approximately two-thirds of the world's uranium deposits have formed below 50°C in near-surface sedimentary environments (Nash et al. 1981). Generally, the U has been leached from a soil or rock that is somewhat enriched in U, such as a granite or rhyolitic tuff. It was then transported in neutral to alkaline, oxidized waters, chiefly as U(VI)-carbonate complexes. When the groundwater encountered a redox interface (see Chap. 11), U(VI) was reduced to U(IV) and precipitated as insoluble uraninite or coffinite, or as a less-crystalline material, such as pitchblende. Reducing conditions at the redox interface were created by H_2S, or by a zone in the sediment, rich in organic matter and/or sulfide minerals such as pyrite or marcasite (FeS_2) (Miller et al. 1984). Most sediment-hosted U deposits formed in this way and have been preserved over geologically significant times by the persistence of a locally reducing environment. Perhaps because of this last requirement, sandstone-hosted U deposits are found almost exclusively in younger rocks (Nash et al. 1981).

Shown in Fig. 13.15 is a schematic cross-section of a sedimentary U deposit described as a roll front. In the United States such deposits are found in Colorado, Texas, and Wyoming, for example. They are described as "roll fronts" because over time they "roll" downgradient as the upgradient edge of the deposit dissolves and is redeposited near the downgradient edge. Depending on the geology and hydrology of the host sandstone, roll-front deposits may range from a few meters to hundreds of meters across.

Redox-sensitive elements such as As, Mo, Se, and V, are soluble along with U(VI) in oxidized groundwaters where they occur as oxyanions. However, in the reduced waters at the redox interface of a roll front or other sedimentary U deposit, they are precipitated nearby along with U(IV) in insoluble minerals (cf. Wanty et al. 1987).

13.3.2 Uranium Ore Deposits as Analogs for a Nuclear Waste Repository

Since the nuclear accidents at the Three Mile Island power plant in the U.S. in 1979, and the near meltdown at Chernobyl, Ukraine in 1986 (cf. Stone 1996; Bradley et al. 1996), the value of uranium

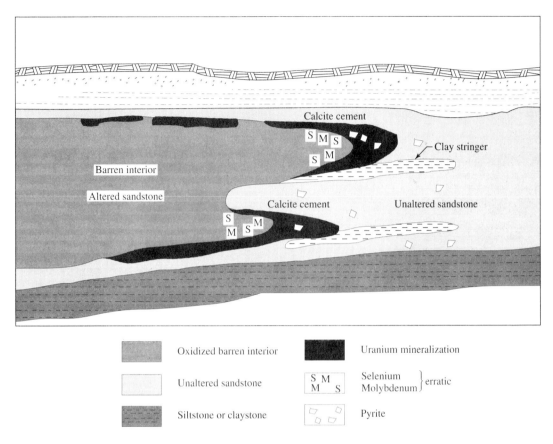

Figure 13.15   Schematic cross-section of an idealized uranium roll-front orebody showing the zonation of elements and primary hydrologic and geochemical features. Oxidized groundwaters flow from left to right. The roll front and associated redox interface moves in the same direction. After Larson (1978).

ore deposits and uranium exploration and mining activity, have both decreased dramatically in the United States and elsewhere. However, numerous projections forecast a resurgence of nuclear power in the first half of the next century, if other energy sources are unable to satisfy burgeoning global energy demands (Nakicenovic 1993). In the meantime, much of the current interest in U deposits is focused on examining their behavior as possible analogs for geologic repositories in which to bury nuclear wastes. In most analog deposits the U occurs as uraninite or pitchblende, similar to spent fuel UO_2. Analog studies have examined the occurrence and mobility of radionuclides from U deposits located both above and below the groundwater table (cf. Curtis et al. 1994). The goal is to learn if and how various radionuclides or analog elements might migrate from nuclear waste repositories in similar settings over long times. Pearcy et al. (1994) have summarized the geologic and mineralogic characteristics of a number of analog U deposits.

The Pena Blanca deposit in northern Mexico occurs in unsaturated and oxidized rhyolitic tuffs, in a geologic and climatic setting similar to that of the proposed U.S. repository at Yucca Mountain, Nevada. Much of the original UO_2 ore has been oxidized and altered, sometimes first to form U(VI) oxide hydrates such as schoepite, and later to precipitate as more stable and abundant U(VI) silicate

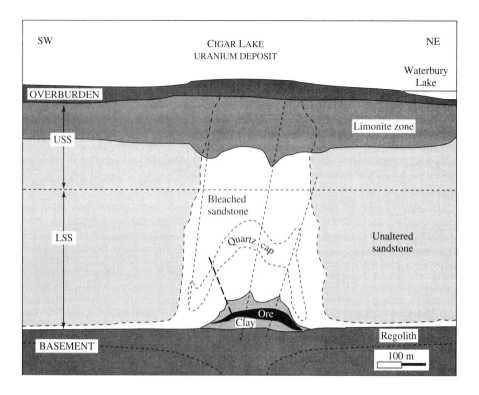

Figure 13.16   Schematic cross-section through the Cigar Lake uranium deposit, showing the U ore and host rocks, including lithologic characteristics related to hydrothermal alteration and weathering. USS and LSS denote upper and lower sandstone. Near-vertical dashed lines denote faults. Modified after J. J. Cramer and J. A. T. Smellie, eds., *Final report of the AECL/SKB Cigar Lake analog study.* Report AECL-10851. Copyright 1994 by Whiteshell Laboratories. Used by permission.

minerals that include chiefly uranophane, with smaller amounts of soddyite (Pearcy et al. 1994). Wilson (1990) produced similar Ca-U(VI) silicate solids by leaching spent fuel with oxidized J-13 well water.

The Cigar Lake uranium deposit in Saskatchewan, Canada, is an analog for a repository located below the groundwater table (cf. Cramer and Sargent 1994; Cramer and Smellie 1994). The deposit, which was formed at elevated temperatures, is currently about 450 m below the land surface. The ore is in sandstones that overlay granitic basement rocks (Fig. 13.16). Primary U ore minerals include uraninite, pitchblende, and minor coffinite. Features that make the Cigar Lake deposit a useful, natural analog site include: (1) the great age of the ore (>1.2 billion y); (2) high ore grades (average 8% up to 55% U); (3) the reducing chemical environment in and adjacent to the orebody; and (4) the relatively high amounts of clay found in the matrix of sandstones surrounding the deposit. Taken together, the age, ore grade, mineralogy, and chemical environment show that accumulations of U can be stable and practically immobile in low-Eh groundwaters for geologically significant time periods. The clay content of the surrounding rock has been used to demonstrate that clay minerals are effective sorbents for any U that is mobilized, preventing large-scale migration of U from the orebody.

13.4 NUCLEAR POWER AND HIGH-LEVEL NUCLEAR WASTES

13.4.1 Composition of Nuclear Fuel and High-Level Nuclear Wastes

Natural uranium from the mining and milling of U ore, has an average composition of 0.0057% ^{234}U, 0.719% ^{235}U, and 99.275% ^{238}U (Rösler and Lange 1972). The ^{235}U isotope is unique, in that it can be split into two atoms (fission fragments) of roughly equal size by the impact of a slowly moving neutron.

$$^{235}U + {}^1n_o \rightarrow \text{fission fragments} + 2\text{--}3 \text{ neutrons} + \text{energy} \qquad (13.12)$$

The reaction, which involves nuclear fission, also produces more neutrons and enormous amounts of energy. Some nuclear power plants use fuel rods made of natural U, unenriched in ^{235}U (e.g., Canadian CANDU fuel; cf. Kathren 1984). In most countries the fuel rods are fabricated from U enriched in ^{235}U to 1.8 to 3.7% of total U (Adloff and Guillaumont 1993). This increases the radioactivity and, therefore, the heat output of the fuel, which is derived from the kinetic energy of the fission fragments and radioactivity of fission products. The heat is used to produce steam to run turbines for generating electricity.

Some of the neutrons from fission of ^{235}U combine with ^{238}U to produce ^{239}Pu (Fig. 13.17)

$$^{238}U + {}^1n_o \rightarrow {}^{239}U \rightarrow {}^{239}Np \rightarrow {}^{239}Pu \qquad (13.13)$$

(Kathren 1984). The ^{239}Pu is also fissile, itself producing neutrons and fission products, along with transuranic (TRU) elements (TRU elements have atomic numbers higher than 92) (see Table 13.5). In fact, about one-third of the heat in the reactor results from the fission of ^{239}Pu. In a nuclear reactor, 1 g of fissile material (^{235}U, ^{239}Pu) ($\sim$10^{21} atoms) can produce an incredible 8.2×10^7 kJ of energy (Kathren 1984; Adloff and Guillaumont 1993). One must burn more than 2.6 metric tons of high-grade coal, yielding $\leq$31 kJ/g of heat (Moore and Moore 1976) to produce the same heat output from coal as obtained from 1 g of fissile nuclear fuel.

After 1 to 3 years the fission products accumulated in nuclear fuel, interfere with neutrons (adsorb or deflect them) so that the fuel rods become inefficient. Once removed from the reactor the rods are referred to as *spent fuel*. The general composition of spent fuel, which is chiefly UO$_2$, is given in Table 13.6. In some countries, but not in the United States, the spent fuel from commercial reactors is reprocessed to remove most of the U and Pu isotopes, which can then be reused to make additional nuclear fuel. Remaining radionuclides, which are mainly short-lived, and include most of the fission products (e.g., ^{135}Cs, ^{137}Cs, ^{90}Sr, ^{99}Tc) and small amounts of the actinides, are usually incorporated in borosilicate glass for disposal. The spent fuel and borosilicate glass with incorporated radionuclides are the chief materials in high level nuclear wastes (cf. Krauskopf 1988; Berlin and Stanton 1989; Wiltshire 1993). High-level nuclear waste generates substantial heat and has a typical radioactivity of 10^{16}–10^{18} Bq/ton (cf. Savage 1995).

The radioactivity of important radionuclides in spent fuel as a function of time after the fuel is removed from a reactor is indicated in Table 13.7. Figures 13.18 and 13.19 describe the water dilution volumes for the radionuclides in spent fuel and reprocessed waste as a function of time. (The water dilution volume is the volume of water needed to dilute the amount of a given isotope in the waste to a concentration safe for ordinary use.) Ordinant values in Figs. 13.18 and 13.19 can also be viewed as a measure of the radioactivities of the isotopes as a function of time. For both spent fuel and reprocessed waste, the chief source of radioactivity for the first 10 to 100 y is the fission products ^{90}Sr and ^{137}Cs. Thereafter, up to about 10,000 y, Am and (briefly) Pu isotopes are the dominant

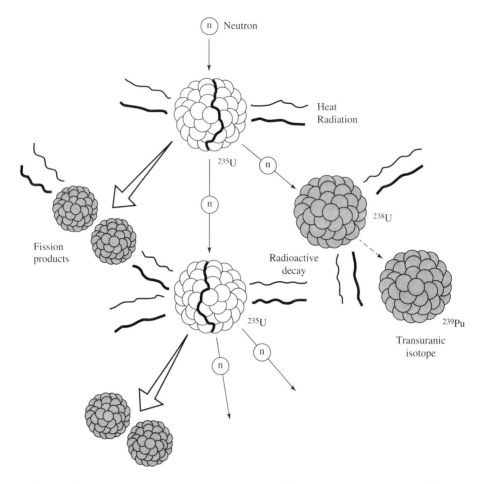

Figure 13.17 Schematic diagram showing fission of ^{235}U and addition of neutrons to ^{238}U to produce ^{239}Pu. The top of the diagram shows the splitting of a ^{235}U nucleus by slow neutrons, resulting in a release of energy and the generation of more neutrons. A fast neutron added to the ^{238}U nucleus produces a nucleus of ^{239}Pu. The diagram suggests direct production of Pu, which is a simplification. (Office of Technology Assessment, U.S. Congress.) After Krauskopf (1988).

TABLE 13.5 Important transuranic (TRU) elements produced in nuclear reactors

Nuclide	Half-life	MCi in 1000 MWe reactor
U-239	6.75 d	1,708
Pu-238	86.4 y	0.138
Pu-239 (fissile)	24,360 y	0.032
Pu-240	6,580 y	0.050
Pu-241	13.2 y	12.4
Pu-243	5 y	22.4
Am and Cm isotopes		1.14
All actinides		3,614

Source: From DOE (1987).

TABLE 13.6   The composition of typical
spent fuel from a commercial nuclear reactor

Nuclide	Wt percent
U-235	0.8
U-236	0.4
U-238	95
Fissile Pu	0.65
Nonfissile Pu	0.25
Fission products	2.9

Source: From Glasstone (1987).

contributors to radioactivity. From roughly 10^4 to 10^7 y, ^{237}Np contributes the most to radioactivity of the waste. Because the reprocessed waste contains smaller amounts of long-lived actinides than does the spent fuel, at later times its radioactivity drops substantially lower than that of spent fuel.

13.4.2 The Long-Term Health Risk of High-Level Nuclear Wastes in a Geologic Repository

Currently the spent-fuel rods from nuclear power plants are being stored in pools of water at power plants. Some of the spent fuel has also been placed in large casks in dry storage. Although pool and cask storage is probably safe for decades, such storage becomes impractical when the power plants are closed down and decommissioned as planned (cf. NWTRB 1996). In most countries there is consensus that the safest long-term disposition of high-level nuclear waste is to isolate the waste in a deep, geologic repository. A variety of geological settings are being considered as potential repositories, including crystalline rocks such as granites, and shales, salt domes, and volcanic rocks. Most national repositories are to be emplaced below the groundwater table. To ensure long-term isolation

TABLE 13.7   The radioactivity of selected radionuclides in spent fuel[†] as a function of fuel age after removal from the reactor in curies per metric ton of initial heavy metal (Ci/MTIHM)

Nuclide	Half-life (y)	Age of Spent Fuel		
		10^2 y	10^4 y	10^6 y
C-14	5.7×10^3	1.5	4.6×10^{-1}	
Sr-90	2.8×10^1	6.7×10^3		
Tc-99	2.1×10^5	1.3×10^1	1.3×10^1	0.504
Cs-137	3.0×10^1	1.0×10^4		
Th-229	7.3×10^3	9.7×10^{-7}	1.6×10^{-2}	0.911
U-233	1.6×10^5	1.7×10^{-4}	4.8×10^{-2}	0.906
U-238	4.5×10^9	3.2×10^{-1}	3.2×10^{-1}	0.318
Np-237	2.1×10^6	4.2×10^{-1}	1.2	0.854
Pu-239	2.4×10^4	3.1×10^2	2.4×10^2	$\sim 10^{-7}$
Pu-240	6.5×10^3	5.3×10^2	1.8×10^2	$\sim 10^{-7}$
Am-241	4.6×10^2	3.7×10^3	6.5×10^{-2}	
Total[‡]		4.1×10^4	4.7×10^2	20.2

[†]Basis is pressurized water reactor fuel with 33,000 megawatts per day/MTIHM.

[‡]Total is of all radionuclides, both selected and unselected.

Source: From the Oak Ridge National Laboratory Spent Fuel Repository Characteristics Data Base (Notz 1990).

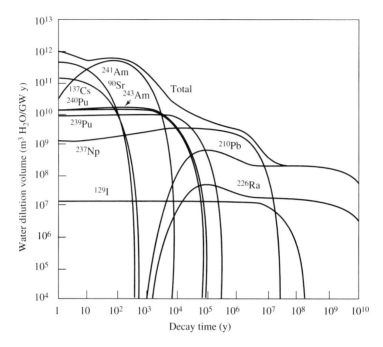

Figure 13.18   Water-dilution volumes for radionuclides in spent fuel discharged from a 1-GW(e) pressurized-water reactor as a function of decay time. After J. Choi and H. Pigford, Water dilution volumes for high-level wastes, *ANS Transactions* 39, p. 176. Copyright 1981 by the American Nuclear Society, Inc., LaGrange Park, Illinois. Used by permission.

of the wastes, groundwater flow rates from the waste to the accessible environment[†] are required to be very slow. In the U.S. program, the repository may be located in unsaturated rock above the water table. The heat of the spent fuel is then expected to boil away any infiltrating water, and thereby prevent corrosion and waste-package failure for one thousand years or more. Most national programs intend that the engineered barriers surrounding the waste and the natural geological system together will prevent dangerous releases of radionuclides to groundwater supplies or to surface environments for tens of thousands to millions of years.

It seems appropriate to ask whether we should be concerned about the nuclear wastes in a repository for such long times. Certainly, spent fuel fresh from a reactor is very hot and dangerously radioactive if unshielded. However, this radioactivity declines exponentially with time (Figs. 13.18 and 13.19). Shown in Fig. 13.20 is the relative toxicity hazard of spent fuel and reprocessed waste as a function of time. (This is the toxicity produced by ingesting a given weight of nuclear waste, spent fuel, or metal ore.) Also plotted is the time-invariant toxicity hazard for mercury, lead, silver, and uranium ore. As shown by the figure, after 10,000 and 100,000 y the respective toxicities of reprocessed waste and spent fuel are the same as and less than that of 0.2% U ore (DOE 1980—see also Wick and Cloninger 1980; OTA 1985; Savage 1995). As noted by Goodwin et al. (1994), "At very long times, a waste disposal vault resembles a high-grade uranium ore deposit." It is logical to question why we should be concerned about the nuclear waste in a repository once the health risk of the waste is less than that of the ore deposit from which the uranium began its journey in the nuclear

---

[†]In the U.S. program the accessible environment is assumed to be at a distance of 5 km from the repository.

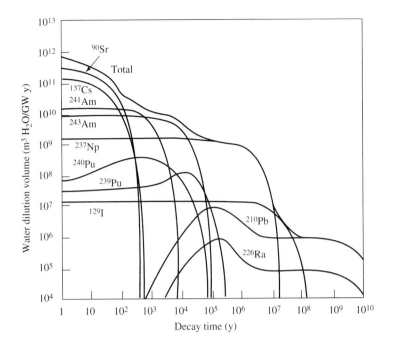

Figure 13.19 Water-dilution volumes for radionuclides in spent-fuel reprocessing wastes formed by operating a 1-GW(e) pressurized-water reactor for one year, plotted as a function of decay time. After J. Choi and H. Pigford, Water dilution volumes for high-level wastes, *ANS Transactions* 39, p. 176. Copyright 1981 by the American Nuclear Society, Inc., LaGrange Park, Illinois. Used by permission.

fuel cycle. Based on Fig. 13.20 one could also argue in terms of relative risk that we should somehow eliminate the risk due to unmined Hg, Pb and rich Ag ores, before concerning ourselves with nuclear wastes beyond 10,000 to 100,000 years.

13.5 GEOCHEMISTRY OF IMPORTANT RADIONUCLIDES IN A GEOLOGICAL REPOSITORY

13.5.1 Thermodynamic Stability and Geochemistry of I, Tc, Am, Np, and Pu Aqueous Species and Solids

Previous discussion indicated that for both spent fuel and reprocessed waste, the chief sources of radioactivity for the first 10 to 100 y are the fission products ^{90}Sr and ^{137}Cs. For later times out to 10^7 y, isotopes of Am, Pu, and Np are successively the dominant contributors to radioactivity. Assuming, as is likely, that corrosion failure of metal-waste packages in a geologic repository will not be significant before 10^3 to 10^4 y, both ^{90}Sr and ^{137}Cs with ~30 y half-lives, will have decayed to insignificance. The actinides and long-lived fission products ^{129}I and ^{99}Tc remain important at later times (cf. Krauskopf 1988). The U.S. program assumes that ^{129}I, ^{99}Tc, and the actinides ^{233}U, ^{234}U, and ^{233}Np are soluble and poorly adsorbed by Yucca Mountain tuffs and, therefore, could reach the accessible environment before 50,000 y after disposal, assuming containment failure (Rechard 1995).

The probability of release and transport of radionuclides from the vicinity of the waste is largely determined by: (1) their solubilities in the groundwater; (2) their tendency to be adsorbed

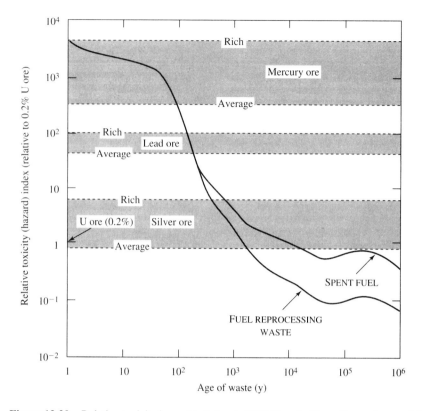

Figure 13.20   Relative toxicity hazard (relative to 0.2% U ore) of ingesting a given weight of high-level nuclear waste or spent fuel and the same weight of different metal ores or 0.2% U ore, as a function of the age of the waste (DOE 1980). The figure shows that after about 1,000 y and 10,000 y, respectively, the reprocessed waste and the spent fuel are less toxic than 0.2% U ore.

onto minerals in any clay backfill surrounding the waste or by minerals in the rock, and (3) by the likelihood that their colloidal forms will be filtered out by the backfill or rock, become unstable and dissolve, or adsorb or settle out during transport. Solubilities of the radionuclides can only be predicted if reliable thermodynamic data are available for the probable aqueous species and solids involved. Adsorption modeling using the electrostatic adsorption models also requires knowledge of the stabilities of radionuclide aquocomplexes, as does a qualitative understanding and application of measured distribution coefficient (K_d) values for adsorption of the actinides.

Thermodynamic data for U and its adsorption reactions have already been discussed. Among other important radioisotopes, the behavior of iodine in ^{129}I is relatively simple and well understood. Iodine occurs as iodate (IO_3^-) in highly oxidized waters and iodide (I^-) under more reducing conditions, including in most groundwaters (cf. Pourbaix 1966). At 25°C and below pH = 5, the redox boundary defining equal concentrations of these species for $\Sigma I = 10^{-6}$ M is given by

$$Eh = 1.097 - 0.0592 \text{ pH} \tag{13.14}$$

where $E°$ has been computed from $\Delta G_f°$ data in Wagman et al. (1982). Iodide salts are less soluble than iodate salts, but are still too soluble to limit maximum possible I concentrations in groundwater.

Critically evaluated thermodynamic data for U (Grenthe et al. 1992) and Am (Silva et al. 1995) (see Table A13.5) have recently been published by international scientists collaborating through the

Nuclear Energy Agency (NEA). The NEA expects a similar critical review for Tc to be completed in 1996. Publication of reviews for Pu and Np is expected in 1997.[†] The thermodynamic data produced under NEA guidance are internally consistent with CODATA key values (Cox et al. 1989). Unfortunately, currently available thermodynamic data for Tc, Np, and Pu is of mixed reliability. Preliminary lists of internally consistent ΔG_f° data for Tc and Np are given in Tables A13.5 and A13.6, with data selections explained in table footnotes. Puigdomenech and Bruno (1991) have critically reviewed the published ΔG_f° and ΔH_f° data for Pu (see also Guillaumont and Adloff 1992; Adloff and Guillaumont 1993). Their data selections for Pu are given in Table A13.8. Unfortunately, enthalpy and entropy data for Tc and the actinides are largely unavailable. This makes prediction of their reactions near 100°C, such as might be expected near nuclear wastes, a highly uncertain exercise.

The tabulated ΔG_f° data for Tc has been used to construct the Eh-pH diagram in Fig. 13.21 and the $TcO_2 \cdot 2H_2O(s)$ solubility diagram in Fig. 13.22. Figure 13.21 shows that in oxidized waters, such as those at the potential U.S. Yucca Mountain repository, Tc(VII), as pertechnetate ion (TcO_4^-), is the dominant Tc species. A large anion, TcO_4^-, is highly soluble and mobile in water. However, in low-Eh groundwaters TcO_4^- is readily reduced and precipitated in $TcO_2 \cdot 2H_2O(s)$, which has a solubility of about 10^{-8} M in systems of near-neutral pH, under which conditions the predominant species is $TcO(OH)_2^\circ$ (Fig. 13.21). As indicated by Fig. 13.22, the amphoteric solubility of $TcO_2 \cdot 2H_2O(s)$ is increased by carbonate complexing at high pH. Little is known about possible Tc(IV) sulfate, phosphate, or fluoride complexing for example, which may further increase Tc(IV) solubility, particularly in acid waters (cf. Rard 1983, 1990).

Example 13.7

Smith and Martell (1976) give $K = 10^{2.44}$ and $K = 10^{2.5}$ for the association constants of $VOSO_4^\circ$ and $TiOSO_4^\circ$ at 25°C. Assuming as an estimate that $K = 10^{2.5}$ for the reaction $TcO^{2+} + SO_4^{2-} = TcOSO_4^\circ$, how much would the sulfate complex increase the solubility of $TcO_2 \cdot 2H_2O(s)$ at pH = 2 for $[SO_4^{2-}] = 10^{-2}$ M? Ignore activity coefficients in your calculation. In the absence of sulfate complexing the solubility of $TcO_2 \cdot 2H_2O(s)$ is 3.7×10^{-8} M at pH = 2.

Free-energy data from Tables A13.2 and A13.6 may be used to estimate $\Delta G_f^\circ(TcOSO_4^\circ) = -205.27$ kcal/mol from the association constant. Dissolution of $TcO_2 \cdot 2H_2O(s)$ may then be written

$$TcO_2 \cdot 2H_2O(s) + SO_4^{2-} + 2H^+ = TcOSO_4^\circ + 3H_2O \tag{13.15}$$

Tabulated ΔG_f° data then lead to $K_{eq} = 10^{-1.94}$, and the expression

$$[TcOSO_4^\circ] = 10^{-1.94} [SO_4^{2-}] [H^+]^2 \tag{13.16}$$

Substituting, we find $[TcOSO_4^\circ] = 1.15 \times 10^{-8}$ M, and so the complex increases Tc concentrations by about 30%.

Many researchers have noted similarities in the thermodynamic properties of lanthanide and actinide solids and aqueous complexes when different metal cations and oxy-cations of the same charge and structure and of the same approximate size are compared (cf. Hobart 1990; Nitsche 1991a; Seaborg 1994; Seaborg and Hobart 1996; Silva and Nitsche 1996). Nitsche (1991) and Palmer et al. (1992) used the analog actinide approach to estimate unknown or poorly known stabilities of

---

[†]Thermodynamic data bases are available from OECD Nuclear Energy Agency, 12 Boulevard des Iles, 92130 Issy-les-Moulineaux, France.

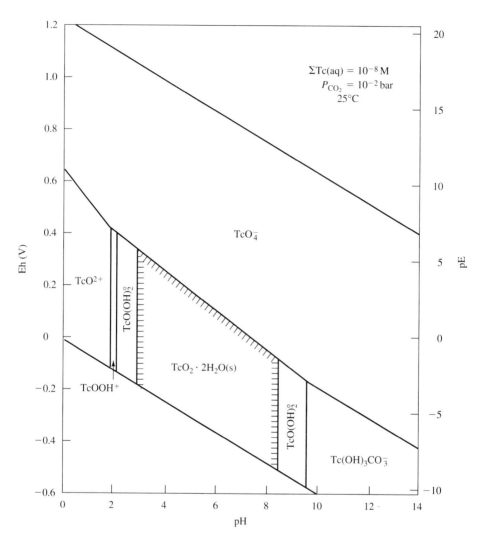

Figure 13.21 Eh-pH diagram for the system Tc-O_2-CO_2-H_2O at 25°C and 1 bar pressure for $P_{CO_2} = 10^{-2.0}$ bar and $\Sigma Tc(aq) = 10^{-8.0}$ M. Boundaries of the $TcO_2 \cdot 2H_2O(s)$ stability field at $\Sigma Tc(aq) = 10^{-8.0}$ M have been stippled.

a number of aqueous complexes and solids of Am, Np, Pu, and U from better known values for use in the EQ3/6 thermodynamic data base. Shown in Table 13.8 are the important oxidation states of Th, U, Np, Pu, and Am in natural systems. Chemistries of Th (cf. Langmuir and Herman 1980; Wagman et al. 1982) and Am (cf. Silva et al. 1995) are relatively simple in that only Th(IV) and Am(III) are observed in nature. In contrast, the multiple oxidation states of Pu in particular greatly complicate its behavior. The actinide cations are hard acid cations (see Chap. 3) and so form their strongest complexes with ligands that are hard bases. The relative strengths of actinide complexes with hard bases generally decrease in the order

$$CO_3^{2-} > OH^- > F^-, HPO_4^{2-} > SO_4^{2-} > Cl^-, NO_3^- \qquad (13.17)$$

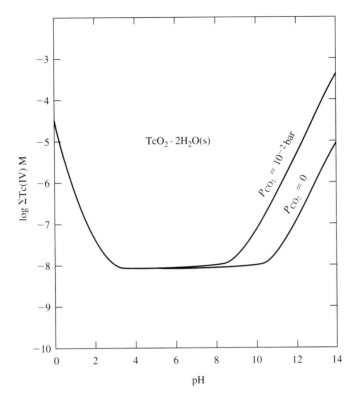

Figure 13.22 Solubility of $TcO_2 \cdot 2H_2O(s)$ as Tc(IV)(aq) species at 25°C and 1 bar total pressure as a function of pH in the absence of CO_2 and for a CO_2 pressure of $10^{-2.0}$ bar.

(Fig. 13.23). Kim (1993) and Silva and Nitsche (1996), among others, have noted that the stabilities of actinide (An) cation complexes with the same ligand most often decrease in the order

$$An^{4+} > An^{3+} \approx AnO_2^{2+} > AnO_2^+ \tag{13.18}$$

This order also roughly describes the increasing solubility (decreasing stability) of actinide solids formed with a given ligand, as well as the decreasing tendency for the actinide cations to be adsorbed (see below).

The actinide cations can form strong complexes with humic and fulvic acid ligands. This may reflect the presence of phenolic, amine, and alcoholic OH-groups, and also carbonate groups in

TABLE 13.8 Important (■) and usually unimportant (□) oxidation states of some actinides (An) in natural water/rock systems

Actinide Charge	An Cation	Element/Atomic Number				
		Th/90	U/92	Np/93	Pu/94	Am/95
6+	AnO_2^{2+}		■	□	■	
5+	AnO_2^+		□	■	■	
4+	An^{4+}	■	■	■	■	
3+	An^{3+}			□	■	■

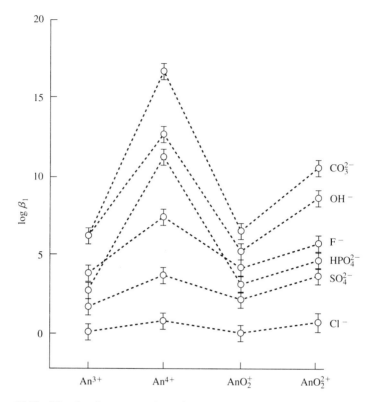

Figure 13.23   Plot showing average formation constants ($\log \beta_1$ values) for the formation of 1:1 complexes of actinide (An) cations An^{3+}, An^{4+}, AnO_2^+, and AnO_2^{2+} with different ligands. Modified after Lieser and Mohlenweg, Neptunium in the hydrosphere and in the geosphere. *Radiochim. Acta* 43:27–35. Copyright 1988 by Oldenbourg Verlag GmbH. Used by permission.

humic-fulvic substances (Silva and Nitsche 1996). Stabilities of actinide complexes with humic-fulvic ligands generally increase with pH and have respective $\log K_{\text{assoc}}$ values of about 5, 8, 12, and 16, with AnO_2^+, AnO_2^{2+}, An^{3+}, and An^{4+} ions (Silva and Nitsche 1996). Choppin and Allard (1985) and Moulin and Moulin (1995) point out that because OH^- forms stronger complexes with An^{4+} species and CO_3^{2-} stronger complexes with AnO_2^+ and AnO_2^{2+} species than with humic and fulvic acid ligands, complexes involving these actinide cations and the humics and fulvics can be ignored. Important actinide/humic-fulvic complexing does, however, occur with An(III) cations Am^{3+} and Pu^{3+}.

Americium (III) forms strong hydroxyl and carbonate complexes (Fig. 13.24). The carbonate complexes predominate above roughly pH 5.5 to 7.5, depending on total dissolved carbonate (C_T), which is higher in Fig. 13.24(b) ($C_T = 0.01$ M) than in Fig. 13.24(a) ($P_{CO_2} = 10^{-3.5}$ bar). In the usual range of surface and subsurface CO_2 pressures, the least soluble Am solid in natural waters is $AmOHCO_3(c)$ (Fig. 13.25). Its solubility, which is amphoteric, is less than 10^{-8} M between pH 7 and 9, but may increase substantially above pH 9 due to carbonate complexing (Fig. 13.26).

Example 13.8

Sulfate complexing can substantially increase the solubility of $AmOHCO_3(c)$. Assuming $P_{CO_2} = 10^{-3.0}$ bar, $[SO_4^{2-}] = 10^{-2.0}$ M, and pH = 7.0, and ignoring ionic strength, compute the

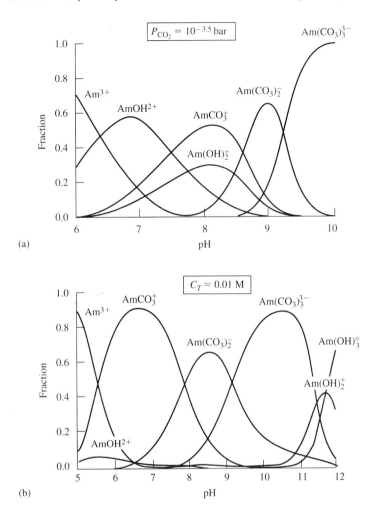

Figure 13.24 Plots showing the fractional distribution of Am(III) complexes as a function of pH (a) for $P_{CO_2} = 10^{-3.5}$ bar and (b) for C_T (total carbonate) = 0.01 M. Reprinted from *Chem. Thermo.* 2, R. J. Silva et al., Actinide environmental chemistry, copyright 1995 with kind permission of Elsevier Science-NL, Sara Burgerhartstraat 25, 1055 KV Amsterdam, The Netherlands.

solubility of Am(III) as the $AmSO_4^+$ complex, and compare it to the solubility of $AmOHCO_3(c)$ in pure water at $P_{CO_2} = 10^{-3.0}$ bar and pH = 7, which is about 1.6×10^{-8} M as shown in Fig. 13.26.

The dissolution reaction may be written

$$AmOHCO_3(c) + SO_4^{2-} + 3H^+ = AmSO_4^+ + CO_2(g) + 2H_2O \qquad (13.19)$$

Tabulated free energies lead to $K_{eq} = 10^{14.81}$. Substituting into the K_{eq} expression yields $[AmSO_4^+] = 6.5 \times 10^{-6}$ M. Sulfate complexing increases Am solubility by more than 400 times over its solubility in the sulfate-free system.

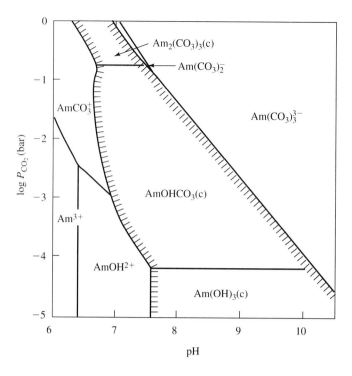

Figure 13.25 Stability fields of solids and aqueous complexes in the system Am(III)-CO$_2$-H$_2$O as a function of CO$_2$ pressure and pH for ΣAm(III)(aq) = 10^{-6} M. Solid/aqueous boundaries are stippled. Reprinted from *Chem. Thermo.* 2, R. J. Silva et al., Actinide environmental chemistry, copyright 1995 with kind permission of Elsevier Science-NL, Sara Burgerhartstraat 25, 1055 KV Amsterdam, The Netherlands.

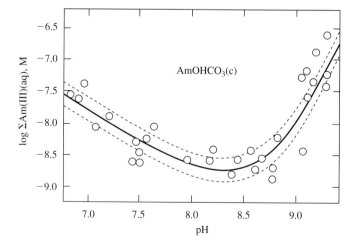

Figure 13.26 Solubility of AmOHCO$_3$(c) at 25°C and 1 bar total pressure as a function of pH for P_{CO_2} = 10$^{-3.0}$ bar. Data points are measurements of Felmy et al. (1990). Curve has been drawn by Silva et al. (1995) using the Am thermodynamic data given in Table A13.5. Dotted curves show the associated uncertainty. Reprinted from *Chem. Thermo.* 2, R. J. Silva et al., Actinide environmental chemistry, copyright 1995 with kind permission of Elsevier Science-NL, Sara Burgerhartstraat 25, 1055 KV Amsterdam, The Netherlands.

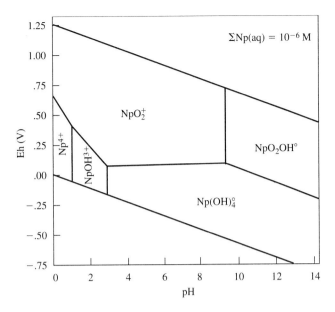

Figure 13.27   Eh-pH diagram for the system Np-O$_2$-H$_2$O at 25°C and 1 bar total pressure for ΣNp(aq) = 10^{-6} M, showing the stability fields of predominant aqueous species.

Neptunium occurs in Np(IV) and Np(V) oxidation states in natural waters. As NpO$_2^+$ it is quite soluble and mobile. The immobility of Np(IV) in low-Eh groundwaters reflects its occurrence in highly insoluble oxide and hydroxide solids. Assuming the Np(IV) concentration is bounded by the solubilities of NpO$_2$(c) and Np(OH)$_4$(am) as Np(OH)$_4^o$, Np(OH)$_4^o$ at saturation ranges from 10$^{-8.3}$ to 10$^{-17.5}$ M. Both Np(IV) and Np(V) form strong CO$_3^{2-}$ complexes, which may predominate over the OH$^-$ complexes above roughly pH 6 and 8, respectively (Figs. 13.27 and 13.28). Fluoride complexing of Np^{4+} below pH 4 to 5 can substantially increase the solubility of Np(IV) in reducing groundwaters. Nevertheless, NpO$_2$(c) is quite insoluble above pH 2, even in oxidizing waters with important carbonate complexing (Fig. 13.29).

Preliminary thermodynamic data for plutonium from Puigdomenech and Bruno (1991) are given in Table A13.8. Unfortunately, the identities and stabilities of the important carbonate complexes of Pu(III) and Pu(IV) remain in dispute. Puigdomenech and Bruno (1991) suggest 1:1 Pu^{3+}-CO$_3^{2-}$ and 1:1 and 1:5 Pu^{4+}-CO$_3^{2-}$ complexes, whereas Nitsche (1991) proposes, in addition, 1:2 and 1:3, Pu^{3+}-CO$_3^{2-}$ complexes and 1:2, 1:3, and 1:4 Pu^{4+}-CO$_3^{2-}$ complexes (see also Adloff and Guillaumont 1993). Stability constants for the complexes recognized by both sources, half of which have been estimated, are in fair agreement.

A number of important and unique features of Pu geochemical behavior should be noted. First, Pu occurs in natural water/rock systems in 3+, 4+, 5+, and 6+ oxidation states (Fig. 13.30). Generally Pu exists in solution as PuO$_2^+$ and PuO$_2^{2+}$ aqueous species with PuO$_2^+$ predominating in oxidized natural waters (Choppin 1990). For the most part, Pu^{3+} and Pu^{4+} are present in solids. Examination of Fig. 13.30 shows that there are four Eh-pH triple points at which Pu species with three different oxidation states may coexist at equilibrium. Defined by the species involved, these are: (A) PuO$_2^{2+}$/Pu^{4+}/Pu^{3+}; (B) PuO$_2^{2+}$/PuO$_2^+$/Pu^{3+}; (C) PuO$_2^+$/Pu(OH)$_4^o$/Pu^{3+}; and (D) PuO$_2$(OH)$_2^o$/PuO$_2^+$/Pu(OH)$_4^o$.

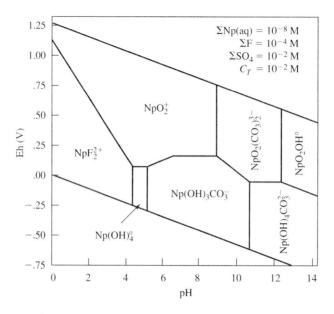

Figure 13.28 Eh-pH diagram for the system Np-O$_2$-F-SO$_4$-CO$_2$-H$_2$O at 25°C and 1 bar total pressure for ΣNp(aq) = 10^{-8} M, showing the stability fields of predominant aqueous species, with ΣF = 10^{-4} M, ΣSO$_4^{2-}$ = 10^{-2} M, and C_T (total carbonate) = 10$^{-2.0}$ M.

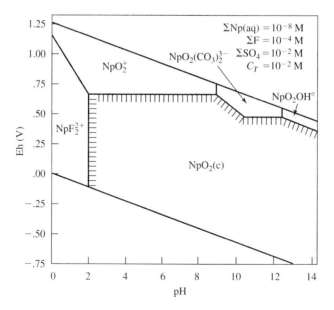

Figure 13.29 Eh-pH diagram for the system Np-O$_2$-F-SO$_4$-CO$_2$-H$_2$O at 25°C and 1 bar total pressure for ΣNp(aq) = 10^{-8} M, showing the stability fields of NpO$_2$(c) and of predominant aqueous species, with ΣF = 10^{-4} M, ΣSO$_4^{2-}$ = 10^{-2} M, and C_T (total carbonate) = 10$^{-2.0}$ M.

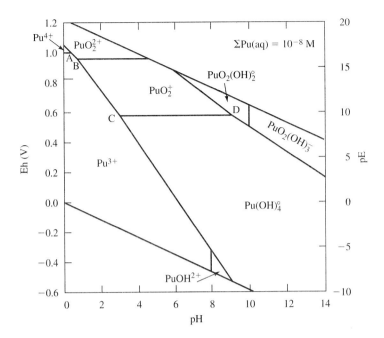

Figure 13.30   Eh-pH diagram for the system Pu-O$_2$-H$_2$O at 25°C and 1 bar total pressure for ΣPu(aq) = 10^{-8} M, showing the stability fields of predominant aqueous species. Letters A, B, C, and D identify triple points where species with three different Pu oxidation states exist at equilibrium. Species disproportionation may occur at and near these points.

In 3+, 5+, and 6+ oxidation states, Pu forms strong carbonate complexes above pH 5 (Fig. 13.31). Nevertheless, PuO$_2$(c) is so insoluble that its stability field in Fig. 13.32, which defines conditions where Pu concentrations are less than 10^{-8} M, covers most of the Eh-pH diagram for C_T = 10^{-2} M. The stability fields of Pu(OH)$_4$(am) (not shown) and PuO$_2$(c) overlap part or all of the fields of Pu(III), Pu(V), and Pu(VI) aqueous species in Fig. 13.32. This fact and the existence of the four triple points among the aqueous species, leads to the likelihood that disproportionation of these species will occur near the triple points because of shifts in Eh, pH, or species concentrations (cf. Adloff and Guillaumont 1993; Seaborg and Hobart 1996). For example, disproportionation of PuO$_2^+$ may result in PuO$_2^{2+}$ and Pu^{4+} species. Also, disproportionation of Pu^{4+} may take place by a reaction such as

$$3Pu^{4+} + 2H_2O = 2Pu^{3+} + PuO_2^{2+} + 4H^+ \tag{13.20}$$

(Nitsche et al. 1987). Humic acid has been shown to facilitate disproportionation of Pu species (Guillaumont and Adloff 1992).

Radioactive decay of ^{239}Pu releases alpha particles (He nuclei), which can lead to alpha radiolysis. In alpha radiolysis, the alpha particles react with water to produce H$_2$O$_2$ and free radicals, which are highly reactive, such as H• and HO$_2$•. These species, in turn, may cause the reduction of aqueous Pu(VI)(Cleveland 1979a, 1979b). Disproportionation or reduction of oxidized Pu species or an increase in pH may lead to the formation of colloidal Pu(OH)$_4$(am) (also termed Pu(IV) polymer, cf. Nitsche et al. 1995). Pu(IV) colloids are often kinetically quite stable and can be transported long distances in groundwater (cf. Penrose et al. 1990; Triay et al. 1995), although they are likely to

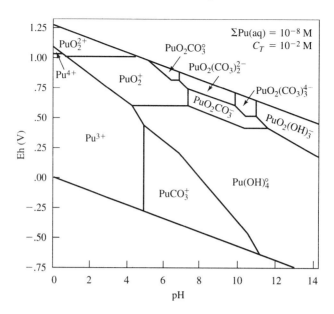

Figure 13.31 Eh-pH diagram for the system Pu-O_2-CO_2-H_2O at 25°C and 1 bar total pressure for $\Sigma Pu(aq) = 10^{-8}$ M, and $C_T = 10^{-2.0}$ M, showing the stability fields of predominant aqueous species.

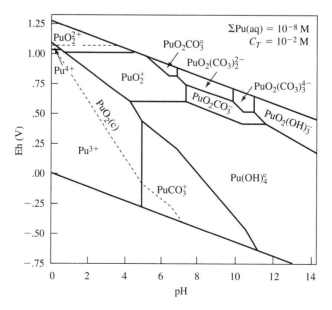

Figure 13.32 Eh-pH diagram for the system Pu-O_2-CO_2-H_2O at 25°C and 1 bar total pressure for $\Sigma Pu(aq) = 10^{-8}$ M, and $C_T = 10^{-2.0}$ M, showing the stability fields of predominant aqueous species. Boundaries of the $PuO_2(c)$ stability field for $\Sigma Pu(aq) = 10^{-8}$ M are shown by the dashed line.

eventually be dissolved, adsorbed, or filtered out. Gradual crystallization of Pu(IV) colloids increases their stability, making them less soluble. The tabulated ΔG_f° data for $PuO_2(c)$ and $Pu(OH)_4(am)$ indicate that for the same solution conditions, $PuO_2(c)$ should be about $10^{6.6}$ times less soluble than $Pu(OH)_4(am)$. However, experimental measurements by Rai et al. (1980) suggest that this difference is roughly 10^2 or less, with $PuO_2(c)$ much more soluble than predicted. Alpha radiolysis evidently damages the crystal structure of $PuO_2(c)$ (Puigdomenech and Bruno 1991) making it considerably more soluble than predicted from its tabulated ΔG_f° value.

13.5.2 Solubility Controls on Releases

The perfect high-level waste repository (or toxic-waste site) is one in which individual waste components are at thermodynamic equilibrium with the host water-rock system. For such conditions there is no tendency for the waste components to dissolve and be transported from the site to the accessible environment. It was shown earlier that low-Eh crystalline-rock groundwaters are often near saturation with respect to UO_2 (Section 13.2.3). Spent fuel UO_2 in such a system should have little or no tendency to dissolve and release other radionuclides to the groundwater. An appropriate first task in characterizing a potential repository site should, therefore, be to obtain accurate groundwater analyses to determine if the groundwater is saturated with respect to UO_2.

A related primary goal of geochemists in high-level waste programs has been to predict maximum concentrations of important radionuclides that could escape to the environment once the metal waste containers holding the waste have corroded and failed.[†] The radionuclides would be released by water leaching the spent fuel or radionuclide-containing waste glass. The usual approach to such prediction has been to calculate or measure the solubility of the least-soluble actinide solid phases that might precipitate under conditions anticipated in and near a HLW repository (cf. Nitsche 1991a; Nitsche et al. 1995).[‡] This is a conservative approach in that other than U and Th, which may occur in pure minerals in natural systems, the concentrations of TRU elements (and Ra) are likely to be so low relative to concentrations of the major elements, that if they precipitate, they do so as coprecipitates within solid solutions of the major elements (cf. Gnanapragasam and Lewis 1995) rather than as pure phases, which are usually more soluble.

McKinley and Savage (1994) discuss the solubilities of radionuclides in groundwater compositions chosen as representative by the different repository programs. The Swiss, Swedish, Finnish, Canadian, and Japanese programs assume that geologic disposal will be below the groundwater table at a groundwater pH of 7.0 to 8.7, and Eh of -250 ± 100 mV. For these conditions there is general agreement on the least soluble pure solids (Table 13.9). However, some programs choose the crystalline oxides of Np, Pu, Tc, Th, and U as solubility controls, whereas others choose their amorphous oxides or hydroxides. Because the solubility difference between crystalline and amorphous forms is typically 10^8 to 10^9 times for (4+) cations, this is obviously a critical choice and explains some of the disparities in Table 13.9. Other reasons for disparity are disagreements in the thermodynamic data being used. The ranges and probable concentrations for Am, Np, and Pu adapted in the U.S. program (Andrews et al. 1995) are largely based on the solubility measurements of Nitsche et al. (1993, 1995) and Torretto et al. (1995).

Using MINTEQA2, it is interesting to compute the solubilities of Am, Np, Pu, Tc, Th, and U in low-Eh Swedish groundwater (SKI-90) and in the high-Eh waters from Yucca Mountain in the

---

[†]These maximum possible concentrations must be known to help define the "source term" or concentrations at a failed waste package, to permit modeling and prediction of the effects of radionuclide releases on future repository performance.

[‡]It is important to note that the different radioisotopes of an element such as U or Pu share its solubility in proportion to their relative concentrations.

TABLE 13.9 Assumed solubility-limiting solids and concentrations of radioelements in low-Eh reference ground-waters, in geological formation types proposed for repository development by different repository programs (McKinley and Savage 1994), and in oxidized (high-Eh) unsaturated zone waters similar to those in the proposed repository horizon at Yucca Mountain, NV

	Solubility-limiting solids in low-Eh groundwaters[†]	General range of assumed concentrations (M)[‡]	Solubility-limiting solids for high-Eh unsaturated conditions [§]	Measured steady-state concentrations at 25°C and pH = 7.0 (M)[‖]	Assumed range and probable concentrations (M)[#]
Am	$AmOH(CO_3)(c)$	2×10^{-9} to 10^{-5}	$AmOH(CO_3)(c)$	1.2×10^{-9}	10^{-10} to 10^{-6} (5×10^{-7})
Np	$NpO_2(c)/$ $Np(OH)_4(am)$	10^{-10} to 10^{-8}	$NaNpO_2(CO_3) \cdot 2H_2O(c)$	1.3×10^{-4}	5×10^{-6} to 10^{-2} (1.4×10^{-4})
Pu	$PuO_2(c)/$ $Pu(OH)_4(am)$	3×10^{-11} to 4×10^{-4} (esp. ~10^{-8})	Amorphous Pu oxyhydroxides and carbonate; mixed V > VI > IV > III oxidation states	2.3×10^{-7}	10^{-8} to 10^{-6} (5.1×10^{-7})
Ra	Coprecip. with Ca or $RaSO_4$	10^{-10} to 10^{-4}	Coprecip., esp. in sulfates		10^{-9} to 10^{-5} (10^{-7})
Tc	$TcO_2(c)/$ $TcO_2 \cdot nH_2O(am)$	10^{-12} to 10^{-7}	Not solubility limited		3.6×10^{-7} to 10 (10^{-3})
Th	$ThO_2(c)/$ $Th(OH)_4(am)$	10^{-10} to 5×10^{-4} (esp. ~10^{-10})	$ThO_2(c)/Th(OH)_4(am)$		10^{-10} to 10^{-7}
U	$UO2(c)/UO2(am)/$ $UO_2(fuel)$	10^{-10} to 4×10^{-5} (esp. 10^{-8})	Uranophane[††]		10^{-8} to 10^{-2} (3.2×10^{-5})

[†]Summary of the solids chosen in Swiss, Finnish, Japanese, Swedish and Canadian programs (McKinley and Savage 1994).

[‡]Concentration range that includes the concentrations chosen by most or all of the programs.

[§]Solids of Am, Np, and Pu have been identified in solubility experiments run from supersaturation in J-13 groundwater by Nitsche et al. (1993). $Na_{0.6}NpO_2(CO_3)_{0.8} \cdot 2.5H_2O$ was identified in pH 5.9 and 8.5 experiments at 25°C and in general at 60°C. Other solubility-limiting solids have been chosen in this study.

[‖]Measured in solubility expriments with J-13 groundwater by Nitsche et al. (1993).

[#]Assumed range and probable concentrations (in parentheses), except for Tc, have been obtained by expert elicitation (TSPA-95 1995).

[††]Uranophane is $Ca(H_3O)_2(UO_2)_2(SiO_4)_2 \cdot 3H_2O$.

U.S. (J-13 and UZ4-TP-7) (chemical analyses in Example 13.3 and Table 13.10). Shown in Table 13.11 are the solid phases assumed to control element concentrations, and their computed solubilities obtained with MINTEQA2. Solubilities at low Eh are probably maximum possible values for quadrivalent Np, Pu, Th and U, for which the solid chosen is the amorphous oxide or hydroxide. These elements are more likely to equilibrate with partially crystalline solids and so be at concentrations several orders of magnitude lower than given in Table 13.11.[†] Above pH 4 to 6, solubilities of the (4+) oxides and hydroxides of Np, Pu, Th and U are generally as their neutral hydroxide complexes and so are independent of pH. The MINTEQA2 calculation indicates that all of the elements listed are highly insoluble in low-Eh near neutral groundwaters. Comparison with concentrations in Table 13.9, which generalizes the different repository programs, suggests that maximum concentrations of Am of 10^{-5} M, Th of 5×10^{-4} M and U(VI) of 4×10^{-5} and 3.2×10^{-5} M are probably overly

---

[†]Perhaps only 10^2 times lower for Pu. See earlier discussion of Pu(IV) oxide and hydroxide solubilities.

TABLE 13.10   Composition of a representative reference groundwater from crystalline rocks (SKI-90) and of unsaturated zone water from near the proposed U.S. repository at Yucca Mountain, NV (UZ4-TP-7)

Species/ parameter	SKI-90 (nonsaline Swedish reference groundwater, mM)[†]	UZ4-TP-7 (Unsaturated zone water, Yucca Mountain, NV, mM)[‡]
Na^+	1.39	2.09
K^+	0.0256	0.38
Ca^{2+}	0.50	2.00
Mg^{2+}	0.0823	0.62
Fe (total)	0.00179	0.00030
SiO_2	0.0682	1.48
HCO_3^-	2.00	2.82
Cl^-	0.423	2.40
SO_4^{2-}	0.0417	1.28
F^-	0.142	—
PO_4^{3-}	3.75×10^{-5}	—
pH	8.2	7.4
Eh (mV)	−300	+700 (assumed)

[†]From SKI (1991).

[‡]From near the proposed repository horizon at Yucca Mountain and presumably of similar composition to waters from that horizon. Bicarbonate has been computed from the charge balance. The water also contains 0.066 mg/L Mn, 0.552 mg/L Zn, and 0.983 mg/L Sr (Yang 1992; see also Yang et al. 1996).

conservative. It seems likely that the low solubility of uranophane will limit maximum U concentrations to values near 10^{-7} M at the proposed Yucca Mountain repository. The computed solubility of $NaNpO_2CO_3 \cdot 3.5H_2O$ (actually of variable H_2O) is in fair agreement with its measured solubility in J-13 well water (Nitsche et al. 1993, 1995; Torretto et al. 1995)

The solubility calculations indicate that some radionuclides are highly insoluble in any proposed repository environment of near neutral pH, whether oxidizing or reducing (e.g., Am, Pu, Th). Other radioelements are insoluble in low-Eh environments and may be quite soluble at high Eh values (e.g. Np, Tc, and U). Several words of caution are appropriate here, however. Thermodynamic data for potentially important aqueous complexes of Np, Pu, and Tc formed with silicate and phosphate ligands, for example, is often poorly known or lacking. Such complexes could increase the computed solubilities of the (4+) metal oxides, hydroxides and other solids by orders of magnitude. Data are also largely lacking on the effect of increased temperature on actinide complexation constants and solid solubilities. Until these data are generated, it would be very useful to see the results of laboratory solubility measurements for these elements in reference-type low-Eh groundwaters.

In their solubility experiments, Nitsche et al. (1993, 1995) and Torretto et al. (1995) assume that Np will occur only as NpO_2^+ in the oxidized groundwaters at Yucca Mountain. With this assumption, the soluble phase $NaNpO_2CO_3 \cdot 3.5H_2O$ limits Np concentrations to between 2.5×10^{-3} and 2.7×10^{-5} M. However, experimental and EQ3/6-modeled leaching of spent fuel and waste glass suggests that Np concentrations are unlikely to exceed 10^{-6} M at Yucca Mountain, and may be far lower. Using the EQ3/6 geochemical code to model leaching by oxidized J-13 well water at 25 and 90°C, of spent fuel (Bruton and Shaw 1988), and waste glass (Bruton 1988), Np concentrations did not exceed about 10^{-6} M from either waste form. In a follow-up experimental study of spent-fuel leaching by oxidized J-13 water at 25 and 85°C, Wilson and Bruton (1990) found that after six

TABLE 13.11   Total concentrations of radioelements at saturation with possible solubility-limiting solids in Swedish SKI-90 reference groundwater (SKI 1991) and in U.S. Yucca Mountain J-13 groundwater (Ogard and Kerrisk 1984) and UZ4-TP-7 unsaturated zone water (Yang 1992)

Element	Solid assumed at saturation in low-Eh waters	SKI-90 (low-Eh groundwater)
Am	$AmOH(CO_3)(c)$	1.4×10^{-7}
Np	$Np(OH)_4(am)$	$\leq 1.6 \times 10^{-9}$
Pu	$Pu(OH)_4(am)$	$\leq 1.7 \times 10^{-9}$
Tc	$TcO_2 \cdot 2H_2O(s)$	$\leq 3.3 \times 10^{-8}$
Th	$Th(OH)_4(am)$	$\leq 5.7 \times 10^{-7}$
U	$U(OH)_4(am) \equiv UO_2(am)$	$\leq 1.4 \times 10^{-8}$

Element	Solid assumed at saturation in high-Eh waters	J-13 (high-Eh groundwater)	UZ4-TP-7 (high-Eh unsaturated zone)
Am	$AmOH(CO_3)(c)$	5.6×10^{-8}	8.9×10^{-7}
Np	$NaNpO_2CO_3 \cdot 3.5H_2O(c)$	8.9×10^{-4}	4.8×10^{-4}
Pu	$Pu(OH)_4(am)$	$\leq 6.6 \times 10^{-8}$	$\leq 1.7 \times 10^{-9}$
Tc	None		
Th	$Th(OH)_4(am)$	$\leq 6.0 \times 10^{-7}$	$\leq 5.7 \times 10^{-7}$
U	$Ca(H_3O)_2(UO_2)_2(SiO_4)_2 \cdot 3H_2O(c)$ uranophane	5.4×10^{-7}	1.6×10^{-7}

Note: Calculations have been made using the MINTEQA2 thermodynamic data base modified by Turner et al. (1993), with revised thermodynamic data for Am from Silva et al. (1995) and revised thermodynamic data for U from this chapter. Log K for formation of $AmOH(CO_3)(c)$ in MINTEQA2 has been changed from 8.605 to 7.20 (Silva et al. 1995). $Np(OH)_5^-$ and $Pu(OH)_5^-$ complexes have been eliminated from the data base by setting their log K values to -30. The hydroxides of Np, Pu, Th, and U are their most soluble, relatively amorphous forms.

months, steady-state Np(aq) concentrations leveled out at values from below the detection limit of 6×10^{-10} M, up to 3×10^{-9} M. The Np oxidation state was not reported.

Separately, based on geochemical modeling, a number of researchers have concluded that the least soluble Np phase in near-neutral, oxidized groundwaters such as occur at Yucca Mountain, is Np(IV) oxide or hydroxide (Wilson and Bruton 1990; Hakanen and Lindberg 1991; Wolery et al. 1995), not a more soluble Np(V) phase such as $NaNpO_2CO_3 \cdot 3.5H_2O$.

Example 13.9

Compute the Np concentration in J-13 groundwater from Yucca Mountain (analysis in Example 13.3) if Np(aq) is controlled by the solubility of $Np(OH)_4(am)$. Assume, conservatively, that the Eh is controlled by atmospheric O_2 ($P_{O_2} = 0.21$ bar; Eh = 0.826 mV). Hint: In PRODEFA2 assign $Np(OH)_4(am)$ as a "finite solid" and specify the redox couple NpO_2^+/Np^{4+}.

At saturation with $Np(OH)_4(am)$, computed groundwater concentrations are: $\Sigma Np(V) = 5 \times 10^{-10}$ M and $\Sigma Np(IV) = 9 \times 10^{-8}$ M, in reasonable agreement with the Np(aq) values reported from the leaching of spent fuel (Wilson and Bruton 1990). In the same water $NaNpO_2CO_3 \cdot 3.5H_2O$ is unstable and greatly undersaturated, with a saturation index of -6.5.

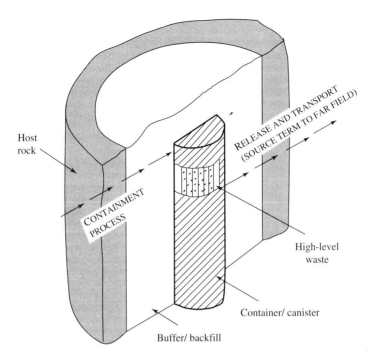

Figure 13.33   Schematic diagram of the engineered barrier system (EBS) showing the high-level nuclear waste in its metal container, surrounded by a buffer or backfill (usually of compacted bentonite clay), in contact with the host rock. The EBS and rock affected thermally by the waste are sometimes termed the near field, with more distant surrounding rock termed the far field. After *The status of near field modeling.* Proc. Technical Workshop, copyright 1995 by OECD. Used by permission.

Because of its high solubility as NpO_2^+, the U.S. program considers Np the most hazardous radionuclide for repository times beyond about 10^4 y (TSPA-95). If, instead, Np(IV) solids limit Np solubility at Yucca Mountain as suggested by Example 13.9, or the steady-state release of Np(aq) from spent fuel is at or below 10^{-9} M (Wilson and Bruton 1989), the threat of Np releases may largely disappear.

Although generally insoluble, Am and Pu tend to form stable colloids (cf. Bates et al. 1992). National programs considering waste disposal below the water table in crystalline rock generally plan to surround the waste with a backfill of compacted bentonite clay (Fig. 13.33) (cf. Langmuir and Apted 1992; OECD 1993). The clay should filter out and prevent the escape of colloidal Am, Pu, or other radionuclides. Colloid filtration by clays has been an effective barrier to radionuclide migration in groundwaters at the Cigar Lake uranium deposit (Vilks et al. 1993).[†] Colloidal transport of radionuclides through the matrix of unsaturated rock, such as the volcanic tuff at Yucca Mountain is also unlikely and could only occur if such colloids were carried by rapid fracture flow.

Example 13.10

Using MINTEQA2, compute the maximum possible concentrations of Am, Np, Pu, Th, Tc, and U in well water no. 3 from Table A13.4, which is from granites at Äspö, Sweden. Assume

---

[†]McCarthy and Degueldre (1993), Degueldre (1993), and Triay et al. (1995) discuss colloidal transport in groundwater as it relates to the transport of contaminants, including radionuclides.

the same solids limit concentrations in low-Eh groundwaters as are listed in Table 13.11. Note: The easiest way to solve this problem is to enter each solid as a "finite solid" in PRODEFA2.

Total concentrations (M) resulting from the computation are Am = 2.7×10^{-7}, Np = 1.6×10^{-9}, Pu = 1.7×10^{-9}, Tc = 3.2×10^{-8}, Th = 5.5×10^{-7}, and U = 1.4×10^{-8}. If partially crystalline to crystalline oxides of Np, Pu, Th, and U had been selected, solubilities of these elements would have been orders of magnitude lower.

13.5.3 Adsorption Controls on Releases

Preceding discussion of solubility controls suggests that concentrations of ^{129}I generally, and of ^{99}Tc in oxidized waters, will not be usefully limited by mineral solubilities. Other nuclides of potentially high solubility, and thus mobility in groundwater, are Ra and, perhaps, Np and U in oxidized repositories. Adsorption reactions can retard the migration of radionuclides and can prevent those having short half-lives (e.g. ^{137}Cs and ^{90}Sr) from escaping to the accessible environment for times that assure their decay to insignificance. Adsorption can also dampen radionuclide peak releases that might otherwise exceed health standards. Most nuclear-waste programs are considering installing a clay backfill around waste packages. Radionuclides would be adsorbed by the backfill, as well as by surrounding rock.

Silva and Nitsche (1996) note that distribution coefficient (K_d) values[‡] for adsorption of the actinides are roughly equal for actinide cations of similar charge and structure. Based on earlier work, they report average actinide (An) cation K_d's for 12 different minerals and 4 rock types of 500, 50, 5, and 1 respectively, for An^{4+}, An^{3+}, AnO$_2^{2+}$, and AnO$_2^+$ ions. The sequence of adsorption edges with increasing pH for Th(IV), Am(III), Np(V), and Pu(V) adsorption by γ-Al$_2$O$_3$ in Fig. 13.34 (Bidoglio et al. 1989) is consistent with this order, with pH values of 2.5, 5.8, and 7.3, respectively, at 50% ad-

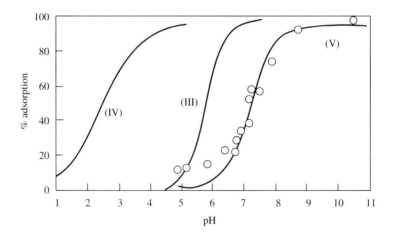

Figure 13.34 Comparison of the adsorption of actinide cations of different oxidation states onto γ-Al$_2$O$_3$. The solid lines refer to Th(IV), Am(III), and Np(V). Open circles are adsorption data for Pu(V), with ΣPu(V) = 2×10^{-10} M. γ-Al$_2$O$_3$ concentrations are 10 mg/L for Th and Am and 200 mg/L for Np and Pu. Modified after Bidoglio et al., *Interactions and transport of plutonium-humic acid particles in groundwater environments,* Mat. Res. Soc. Symp. Proc. 1989, 127:823–30. Used by permission.

[‡]K_d = [(wt adsorbed/wt sorbent)/(solute concentration)], usually in units of ml or cm^3/g,or L/kg. See Chap. 10.

sorbed. Trace concentrations of Pu(IV), U(VI), and Np(V) are 50% adsorbed by α-FeOOH at about pH 3.2, 4.2, and 7.0, respectively (Turner 1995), also in agreement with the proposed order. However, individual K_d values for actinide ions often grossly differ from the average values reported by Silva and Nitsche (1996). This is because K_d not only depends on the oxidation state of the actinide, but also depends on pH and competitive adsorption, the nature and extent of actinide complexing, and the characteristics of sorbent phases.

The adsorption edge plots in Fig. 13.34 show that adsorption of actinide cations increases with pH and has a strong pH dependence. This reflects that the actinide cations must compete with protons for adsorption sites. Cations whose adsorption edges occur at lowest pH values must, therefore, form the strongest bonds with sorbent surface sites. The adsorption edges of weaker-bonding cations then occur at successively higher pH values.

Example 13.11

Based on Fig. 13.34, with 0.2 g/L of sorbent γ-Al$_2$O$_3$ at pH = 6.0, 6% of Np(V) is adsorbed, whereas at pH = 9.0, 95% is adsorbed. From this information calculate K_d (ml/g) for Np(V) adsorption by γ-Al$_2$O$_3$ at the two pH values.

Because units of weight adsorbed and dissolved cancel out in the K_d expression, we can use any weight units we choose. At pH = 6, assuming 6% adsorbed is 6 μg on 0.2 g/L sorbent, then 94% dissolved in 1 L is equivalent to 94 μg/L or 0.094 μg/ml. K_d then equals

$$K_d = \frac{6 \ \mu\text{g}/0.2 \ \text{g}}{0.094 \ \mu\text{g/ml}} = 320 \ \text{ml/g} \qquad (13.21)$$

The same approach at 95% of Np(V) adsorbed gives $K_d = 9.5 \times 10^4$ ml/g.

The pH-dependence of K_d values is also apparent from Fig. 13.13, which describes U(VI) adsorption by HFO. Based on this figure, K_d ranges from about 10^2 to $>10^7$ between pH 4 and 9. Finally, K_d values for Th(IV) adsorption by quartz shown in Fig. 10.24 are seen to increase from 10^2 to $>10^5$ between pH 3 and 6.

The pH-dependence of radionuclide cation adsorption partly reflects competition with H$^+$ ions for adsorption sites. Competition for adsorption sites by major electrolyte ions also limits radionuclide adsorption in saline waters (cf. Mahoney and Langmuir 1991). Also, competition for sites with Ca^{2+} can significantly reduce Ra^{2+} adsorption by quartz and kaolinite (Riese 1982), as well as NpO$_2^+$ adsorption by smectite clays (Kozai et al. 1995).

The oxidation state of an actinide is obviously key to its adsorption behaviour. Lieser and Mohlenweg (1988) measured the apparent K_d for Np adsorption at pH = 7.0 as a function of Eh. Their results in Fig. 13.35 indicate that $K_d \approx 1$ to 6 in oxidized systems where NpO$_2^+$(aq) dominates, but increases to about 10^3 for Eh < 200 mV, when relatively insoluble Np(OH)$_4$(am) colloid is likely to have precipitated. In the latter case, the K_d value is, of course, a measure of Np solubility not its adsorption.

Complexing of actinide cations also drastically changes their adsorption behavior. The adsorption of actinide cations by oxide, hydroxide, and aluminosilicate minerals is limited in waters that contain abundant dissolved carbonate. This has been thought to reflect either: (1) competition for adsorption sites by carbonate species (cf. Murray and Coughlin 1992; van Geen et al. 1994) or (2) weaker adsorption of neutral or negatively charged carbonate complexes than of actinide-hydroxide complexes (Hsi and Langmuir 1985; Bidoglio et al. 1987; Waite et al. 1994; Turner 1995).

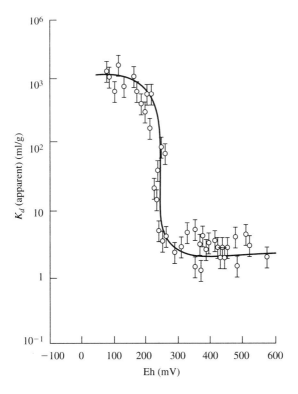

Figure 13.35   Apparent K_d values for Np(V) as a function of Eh with $\Sigma Np = 10^{-7}$ M. Sorbent sediments contained quartz, K-feldspar, plagioclase feldspar, kaolinite, muscovite, calcite, chlorite, and dolomite. Apparent K_d values from >10 to $>10^3$, measured for Eh values below about 250 mV, reflect precipitation of Np(IV) solids rather than adsorption. Modified after Lieser and Mohlenweg, Neptunium in the hydrosphere and in the geosphere. *Radiochim. Acta* 43:27–35. Copyright 1988 by Oldenbourg Verlag GmbH. Used by permission.

Under acid conditions, Th(IV)-SO_4^{2-} complexing lowers Th(IV) adsorption by quartz (Fig. 10.24). Complexing by SO_4^{2-} and F^- probably reduces An^{4+}, An^{3+}, and AnO_2^{2+} cation adsorption by oxide, hydroxide, and aluminosilicate minerals, generally. Presumably in all of these cases, the actinide complexes are poorly adsorbed. At the elevated Cl^- level of brines (e.g., $Cl^- = 5.48$ M) Cl^- complexing similarly limits the adsorption of An^{3+} cations (cf. Bidoglio et al. 1984). Humic-acid complexing has little or no effect on the adsorption of most actinide cations, but can reduce the adsorption of species such as Am^{3+} (Fig. 13.36) (Kung and Triay 1994; Tanaka and Senoo 1995).

As indicated, other than the actinide-hydroxyl complexes, most complexes are poorly adsorbed by oxyhydroxide minerals. However, PO_4^{3-} is itself strongly adsorbed, particularly by the Fe(III) oxyhydroxides, and so creates negatively charged surface sites that enhance the adsorption of multivalent actinide cations (cf. Hsi 1981). This behavior probably reflects the tendency of PO_4^{3-} to bond with multivalent actinide cations, forming strong aquocomplexes, as well as forming insoluble phosphate solids and solid solutions (cf. Langmuir and Apted 1992).

Among common minerals, the strongest sorbents for most actinide cations are the ferric oxyhydroxides and especially HFO. Quartz and the feldspars are among the weakest sorbent phases (Fig. 13.37) (cf. Beall and Allard 1981). Apatite, a common accessory phosphate mineral, is also a

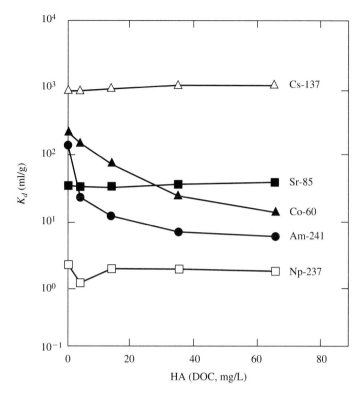

Figure 13.36   Effect of increasing humic acid (HA) concentrations as dissolved organic carbon (DOC) on K_d for the adsorption of various radionuclides by a sandy soil. After Tanaka and Senoo, *Sorption of* ^{60}Co, ^{85}Sr, ^{137}Cs, ^{237}Np *and* ^{241}Am *on soil under coexistence of humic acid: Effects of molecular size of humic acid,* Mat. Res. Soc. Symp. Proc. 1995, 353:1013–20. Used by permission.

strong sorbent for multivalent actinide cations (Andersson et al. 1982). A few percent of strongly adsorbing phases in a rock or sediment will often dominate its adsorption behavior toward trace species such as the actinides (see Chap. 10).

A primary goal of adsorption measurements and models is to predict adsorption in a variety of possible water/rock systems for a range of conditions. For the actinides, such prediction is necessary to develop confidence that radionuclide releases from a nuclear waste repository will not exceed health standards at some time and distance from the repository. Most repository programs use the K_d approach to describe and model radionuclide adsorption (cf. Meijer 1992; OECD 1993). In part, this surely reflects the simplicity of incorporating K_d values in transport codes (Freeze and Cherry 1979; Javandel et al. 1984) (see Chap. 10). The obvious danger is that a single K_d value may be grossly in error. The U.S. program takes the conservative approach of considering minimum and maximum possible K_d values for adsorption of each radionuclide by each general rock type at Yucca Mountain (TSPA-95). Concern is then focused on nuclides that have minumum K_d values near zero ml/g (e.g., ^{99}Tc and ^{129}I), and greater than zero but less than 100 ml/g (e.g., Np(V) and U(VI)) for adsorption by tuff rock and tuff minerals (cf. Meijer 1990, 1992).[†]

[†]If $K_d = 0$ ml/g, there is no retardation of a contaminant. If $K_d = 100$ ml/g for a contaminant, equations given by Freeze and Cherry (1979) indicate that the groundwater moves about 400 to 1000 times faster than the contaminant (see Chap. 10).

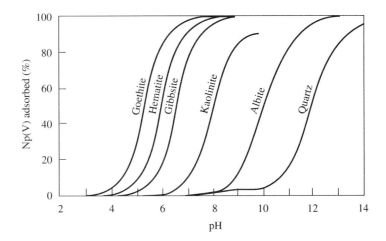

Figure 13.37   Adsorption of Np(V) by different minerals, for ΣNp(V) = (1.1 to 1.3) $\times$ 10^{-7} M, in I = 0.01 M $NaClO_4$ solutions. Total site concentration [SOH] = 5×10^{-4} M. From Kohler et al. (1992). Reprinted from: Kharaka, Y. K. & A. S. Maest (eds.), *Water-rock interaction—Proceedings of the 7th international symposium, WRI-7, Park City, Utah, 13–18 July 1992.* 1992. Volume 1. Set of two volumes, 1730 pp., Hfl. 325/US$190.00. Please order from: A. A. Balkema, Old Post Road, Brookfield, Vermont 05036 (telephone: 802-276-3162; telefax: 802-276-3837).

Many national programs plan to surround containers of their nuclear waste in a geologic repository, with a backfill of compacted bentonite clay (Fig. 13.33). A chief function of the clay backfill is to adsorb radionuclides and so retard their release from the engineered barrier system. Conca (1992) measured the apparent diffusion coefficient (D_a) and apparent distribution coefficient (K_d [ml/g]) of some radionuclides in bentonite clay as a function of clay moisture content and compaction density. Measurements were made for clay densities from 0.2 to 2.0 g/cm^3, which correspond to porosities of 93 to 25%, respectively. With decreasing porosity, D_a values declined by roughly 10 to 10^2-fold. However, for the same porosity reduction, K_d values were usually lowered by 10-fold and more, indicating less adsorption with compaction (Fig. 13.38).

The retardation of radionuclides by backfill only slows their eventual release. Thus, actinides with $K_d > 10^4$ ml/g are released from a 1-m-thick backfill at times beyond 10,000 y (Langmuir and Apted 1992). This is sufficient time for the decay of short-lived radioisotopes such as ^{90}Sr, ^{137}Cs, and [241]Am, but not of long-lived [237]Np, ^{99}Tc, or [129]I, for example. (Half-lives are given in Table 13.7. For [129]I, $t_{1/2}$ = 1.7 $\times$ 10^7 y.) Further, I, Tc, and Np have K_d values in bentonite of roughly 10^{-1} to 1 ml/g for IO_3^-, about 1 ml/g for TcO_4^-, and 10^2 to 10^3 ml/g for Np(V) (Conca 1992); thus all would presumably begin to escape from the backfill before 10,000 y had passed. The extent of attenuation of radionuclide releases by adsorption increases in direct proportion to the distance of groundwater flow, but only with the square root of increasing K_d. This suggests that adsorption by the host rock is potentially more important to waste isolation than adsorption by a clay backfill (Langmuir and Apted 1992).

As detailed above, the adsorption behavior of most actinides varies widely with solution pH, Eh, complexation, competitive adsorption and ionic strength, and the surface properties of sorbent phases. For this reason, many researchers have modeled actinide adsorption using surface complexation (SC) models that can quantitatively account for such variables. These models include the constant capacitance (CC), diffuse-layer (DL), and triple-layer (TL) models (Chap. 10). Much of the ra-

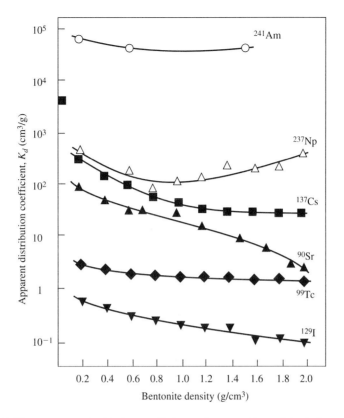

Figure 13.38   Apparent distribution coefficients of some radionuclides in bentonite as a function of bentonite compaction density. Except for ^{237}Np, adsorption of the radionuclides decreases with increasing compaction (decreasing porosity). After Conca (1992).

dionuclide adsorption work that has used SC models is critiqued and systematized by Kent et al. (1986) and by Turner (1991, 1993, 1995). These authors point out that the specific K^{int} values chosen to model-fit a set of empirical adsorption data are only valid when the aqueous solution model employed in their derivation is also used in their application. For example, Turner (1995) has derived CC, DL, and TL model K^{int} values for U(VI) adsorption by HFO, assuming log K_{assoc} = 17.7 for $UO_2(OH)_2^\circ$ from Grenthe et al. (1992). If, instead, log K_{assoc} = 16.0 or less for $UO_2(OH)_2^\circ$ (Table A13.1), with this revised constant a different set of K^{int} values would be derived from the experimental adsorption data.

Turner (1995) lists and reviews most of the published work on SC modeling of actinide cation adsorption. Further SC-modeling references include: Np(V) adsorption by α-FeOOH and Fe_3O_4 (Fujita et al. 1995); Pu(IV) and Am(III) adsorption by HFO, CaSi concrete, mortar, sand/bentonite, tuff, and sandstone (Baston et al. 1995a); and U(IV), U(VI), Tc(IV), and Tc(VII) adsorption by bentonite and tuff, silica, alumina, and goethite (Baston et al. 1995b).

An advantage of the SC models is that they provide a mechanistic and scientifically defensible approach for predicting adsorption behavior outside the range of laboratory conditions used to parameterize the models. The models are capable of relatively accurate predictions of actinide adsorption behavior in complex systems. By comparison, K_d values are largely restricted in application to systems similar to those used for the K_d measurement. Alternatively, highly conservative (often unnecessarily conservative) K_d values may be selected to assure that the expected adsorption behavior is bounded.

As a disadvantage, the SC models require considerable experimental work to obtain the model parameters needed for their application to natural water/rock systems. However, because any of the SC models have been shown to fit the same adsorption data equally well in most cases (cf. Westall and Hohl 1980; Turner 1995), many recent workers have suggested that modeling efforts be limited to the DL model, which is the simplest model to parameterize and apply (Dzombak and Morel 1990). As a further simplification, Turner (1995) has shown that most actinide adsorption data can be modeled with reasonable accuracy assuming that adsorption is only of a single monodentate and mononuclear surface complex, which is usually the same complex regardless of the model. For example, the data of Sanchez et al. (1985) for Pu(V) adsorption by goethite is closely fit by DL, CC, and TL models assuming a single-surface complex $FeOH_2$-PuO_2OH^+ is formed by the adsorption reaction

$$FeOH + PuO_2^+ + H_2O = Fe\ OH_2\text{-}PuO_2OH^+ \qquad (13.22)$$

(Fig. 13.39). The accumulation of carefully measured empirical adsorption data and K^{int} values for actinide adsorption, have encouraged a number of researchers to estimate SC parameters for

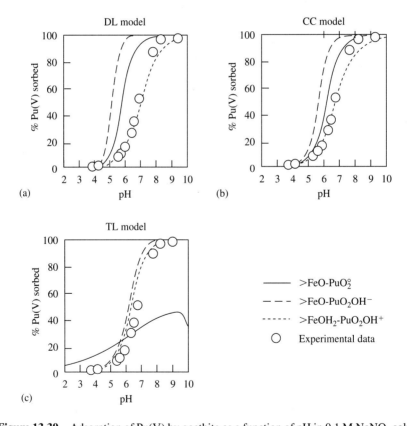

Figure 13.39 Adsorption of Pu(V) by goethite as a function of pH in 0.1 M $NaNO_3$ solutions with $\Sigma Pu(V) = 1 \times 10^{-11}$ M, and a sorbent/liquid ratio of 0.55 g/L. The experimental data (open circles) are from Sanchez et al. (1985). The various curves are model fits to the experimental data assuming that adsorption is as one of three possible mononuclear surface species. The best fit of the data with DL, CC, and TL models is obtained assuming that $FeOH_2$-PuO_2OH^+ is the adsorbed species. From Turner (1995).

water/rock systems from literature values for individual minerals and component oxides (cf. Baston et al. 1995a, 1995b). Such an approach is probably more reliable than using single K_d values and may be sufficiently accurate for most applications.

STUDY QUESTIONS

1. Discuss factors governing the attainment of radioactive and chemical equilibrium in a water-rock system. What chemical and physical processes aid or hinder the attainment of radioactive (secular) equilibrium in a groundwater?

2. The U.S. EPA is about to institute a primary drinking water standard for U. Discuss the merits of setting the limit in terms of radioactivity or mass-based concentration units given the following arguments: (1) the known adverse health effects of ingesting U derive from both its heavy-metal behavior (i.e., a mass-dependent effect) and its radioactivity; (2) it is much easier, faster, and cheaper to measure total U in mass units than to measure the radioactivity of individual U isotopes; and (3) the conversion from mass to radio-activity requires some assumptions (what are they?), whereas the conversion the other way is direct (why?).

3. Typically, ^{214}Bi is measured in a solid sample as a proxy for ^{238}U. For example, airborne radiometric surveys use a gamma spectrometer that is calibrated for ^{214}Bi. The resultant measured concentration of ^{214}Bi is then converted to effective U concentration (eU). What assumptions make this conversion calculation possible? Under what conditions would these assumptions invalid? Suppose ^{222}Rn emanation was very high for a particular soil. How would the measured eU compare to a direct measurement of U in such a sample?

4. In this chapter we noted that about 10,000 y after high-level nuclear waste has been removed from a nuclear reactor, the health risk from ingesting a given weight of the aged spent fuel is less than that from ingesting the same weight of ore from a high-grade uranium deposit. What design characteristics and conditions in a repository and surrounding rock might make the nuclear waste more dangerous than a particular uranium ore deposit at later times?

5. What geologic conditions lead to the formation of sedimentary uranium ore deposits and to their preservation for geologic times?

6. What are the important oxidation states of ^{129}I, Tc, Am, Np, Pu, and U in the environment, and in what forms do these radioelements chiefly occur in natural waters?

7. **(a)** What is the general order of increasing stability of actinide cation (An) complexes of different An oxidation states?
   (b) What is the general order of increasing tendency of actinide cations to be adsorbed by a given sorbent mineral?

8. Which actinide cations form important organic complexes in natural waters and which do not. Why?

9. A wastewater contains 10^{-6} M each of dissolved Am, Np, and U. How could you separate these actinides from their mixture by adjustments in solution composition?

10. What are some characteristics of Pu that makes its behavior and mobility so difficult to understand and predict in natural waters?

11. What kinds of aqueous complexes make Tc and the actinide elements more soluble at high and low pH's than at intermediate pH values?

12. Discuss some of the important similarities and differences in radionuclide behavior in uranium ore deposits and in geologic repositories for high level nuclear waste, as discussed in this chapter.

13. Compare the application of the distribution coefficient (K_d) concept and surface complexation adsorption models for predicting the adsorption behavior of radionuclides. Discuss potential advantages and pitfalls of assuming bounding K_d values for a given radionuclide in groundwater.

14. Examine some of the Eh-pH diagrams in this chapter and explain:
    (a) why there are breaks in slope of some of the stability boundaries, even though the predominant species on either side of the boundary are the same; and

(b) why some of the triple points are formed by the intersection of three nonvertical lines, whereas in other cases one of the three lines is vertical.

15. In groundwaters distant from the engineered portions of a high-level waste repository the mobilities of U and Th may be limited by adsorption at low U and Th concentrations and by the solubilities of U and Th solids at higher concentration. In the same groundwaters, the mobilities of Ra, I, Tc, Np, and Pu are more likely to be limited (if they are limited) by adsorption and/or by coprecipitation in the structures of major mineral precipitates, or by the instability or filtration of colloids. Explain these statements and discuss the possible detailed behavior of each element.

PROBLEMS

1. From the information in Table 13.2 show the radioactive decay series for ^{235}U and ^{232}Th graphically, in a form analogous to that used in Fig. 13.2.

2. It is often argued that after seven half-lives have passed the radioactivity of a radioisotope becomes insignificant (assuming no input of new material).
   (a) What is the significance of seven half-lives?
   (b) Suppose the original concentration of the radionuclide is 150 times the safety standard. Will the radionuclide still present a health hazard after seven half-lives?

3. The Oklo natural reactor in Gabon, Africa, is a high-grade sandstone-hosted U deposit. It was formed about 1.2 billion years ago and reached sufficient U ore grade that nuclear fission occurred spontaneously in the deposit. Although there are a number of factors that control the spontaneity of fission reactions, an important control is the ratio of fissionable ^{235}U to nonfissionable ^{238}U. The present-day atomic $^{235}U/^{238}U$ ratio is 0.719/99.275. What was this ratio when the ore deposit formed?

4. A granite rock sample contains 6 $\mu g/g$ (ppm) total U. Assuming that all daughters in the ^{238}U decay series are in secular equilibrium in this sample, calculate the concentrations of ^{238}U, ^{234}U, and ^{226}Ra, and express the results in terms of Bq/g and pCi/g. Speculate on the mineralogic location of the U in the rock and the geologic and hydrogeologic history of the sample.

5. The radioactivity of ^{237}Np leached from some spent nuclear fuel by an oxidized water ranges from 0.1 to 0.5 pCi/ml. Calculate the concentration range of Np in mol/L from this information, given that $t_{1/2} = 2.14 \times 10^6$ y for ^{237}Np, and in general for actinide An

$$An \text{ (mol/L)} = 10^{-25.05} D \text{ (pCi/L)} \cdot t_{1/2}$$

(Langmuir and Herman 1980), where D is the decay rate in pCi/L, and $t_{1/2}$ is the half-life of the radionuclide in seconds.

6. Calcite ($CaCO_3$) and cristobalite (SiO_2) are common minerals in the host rock of the proposed repository at Yucca Mountain, Nevada. Assuming uranophane solubility will limit U(VI) concentrations in the subsurface at Yucca Mountain, use MINTEQA2 to compute the pH and U(VI) concentration at equilibrium with all three minerals at 25°C. Before running the calculation, revise K_{eq} for uranophane in the file *type6.dbs* to agree with its value given in this chapter. Assume $P_{CO_2} = 10^{-2.0}$ bar and $Na^+ = HCO_3^- = 10^{-2}$ M.

7. In many high-level waste programs concrete will be used in repository underground tunnel construction, to fabricate containers for holding the nuclear wastes, and/or as backfill (Savage 1995).
   (a) Assume $Na^+ = Cl^- = 0.1$ M in the groundwater. Also assume that portlandite [$Ca(OH)_2(c)$] in the concrete and calcite in the rock control the pH and CO_2 pressure near the high-level waste and that the groundwater Eh = −300 mV. Use the Turner (1996) version of MINTEQA2 with its thermodynamic data base for Tc, to compute the concentration of $\Sigma Tc(aq)$ as ^{99}Tc in the groundwater, assuming it is controlled by the solubility of $TcO_2 \cdot 2H_2O(s)$.
   (b) Rerun the calculation performed in part (a) after revising the stability constants and $E°$ values for Tc in MINTEQA2, using the $\Delta G_f°$ data for Tc given in Table A13.6. You will need to add several

Tc(IV) complexes to the MINTEQA2 data base. Compare your results to the results obtained in part (a) and discuss.

(c) With the same assumptions and the revised thermodynamic data base for Tc from part (b), compute the solubility of $TcO_2 \cdot 2H_2O(s)$ in the groundwater out of contact with the portlandite. As before assume $Eh = -300$ mV and equilibrium with calcite, but with pH = 8.0.

(d) Given the results of parts (b) and (c), explain the likely fate of dissolved Tc released from wastes in contact with concrete as the Tc moves away in low-Eh groundwaters.

8. Borehole BH12, adjacent to the Nopal I uranium deposit in Mexico, contains water with the composition tabulated here (Pearcy et al. 1994). The pH is 7.5.

Species	Concentration (mg/L)	Species	Concentration (mg/L)
Al	0.17	Ca	57.7
Cl	3.0	Mg	<1.0
F	1.6	K	1.5
NO_3	2.5	Na	13.8
SO_4	39	SiO_2	81
HCO_3	161.5		

Alteration of uraninite [$UO_2(c)$] in the deposit produces U(VI) oxyhydroxides and subsequently uranophane. The local geology is welded silicic tuffs, similar to the geology of Yucca Mountain, Nevada. Assuming 25°C and oxidizing conditions, and assuming that the solubility of uranophane limits U concentrations, what U(aq) concentration would you expect to find in the borehole water?

9. (a) With the sweep option of MINTEQA2, separately compute the solubilities of $NaNpO_2CO_3 \cdot 3H_2O$, $Np(OH)_4(am)$, and $NpO_2(c)$ at 0.5 pH-unit increments from pH 4 to 9.5. Assume $P_{CO_2} = 10^{-3}$ bar and that the water also contains $Na^+ = Cl^- = 0.01$ M. Plot your results and compare the three solubility curves.

(b) What aqueous complex or complexes increase the solubilities of $Np(OH)_4(s)$ and $NpO_2(s)$ at pH values above those of their minimum solubilities?

(c) Would the computed solubilities of the Np(IV) solids have been greater or less than plotted if instead you had used the Np thermodynamic data from Table A13.7 to determine the stabilities of the aqueous Np(IV) species? Explain. Estimate the solubilities of $NpO_2(s)$ and $Np(OH)_4(s)$ above pH 5 using the ΔG_f° data for Np species from Table A13.7.

10. Compute the adsorption of Am(III) by 20 g/L of goethite (α-FeOOH) from pH 2 to 10 at 25°C using the constant capacitance (CC) model, given that $\Sigma Am(III) = 10^{-7}$ M and that the solution contains 0.1 M $NaNO_3$. Other assumptions are that $C_1 = 1.06$ F/m^2, $N_S = 3.0 \times 10^{-5}$ mol sites/m^2, and $S_A = 45$ m^2/g. Intrinsic constant expressions required for the calculation are given here. Tabulate percent Am(III) adsorbed and total dissolved Am(III) values (mol/L) from pH 2 to 10, and discuss. Contrast the merits of presenting the adsorption results in percent adsorbed versus as the concentration of total dissolved Am(III).

Adsorption reaction	Intrinsic constant
$FeOH + H_s^+ = FeOH_2^+$	$\log K_+^{int} = 7.31$
$FeOH - H_s^+ = FeO^-$	$\log K_-^{int} = -8.80$
$FeOH + Am^{3+} = H^+ + FeOAm^{2+}$	$\log K^{int} = 4.02^\dagger$

†Assumed to be the same as for Am(III) adsorption by γ-Al_2O_3 with the CC model (Turner 1995).

11. Granitic rocks have a relatively high average U content of 4.4 mg/kg (Table 13.1). Because rock surface permits Rn releases to soil gas or moisture, the ability of a mass of granite to release ^{222}Rn gas from ^{238}U decay is proportional to the internal plus external surface area/mass (A/M) ratio of the rock. The A/M ratio is thus proportional to the porosity and degree of fracturing and weathering of the rock. A polished surface exposes less area than an unpolished one.

Would a polished granite countertop in your kitchen be a source of dangerous levels of Rn in your house? The worksheet below permits a calculation of the Rn released to household air by a granite countertop. Follow the instructions in the second column and fill in the blank spaces on the right. Typical values have been suggested in rows 1b, 2a, and 2b. Assume the area of the countertop is 70 ft^2 and that it is in a house of 3000 ft^2 area with 8-foot-high ceilings. Compute the ^{222}Rn (pCi/L) in household air assuming no exchange of indoor and outdoor air. (Indoor and outdoor air are exchanged 0.5 to 1.5 times/hour in the average house.) Compare your result to the U.S. EPA standard of 4 pCi/L, which is not to be exceeded in indoor air, and comment on the health risk posed by the granite countertop. (This calculation, which was suggested by Richard Wanty, is reproduced with permission from the Marble Institute of America, Columbus, OH (Langmuir 1995).) Eisenbud (1987) reports that the average contributions of radon from various sources to indoor air are 1.5 pCi/L from the soil under and around the house, 0.01 pCi/L from public water supplies, 0.4 pCi/L from private wells, 0.05 pCi/L from building materials, and 0.2 pCi/L from outdoor air.

1a	Dimensions of countertop (sq ft).	
1b	Reactive thickness of countertop (in).[†]	10^{-6} in
1c	Multiply 1a by 1b, then multiply the result by 2360 to get the effective volume of the countertop in cm^3 (i.e., the volume that can release Rn).	
1d	Multiply 1c by 2.6 g/cm^3 (the average density of granite) to get the total mass of the reactive portion of the countertop in grams.	
2a	The average concentration of U in the countertop material in mg/kg.	4 mg/kg
2b	Multiply 2a by 1.47 to get the U concentration in pCi/g.	5.88 pCi/g
3a	Multiply 1d by 2b to get the total number of pCi of U in the reactive thickness of the countertop. Assuming radioactive equilibrium in the rock (a good assumption),[‡] this is also the total number of pCi or ^{222}Rn produced within the reactive thickness of the countertop.	
3b	Multiply 3a by the emanation coefficient—the proportion of Rn produced by the rock which escapes to the air. The highest possible value is 1, the lowest 0. For most dry rocks 0.2 is a good general figure.[†,‡]	
4a	Square footage of house in which countertop is installed.	
4b	Average ceiling height in ft.	
4c	Multiply 4a by 4b, then multiply that result by 28.32 to get the total volume of air in the house in liters.	
5	Divide 3b by 4c to get pCi/L of Rn in the indoor air of the house, which is contributed by the countertop only.	

Source: [†]Emanation of Rn is limited to an extremely thin veneer of about 30 to 40 nm or about 10^{-6} in from the external surface of the material (Flugge and Zimens 1939; Bossus 1984). [‡]Wanty et al. (1992), Thamer et al. (1981).

Comment on the importance of Rn emanation from the granite countertop versus amounts from these other Rn sources.

CHAPTER 13 APPENDIX

The tables in this section present thermodynamic data for uranium, americium, technecium, neptunium, plutonium, and auxiliary species. Also presented are chemical analyses of some groundwaters from uranium deposits and granites.

TABLE A13.1 Gibbs free energy and enthalpy of formation from the elements at 25°C and 1 bar total pressure of some uranium aqueous species and solids of geochemical interest

Aqueous species or solid	ΔG_f° (kcal/mol)	ΔH_f° (kcal/mol)	Source
U^{4+}	-126.45	-141.3	†,‡
UOH^{3+}	-182.25	-198.4	†,‡
$U(OH)_4^\circ$	-336.6		§
$UO_2(am)$	-234.15		§
$UO_2(c)$ uraninite	-246.61	-259.3	‡
$\beta\text{-}U_4O_9(c)$	-1022.4	-1078.1	‡
$\beta\text{-}U_3O_7(c)$	-774.8	-818.8	‡
$U_3O_8(c)$	-805.35	-854.4	‡
$USiO_4(am)$ coffinite	-438.63		‖
$USiO_4(c)$ coffinite	-450.76		#
UF^{3+}	-206.4	-222.8	‡
UF_2^{2+}	-283.1	-302.4	‡
UF_3^+	-357.8	-381.6	‡
UF_4°	-430.4	-462.6	‡
UF_5^-	-499.7		‡
UF_6^{2-}	-569.8		‡
$UF_4(c)$	-435.8	-457.5	‡
$UF_4 \cdot 2.5H_2O(c)$	-583.2	-638.5	‡
UCl^{3+}	-160.2	-185.8	‡
USO_4^{2+}	-313.2	-356.7	‡
$U(SO_4)_2^\circ$	-496.4	-568.2	‡
$U(HPO_4)_2 \cdot 4H_2O(c)$	-918.7	-1035.9	‡
$CaU(PO_4)_2 \cdot 2H_2O(c)$ ningyoite	-939.0		††
$U(CO_3)_4^{4-}$	-676.0		‡
$U(CO_3)_5^{6-}$	-803.5	-953.0	‡
UO_2^+	-229.69	-245.0	‡
UO_2^{2+}	-227.68	-243.5	‡
UO_2OH^+	-277.25	-301.6	‡
$UO_2(OH)_2^\circ$	-324.6		‡‡
$UO_2(OH)_3^-$	-371.5		‡

Aqueous species or solid	ΔG_f° (kcal/mol)	ΔH_f° (kcal/mol)	Source
$(UO_2)_2(OH)_2^{2+}$	-561.0	-614.7	‡
$(UO_2)_3(OH)_4^{2+}$	-893.5		‡
$(UO_2)_3(OH)_5^+$	-945.2	-1049.0	‡
$(UO_2)_4(OH)_7^+$	-1277.5		‡
$(UO_2)_3(OH)_7^-$	-1037.4		‡
$\gamma\text{-}UO_3(c)$	-273.8	-292.5	‡
$UO_3(am)$ gummite	-270.1	-288.7	‡
$\beta\text{-}UO_3 \cdot 2H_2O(c)$ schoepite	-391.1	-436.4	‡
$UO_2CO_3^\circ$	-367.03	-403.7	‡
$UO_2CO_3(c)$ rutherfordine	-373.60	-403.8	‡
$UO_2(CO_3)_2^{2-}$	-503.2	-561.9	‡
$UO_2(CO_3)_3^{4-}$	-635.7	-736.8	‡
$UO_2(CO_3)_3^{5-}$	-618.3		‡
$(UO_2)_2CO_3(OH)_3^-$	-750.4		‡
UO_2F^+	-301.9	-323.3	‡
$UO_2F_2^\circ$	-374.0	-403.3	‡
$UO_2F_3^-$	-444.4	-483.4	‡
$UO_2F_4^{2-}$	-512.8	-564.1	‡
UO_2Cl^+	-259.3	-281.6	‡
$UO_2(Cl)_2^\circ$	-288.9	-319.8	‡
$UO_2SO_4^\circ$	-409.8	-456.2	‡
$UO_2(SO_4)_2^{2-}$	-589.0	-669.8	‡
$UO_2PO_4^-$	-491.4		‡
$(UO_2)_3(PO_4)_2(c)$	-1222.8	-1312.5	‡
$(UO_2)_3(PO_4)_2 \cdot 4H_2O(c)$	-1467.3	-1610.7	‡
$(UO_2)_3(PO_4)_2 \cdot 6H_2O(c)$	-1581.7		‡
$UO_2HPO_4^\circ$	-500.1		‡
$UO_2HPO_4 \cdot 4H_2O(c)$ H-autunite	-732.5	-829.3	‡
$UO_2H_2PO_4^+$	-503.9		‡
$UO_2(H_2PO_4)_2^\circ$	-778.3		‡

TABLE A13.1 (continued)

Aqueous species or solid	ΔG_f° (kcal/mol)	ΔH_f° (kcal/mol)	Source
$H_2(UO_2)_2(PO_4)_2$(c) H-autunite	−1011.6		§§
$Na_2(UO_2)_2(PO_4)_2$(c) Na-autunite	−1135.7		‖
$K_2(UO_2)_2(PO_4)_2$(c) K-autunite	−1145.6		‖
$(NH_4)_2(UO_2)_2(PO_4)_2$(c) uramphite	−1054.3		##
$Mg(UO_2)_2(PO_4)_2$(c) saleeite	−1115.1		‖
$Ca(UO_2)_2(PO_4)_2$(c) autunite	−1138.6		‖
$Sr(UO_2)_2(PO_4)_2$(c) Sr-autunite	−1140.1		‖
$Ba(UO_2)_2(PO_4)_2$(c) uranocircite	−1141		‖
$Fe(UO_2)_2(PO_4)_2$(c) bassetite	−1028.7		‖
$Cu(UO_2)_2(PO_4)_2$(c) torbernite	−991.7		‖
$Pb(UO_2)_2(PO_4)_2$(c) przhevalskite	−1032.9		‖

Aqueous species or solid	ΔG_f° (kcal/mol)	ΔH_f° (kcal/mol)	Source
$K_2(UO_2)_2(VO_4)_2$(c) carnotite	−1097		†††
$Ca(UO_2)_2(VO_4)_2$(c) tyuyamunite	−1090		‡‡‡
$(UO_2)_2SiO_4 \cdot 2H_2O$(c) soddyite	−873.47		§§§
$Ca(H_2O)_2(UO_2)_2(SiO_4)_2 \cdot 3H_2O$(c) uranophane	−1483.18		§§§
$Na(H_3O)(UO_2)SiO_4 \cdot H_2O$(c) Na-boltwoodite	≥−708.28		§§§
$Na_2(UO_2)_2(Si_2O_5)_3 \cdot 7H_2O$(c) Na-weeksite	−2170.64		§§§
$Ca(UO_2)_2(Si_2O_5)_3 \cdot 5H_2O$(c) haiweeite	>−2239.8		‖‖‖

Note: ΔG_f° and ΔH_f° values for U(IV) aqueous species and solids, attributed to Grenthe et al. (1992), have been adjusted as necessary to be consistent with $\Delta G_f^\circ(U^{4+}) = -126.45$ kcal/mol in the table from Giridhar and Langmuir (1991).

Source: *Based on $E^\circ = 0.263 \pm 0.004$ V for the UO_2^{2+}/U^{4+} couple. log $K = -0.65$ for the reaction $U^{4+} + H_2O = U(OH)^{3+} + H^+$ at 25°C and 1 bar pressure. (Giridhar and Langmuir 1991). $E^\circ = 0.2674 \pm 0.0015$ V (Grenthe et al. 1992) was obtained neglecting important U^{4+} complexing and is based on a faulty extrapolation to $I = 0$.

‡Grenthe et al. (1992).

§Computed in this study from the UO_2(am) solubility measurements of Rai et al. (1990). See Table 13.4, footnote †. $\Delta G_f^\circ[U(OH)_4^\circ] = -347.18$ kcal/mol proposed by Grenthe et al. (1992) is in serious error (cf. Grenthe et al. 1995).

‖Based on the average maximum ion activity product of coffinite in three groundwaters from coffinite-bearing formations, which gives log $K = 0.50 \pm 0.03$ for the reaction $USiO_4$(am) $+ 4H^+ = U^{4+} + H_4SiO_4^\circ$.

#Estimated by Hemingway (1982).

††Computed from solubility data reported by Muto (1965), which makes ningyoite metastable relative to UO_2(c) by about 10^5 times and with a solubility roughly equal to that of $U(HPO_4)_2 \cdot 4H_2O$(c).

‡‡$\Delta G_f^\circ = \geq -327.0$ kcal/mol for $UO_2(OH)_2^\circ$ (Grenthe et al. 1992) is based on log $K \leq -10.3$ for $UO_2^{2+} + 2H_2O = UO_2(OH)_2^\circ + 2H^+$. Silva (1992) found that log $K = -10.3$ gave poor agreement between his predicted and measured schoepite solubility. Agreement was improved with log $K \leq -11.5$. Choppin and Mathur (1991) found log $K = -12.43$ in 0.1 M $NaClO_4$ solution. Corrected to $I = 0$ this gives log $K = -12.0$, which is the basis for the tabulated ΔG_f° value. Fuger (1992) has proposed log $K = -13.0 \pm 0.25$, making the complex even less important. §§Based on $\Delta G_f^\circ(UO_2HPO_4 \cdot 4H_2O$, H-autunite) from this table and the assumption that $\Delta G_r^\circ = 0$ for the reaction $2[UO_2HPO_4 \cdot 4H_2O]$(c) = $H_2(UO_2)_2(PO_4)_2$(c) $+ 8H_2O$.

‖‖Most published solubility studies of the cation-uranyl phosphates are inconsistent and/or poorly documented (cf. Grenthe et al. 1992). To avoid dealing with the questionable solubility data, tabulated ΔG_f° values for the autunites are based on ΔG_f°(H-autunite) from this table and ΔG_r° for the general exchange reaction $H_2(UO_2)_2(PO_4)_2 + M_n = M_n(UO_2)_2(PO_4)_2 + 2H^+$, where $n = 1$ and 2 for divalent and monovalent cations, respectively. The empirical ion exchange results are from Muto et al. (1968).

##Based on results of the H-autunite/uramphite exchange reaction (see footnote ‖‖) as reported by Vesely et al. (1965). Other exchange reaction measurements by these authors lead to ΔG_f°(Na-autunite) = −1137.6 kcal/mol, and ΔG_f°(K-autunite) = −1149.4 kcal/mol, in good agreement with the tabulated values.

†††Recomputed from the empirical solubility data in Hostetler and Garrels (1962) consistent with the data in this table and with corrections for ion activities and complexes.

‡‡‡ΔG_f° computed assuming ΔG_f° for the tyuyamunite/carnotite exchange reaction equals its value for the autunite/K-autunite exchange reaction.

§§§Nguyen et al. (1991) measured the K_{eq} values for the solution reactions listed in Table A13.3. ΔG_f° values for the minerals have been computed from the K_{eq} values using ΔG_f° data in Tables A13.1 and A13.2.

‖‖‖ΔG_f°(haiweeite) estimated by Hemingway (1982) makes it more stable than uranophane in general, which is unlikely. The listed ΔG_f° value, which is probably a maximum stability, is based upon the assumption that uranophane is more stable than haiweeite when $H_4SiO_4 = 10^{-3.0}$ M.

TABLE A13.2   Thermochemical data for some auxiliary aqueous species and solids at 25°C and 1 bar total pressure

Aqueous species or solid	ΔG_f° (kcal/mol)	ΔH_f° (kcal/mol)	S° (cal/mol K)	Source
$H_2(g)$	0	0	31.233	†
$O_2(g)$	0	0	49.033	†
$H_2O(l)$	−56.678	−68.315	16.72	†
OH^-	−37.57	−54.97	−2.605	†
HF°	−71.63	−77.23	21.03	†
F^-	−67.28	−80.15	−3.30	†
Cl^-	−31.36	−39.94	13.53	†
$CO_2(g)$	−94.26	−94.05	51.10	†
$H_2CO_3^\circ$	−148.93	−167.10	45.3	†
HCO_3^-	−140.25	−164.9	23.5	†
CO_3^{2-}	−126.17	−161.4	−12.0	†
$H_3PO_4^\circ$	−274.7	−309.3	38.7	†
$H_2PO_4^-$	−271.8	−311.3	22.1	†
HPO_4^{2-}	−262.0	−310.5	−8.0	†
PO_4^{3-}	−245.08	−306.6	−51.6	†
$SiO_2(c)$ quartz	−204.66	−217.66	9.92	‡
$SiO_2(am)$	−202.91	−215.33	11.8	§
$H_4SiO_4^\circ$	−312.58	−348.30	45.1	§
$H_3SiO_4^-$	−299.18	−342.18	20.7	‖
$H_2SiO_4^{2-}$	−281.31	−	−	#
HSO_4^-	−180.52	−212.0	31.5	†
SO_4^{2-}	−177.82	−217.3	4.42	†
$H_4VO_4^+$	−253.67	−291.9	23.3	§
$H_3VO_4^\circ$	−249.2	(−289.9)	(44)	§
$H_2VO_4^-$	−244.0	−280.6	29	§
HVO_4^{2-}	−233.0	−277.0	4	††
VO_4^{3-}	−214.9		(−41)	§
VO^{2+}	−106.7	−116.3	−32	††
$VOOH^+$	−155.65	−	−	§
V^{3+}	−57.8	−	−	§
VOH^{2+}	−111.41	−	−	§
$V(OH)_2^+$	(−163.2)	−	−	§
$V(OH)_3^\circ$	(−212.9)	−	−	§
Na^+	−62.62	−57.44	14.0	†
K^+	−67.52	−60.26	24.2	†
NH_4^+	−18.98	−31.85	26.6	†
Mg^{2+}	−108.8	−111.6	−32.7	†
Ca^{2+}	−132.12	−129.8	−13.4	†
Sr^{2+}	−134.8	−131.7	−7.53	†
Ba^{2+}	−132.7	−127.3	2.01	†
Fe^{2+}	−18.85	−21.3	−32.9	††
Cu^{2+}	15.54	15.51	−23.4	†
Pb^{2+}	−5.78	0.22	4.42	†

Note: Values in parentheses are estimates or have been computed using estimated values.

Source: †Cox et al. (1989). ‡CODATA (1976). §Langmuir (1978). ‖Busey and Mesmer (1977). #Baes and Mesmer (1976). ††Wagman et al. (1982).

TABLE A13.3 Stability constants and ΔH_r° values for some reactions involving uranium aqueous species and solids of geochemical interest at 25°C and 1 bar total pressure

Reaction	$\log_{10} K$	ΔH_r° (kcal/mol)
$UO_2^{2+} + 4H^+ + 2e^- = U^{4+} + 2H_2O$	8.89 ($E^\circ = 0.263$ V)	−34.4
$UO_2^{2+} + e^- = UO_2^+$	2.97 ($E^\circ = 0.0879$ V)	−1.46
$U^{4+} + H_2O = UOH^{3+} + H^+$	−0.65	−11.21
$U^{4+} + 4H_2O = U(OH)_4^\circ + 4H^+$	−12.0	
$UO_2^{2+} + H_2O = UO_2OH^+ + H^+$	−5.2	
$UO_2^{2+} + 2H_2O = UO_2(OH)_2^\circ + 2H^+$	−12.0	
$UO_2^{2+} + 3H_2O = UO_2(OH)_3^- + 3H^+$	−19.2	
$2UO_2^{2+} + 2H_2O = (UO_2)_2(OH)_2^{2+} + 2H^+$	−5.62	
$3UO_2^{2+} + 5H_2O = (UO_2)_3(OH)_5^+ + 5H^+$	−15.55	
$3UO_2^{2+} + 7H_2O = (UO_2)_3(OH)_7^- + 7H^+$	−31.0	
$4UO_2^{2+} + 7H_2O = (UO_2)_4(OH)_7^+ + 7H^+$	−21.9	
$\beta\text{-}UO_3 \cdot 2H_2O(c)(\text{schoepite}) + 2H^+ = UO_2^{2+} + 3H_2O$	5.20	−12.0
$UO_2(\text{am}) + 4H^+ = U^{4+} + 2H_2O$	4.16	
$UO_2(c)(\text{uraninite}) + 4H^+ = U^{4+} + 2H_2O$	−4.99	−18.6
$USiO_4(\text{am})(\text{coffinite}) + 4H^+ = U^{4+} + H_4SiO_4^\circ$	0.50	
$U^{4+} + F^- = UF^{3+}$	9.3	−1.34
$U^{4+} + 2F^- = UF_2^{2+}$	16.22	−0.84
$U^{4+} + 3F^- = UF_3^+$	21.6	0.12
$U^{4+} + 4F^- = UF_4^\circ$	25.5	−0.87
$U^{4+} + 5F^- = UF_5^-$	27.01	
$U^{4+} + 6F^- = UF_6^{2-}$	29.1	
$UO_2^{2+} + F^- = UO_2F^+$	5.09	0.41
$UO_2^{2+} + 2F^- = UO_2F_2^\circ$	8.62	0.50
$UO_2^{2+} + 3F^- = UO_2F_3^-$	10.9	0.56
$UO_2^{2+} + 4F^- = UO_2F_4^{2-}$	11.7	0.07
$U^{4+} + Cl^- = UCl^{3+}$	1.72	−4.50
$UO_2^{2+} + Cl^- = UO_2Cl^+$	0.17	1.91
$UO_2^{2+} + 2Cl^- = UO_2Cl_2^\circ$	−1.1	3.59
$U^{4+} + SO_4^{2-} = USO_4^{2+}$	6.58	1.91
$UO_2^{2+} + SO_4^{2-} = UO_2SO_4^\circ$	3.15	4.66
$UO_2^{2+} + 2SO_4^{2-} = UO_2(SO_4)_2^{2-}$	4.14	8.39

TABLE A13.3   (continued)

Reaction	$\log_{10} K$	ΔH_r° (kcal/mol)
$UO_2^{2+} + PO_4^{3-} = UO_2PO_4^-$	13.69	
$UO_2^{2+} + HPO_4^{2-} = UO_2HPO_4^\circ$	7.71	
$UO_2^{2+} + H_3PO_4^\circ = H^+ + UO_2H_2PO_4^+$	1.12	
$UO_2^{2+} + H_3PO_4^\circ = UO_2H_3PO_4^{2+}$	0.76	
$UO_2^{2+} + 2H_3PO_4^\circ = 2H^+ + UO_2(H_2PO_4)_4^\circ$	0.87	
$UO_2^{2+} + H_3PO_4^\circ + 4H_2O = 2H^+ + UO_2HPO_4 \cdot 4H_2O(c)(\text{H-autunite})$	2.50	
$U^{4+} + 2H_3PO_4^\circ + 4H_2O = 4H^+ + U(HPO_4)_2 \cdot 4H_2O(c)$	11.79	
$3UO_2^{2+} + 2H_3PO_4^\circ + 4H_2O = 6H^+ + (UO_2)_3(PO_4)_2 \cdot 4H_2O(c)$	6.0	
$Ca^{2+} + 2UO_2^{2+} + 2PO_4^{3-} = Ca(UO_2)_2(PO_4)_2(c)(\text{autunite})$	44.7	
$2K^+ + 2UO_2^{2+} + 2VO_2^+ + 4H_2O = 8H^+ + K_2(UO_2)_2(VO_4)_2(c)(\text{carnotite})$	−0.56	
$2K^+ + 2UO_2^{2+} + 2VO_4^{3-} = K_2(UO_2)_2(VO_4)_2(c)(\text{carnotite})$	56.3	
$Ca^{2+} + 2UO_2^{2+} + 2VO_2^+ + 4H_2O = 8H^+ + Ca(UO_2)_2(VO_4)_2(c)(\text{tyuyamunite})$	3.56	
$Ca^{2+} + 2UO_2^{2+} + 2VO_4^{3-} = Ca(UO_2)_2(VO_4)_2(c)(\text{tyuyamunite})$	53.3	
$UO_2^{2+} + CO_3^{2-} = UO_2CO_3^\circ$	9.67	1.2
$UO_2^{2+} + CO_3^{2-} = UO_2CO_3(c)(\text{rutherfordine})$	14.49	1.1
$UO_2^{2+} + 2CO_3^{2-} = UO_2(CO_3)_2^{2-}$	17.0	4.4
$UO_2^{2+} + 3CO_3^{2-} = UO_2(CO_3)_3^{4-}$	21.63	−9.1
$U^{4+} + 5CO_3^{2-} = U(CO_3)_5^{6-}$	33.9	−4.8
$2UO_2^{2+} + CO_2(g) + 4H_2O = 5H^+ + (UO_2)_2CO_3(OH)_3^-$	−19.0	
$2UO_2^{2+} + H_4SiO_4^\circ + 2H_2O = 4H^+ + (UO_2)_2SiO_4 \cdot 2H_2O(c)(\text{soddyite})$	−5.74	
$Ca^{2+} + 2H_4SiO_4^\circ + 2UO_2^{2+} + 5H_2O = 6H^+ +$ $Ca(H_3O)_2(UO_2)_2(SiO_4)_2 \cdot 3H_2O(c)(\text{uranophane})$	−9.42	
$2Na^+ + 2UO_2^{2+} + 6H_4SiO_4^\circ = 6H^+ + 5H_2O +$ $Na_2(UO_2)_2(Si_2O_5)_3 \cdot 4H_2O(c)(\text{Na-weeksite})$	−1.50	
$Ca^{2+} + 2UO_2^{2+} + 6H_4SiO_4^\circ = 6H^+ + 4H_2O +$ $Ca(UO_2)_2(Si_2O_5)_3 \cdot 5H_2O(c)(\text{haiweeite})$	<2.60	

Note: For the redox couples, $\log_{10} K = nE^\circ/0.05916$. Tabulated values have been computed using thermodynamic data in Tables A13.1 and A13.2.

TABLE A13.4 Chemical analyses and related information for some well waters from uranium deposits and two unmineralized granites (Äspö and Stripa, Sweden)

	Source										
	1	2	3	4	5	6	7	8	9	10	11
Species (mg/L)											
Na^+	2.70	28.9	1180	430	330	391	22.3	1.24	1.49	218	137
K^+	1.23	8.5	6.3	31	19	4.3	1.2	11.8	34.2	0.44	2.8
Mg^{2+}	0.94	3.4	30	29	21	4.6	4.4	0.3	3.0	0.03	1.9
Ca^{2+}	1.86	10.9	741	210	210	26.1	18.0	10.8	74	94	19
Sr^{2+}	—	0.198	12.3	—	—	—	—	0.26	1.5	0.89	—
SO_4^{2-}	0.55	26.4	265	570	460	768	30.0	33.4	360	57	224
F^-	0.67	0.69	2.7	0.4	0.4	1.14	0.36	5.4	4.3	5.3	0.34
Cl^-	1.23	38.8	3030	650	520	74.5	1.4	—	—	460	12
Br^-	<0.01	0.56	15.9	—	—	—	—	—	—	4.5	—
HCO_3^-	12.3	31.1	69	275	289	61.0	85.4	31.4	18.4	22	128
NO_3^-	<0.065	0.54	—	6.1	5.1	—	—	—	—	—	0.05
Al	<0.10	<0.15	—	—	—	—	—	0.6	2.5	0.009	0.41
$Fe(II)$	—	—	—	—	—	—	—	9.96	53.4	—	—
Fe (total)	0.62	<.012	0.33	0.03	0.06	<0.05	<0.05	10.6	53.5	—	0.02
$Mn(II)$	0.05	0.08	0.31	0.09	0.07	—	—	2.95	12	—	0.005
$SiO_2(aq)$	18.6	9.4	4.1	45	48	7.8	9.6	37.3	—	18	—
Species (µg/L)											
B	<10	750	—	—	—	—	—	20	—	—	100
Ba^{2+}	184	33	—	—	—	—	—	110	50	9.0	300
Cd	<1	<1	—	—	—	—	—	—	—	—	5
Cu	<2	<2	—	—	—	—	—	<5	0.8	7.0	5
Mo	<6.5	150	—	2	2	—	—	<35	1.4	—	100
Ni	<15	—	—	—	—	—	—	<25	—	—	10
Th	<0.02	<0.02	—	—	—	—	—	0.13	4.0	—	—
ΣU	0.36	7.8	0.459	0.4	0.2	—	—	11.0	0.06	0.105	215
$U(IV)$	—	—	—	—	—	1.43	5.3	—	25	—	—
$U(VI)$	—	—	—	—	—	0.11	—	—	—	—	—
V	<10	—	—	9	4	—	185	170	—	—	42
Zn	5.7	2.2	—	—	—	—	—	—	2020	—	40

TABLE A13.4 (continued)

					Source						
	1	2	3	4	5	6	7	8	9	10	11
Other Information											
T (°C)	12.3	7.4	(25)	(25)	(25)	(25)	(25)	24	22	8.6	(25)
pH	5.85	7.37	7.9	6.9	7.0	9.05	8.42	6.30	5.45	8.50	8.5
Eh (mV)	130	−242	−280	(−200)	(−223)	−92	−55	155	420	(−400)	(+600)
DO (mg/L)	0.002	0.003	—	0.4	1.2	—	—	<0.008	<0.008	(0)	—
SpC (μmhos/cm)	—	—	—	3300	2650	—	—	—	—	—	—
TDS (mg/L)	78	158	—	—	—	—	—	—	—	—	452
He (mg/L)	—	—	—	10.4	11.6	—	—	—	—	—	—
Ra-226 (Bq/L)	0.59	4.28	—	6.1	0.80	—	—	—	—	2.81	5.2
Rn-222 (Bq/L)	1630	—	—	92.2	87.2	—	—	—	—	20300	—
Date collected	9-12-87	6-21-89	1991?	8-17-79	8-21-79	1992–93	1992–93	12-20-88	2-1-88	1983?	1978?
Sample depth (m)	423 to 427	432 to 440	334-343	?	?	240 to 246	90 to 95	275 to 300	50 to 78	908 to 969	?

Note: Values in parentheses are estimated or assumed.

Source: (1) and (2) ore zone of Cigar Lake uranium deposit, Athabaska Sandstone Formation, Saskatchewan, Canada. Ore minerals: uraninite (UO_2 to $UO_{2.33}$) and coffinite. In (2), $H_2(aq) =$ 1.4 mg/L. (Cramer and Smellie 1994). (3) Groundwater from granite at Äspö, Sweden. Water also contains 0.32 mg/L NH_4^+ and 3.6 μg/L PO_4^{3-} (Wikberg et al. 1991). (4) and (5) Oakville (fluvial sandstone) Formation, Texas. Ore minerals uraninite and coffinite. $H_2S > 10$ mg/L in both waters (Chatham et al. 1981; Wanty et al. 1987). (6) and (7) Palmottu uranium deposit in fractured crystalline rock, Finland (Ahonen et al. 1993; Ervanne et al. 1994). Ore is uraninite with 7 to 10% ThO_2, and 10 to 17% PbO by wt. Authors assume UO_2 oxidation up to $U_{2.33}$. (8) and (9) Osamu Utsumi Mine, Pocos de Caldas, Brazil (Nordstrom et al. 1990; Miekeley et al. 1991). Ore is pitchblende (relatively amorphous UO_2). (10) Well water from granite at Stripa, Sweden (Nordstrom et al. 1989; Andrews et al. 1989). The reported pH was 10.06, which makes the water supesaturated with calcite by more than 10-fold. Readjustment to calcite saturation gives pH = 8.50. (11) Well in Wasatch (fluvial) Formation, on property of the Irigaray Uranium Solution Mining Project, Wyoming, U.S.A. (NRC 1978). Ore is coffinite with lesser amounts of pitchblende (cf. Dahl and Hagmaier 1974).

TABLE A13.5 Thermodynamic data for selected aqueous species and solids of americium at 25°C and 1 bar total pressure

Aqueous species or solid	ΔG_f° (kcal/mol)	ΔH_f° (kcal/mol)	S° (cal/mol K)
Am^{3+}	−143.09	−147.39	−48.04
$Am_2O_3(c)$	−385.59	−404.02	38.24
$AmOH^{2+}$	−191.04		
$Am(OH)_2^+$	−237.21		
$Am(OH)_3(am)$	−287.54		
$Am(OH)_3^\circ$	−278.06		
$Am(OH)_3(c)$	−292.39		
AmF^{2+}	−215.02		
AmF_2^+	−285.58		
$AmF_3(c)$	−363.01	−379.54	30.50
$AmCl^{2+}$	−175.89		
$AmSO_4^+$	−326.17		
$Am(SO_4)_2^-$	−506.10		
$AmPO_4(am,hydr)$	−422.01		
$AmH_2PO_4^{2+}$	−418.97		
$AmCO_3^+$	−279.90		
$Am(CO_3)_2^-$	−412.21		
$Am(CO_3)_3^{3-}$	−542.34		
$Am(CO_3)_5^{6-}$	−767.26		
$Am_2(CO_3)_3(c)$	−710.26		
$AmOHCO_3(c)$	−335.76		

Source: From Silva et al. (1995).

TABLE A13.6 Gibbs free energies of formation of some aqueous species and solids of technetium of geochemical interest at 25°C and 1 bar total pressure

Aqueous species or solid	ΔG_f° (kcal/mol)	Source
TcO_4^-	−148.45	†
TcO^{2+}	−24.04	†
$TcOOH^+$	−79.18	†
$TcO(OH)_2^\circ$	−132.89	†
$[TcO(OH)_2^\circ]_2$	−274.67	†
$TcO_2 \cdot 2H_2O(s)$	−200.13	‡
$TcO(OH)_3^-$	−174.08	§
$Tc(OH)_2CO_3^\circ$	−228.04	§
$Tc(OH)_3CO_3^-$	−273.45	§

Source: †Rard (1983). ‡Based on $E^\circ = 0.747$ V for the reaction $TcO_4^- + 4H^+ + 3e^- = TcO_2 \cdot 2H_2O(s)$ (Meyer et al. 1989) and ΔG_f° (TcO_4^-) from this table. §Based on measured stability constants for dissolution of $TcO_2 \cdot 2H_2O(s)$ to form the complexes from Eriksen et al. (1993).

TABLE A13.7 Gibbs free energies of formation from the elements of some geochemically important aqueous species and solids of neptunium at 25°C and 1 bar total pressure

Aqueous species or solid	ΔG_f° (kcal/mol)	Source
NpO_2^{2+}	−190.2	†
NpO_2^+	−218.7	†
NpO_2OH°	−262.90	‡
$NpO_2(OH)_2^-$	−299.73	§
$NpO_2CO_3^-$	−338.19	‖
$NpO_2(CO_3)_2^{3-}$	−480.66	‖
$NpO_2(CO_3)_3^{5-}$	−607.22	#
$NaNpO_2CO_3(s)$	−423.26	††
$NpO_2OH(am)$	−269.60	††
$Np_2O_5(s)$	−481.12	††
Np^{4+}	−120.2	†
$NpOH^{3+}$	−175.45	‡
$Np(OH)_4^\circ$	−333.57	‡‡
$Np(OH)_4(s)$	−344.87	§§
$NpO_2(c)$	−244.06	§§
$Np(OH)_3CO_3^-$	−421.55	‡‡
$Np(OH)_4CO_3^{2-}$	−463.83	‡‡
$NpCO_3^{2+}$	(−263.15)	‡
$Np(CO_3)_2^\circ$	(−404.39)	‡
$Np(CO_3)_3^{2-}$	(−539.64)	‡
$Np(CO_3)_4^{4-}$	(−669.90)	‡
NpF^{3+}	−199.32	‡
NpF_2^{2+}	−276.18	‡
$NpSO_4^{2+}$	−307.00	‡
$Np(SO_4)_2^\circ$	−489.20	‡
$Np(HPO_4)_2(s)$	−784.24	††

Note: Values in parentheses are estimates.

Source: †ΔG_f° values from Fuger and Oetting (1976). ‡Computed from stability constants given by Nitsche (1991). Nitsche (1991) gives estimated constants for $Np(OH)_2^{2+}$, $Np(OH)_4^+$, $Np(OH)_4^\circ$, and $Np(OH)_5^-$. $Np(OH)_5^-$ has been discredited (cf. Lieser and Mohlenweg 1988), and Nitsche's stability for $Np(OH)_4^\circ$ is entirely inconsistent with the measured solubility of $Np(OH)_4(s)$ (cf. Eriksen et al. 1993). Consistent with U^{4+} hydrolysis, this review limits the hydrolysis species to $NpOH^{3+}$ and $Np(OH)_4^\circ$. §Computed from the average of stability constants given by Fujita et al. (1995) and Novak and Roberts (1995). ‖Fuger et al. (1992). #Computed from the stability constant given by Fujita et al. (1995). ††ΔG_f° based on Lemire (1984). Lemire (1984) gives $K_{sp} = 10^{-11.56}$ for $NaNpO_2CO_3(s)$ (actually a hydrate of variable H_2O content). In good agreement Novak and Roberts (1995) suggest $K_{sp} = 10^{-11.28}$. ‡‡Computed from stability constants given by Eriksen et al. (1993). When modeling Np^{4+} carbonate complexing, the assumption of these two hydroxyl-carbonate complexes should be made, or the assumption of the four Np^{4+}-carbonate complexes with estimated ΔG_f° values from Nitsche (1991b), but not both. §§Based on data given by Rai et al. (1987).

TABLE A13.8 Gibbs free energies and enthalpies of formation from the elements of some geochemically important aqueous species and solids of plutonium at 25°C and 1 bar total pressure

Aqueous species or solid	ΔG_f° (kcal/mol)	ΔH_f° (kcal/mol)	Data quality[†]
Pu^{3+}	−138.29	−141.52	g
Pu^{4+}	−115.11	−128.22	g
PuO_2^+	−203.15	−218.61	f
PuO_2^{2+}	−180.90	−196.48	f
$PuOH^{2+}$	−184.11		g
$PuCO_3^+$	−276.74		s
$Pu(H_2PO_4)^{2+}$	−411.81		g
$PuOH^{3+}$	−171.10		f
$Pu(OH)_2^{2+}$	−225.31		f
$Pu(OH)_3^+$	−277.96		f
$Pu(OH)_4^\circ$	−329.35	−375.835	f
$PuCO_3^{2+}$	−259.02		s
$Pu(CO_3)_5^{6-}$	−799.24		s
$Pu(HPO_4)^{2+}$	−393.16		f
$Pu(HPO_4)_2^\circ$	−668.26	−739.59	f
$Pu(HPO_4)_3^{2-}$	−941.68		f
$Pu(HPO_4)_4^{4-}$	−1215.34		f
PuO_2OH°	−246.65	−270.39	f
$PuO_2CO_3^-$	−336.28		f
$PuO_2(CO_3)_2^{3-}$	−465.11		s
$(PuO_2)_2(OH)_2^{2+}$	−463.91		f
$(PuO_2)_3(OH)_5^+$	−796.61		f
PuO_2OH^+	−229.90		f
$PuO_2(OH)_2^\circ$	−278.99	−313.21	f
$PuO_2(OH)_3^-$	−322.04		f
$PuO_2CO_3^\circ$	−319.62	−352.99	g
$PuO_2(CO_3)_2^{2-}$	−453.43		g
$PuO_2(CO_3)_3^{4-}$	−583.15		g
$PuO_2H_2PO_4^+$	−456.50		g
$PuO_2(OH)_2HCO_3^-$	−423.67		f
β-$Pu_2O_3(c)$	−380.98		f
$Pu(OH)_3(s)$	−277.72		f
$PuO_2(c)$	−238.53		g
$Pu(OH)_4(am)$	−342.97		f
$PuO_2OH(am)$	−252.39		f
$PuO_2(OH)_2(c)$	−288.31		f
$PuOHCO_3(s)$	−332.06		s
$PuO_2CO_3(s)$	−326.45		f
$Pu(HPO_4)_2(c)$	−673.52		f
$PuO_2HPO_4(c)$	−458.41		f

[†]g, f, and s denote good, fair, and speculative ΔG_f° and/or ΔH_f° data, respectively.

Source: Data from Puigdomenech and Bruno (1991), whose sources were chiefly Lemire and Tremaine (1980) and Lemire and Garisto (1989).

Geochemical
Computer Models

SOME EXAMPLE GEOCHEMICAL COMPUTER MODELS[†,‡]

Mass-Balance Models

Examples: BALANCE (Parkhurst et al. 1982), NETPATH (Plummer et al. 1991, 1994).

Purposes: Used to define net masses of minerals (or gases) dissolved and/or precipitated along a flow path between two wells. Also considers ion exchange and other user-definable mass transfer processes. Can account for the mixing of two waters to produce a third, final water. NETPATH can also be used to age-date groundwaters (using ^{14}C) and can solve isotope evolution problems.

User input: Two complete chemical analyses are required (or three in the case of a mixing problem). Isotopic ($\delta^{34}S$, $\delta^{13}C$, $\delta^{18}O$, $\delta^{2}H$, ^{14}C, $^{87}Sr/^{86}Sr$, etc.) analyses are strongly recommended. Mineralogical information is recommended. Guesses of possible mass transfer reactions are required.

Interpretation of results: Usually several plausible models are found. The user must apply his knowledge of aquifer mineralogy and of the isotopic compositions of the water and reacting phases, as well as his knowledge of the probability that given reactions could occur, to eliminate as many models as possible.

Limitations and strengths: The models found are not dependent on thermodynamic data or on the assumption of thermodynamic equilibrium. The user must have the experience and judgment to be able to eliminate thermodynamically and kinetically unrealistic models. Some knowledge of aquifer/soil mineralogy between the two well points is required. The user must have knowledge of the hydrogeologic flow paths. A steady-state flow system is assumed. Dispersion of chemical constituents along the flow path is assumed insignificant or can be characterized as the mixing of two initial waters.

[†]This discussion is largely after Glynn et al. (1992).

[‡]For a discussion and comparison of most of the codes described below and many other similar and/or related codes see Mangold and Tsang (1991) and van der Heijde and Elnawawy (1993).

Observations: the mass-balance models can help the user to determine and quantify the geochemical processes that are most important to the chemical evolution of a water, whether natural or contaminated. With sufficient information (e.g., isotopic data) the models can provide information on the hydrogeology, including groundwater flow paths and velocities. If groundwater velocities are known, the models can determine reaction rates. Use of these models can help to identify critically needed data or measurements, such as identification of the mineralogy or isotopic compositions, etc.

Speciation Models

Examples: WATEQF (Plummer et al. 1984), WATEQ4F (Ball and Nordstrom 1991).

Purposes: Can calculate the partitioning of an element between different aqueous species and complexes (inorganic and organic) and determine whether a water is supersaturated or undersaturated with respect to various minerals or gas phases (useful in testing mass-balance models).

User input: Requires a complete chemical analysis of a water, including its pH, and some knowledge of pE or Eh (redox potential) or of the concentrations of redox couples (e.g., Fe^{3+}/Fe^{2+}).

Limitations and problems: For reliable results, the thermodynamic data base should be internally consistent and of high quality (cf. Nordstrom and Munoz 1994). The data bases often differ among the different models. The aqueous species and minerals considered also differ among the different models. The number of organic species considered is generally small. The models are usually limited to modeling speciation of dilute waters with ionic strengths less than seawater (< 0.7 M). Nonequilibrium conditions among redox couples or species forming aqueous complexes, etc., cannot be addressed. The quality of input water analyses is very important. An inaccurate water analysis, with a poor charge balance, for example, or with inaccurate pH or Eh values, can lead to meaningless model results.

Observations: Speciation models are useful for determining the relative importance of individual aqueous complexes and the toxicity of contaminated waters. They can establish whether a water has the potential to precipitate or dissolve a mineral or gas phase and whether or not various mass transfer processes such as ion exchange have the potential to affect the concentrations of various constituents. They are often used in conjunction with mass-balance models.

Mass Transfer Codes

Examples: PHREEQE (Parkhurst et al. 1990), PHREEQC (Parkhurst 1995), PHRQPITZ (Plummer et al. 1988), SOLMINEQ.88 (Kharaka et al. 1988), MINTEQA2 (Allison et al. 1991), MINTEQ(4.00) (Eary and Jenne 1992), MINEQL+ (Schecher and McAvoy 1991), EQ3/6 (Wolery 1992a, 1992b), Geochemist's Workbench (Bethke 1994, 1996).

Purposes: These models can speciate an aqueous solution, just as the speciation models do. They can also simulate changes in solution chemistry caused by mass transfer processes, such as dissolution/precipitation, ingassing/outgassing, ion exchange/adsorption, evaporation, boiling temperature and pressure changes, and mixing of two waters.

Limitations and problems additional to those of the speciation codes: The models do not consider solid-solution mass transfer and provide only limited information on ion exchange/adsorption mass transfer. All the programs except PHREEQE, PHRQPITZ, and MINTEQA2 keep track of water mass. Except for EQ3/6 and the Geochemist's Workbench, rate laws for mass transfer kinetics cannot be specified. Convergence problems occur more often than for the speciation codes.

Observations: The mass transfer models can predict the overall geochemical behavior of a contaminant and whether reactions go to equilibrium within a system. They are often used in reaction path modeling.

Chemical Mass Transport Codes

Examples: PHREEQM-2D (Nienhuis et al. 1994), PHREEQC (Parkhurst 1995), CHMTRNS (Noorishad et al. 1987).

Purposes: These codes can speciate an aqueous solution and allow for chemical mass transfer processes. They can also simulate hydrodynamic advection and dispersion of chemical constituents in a porous medium.

Limitations and special properties: A system of partial differential equations has to be solved in addition to the system of algebraic chemical equations typical of speciation and mass transfer codes. The local equilibrium assumption is generally made. Problems of scale in using the advection-dispersion equation are generally ignored. Flow modeling is often kept simple, usually with steady-state 1-D flow fields in homogeneous geological media and with simple boundary conditions. Numerical dispersion and other problems pertaining to the numerical approximation of the transport equations can be significant.

Observations: The mass transport models can be used to predict "best-case" and "worst-case" scenarios of contaminant transport, but in most cases are not exact predictive tools. Both mass transfer and mass transport models are useful to help establish possible contaminant cleanup strategies and, more generally, to help understand the processes that affect the chemical evolution of groundwaters.

OBTAINING GEOCHEMICAL SOFTWARE

Software for current PC DOS versions of geochemical models supported by the U.S. Environmental Protection Agency (EPA) and the software documentation may be obtained at no charge via the Internet. Go to: **ftp://ftp.epa.gov/epa_ceam/wwwhtml/software.htm** to obtain downloadable, compressed software programs, including MINTEQA2, CHEMFLO, MT3D (MODFLOW), and VLEACH. Alternatively, the software and documentation may be ordered from The Center for Exposure Assessment Modeling (CEAM), U.S. Environmental Protection Agency, Office of Research and Development, Environmental Research Laboratory, College Station Road, Athens, GA 30613-0801. Telephone and e-mail addresses are (706) 546-3549 and **ceam@athens.ath.epa.gov**.

The latest UNIX versions of U.S. Geological Survey (USGS) geochemical software and the documentation manuals may be downloaded free of charge from the Internet address **http://h2o.usgs.gov/software**. PC DOS versions of the USGS software should be available from this same address in late 1996. Until then, the PC DOS versions may be downloaded by anonymous ftp from **ftp://brrcrftp.cr.usgs.gov/geochem**. The files for PHREEQC, for example, reside in **/pc/phreeqc**. PC DOS versions of other USGS geochemical software are similarly available for BALNINPT, NETPATH, PHREEQE, PHRQPITZ, and WATEQ4F. The software and documentation for these programs may also be purchased from U.S. Geological Survey, NWIS Program Office, 437 National Center, Reston, VA 22092 (telephone 703-648-5695). For further information on USGS water resources applications software contact U.S. Geological Survey, Hydrologic Analysis Software Support Team, 437 National Center, Reston, VA 20192. The e-mail address is **h2osoft@usgs.gov**.

A practical difficulty in using many of the USGS codes has been the unfriendly nature of their data entry files. The program CHEMFORM (Toran and Sjoreen 1996) simplifies this problem. CHEMFORM reads existing ASCII chemical data files and formats the data correctly for use by USGS programs, including NETPATH (Plummer et al. 1994), PHREEQE (Parkhurst et al. 1990), and WATEQ4F (Ball and Nordstrom 1991). CHEMFORM is available at no charge by anonymous ftp to **ftp://ftp.esd.ornl.gov/pub**. It can also be downloaded along with the USGS geochemical codes by anonymous ftp to **ftp://brrcrftp.cr.usgs.gov/contrib**.

The thermodynamic data bases of most geochemical models are only occasionally corrected and updated. Updated data bases for the Geochemist's Workbench (Bethke 1994, 1996) and EQ3/6 (Wolery 1992a, 1992b) may be obtained by anonymous ftp from **ftp://s32.es.llnl.gov/johnson** where they are located in the files **gwb** and **eq36**. For further information contact Jim Johnson at his e-mail address **jwjohnson@llnl.gov**.

The current version of MINEQL$^+$ (Schecher and McAvoy 1991) may be downloaded from the Internet at web site **http://www.agate.net/~ersoftwr/mineql.html**. For further information contact William Schecher at his e-mail address, **ersoftwr@agate.net**.

References

AAGAARD, P., and H. C. HELGESON. 1982. Thermodynamic and kinetic constraints on reaction rates among minerals and aqueous solutions. *Am. J. Sci.* 282:237–85.

AAGAARD, P., and H. C. HELGESON. 1983. Activity/composition relations among silicates and aqueous solutions: II. Chemical and thermodynamic consequences of ideal mixing of atoms on homological sites in montmorillonites, illites, and mixed-layer clays. *Clays Clay Minerals* 31(3):207–17.

ACKER, J. G., and O. P. BRICKER. 1992. The influence of pH on biotite dissolution and alteration kinetics at low temperatures. *Geochim. Cosmochim. Acta* 56:3073–92.

ADAMSON, A. W. 1990. *Physical chemistry of surfaces.* 5th ed. New York: John Wiley & Sons.

ADLOFF, J-P., and R. GUILLAUMONT. 1993. *Fundamentals of radiochemistry.* Boca Raton, FL: CRC Press.

AGGARWAL, P. K., W. D. GUNTER, and Y. K. KHARAKA. 1990. Effect of pressure on aqueous equilibria. In *Chemical modeling of aqueous systems II,* ed D. C. Melchior and R. L. Bassett. Am. Chem. Soc. Symp. Ser. 416, pp. 87–101. Washington, DC: Am. Chem. Soc.

AHONEN, L., H. ERVANNE, T. RUSKEENIEMI, T. JAAKKOLA, and R. BLOMQVIST. 1993. Uranium mineral-groundwater equilibration at the Palmottu natural analogue study site, Finland. *Mat. Res. Soc. symp. proc.* 294, pp. 497–504.

AHRENS, L. H. 1952. The use of ionization potentials. Part I. Ionic radii of the elements. *Geochim. Cosmochim. Acta* 2:155–69.

AHRLAND, S. 1973. Thermodynamics of the stepwise formation of metal-ion complexes in aqueous solution. *Structure and Bonding* 15:16–88.

AJA, S. U., and P. E. ROSENBERG. 1992. The thermodynamic status of compositionally-variable clay minerals: A discussion. *Clays Clay Minerals* 40(3):292–99.

AJA, S. U., P. E. ROSENBERG, and J. A. KITTRICK. 1991a. Illite equilibria in solutions: I. Phase relationships in the system $K_2O-Al_2O_3-SiO_2-H_2O$ between 25 and 250°C. *Geochim. Cosmochim. Acta* 55:1353–64.

AJA, S. U., P. E. ROSENBERG, and J. A. KITTRICK. 1991b. Illite equilibria in solutions: II. Phase relationships in the system $K_2O-MgO-Al_2O_3-SiO_2-H_2O$. *Geochim. Cosmochim. Acta* 55:1363–74.

AL-BORNO, A., and M. B. TOMSON. 1994. The temperature dependence of the solubility product constant of vivianite. *Geochim. Cosmochim. Acta* 58(24):5373–78.

ALLISON, J. D., D. S. BROWN, and K. J. NOVO-GRADAC. 1991. *MINTEQA2, A geochemical assessment data base and test cases for environmental systems: Vers. 3.0 user's manual.* Report EPA/600/3-91/-21. Athens, GA: U.S. EPA.

ALLRED, A. L. 1961. Electronegativity values from thermochemical data. *J. Inorg. Nucl. Chem.* 17:215–21.

ALPERS, C. N., D. K. NORDSTROM, and J. W. BALL. 1989. Solubility of jarosite solid solutions precipitated from acid mine waters, Iron Mountain, California, U.S.A. *Sci. Geol. Bull. (Strasbourg)* 42:281–98.

ALPERS, C. N., R. O. RYE, D. K. NORDSTROM, L. D. WHITE, and B-S. KING. 1992. Chemical, crystallographic and stable isotopic properties of alunite and jarosite from acid-hypersaline Australian lakes. *Chem. Geol.* 96:203–26.

ALVETEG, M., H. SVERDRUP, and P. VARFVINGE. 1995. Developing a kinetic alternative in modeling soil aluminum. *Water, Air and Soil Pollution* 79:377–89.

ANDERSON, B. J., E. A. JENNE, and T. T. CHAO. 1973. The sorption of silver by poorly crystallized manganese oxides. *Geochim. Cosmochim. Acta* 37:611–22.

ANDERSON, G. M. 1972. Silica solubility. In *Encyclopedia of geochem. and envir. sciences,* ed R. W. Fairbridge. New York: Van Nostrand Reinhold Co.

ANDERSON, G. M., and D. A. CRERAR. 1993. *Thermodynamics in geochemistry, the equilibrium model.* New York: Oxford Univ. Press.

ANDERSON, L. D., D. B. KENT, and J. A. DAVIS. 1994. Batch experiments characterizing the reduction of Cr(VI) using suboxic material from a mildly reducing sand and gravel aquifer. *Envir. Sci. & Technol.* 28:178–85.

ANDERSON, P. R., and M. M. BENJAMIN. 1990a. Constant-capacitance surface complexation model. In *Chemical modeling of aqueous systems II*, ed D. C. Melchior and R. L. Bassett. Am. Chem. Soc. Symp. Ser. 416, pp. 272–81. Washington DC: Am. Chem. Soc.

ANDERSON, P. R., and M. M. BENJAMIN. 1990b. Modeling adsorption in aluminum-iron binary oxide suspensions. *Envir. Sci. & Technol.* 24:1586–92.

ANDERSON, P. R., and M. M. BENJAMIN. 1990c. Surface and bulk characteristics of binary oxide suspensions. *Envir. Sci. & Technol.* 24:692–98.

ANDERSON, W. C., and M. P. YOUNGSTROM. 1976. Coal pile leachate-quantity and quality characteristics. *Proc. 6th symp. coal mine drainage research*, pp. 17–33. Washington DC: N. H. Coal Assoc.

ANDERSSON, K., B. TORSTENFELT, and B. ALLARD. 1982. *Sorption behavior of long-lived radionuclides in igneous rock.* Intl. Atomic Energy Agency Report IAEA-SM-257/20:111–31.

ANDREWS, J. N. et al. 1982. Radioelements, radiogenic helium and age relationships for groundwaters from the granites at Stripa, Sweden. *Geochim. Cosmochim. Acta* 46:1533–43.

ANDREWS, R. A. et al. 1995. *Total system performance assessment-1995: An evaluation of the potential Yucca Mountain repository.* Las Vegas, NV: Civilian Radioactive Waste Management System Management and Operating Contractor.

ANSTO. 1992. *Alligator Rivers Analogue Project final report.* DOE/HMIP/RR/92/072, SKI TR 92:20-2. Australia Nuclear Science and Technology Organization.

APGAR, M. A., and D. LANGMUIR.1971. Ground water pollution potential of a landfill above the water table. *Ground Water* 9:76–96.

APPELO, C. A. J., and D. POSTMA. 1993. *Geochemistry, groundwater and pollution.* Rotterdam: A. A. Balkema.

APPS, J. A. 1983. Hydrothermal evolution of repository groundwaters in basalt. In *NRC nuclear waste geochemistry '83*, ed D. H. Alexander and G. F. Berchard. Report NUREG/CP-0052, pp. 14–51. U.S. Nuclear Regulatory Commission.

APRIL, R., R. NEWTON, and L. COLES. 1986. Chemical weathering in two Adirondack watersheds; Past and present day rates. *Bull. Geol. Soc. Am.* 97:1232–38.

ATKINS, P. W. 1978. *Physical chemistry.* San Francisco: W. H. Freeman & Co.

BAAS BECKING, L. G. M., I. R. KAPLAN, and D. MOORE. 1960. Limits of the natural environment in terms of pH and oxidation-reduction potentials. *J. Geol.* 68:243–84.

BACK, W., and B. B. HANSHAW. 1970. Comparison of chemical hydrogeology of the carbonate peninsulas of Florida and Yucatan. *J. Hydrol.*, special issue 10(4):77–93.

BACK, W., B. B. HANSHAW, J. S. HERMAN, and J. N. VAN DRIEL. 1986. Differential dissolution of a Pleistocene reef in the ground-water mixing zone of coastal Yucatan, Mexico. *Geology* 14:137–40.

BAEDECKER, M. J., and W. BACK. 1979. Modern marine sediments as a natural analog to the chemically stressed environment of a landfill. *J. Hydrol.* 43:393–414.

BAEDECKER, P. A., et al. 1990. Acidic deposition, Rept. 19. In *Effects of acidic deposition on material.* Washington, DC: State of Science & Technol., Natl. Acid Precip. Assessment Program.

BAES, C. F., JR., and R. E. MESMER. 1976. *The hydrolysis of cations.* New York: Wiley-Interscience.

BAES, C. F., JR., and R. E. MESMER. 1981. The thermodynamics of cation hydrolysis. *Am. J. Sci.* 281:935–62.

BAGANDER, L. E., and R. CARMAN. 1994. In situ determination of the apparent solubility product of amorphous iron sulphide. *Applied Geochem.* 9:379–86.

BAIN, D. C., A. MELLOR, M. S. E. ROBERTSON, and S. T. BUCKLAND. 1991. Variations in weathering processes and rates with time in a chronosequence of soils from Glen Feshie, Scotland. *Proc. 2nd. int. symp. envir. geochem.*, ed O. Selinus, Uppsala.

BAIN, D. C., A. MELLOR, and M. WILSSON. 1990. Nature and origin of an aluminous vermiculite weathering product in acid soils from upland catchments in Scotland. *Clay Minerals* 25:467–75.

BAKER, J. P., W. J. WARREN-HICKS, J. GALLAGHER, and S. W. CHRISTENSEN. 1993. Fish population losses from Adirondack lakes: The role of surface water acidity and acidification. *Water Resources Research* 29(4):861–74.

BALL, J. W., and D. K. NORDSTROM. 1991. *User's manual for WATEQ4F, with revised thermodynamic database and test cases for calculating speciation of major, trace, and redox elements in natural waters.* U.S. Geol. Survey Open File Rept. 91-183.

BALL, J. W., D. K. NORDSTROM, and E. A. JENNE. 1980. *Additional and revised thermochemical data and computer code for WATEQ2. A computerized chemical model for trace and major element speciation and mineral equilibria of natural waters.* U.S. Geol. Survey Water-Resources Inv. 78-116.

BALTPURVINS, K. A., R. C. BURNS, G. A. LAWRANCE, and A. D. STUART. 1996. Effect of pH and anion type on the aging of freshly precipitated iron (III) hydroxide sludges. *Environ. Sci. Technol.* 30:939–44.

BANKS, D., C. REIMANN, O. ROYSET, H. SKARPHAGEN, and O. M. SAETHER. 1995. Natural concentrations of major and trace elements in some Norwegian bedrock groundwaters. *Applied Geochem.* 10:1–16.

BARBERO, J. A., K. G. MCCURDY, and P. R. TREMAINE. 1982. Apparent molal heat capacities and volumes of aqueous hydrogen sulfide and sodium hydrogen sulfide near 25°C: The temperature dependence of H_2S ionization. *Canadian J. Chem.* 60:1872–80.

BARNES, H. L., ed. 1979. *Geochemistry of hydrothermal ore deposits.* New York: Wiley-Interscience.

BARNES, H. L., and D. LANGMUIR. 1978. *Geochemical prospecting handbook for metals and associated elements.* Natl. Science Foundation Grant No. AER77-06511 AO2, Annual Report.

BARNES, I. 1964. *Field measurement of alkalinity and pH.* U.S. Geol. Survey Water Supply Paper 1535-H.

BARNES, I. 1970. Metamorphic waters from the Pacific tectonic belt of the West Coast of the United States. *Science* 168:973–75.

BARNES, I., W. T. STUART, and D. W. FISHER. 1964. *Field investigations of mine waters in the northern anthracite field, Pennsylvania.* U.S. Geol. Survey Prof. Paper 473-B.

BARON, D., and C. D. PALMER. 1996. Solubility of jarosite at 4–35°C. *Geochim. Cosmochim. Acta* 60(2):185–95.

BARSHAD, I. 1966. The effect of variation in precipitation on the nature of clay mineral formation in soils from acid and basic igneous rocks. *Proc. int. clay conf. Jerusalem* 1, pp. 167–73.

BARTON, A. F. M., and N. M. WILDE. 1971. Dissolution rates of polycrystalline samples of gypsum and orthorhombic forms of $CaSO_4$ by a rotating disk method. *Trans. Farad. Soc.* 67:3590–97.

BARTON, P. B., JR., and B. J. SKINNER. 1979. Sulfide mineral stabilities.In *Geochemistry of hydrothermal ore deposits*, 2d ed., ed H. L. Barnes, pp. 278–390. New York: Wiley-Interscience.

BASSETT, R. L., Y. K. KHARAKA, and D. LANGMUIR. 1979. Critical review of the equilibrium constants for kaolinite and sepiolite. In *Chemical modeling in aqueous systems*, ed E. A. Jenne. Am. Chem. Soc. Symp. Ser. 93, pp. 389–400. Washington, DC: Am. Chem. Soc.

BASTON, G. M. N. et al. 1995a. Sorption of Pu and Am on repository backfill and geological materials relevant to the JNFL low-level radioactive waste repository at Rakkasho-Mura. *Mat. Res. Soc. symp. proc.* 353, pp. 957–64.

BASTON, G. M. N. et al. 1995b. The sorption of U and Tc on bentonite, tuff and granodiorite. *Mat. Res. Soc. symp. proc.* 353, pp. 989–96.

BATEMAN, H. 1910. Solution of a system of differential equations occurring in the theory of radioactive transformations. *Proc. Cambridge Phil. Soc.* 15, pp. 423–27.

BATES, J. K., J. P. BRADLEY, A. TEETSOV, C. R. BRADLEY, and M. BUCHHOLTZ TEN BRINK. 1992. Colloid formation during waste form reaction: Implications for nuclear waste disposal. *Science* 256: 649–52.

BATES, R. G. 1964 *Determination of pH, theory and practice.* New York: John Wiley & Sons.

BATH, A. H. et al. 1987. *Trace element and microbiological studies of alkaline groundwaters in Oman, Arabian Gulf: A natural analogue for cement.* Tech. Rept. 87-16 NAGRA, Swiss Fed. Inst. for Reactor Research, Wurenlingen.

BEALL, G. W., and B. ALLARD. 1981. Sorption of actinides from aqueous solutions under environmental conditions. In *Adsorption from aqueous solutions,* ed P. H. Tewari, pp. 193–212. New York: Plenum Press.

BECK, M. T. 1970. *Chemistry of complex equilibria.* New York: Van Nostrand Reinhold.

BENJAMIN, M. M., and J. O. LECKIE. 1981. Competitive adsorption of Cd, Cu, Zn and Pb on amorphous iron hydroxide. *J. Colloid Interface Sci.* 83:410–19.

BENSON, S. W. 1960. *The foundations of chemical kinetics.* New York: McGraw-Hill.

BERGER, G., E. CADORE, J. SCHOTT, and P. M. DOVE. 1994. Dissolution rate of quartz in lead and sodium electrolyte solutions between 25 and 300°C: Effect of the nature of surface complexes and reaction affinity. *Geochim. Cosmochim. Acta* 58:541–52.

BERLIN, R. E., and C. C. STANTON. 1989. *Radioactive waste management.* New York: John Wiley & Sons.

BERNER, E. K., and R. A. BERNER. 1987. *The global water cycle, geochemistry and environment.* Englewood Cliffs, NJ: Prentice-Hall, Inc.

BERNER, E. K., and R. A. BERNER. 1996. *Global environment: water, air, and geochemical cycles.* Upper Saddle River, NJ: Prentice Hall, Inc.

BERNER, R. A. 1963 Electrode studies of hydrogen sulfide in marine sediments. *Geochim. Cosmochim. Acta* 27:563–75.

BERNER, R. A. 1967. Thermodynamic stability of sedimentary iron sulfides. *Am. J. Sci.* 265:773–85.

BERNER, R. A. 1971. *Principles of chemical sedimentology.* New York: McGraw-Hill.

BERNER, R. A. 1978. Rate control of mineral dissolution under earth surface conditions. *Am. J. Sci.,* 278:1235–52.

BERNER, R. A. 1980. *Early diagenesis—A theoretical approach.* Princeton: Princeton Univ. Press.

BERNER, R. A. 1981a. Kinetics of weathering and diagenesis. In *Kinetics of geochemical processes,* ed A. C. Lasaga and R. S. Kirkpatrick. Reviews in Mineralogy 8, pp. 111–34. Min. Soc. Am.

BERNER, R. A. 1981b. A new geochemical classification of sedimentary environments. *J. Sed. Petrology* 51:359–65.

BETHKE, C. M. 1994. *The geochemist's workbench, version 2.0, A users guide to Rxn, Act2, Tact, React, and Gtplot.* Hydrogeology Program. Urbana, IL: Univ. of Illinois.

BETHKE, C. M. 1996. *Geochemical reaction modeling.* New York: Oxford University Press.

BEVEN, K., and GERMANN, P. 1982. Macropores and water flow in soils. *Water Resources Res.* 18:1311–25.

BEZMEN, N. I., and T. A. SMOLYAROVA. 1977. Enthalpy of formation of monoclinic pyrrhotite and its stability in the system Fe-S. *Intl. Geol. Rev.* 19:761–65.

BIDOGLIO, G., A. AVOGADRO, A. DE PLANO, and G. P. LAZZARI. 1984. Geochemical pattern of americium in the saline area surrounding a rock salt formation. *Radioactive Waste Management and the Nuclear Fuel Cycle* 5(4):311–326.

BIDOGLIO, G., A. DE PLANO, and L. RIGHETTO, L. 1989. Interactions and transport of plutonium-humic acid particles in groundwater environments. *Mat. Res. Soc. symp. proc.* 127, pp. 823–30.

BIDOGLIO, G., P. OFFERMANN, and A. SALTELLI. 1987. Neptunium migration in oxidizing clayey sand. *Applied Geochem.* 2:275–284.

BIRCH, F. 1966. Compressibility; elastic constants. In *Handbook of physical constants,* rev. ed., ed S. P. Clark, Jr. Geol. Soc. Am. Mem. 97, pp. 97–173.

BITTELL, J. E., and R. J. MILLER. 1974. Lead, cadmium, and calcium selectivity coefficients on a montmorillonite, illite, and kaolinite. *J. Envir. Quality* 3(3):250–53.

BLOUNT, C. W. 1977. Barite solubilities and thermodynamic quantities up to 300°C and 1400 bars. *Am. Mineral.* 62:942–57.

BOCKRIS, J. O., and A. K. N. REDDY. 1973. *Modern electrochemistry,* vol. 1. New York: Plenum Press.

BODEK, I. B., W. J. LYMAN, W. F. REEHL, and D. H. ROSENBLATT, eds. 1988. *Environmental inorganic chemistry; properties, processes, and estimation methods.* SETAC Spec. Publ. Ser. New York: Pergamon Press.

BOHN, H. L., B. L. MCNEAL, and G. A. O'CONNOR. 1985. *Soil chemistry.* 2d ed. New York: Wiley-Interscience.

BOLT, G. H., ed. 1979. *Soil chemistry, B. Physico-chemical models. Developments in soil science 5B.* Amsterdam: Elsevier Sci. Publ. Co.

BOLT, G. H., and W. H. VAN RIEMSDIJK. 1987. Surface chemical processes in soil. In *Aquatic surface chemistry,* ed W. Stumm, pp. 127–64. New York: Wiley-Interscience.

BOSSUS, D. A. W. 1984. Emanating power and specific surface area. *Radiation protection dosimetry* 7(1–4):73–76.

BÖTTCHER, J. O. STREBEL, and H. M. DUYNISVELD. 1985. Vertikale Stoffkonzentrationsprofile im Grundwasser eines Lockergesteins-Aquifers und deren Interpretation (Beispiel Fuhrberger Feld). *Z. dt. geol Ges.* 136:543–52.

BOULEGUE, J. 1978. Electrochemistry of reduced sulfur species in natural waters. I. The H_2S-H_2O system. *Geochim. Cosmochim. Acta* 42:1439–45.

BOULEGUE, J., C. J. LORD, III, and T. M. CHURCH. 1982. Sulfur speciation and associated trace metals (Fe, Cu) in the pore waters of Great Marsh, Delaware. *Geochim. Cosmochim. Acta* 46:453–64.

BOULEGUE, J., and G. MICHARD. 1979. Sulfur speciations and redox processes in reducing environments. In *Chemical modeling in aqueous systems,* ed E. A. Jenne. Am. Chem. Soc. Symp. Ser. 93, pp. 25–50. Washington, DC: Am. Chem. Soc.

BOUTRON, C. F., U. GORLACH, J. P. CANDELONE, M. BOL'SHOV, and R. DELMAS. 1991. Decrease in anthropogenic lead, cadmium and zinc in Greenland snows since the late 1960s. *Nature* 353:153–56.

BOWERS, T. S., K. J. JACKSON, and H. C. HELGESON. 1984. *Equilibrium activity diagrams.* New York: Springer-Verlag.

BOWIE, G. L., W. MILLS, D. B. PORCELLA, C. L. CAMPBELL, J. R. PAGENKOPF, et al. 1985. *Rates, constants, and kinetics formulations in surface water quality modeling,* 2d ed. Report EPA/600/3-85/04. Athens, GA: U.S. EPA.

BRADLEY, D. J., C. W. FRANK, and Y. MIKERIN. 1996. Nuclear contamination from weapons complexes in the former Soviet Union and the United States. *Physics Today* (April):40–45.

BRADY, N. C. 1974. *The nature and properties of soils.* 8th ed. New York: Macmillan Publ. Co.

BRADY, P. V. 1992. Silica surface chemistry at elevated temperatures. *Geochim. Cosmochim. Acta* 56:2941–46.

BRADY, P. V., and J. V. WALTHER. 1989. Controls on silicate dissolution rates in neutral and basic solutions at 25°C. *Geochim. Cosmochim. Acta* 53:2823–30.

BRENDEL, P. J., and G. W. LUTHER, III. 1995. Development of a gold amalgam voltametric microelectrode for the de-

termination of dissolved Fe, Mn, O_2, and S(-II) in porewaters of marine and freshwater sediments. *Envir. Sci. & Technol.* 29:751–61.

BRICKER, O. P. 1965. Some stability relations in the system Mn-O_2-H_2O at 25°C and one atmosphere total pressure. *Am. Mineral.* 50:1296–354.

BRICKER, O. P., and R. M. GARRELS. 1965. Mineralogic factors in natural water equilibria. In *Principles and applications of natural water chemistry,* ed S. Faust and J. V. Hunter, 449–69. New York: John Wiley & Sons.

BRICKER, O. P., A. E. GODFREY, and E. T. CLEAVES. 1968. *Mineral-water interaction during the chemical weathering of silicates.* Am. Chem. Soc. Symp. Ser. 73, pp. 128–42. Washington, DC: Am. Chem. Soc.

BRIMHALL, G. H., and D. A. CRERAR. 1987. Ore fluids: magmatic to supergene. In *Thermodynamic modeling of geological materials: Minerals, fluids and melts,* ed I. S. E. Carmichael and H. P. Eugster, pp. 235–311. Min. Soc. Am.

BRINDLEY, G. W., and G. BROWN. 1980. *Crystal structures of clay minerals and their X-ray identification.* Monograph No. 5. London: Min. Soc.

BROWN, D. S., and J. D. ALLISON. 1987. *MINTEQA1, An equilibrium metal speciation model: User's manual.* Report EPA/600/3-87/012. U.S. EPA.

BROWN, G., A. C. D. NEWMAN, J. H. RAYNER, and A. H. WEIR. 1978. The structures and chemistry of soil clay minerals. In *The chemistry of soil constituents,* ed D. J. Greenland and M. H. B. Hayes, pp. 29–178. New York: John Wiley & Sons.

BROWN, P. L., and R. N. SYLVA. 1987. Unified theory of metal ion complex formation constants. *J. Chem. Res.* 4–5(M):110–81.

BROWNLOW, A. H. 1979. *Geochemistry.* Englewood Cliffs, NJ: Prentice-Hall, Inc.

BRUNO, J., I. CASAS, B. LAGERMAN, and M. MUNOZ. 1987. The determination of the solubility of amorphous $UO_2(s)$ and the mononuclear hydrolysis constants of uranium(IV) at 25°C. *Mat. Res. Soc. symp. proc.* 84, pp. 153–60.

BRUNO, J., P. WERSIN, and W. STUMM. 1992. On the influence of carbonate in mineral dissolution: II. The solubility of $FeCO_3(s)$ at 25°C and 1 atm total pressure. *Geochim. Cosmochim. Acta* 56:1149–55.

BRUTON, C. J. 1988. Geochemical simulation of dissolution of West Valley and DWPF glasses in J-13 water at 90°C. *Mat. Res. Soc. symp. proc.* 112, pp. 607–19.

BRUTON, C. J., and H. J. SHAW. 1988. Geochemical simulation of reaction between spent fuel waste form and J-13 water at 25°C and 90°C. *Mat. Res. Soc. symp. proc.* 112, pp. 485–94.

BRUUN HANSEN, H. C., O. K. BORGGAARD, and J. SORENSEN. 1994. Evaluation of the free energy of formation of Fe(II)-Fe(III) hydroxide-sulphate (green rust) and its reduction of nitrite. *Geochim. Cosmochim. Acta* 58(12):2599–608.

BUNTING, B. T. 1965. *The geography of soil.* London: Hutchinson Univ. Press.

BUSENBERG, E., and L. N. PLUMMER. 1985. Kinetic and thermodynamic factors controlling the distribution of SO_4^{2-}

and Na$^+$ in calcites and selected aragonites. *Geochim. Cosmochim. Acta* 9:713–25.

BUSENBERG, E., and L. N. PLUMMER. 1986. The solubility of BaCO$_3$(cr) (witherite) in CO$_2$-H$_2$O solutions between 5 and 90°C, evaluation of the association constants of BaHCO$_3^+$(aq) and BaCO$_3^0$(aq) between 5 and 80°C, and a preliminary evaluation of the thermodynamic properties of Ba^{2+}(aq). *Geochim. Cosmochim. Acta* 50:2225–33.

BUSENBERG, E., and L. N. PLUMMER. 1989. Thermodynamics of magnesian calcite solid-solutions at 25°C and 1 atm total pressure. *Geochim. Cosmochim. Acta* 53(6):1189–1208.

BUSENBERG, E., L. N. PLUMMER, and V. B. PARKER. 1984. The solubility of strontianite (SrCO$_3$) in CO$_2$-H$_2$O solutions between 2 and 91°C, the association constants of SrHCO$_3^+$(aq) and SrCO$_3^0$(aq) between 5 and 80°C, and evaluation of the thermodynamic properties of Sr^{2+}(aq) and SrCO$_3$(cr) at 25°C and 1 atm total pressure. *Geochim. Cosmochim. Acta* 48:2012–35.

BUSEY, R. H., and R. E. MESMER. 1977. Ionization equilibria of silicic acid and polysilicate formation in aqueous sodium chloride solutions to 300°C. *Inorg. Chem.* 16:2444–50.

BUTLER, J. N. 1964. *Ionic equilibrium, a mathematical approach.* Reading, MA: Addison-Wesley Publ. Co. Inc.

CANFIELD, D. E., and D. DES MARAIS. 1991. Aerobic sulfate reduction in microbial mats. *Science* 251:1471–73.

CARROLL, S. A., and J. V. WALTHER. 1990. Kaolinite dissolution at 25°, 60°, and 80°C. *Am. J. Sci.* 290:797–810.

CARROLL-WEBB, S. A., and J. V. WALTHER. 1988. A surface complex reaction model for the pH-dependence of corundum and kaolinite dissolution rates. *Geochim. Cosmochim. Acta* 52: 2609–23.

CARUCCIO, F. T., and J. C. FERM. 1974. Paleoenvironment-predictor of acid mine drainage problems. *Proc. 5th symp. on coal mine drainage*, pp. 5–10. Washington, DC: Natl. Coal Assoc.

CARUCCIO, F. T., G. GEIDEL, and J. M. SEWELL. 1976. The character of drainage as a function of the occurrence of framboidal pyrite and ground water quality in eastern Kentucky. *Proc. 6th symp. on coal mine drainage*, pp. 1–16. Washington, DC: Natl. Coal Assoc.

CASAGRANDE, D. J., R. B. FINKELMAN, and F. T. CARUCCIO. 1990. The non-participation of organic sulphur in acid mine drainage production. *Envir. Geochem. & Health* 11(3–4):187–92.

CATTS, J. G. 1982. Adsorption of Cu, Pb, Zn onto birnessite. PhD thesis T-2538, Colorado School of Mines, Golden, CO.

CATTS, J. G., and D. LANGMUIR. 1986. Adsorption of Cu, Pb and Zn by δ-MnO$_2$: Applicability of the site binding-surface complexation model. *J. Applied Geochem.* 1:255–64.

CATTS, J. G., M. J. PAVELICH, and D. LANGMUIR. 1981. *Dissolution and precipitation kinetics of calcite and gypsum, some ferric oxyhydroxides and silica polymorphs: A literature survey and critique.* Battelle Pacific Northwest Laboratories.

CDM FEDERAL PROGRAMS CORP. 1990. Remedial investigation addendum for Sharon Steel/Midvale Tailings Site, Midvale UT. 1989–1990 *Ground water/geochemistry data report*, vol. II—*Appendices*. Final Report, U.S. EPA Contract No. 68-W9-0021.

CHAMP, D. R., and GULENS, J. 1979. Oxidation-reduction sequences in ground water flow systems. *Can. J. Earth Sci.* 16:12–23.

CHAO, T. T., and P. K. J. THEOBALD. 1976. The significance of secondary iron and manganese oxides in geochemical exploration. *Econ. Geol.* 71:1560–69.

CHAPELLE, F. H. 1983. Groundwater geochemistry and calcite cementation of the Aquia Aquifer in southern Maryland. *Water Resources Research* 19(2):545–58.

CHAPELLE, F. H., and L. L. KNOBEL. 1983. Aqueous geochemistry and exchangeable cation composition of glauconite in the Aquia Aquifer, Maryland. *Ground Water* 21(3):343–52.

CHAPELLE, F. H., and D. R. LOVLEY. 1992. Competitive exclusion of sulfate reduction by Fe(III)-reducing bacteria: A mechanism for producing discrete zones of high-iron ground water. *Ground Water* 30(1): 29–36.

CHATHAM, J. R., R. B. WANTY, and D. LANGMUIR. 1981. *Groundwater prospecting for sandstone-type uranium deposits: The merits of mineral-solution equilibria versus single element tracer methods.* U.S. Dept. of Energy, Report GJO 79-360-E. 197.

CHERMAK, J. A., and J. D. RIMSTIDT. 1989. Estimating the thermodynamic properties (ΔG_f^o and ΔH_f^o) of silicate minerals at 298 K from the sum of polyhedral contributions. *Am. Mineral.* 74:1023–31.

CHERMAK, J. A., and J. D. RIMSTIDT. 1990. Estimating the free energy of formation of silicate minerals at high temperatures from the sum of polyhedral contributions. *Am. Mineral.* 75:1376–80.

CHIOU, C. T., V. H. FREED, D. W. SCHMEDDING, and R. L. KOHNERT. 1977. Partition coefficient and bioaccumulation of selected organic chemicals. *Envir. Sci. & Technol.* 11:475–77.

CHIOU, C. T., L. J. PETERS, and V. H. FREED. 1979. A physical concept of soil-water equilibria for nonionic organic compounds. *Science* 206:831–32.

CHIOU, C. T., P. E. PORTER, and D. W. SCHMEDDING. 1983. Partition equilibria of nonionic organic compounds between soil organic matter and water. *Envir. Sci. & Technol.* 17:227–31.

CHISHOLM-BRAUSE, C. J., and D. E. MORRIS. 1992. Speciation of uranium(VI) sorption complexes on montmorillonite. *Proc. 7th intl. symp. on water-rock interaction*, ed Y. K. Kharaka and A. S. Maest, pp. 137–40. Rotterdam: A. A. Balkema Publ.

CHOI, J.-S., and T. H. PIGFORD. 1981. Water-dilution volumes for high-level wastes. *Trans. Am. Nucl. Soc.* 39:176–77.

CHOPPIN, G. R. 1990. Actinide speciation in spent fuel leaching studies. *Mat. Res. Soc. symp. proc.* 176, pp. 449–56.

CHOPPIN, G. R., and B. ALLARD. 1985. In *Handbook of the chemistry and physics of the actinides*, ed A. J. Freeman and C. Keller, vol. 3, pp. 407–29. Amsterdam: Elsevier Publ.

CHOPPIN, G. R., and J. N. MATHUR. 1991. Hydrolysis of actinyl(VI) cations. *Radiochim. Acta* 52/53: 25–28.

CHOU, L., and R. WOLLAST. 1985. Steady-state kinetics and dissolution mechanisms of albite. *Am. J. Sci.* 285:963–95.

CHRISTENSEN, J. J., D. J. EATOUGH, and R. M. IZATT. 1975. *Handbook of metal ligand heats and related thermodynamic quantities.* 2d ed. New York: Marcel Dekker.

CHRISTOFFERSON, J., and M. R. CHRISTOFFERSON. 1976. The kinetics of dissolution of calcium sulfate dihydrate in water. *J. Crystal Growth* 35:29–88.

CLAASSEN, H. C. 1981. Estimation of calcium sulfate solution rate and effective aquifer surface area in a groundwater system near Carlsbad, New Mexico. *Ground Water* 19:287–97.

CLEVELAND, J. M. 1979a. Critical review of plutonium equilibria of environmental concern. In *Chemical. modeling in aqueous systems*, ed E. A. Jenne pp. 321–38. Washington, DC: Am. Chem. Soc.

CLEVELAND, J. M. 1979b. *The chemistry of plutonium.* La Grange Park, IL: Am. Nucl. Soc.

CLIFFORD, A. F. 1959. The electronegativity of groups. *J. Phys. Chem.* 63:1227–31.

CODATA TASK GROUP ON KEY VALUES FOR THERMODYNAMICS. 1976. Recommended key values for thermodynamics 1975. *J. Chem. Thermo.* 8:603–5.

CODATA TASK GROUP ON KEY VALUES FOR THERMODYNAMICS. 1977. Recommended key values for thermodynamics 1976. *J. Chem. Thermo.* 9:705–6.

COLE, J. J., N. F. CARACO, G. W. KLING, and T. K. KRATZ. 1994. Carbon dioxide supersaturation in the surface waters of lakes. *Science* 265:1568–70.

COMBES, J. M., A. MANCEAU, and G. CALAS. 1990. Formation of ferric oxides from aqueous solutions: A polyhedral approach by X-ray absorption spectroscopy: II. Hematite formation from ferric gels. *Geochim. Cosmochim. Acta* 54:1083–91.

CONCA, J. 1992. Transport in unsaturated flow systems using centrifuge techniques. *Proc. DOE/Yucca Mountain site characterization project radionuclide adsorption workshop at Los Alamos National Laboratory.* Report LA-12325-C, pp. 125–52.

COTTON, F., and G. WILKINSON. 1988. *Advanced inorganic chemistry, a comprehensive text.* 5th ed. New York: John Wiley & Sons.

COX, J. D., D. D. WAGMAN, and V. A. MEDVEDEV. 1989. *CODATA key values for thermodynamics.* New York: Hemisphere Publ. Corp.

COZZARELLI, I. M, J. S. HERMAN, and R. A. PARNELL, JR. 1987. The mobilization of aluminum in a natural soil system: Effects of hydrological pathways. *Water Resources Research* 23(5):859–74.

CRAMER, J. J., and F. P. SARGENT. 1994. The Cigar Lake analog study: An international R&D project. *Proc. 5th annual intl. conf. Am. Nucl. Soc.* 4, pp. 2237–42.

CRAMER, J. J., and J. A. T. SMELLIE, eds. 1994. *Final report of the AECL/SKB Cigar Lake analog study.* Report AECL-10851. Whiteshell Laboratories, Pinawa, Manitoba, Canada ROE 1L0.

CRONAN, C. S. 1985. Chemical weathering and solution chemistry in acid forest soils: Differential influence of soil type, biotic processes and H+ deposition. In *The chemistry of weathering,* ed J. I. Drever, pp. 175–95. Dordrecht, Holland: D. Reidel Publ. Co.

CUADROS, J., and J. LINARES. 1996. Experimental kinetic study of the smectite-to-illite transformation. *Geochim. Cosmochim. Acta* 60(3):439–53.

CULLIMORE, D. R. 1991. *Practical manual for groundwater microbiology.* Chelsea, MI: Lewis Publ.

CURTIS, D. B., J. FABRYKA-MARTIN, R. AGUILAR, M. ATTREP, and F. ROENSCH. 1992. Plutonium in uranium deposits: Natural analogues of geologic repositories for plutonium-bearing nuclear wastes. *Proc. 3d annual intl. conf. Am. Nucl. Soc.* 1, pp. 338–94.

CURTIS, D. B., J. FABRYKA-MARTIN, P. DIXON, R. AGUILAR, and J. CRAMER. 1994. Radionuclide release rates from natural analogues of spent nuclear fuel. *Proc. 5th annual intl. conf. Am. Nucl. Soc.* 4, pp. 2228–36.

DAHL, A. R., and J. L. HAGMAIER. 1974. Genesis and characteristics of the southern Powder River Basin uranium deposits, Wyoming, U.S.A. *Formation of uranium deposits. Proc. symp. Intl. Atomic Energy Agency,* pp. 201–18, Vienna.

DALL'AGLIO, M., R. GRAGNANI, and E. LOCARDI. 1974. Geochemical factors controlling the formation of secondary minerals of uranium. *Formation of uranium deposits. Proc. symp. Intl. Atomic Energy Agency,* pp. 33–47, Vienna.

DARKEN, L. S., and R. G. GURRY. 1953. *Physical chemistry of metals.* New York: McGraw-Hill.

DASKALAKIS, D. K., and G. R. HELZ. 1992. Solubility of CdS (greenockite) in sulfidic waters at 25°C. *Envir. Sci. & Technol.* 26:2462–68.

DAVIDSON, P. M., G. H. SYMMES, B. A. COHEN, R. J. REEDER, and D. H. LINDSLEY. 1994. Synthesis of the new compound $CaFe(CO_3)_2$ and experimental constraints on the $(Ca,Fe)CO_3$ join. *Geochim. Cosmochim. Acta* 58:5105–9.

DAVIS, A., and D. D. RUNNELLS. 1987. Geochemical interactions between acidic tailings fluid and bedrock: use of the computer model MINTEQ. *Applied Geochem.* 2:231–41.

DAVIS, J. A. 1984. Complexation of trace metals by adsorbed natural organic matter. *Geochim. Cosmochim. Acta* 48:679–91.

DAVIS, J. A., R. O. JAMES, and J. O. LECKIE. 1978. Surface ionization and complexation at the oxide/water interface. I. Computation of electrical double layer properties in simple electrolytes. *J. Colloid Interface Sci.* 63:480–99.

DAVIS, J. A., and D. B. KENT. 1990. Surface complexation models in aqueous geochemistry. In *Mineral-water interface geochemistry,* ed M. F. Hochella and A. F. White. Reviews in Mineralogy 23, pp. 177–260. Min. Soc. Am.

DAVIS, J. A., and J. O. LECKIE. 1978. Surface ionization and complexation at the oxide/water interface. II. Surface properties of amorphous iron oxyhydroxide and adsorption of metal ions. *J. Colloid Interface Sci.* 67:90–107.

DAVIS, J. A., and J. O. LECKIE. 1980. Surface ionization and complexation at the oxide/water interface. III. Adsorption of anions. *J. Colloid Interface Sci.* 74:32–42.

DAVIS, S. N. 1964. Silica in streams and ground waters. *Am. J. Sci.* 262:870–91.

DAVIS, S. N., and R. J. M. DEWIEST. 1966. *Hydrogeology.* New York: John Wiley & Sons.

DEAN, J. A., ed. 1985. *Lange's handbook of chemistry*. 13th ed. New York: McGraw-Hill.

DEBRAAL, J. D., and Y. K. KHARAKA. 1989. *SOLINPUT: A Computer code to create and modify input files for the geochemical program SOLMINEQ.88.* Open-File Report 89-616. Menlo Park, CA: U.S. Geol. Survey.

DEER, W. A., R. A. HOWIE, and J. ZUSSMAN. 1992. *An introduction to the rock-forming minerals.* New York: John Wiley & Sons.

DEGUELDRE, C. 1993. Colloid properties in granitic groundwater systems, with emphasis on the impact on safety assessment. *Mat. Res. Soc. symp. proc.* 294, pp. 817–23.

DEINES, P., D. LANGMUIR, and R. S. HARMON. 1974. Stable carbon isotopes to indicate the presence or absence of a gas phase in the evolution of carbonate ground waters. *Geochim. Cosmochim. Acta* 38:1147–64.

DESNOYERS, J. E. 1979. Experimental techniques: Cryoscopy and other methods. In *Activity coefficients in electrolyte solutions*, ed R. M. Pytkowicz, pp. 139–55. Boca Raton, FL: CRC Press, Inc.

DEUTSCH, W. J., W. J. MARTIN, L. E. EARLY, and R. J. SERNE. 1985. *Methods of minimizing ground-water contamination from in situ leach uranium mining.* Pacific Northwest Laboratory, Final Report NUREG/CR-3709.

DEVIDAL, J.-L., J.-L. DANDURAND, and R. GOUT. 1996. Gibbs free energy of formation of kaolinite from solubility measurement in basic solution between 60 and 170°C. *Geochim. Cosmochim. Acta* 60:553–64.

DOE. 1980. *Project review: uranium mill tailings remedial action program.* U. S. Dept. of Energy.

DOE. 1987. *Integrated data base for 1987. Spent fuel and radioactive waste inventories, projections and characteristics.* DOE/RW-0006, rev. 3. U.S. Dept. of Energy.

DOEHRING, D. O., and J. H. BUTLER. 1974. Hydrogeologic constraints on Yucatan's development. *Science* 186:519–95.

DOMENICO, P. A., and F. W. SCHWARTZ. 1990. *Physical and chemical hydrogeology.* New York: John Wiley & Sons.

DOVE, P. M. 1994. The dissolution kinetics of quartz in sodium chloride solutions at 25°C to 300°C. *Am. J. Sci.* 294:665–712.

DOVE, P. M., and J. D. RIMSTIDT. 1994. Silica-water interactions. In *Silica, physical behavior, geochemistry and materials applications*, ed P. J. Heaney, C. T. Prewitt, and G. V. Gibbs. Reviews in Mineralogy 29, pp. 259–308. Min. Soc. Am.

DREVER, J. I., ed. 1985. *NATO ASI series c: Mathematical and physical sciences*, vol. 149. *The chemistry of weathering.* Dordrecht, Holland: D. Reidel.

DREVER, J. I. 1988. *The geochemistry of natural waters.* 2d ed. Englewood Cliffs, NJ: Prentice Hall, Inc.

DREVER, J. I. 1994. The effect of land plants on weathering rates of silicate minerals. *Geochim. Cosmochim. Acta* 58(10):2325–32.

DRISCOLL, C. T., S. W. EFFLER, and S. M. DOERR. 1994. Changes in inorganic carbon chemistry and deposition of Onondaga Lake, New York. *Envir. Sci. & Technol.* 28:1211–16.

DUBROVSKY, N. M., J. A. CHERRY, E. J. REARDON, T. P. LIM, and A. J. VIVYURKA. 1985. Geochemical evolution of inactive pyritic tailings in the Elliot Lake Uranium District: 1. The groundwater zone. *Canadian Geotech. J.* 22(1):110–28.

DUBY, P. 1977. The thermodynamic properties of aqueous inorganic copper systems. In *The metallurgy of copper.* INCRA Monograph IV.

DURRANCE, E. M. 1986. *Radioactivity in geology.* New York: John Wiley & Sons.

DZOMBAK, D. A., and F. M. M. MOREL. 1987. Adsorption of inorganic pollutants in aquatic systems. *J. Hydraulic Eng.* 113:430–75.

DZOMBAK, D. A., and F. M. M. MOREL. 1990. *Surface complexation modeling. Hydrous ferric oxide.* New York: Wiley-Interscience.

EARY, L. E., and E. A. JENNE. 1992. *Version 4.00 of the MINTEQ code.* Report PNL-8190/UC-204. Richland, WA: Pacific Northwest Laboratory.

EARY, L. E., and D. RAI. 1988. Chromate removal from aqueous wastes by reduction with ferrous ion. *Envir. Sci. & Technol.* 22(8): 972–77.

EARY, L. E., and J. A. SCHRAMKE. 1990. Rates of inorganic oxidation reactions involving dissolved oxygen. In *Chemical modeling of aqueous systems II*, ed D. C. Melchior and R. L. Bassett. Am. Chem. Soc. Adv. Chem. Ser. 416, pp. 379–96. Washington, DC: Am. Chem. Soc.

EDMUNDS, W. M. et al. 1987. Baseline geochemical conditions in the Chalk Aquifer, Berkshire, U.K.: A basis for groundwater quality management. *Applied Geochem.* 2:251–74.

EISENBUD, M. 1987. *Environmental radioactivity.* 3d ed. New York: Academic Press, Inc.

ELBERLING, B. and R. V. NICHOLSON. 1996. Field determination of sulphide oxidation rates in mine tailings. *Water Resources Research* 32(6):1773–84.

ELLIS, A. J., and W. A. J. MAHON. 1964. Natural hydrothermal systems and experimental hot-water/rock interaction. *Geochim. Cosmochim. Acta* 28:1323–57.

EMBER, L. R. 1981. Acid pollutants: Hitchhkers ride with wind. *Chem. & Eng. News* (Sept. 14): 20–31.

ENVIRONMENTAL PROTECTION AGENCY. 1986. *A citizen's guide to radon.* U.S. EPA Report EPA-86-004.

ENVIRONMENTAL PROTECTION AGENCY. 1990a. *Assessing the geochemical fate of deep-well-injected hazardous waste. A reference guide.* Report EPA/625/6-89/025a. Cincinnati, OH: U.S. EPA.

ENVIRONMENTAL PROTECTION AGENCY. 1990b. *Assessing the geochemical fate of deep-well-injected hazardous waste. Summaries of recent research.* Report EPA/625/6-89/025b. Cincinnati, OH: U.S. EPA.

ENVIRONMENTAL PROTECTION AGENCY. 1991. *National primary drinking water standards for radionuclides, proposed rule, June, 1991.* EPA Fact Sheet, Radionuclides in Drinking Water, 570/9-91-700.

ERIKSEN, T. E. et al. 1993. *Solubility of the redox-sensitive radionclides ^{99}Tc and ^{237}Np under reducing conditions in neutral to alkaline solutions. Effect of carbonate.* Swedish Nuclear Fuel and Waste Mgmt. Co. Tech. Report 93-18.

ERVANNE, H., L. AHONEN, T. JAAKKOLA, and R. BLOMQVIST. 1994. Redox chemistry of uranium in groundwater of Palmottu uranium deposit, Finland. *Third intl. conf. on migration behavior of actinides and fission products in the geosphere, 1993*. Brussels, Belgium: Commission of the European Communities.

EVANS, R. 1952. *An introduction to crystal chemistry*. Cambridge: Cambridge Univ. Press.

EVERETT, D. H. 1988. *Basic principles of colloid science*. London: Royal Soc. of Chem. Paperbacks.

FALINI, G., S. ALBECK, S. WEINER, and L. ADDADI. 1996. Control on aragonite or calcite polymorphism by mollusk shell macromolecules. *Science* 271:67–69.

FARRAH, H., D. HATTON, and W. F. PICKERING. 1980. The affinity of metal ions for clay surfaces. *Chem. Geol.* 28:55–68.

FAURE, G. 1991. *Principles and applications of inorganic geochemistry*. New York: Macmillan Publ. Co.

FAUST, S. D., and O. ALY. 1981. *Chemistry of natural waters*. Boston, MA: Butterworths.

FELMY, A. R., S. A. PETERSON, and R. J. SERNE. 1987. Interactions of acidic uranium mill tailings solution with sediments: Predictive modeling of precipitation/dissolution reactions. *Uranium* 4:25–41.

FELMY, A. R., D. RAI, and R. W. FULTON. 1990. The solubility of $AmOHCO_3(c)$ and the aqueous thermodynamics of the system Na^+-Am^{3+}-HCO_3^--CO_3^{2-}-OH^--H_2O. *Radiochim. Acta* 50:193–204.

FERNANDEZ, M. A., L MARTINEZ, M. SEGARRA, J. C. GARCIA, and F. ESPIELL. 1992. Behavior of heavy metals in the combustion gases of urban waste incinerators. *Envir. Sci. Technol.* 26(5):1040–47.

FETH, J. H., C. E. ROBERSON, and W. L. POLZER. 1964. *Sources of mineral constituents in water from granitic rocks, Sierra Nevada, California and Nevada*. U.S. Geol. Survey Water Supply Paper 1535-I.

FETTER, C. W. 1988. *Applied hydrogeology*. 2d ed. New York: Macmillan Publ. Co.

FILIPEK, L. H., D. K. NORDSTROM, and W. H. FICKLIN. 1987. Interaction of acid mine drainage with waters and sediments of West Squaw Creek in the West Shasta Mining District, California. *Envir. Sci. & Technol.* 21(4): 388–96.

FIX, P. G. 1956. *Hydrochemical exploration for uranium*. U.S. Geol. Survey Prof. Paper 300, 667–71.

FLEISCHER, R. L. 1988. Alpha-recoil damage: Relation to isotopic disequilibrium and leaching of radionuclides. *Geochim. Cosmochim. Acta* 52(6):1459–66.

FLUGGE, S., and K. E. ZIMENS. 1939. Die Bestimmung von Korngrossen und Kiffusionskonstanten aus dem Emaniervermögen (Die Theorie der Emaniermethode). *Zeitschrift für physikalische Chemie* (Leipzig) B42:179–220.

FOLGER, P. F. 1995. A multidisciplinary study of the variability of dissolved ^{222}Rn in ground water in a fractured crystalline rock aquifer and its impact on indoor air. PhD thesis, T-4696, Colorado School of Mines, Golden, CO.

FOLGER, P. F., P. NYBERG, R. B. WANTY, and E. POETER. 1994. Relationships between ^{222}Rn dissolved in ground water supplies and indoor ^{222}Rn concentrations in some Colorado Front Range houses. *Health Phys.* 67(3):244–52.

FOLGER, P. F., E. POETER, R. B. WANTY, D. FRISHMAN, and W. DAY. 1996. Controls on ^{222}Rn variations in a fractured crystalline rock aquifer evaluated using aquifer tests and geophysical logging. *Ground Water* 34(2):250–61.

FOLK, R. L., and L. S. LAND. 1975. Mg/Ca ratio and salinity: Two controls over crystallization of dolomite. *Am. Assoc. Petrol. Geol. Bull.* 59(1):60–68.

FOTH, H. D. 1984. *Fundamentals of soil science*. 7th ed. New York: John Wiley & Sons.

FOURNIER, R. O. 1985. In *Geology and geochemistry of epithermal systems*, ed B. R. Berger and P. M Bethke. Reviews in Econ. Geol. 2. Chelsea, MI: Soc. Econ. Geologists.

FOURNIER, R. O., and J. J. ROWE. 1962. The solubility of cristobalite along the three-phase curve, gas plus liquid plus cristobalite. *Am. Mineral.* 47:897–902.

FOX, L. E. 1988. The solubility of colloidal ferric hydroxide and its relevance to iron concentrations in river water. *Geochim. Cosmochim. Acta* 52:771–77.

FREEZE, R. A., and J. A. CHERRY. 1979. *Groundwater*. Englewood Cliffs, NJ: Prentice-Hall, Inc.

FRIEND, J. P. 1973. The global sulfur cycle. In *Chemistry of the lower atmosphere*, ed S. I. Rasool, pp. 177–201. New York: Plenum Press.

FRIZADO, J. P. 1979. Ion exchange on humic materials—a regular solution approach. In *Chemical modeling in aqueous systems*, ed E. A. Jenne. Am. Chem. Soc. Symp. Ser. 93:133–45. Washington, DC: Am. Chem. Soc.

FRUCHTER, J. S. et al. 1985. *Radionuclide migration in ground water*. Final Report NUREG/CR-4030 PNL-5299, Battelle Pacific Northwest Laboratory.

FUGER, J. 1992. Thermodynamic properties of actinide aqueous species relevant to geochemical problems. *Radiochim. Acta* 58/59:81–91.

FUGER, J. and F. L. OETTING. 1976. *The chemical thermodynamics of actinide elements and compounds. 2. The actinide aqueous ions*. Vienna: Intl. Atomic Energy Agency.

FUGER, J. et al. 1992. *The chemical thermodynamics of actinide elements and compounds. 12. The actinide aqueous inorganic complexes*. Vienna: Intl. Atomic Energy Agency.

FUJITA, T., M. TSUKAMOTO, T. OHE, S. NAKAYAMA, and Y. SAKAMOTO. 1995. Modeling of neptunium(V) sorption behavior onto iron-containing minerals. *Mat. Res. Soc. symp. proc.* 353, pp. 965–72.

FYFE, W. S. 1964. *Geochemistry of solids*. New York: McGraw-Hill.

GALKIN, N. P., and M. A. STEPANOV. 1961. Solubility of uranium(IV) hydroxide. *Soviet J. Atomic Energy* 2:231–33.

GANG, M. A., and D. LANGMUIR. 1974. Controls on heavy metals in surface and ground waters affected by coal mine drainage. *Proc. 5th symp. coal mine drainage research*, pp. 39–69. Washington, DC: Natl. Coal Assoc.

GARCIA-MIRAGAYA, J., and A. L. PAGE. 1976. Influence of ionic strength and inorganic complex formation on the sorption of trace amounts of Cd by montmorillonite. *Soil Sci. Soc. Am. J.* 40(5):658–63.

GARDINER, W. C. J. 1972. *Rates and mechanisms of chemical reactions*. Menlo Park, CA: W. A. Benjamin, Inc.

GARRELS, R. M. 1960. *Mineral equilibria.* New York: Harper & Bros.

GARRELS, R. M. 1984. Montmorillonite/illite stability diagrams. *Clays Clay Minerals* 32:161–66.

GARRELS, R. M., and C. L. CHRIST. 1965. *Solutions, minerals and equilibria.* New York: Harper & Row Publ. Co.

GARRELS, R. M., and F. T. MACKENZIE. 1971. *Evolution of sedimentary rocks.* New York: W. W. Norton & Co., Inc.

GARRELS, R. M., F. T. MACKENZIE, and C. HUNT. 1975. *Chemical cycles and the global environment: Assessing human influences.* Los Altos, CA: William Kaufmann.

GARRELS, R. M., and C. R. NAESER. 1958. Equilibrium distribution of dissolved sulfur species in water at 25°C and 1 atm total pressure. *Geochim. Cosmochim. Acta* 15:113–30.

GARRELS, R. M., and M. E. THOMPSON. 1960. Oxidation of pyrite by iron sulfate solutions. *Am. J. Sci.* 258A:57–67.

GARRELS, R. M., M. E. THOMPSON, and R. SIEVER. 1960. Stability of some carbonates at 25°C and 1 atmosphere total pressure. *Am. J. Sci.* 258:402–18.

GASCOYNE, M. 1989. High levels of uranium in groundwaters at Canada's Underground Research Laboratory, Lac du Bonnet, Manitoba, Canada. *Applied Geochem.* 4:577–91.

GAYER, K. H., and H. LEIDER. 1955. The solubility of uranium trioxide, $UO_3 \cdot H_2O$ in solutions of sodium hydroxide and perchloric acid at 25°C. *J. Am. Chem. Soc.* 77:1448–50.

GAYER, K. H., and H. LEIDER. 1957. The solubility of uranium(IV) hydroxide in solutions of sodium hydroxide and perchloric acid at 25°C. *Can. J. Chem.* 35:5–7.

GESELL, T. F., and H. M. PRICHARD. 1980. *The contribution of radon in tap water to indoor radon concentrations.* U.S. Dept. of Energy Report Conf-780422, ed T. F. Gesell and W. M. Lowder, pp. 5–56. Houston, TX: Natl. Tech. Info. Service.

GIBBS, R. J. 1970. Mechanisms controlling world water chemistry. *Science* 170:1088–90.

GIBLIN, A. M. 1987. Applications of groundwater geochemistry to genetic theories and exploration methods for early Tertiary sediment-hosted uranium deposits in Australia. *Uranium* 3:165–86.

GIGGENBACH, W. 1974. Kinetics of polysulfide-thiosulfate disproportionation up to 240°C. *Inorg. Chem.* 13:1730–33.

GIRIDHAR, J., and D. LANGMUIR. 1991. Determination of E° for the UO_2^{2+}/U^{4+} couple from measurement of the equilibrium: $UO_2^{2+} + Cu(s) + 4H^+ = U^{4+} + Cu^{2+} + 2H_2O$ at 25°C and some geochemical implications. *Radiochim. Acta* 54:133–38.

GLASSTONE, S. 1946. *Physical chemistry.* 2d ed. New York: D. Van Nostrand Co., Inc.

GLASSTONE, S. 1982. *Energy deskbook.* U.S. Dept. of Energy Rept. DOE/IR/05114-1.

GLYNN, P. D. 1990. Modeling solid-solution reactions in low-temperature aqueous systems. In *Chemical modeling of aqueous systems II,* ed D. C. Melchior and R. L. Bassett, Washington, DC: Am. Chem. Soc. Symp. Ser. 416, pp. 74–86.

GLYNN, P. D., D. L. PARKHURST, and L. N. PLUMMER. 1992. Principles and applications of modeling chemical reactions in ground water. IGWMC Short Course. Colorado School of Mines, Golden, CO.

GNANAPRAGASAM, E. K. and B. G. LEWIS. 1995. Elastic strain energy and the distribution coefficient of radium in solid solutions with calcium salts. *Geochim. Cosmochim. Acta* 59(24):5103–11.

GOLDBERG, E. D., W. S. BROECKER, M. G. GROSS, and K. K. TUREKIAN. 1971. Marine chemistry. In *Radioactivity in the marine environment,* pp. 137–46. Washington DC: Natl. Acad. Sciences.

GOLDBERG, R. N. 1979. Evaluated activity and osmotic coefficients for aqueous solutions: Bi-univalent compounds of lead, copper, manganese and uranium. *J. Phys. Chem. Ref. Data* 8(4):1005–50.

GOLDBERG, R. N., and R. L. NUTTALL. 1978. Evaluated activity and osmotic coefficients for aqueous solutions: The alkaline earth metal halides. *J. Phys. Chem. Ref. Data* 7(1):263–310.

GOLDBERG, S., and R. A. GLAUBIG. 1985. Boron adsorption on aluminum and iron oxide minerals. *Soil Sci. Soc. Am. J.* 49:1374–79.

GOLDBERG, S., and G. SPOSITO. 1984. A chemical model of phosphate adsorption by soils. I. Reference oxide minerals. *Soil. Sci. Soc. Am. J.* 48:772–78.

GOLDHABER, M. B., B. S. HEMINGWAY, A. MOHAGHEGHI, R. L. REYNOLDS, and H. R. NORTHROP. 1987. Origin of coffinite in sedimentary rocks by a sequential adsorption-reduction mechanism. *Bull. Mineral.* 110:131–44.

GOLDHABER, M. B., and I. R. KAPLAN. 1974. The sulfur cycle. In *The sea.* 5. *Marine chemistry,* ed E. D. Goldberg, pp. 569–655. New York: John Wiley & Sons.

GOLDICH, S. S. 1938. A study in rock weathering. *Geology* 46:17–58.

GOLEVA, G. A. 1968. *Hydrogeochemical prospecting of hidden ore deposits* (in Russian). Moscow: Nedra Publ. House. English ranslation available from the Geological Survey of Canada (1970).

GOODWIN, B. W. et al. 1994. *The disposal of Canada's nuclear fuel waste: Postclosure assessment of a reference system.* Report AECL-10717. Pinawa, Manitoba: Whiteshell Laboratories.

GORDY, W., and W. J. O. THOMAS. 1956. Electronegativities of the elements. *J. Chem. Phys.* 24(2):439–44.

GRANAT, L. 1972. On the relation between pH and the chemical composition of atmospheric precipitation. *Tellus* 24:550–60.

GRANAT, L., H. RODHE, and R. O. HALLBERG. 1976. The global sulfur cycle. In *Nitrogen, phosphorus, and sulfur global cycles,* ed B. H. Svensson and R. Sjoderlund. SCOPE Report no. 7. *Ecol. Bull.* (Stockholm) 22:89–134.

GRANDSTAFF, D. E. 1986. The dissolution rate of forsteritic olivine from Hawaiian Beach Sand. In *Rates of chemical weathering of rocks and minerals,* ed M. Colman and D. P. Dethier, pp. 41–59. New York: Academic Press, Inc.

GRANGER, H. C., and C. G. WARREN. 1969. Unstable sulfur compounds and the origin of roll-type uranium deposits. *Econ. Geol.* 64: 160–71.

GREENLAND, D. J., and M. H. B. HAYES, eds. 1978. *The chemistry of soil constituents*. New York: John Wiley & Sons.

GREENLAND, D. J., and C. J. B. MOTT. 1978. Surfaces of soil particles. In *The chemistry of soil constituents*, ed D. J. Greenland and M. H. B. Hayes, pp. 321–53. New York: John Wiley & Sons.

GRENTHE, I. et al., eds. 1992. *Chemical thermodynamics of uranium*. Chem. Thermo. Ser. Amsterdam: Elsevier Science Publ.

GRENTHE, I., I. PUIGDOMENECH, M. C. A. SANDINO, and M. H. RAND. 1995. Appdx. D, Chemical Thermodynamics of Americium. *Chemical thermodynamics of uranium*. Chem. Thermo. 2, Nuclear Energy Agency/OECD, pp. 347–74. Amsterdam: Elsevier Science Publ.

GRENTHE I., and H. WANNER. 1989. *Guidelines for the extrapolation to zero ionic strength,* Report NEA-TDB-2.1, F-91191. Giv-sur Yvette, France: OECD, Nuclear Energy Agency Data Bank.

GRIFFIN, R. A., and N. F. SHIMP. 1976. Effect of pH in exchange-adsorption or precipitation of Pb from landfill leachates by clay minerals. *Envir. Sci. Technol.* 10(13):1256–61.

GRIM, R. E. 1968. *Clay mineralogy.* New York: McGraw Hill.

GU, Y., C. H. GAMMONS, and M. S. BLOOM. 1994. A one-term extrapolation method for estimating equilibrium constants of aqueous reactions at elevated temperatures. *Geochim. Cosmochim. Acta* 58(17):3545–60.

GUILLAUMONT, R., and J. P. ADLOFF. 1992. Behavior of environmental pollution at very low concentration. *Radiochim. Acta* 58/59:53–60.

GUNNERIUSSON, L., L. LOVGREN, and S. SJOBERG. 1994. Complexation of Pb(II) at the goethite (α-FeOOH)/water interface: The influence of chloride. *Geochim. Cosmochim. Acta* 58:4973–83.

HAAN, C. T. 1977. *Statistical methods in hydrology.* Ames, IA: Iowa Univ. Press.

HAAR, L., J. S. GALLAGHER and G. S. KELL. 1984. Natl. Bur. Stds./Natl. Res. Council Steam Tables. Washington DC: Hemisphere Publ. Co.

HACKETT, G. 1987. A review of chemical treatment strategies for iron bacteria in wells. *Water Well J.* 41(2):37–42.

HAKANEN, M., and A. LINDBERG. 1991. Sorption of neptunium under oxidizing and reducing groundwater conditions. *Radiochim. Acta* 52/53:147–51.

HALL, J. C., and R. L. RAIDER. 1993. A reflection on metals criteria. *Water Envir. & Technol.* (June 1993):63–64.

HAMER, W., and Y. WU. 1972. Osmotic coefficients and mean activity coefficients of uni-univalent electrolytes in water at 25°C. *J. Phys. Chem. Ref. Data* 1(4):1047–1100.

HANCOCK, R. D., N. P. FINKELSTEIN, and A. EVERS. 1974. Linear free energy relationships in aqueous complex-formation reactions of the d^{10} metal ions. *J. Inorg. Nucl. Chem.* 36:2539–43.

HANCOCK, R. D., and F. MARSICANO. 1978. Parametric correlation of formation constants in aqueous solution. 1. Ligands with small donor atoms. *Inorg. Chem.* 17(3):560–64.

HANSON, B. K., and D. POSTMA. 1995. Acidification, buffering, and salt effects in the unsaturated zone of a sandy aquifer, Klosterhede, Denmark. *Water Resources Research* 31(11):2795–809.

HANSON, D. 1992. Drinking water limits set for 23 more compounds. *Chem. & Eng. News* (June 1):19.

HANSON, W. C., ed. 1980. *Transuranic elements in the environment.* Tech. Info. Center/U.S. Dept. of Energy. DOE/TIC-22800, NTIS. Springfield VA: U.S. Dept. of Commerce.

HARNED, H. S., and B. B. OWEN. 1958. T*he physical chemistry of electrolyte solutions.* 3d ed. Am. Chem. Soc. Monograph Series. New York: Reinhold Publ. Corp.

HART, J. 1993. Designing chelates for selective waste treatment processes. *Amer. Laboratory* (July):23–24.

HARVIE, C. E., N. MOLLER, and J. H. WEARE. 1984. The prediction of mineral solubilities in natural waters: The Na-K-Mg-Ca-H-Cl-SO_4-OH-HCO_3-CO_3-CO_2-H_2O system to high ionic strengths at 25°C. *Geochim. Cosmochim. Acta* 48:723–51.

HARVEY, B. G. 1962. *Introduction to nuclear physics and chemistry.* Englewood Cliffs, NJ: Prentice-Hall.

HEALY, T. W. 1974. Principles of adsorption of organics at solid-solution interfaces. *J. Macromol. Sci.-Chem.* A8(3):603–19.

HEATH, R. C. 1989. *Basic ground-water hydrology.* U.S. Geol. Survey Water Supply Paper 2220.

HELGESON, H. C. 1967. Thermodynamics of complex dissociation in aqueous solutions at elevated temperatures. *J. Phys. Chem.* 71(10):3121–36.

HELGESON, H. C. 1969. Thermodynamics of hydrothermal systems at elevated temperatures and pressures. *Am. J. Sci.* 267:729–804.

HELGESON, H. C. 1985. Activity/composition relations among silicates and aqueous solutions. I. Thermodynamics of intrasite mixing and substitutional order/disorder in minerals. *Am. J. Sci.* 285:769–844.

HELGESON, H. C., T. H. BROWN, and R. H. LEEPER. 1969. *Handbook of theoretical activity diagrams depicting chemical equilibria in geological systems involving an aqueous phase at one atm and 0° to 300°C.* San Francisco: Freeman, Cooper & Co.

HELGESON, H. C., J. M. DELANY, H. W. NESBITT, and D. K. BIRD. 1978. Summary and critique of the thermodynamic properties of rock-forming minerals. *Am. J. Sci.* 278A.

HELGESON, H. C., and D. H. KIRKHAM. 1974. Theoretical prediction of the thermodynamic behavior of aqueous electrolytes at high pressures and temperatures: I. Summary of the thermodynamic/ electrostatic properties of the solvent. *Am. J. Sci.* 274:1089–198.

HELGESON, H. C., D. H. KIRKHAM, and G. C. FLOWERS. 1981. Theoretical prediction of the thermodynamic behavior of aqueous electrolytes at high temperatures and pressures: IV. Calculation of activity coefficients, osmotic coefficients, and apparent molal properties to 600°C and 5 kb. *Am. J. Sci.* 281:1249–516.

HELGESON, H., W. MURPHY, and P. AAGAARD. 1984. Thermodynamic and kinetic constraints on reaction rates among minerals and aqueous solutions. I. Rate constants, effective surface area and the hydrolysis of feldspar. *Geochim. Cosmochim. Acta* 48:2405–32.

HELLMAN, R. 1994. The albite-water system: Part I. The kinetics of dissolution as a function of pH at 100, 200, and 300°C. *Geochim. Cosmochim. Acta* 58:595–612.

HELZ, G. R., J. H. DAI, P. J. KIJAK, and N. J. FENDINGER. 1987. Processes controlling the composition of acid sulfate solutions evolved from coal. *Applied Geochem.* 2:427–36.

HELZ, G. R., and S. A. SINEX. 1974. Chemical equilibria in the thermal spring waters of Virginia. *Geochim. Cosmochim. Acta* 38:1807–20.

HEM, J. D. 1980. Redox coprecipitation mechanisms of manganese oxides. In *Particulates and water*, ed M. C. Kavanaugh and J. O. Leckie. Adv. in Chem. Ser. 189, pp. 45–72. Washington, DC: Am. Chem. Soc.

HEM, J. D. 1985. *Study and interpretation of the chemical characteristics of natural waters*. 3d ed. U.S. Geol. Survey Water-Supply Paper 2254.

HEM, J. D., and W. H. DURUM. 1973. Solubility and occurrence of lead in surface water. *J. Am. Water Works Assoc.* 65:562–8.

HEM, J. D., C. J. LIND, and C. E. ROBERSON. 1989. Coprecipitation and redox reactions of manganese oxides with copper and nickel. *Geochim. Cosmochim. Acta* 53:2811–22.

HEMINGWAY, B. S. 1982. *Thermodynamic properties of selected uranium compounds and aqueous species at 298.15 K and 1 bar at higher temperatures—Preliminary models for the origin of coffinite deposits*. Open File Report 82-619, U.S. Geol. Survey.

HEMINGWAY, B. S., and R. A. ROBIE. 1973. A calorimetric determination of the standard enthalpies of formation of huntite, $CaMg_3(CO_3)_4$ and artinite, $Mg_2(OH)_2CO_3 \cdot 3H_2O$, and their standard Gibbs free energies of formation. *J. Res. U.S. Geol. Survey* 1(5):535–41.

HENDERSON, P. 1982. *Inorganic geochemistry*. London: Pergamon Press.

HENLEY, R.W. 1984. pH calculations for hydrothermal fluids. In *Fluid-mineral equilibria in hydrothermal systems*. Rev. in Econ. Geology 1, pp. 93–98.

HERMAN, J. S., and M. M. LORAH. 1987. CO_2 outgassing and calcite precipitation in Falling Spring Creek, Virginia. *Chemical Geol.* 62(3–4):251–62.

HERMAN, J. S., and W. B. WHITE. 1985. Dissolution kinetics of dolomite: Effects of lithology and fluid flow velocity. *Geochim. Cosmochim. Acta* 49:2017–26.

HERON, G., and T. H. CHRISTENSEN. 1995. Impact of sediment-bound iron on redox buffering in a landfill leachate polluted aquifer (Vejen, Denmark). *Envir. Sci. & Technol.* 29:187–92.

HERON, G., T. H. CHRISTENSEN, and J. C. TJELL. 1994. Oxidation capacity of aquifer sediments. *Envir. Sci. & Technol.* 28:153–58.

HERSEY, J. P., T. PLESE, and F. J. MILLERO. 1988. The pK_1^* for the dissociation of H_2S in various ionic media. *Geochim. Cosmochim. Acta* 52:2047–51.

HESS, C. T., J. K. KORSAH, and C. J. EINLOH. 1987. Radon in houses due to radon in potable water. In *Radon and its de-*

cay products, ed P. K. Hopke, pp. 30–41. New York: Am. Chem. Soc.

HESS, P. 1966. Phase equilibria of some minerals in the K_2O-Na_2O-Al_2O_3-SiO_2-H_2O system at 25°C and 1 atmosphere. *Am. J. Sci.* 264:289–309.

HILDEBRAND, J. H., J. M. PRAUSINITZ, and R. L. SCOTT. 1970. *Regular and related solutions; The solubility of gases, liquids, and solids*. New York: Van Nostrand Reinhold.

HINGSTON, F. J. 1981. A review of anion adsorption. In *Adsorption of inorganics at solid-liquid interfaces,* ed M. A. Anderson and A. J. Rubin, pp. 51–90. Ann Arbor, MI: Ann Arbor Science.

HITCH, B. F., and R. E. MESMER. 1976. The ionization of aqueous ammonia to 300°C in KCl media. *J. Soln. Chem.* 5:667–80.

HOBART, D. E. 1990. Actinides in the environment. In *Fifty years with transuranium elements. Proc. Robert Welch Foundation conf. on chem. research* 34, pp. 379–436.

HOFFMAN, M. R. 1981. Thermodynamic, kinetic, and extra-thermodynamic considerations in the development of equilibrium models for aqueous systems. *Envir. Sci. Technol.* 15:345–53.

HOFFMAN, M. R. 1990. Catalysis in aquatic environments. In *Aquatic chemical kinetics,* ed W. Stumm, pp. 71–111. New York: Wiley-Interscience.

HOHL, H., L. SIGG, and W. STUMM. 1980. Characterization of surface chemical properties of oxides in natural waters. In *Particulates in water,* ed M. C. Kavanaugh and J. O. Leckie. Am. Chem. Soc. Symp. Ser. 189, pp. 1–31. Washington DC: Am. Chem. Soc.

HOLLAND, H. D. 1978. *The chemistry of the atmosphere and oceans*. New York: John Wiley & Sons.

HOLLAND, T. J. B. 1989. Dependence of entropy on volume for silicate and oxide minerals: A review and a predictive model. *Am. Mineral.* 74:5–13.

HONEYMAN, B. D. 1984. Cation and anion adsorption at the oxide/solution interface in systems containing binary mixtures of adsorbents: An investigation of the concept. PhD thesis, Stanford Univ., Stanford, CA.

HOOGART, J. C., and C. W. S. POSTHUMUS, eds. 1990. *Hydrochemistry and energy storage in aquifers*. Proc. and Info. No. 43, Tech. Meeting 48, Ede, The Hague, The Netherlands.

HOPKE, P. K. 1987. *Radon and its decay products*. New York: Am. Chem. Soc.

HORNE, R. A. 1969. *Marine chemistry*. New York: John Wiley & Sons.

HOSTETLER, P. B. 1964. The degree of saturation of magnesium and calcium carbonate minerals in natural waters. *I.A.S.H. Comm. Subterranean Waters* 64:34–49.

HOSTETLER, P. B., and R. M. GARRELS. 1962. Transportation and precipitation of uranium and vanadium at low temperatures, with special reference to sandstone type uranium deposits. *Econ. Geol.* 57:137–67.

HOSTETTLER, J. D. 1984. Electrode electrons, aqueous electrons, and redox potential in natural waters. *Am. J. Sci.* 284:734–59.

HOWARD, P. H. 1990a. *Handbook of environmental fate and exposure data for organic chemicals*. Vol. I, *Large pro-*

duction and priority pollutants. Chelsea, MI: Lewis Publ. Inc.

HOWARD, P. H. 1990b. *Handbook of environmental fate and exposure data for organic chemicals.* Vol II, *Solvents.* Chelsea, MI: Lewis Publ. Inc.

HSI, C-K. D. 1981. Sorption of uranium (VI) by iron oxides. PhD thesis, T-2514. Colorado School of Mines, Golden, CO.

HSI, C-K. D., and D. LANGMUIR. 1985. Adsorption of uranyl onto ferric oxyhydroxides: Application of the surface complexation site binding model. *Geochim. Cosmochim. Acta* 49(11):2423–32.

HUANG, C. P. 1981. The surface acidity of hydrous solids. In *Adsorption of inorganics and solid-liquid interfaces*, ed M. A. Anderson and A. J. Rubin, pp. 183–217. Ann Arbor, MI: Ann Arbor Science.

HUANG, W-L., J. M. LONGO, and D. R. PEVEAR. 1993. An experimentally derived kinetic model for smectite-to-illite conversion and its use as a geothermometer. *Clays Clay Minerals* 41(2):162–77.

HUTCHEON, I., and M. SHEVALIER, and H. J. ABERCROMBIE. 1993. pH buffering by metastable mineral-fluid equilibria and evolution of carbon dioxide fugacity during burial diagenesis. *Geochim. Cosmochim. Acta* 57:1017–27.

HUTCHINSON, G. E. 1957. *Geography, physics & chemistry.* Vol. 1. *A treatise on limnology.* New York: John Wiley & Sons.

INSKEEP, W. P., and P. R. BLOOM. 1986. Kinetics of calcite precipitation in the presence of water soluble organic ligands. *Soil Sci. Soc. Am. J.* 50:1167–72.

IVANOVICH, M. 1992. The phenomenon of radioactivity. In *Uranium-series disequilibrium: Applications to earth, marine and environmental systems*, ed M. Ivanovich and R. S. Harmon, pp. 1–33. Oxford: Oxford Univ. Press.

JACOBSON, R. L. 1973. Controls on the quality of some carbonate ground waters: Dissolution constants of calcite and CaHCO$_3^+$ from 0 to 50°C. PhD thesis, The Pennsylvania State Univ., University Park, PA.

JACOBSON, R. L., and D. LANGMUIR. 1974a. Controls on the quality variations of some carbonate spring waters. *J. Hydrol.* 23:247–65.

JACOBSON, R. L., and D. LANGMUIR. 1974b. Unpublished data. Dept. of Geosciences, The Pennsylvania State Univ., University Park, PA.

JACOBSON, R. L., D. LANGMUIR, and P. DEINES. 1971. Unpublished manuscript. Dept. of Geosciences, The Pennsylvania State Univ., University Park, PA.

JAMES, A. M., and A. R. R. LUPTON. 1978. Gypsum and anhydrite in foundations of hydraulic structures. *Geotechnique* 28:249–72.

JAMES, R. O., and T. W. HEALY. 1972. Adsorption of hydrolyzable metal ions at the oxide-water interface. *J. Colloid & Interface Sci.* 40(1):65–81.

JAMES, R. O., and G. A. PARKS. 1982. Characterization of aqueous colloids by their electric double-layer and intrinsic surface chemical properties. *Surface Colloid Sci.* 12:119–216.

JAVENDEL, I., C. DOUGHTY, and C.-F. TSANG. 1984. *Groundwater transport: Handbook of mathematical models.* Water Resources Monograph Ser. 10. Washington, DC: Amer. Geophys. Union.

JENNE, E. A. 1990. Aquifer thermal energy storage: The importance of geochemical reactions. In *Hydrochemistry and energy storage in aquifers*, ed J. C. Hooghart and C. W. S. Posthumus. Proc. & Info. No. 43, pp. 19–36, Tech. Mtg. 48, Ede, The Hague, The Netherlands.

JENNE, E. A. 1992. Verbal communication. Pacific Northwest Laboratory, Richland, WA.

JENNE, E. A. 1995. Metal adsorption onto and desorption from sediments. I. Rates. In *Metal speciation and contamination of aquatic sediments*, ed H. E. Allen, pp. 81–112. Ann Arbor, MI: Ann Arbor Press.

JENNY, H. 1941. *Factors of soil formation.* New York: McGraw-Hill.

JENNY, H., and C. D. LEONARD. 1934. Functional relationships between soil properties and rainfall. *Soil Sci.* 38:363–81.

JONES, L. H. P., and K. A. HANDRECK. 1963. Effect of iron and aluminum oxides on silica in solution in soils. *Nature* 198:852–53.

JONSSON, C., P. WARFVINGE, and H. SVERDRUP. 1995. Uncertainty in predicting weathering rate and environmental stress factors with the PROFILE model. *Water, Air and Soil Pollution* 81:1–23.

JORGENSEN, B. B. 1990. A thiosulfate shunt in the sulfur cycle of marine sediments. *Science* 249:152–54.

JORGENSEN, B. B., M. F. ISAKSEN, and H. W. JANNASCH. 1992. Bacterial sulfate reduction above 100°C in deep-sea hydrothermal vent sediments. *Science* 258:1756–57.

JUNGE, C. E., and R. T. WERBY. 1958. The concentration of chloride, sodium, potassium, calcium and sulfate in rain water over the United States. *J. Meteorol.* 15:417–25.

KALINOWSKI B. E., and P. SCHWEDA. 1996. Kinetics of muscovite, phlogopite, and biotite dissolution and alteration at pH 1–4, room temperature. *Geochim. Cosmochim. Acta* 60(3):367–85.

KARAMENKO, L. Y. 1969. Biogenic factors in the formation of sedimentary ores. *Intl. Geol. Rev.* 11:271–81.

KARSHIN, V. P., and V. A. GRIGORYAN. 1970. Kinetics of the dissolution of gypsum in water. *Russ. J. Chem.* 44:762–63.

KATHREN, R. L. 1984. *Radioactivity in the environment.* Philadelphia: Harwood Acad. Publ.

KATZ, B. G., O. P. BRICKER, and M. M. KENNEDY. 1985. Geochemical mass balance relationships for selected ions in precipitation and stream water, Catoctin Mountains, Maryland. *Am. J. Sci.* 285:931–62.

KEMPTON, J. H., LINDBERG, R. D., and D. D. RUNNELLS. 1990. Numerical modeling of platinum Eh measurements by using heterogeneous electron-transfer kinetics. In *Chemical modeling of aqueous systems II*, ed D. C. Melchior and R. L. Bassett. Am. Chem. Soc. Symp. Ser. 416, pp. 339–49. Washington, DC: Am. Chem. Soc.

KENNEDY, V. C. 1965. *Mineralogy and cation-exchange capacity of sediments from selected streams.* U.S. Geol. Survey Prof. Paper 433-D.

KENNEDY, V. C. 1971. Silica variations in stream water with time and discharge. In *Nonequilibrium systems in natural-water chemistry*, Adv. Chem. Ser. 106, pp. 94–130. Washington, DC: Am. Chem. Soc.

KENNEDY, V. C., and R. L. MALCOLM. 1978. *Geochemistry of the Mattole River in northern California.* U.S. Geol. Survey Open-File Report 78-205.

KENT, D. B., J. A. DAVIS, C. D. ANDERSON, B. A. REA, and T. D. WAITE. 1994. Transport of chromium and selenium in the suboxic zone of a shallow aquifer: Influence of redox and adsorption reactions. *Water Resources Research* 30(4): 1099–114.

KENT, D. B., V. S. TRIPATHI, N. B. BALL, and J. O. LECKIE. 1986. *Surface-complexation modeling of radionuclide adsorption in sub-surface environments.* Stanford Civil Eng. Tech. Rept. 294, NUREG Rept. CR-4897.

KHARAKA, Y. K., W. D. GUNTER, P. K. AGGARWAL, E. H. PERKINS, and J. D. DEBRAAL. 1988. *SOLMINEQ.88: A computer program for geochemical modeling of water-rock interactions.* U.S. Geol. Survey Water Resources Inv. 88–4227. Menlo Park, CA: U.S. Geol. Survey.

KHODAKOVSKY, I. L. 1993. Verbal communication. Vernadsky Inst. of Geochem. and Analytical Chem., Russian Acad. of Sciences, Moscow.

KIELLAND, J. 1937. Individual activity coefficients of ions in aqueous solutions. *J. Am. Chem. Soc.* 59:1675–735.

KIM, A. G. 1968. An experimental study of ferrous iron oxidation in acid mine water. *Proc. 2d symp. coal mine drainage research,* pp. 40–45. Monroeville, PA: Bituminous Coal Research, Inc.

KIM, J-I. 1993. The chemical behavior of transuranium elements and barrier functions of natural aquifer systems. *Mat. Res. Soc. symp. proc.* 294, pp. 3–21.

KING, D. W., H. A. LOUNSBURY, and F. J. MILLER. 1995. Rates and mechanism of Fe(II) oxidation at nanomolar total iron concentrations. *Envir. Sci. & Technol.* 29:818–24.

KING, T. V. V., ed. 1995. Environmental considerations of active and abandoned mine lands. Lessons from Summitville, Colorado, *U.S. Geol. Survey Bull.* 2220.

KINNIBURGH, D. G., and M. L. JACKSON. 1982. Concentration and pH dependence of calcium and zinc adsorption by iron hydrous oxide gel. *Soil Sci. Soc. Am. J.* 46(1):56–61.

KITTRICK, J. A. 1966. Free energy of formation of kaolinite from solubility measurements. *Am. Mineral.* 51:1457–66.

KLEIBER, P., and W. E. ERLEBACH. 1977. *Limitations of single water samples in representing mean water quality.* III. *Effect of variability in concentration measurements on estimates of nutrient loadings in the Squamish River, B.C.* Tech. Bull. 103, Inland Water Directorate, Pacific & Yukon Region, Water Qual. Branch, Fisheries & Envir. Canada.

KLEINMANN, R. L. P., D. A. CRERAR, and R. R. PACELLI. 1981. Biogeochemistry of acid mine drainage and a method to control acid formation. *Mining Eng.* (March):300–305.

KLOTZ, I. M. 1964. *Chemical thermodynamics: Basic theory and methods.* rev. ed. Menlo Park, CA: W. A. Benjamin, Inc.

KLOTZ, I. M., and R. M. ROSENBERG. 1986. *Chemical thermodynamics. Basic theory and methods.* 4th ed. Menlo Park, CA: The Benjamin/Cummings Publ. Co., Inc.

KLUSMAN, R. W. 1996. Verbal communication. Dept. of Chemistry & Geochemistry, The Colorado School of Mines, Golden, CO.

KOHLER, M., E. WIELAND, and J. O. LECKIE. 1992. Metal-ligand-surface interactions during sorption of uranyl and neptunyl on oxides and silicates. *Proc. 7th intl. symp. water-rock interaction,* ed Y. K. Kharaka and A. S. Maest, 1, pp. 51–54, Rotterdam, The Netherlands: A. A. Balkema.

KONIGSBERGER, E., J. BUGAJSKI, and H. GAMSJAGER. 1989. Solid-solute phase equilibria in aqueous solution. II. A potentiometric study of the aragonite-calcite transition. *Geochim. Cosmochim. Acta* 53:2807–10.

KOZAI, N., T. OHNUKI, and S. MURAOKA. 1995. Sorption behavior of neptunium on bentonite—effect of calcium ion on the sorption. *Mat. Res. Soc. symp. proc.* 353, pp. 1021–28.

KOZIOL, A. M., and R. C. NEWTON. 1995. Experimental determination of the reactions magnesite + quartz = enstatite + CO_2 and magnesite = periclase + CO_2, and the enthalpies of formation of enstatite and magnesite. *Am. Mineral.* 80:1252–60.

KRAUSKOPF, K. B. 1967. *Introduction to geochemistry.* New York: McGraw-Hill.

KRAUSKOPF, K. B. 1988. *Radioactive waste disposal and geology. Topics in earth sci. 1.* New York: Chapman and Hall Publ.

KRAUSKOPF, K. B., and D. K. BIRD. 1995. *Introduction to geochemistry.* 3d ed. New York: McGraw-Hill.

KRIEGER, H. L., and E. L. WHITTAKER. 1980. *Prescribed procedures for measurement of radioactivity in drinking water.* Report EPA-600/4-80-032. U.S. EPA.

KRUPKA, K. M., E. A. JENNE, and W. J. DEUTCH. 1983. *Validation of the WATEQ4 geochemical model for uranium.* Pacific Northwest Laboratory Report PNL-4333.

KUNG, S. K., and I. R. TRIAY. 1994. Effect of natural organics on Cd and Np sorption. *Radiochim. Acta* 66/67:421–26.

LAFON, G. M., and F. T. MACKENZIE. 1974. Early evolution of the oceans; A weathering model. *Soc. Econ. Paleont. & Mineral.*, spec. pub., 20:205–10.

LAHERMO, P., J. MANNIO, and T. TARAVAINEN. 1995. The hydrogeochemical comparison of streams and lakes in Finland. *Applied Geochem.* 10:45–64.

LANGMUIR, D. 1965. Stability of carbonates in the system MgO-CO_2-H_2O. *J. Geol.* 73:730–54.

LANGMUIR, D. 1968. Stability of calcite based on aqueous solubility measurements. *Geochim. Cosmochim. Acta* 32(8):835–51.

LANGMUIR, D. 1969a. *The Gibbs free energies of substances in the system Fe-O_2-H_2O-CO_2 at 25°C.* U.S. Geol. Survey Prof. Paper 650-B.

LANGMUIR, D. 1969b. Iron in ground-water of the Magothy and Raritan Formations in Camden and Burlington Counties, NJ. *New Jersey Water Res. Circ.* 19.

LANGMUIR, D. 1971a. Eh-pH determination. In *Procedures in sedimentary petrology*, ed R. E. Carver, pp. 597–634. New York: John Wiley & Sons.

LANGMUIR, D. 1971b. Particle size effect on the reaction: goethite = hematite + water. *Am J. Sci* 271(2):147–156.

LANGMUIR, D. 1971c. The geochemistry of some carbonate groundwaters in central Pennsylvania. *Geochim. Cosmochim. Acta* 35:1023–45.

LANGMUIR, D. 1972. Correction. Particle size effect on the reaction: goethite = hematite + water. *Am J. Sci* 272(10):972.

LANGMUIR, D. 1978. Uranium solution-mineral equilibria at low temperatures with applications to sedimentary ore deposits. *Geochim. Cosmochim. Acta* 42(6):547–69.

LANGMUIR, D. 1979. Techniques of estimating thermodynamic properties for some aqueous complexes of geochemical interest. In *Chemical modeling in aqueous systems,* ed E. A. Jenne. Am. Chem. Soc. Symp. Ser. 93, pp. 353–87. Washington, DC: Am. Chem. Soc.

LANGMUIR, D. 1981. The power exchange function: A general model for metal adsorption onto geological materials. In *Adsorption from aqueous solutions,* ed D. H. Tewari, pp. 1–17. New York: Plenum Press.

LANGMUIR, D. 1984. Physical and chemical characteristics of carbonate water. In *Guide to the hydrology of carbonate rocks,* ed P. E. Lamoreaux, B. M. Wilson, and B. A. Memeon. Paris: UNESCO, pp. 60–105, 116–30.

LANGMUIR, D. 1987. Overview of coupled processes with emphasis in geochemistry. In *Coupled processes associated with nuclear waste repositories,* ed Chin-Fu Tsang, pp. 67–101. Orlando, FL: Academic Press, Inc.

LANGMUIR, D. 1995. Review of the article "Granite and radon" published in *Solid surface.* In *Granite & radon—the truth.* Special bulletin, pp. 2–4. Columbus, OH: Marble Inst. of Amer.

LANGMUIR, D. 1997. *Worked problems in aqueous environmental geochemistry.* Upper Saddle River, NJ: Prentice Hall, Inc.

LANGMUIR, D., and M. A. APTED. 1992. Backfill modification using geochemical principles to optimize high level nuclear waste isolation in a geological repository. *Mat. Res. Soc. symp. proc.* 257, pp. 13–24.

LANGMUIR, D., and H. L. BARNES. 1972. Water-substance and solvent.In *Encyclopedia of geochemistry and mineralogy,* ed R. W. Fairbridge, pp. 1085–87. New York: Reinhold Publ. Corp.

LANGMUIR, D., and J. R. CHATHAM. 1980. Groundwater prospecting for sandstone-type uranium deposits: A preliminary comparison of the merits of mineral-solution equilibria, and single-element tracer methods. *J. Geochem. Explor.* 13:201–19.

LANGMUIR, D., and J. S. HERMAN. 1980. The mobility of thorium in natural waters at low temperatures. *Geochim. Cosmochim. Acta* 44(11):1753–66.

LANGMUIR, D., and J. MAHONEY. 1985. Chemical equilibrium and kinetics of geochemical processes in ground water studies. In *Practical applications of ground water geochemistry, Proc. 1st Canadian/American conf. on hydrogeology,* ed B. Hitchon and E. I. Wallick, pp. 69–95. Worthington, OH: Natl. Water Well Assoc.

LANGMUIR, D., and D. MELCHIOR. 1985. The geochemistry of Ca, Sr, Ba and Ra sulfates in some deep brines from the Palo Duro Basin, Texas. *Geochim. Cosmochim. Acta* 49(11):2423–32.

LANGMUIR, D., and D. K. NORDSTROM. 1995. The geochemistry of uranium mill tailings: An acid mine drainage problem. Unpublished ms.

LANGMUIR, D., and R. L. SCHMIERMUND. 1986. *Geochemistry of lime mortar and portland cement weathering by acid rain.* Phase I Rept., Part 1, for 7/15/86 to 9/30/86, Coop. Agreement CA-0424-6-8001 between Natl. Park Service and Colorado School of Mines.

LANGMUIR, D., and D. O. WHITTEMORE. 1971. Variations in the stability of precipitated ferric oxyhydroxides. In *Nonequilibrium systems in natural water chemistry,* ed R. F. Gould. Am. Chem. Soc. Symp. Ser. 106, pp. 209–34. Washington, DC: Am. Chem. Soc.

LARSON, W. C. 1978. *Uranium in situ leach mining in the United States.* Info. Circ. 8777. U.S. Dept. of Interior, Bureau of Mines.

LASAGA, A. C. 1981a. Rate laws of chemical reactions. In *Kinetics of geochemical processes,* ed A. C. Lasaga and R. J. Kirkpatrick. Rev. in Mineralogy 8, pp. 1–68. Min. Soc. Am.

LASAGA, A. C. 1981b. Transition state theory. In *Kinetics of geochemical processes,* ed A. C. Lasaga and R. J. Kirkpatrick. Rev. in Mineralogy 8, pp. 135–69. Min. Soc. Am.

LASAGA, A. C., J. M. SOLER, J. GANOR, T. E. BURCH, and K. L. NAGY. 1994. Chemical weathering rate laws and global geochemical cycles. *Geochim. Cosmochim. Acta* 58(10):2361–86.

LATIMER, W. M. 1952. *Oxidation potentials.* 2d ed. Englewood Cliffs, NJ: Prentice-Hall.

LEENHEER, J. A., R. L. MALCOLM, P. W. MCKINLEY, and L. A. ECCLES. 1974. Occurrence of dissolved organic carbon in selected groundwater samples in the United States. *U.S. Geol. Survey J. of Research* 2:361–69.

LEMIRE, R. J. 1984. *An assessment of the thermodynamic behavior of neptunium in water and model groundwater from 25°C to 150°C.* AECL-7817. Pinawa Manitoba: Atomic Energy of Canada Ltd.

LEMIRE, R. J., and F. GARISTO. 1989. *The solubility of U, Np, Pu, Th and Tc in a geological disposal vault for used nuclear fuel.* Report ROE 1LO, AECL-10009. Pinawa, Manitoba: Atomic Energy of Canada Ltd., Whiteshell Nuclear Research Establ.

LEMIRE, R. J., and P. R. TREMAINE. 1980. Uranium and plutonium equilibria in aqueous solutions to 200°C. *J. Chem. Eng. Data* 25:361–70.

LERMAN, A. 1979. *Geochemical processes. Water and sediment environments.* New York: John Wiley & Sons.

LERMAN, A. 1990. Transport and kinetics in surficial processes. In A*quatic chemical kinetics,* ed W. Stumm, pp. 505–34. New York: Wiley-Interscience.

LEVINSON, A. A. 1980. *Introduction to exploration geochemistry.* Wilmette, IL: Applied Publishing Ltd.

LEVINSON, A. A., P. M. D. BRADSHAW, and I. THOMSON. 1987. *Practical exploration geochemistry.* Wilmette, IL: Applied Publishing Ltd.

LEWIS, G. N., and M. RANDALL. 1921. The activity coefficient of strong electrolytes. *J. Am. Chem. Soc.* 43:1112–54.

LEWIS, G. N., and M. RANDALL. 1961. *Thermodynamics.* 2d ed. revised by K. S. Pitzer and L. Brewer. New York: McGraw Hill.

LEWIS-RUSS, A. 1991. Measurement of surface charge of inorganic geological materials: Techniques and their consequences. In *Advances in agronomy,* ed D. L. Sparks, vol. 46, pp. 199–243. New York: Academic Press, Inc.

LIANG, L., and J. J. MORGAN. 1990. Chemical aspects of iron oxide coagulation in water: Laboratory studies and implications for natural systems. *Aquatic Sciences* 52(1):32–55.

LIBÚS, Z., and H. TIALOWSKA. 1975. Stability and nature of complexes of the type MCl$^+$ in aqueous solution (M = Mn, Co, Ni, Zn). *J. Soln. Chem.* 4(12):1011–22.

LICHTNER, P. C., C. I. STEEFEL, and E. H. OELKERS, eds. 1996. *Reactive transport in porous media.* Reviews in Mineralogy 34. Min. Soc. Am.

LIDE, D. R., ed. 1995. *Handbook of chemistry and physics.* 76th ed. Boca Raton, FL: CRC Press.

LIESER, K. H., and U. MOHLENWEG. 1988. Neptunium in the hydrosphere and in the geosphere. *Radiochim. Acta* 43:27–35.

LIKENS, G. E. 1976. Acid precipitation. *Chem. & Eng. News* 54(48):29–44.

LIKENS, G. E. 1979. Analysis of a North American lake ecosystem, III. Sources, loading and fate of nitrogen and phosphorus. *Limnol.* 20(1):568–73.

LIKENS, G. E., F. H. BORMANN, R. S. PIERCE, J. S. EATON, and N. M. JOHNSON. 1977. *Biogeochemistry of a forested ecosystem.* New York: Springer-Verlag.

LIKENS, G. E., C. T. DRISCOLL, and D. C. BUSO. 1996. Long-term effects of acid rain: Response and recovery of a forested watershed. *Science* 272:244–45.

LINDBERG, R. E., and D. D. RUNNELLS. 1984. Ground water redox reactions: An analysis of equilibrium state applied to Eh measurements and geochemical modeling. *Science* 225:925–27.

LINDSAY, W. L. 1979. *Chemical equilibria in soils.* New York: Wiley-Interscience.

LIPFERT, F. W. 1989. Air pollution and materials damage. In *The handbook of environmental chemistry,* ed O. Hutzinger, vol. 4B, pp. 114–86, Berlin: Springer-Verlag.

LIPPMANN, F. 1979. Stabilitätsbeziehungen der Tonminerale (Stability relations of clay minerals). *N. Jb. Miner. Abh.* 136(3):287–309.

LIU, S-T., and G. H. NANCOLLAS. 1971. The kinetics of dissolution of calcium sulfate hydrate. *J. Inorg. Nucl. Chem.* 33:2311–16.

LIVINGSTONE, D. A. 1963. *Chemical composition of rivers and lakes.* 6th ed. U.S. Geol. Survey Prof. Paper 440-G.

LOUGHNAN, F. C. 1969. *Chemical weathering of silicate minerals.* New York: Amer. Elsevier Publ. Co., Inc.

LOUX, N. T., D. S. BROWN, C. R. CHAFIN, J. D. ALLISON, and S. M. HASSAN. 1989. Chemical speciation and competitive cationic partitioning on a sandy aquifer material. *J. Chem. Speciation & Bioavailability* 1:111–25.

LOUX, N. T., D. S. BROWN, C. R. CHAFIN, J. D. ALLISON, and S. M. HASSAN. 1990. Modeling geochemical processes attenuating inorganic contaminant transport in the subsur-

face region: Adsorption on amorphous iron oxide. In *Proc. envir. res. conf. on groundwater qual. and waste disposal,* ed I. Murarka and S. Cordle. Report EPRI EN-6749. Palo Alto, CA: Electrical Power Research Inst.

LOVLEY, D. R., F. H. CHAPELLE, and J. C. WOODWARD. 1994. Use of dissolved H$_2$ concentrations to determine distribution of microbially catalyzed redox reactions in anoxic groundwater. *Envir. Sci. & Technol.* 28(7):1205–10.

LOVLEY, D. R., and S. GOODWIN. 1988. Hydrogen concentrations as an indicator of the predominant terminal electron-accepting reactions in aquatic sediments. *Geochim. Cosmochim. Acta* 52: 2993–3003.

LUNDEGARD, P. D., and Y. K. KHARAKA. 1990. Geochemistry of organic acids in subsurface waters. In *Chemical modeling of aqueous systems II,* ed D. C. Melchior and R. L. Bassett. Am. Chem. Soc. Symp. Ser. 416, pp. 169–89. Washington, DC: Am. Chem. Soc.

LUTHER, G. W., III, D. T. RICKARD, S. THEBERGE, and A. OLROYD. 1996. Determination of metal (bi)sulfide stability constants of Mn^{2+}, Fe^{2+}, Co^{2+}, Ni^{2+}, Cu^{2+}, and Zn^{2+} by voltammetric methods. *Envir. Sci. & Technol.* 30:671–79.

LYMAN, W. J., W. F. REEHL, and D. H. ROSENBLATT. 1982. *Handbook of chemical property estimation methods. Environmental behavior of organic compounds.* New York: McGraw-Hill.

MACALADY, D. L., D. LANGMUIR, T. GRUNDL, and A. ELZERMAN. 1990. Use of model generated Fe^{3+} ion activities to compute Eh and ferric oxyhydroxide solubilities in anaerobic systems. In *Chemical modeling in aqueous ststems II,* ed D. C. Melchior and R. L. Bassett. Am. Chem. Soc. Symp. Ser. 416, pp. 350–67. Washington, DC: Am. Chem. Soc.

MACINNES, D. A. 1919. The activities of the ions of strong electrolyes. *J. Am. Chem. Soc.* 41:1086–92.

MACINNES, D. A. 1961. *The principles of electrochemistry.* New York: Dover Publications, Inc.

MACKENZIE, F. T. et al. 1983. *Magnesian calcites: Low-temperature occurrences, solubility and solid-solution behavior.* Reviews in Mineralogy, pp. 97–144. Min. Soc. Am.

MACKENZIE, F. T., and R. GEES. 1971. Quartz: Synthesis at Earth-surface conditions. *Science* 173:533–34.

MACKENZIE, F. T., and R. WOLLAST. 1977. In *The sea.* Vol. 6, p. 739. New York: Wiley-Interscience.

MACKIN, J. E., and K. T. SWIDER. 1987. Modeling and dissolution behavior of standard clays in seawater. *Geochim. Cosmochim. Acta* 51(11): 2947–64.

MAEST, A. S., S. P. PASILI, L. G. MILLER, and D. K. NORDSTROM. 1992. Redox geochemistry of arsenic and iron in Mono Lake, California, USA. *Proc. 7th intl. symp. water-rock interaction,* ed Y. K. Kharaka and A. S. Maest, 1, pp. 507–71.

MAHONEY, J. J., and D. LANGMUIR. 1991. Adsorption of Sr on kaolinite, illite and montmorillonite at high ionic strengths. *Radiochim. Acta* 54(3):139–44.

MALCOLM, R. L. 1985. Humic substances in rivers and streams. In *Humic substances* I. *Geochemistry, characterization, and isolation,* ed G. R. Aiken, P. MacCarthy, D. McKnight, and R. L. Wershaw. New York: John Wiley & Sons.

MALCOLM, R. L., and V. C. KENNEDY. 1970. Variation of cation exchange capacity and rate with particle size in stream sediments. 2. *Water Pollution Control Fed. J.* 42(5):153–60.

MALEY, K. M. Characterization and modeling of surface water geochemistry at the Rawley Twelve mine and Mill Site, Bonanaza Mining District, Saguache County, Colorado. M.S. thesis, T-4574, Colorado School of Mines, Golden, CO.

MANAHAN, S. E. 1991. *Environmental chemistry.* 5th ed. Chelsea, MI: Lewis Publ. Co.

MANAHAN, S. E. 1994. *Environmental chemistry.* 6th ed. Chelsea, MI: Lewis Publ. Co.

MANGOLD, D. C., and C. TSANG. 1991. A summary of subsurface hydrological and hydrochemical models. *Reviews of Geophysics* 29(1):51–80.

MANN, A. W. 1974. *Chemical ore genesis models for the precipitation of carnotite in calcrete.* Rept. No. FP7, CSIRO Minerals Res. Labs, Div. Mineralogy.

MARSHALL, W. L., and C. A. CHEN. 1982. Amorphous silica solubilities. V. Prediction of solubility behavior in aqueous mixed electrolyte solutions to 300°C. *Geochim. Cosmochim. Acta* 46:289–91.

MARSICANO, F., and R. HANCOCK. 1978. The linear free energy relation in the thermodynamics of complex formation, 2. Analysis of the formation constants of complexes of the large metal ions Ag^+, Hg^{2+}, and Cd^{2+} with ligands having "soft" and nitrogen-donor atoms. *J. Chem. Soc. Dalton* 1978:228–34.

MARTELL, A. E., and R. M. SMITH. 1977. *Critical stability constants. 3, Other organic ligands.* New York: Plenum Press.

MASON, B., and C. B. MOORE. 1982. *Principles of geochemistry.* 4th ed. New York: John Wiley & Sons.

MASSCHELEYN, P. H., R. D. DALAUNE, and W. H. PATRICK, JR. 1991. Effect of redox potential and pH on arsenic speciation and solubility in a contaminated soil. *Envir. Sci. & Technol.* 25:1414–19.

MAST, M. A., and J. I. DREVER. 1987. The effect of oxalate on the dissolution rates of oligoclase and tremolite. *Geochim. Cosmochim. Acta* 51:2559–68.

MAST, M. A., J. I. DREVER, and J. BARON. 1990. Chemical weathering in the Loch Vale Watershed, Rocky Mountain National Park, Colorado. *Water Resources Research* 26(12):2971–78.

MATHIESON, J. G., and B. E. CONWAY. 1974. Partial molal compressibilities of salts in aqueous solution and assignment of ionic contributions. *J. Soln. Chem.* 3:455–77.

MATIJEVIC, E., and P. SCHEINER. 1978. Ferric hydrous oxide sols. III. Preparation of uniform particles by hydrolysis of Fe^{III}-chloride, -nitrate, and -perchlorate solutions. *J. Colloid Interface Sci.* 63:509–24.

MAY, H. M., D. G. KINNIBURGH, P. A. HELMKE, and M. L. JACKSON. 1986. Aqueous dissolution, solubilities and thermodynamic stabilities of common aluminosilicate clay minerals: Kaolinite and smectites. *Geochim. Cosmochim. Acta* 50:1667–77.

MAYER, L. M. 1982a. Retention of riverine iron in estuaries. *Geochim. Cosmochim. Acta* 46:1003–9.

MAYER, L. M. 1982b. Aggregation of colloidal iron during estuarine mixing: Kinetics, mechanism and seasonality. *Geochim. Cosmochim. Acta* 46:2527–35.

MCAULEY, A., and G. H. NANCOLLAS. 1961. Thermodynamics of ion association. VII. Some transition metal oxalates. *J. Chem. Soc.* 1961:2215–21.

MCCARTHY, J. F., and C. DEGUELDRE. 1993. Sampling and characterization of colloids and particles in ground water for studying their role in contaminant transport. In *Characterization of environmental particles*, ed H. P. van Leeuwen and J. Buffle. IUPAC Envir. Analytical and Physical Chem. Ser. II. Chelsea, MI: Lewis Publ.

MCKIBBEN, M. A., and H. L. BARNES. 1986. Oxidation of pyrite in low temperature acidic solutions: Rate laws and surface textures. *Geochim. Cosmochim. Acta* 50:1509–20.

MCKINLEY, I. G., and D. SAVAGE. 1994. Comparison of solubility databases used for HLW performance assessment. *Fourth intl. conf. on the chemistry and migration behavior of actinides and fission products in the geosphere,* pp. 657–65. München: R. Oldenbourg Verlag Publ.

MCKNIGHT, D. M., B. A. KIMBALL, and K. E. BENCALA. 1988. Iron photoreduction and oxidation in an acidic mountain stream. *Science* 240:637–40.

MCWHORTER, D. B., R. K. SKOGERBOE, and G. V. SKOGERBOE. 1974. Water pollution potential of mine spoils in the Rocky Mountain region. *Proc. 5th symp. coal mine drainage research,* pp. 25–38. Monroeville, PA: Bituminous Coal Research, Inc.

MEIJER, A. 1990. *Yucca Mountain far-field soption studies and data needs.* Los Alamos Natl. Lab. Rept. LA-11671-MS. Los Alamos, NM: Los Alamos Natl. Lab.

MEIJER, A. 1992. A strategy for the derivation and use of sorption coefficients in performance assessment calculations for the Yucca Mountain Site. In *Proc. DOE/Yucca Mountain Site characterization project radionuclide adsorption.* Workshop at Los Alamos Natl. Lab., Rept. LA-12325-C, UC-510, pp. 9–40. Los Alamos, NM.

MENG, X., and R. D. LETTERMAN. 1993. Effect of component oxide interaction on the adsorption properties of mixed oxides. *Envir. Sci. & Technol.* 27:970–75.

MESMER, R. E., C. F. BAES, JR., and F. H. SWEETON. 1972. Acidity measurements at elevated temperatures. VI. Boric acid equilibria. *Inorg. Chem.* 11:537.

MESUERE, K., and W. FISH. 1992a. Chromate and oxalate adsorption on goethite. 1. Calibration of surface complexation models. *Envir. Sci. & Technol.* 26(12):2357–64.

MESUERE, K., and W. FISH. 1992b. Chromate and oxalate adsorption on goethite. 2. Surface complex modeling of competitive adsorption. *Envir. Sci. & Technol.* 26(12):2365–70.

MEUNIER, J-D., P. LANDAIS, M. MONTHIOUX, and M. PAGEL. 1987. Oxidation-reduction processes in the genesis of the uranium-vanadium tabular deposits of the Cottonwood Wash mining area (Utah, USA): Evidence from petrological study and organic matter analysis. *Bull. Mineral.* 110:145–56.

MEYER, R. E., W. D. ARNOLD. F. I. CASE, and G. D. O'KELLEY. 1989. *Thermodynamics of Tc related to nuclear waste disposal.* Oak Ridge Natl. Lab. Report NUREG/CR-5235.

MIDDLEBURG, J. J. 1989. A simple rate model for organic matter decomposition in marine sediments. *Geochim. Cosmochim. Acta* 53:1577–81.

MIEKELEY, N. et al. 1991. *Natural series radionuclide and rare-earth element geochemistry of waters from the Osamu Utsumi mine and Morro do Ferro analogue study sites, Pocos de Caldas, Brazil.* Pocos de Caldas Report No. 8., NAGRA NTB 90-26, SKB TR 90-17.

MILLAR, C. E., L. M. TURK, and H. D. FOTH. 1958. *Fundamentals of soil science.* 3d ed. New York: John Wiley & Sons.

MILLER, G. C., W. B. LYONS, and A. DAVIS. 1996. Understanding the water quality of pit lakes. *Envir. Sci. & Technol.* 30(3):118A–123A.

MILLER, W. R., R. B. WANTY, and J. B. McHUGH. 1984. Application of mineral-solution equilibria to geochemical exploration for sandstone-hosted uranium deposits in two basins in west-central Utah. *Econ. Geol.* 79(2):266–83.

MILLERO, F. J. 1982. The effect of pressure on the solubility of minerals in water and sea water. *Geochim. Cosmochim. Acta* 46:11–22.

MILLERO, F. J. 1983. The estimation of the pK^*_{HA} of acids in sea water using Pitzer equations. *Geochim. Cosmochim. Acta* 48:571–81.

MILLERO, F. J., and D. R. SCHREIBER. 1982. Use of the ion pairing model to estimate activity coefficients of the ionic components of natural waters. *Am. J. Sci.* 282:1508–40.

MONCURE, G., P. A. JANKOWSKI, and J. I. DREVER. 1992. The hydrochemistry of arsenic in reservoir sediments, Milltown, Montana. *Proc. 7th intl. symp. water-rock interaction,* ed Y. K. Kharaka and A. S. Maest, 1, pp. 513–16. Rotterdam, The Netherlands: A. A. Balkema.

MOORE, J. W., and E. A. MOORE. 1976. *Environmental chemistry.* New York: Academic Press.

MOREL, F. M. M. 1983. *Principles of aquatic chemistry.* New York: John Wiley & Sons.

MOREL, F. M. M., and J. G. HERING. 1993. *Principles and applications of aquatic chemistry.* New York: John Wiley & Sons.

MOREY, G. W., R. O. FOURNIER, and J. J. ROWE. 1964. The solubility of amorphous silica at 25°C. *J. Geophys. Res.* 69(10):1995–2002.

MORSE, J. W., F. J. MILLERO, J. C. CORNWELL, and D. RIKARD. 1987. The chemistry of the hydrogen sulfide and iron sulfide systems in natural waters. *Earth Sci. Rev.* 24:1–42.

MORTH, A. H., E. E. SMITH, and K. S. SHUMATE. 1972. *Pyritic systems: A mathematical model.* Envir. Protection Technol. Ser. Rept. EPA-R2-72-002. Washington, DC: U.S. EPA.

MOSES, C. O., and J. S. HERMAN. 1991. Pyrite oxidation at circumneutral pH. *Geochim. Cosmochim. Acta* 55:471–82.

MOSES, C. O., K. K. NORDSTROM, J. S. HERMAN, and A. L. MILLS. 1987. Aqueous pyrite oxidation by dissolved oxygen and by ferric iron. *Geochim. Cosmochim. Acta* 51:1561–71.

MOSKVIN, A. I. 1975. Mechanism of the interchange of ligands in actinide coordination compounds. *Koord. Khim.* 1:83–92.

MOULIN, V., and C. MOULIN. 1995. Fate of the actinides in the presence of humic substances under conditions relevant to nuclear waste disposal. *Applied Geochim.* 10:573–80.

MUCCI, A. 1991. The solubility and free energy of formation of natural kutnahorite. *Canadian Mineral.* 29:113–21

MUMPTON, F. A., ed. 1977. *Mineralogy and geology of natural zeolites.* Reviews in Mineralogy 4. Min. Soc. Am.

MURAKAMI, T., and R. C. EWING, eds. 1995. *18th symp. scientific basis for nuclear waste management, 1994.* Pittsburgh, PA: Materials Research Soc.

MUROWCHICK, J. B., and H. L. BARNES. 1986. Marcasite precipitation from hydrothermal solutions. *Geochim. Cosmochim. Acta* 50:2615–29.

MURPHY, W. M., and H. C. HELGESON. 1989. Thermodynamic and kinetic constraints on reaction rates among minerals and aqueous solutions. IV. Retrieval of rate constants and activation parameters for the hydrolysis constants of pyroxene, wollastonite, olivine, andalusite, quartz, and nepheline. *Am. J. Sci.* 289:17–101.

MURPHY, W. M., and PALABAN, R. T. 1994. *Geochemical investigations related to the Yucca Mountain Environment and Potential Nuclear Waste Repository.* NUREG/CR-6288. San Antonio, TX: Southwest Research Inst., Center for Nuclear Waste Analysis.

MURRAY, J. W., and B. R. COUGHLIN. 1992. Competitive adsorption: The effect of carbonate alkalinity on the adsorption of Pb, Th, and Pu by goethite (α-FeOOH). *Proc. 7th intl. symp. water-rock interaction,* ed Y. K. Kharaka and A. S. Maest, pp. 55–57. Rotterdam: A. A. Balkema.

MURRAY, R. C., and J. W. COBBLE. 1980. Chemical equilibria in aqueous systems at high temperatures. *Proc. 41st annual intl. water conf.* IWC-80-25.

MUSGROVE, M., and J. L. BANNER. 1993. Regional groundwater mixing and the origin of saline fluids: Midcontinent, United States. *Science* 259:1877–82.

MUTO, T. 1965. Thermochemical stability of ningyoite. *Min. Jour.* 4:245–74.

MUTO, T., S. HIRONO, and H. KURATA. 1968. Some aspects of fixation of uranium from natural waters. Japan Atomic Energy Research Inst. Report NSJ Transl. No. 91. From *Mining Geol.* (Japan) 15:287–98, 1965.

NAGY, K. L., A. E. BLUM, and A. C. LASAGA. 1991. Dissolution and precipitation kinetics of kaolinite at 80°C and pH 3: The dependence on solution saturation state. *Am. J. Sci.* 291(7):649–86.

NAKICENOVIC, N. 1993. Energy gases—The methane age and beyond. In *The future of energy gases,* ed D. G. Howell, pp. 661–75. U.S. Geol. Survey Prof. Paper 1570.

NANCOLLAS, G. H. 1966. *Interactions in electrolyte solutions.* New York: Elsevier Publ. Co.

NASH, J. T., H. C. GRANGER, and S. S. ADAMS. 1981. Geology and concepts of genesis of important types of uranium deposits. *Econ. Geol.* (75th ann. vol): 63–116.

NATIONAL ATMOSPHERIC DEPOSITION PROGRAM (NRSP-3)/ NATIONAL TRENDS NETWORK. 1996. NADP/NTN Coordination Office, Natl. Resource Ecology Laboratory, Colorado State Univ., Fort Collins, CO.

NAUMOV, G. B., B. N. RYZHENKO, and I. L. KHODAKOVSKY. 1974. *Handbook of thermodynamic data.* PB-226 722. Springfield, VA: Natl. Tech. Info. Serv., U.S. Dept. Commerce.

NAZAROFF, W. W., and A. V. NERO, JR. 1988. *Radon and its decay products in indoor air.* New York: John Wiley & Sons.

NEAL, C., and G. STANGER. 1985. Past and present serpentinization of ultramafic rocks; An example from the Semail Ophiolite Nappe of northern Oman. In *The chemistry of weathering,* ed J. I. Drever. NATO ASI ser. C, Mathematical and Physical Sciences 149, pp. 249–75. Boston: D. Reidel Publ. Co.

NESBITT, H. W., and R. E. WILSON. 1992. Recent chemical weathering of basalts. *Am. J. Sci.* 292: 740–77.

NGUYEN, S. N., R. J. SILVA, H. C. WEED, and J. E. ANDREWS, JR. 1991. Standard Gibbs free energies of formation at the temperature 303.15 K of four uranyl silicates: soddyite, uranophane, sodium boltwoodite and sodium weeksite. *J. Chem. Thermo.* 24:359–76.

NICHOLSON, R. V., R. W. GILLHAM, and E. J. REARDON. 1988. Pyrite oxidation in carbonate-buffered solution: I. Experimental kinetics. *Geochim. Cosmochim. Acta* 52:1077–85.

NIEBOER, E., and W. A. E. MCBRYDE. 1973. Free energy relationships in coordination chemistry. III. A comprehensive index to complex stability. *Can. J. Chem.* 51:2512–24.

NIENHUIS, P., T. APPELO, and G. WILLEMSEN. 1994. *User guide PHREEQM-2D: A multi-component mass transport model.* Ver. 2.01 (5/94). IT Technology, Amhem, The Netherlands.

NIGHTINGALE, E. R., JR. 1958. Poised oxidation-reduction systems. *Anal. Chem.* 30:267–72.

NITSCHE, H. 1991a. Solubility studies of transuranium elements for nuclear waste disposal: Principles and overview. *Radiochim. Acta* 52/53:3–8.

NITSCHE, H. 1991b. Basic research for assessment of geologic nuclear waste repositories: What solubility and speciation studies of transuranium elements can tell us. *Mat. Res. Soc. symp. proc.* 212, pp. 517–28.

NITSCHE, H. et al. 1993. *Measured solubilities and speciations of Np, Pu, and Am in a typical groundwater (J-13) from the Yucca Mountain Region.* Report LA-12562-MS. Los Alamos, NM: Los Alamos Natl. Lab.

NITSCHE, H. et al. 1995. *Solubility and speciation results from over- and undersaturation experiments on Np, Pu, and Am in water from Yucca Mountain Region Well UE-25p#1.* Report LA-13017-MS. Los Alamos, NM: Los Alamos Natl. Lab.

NITSCHE, H., S. C. LEE, and R. C. GATTI. 1987. Determination of Pu oxidation states at trace levels pertinent to nuclear waste disposal. Lawrence Berkeley Laboratory Report LBL-23158.

NOORISHAD, J. C. L. CARNAHAN, and L. V. BENSON. 1987. *CHMTRNS: A temperature-dependent non-equilibrium reactive transport code.* Earth Sciences Div., Lawrence Berkeley Laboratory.

NORD, G. L. 1977. Characteristics of fine-grained black uranium ores by transmission electron microscopy. *Short Papers of the U.S. Geol. Survey, uranium-thorium symp.,* ed J. A. Campbell. *U.S. Geol. Survey Circ.* 753:29–31.

NORDSTROM, D. K. 1982. The effect of sulfate on aluminum concentrations in natural waters: Some stability relations in the system Al_2O_3-SO_3-H_2O at 298 K. *Geochim. Cosmochim. Acta* 46:681–92.

NORDSTROM, D. K. 1985. *The rate of ferrous iron oxidation in a stream receiving acid mine effluent. Selected papers in the hydrologic sciences.* U.S. Geol. Survey Water Supply Paper 2270, pp. 113–119.

NORDSTROM, D. K., and J. W. BALL. 1984. Chemical models, computer programs and metal complexation in natural waters.In *Complexation of trace metals in natural waters,* pp. 149–64. The Hague: Dr. W. Junk Publ.

NORDSTROM, D. K., and J. W. BALL. 1986. The geochemical behavior of aluminum in acidified surface waters. *Science* 232:54–56.

NORDSTROM, D. K., E. A. JENNE, and J. W. BALL. 1979. Redox equilibria of iron in acid mine waters. In *Chemical modeling in aqueous systems,* ed E. A. Jenne. Am. Chem. Soc. Symp. Ser. 93, pp. 51–79. Washington, DC: Am. Chem. Soc.

NORDSTROM, D. K., and H. M. MAY, 1989. Aqueous equilibrium data for mononuclear aluminum species. In *The environmental chemistry of aluminum,* ed G. Sposito, pp. 29–53. Boca Raton, FL: CRC Press.

NORDSTROM, D. K., and J. L. MUNOZ. 1985. *Geochemical thermodynamics.* Menlo Park, CA: The Benjamin/Cummings Publ. Co., Inc.

NORDSTROM, D. K., and J. L. MUNOZ. 1994. *Geochemical thermodynamics.* 2d ed. Cambridge, MA: Blackwell Scientific Publications, Inc.

NORDSTROM, D. K., T. OLSSON, L. CARLSSON, and P. FRITZ. 1989. Introduction to the hydrogeochemical investigations within the International Stripa Project. *Geochim. Cosmochim. Acta* 53:1717–26.

NORDSTROM, D. K., L. N. PLUMMER, D. LANGMUIR, E. BUSENBERG, and H. M. MAY. 1990. Revised chemical equilibrium data for major water-mineral reactions and their limitations. In *Chemical modeling of aqueous systems II,* ed D. C. Melchior and R. L. Bassett. Am. Chem. Soc. Ser. 416, pp. 398–413. Washington, DC: Am. Chem. Soc.

NORDSTROM, D. K., L. N. PLUMMER, T. M. L. WIGLEY, T. J. WOLERY, and J. W. BALL. 1979. A comparison of computerized chemical models for equilibrium calculations in aqueous systems. In *Chemical modeling in aqueous systems,* ed E. A. Jenne. Am. Chem. Soc. Symp. Ser. 93, pp. 857–92. Washington, DC: Am. Chem. Soc.

NORDSTROM, D. K., J. A. T. SMELLIE, and M. WOLF. 1990. *Chemical and isotopic composition of groundwaters and their seasonal variability at the Osamu Utsumi and Morro do Ferro Analogue Study Sites, Pocos de Caldas, Brazil.* Pocos de Caldas Report No. 8. NAGRA NTB 90-24, SKB TR 90-15.

NOTZ, K. J. 1990. *Characteristics data base, systems integration program.* Oak Ridge, TN: Oak Ridge Natl. Lab.

NOVAK, C. F., and K. E. ROBERTS. 1995. Thermodynamic modeling of Np(V) solubility in concentrated Na-CO$_3$-HCO$_3$-Cl-ClO$_4$-H-H$_2$O systems. *Mat. Res. Soc. symp. proc.* 353, pp. 1119–28.

NRC. 1978. *Final environmental impact statement related to the Wyoming Mineral Corp. Irigaray Uranium Solution Mining Project (Wyoming, U.S.A.).* Report NUREG-0481. U.S. Nuclear Regulatory Commission.

NRIAGU, O. J. 1972. Stability of vivianite and ion-pair formation in the system Fe$_3$(PO$_4$)$_2$-H$_3$PO$_4$-H$_2$O. *Geochim. Cosmochim. Acta* 36:459–70.

NRIAGU, O. J. 1996. A history of global metal pollution. *Science* 272:223–24.

NWTRB. 1996. *Disposal and storage of spent nuclear fuel—Finding the right balance.* A Report to Congress and the Secretary of Energy. Arlington, VA: Nuclear Waste Tech. Review Board.

O'BRIEN, D. J., and F. B. BIRKNER. 1977. Kinetics of oxygenation of reduced sulfur species in aqueous solution. *Envir. Science & Technol.* 11(12):1114–20.

O'CONNOR, J. J., and C. E. RENN. 1964. Soluble-adsorbed zinc equilibrium in natural waters. *J. Am. Water Works Assoc.* 56:1055–61.

OECD. 1993. *The status of near-field modeling. Proc. technical workshop, Cadarache, France.* Paris: Organisation for Econ. Co-operation & Devel.

ODOM, J. M., and SINGLETON, R., JR., eds. 1993. *The sulfate reducing bacteria: Contemporary perspectives.* New York: Springer-Verlag.

OGARD, A. E., and J. F. KERRISK. 1984. *Groundwater chemistry along the flow path between a proposed repository site and the accessible environment.* Report LA-10188-MS. Los Alamos, NM: Los Alamos Natl. Lab.

OGINO, T., T. SUZUKI, and K. SAWADA. 1987. The formation and transformation mechanism of calcium carbonate in water. *Geochim. Cosmochim. Acta* 51:2757–67.

OTA. 1985. *Managing the nation's commercial high-level radioactive waste.* Office of Technol. Assessment, Congress of the United States. UNIPUB, InfoSource Intl.

OXBURGH, R., J. I. DREVER, and Y.-T. SUN. 1994. Mechanism of plagioclase dissolution at 25°C. *Geochim. Cosmochim. Acta* 58:661–69.

OZSVATH, D. 1979. Modeling heavy metal sorption from subsurface waters with the n-power exchange function. M.S. thesis, The Pennsylvania State Univ., University Park, PA.

PACES, T. 1969. Chemical equilibria and zoning in subsurface water from the Jachymov ore deposit, Czechoslovakia. *Geochim. Cosmochim. Acta* 33:591–609.

PACES, T. 1978. Reversible control of aqueous aluminum and silica during the irreversible evolution of natural waters. *Geochim. Cosmochim. Acta* 42:1487–93.

PACES, T. 1983. Rate constants of dissolution derived from the measurements of mass balance in hydrological catchments. *Geochim. Cosmochim. Acta* 47:1855–63.

PACK, D. H. 1980. Precipitation chemistry patterns: A two-network data set. *Science* 208:1143–45.

PACQUET, A., G. CAPUS, R. MAGNE, and A. OBELLIANNE. 1987. Formation processes of uranium deposits in Tertiary sediments of the Limagnes, Central Massif, France. *Uranium* 3:261–83.

PAGENKOPF, G. K. 1978. *Introduction to natural water chemistry.* New York: Marcel Dekker, Inc.

PAIGE, C. R., W. A. KORNICKER, O. E. HILEMAN, JR., and W. J. SNODGRASS. 1992. Modeling solution equilibria for uranium ore processing: The PbSO$_4$-H$_2$SO$_4$-H$_2$O and PbSO$_4$-Na$_2$SO$_4$-H$_2$O systems. *Geochim. Cosmochim. Acta* 56:1165–73.

PALACHE, C., H. BERMAN, and C. FRONDEL. 1944. *The system of mineralogy.* Vol I. 7th ed. New York: John Wiley & Sons.

PALACHE, C., H. BERMAN, and C. FRONDEL. 1951. *The system of mineralogy.* Vol II. 7th ed. New York: John Wiley & Sons.

PALMER, C. E. A., R. J. SILVA, and C. W. MILLER. 1992. Speciation calculations of Pu, Np, Am, and U in J-13 well water. Effects of anions concentration and pH. Unpublished ms, Lawrence Livermore Laboratory.

PALMER, D. A., and K. E. HYDE. 1993. An experimental determination of ferrous chloride and acetate complexation in aqueous solutions to 300°C. *Geochim. Cosmochim. Acta* 57(7):1393–408.

PANKOW, J. F. 1991. *Aquatic chemistry concepts.* Chelsea, MI: Lewis Publ. Inc.

PANKOW, J. F. 1992. *Aquatic chemistry problems.* Portland, OR: Titan Press.

PANKOW, J. F., and J. J. MORGAN. 1981. Kinetics for the aquatic environment. *Envir. Science & Technol.* 15(10):1155–64.

PARIZEK, R. R., W. B. WHITE, and D. LANGMUIR. 1971. Hydrogeology and geochemistry of folded and faulted carbonate rocks of central Appalachian type and related land use problems. In *Field trip guidebook.* Washington, DC: Geol. Soc. Am.

PARKHURST, D. L. 1990. Ion-association models and mean activity coefficients of various salts. In *Chemical modeling of aqueous systems II*, ed D. C. Melchior and R. L. Bassett, Am. Chem. Soc. Symp. Ser. 416, pp. 30–43. Washington DC: Am. Chem. Soc.

PARKHURST, D. L. 1995. *Users guide to PHREEQC—A computer program for speciation, reaction-path, advective-transport, and inverse geochemical calculations.* U.S. Geol. Survey Water Resources Inv. Rept. 95-4227.

PARKHURST, D. L., L. N. PLUMMER, and D. C. THROSTENSON. 1982. *BALANCE—A computer program for calculating mass transfer for geochemical reactions in ground water.* U.S. Geol. Survey Water Resources Inv. Rept. 82-14.

PARKHURST, D. L., D. C. THORSTENSON, and L. N. PLUMMER. 1990. *PHREEQE—A computer program for geochemical calculations.* Rev. U.S. Geol. Survey Water Resources Inv. Rept. 80-96.

PARKS, G. A. 1965. The isoelectric points of solid oxides, solid hydroxides, and aqueous hydroxo complex systems. *Chem. Rev.* 65:177–98.

PARKS, G. A., and D. C. POHL. 1988. Hydrothermal solubility of uraninite. *Geochim. Cosmochim. Acta* 52:863–75.

PARSONS, R. 1985. Standard electrode potential, units, conventions and methods of determination. In *Standard po-

tentials in aqueous solutions IUPAC, ed A. J. Bard, R. Parsons, and J. Jordan. New York: Marcel Dekker.

PATEL, S. B. 1991. *Nuclear physics: An introduction.* New York: John Wiley & Sons.

PATTERSON, C. G., and D. D. RUNNELLS. 1992. Dissolved gases in ground water as indicators of redox conditions. *Proc. 7th intl. symp. water-rock interaction,* ed Y. K. Kharaka and A. S. Maest, 1, pp. 517–20, Rotterdam: A. A. Balkema.

PAULING, L. 1960. *The nature of the chemical bond.* 3d ed. Ithaca, NY: Cornell Univ. Press.

PAVLIK, H. F., F. G. BAKER, and D. D. RUNNELLS. 1992. Use of batch equilibration data in the retardation equation. In *Proc. 7th intl. symp. water-rock interaction*, ed Y. K. Kharaka and A. S. Maest, 1, pp. 59–62, Rotterdam: A. A. Balkema.

PEARCY, E. C., J. D. PRIKRYL, W. M. MURPHY, and B. W. LESLIE. 1994. Alteration of uraninite from the Nopal I deposit, Pena Blanca District, Chihuahua, Mexico, compared to degradation of spent nuclear fuel in the proposed U.S. high-level nuclear waste repository at Yucca Mountain, Nevada. *Applied Geochem.* 9:713–32.

PEARSON, F. J., D. W. FISHER, and L. N. PLUMMER. 1978. Correction of ground-water chemistry and carbon isotopic composition for effects of CO_2 outgassing. *Geochim. Cosmochim. Acta* 42:1799–1807.

PEARSON, F. J., J. L. LOLCAMA, and A. SCHOLTIS. 1989. *Chemistry of waters in the Bottstein, Weiach, Riniken, Schafisheim, Kaisten and Leuggern Boreholes: A hydrochemically consistent data set.* NAGRA Tech. Report 86-19.

PEARSON, F. J., and A. SCHOLTIS. 1992. *Reference waters for the crystalline basement of Northern Switzerland.* NAGRA Report NIB 92-59.

PEARSON, F. J. et al. 1991. *Applied isotope hydrology. A case study in northern Switzerland.* Studies in Envir. Science 43. Amsterdam: Elsevier Publ. Co.

PEARSON, R. G. 1968. Hard and soft acids and bases, I. *J. Chem. Educ.* 45:581–87.

PEARSON, R. G. 1973. *Hard and soft acids and bases.* Stroudsburg, PA: Dowden, Hutchinson & Ross.

PENROSE, W. R., W. L. POLZER, E. H. ESSINGTON, D. M. NELSON, and K. A. ORLANDINI. 1990. Mobility of plutonium and americium through a shallow aquifer in a semiarid region. *Envir. Sci. & Technol.* 24:228–34.

PETERSON, S. R., A. R. FELMY, R. J. SERNE, and G. W. GEE. 1983. *Predictive geochemical modeling of interactions between uranium mill tailings solutions and sediments in flow-through system.* NUREG/CR-3404, PNL-4782. U.S. Nuclear Regulatory Commission.

PHILLIPS C. S. G., and R. J. P. WILLIAMS. 1965. *Inorganic chemistry, 1. Principles and non-metals.* New York: Oxford Univ. Press.

PHILLIPS, C. S. H., and R. J. P. WILLIAMS. 1966. *Inorganic chemistry, 2. Metals.* New York: Oxford Univ. Press.

PITZER, K. S. 1987. Thermodynamic model for aqueous solutions of liquid-like density. In *Thermodynamic modeling of geological materials: Minerals, fluids and melts,* ed I. S. E. Carmichael and H. P. Eugster, Reviews in Mineralogy 17, pp. 97–142. Min. Soc. Am.

PLUMMER, L. N. 1975. Mixing of sea water with calcium carbonate water. *Geol. Soc. Am. Memoir* 142:219–36.

PLUMMER, L. N. 1977. Defining reactions and mass transfer in part of the Floridan Aquifer. *Water Resources Res.* 13:801–12.

PLUMMER, L. N., and E. BUSENBERG, 1982. The solubilities of calcite, aragonite and vaterite in CO_2 solutions between 0 and 90°C, and an evaluation of the aqueous model for the system $CaCO_3$-CO_2-H_2O. *Geochim. Cosmochim. Acta* 46(6):1011–40.

PLUMMER, L., B. F. JONES, and A. H. TRUESDELL. 1984. *WATEQF—A FORTRAN IV version of WATEQ, a computer program for calculating chemical equilibrium of natural waters.* Rev. U.S. Geol. Survey Water Resources Inv. Rept. 76-13. Reston, VA: U.S. Geol. Survey.

PLUMMER, L. N., D. L. PARKHURST, G. W. FLEMING, and S. A. DUNKLE. 1988. *A computer program (PHRQPITZ) incorporating Pitzer's equations for calculation of geochemical reactions in brines.* U.S. Geol. Survey Water Resources Inv. Rept. 88-4153. Reston, VA: U.S. Geol. Survey.

PLUMMER, L. N., D. L. PARKHURST, and D. C. THORSTENSON. 1983. Development of reaction models for ground-water systems. *Geochim. Cosmochim. Acta* 47:665–86.

PLUMMER, L. N., D. L. PARKHURST, and T. M. L. WIGLEY. 1979. Critical review of the kinetics of calcite dissolution and precipitation. In *Chemical modeling in aqueous systems,* ed E. A. Jenne. Am. Chem. Soc. Symp. 93, pp. 537–73. Washington, DC: Am. Chem. Soc.

PLUMMER, L. N., E. C. PRESTEMON, and D. L. PARKHURST. 1991. *An interactive code (NETPATH) for modeling net geochemical reactions along a flow path.* U.S. Geol. Survey Water Resources Inv. Rept. 91-4078. Reston, VA: U.S. Geol. Survey.

PLUMMER, L. N., E. C. PRESTEMON, and D. L. PARKHURST. 1994. *An interactive code (NETPATH) for modeling net geochemical reactions along a flow path ,ver 2.0.* U.S. Geol. Survey Water Resources Inv. Rept. 94-4169, Reston, VA: U.S. Geol. Survey.

PLUMMER, L. N., and T. M. L. WIGLEY. 1976. The dissolution of calcite in CO_2-saturated solutions at 25°C and 1 atmosphere total pressure. *Geochim. Cosmochim. Acta* 40:191–202.

PLYASUNOV, A. V., and I. GRENTHE. 1994. The temperature dependence of stability constants for the formation of polynuclear cationic species. *Geochim. Cosmochim. Acta* 58(17):3561–82.

POKROVSKII, V. A., and H. C. HELGESON. 1995. Thermodynamic properties of aqueous species and the solubilities of minerals at high pressures and temperatures: The system Al_2O_3-H_2O-NaCl. *Am. J. Sci.* 295(10):1255–342.

POLEMO, M., S. BUFO, and S. PAOLETTI. 1980. Evaluation of ionic strength and salinity of groundwaters: Effect of the ionic composition. *Geochim. Cosmochim. Acta* 44(6):809–14.

PORTER, R. A., and W. J. WEBER, JR. 1971. The interaction of silicic acid with iron (III) and uranyl ions in dilute aqueous solution. *J. Inorg. Nucl. Chem.* 33:2443–49.

POURBAIX, M. 1966. *Atlas of electrochemical equilibria in aqueous solutions.* New York: Pergamon Press.

PUIGDOMENECH, I., and J. BRUNO. 1991. *Plutonium solubilities.* Swedich Nuclear Fuel and Waste Mgmt. Co., SKB Tech. Rept. 91-04.

PYTKOWICZ, R. M., ed. 1979. *Activity coefficients in electrolyte solution,* vols. 1 and 2. Boca Raton, FL: CRC Press, Inc.

PYTKOWICZ, R. M. 1983. *Equilibria, nonequilibria, & natural waters,* vols. 1 and 2. New York: John Wiley & Sons.

PYTKOWICZ, R. M., and J. E. HAWLEY. 1974. *Limnol. Oceanogr.* 19:223.

PYZIK, A. J., and S. E. SOMMER. 1981. Sedimentary iron monosulfides: Kinetics and mechanisms of formation. *Geochim. Cosmochim. Acta* 45:687–98.

RAAB, G. M., et al. 1993. The influence of pH and household plumbing on water lead concentration. *Envir. Geochem. & Health* 15(4):191–200.

RAI, D., et al. 1996. The solubility of Th(IV) and U(IV) hydrous oxides in concentrated NaCl and MgCl$_2$ solutions. *Fifth intl. conf. on the chemistry and migration behavior of actinides and fission products in the geosphere, 1995.* (In press.)

RAI, D., A. R. FELMY, and J. L. RYAN. 1990. Uranium(IV) hydrolysis constants and solubility product of UO$_2$ · xH$_2$O(am). *Inorg. Chem.* 29:260–64.

RAI, D., R. J. SERNE, and D. A. MOORE. 1980. Solubility of plutonium compounds and their behavior in soils. *Soil Sci. Soc. Am. J.* 44(3):490–95.

RAI, D., J. L. SWANSON, and J. L. RYAN. 1987. Solubility of hydrous NpO$_2$(am) in the presence of Cu(I)/Cu(II) redox buffers. *Radiochim. Acta* 42:35–41.

RAISWELL, R. W., P. BRIMBLECOMBE, D. L. DENT, and P. S. LISS. 1980. *Environmental chemistry, the earth-air-water-factory.* Resource & Envir. Sci. Ser. New York: John Wiley & Sons.

RANSOM, B., and H. C. HELGESON. 1993. Compositional end members and thermodynamic components of illite and dioctahedral aluminous smectite solid solutions. *Clays Clay Minerals* 41(5):537–50.

RANVILLE, R. J. 1988. The application of an ion-interaction model to solubilities of some alkaline-earth carbonates, ThO$_2$ and UO$_2$ in brines. MS thesis, T-3232, Colorado School of Mines, Golden, CO.

RARD, J. A. 1983. *Critical review of the chemistry and thermodynamics of technetium and some of its inorganic compounds and aqueous species.* Lawrence Livermore Laboratory, Report UCRL-53440.

RARD, J. A. 1990. *Chemical thermodynamics of technetium (III).* Lawrence Livermore Laboratory, Report UCRL-103728.

REARDON, E. J. 1975. Dissolution constants of some monovalent sulfate ion pairs at 25°C from stoichiometric activity coefficients. *J. Phys. Chem.* 79:422.

REARDON, E. J. 1981. K_d's—Can they be used to describe reversible ion sorption reactions in contaminant transport? *Ground Water* 19:279–86.

REARDON, E. J., and R. D. BECKIE. 1987. Modeling chemical equilibria of acid mine-drainage: The FeSO$_4$-H$_2$SO$_4$-H$_2$O system. *Geochim. Cosmochim. Acta* 51:2355–68.

RECHARD, R. P., ed. 1995. *Performance assessment of the direct disposal in unsaturated tuff of spent nuclear fuel and high-level waste owned by U.S. Department of Energy,* 1. *Executive summary.* Sandia National Laboratories Report SAND94-2563/1.

REEDER, R. J., ed. 1983. *Carbonates: Mineralogy and chemistry.* Reviews in Mineralogy 11. Min. Soc. Am.

REID, C. E. 1990. *Chemical thermodynamics.* New York: McGraw Hill.

RETALLACK, G. J. 1990. *Soils of the past.* London: Harper-Collins.

RHEINHEIMER, G. 1981. *Aquatic microbiology.* New York: Wiley-Interscience.

RICHET, P., K. BOTTINGA, L. DENIELOU, J. P. PETITET, and C. TEQUI. 1982. Thermodynamic properties of quartz, cristobalite and amorphous SiO$_2$: Drop calorimetry measurements between 1000 and 1800 K and a review from 0 to 2000 K. *Geochim. Cosmochim. Acta* 46(12):2639–58.

RIESE, A. C. 1982. Adsorption of radium and thorium onto quartz and kaolinite: A comparison of solution/surface equilibria models. PhD thesis, T-2625, Colorado School of Mines, Golden, CO.

RIMSTIDT, J. D., and H. L. BARNES. 1980. The kinetics of silica water reactions. *Geochim. Cosmochim. Acta* 44:1683–99.

RINGBOM, A. 1963. *Complexation in analytical chemistry.* New York: John Wiley & Sons.

ROBERSON, C. E., and R. B. BARNES. 1978. Stability of fluoride complex with silica and its distribution in natural water systems. *Chem. Geol.* 21:239–56.

ROBIE, R. A., H. T. HASELTON, JR., and B. S. HEMINGWAY. 1984. Heat capacities and entropies of rhodocrosite (MnCO$_3$) and siderite (FeCO$_3$) between 5 and 600 K. *Am. Mineral.* 69:349–57.

ROBIE, R. A., and B. S. HEMINGWAY. 1973. The enthalpies of formation of nesquehonite and hydromagnesite. *U.S. Geol. Survey J. Research* 1(5):543–47.

ROBIE, R. A., B. S. HEMINGWAY, and J. R. FISHER. 1978. *Thermodynamic properties of minerals and related substances at 298.15 K and 1 bar (10^5 pascal) pressure and at higher temperatures.* U.S. Geol. Survey Bull. 1452.

ROBIE, R. A., B. S. HEMINGWAY, and W. H. WILSON. 1976. The heat capacities of Calorimetry Conference copper and of muscovite KAl$_2$(AlSi$_3$)O$_{10}$(OH)$_2$, pyrophyllite Al$_2$Si$_4$O$_{10}$(OH)$_2$, and illite K$_3$(Al$_7$Mg)(Si$_{14}$Al$_2$)O$_{40}$(OH)$_8$ between 15 and 375 K and their standard entropies at 298.15 K. *U.S. Geol. Survey J. Research,* 4:631–44.

ROBINSON, G. R., and J. L. HAAS. 1983. Heat capacity, relative enthalpy, and calorimetric entropy of silicate minerals: an empirical method of prediction. *Am. Mineral.* 68:541–53.

ROBINSON, R. A., and R. H. STOKES. 1970. *Electrolyte solutions.* 2d ed. rev. London: Butterworths.

ROCK, P. A., W. H. CASEY, M. K. MCBEATH, and E. M. WALLING. 1994. A new method for determining Gibbs energies of formation of metal-carbonate solid solutions: 1. The Ca$_x$Cd$_{1-x}$CO$_3$(s) system at 298 K and 1 bar. *Geochim. Cosmochim. Acta* 58(20):4281–91.

ROGERS, J. J. W., and J. A. S. ADAMS. 1970. Geochemistry of uranium. In *Handbook of geochemistry,* ed K. H. Wedephol, vol. II-2, pp. 92E1–92E5. Berlin: Springer-Verlag.

ROSE, A. W., H. E. HAWKES, and J. S. WEBB. 1979. *Geochemistry in mineral exploration.* 2d ed. New York: Academic Press.

ROSENBERG, P. E., J. A. KITTRICK, and S. U. AJA. 1990. Mixed-layer illite/smectite: A multiphase model. *Am. Mineral.* 75:1182–85.

RÖSLER, H., and H. LANGE. 1972. *Geochemical tables.* New York: Elsevier Publ. Co.

ROUSH, W. 1995. Building a wall against toxic waste. *Chemical & Engineering News* (July 28):473.

ROUTSON, R. C., and J. A. KITTRICK. 1971. Illite solubility. *Soil. Sci. Soc. Am.* 35(5):714–18.

ROY, R. N. et al. 1983. The first ionization of carbonic acid in aqueous solutions of potassium chloride including the activity coefficients of potassium bicarbonate. *J. Chem. Thermo.* 15:37–47.

RUBIN, A. J., and D. L. MERCER. 1981. Adsorption of free and complexed metals from solution by activated carbon. In *Adsorption of inorganics at solid-liquid interfaces*, ed M. A. Anderson and A. J. Rubin, pp. 295–325. Ann Arbor, MI: Ann Arbor Science.

RUNNELLS, D. D., T. A. SHEPHERD, and E. E. ANGINO. 1992. Metals in water. *Envir. Sci. & Technol.* 26(12):2316–22.

RYAN, J. L., and D. RAI. 1983. The solubility of uranium(IV) hydrous oxide in sodium hydroxide solutions under reducing conditions. *Polyhedron* 2(9):947–52.

SANCHEZ, A. L., J. W. MURRAY, and T. H. SIBLEY. 1985. The adsorption of plutonium IV and V on goethite. *Geochim. Cosmochim. Acta* 49(11):2297–308.

SATO, M., and H. M. MOONEY. 1960. The electrochemical mechanism of sulfide self-potentials. *Geophysics* 25(1):226–49.

SAVAGE, D., ed. 1995. *The scientific and regulatory basis for the geological disposal of radioactive waste.* New York: John Wiley & Sons.

SCHECHER, W. D., and C. T. DRISCOLL. 1987. An evaluation of uncertainty associated with aluminum equilibrium calculations. *Water Resources Research* 23(4):525–34.

SCHECHER, W. D., and D. C. MCAVOY. 1991. *MINEQL⁺: A chemical equilibrium program for personal computers. User's manual ver. 2.1.* Edgewater, MD: Envir. Research Software.

SCHINDLER, D. W. 1988. Effects of acid rain on freshwater ecosystems. *Science* 239:149–57.

SCHINDLER, P. W. 1967. Heterogeneous equilibria involving oxides, hydroxides, carbonates, and hydroxide carbonates. In *Equilibrium concepts in natural water systems.* Am. Chem. Soc. Adv. Chem. Ser. 67, pp. 196–221. Washington, DC: Am. Chem. Soc.

SCHINDLER, P. W., B. FURST, R. DICK, and P. U. WOLF. 1976. Ligand properties of surface silanol groups. I. Surface complex formation with Fe^{3+}, Cu^{2+}, Cd^{2+}, and Pb^{2+}. *J. Colloid Interface Sci.* 55:469–75.

SCHINDLER, P. W., and W. STUMM. 1987. The surface chemistry of oxides, hydroxides, and oxide minerals. In *Aquatic surface chemistry,* ed W. Stumm, pp. 83–110. New York: Wiley-Interscience.

SCHLAUTMAN, M. A., and J. J. MORGAN. 1994. Adsorption of aquatic humic substances on colloidal-sized aluminum

oxide particles: Influence of solution chemistry. *Geochim. Cosmochim. Acta* 58:4293–303.

SCHMIDT-COLLERUS, J. J. 1967. *Research in uranium geochemistry. Investigations of the relationship between organic matter and uranium deposits—I.* (USEAC Contract No. AT (05-1)-933). Denver Research Inst.

SCHMIERMUND, R. L. 1991. Steady-state weathering of limestone and marble by acidic precipitation—a quantitative laboratory simulation. Ph.D. thesis, T-3963, Colorado School of Mines, Golden, CO.

SCHNITZER, M. 1971. Metal-organic matter interactions in soils and waters. In *Organic compounds in aquatic environments*, ed S. D. Faust and J. V. Hunter, pp. 297–315. New York: Marcel Dekker.

SCHNITZER, M., and S. U. KAHN, eds. 1978. *Soil organic matter.* Developments in soil science 8. New York: Elsevier Sci. Publ. Co.

SCHOONEN, M. A. A. 1994. Calculation of the point of zero charge of metal oxides between 0 and 350°C. *Geochim. Cosmochim. Acta* 58:2845–51.

SCHOONEN, M. A. A., and H. L. BARNES. 1988. An approximation of the second dissociation constant for H_2S. *Geochim. Cosmochim. Acta* 52(3):649–54.

SCHOONEN, M. A. A., and H. L. BARNES. 1991a. Reactions forming pyrite and marcasite from solution: II. Via FeS precursors below 100°C. *Geochim. Cosmochim. Acta* 55:1505–14.

SCHOONEN, M. A. A., and H. L. BARNES. 1991b. Mechanisms of pyrite and marcasite formation from solution: III. Hydrothermal processes. *Geochim. Cosmochim. Acta* 55:3491–504.

SCHNOOR, J. L. 1990. Kinetics of chemical weathering: A comparison of laboratory and field weathering rates. In *Aquatic chemical kinetics*, ed W. Stumm, pp. 475–504. New York: John Wiley & Sons.

SCHWARZENBACH, G. 1961. The general, selective and specific formation of complexes by metallic cations. *Adv. Inorg. Radiochem.* 3:257, 265–70.

SCHWARZENBACH, R. P., P. M. GSCHWEND, and D. M. IMBODEN. 1993. *Environmental organic chemistry.* New York: Wiley-Interscience.

SCHWERTMANN, U., and R. M. TAYLOR. 1977. Iron oxides. In *Minerals in soil environments*, ed J. B. Dixon and S. B. Weed, pp. 379–438. Soil Sci. Soc. Am.

SCOTT, K. M. 1987. Solid solution in, and classification of, gossan-derived members of the alunite-jarosite family, northwest Queensland, Australia. *Am. Mineral.* 72:178–87.

SCOTT, M. J., and J. J. MORGAN. 1990. Energetics and conservative properties of redox systems. In *Chemical modeling of aqueous systems II*, ed D. C. Melchior and R. L. Bassett. Am. Chem. Soc. Symp. Ser 416, pp. 368–78. Washington, DC: Am. Chem. Soc.

SEABORG, G. T. 1994. Origin of the actinide concept. In *Handbook on the physics and chemistry of rare earths.* Vol. 18, *Lanthanides/actinides chemistry*, ed K. A. Gschneider, L. Eyring, G. R. Choppin, and G. H. Lander, pp. 1–27. New York: Elsevier Science.

SEABORG, G. T., and D. E. HOBART. 1996. Summary of the properties of lanthanide and actinide elements. In

IANCAS' frontiers in nuclear chemistry commemorating the centennial of Becquerel's discovery of radioactivity. Berkeley, CA: Lawrence Berkeley Laboratory. (In press.)

SEARS, S. O. 1976. Inorganic and isotopic geochemistry of the unsaturated zone in a carbonate terrane. Ph.D. thesis, The Pennsylvania State Univ., University Park, PA.

SEARS, S. O., and D. LANGMUIR. 1982. Sorption and clay-mineral equilibria controls on moisture chemistry in a C-horizon soil. *J. Hydrol.* 56:287–308.

SEMKOW, T. M. 1990. Recoil-emanation theory applied to radon release from mineral grains. *Geochim. Cosmochim. Acta* 54:425–40.

SEMKOW, T. M., and P. P. PAREKH. 1990. The role of radium distribution and porosity in radon emanation from solids. *Geophys. Res. Letters* 17(6):837–40.

SHANNON, R. D. 1976. Revised effective ionic radii and systematic studies of interatomic distances in halides and chalcogenides. *Acta Cryst.* A32:751–67.

SHANNON, R. D., and C. T. PREWITT. 1969. Effective ionic radii in oxides and fluorides. *Acta Cryst.* B25:925–46.

SHARON STEEL/MIDLALE (UT) PROJECT. 1990. Groundwater/geochemistry RI addendum. EPA ARCS Region VI, VII, VIII Contract No. 68-W9-0021.

SHAW, R. W., T. B. BRILL, A. A. CLIFFORD, C. A. ECKERT, and E. U. FRANCK. 1991. Supercritical water, a medium for chemistry. Special report. *Chem. & Eng. News* (Dec. 23):26–39.

SHOCK, E. L., and H. C. HELGESON. 1988. Calculation of the thermodynamic and transport properties of aqueous species at high pressures and temperatures: Correlation algorithms for ionic species and equation of state predictions to 5 kb and 1000°C. *Geochim. Cosmochim. Acta* 52(8):2009–36.

SHUMATE, K. S., and E. E. SMITH. 1968. Development of a natural laboratory for the study of acid mine drainage production. *Proc. 2d symp. coal mine drainage research*, pp. 223–45. Monroeville, PA: Coal Research, Inc.

SIEBERT, R. M., and P. B. HOSTETLER. 1977a. The stability of the magnesium bicarbonate ion pair from 10° to 90°C. *Am. Jour. Sci.* 277:697–715.

SIEBERT, R. M., and P. B. HOSTETLER. 1977b. The stability of the magnesium carbonate ion pair from 10° to 90°C. *Am. Jour. Sci.* 277: 716–34.

SIEGEL, M. D., V. S. TRIPATHI, M. G. RAO, and D. B. WARD. 1992. Development and validation of a multi-site model for adsorption of metals by mixtures of minerals: 1 Overview and preliminary results. *Proc. 7th intl. symp. water rock interaction*, ed Y. K. Kharaka and A. S. Maest, 1, pp. 63–67. Rotterdam: A. A. Balkema.

SIEVER, R. 1962. Silica solubility, 0–200°C, and the diagenesis of siliceous sediments. *J. Geol.* 70(2):127–50.

SIGG, L. 1987. Surface chemical aspects of the distribution and fate of metal ions in lakes. In *Aquatic surface chemistry. Chemical processes at the particle-water interface*, ed. W. Stumm, pp. 319–49 New York: Wiley-Interscience.

SILLEN, L. G., and A. E. MARTELL. 1964. *Stability constants of metal-ion complexes.* Special publ. no. 17. London: The Chemical Society.

SILMAN, J. F. B. 1958. Ph.D. thesis, Harvard Univ., Cambridge, MA.

SILVA, R. J. 1992. Mechanisms for the retardation of uranium(VI) migration. *Mat. Res. Soc. symp. proc.* 257, pp. 323–30.

SILVA, R. J. et al. 1995. *Chemical thermodynamics of americium.* Chem. Thermo. 2. Nuclear Energy Agency, OECD. Amsterdam: Elsevier Publ. Co.

SILVA, R. J., and H. NITSCHE. 1996. Actinide environmental chemistry. *Radiochim. Acta.* (In press.)

SILVA, R. J., and A. W. YEE. 1981. *Uranium (VI) retardation mechanisms.* Earth Sci. Div. Annual Rept. (LBL-13600). Berkeley CA: Lawrence Berkeley Lab.

SJOBERG, E. L., and D. T. RICKARD. 1984. Temperature dependence of calcite dissolution kinetics between 1 and 62°C and pH 2.7 to 8.4 in aqueous solutions. *Geochim. Cosmochim. Acta* 48:485–93.

SKAGIUS, K., and I. NERETNIEKS. 1988. Measurements of cesium and strontium diffusion in biotite gneiss. *Water Resources Research* 24(1):75–84.

SKI. 1991. *SKI Project 90 Summary.* SKI Tech. Rept. 91-23. Stockholm.

SLAUGHTER, M. 1992. Written communication. Dept. of Chemistry and Geochemistry, Colorado School of Mines, Golden, CO.

SMITH, E. E. et. al. 1970. *Sulfide to sulfate reaction mechanism.* Fed. Water Qual. Admin. Water Poll. Control Res. Study #140-FPS-02–70.

SMITH, H. J. 1918. On equilibrium in the system ferrous carbonate, carbon dioxide, and water. *J. Am. Chem. Soc.* 40:879.

SMITH, K. S. 1986. Adsorption of Cu and Pb onto goethite as a function of pH, ionic strength, and metal and total carbonate concentrations. MS thesis, T-3057, Colorado School of Mines, Golden, CO.

SMITH, K. S. 1991. Factors influencing metal sorption onto iron-rich sediment in acid-mine drainage. PhD thesis, Colorado School of Mines, Golden, CO.

SMITH, K. S., J. F. RANVILLE, and D. L. MACALADY. 1991. *Predictive modeling of Cu, Cd and Zn partitioning between stream water and bed sediment from a stream receiving acid mine drainage, St. Kevin Gulch, Colorado.* U.S. Geol.Survey Water Resources Inv. Rept. 91-4034, pp. 380–86.

SMITH, R. A., R. B. ALEXANDER, and M. G. WOLMAN. 1987. Water-quality trends in the nation's rivers. *Science* 235:1607–15.

SMITH, R. M., and A. E. MARTELL. 1976. *Critical stability constants.* 4, *Inorganic complexes.* New York: Plenum Press.

SMITH, R. W., and E. A. JENNE. 1988. *Compilation, evaluation, and prediction of triple-layer model constants for ions on Fe(III) and Mn(IV) hydrous oxides.* Report PNL-6754. Richland, WA: Battelle Pacific Northwest Laboratory.

SMITH, R. W., and E. A. JENNE. 1991. Recalculation, evaluation, and prediction of surface complexation constants for metal adsorption on iron and manganese oxides. *Envir. Sci. & Technol.* 25:525–31.

SMITH, R. W., C. J. POPP, and D. I. NORMAN. 1986. The dissociation of oxy-acids at elevated temperatures. *Geochim. Cosmochim. Acta* 50(1):137–42.

SNOEYINK, V. L., and D. JENKINS. 1980. *Water chemistry.* New York: John Wiley & Sons.

SOIL SURVEY STAFF. 1975. *Soil taxonomy.* U.S. Dept. of Agriculture Handbook 436. Washington, DC: U.S. Dept. of Agriculture.

SOLOMON, D. K., and T. E. CERLING. 1987. The annual carbon dioxide cycle in a montane soil: observations, modeling, and implications for weathering. *Water Resources Research* 23(12): 2257–65.

SPOSITO, G. 1981. Cation exchange in soils: An historical and theoretical perspective. In *Chemistry in the soil environment*, pp. 13–30. Madison, WI: Am. Soc. of Agronomy.

SPOSITO, G. 1985. Chemical models of weathering in soils. In *The chemistry of weathering*, ed J. I. Drever, pp. 1–18. NATO ASI Ser. C: Mathematical & Physical Sciences. Boston: D. Reidel Publ. Co.

SPOSITO, G. 1989. *The chemistry of soils.* New York: Oxford Univ. Press.

SPOSITO, G. 1994. *Chemical equilibrium and kinetics in soils.* New York: Oxford Univ. Press.

STALLAND, R. F., and J. M. EDMOND. 1981. Geochemistry of the Amazon. 1. Precipitation chemistry and the marine contribution to the dissolved load at the time of peak discharge. *J. Geophys. Res.* 86(C10):9844–58.

STARKEY, R. L. 1945. Transformations of iron by bacteria in water. *J. Am. Water Works Assoc.* 37(10):963–84.

STEEFEL, C. I., and P. V. VAN CAPPELLEN. 1990. A new kinetic approach to modeling water-rock interaction: The role of nucleation, precursors, and Ostwald ripening. *Geochim. Cosmochim. Acta* 54(10):2657–77.

STEINMANN, P., P. C. LICHTNER, and W. SHOTYK. 1994. Reaction path approach to mineral weathering reactions. *Clays Clay Minerals* 42(2):197–206.

STEPANOV, M. A., and N. P. GALKIN. 1960. The solubility product of the hydroxide of tetravalent uranium. *Soviet Atomic Energy* 9:817–21.

STIPP, S. L., M. HOCHELLA, JR., G. A. PARKS, and J. O. LECKIE. 1992. Cd^{2+} uptake by calcite, solid-state diffusion, and the formation of solid-solution: Interface processes observed with near-surface sensitive techniques (XPS, LEED, and AES). *Geochim. Cosmochim. Acta* 56:1941–54.

STOLLENWERK, K. G. 1994. Geochemical interactions between constituents in acidic groundwater and alluvium in an aquifer near Globe, Arizona. *Applied Geochem.* 9:353–69.

STONE, A. T., and J. J. MORGAN. 1990. Kinetics of chemical transformations in the environment. In *Aquatic chemical kinetics*, ed W. Stumm, pp. 1–42. New York: Wiley-Interscience.

STONE, R. 1996. The explosions that shook the world. *Science* 272:352–54.

STRAHLER, A. H., and A. N. STRAHLER. 1992. *Modern physical geography.* 4th ed. New York: John Wiley & Sons.

STUMM, W., ed. 1987. *Aquatic surface chemistry. Chemical processes at the particle-water interface.* New York: John Wiley & Sons.

STUMM, W., ed. 1990. *Aquatic chemical kinetics.* New York: Wiley-Interscience.

STUMM, W. 1992. *Chemistry of the solid-water interface.* New York: Wiley-Interscience.

STUMM, W. 1993. Aquatic colloids as chemical reactants: surface structure and reactivity. Colloids and surfaces A. *Physicochemical and Engineering Aspects* 73:1–18.

STUMM, W., and J. J. MORGAN. 1981. *Aquatic chemistry.* 2d ed. New York: John Wiley & Sons.

STUMM, W., and J. J. MORGAN. 1985. Comment on the conceptual significance of pE. A comment to John D. Hostettler's paper "Electrode electrons, aqueous electrons, and redox potentials in natural waters." *Am. J. Sci.* 285(9):856–63.

STUMM, W., and J. J. MORGAN. 1996. *Aquatic chemistry.* 3d ed. New York: Wiley-Interscience.

SUAREZ, D. L., and D. LANGMUIR. 1975. Heavy metal relationships in a Pennsylvania soil. *Geochim. Cosmochim. Acta* 40:589–98.

SULLIVAN, P. J., A. A. SOBEK, and J. RYBARCZYK. 1986. Evaluating mineral dissolution in laboratory weathering experiments using leachate data. *Soil Sci. Soc. Am. J.* 50:251–54.

SUNDER, S., N. H. MILLER, and A. M. DUCLOS. 1994. XPS and XRD studies of samples from the natural fission reactors in the Oklo uranium deposits. *Mat. Res. Soc. symp. proc.* 333, pp. 631–38.

SUTHERLAND, J. C. 1970. Silicate mineral stability and equilibria in the Great Lakes. *Envir. Sci. Technol.* 4:826–33.

SVERDRUP, H. 1989. Modeling base cation release due to chemical weathering of primary minerals in the field. Unpublished paper, 28th Int. Geological Congress Acid Rain Workshop. Washington, DC.

SVERDRUP, H. 1990. *The kinetics of base cation release due to chemical weathering.* Lund: Lund Univ. Press.

SVERDRUP, H., and P. WARFVINGE. 1988. Weathering of primary silicate minerals in the natural soil environment in relation to a chemical weathering model. *Water Air Soil Pollution* 38:387–408.

SVERDRUP, H., and P. WARFVINGE. 1993. Calculating field weathering rates using a mechanistic geochemical model PROFILE. *Applied Geochem.* 8:273–83.

SVERDRUP, H., and P. WARFVINGE. 1995. Estimating field weathering rates using laboratory kinetics. In *Chemical weathering rates of silicate minerals*, ed A. F. White and S. L. Brantley. Reviews in Mineralogy 31, pp. 485–541. Min. Soc. Am.

SVERJENSKY, D. A. 1993. Physical surface-complexation models for sorption at the mineral-water interface. *Nature* 364:776–80.

SVERJENSKY, D. A. 1994. Zero-point-of-charge prediction from crystal chemistry and solvation theory. *Geochim. Cosmochim. Acta* 58(14):3123–29.

SWAIN, E., D. ENGSTROM, M. E. BRIGHAM, T. HENNING, and P. L. BREZONIC. 1992. Increasing rates of atmospheric mercury deposition in midcontinental North America. *Science* 257:784–87.

Sweeton, F. H., R. E. Mesmer, and C. F. Baes, Jr., 1974. Acidity measurements at elevated temperatures. VII. Dissociation of water. *Solution Chem.* 3:191–214.

Swoboda-Colberg, N. G., and J. I. Drever. 1992. Mineral dissolution rates: A comparison of laboratory and field studies. *Proc. 7th intl. symp. water-rock interaction*, ed J. Kharaka and A. Maest, pp. 115–18. Rotterdam: A. A. Balkema.

Swoboda-Colberg, N. G., and J. I. Drever. 1993. Mineral dissolution rates in plot-scale field and laboratory experiments. *Chem. Geol.* 105:51–69.

Takano, B. 1987. Correlation of volcanic activity with sulfur oxyanion speciation in a crater lake. *Science* 235:1633–35.

Tanaka, T., and M. Senoo. 1995. Sorption of ^{60}Co, ^{85}Sr, ^{137}Cs, ^{237}Np and ^{241}Am on soil under coexistence of humic acid: effects of molecular size of humic acid. *Mat. Res. Soc. symp. proc.* 353, pp. 1013–20.

Tanner, A. B. 1964. Physical and chemical controls on distribution of Ra-226 and Rn-222 in ground water near Great Salt Lake, Utah. In *The natural radiation environment*, ed J. A. Adams and W. M. Lowder, pp. 253–76. Chicago: Univ. of Chicago Press.

Tanger, J. C., IV, and H. C. Helgeson. 1988. Calculation of the thermodynamic and transport properties of aqueous species at high pressures and temperatures: Revised equations of state for the standard partial molal properties of ions and electrolytes. *Am. J. Sci.* 288:19–98.

Tardy, Y., and D. Nahon. 1985. Geochemistry of laterites, stability of Al-goethite, Al-hematite, and Fe^{3+}-kaolinite in bauxites and ferricretes: An approach to the mechanism of concretion formation. *Am. J. Sci.* 285:865–903.

Tarutis, W. J., Jr. 1993. On the equivalence of the power and reactive continuum models of organic matter diagenesis. *Geochim. Cosmochim. Acta* 57:1349–50.

Taylor, B. E., M. C. Wheeler, and D. K. Nordstrom. 1984a. Stable isotope geochemistry of acid mine drainage: Experimental oxidation of pyrite. *Geochim. Cosmochim. Acta* 48:2669–78.

Taylor, B. E., M. C. Wheeler, and D. K. Nordstrom. 1984b. Isotope composition of sulphate in acid mine drainage as measure of bacterial oxidation. *Nature* 308:539–41.

Taylor, M. J. 1979. *Practical study requirements groundwater and seepage uranium mill waste disposal systems*. Written testimony for Royal Commission of Inquiry Health and Envir. Protection, Vancouver, BC. Denver, CO: D'Appolonia Consulting Engineers, Inc.

Thamer, B. J., K. K. Nielson, and K. Felthauser. 1981. *The effects of moisture on radon emanation*. U.S. Bureau of Mines Open-file Rept. 184-82.

Thorstenson, D. C. 1970. Equilibrium distribution of small organic molecules in natural waters. *Geochim. Cosmochim. Acta* 34(7):745–70.

Thorstenson, D. C. 1984. *The concept of electron activity and its relation to redox potentials in aqueous geochemical systems*. U.S. Geol. Survey Open File Rept. 84-72.

Thurman, E. M. 1985. *Organic geochemistry of natural waters*. Boston: Martinus Nijhoff/Dr W. Junk Publ.

Titayeva, N. A. 1994. *Nuclear geochemistry*. Boca Raton, FL: CRC Press.

Toran L., and A. Sjoreen. 1996. CHEMFORM: A formatting program for geochemical data. *Ground Water* 34(3):552–53.

Torrero, M. E. et al. 1991. The solubility of unirradiated UO$_2$ in both perchlorate and chloride test solutions. Influence of the ionic medium. *Mat. Res. Soc. symp. proc.* 212, pp. 229–234.

Torretto, P. et al. 1995. *Solubility and speciation results from oversaturation experiments on Np, Pu, and Am in a neutral elctrolyte with a total carbonate similar to water from Yucca Mountain Region well UE-25p#1*. Los Alamos Natl. Lab., Report LA-13018-MS.

Toth, S. J., and A. N. Ott. 1970. Characterization of bottom sediments: cation exchange capacity and exchangeable cation status. *Envir. Sci. & Technol.* 4(11):935–39.

Tremaine, P. R., R. von Massow, and G. R. Shierman. 1977. A calculation of the Gibbs free energy for ferrous ion and the solubility of magnetite in H$_2$O and D$_2$O to 300°C. *Thermochim. Acta* 19:287–300.

Triay, I. et al. 1995. *Colloid-facilitated radionuclide transport at Yucca Mountain*. Report LA-12779-MS. Los Alamos, NM: Los Alamos Natl. Lab.

Trolard, F., and Y. Tardy. 1987. The stabilities of gibbsite, boehmite, aluminous goethites and aluminous hematites in bauxites, ferricretes and laterites as a function of water activity, temperature and particle size. *Geochim. Cosmochim. Acta* 51:945–57.

Truesdell, A. H., and B. F. Jones. 1974. WATEQ, a computer program for calculating chemical equilibria of natural waters. *U.S. Geol. Survey J. Research* 2:233–48.

Turekian, K. K. 1977. The fate of metals in the oceans. *Geochim. Cosmochim. Acta* 41(8):1139.

Turner, D. R. 1991. *Sorption modeling for high-level waste performance assessment: A literature review*. Report CN-WRA 91-011. San Antonio, TX: Center for Nuclear Waste Regulatory Anal.

Turner, D. R. 1993. *Mechanistic approaches to radionuclide sorption modeling*. Report CNWRA 93-019. San Antonio, TX: Center for Nuclear Waste Regulatory Anal.

Turner, D. R. 1995. *A uniform approach to surface complexation modeling of radionuclide sorption*. Report CN-WRA 95-001. San Antonio, TX: Center for Nuclear Waste Regulatory Anal.

Turner, D. R., T. Griffin, and T. B. Dietrich. 1993. Radionuclide sorption modeling using the MINTEQA2 speciation code. *Mat res. soc. symp. proc.* 294, pp. 783–89.

Usdowski, E. 1994. Synthesis of dolomite and geochemical implications. *Spec. Pubs. Intl. Assoc. Sediment.* 21:345–60.

Van Breemen, N. 1975. Acidification and deacidification of coastal plain soils as a result of periodic flooding. *Proc. Soil Sci. Soc. Am.* 39, pp. 1153–57.

Van Breemen, N., and W. G. Wielemaker. 1974a. Buffer intensities and equilibrium pHs of minerals and soils: I. The contribution of minerals and aqueous carbonate to pH buffering. *Proc. Soil Sci. Soc. Am.* 38, pp. 55–60.

Van Breemen, N., and W. G. Wielemaker. 1974b. Buffer intensities and equilibrium pHs of minerals and soils: II.

Theoretical and actual pH of minerals and soils. *Proc. Soil Sci. Soc. Am.* 38, pp. 61–66.

VAN DER HEIJDE, P. K. M, and O. A. ELNAWAWY, 1993. *Compilation of ground-water models.* U.S. Envir. Protection Agency Report EPA/600/R-93/118. U.S. Office of Research and Devel.

VAN DER WEIJDEN, C. H., R. C. ARTHUR, and D. LANGMUIR. 1976. Sorption of uranyl by hematite: theoretical and geochemical implications, *Geol. Soc. Am. abstracts with programs,* Annual Meeting, Denver, p. 1152.

VAN DER WEIJDEN, C. H., and M. VAN LEEUWEN, The effect of the pH on the adsorption of uranyl onto peat. *Uranium* 2(1):59–66.

VAN GEEN, A., A. P. ROBERTSON, and J. O. LECKIE. 1994. Complexation of carbonate species at the goethite surface: Implications for adsorption of metal ions in natural waters. *Geochim. Cosmochim. Acta* 58(9):2073–86.

VAN SCHUYLENBORG, J. 1973. Report on topic 1.1; Sesquioxide formation and transformation. *Intl. Soc. Soil Sci., Comm. V-VI, Trans. Pseudogley and Gley,* pp. 93–102.

VAN SLYKE, D. D. 1922. On the measurement of buffer values and on the relationship of buffer values to the dissociation constant of the buffer and the concentration and reaction of the buffer solution. *J. Biol. Chem.* 52:525–70.

VARADACHARI, C., M. KUDRAT, and K. GHOSH. 1994. Evaluation of standard free energies of formation of clay minerals by an improved regression method. *Clays Clay Minerals* 42(3):298–307.

VELBEL, M. A. 1985. Geochemical mass balances and weathering rates in forested watersheds of the southern Blue Ridge. *Am. J. Sci.* 285:904–30.

VELDE, B., and G. VASSEUR. 1992. Estimation of the diagenetic smectite to illite transformation in time-temperature space. *Am. Mineral.* 77:967–76.

VESELY, V., V. PEKAREK, and M. ABBRENT. 1965. A study of uranyl phosphates-III. Solubility products of uranyl hydrogen phosphate. Uranyl orthophosphate and some alkali uranyl phosphates. *J. Inorg. Nucl. Chem.* 27:1159–66.

VIEILLARD, P., and Y. TARDY. 1988. Estimation of enthalpies of formation of minerals based on their refined crystal structures. *Am. J. Sci.* 288:997–1040.

VILKS, P., J. J. CRAMER, D. B. BACHINSKI, D. C. DOERN, and H. G. MILLER. 1993. Studies of colloids and suspended particles, Cigar Lake uranium deposit, Saskatchewan, Canada. *Applied Geochem.* 8:605–16.

VOLOSOV, A. G., I. L. KHODAKOWSKIY, and B. N. RYZHENKO. 1972. Equilibria in the system SiO_2-H_2O at elevated temperatures along the lower three-phase curve. *Geochem. Intl.* 9(3):362–77.

WADSWORTH, M. E. 1979. Hydrometallurgical processes. In *Rate processes of extractive metallurgy,* ed H. Y. Sohn and M. E. Wadsworth, pp. 133–86. New York: Plenum Press.

WAGMAN, D. et al. 1982. The NBS tables of chemical thermodynamic properties. *J. Phys. Chem. Ref. Data* 11, sup. no. 2.

WAITE, T. D., J. A. DAVIS, T. E. PAYNE, G. A. WAYCHUNAS, and N. XU. 1994. Uranium(VI) adsorption to ferrihydrite:

Application of a surface complexation model. *Geochim. Cosmochim. Acta* 58:5465–78.

WALLHAUSER, K. H., and H. PUCHELT. 1966. Sulfate-reducing bacteria in sulfur springs and connate waters of Germany and Austria (in German). *Contrib. Min. Petrol.-Beitr. Mineral. Petrologie* 13(1):12–30.

WALLING, E. M., P. A. ROCK, and W. H. CASEY. 1995. The Gibbs energy of formation of huntite, $CaMg_3(CO_3)_4$ at 298 K and 1 bar from electrochemical cell measurements. *Am. Mineral.* 80:355–60.

WANTY, R. B. et al. 1992. Weathering of Pikes Peak granite: Field, experimental and modeling observations. *Proc. 7th intl. conf. water-rock interaction,* ed Y. K. Kharaka and A. S. Maest, pp. 599–602. Rotterdam: A. A. Balkema.

WANTY, R. B., J. R. CHATHAM, and D. LANGMUIR. 1987. The solubilities of some major and minor element minerals in ground waters associated with a sandstone-hosted uranium deposit. *Bull. de Mineralogie* (France) 110(2–3):209–26.

WANTY, R. B., and D. K. NORDSTROM. 1993. Natural radionuclides. In *Regional ground-water quality,* ed W. M. Alley. New York: Van Nostrand Reinhold.

WANTY, R. B., C. A. RICE, D. LANGMUIR, P. H. BRIGGS, and E. P. LAWRENCE. 1991. Prediction of uranium adsorption by crystalline rocks: The key role of reactive surface area. *Mat. Res. Soc. symp. proc.* 212, pp. 695–702.

WARFRINGE, P., and H. SVERDRUP. 1992. Calculating critical loads of acid deposition with PROFILE-a steady-state soil chemistry model. *Water, Air and Soil Pollution* 63:119–43.

WARREN, L. A., and A. P. ZIMMERMAN. 1994. The importance of surface area in metal sorption by oxides and organic matter in a heterogeneous natural sediment. *Applied Geochem.* 9:245–54.

WEARE, J. H. 1987. Models of mineral solubility in concentrated brines with application to field observations. In *Thermodynamic modeling of geological materials: Minerals, fluids and melts.* Reviews in Mineralogy 17, pp. 143–76. Min. Soc. Am.

WEHRLI, B. 1990. Redox reactions of metal ions at mineral surfaces. In *Aquatic chemical kinetics. Reaction rates of processes in natural waters,* ed W. Stumm. New York: Wiley-Interscience.

WELCH, A. H., Z. SZABO, D. L. PARKHURST, P. C. VAN METRE, and A. H. MULLIN. 1995. Gross-beta activity in groundwater: natural sources and artifacts of sampling and laboratory analysis. *Applied Geochem.* 10:491–503.

WELLS, A. F. 1962. *Structural inorganic chemistry.* 3d ed. Oxford: Clarendon Press.

WESTALL, J. 1980. Chemical equilibrium including adsorption on charged surfaces. In *Particulates in water,* ed M. C. Kavanaugh and J. O. Leckie. Adv. Chem. Ser. 189, pp. 34–44. Washington, DC: Am. Chem. Soc.

WESTALL, J. C. 1986. Reactions at the oxide-solution interface: Chemical and electrostatic models. In *Geochemical processes and mineral surfaces,* ed J. A. Davis and K. F. Hayes. Am. Chem. Soc. Symp. Ser. 323, pp. 54–78. Washington DC: Am. Chem. Soc.

WESTALL, J. C., and H. HOHL. 1980. A comparison of electrostatic models for the oxide/solution interface. *Adv. Colloid Interface Sci.* 12:265–94.

WESTALL, J. C., J. L. ZACHARY, and F. M. M. MOREL. 1976. *MINEQL, A computer program for the calculation of chemical equilibrium composition of aqueous systems.* Tech. Note 18, Dept. Civil Eng., MIT, Cambridge, MA.

WESTRICH, J. T., and R. A. BERNER. 1984. The role of sedimentary organic matter in bacterial sulfate reduction: the G model tested. *Limnol. Oceanogr.* 29:236–49.

WHITE, A. F., and H. C. CLAASSEN. 1979. Dissolution kinetics of silicate rocks-application to solute modeling. In *Chemical modeling in aqueous systems*, ed E. A. Jenne. Am. Chem. Soc. Symp. 93, pp. 447–73. Washington, DC: Am. Chem. Soc.

WHITE, A. F., and M. L. PETERSON. 1990. Role of reactive-surface-area characterization in geochemical kinetic models In *Chemical modeling of aqueous systems II*, ed. D. C. Melchior and R. L. Bassett. ACS Symposium Series 416, pp. 461–75. Washington, DC: Am. Chem. Soc.

WHITE, D. E., J. D. HEM, and G. A. WARING. 1963. Chemical composition of subsurface waters. In *Data of geochemistry*, 6th ed. U.S. Geol Survey Prof. Paper 440-F.

WHITFIELD, M. 1974. Thermodynamic limitations on the use of the platinum electrode in Eh measurements. *Limnol. & Oceanogr.* 19:857–65.

WHITTEMORE, D. O., and D. LANGMUIR. 1975. The solubility of ferric oxyhydroxides in natural waters. *Ground Water* 13(4):360–65.

WICK, O. J., and M. O. CLONINGER. 1980. *Comparison of potential radiological consequences from a spent-fuel repository and natural uranium deposits.* Pacific Northwest Laboratory Report PNL-3540.

WICKBERG, P. E. GUSTAFSON, I RHÉN, and R. STANFORS. 1991. *Äspö hard rock laboratory, evaluation and conceptual modelling based on the pre-investigations 1986–1990.* Report SKB TR 91-22. Stockholm, Sweden: Swedish Nuclear Fuel and Waste Mgmt. Co.

WICKS, C. M., and J. S. HERMAN. 1996. Regional hydrogeochemistry of a modern coastal mixing zone. *Water Resources Research* 32(2):401–7.

WIELAND, E., B. WEHRLI, and W. STUMM. 1988. The coordination chemistry of weathering; III. A generalization on the dissolution rates of minerals. *Geochim. Cosmochim. Acta* 52:1969–81.

WIERSMA, C. L., and J. D. RIMSTIDT. 1984. Rates of reaction of pyrite and marcasite with ferric ion at pH = 2. *Geochim. Cosmochim. Acta* 48:85–92.

WIKLANDER, L. 1964. Cation and anion exchange phenomena. In *Chemistry of the soil*, ed F. E. Bear. New York: Reinhold Publ. Co.

WILLIAMSON, M. A. 1996. Personal communication. Adrian Brown Consultants, Inc., Denver, CO.

WILLIAMSON, M. A., and J. D. RIMSTIDT. 1992. Correlation between structure and thermodynamic properties of aqueous sulfur species. *Geochim. Cosmochim. Acta* 56:3867–80.

WILLIAMSON, M. A., and J. D. RIMSTIDT. 1994. The kinetics and electrochemical rate-determining step of aqueous pyrite oxidation. *Geochim. Cosmochim. Acta* 58(24):5443–54.

WILSON, C. N. 1990. *Results from NNWSI series 3 spent fuel dissolution tests.* Pacific Northwest Laboratory, Report PNL-7170.

WILSON, C. N., and C. J. BRUTON. 1990. Studies on spent fuel dissolution behavior under Yucca Mountain repository conditions. *Nuclear Waste Mgmt. III. Ceramic Trans.* 9:423–41.

WILSON, E. K. 1995. Zero-valent metals provide possible solution to groundwater problems. *Chemical & Engineering News* 269:19–22.

WILSON, M. J., and P. H. NADEAU. 1985. Interstratified clay minerals and weathering processes. In *The chemistry of weathering*, ed J. I. Drever. NATO ASI Ser. C: Mathematical & Physical Sciences, 149, pp. 97–118. Dordrecht, Holland: D. Reidel Publ. Co.

WILTSHIRE, S. D. 1993. The nuclear waste primer, a handbook for citizens. rev. ed. The League of Women Voters Educ. Fund. New York: Lyons & Burford Publ. Co.

WOGELIUS, R. A, and J. V. WALTHER. 1991. Olivine dissolution at 25°C: Effects of pH, CO_2, and organic acids. *Geochim. Cosmochim. Acta* 55:943–54.

WOLERY, T. J. 1992. *EQ3/6, A software package for geochemical modeling of aqueous systems: Package overview and installation guide (Ver. 7.0).* UCRL-MA-110662 Pt I. Lawrence Livermore Natl. Lab.

WOLERY, T. J. 1992. *EQ3NR, A computer program for geochemical aqueous speciation-solubility calculations: Theoretical manual, user's guide, and related documentation (Ver. 7.0).* UCRL-MA-110662 Pt III. Lawrence Livermore Natl. Lab.

WOLERY, T. J., and S. A. DAVELER. 1992. *EQ6, A computer program for reaction path modeling of aqueous geochemical systems: Theoretical manual, user's guide, and related documentation (Ver. 7.0).* UCRL-MA-11066 Pt IV. Lawrence Livermore Natl. Lab.

WOLERY, T. J. et al. 1990. Current status of the EQ3/6 software package for geochemical modeling. In *Chemical modeling of aqueous systems II*, ed D. C. Melchior and R. L. Bassett. Am. Chem. Soc. Symp. Series 416, pp. 104–16. Washington, DC: Am. Chem. Soc.

WOLERY, T. J., C. E. PALMER, and K. G. KNAUSS. 1995. The neptunium solubility problem in repository performance assessment: A white paper. Unpublished paper. Yucca Mtn. Site Characterization Project, Lawrence Livermore Natl. Lab.

WOOD, B. J., and D. G. FRASER. 1976. *Elementary thermodynamics for geologists.* Oxford: Oxford Univ. Press.

WOOD, R. J., and J. V. WALTHER. 1983. Rates of hydrothermal reactions. *Science* 222:413–15.

WOOD, W. W. 1985. Origin of caves and other solution openings in the unsaturated (vadose) zone of carbonate rocks: A model for CO_2 generation. *Geology* 13(11):753–832.

WOOD, W. W., and M. J. PETRAITIS. 1984. Origin and distribution of carbon dioxide in the unsaturated zone of the southern high plains. *Water Resources Research* 20(9):1193–208.

WOODS, T. L., and R. M. GARRELS. 1992. Calculated aqueous-solution-solid-solution relations in the low-temperature system $CaO-MgO-FeO-CO_2-H_2O$. *Geochim. Cosmochim. Acta* 56:3031–43.

XIE, Z., and J. V. WALTHER. 1992. Incongruent dissolution and surface area of kaolinite. *Geochim. Cosmochim. Acta* 56: 3357–63.

YAJIMA, T., Y. KAWAMURA, and S. UETA. 1995. Uranium (IV) solubility and hydrolysis constants under reduced conditions. *Mat. Res. Soc. symp. proc.* 353, pp. 1137–42.

YANG, I. C. 1992. Flow and transport through unsaturated rock—data from two test holes, Yucca Mountain, Nevada. *Proc. 3d intl. conf. on high level radioactive waste management* 1, pp. 732–37. La Grange Park, IL: Am. Nucl. Soc. Inc.

YANG, I. C., G. W. RATTRAY, and P. YU. 1996. *Interpretations of chemical and isotopic data from boreholes in the unsaturated-zone at Yucca Mountain, Nevada*. U.S. Geol. Survey Water-Resources Inv. Rept. WRIR 96–4058. (In press.)

YARIV, S., and H. CROSS. 1979. *Geochemistry of colloid systems*. Berlin: Springer-Verlag.

YATSIMIRSKII, K., and V. P. VASIL'EV. 1966. Instability constants of complex compounds. New York: D. Van Nostrand, Inc.

ZACHARA, J. M., C. E. COWAN, and C. T. RESCH. 1993. Metal cation/anion adsorption on calcium carbonate: Implications to metal ion concentrations in ground water. In *Metals in groundwater*, ed. H. E. Allen, E. M. Perdue, and D. S. Brown, pp. 38–71. Boca Raton, FL: Lewis Publishers.

ZACHARA, J. M., J. A. KITTRICK, and J. B. HARSH. 1988. The mechanism of Zn^{2+} adsorption on calcite. *Geochim. Cosmochim. Acta* 52:2281–91.

ZEHNDER, A. J. B., ed. 1988. *Biology of anaerobic microorganisms*. New York: John Wiley & Sons.

ZIELINSKI, R. A., S. ASHER-BOLINDER, and A. L. MEIER. 1995. Uraniferous waters of the Arkansas River Valley, Colorado, U.S.A.: A function of geology and land use. *Applied Geochem.* 10:133–44.

ZOBELL, C. E. 1946. Studies on redox potential of marine sediments. *Bull. Am. Assoc. Petrol. Geologists* 30:477–513.

ZOBRIST, J., and W. STUMM. 1979. Chemical dynamics of the Rhine catchment area in Switzerland; Extrapolation to the "pristine" Rhine River input into the ocean. *Proc. review and workshop on river inputs to ocean systems (RIOS)*. FAO, Rome.

Index

Boldface numbers indicate tables and figures.

carbon dioxide *(continued)* consumption with groundwater flow, 158; content of streams and lakes, 156; controls of content in surface and groundwaters, **157**; dissolution of, 202–204; effect on acid rain pH, 288; effect of snow cover, 158; effect on total carbonate in oceans, 309; effect on weathering rate, 261; exsolution of, 202–204; from fossil fuels, 309; increase since 1850, 275–276; at Mauna Loa, Hawaii, **276**; from organic matter, 158; production with groundwater flow, 160–161; respiratory coefficient, 158; respired, 223

carbon dioxide–acidity titration, 171, 172

carbon dioxide pressure, 160, **225**; trends in surface and groundwater, 160

carbon isotopic analyses, Slab Cabin Run, **224**

carbon isotopic composition, 223, 224; typical values, 223

carbonate complexes, K_{dissoc}-temperature functions, **25**

carbonate mineral saturation state, 225–**227**, 228

carbonate mineral supersaturation: explanation for, 219–223

carbonate minerals; dissolution rates, 60–61, 73–74, 78; effects on solubility of, 202–207; K_{sp}-temperature functions, **26, 206**; and molar Ca^{2+}/Mg^{2+} ratio, 208–211; solubility products of, **194, 218**; stability of Ca carbonates, **193–195**; stability of Mg carbonates, **194–196**; surface charge, 349; wind-blown, 283

carbonate speciation, Slab Cabin Run, **225**

carbonate species: concentrations as function of pH, **156, 212, 213**

carbonate stone, damage due to acid rain, 286, 311

carbonic acid, 28; buffer capacity, 183–185, **186**; concentrations of, 293; dissociation, 154; relation to pH in natural water, 155–161; and rock weathering, 153; true concentration of, 154

carbonic acid species, 153–161

carnotite, 506; solubility of, with pH, **500**; uranium speciation solubility of, 501

cation exchange capacity, 346–353; of natural materials, 351–353; of organic matter, 352; and sediment particle size, 352–353; of soils, 352; of stream sediments versus soils, 353; zeolites, 349

cation export, rates of, **259**

cations: adsorbed, 352; behavior in water, 95–99; divalent complexes, **106, 108**; electronegativity, 99–103; entropy of dehydration of, 111; hard-acid, 103–105; hydrolysis in water, 95–99; ionic potentials, **86**; monovalent complexes, **106**; radii of, **86, 98**; soft-acid, 103–106, 118, 119; transition element, 106–109

celestite, 31, 32; solubility of, 32

cerargyrite, 28

^{135}Cs, as nuclear fission product, 515

^{137}Cs, 536; as nuclear fission product, 515, 519

chalcedony, 244; solubility as a function of temperature, 244

Chalk (limestone) aquifer, U.K., 425, **426**

characteristic solid phases in redox environments, **421**

charge-balance equation, 3, 274, 310

charge densities, 86

charge-potential relationships, 376

chelation, entropy of, 111

chemical equilibrium, 1, 50–56

chemical oxygen demand (COD), 421

chemical potential, 4, 7–9, 123

chemical tracers, 292

chemical wastewaters, 299

chemical weathering, 97, 231–265, 292. *See also* weathering, weathering rates

Chernobyl, Ukraine, 512

chloride, mass-balance calculation, 257

chlorinated hydrocarbons, 356, 358–360, 430

Cl^-, in rain over United States, 276, **280**

chlorite group minerals, **314**

chromate competition, triple-layer model, 385

chromate reduction, 430

chromium, in groundwater, 430

Cigar Lake uranium deposit, Canada, **337**, 504, 506, 508, **514**, 535

cigarette smoking, 493

clay backfill, at nuclear waste repository, 520

clay formation, 293

clay mineral phase diagrams, 324–342; assumptions of, 324–325; effect of temperature, **333–334**

clay mineral saturation, calculation, 342

clay minerals. *See also individual mineral sources*: crystal chemistry of, 312; equilibrium concepts, 322–337; occurrence of, 319–322; structure, **315**; thermodynamic stability of, 337–339

clay moisture content, 540

clays, 509; base-poor, **241**; base-rich, **241**; buffer capacity, 187; formation of, 232; importance of small amounts, 187; kinetics of transformation, 324; metastability, 322–324; mixed-layer, 316; particle size of, 312, 347; sequence of forming, **320**; solubility in river water, 339; surface charge, 346, **347, 348**

climate, related to soil group formation, **240**

closed carbonate system, 211–216

closed system, 51, 174, 492; defined, 1–2; and low groundwater velocities, 270

coagulation, rate of, 439

coal deposits, 457

coal leachate, pyrite oxidation in, 476

coal mine drainage, **305**

coal mines, flooded, 464

coal; carbon isotopic composition, 223; eastern U.S., 467; high-sulfur, 277; of marine versus fresh water origin, 457

$^{90}Co^{2+}$, 27

cobalt oxidation, 430

CODATA key values, 521, **550**

coefficients of variation, of groundwater, 273

coffinite, 495, 497, 502, 512, 514; partial oxidation of, 505; saturation in groundwaters from Cigar Lake, 508; saturation in groundwaters from crystalline rocks, 507; thermodynamic stability, 502, 504, 548–**549, 551**

coions, defined, 372

colloidal filtration, 520, 535

colloidal $Pu(OH)_4$(am), groundwater transport of, 529

colloidal suspensions, stability of, 343, 438–441

colluvium, 236

common ion effect, 205

compaction density, 540

complex composite rate law, 60

complex formation, extent of, 90

complexation: mass balance equation, 90–95; thermodynamics of, 111–112

complexes, aqueous, 82–122; divalent metal-sulfate, 87; inner-sphere, 84–88, 99–103; metal cation–ligand relations in, 88–90; outer-sphere, 84–88; predicting stability of, 109–111

components, defined, 2

Comprehensive Soil Classification System, 239, **241**

compressibility, 32; isothermal, 38

computer models. *See specific models*

concentrations versus stream discharge, 301, **302, 303–305**

concrete, 163; effect in repositories on technetium solubility, 544

congruent reaction, defined, 4

constant capacitance model, 371, **372**, 374, 376–381, 400, 509, 545; adsorption parameters, 381, **382**; applicable conditions for, 381; hydrolysis constants of metal cations versus intrinsic surface complex constants, **389**; input values, 381; in MINTEQA2, 399

coordination numbers, 89

copper: adsorption by organic matter, 393; frequency distribution in shale groundwater, **307**

core cations, charge of, **98**

corundum, laboratory weathering rates of, **259**

coulombic function, 98–99

counterions, defined, 372